Kernel-based Approximation Methods using MATLAB

INTERDISCIPLINARY MATHEMATICAL SCIENCES*

Published

Vol. 11 Advances in Interdisciplinary Applied Discrete Mathematics
eds. Hemanshu Kaul & Henry Martyn Mulder

Vol. 12 New Trends in Stochastic Analysis and Related Topics: A Volume in Honour of Professor K D Elworthy
eds. Huaizhong Zhao & Aubrey Truman

Vol. 13 Stochastic Analysis and Applications to Finance: Essays in Honour of Jia-an Yan
eds. Tusheng Zhang & Xunyu Zhou

Vol. 14 Recent Developments in Computational Finance: Foundations, Algorithms and Applications
eds. Thomas Gerstner & Peter Kloeden

Vol. 15 Recent Advances in Applied Nonlinear Dynamics with Numerical Analysis: Fractional Dynamics, Network Dynamics, Classical Dynamics and Fractal Dynamics with Their Numerical Simulations
eds. Changpin Li, Yujiang Wu & Ruisong Ye

Vol. 16 Hilbert–Huang Transform and Its Applications (2nd Edition)
eds. Norden E Huang & Samuel S P Shen

Vol. 17 Festschrift Masatoshi Fukushima In Honor of Masatoshi Fukushima's Sanju
eds. Zhen-Qing Chen, Niels Jacob, Masayoshi Takeda & Toshihiro Uemura

Vol. 18 Global Attractors of Non-Autonomous Dynamical and Control Systems (2nd Edition)
by David N Cheban

Vol. 19 Kernel-based Approximation Methods using MATLAB
by Gregory Fasshauer & Michael McCourt

*For the complete list of titles in this series, please go to
http://www.worldscientific.com/series/ims

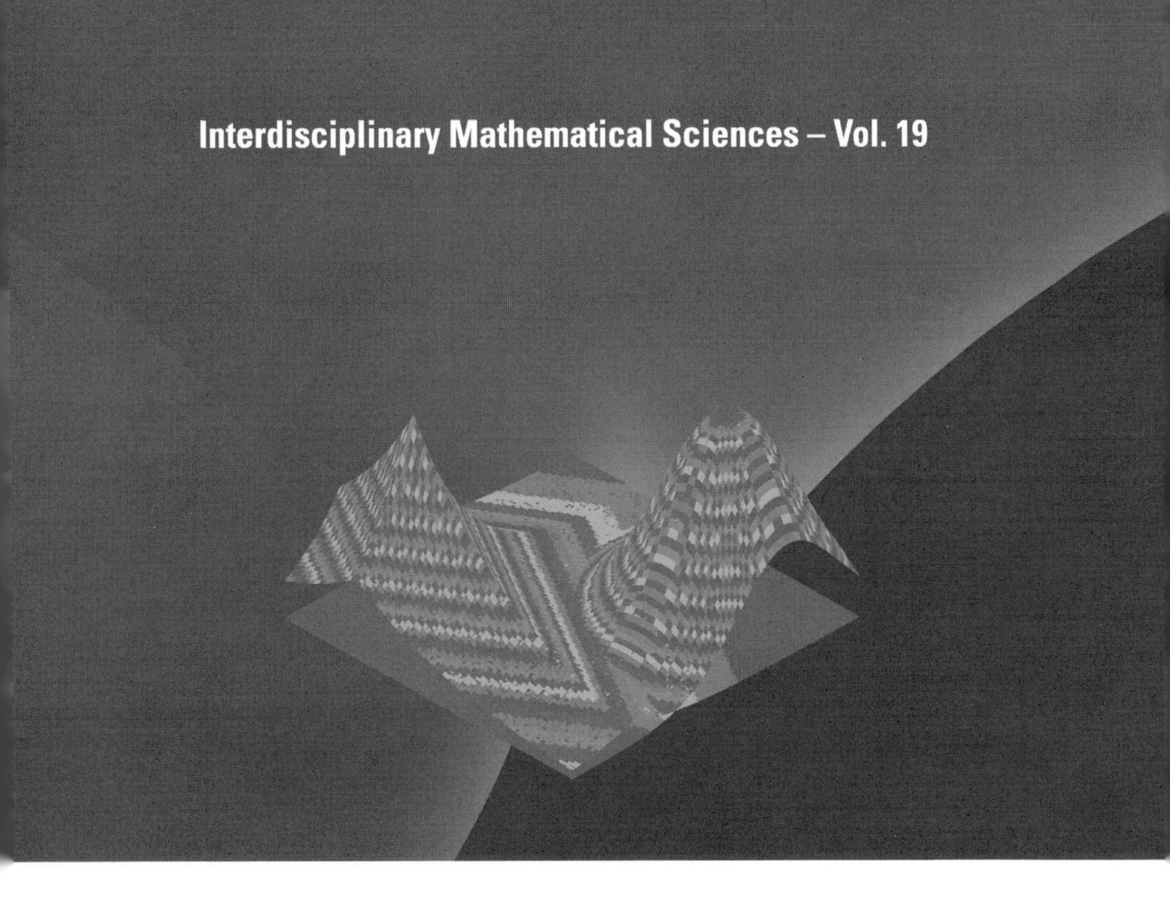

Kernel-based Approximation Methods using MATLAB

Gregory Fasshauer
Illinois Institute of Technology, USA

Michael McCourt
University of Colorado Denver, USA

NEW JERSEY • LONDON • SINGAPORE • BEIJING • SHANGHAI • HONG KONG • TAIPEI • CHENNAI

Published by

World Scientific Publishing Co. Pte. Ltd.
5 Toh Tuck Link, Singapore 596224
USA office: 27 Warren Street, Suite 401-402, Hackensack, NJ 07601
UK office: 57 Shelton Street, Covent Garden, London WC2H 9HE

Library of Congress Cataloging-in-Publication Data
Fasshauer, Gregory E.
Kernel-based approximation methods using MATLAB / by Gregory Fasshauer (Illinois Institute of Technology, USA), Michael McCourt (University of Colorado Denver, USA).
pages cm. -- (Interdisciplinary mathematical sciences ; vume 19)
Includes bibliographical references and index.
ISBN 978-9814630139 (hardcover : alk. paper)
1. Approximation theory--Mathematics. 2. Kernel functions. 3. MATLAB. I. McCourt, Michael J. II. Title.
QA221.F37 2015
511'.42--dc23

2015014679

British Library Cataloguing-in-Publication Data
A catalogue record for this book is available from the British Library.

Printed by FuIsland Offset Printing (S) Pte Ltd Singapore

This book is dedicated to
Inge, Conny, Marc and Pat

Jason, Laurie and Kevin

---

and to Kaya, who kept us company
during some of the hardest work.

Preface

This book focuses on recent developments in the field of meshfree approximation using positive definite reproducing kernels, both from a theoretical and practical point of view. Since positive definite kernels play an important role in many different areas of mathematics, science and engineering, we hope to provide a broad view of this field of research. On the one hand, we aim to speak to graduate students and researchers in these diverse fields by providing each of them with at least some content they are familiar with. And on the other hand, we hope that our presentation of ideas from areas such as approximation theory, numerical analysis, spatial statistics, machine learning, finance, and computing with MATLAB[1] will enable our readers to take some of those ideas and apply them successfully to their own work which may be housed in only one of these areas, or even an entirely different area). Thus, we view this book as a hands-on guide for graduate students in applied mathematics and engineering interested in understanding and applying some of the most recent advances in kernel-based approximation.

Encouraged by the success of the format used for "Meshfree Approximation Methods with MATLAB" (Volume 6 in this book series [Fasshauer (2007)]), where a gentle introduction to the underlying theory of radial basis functions and moving least squares methods was combined with many — relatively short — MATLAB scripts that illustrated those concepts, we have decided on a similar format for this book. We need to emphasize that this book should in no way be considered a second (or revised and enlarged) edition of "Meshfree Approximation Methods with MATLAB". If anything, the present volume can be considered as a "Volume 2". There is very little theory overlap between the two books, and the use of MATLAB is considerably more sophisticated — due in part to some relatively recent additions to the MATLAB software such as `bsxfun`, `cellfun`, and other related functionality. Therefore, books developing the theoretical foundation for the use of kernels in approximation theory (such as, e.g., [Buhmann (2003); Wendland (2005)]), spatial

[1]MATLAB® is a registered trademark of MathWorks® and is used with permission. MathWorks does not warrant the accuracy of the text or exercises in this book. This book's use or discussion of MATLAB software or related products does not constitute endorsement or sponsorship by MathWorks of a particular pedagogical approach or particular use of the MATLAB software.

statistics (e.g., [Stein (1999)]), statistical learning (e.g., [Steinwart and Christmann (2008)]) or statistics in general [Wahba (1990)] remain of great importance.

The book is divided into two parts: Part I focusses mostly on background information (and thus is probably a bit more theory-heavy), while Part II aims mainly to illustrate the basic concepts in the context of applications.

We begin the first part of the book with an chapter that puts positive definite kernels into perspective — both historically, as well as scientifically by pointing out connections to and important resources in related fields such as analysis, approximation theory, the theory of integral equations, mathematical physics, probability theory and statistics, geostatistics, statistical or machine learning, and various kinds of engineering or physics applications. Many of the fundamental concepts we develop to support our view of positive definite kernels should be familiar to beginning graduate students in applied mathematics and engineering: eigenvalues, eigenfunctions, orthogonality, change of basis, Sturm–Liouville theory, Green's kernels, maximum likelihood estimation, Bayesian statistics, convex optimization, etc. We do not aim to give any of these topics a thorough theoretical treatment — that is what we provide many pointers and references for. Instead, it is our goal to present these concepts as they relate to our work on positive definite kernels.

However, there are also new developments — or new interpretations of old ideas — that we believe to be important for the field of meshfree approximation. The first of these ideas is the use of different kinds of basis transformations to map the standard kernel basis to an alternate basis that is more advantageous for the application at hand. In particular, this leads to the so-called Hilbert–Schmidt SVD introduced in Chapter 13, a framework developed by the authors (motivated, however, by the ground-breaking work of Bengt Fornberg and his collaborators) which opens the door to stable computation with kernels in their often most accurate close-to-flat state. By stable computation we mean not only the solution and evaluation of interpolation problems, but also corresponding tasks for partial differential equations, as well as estimation of "optimal" kernel parametrizations via criteria such a maximum likelihood estimation, cross-validation and minimization of the kriging variance (or power function).

Another relatively new idea presented in this book is the notion of a designer kernel, i.e., the construction of a kernel that is ideally suited for a certain kind of application. We investigate several such approaches such as the use of the generalized Sobolev spaces of Chapter 8 (which are closely related to the Green's kernels of Chapter 6), or building a kernel from its eigenfunctions and eigenvalues (see, e.g., Section 3.9 and Chapter 7). This latter idea of working with a kernel in series form — rather than in closed form — ties in nicely with the Hilbert–Schmidt framework.

A third important topic is the implementation of various types of kernels in MATLAB. Chapter 4 lays the groundwork upon which later applications build upon. We go considerably beyond the use of basic radial kernels (RBFs) in this book. In particular, we consider the use of anisotropic kernels (both in radial and in tensor product form), zonal kernels, compactly supported kernels, space-time kernels, and kernels given only in series form.

In Part II we apply the fundamental concepts developed in the first part to applications in data fitting, machine learning, the numerical solution of partial differential equations, and finance. These concepts are illustrated in the context of kernel methods with examples based on accessible "textbook" versions of MATLAB code. However, our `GaussQR` library (discussed in Appendix D) contains "production" versions of the code which is also accessible to the interested reader. Discussion of context that deals with computational cost is spread throughout the book and can be found by looking up this topic in the index.

This book contains 72 MATLAB programs, 76 figures, 11 tables, and more than 650 references). Some code requires one or more of the following MATLAB toolboxes: Curve Fitting Toolbox™, Optimization Toolbox™, Statistics and Machine Learning Toolbox™. All MATLAB programs printed in the text are contained — in an extended version — in the `GaussQR` library. The code was developed and tested on MATLAB R2015a, the most recent update prior to publication; however, earlier versions of the code were also used on older MATLAB releases.

The manuscript for this book is based on the lecture notes for an evolving course on Meshfree Methods taught by the first author every two years at the Illinois Institute of Technology. Moreover, many of the ideas presented in this book emerged from discussions and research performed in IIT's Meshfree Methods seminar, a group co-organized by the first author together with Fred Hickernell. We thank all the participants of this group for their contributions, discussions, and willingness to be a sounding board for our ideas. However, special thanks are due to Qi Ye, whose Ph.D. research on generalized Sobolev spaces formed the basis for Chapter 8, and to Fred Hickernell, who inspired us with many insightful comments and ideas, some of which found their way into this book (such as the Chebyshev kernels of Section 3.9.2 and some of the fundamental insights on kernel parametrizations presented in Chapter 14).

Thanks are also due to Bengt Fornberg and Natasha Flyer for welcoming the second author to their research seminars at CU–Boulder, as well as his colleagues at UC–Denver, most notably Jan Mandel, Loren Cobb and Troy Butler. We thank MathWorks and Naomi Fernandes at their book program for providing us with a complimentary copy of their software and for helping us with some issues that arose during beta-testing of the latest release. Finally, many thanks are due to Rajesh Babu, Rok Ting Tan, and Elena Nash (as well as all the unnamed people working in the background) at World Scientific Publishing Co. who helped make this project a success.

Greg Fasshauer and Mike McCourt
Chicago, IL, and Denver, CO, February 2015

Contents

PART 1

An Introduction to Kernel-Based Approximation Methods and Their Stable Computation

Chapter 1

Introduction

1.1 Positive Definite Kernels: Where Do They Fit in the Mathematical Landscape?

Let us begin by roughly outlining three general areas of study that are involved in our treatment of positive definite kernels.

Theory: The foundation for our work lies mostly in *functional, numerical and stochastic analysis* and deals with concepts such as reproducing kernel Hilbert spaces, Sobolev spaces, positive definite kernels, Hilbert–Schmidt integral operators, Mercer's theorem, Green's kernels, Sturm–Liouville eigenvalue problems, convergence analysis, alternate basis representations, and (Gaussian) random fields.

Computation: This component of our work reaches into *numerical linear algebra, computational statistics and computer science* and is concerned with issues such as parameter selection, stable, fast and efficient algorithms, regularization techniques, and the use of appropriate data structures — especially within MATLAB.

Applications: This segment is arguably the largest of these three areas and therefore we dedicate the entire second part of this book to applications. It covers problems such as basic data fitting (in both the deterministic and stochastic settings), the numerical solution of partial differential equations (both deterministic and stochastic), statistical or machine learning and classification, multivariate integration, multivariate optimization, engineering design (both in the sense of geometric design as well as in the sense of design of experiments), computer graphics, signal processing, finance, and many more. Our examples in Part II cover many, but certainly not all, of these topics.

Even though we have decided to describe the field of positive definite kernels using these three categories, the boundaries separating the areas are rather soft. In order for someone to make significant progress on any of the topics listed above, that person will almost certainly require at least some expertise in all three categories.

All of this literature provides excellent resources for additional background reading or exposure to alternative discussions of closely related material.

An indication of the important role positive definite kernels play in many different fields is provided by the following rapidly growing, and almost certainly incomplete, list of monographs. All of these books contain at least a significant portion that is concerned with positive definite kernels and/or reproducing kernel Hilbert spaces.

Analysis: Agler and McCarthy (2002), Akhiezer and Glazman (1993), Aubin (2000), Berg, Christensen and Ressel (1984), Bergman (1950), Bochner (1932), Meschkowski (1962), Moore (1935), Riesz and Sz.-Nagy (1955), Ritter (2000), Saitoh (1988, 1997), Wells and Williams (1976)

Approximation Theory: Buhmann (2003), Cheney and Light (1999), Dick and Pillichshammer (2010), Fasshauer (2007), Freeden, Gervens and Schreiner (1998), Golberg and Chen (1998), Iske (2004), Michel (2013), Novak and Woźniakowski (2008), Shapiro (1971), Wendland (2005)

Integral Equations: Cochran (1972), Hackbusch (1989), Hochstadt (1973), Kress (1999), Pogorzelski (1966), Smithies (1958), Stakgold (1979)

Mathematical Physics: Bergman and Schiffer (1953), Courant and Hilbert (1953), Morse and Feshbach (1953)

Engineering and Physics Applications: Ali, Antoine and Gazeau (2000), Atluri (2004), Atluri and Shen (2002), Belytschko and Chen (2007), Chen, Fu and Chen (2014b), Chen, Lee and Eskandarian (2006), Forrester, Sobester and Keane (2008), Ghanem and Spanos (2003), Li and Liu (2007), Liu (2002), Liu and Liu (2003), Sarra and Kansa (2009), Van Trees (2001)

Probability Theory and Statistics: Berlinet and Thomas-Agnan (2004), Gu (2013), Ramsay and Silverman (2005), Wahba (1990)

Geostatistics: Gandin (1963), Cressie (1993), Kitanidis (1997), Matérn (1986), Matheron (1965), Stein (1999), Wackernagel (2003)

Statistical/Machine Learning: Alpaydin (2009), Catoni (2004), Cristianini and Shawe-Taylor (2000), Cucker and Zhou (2007), Hastie, Tibshirani and Friedman (2009), Herbrich (2002), Joachims (2002), Rasmussen and Williams (2006), Schölkopf and Smola (2002), Shawe-Taylor and Cristianini (2004), Steinwart and Christmann (2008), Suykens, Van Gestel, De Brabanter, De Moor and Vandewalle (2002), Vapnik (1998)

In addition to these monographs, the reader should not forget the historical papers by Schmidt (1908), Mercer (1909), and Aronszajn (1950) as well as a few more modern survey articles such as those by Hille (1972), Stewart (1976), Dyn (1987, 1989), Powell (1987, 1992), Schaback (1999, 2000), Buhmann (2000), Cucker and Smale (2002), Schaback and Wendland (2006), Fasshauer (2011b), Ferreira and Menegatto (2013), or Scheuerer, Schaback and Schlather (2013). In the next section we will provide a more complete historical picture and mention more references.

1.2 A Historical Perspective

The study of *positive definite functions*, or — more generally — *positive definite kernels*, in the field of analysis began either with the work of Maximilian Mathias (1923), a student of Erhard Schmidt's at the University of Berlin or with that of James Mercer (1909), who was a Fellow of Trinity College at Cambridge University at that time. As pointed out in the survey by Stewart (1976),

> *Mathias and the other early workers with p.d. functions of a real variable were chiefly concerned with Fourier transforms and apparently did not realize that more than a decade previously Mercer and others had considered the more general concept of positive definite kernels $K(x, y)$ [...] in research on integral equations. I have likewise found that present-day mathematicians working with some of the manifestations of p.d. functions are unaware of other closely related ideas. Thus one of the purposes of this survey is to correlate some of the more important generalizations of p.d. functions with the hope of making them better known.*

Perhaps some of the most fundamental contributions, namely characterizations of positive definite functions in terms of Fourier transforms, were made a few years later by Salomon Bochner (1932), also a former student of Erhard Schmidt's and lecturer at the University of Munich at the time (and later at Princeton and Rice Universities), and by Isaac "Iso" Schoenberg (1938), an assistant professor at Colby College at that time (and later at the Universities of Pennsylvania and Wisconsin–Madison). These early contributions were used by Micchelli (1986) as the starting point for his proofs of the nonsingularity of the system matrices associated with kernel-based interpolation (see Section 1.3). Also in the 1930s, Aleksandr Khintchine (1934), a professor at Moscow University, used Bochner's theorem to establish the foundation for the study of *stationary stochastic processes* in probability theory.

It is difficult to put an exact date on the origins of the theory of *reproducing kernel Hilbert spaces*. Many people point to [Aronszajn (1950)] or [Bergman (1950)]. However, one should probably credit several people whose work over then span of more than 40 years produced the foundations laid out so comprehensively in [Aronszajn (1950); Bergman (1950)]. Reproducing kernels were indeed introduced by Nachman Aronszajn (1943, 1950), who was in England from 1940–45, and at Oklahoma A&M University from 1948 (and a professor at the University of Kansas from 1951). And reproducing kernels were also introduced by Stefan Bergman (1950), at the time at Harvard University (and later a professor at Stanford University). However, E. H. Moore (1916, 1935), a professor at the University of Chicago, referred to reproducing kernels as *positive Hermitian matrices* considerably earlier. In fact, Aronszajn (1950) gives much credit to Bergman, Mercer and Moore. Moreover, the first person to discuss the *reproducing property* — without developing a full theory — seems to have been the Polish mathematician Stanislaw Zaremba (1907, 1909), a professor at Jagiellonian University in Kraków (also referenced in [Aronszajn (1950)]), and Bergman's work on reproducing kernels already began in his Ph.D. thesis [Bergmann (1921, 1922)] at the University of Berlin under the direction of

Richard von Mises.

The *reproducing property* satisfied by a symmetric reproducing kernel K ensures that

$$\langle K(\boldsymbol{x},\cdot), f\rangle_{\mathcal{H}_K(\Omega)} = f(\boldsymbol{x}), \tag{1.1}$$

i.e., the reproducing kernel acts as a point evaluation functional for all functions $f \in \mathcal{H}_K(\Omega)$. Here $\mathcal{H}_K(\Omega)$, the reproducing kernel Hilbert space of K, is a *Hilbert space* of functions with domain Ω and $\langle\cdot,\cdot\rangle_{\mathcal{H}_K(\Omega)}$ denotes the associated inner product. We will provide more details for reproducing kernel Hilbert spaces in Section 2.3. The synonymous term *native space of* K was introduced by Robert Schaback (1999, 2000) in the context of radial basis functions.

Splines — seen as piecewise polynomial functions with prescribed global smoothness — were introduced and named by Schoenberg (1946a,b). However, Schumaker (1981) claims that Runge (1901) already worked on splines, and Schoenberg (1973) as well as Birkhoff and de Boor (1965) even attribute the idea of splines to Leonhard Euler (1755). In the general multivariate setting, possibly the first use of the term "spline" and also the foundations of a variational theory via Green's kernels interpreted as reproducing kernels in the sense of Aronszajn and Bergman are due to Marc Atteia (1966), who was at the Université Joseph Fourier in Grenoble at the time (and later at the Université Paul Sabatier in Toulouse). The connection between Green's kernels and reproducing kernels just mentioned — and discussed in detail in Chapter 6 — suggests that one could also pinpoint the origins of kernel methods to the work of George Green (1828).

While *radial basis functions* (RBFs) do not explicitly appear in [Atteia (1966)], other researcher introduced functions that are now classified as RBFs in their work in the 1970s. For example, Rolland Hardy (1971), a civil engineering professor at Iowa State University, introduced *multiquadrics*, and Harder and Desmarais (1972), worked in the aerospace industry in the United States with functions called *surface splines*. In Europe, Jean Duchon (1976, 1977), who still is a mathematics researcher at the Université Joseph Fourier in Grenoble, introduced similar functions called *thin plate splines* and — more generally — *polyharmonic splines*. The Belgian mathematician Jean Meinguet (1979) (Université Catholique de Louvain) was also an early contributor to the theory of surface splines. The "breakthrough" for radial basis functions came with a comprehensive study comparing many methods for multivariate scattered data interpolation by Richard Franke (1982) (and as a comprehensive report [Franke (1979)]). In this study he recommended the use of thin plate splines and multiquadrics over many other methods and this recommendation in turn led to work on the theoretical foundation for radial basis functions reflected in papers such as [Madych and Nelson (1983)] and [Micchelli (1986)]. It appears that the first use of the term "radial basis function" in a publication may have been made by Dyn and Levin (1983) (with related work in [Dyn and Levin (1981)]).

On the stochastic side, as already mentioned, the foundation for stochastic processes or random fields was provided by Aleksandr Khinchin. An infinite series representation of a stochastic process in terms of orthogonal functions akin to Mercer's theorem now known as the *Karhunen–Loève theorem* was provided by Kari Karhunen (1947), a mathematician at the University of Helsinki (and later a CEO for a Finnish insurance company) and Michel Loève (1977), a mathematician at the University of California–Berkeley. The South African mining engineer Danie Krige (1951) introduced a method for predicting gold ore distributions from a collection of ore samples which was later referred to as *kriging* by the French mathematician and geostatistician Georges Matheron (1965). It was Matheron who also provided the theoretical foundation for that approach. Even though the kriging method carries Krige's name, one should probably also mention the earlier work of Herman Wold (1938), a Norwegian-born statistician at Uppsala University, Andrey Kolmogorov (1962), a mathematician at Moscow State University, and Norbert Wiener (1949), a mathematician at Princeton University, all of whom discussed optimal linear prediction methods — albeit in the context of a one-dimensional time series setting, instead of the spatial setting of Krige (see [Cressie (1993)]). We will discuss kriging and how it fits into the general kernel-based setting in Chapter 5. The duality between reproducing kernel Hilbert spaces and corresponding spaces of random variables that is revealed in Chapter 5 goes back to the work of Emanuel Parzen (1961, 1970), a student of Michel Loève's and professor of statistics at Stanford University at the time. The book by Grace Wahba (1990), a statistician at the University of Wisconsin–Madison and a student of Parzen's, is one of the first monographs discussing reproducing kernels in the stochastic setting.

The use of reproducing kernels in the context of machine learning (as support vector machines or regularization networks) started with [Boser *et al.* (1992)] and is based on work of Vapnik and Lerner (1963).

1.3 The Fundamental Application: Scattered Data Fitting

Scattered data fitting is a basic problem in many mathematical disciplines including approximation theory, statistics and data modeling in general. The first and fundamental mathematical challenge is to ensure that the scattered data fitting problem (to be formulated shortly) has a *well-posed problem formulation* in the sense of Hadamard, i.e., that it not only *has* a solution, but that this solution is also unique and that small changes in the data lead only to small changes in the solution. Moreover, we want this to be true for *arbitrary space dimensions* and for an *arbitrary number of data points* that can be *placed at arbitrary locations.*

Before we give a proper mathematical definition of the problem, let us discuss what we mean in "plain English." We can, e.g., imagine that we are given a set of data (such as measurements along with locations at which these measurements were obtained), and we want to find a rule which allows us to deduce information about the process we are studying also at locations different from those at which we obtained our measurements.

Typical examples could be a series of measurements taken over a certain time period, or weather data collected at weather stations, or data obtained via some physical or computer experiment which involves many different input parameters (i.e., the problem is high-dimensional). For all of these examples our goal is then to produce a function s that faithfully represents the given data and that allows us to make predictions at other times, locations, or parameter settings. In some contexts such a function is referred to as a *surrogate model*, a *simulation metamodel*, or a *response surface*.

We should probably be a bit more specific about what we mean when we state that our model should "faithfully represent the given data". In particular, we may want the function s to exactly match the given measurements at the corresponding locations, or we may be satisfied if the function s merely approximately matches the given measurements at the corresponding locations. In the former case we would refer to the process as *(scattered data) interpolation*, while in the latter case it would be *(scattered data) approximation*.

Most of our discussions will focus on the interpolation setting. Moreover, we will usually assume that the data is provided *without error*, although we will also discuss approximation methods, which are more appropriate for data measurements that contain a certain amount of error or noise.

For a more precise mathematical description we need to set up some notation that we will be using throughout this book. First, we assume that the measurement locations (or *data sites*) are denoted by the set $\mathcal{X} = \{\boldsymbol{x}_1, \ldots, \boldsymbol{x}_N\} \subset \Omega \subset \mathbb{R}^d$. In the statistics literature and the literature on computer experiments the set $\mathcal{X}$ is usually referred to as the *design*. Here Ω is usually a bounded and connected domain whose boundary is at least Lipschitz. We will denote the measurements (or *data values*) corresponding to $\boldsymbol{x}_i$, $i = 1, \ldots, N$, by $y_i \in \mathbb{R}$. Moreover, we will frequently assume that the data values are obtained by sampling some (unknown) function f at the data sites, i.e., $y_i = f(\boldsymbol{x}_i)$, $i = 1, \ldots, N$. We are now ready to formulate the scattered data interpolation problem.

Problem 1.1 (Scattered Data Interpolation). *Given data* $\{(\boldsymbol{x}_i, y_i)\}_{i=1}^N$ *with* $\boldsymbol{x}_i \in \mathbb{R}^d$, $y_i \in \mathbb{R}$, *find a (continuous) function* s *such that* $s(\boldsymbol{x}_i) = y_i$, $i = 1, \ldots, N$.

A typical scattered data fitting result for geographical data (generated with C^2 Matérn kernels) is shown in Figure 1.1. The data locations $\boldsymbol{x}_i$ are $N = 530$ points scattered throughout the unit square, the corresponding y-values are height measurements at these locations indicated by vertical bars. The resulting interpolating surface is shown as well.

A convenient and common approach to obtaining the interpolant s is to assume that it is given by a linear combination of certain *basis functions* B_j, $j = 1, \ldots, N$,

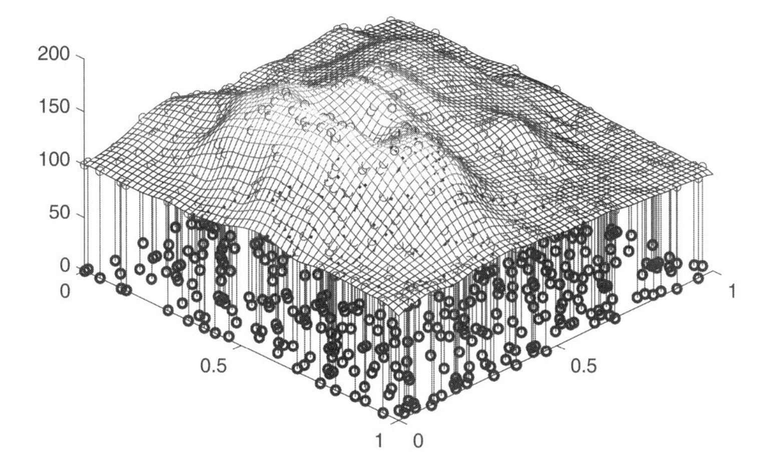

Fig. 1.1 A typical scattered data interpolant.

i.e.,

$$s(\boldsymbol{x}) = \sum_{j=1}^{N} c_j B_j(\boldsymbol{x}), \qquad \boldsymbol{x} \in \mathbb{R}^d. \tag{1.2}$$

Solving the interpolation problem under this assumption leads to a system of linear equations of the form

$$\mathsf{B}\boldsymbol{c} = \boldsymbol{y},$$

where the entries of the *interpolation matrix* B are given by $(\mathsf{B})_{i,j} = B_j(\boldsymbol{x}_i)$, $i, j = 1, \ldots, N$, $\boldsymbol{c} = \begin{pmatrix} c_1 & \cdots & c_N \end{pmatrix}^T$, and $\boldsymbol{y} = \begin{pmatrix} y_1 & \cdots & y_N \end{pmatrix}^T$.

The scattered data fitting problem will have a unique solution if and only if the matrix B is nonsingular. For $d = 1$ it is well known that one can interpolate to arbitrary data at N distinct data sites using a polynomial of degree $N - 1$. As we now show, if $d > 1$ the situation is no longer as straightforward.

1.3.1 *The Haar–Mairhuber–Curtis theorem: Why using kernels is a "natural" approach*

Haar systems (or Chebyshev/Tchebycheff [1] systems) play a fundamental role in the theory of function approximation by interpolation since existence of such a system guarantees uniqueness of an interpolant formed from such a system, and thus the first step toward having a well-posed problem formulation. A detailed introduction

[1]The spelling of Chebyshev's last name varies throughout the literature. We use *Chebyshev* throughout our discussion.

to Haar systems is presented, e.g., in [Schumaker (1981, Chapter 2)]. The following definition is taken from [Fasshauer (2007, Chapter 1)] and applies directly to the scattered data interpolation problem as formulated above.

Definition 1.1 (Haar system). *Let the finite-dimensional linear function space* $\mathcal{B} \subseteq C(\Omega)$ *have a basis* $\{B_1, \ldots, B_N\}$. *Then* $\mathcal{B}$ *is a* Haar space *on* Ω *if*

$$\det \mathsf{B} \neq 0$$

for any set of distinct $\boldsymbol{x}_1, \ldots, \boldsymbol{x}_N$ *in* Ω. *Here* B *is the matrix with entries* $(\mathsf{B})_{i,j} = B_j(\boldsymbol{x}_i)$. *The set* $\{B_1, \ldots, B_N\}$ *is called a* Haar system.

As mentioned above, for $d = 1$ and $\Omega = [a, b]$ the situation is straightforward. The following theorem paraphrases [Schumaker (1981, Theorem 2.21)].

Theorem 1.1. *A set* $\{B_1, \ldots, B_N\}$ *of continuous functions on* $[a, b]$ *forms a Haar system if and only if any nontrivial linear combination of* $B_1, \ldots, B_N$ *has at most* $N - 1$ *zeros in* (a, b).

As soon as $d > 1$, one can no longer ensure a unique interpolant if one chooses the basis independent of the data sites. This fact is implied by the *Haar–Mairhuber–Curtis theorem* [Haar (1918); Mairhuber (1956); Curtis (1959)].

Theorem 1.2 (Haar–Mairhuber–Curtis). *If* $\Omega \subset \mathbb{R}^d$, $d \geq 2$, *contains an interior point, then there exist no Haar spaces of continuous functions except for trivial ones, i.e., spaces spanned by a single function.*

Note that this theorem contains two notions of "dimension": the dimension d of the space in which the data sites lie, and the dimension N of the function space $\mathcal{B}$. The Haar–Mairhuber–Curtis theorem implies that one can not fix a nontrivial basis consisting of more than $N = 1$ function and hope that interpolation from that space at N arbitrary (distinct) points $\boldsymbol{x}_i \in \mathbb{R}^d$, $d \geq 2$, has a unique solution. However, if one selects the basis *after* the data sites are given, then there are no problems suggested by this theorem. This motivates the use of kernel-based methods in higher space dimensions since we can select the set $\mathcal{B} = \{K(\cdot, \boldsymbol{x}_1), \ldots, K(\cdot, \boldsymbol{x}_N)\}$ as a basis that is naturally adapted to the points $\mathcal{X} = \{\boldsymbol{x}_1, \ldots, \boldsymbol{x}_N\}$ so that (1.2) becomes

$$s(\boldsymbol{x}) = \sum_{j=1}^{N} c_j K(\boldsymbol{x}, \boldsymbol{x}_j) = \boldsymbol{k}(\boldsymbol{x})^T \boldsymbol{c}, \qquad \boldsymbol{x} \in \mathbb{R}^d, \tag{1.3a}$$

and the coefficients c_j are obtained by solving the linear system[2]

$$\mathsf{K}\boldsymbol{c} = \boldsymbol{y}. \tag{1.3b}$$

[2]At this point we want to alert the reader to our notational conventions. We use upright fonts (such as K) for matrices as in (1.3b), and we reserve standard italicized math fonts (such as K) for kernels, which are formally introduced in Section 2.1, as in (1.3a). Other quantities associated with the kernel K will use the same letter in a different font. For example, $\mathcal{K}$ will denote an integral operator associated with K, and $\boldsymbol{k}$ will represent a vector of copies of $K(\cdot, \boldsymbol{x}_j)$ as in (1.3c).

Having found the coefficients as $\boldsymbol{c} = \mathsf{K}^{-1}\boldsymbol{y}$, we can now evaluate the kernel interpolant (1.3a) as

$$s(\boldsymbol{x}) = \boldsymbol{k}(\boldsymbol{x})^T \mathsf{K}^{-1}\boldsymbol{y}, \tag{1.3c}$$

where the vector $\boldsymbol{k}(\boldsymbol{x})^T = \big(K(\boldsymbol{x},\boldsymbol{x}_1) \cdots K(\boldsymbol{x},\boldsymbol{x}_N)\big)$. This kernel approach can be placed on a solid theoretical foundation by using (strictly) *positive definite kernels* (defined in Section 2.1), so that the matrix K is positive definite, and therefore invertible.

Another possible choice of basis for well-posed higher-dimensional interpolation is provided by the *least polynomial spaces* of de Boor and Ron (1990, 1992a,b) with recent extensions by Narayan and Xiu (2012) (see Remark 10.4).

1.3.2 *Variations of scattered data fitting*

The reason we refer to the scattered data fitting problem as the *fundamental application* is that it is either directly involved in some of the other topics discussed in this book (such as kriging), or that it can be modified to deal with related problems (such as derivative estimation, Hermite interpolation and the numerical solution of PDEs).

To illustrate those connections, the remainder of Part I contains a summary of some of the theory on positive definite kernels and reproducing kernel Hilbert spaces, which includes Hilbert–Schmidt theory and the Hilbert–Schmidt integral eigenvalue problem whose eigenpairs provide us with the building blocks for the Mercer series representation of a positive definite kernel. Other topics mentioned in Chapter 2 include the Karhunen-Loève expansion and feature maps (topics which are popular in uncertainty quantification and machine learning). Chapters 3 and 4 contain an overview of many different types of kernels and their implementation in MATLAB: radial (both isotropic and anisotropic), translation invariant, in series form, as zonal functions, in tensor product form, with compact support, and in space-time form.

We spend all of Chapter 5 to discuss connections between the deterministic world in which we formulated the scattered data fitting problem above and a stochastic world based on Gaussian random fields. The stochastic interpretation not only allows us to describe the parallels between kernel-based scattered data interpolation and kriging, but it also provides deeper insights into the interpretations of optimality as viewed from the stochastic and deterministic perspectives. These insights will be of great importance to our development of kernel parametrization criteria in Chapter 14, which in turn are put to use in the applications discussed in Part II.

Another interesting theoretical connection exists between reproducing kernels and Green's kernels. We explore and extend this relationship in Chapters 6–8, leading to the concept of so-called *generalized Sobolev spaces.* We employ these spaces in Chapter 9 in the context of error estimates and sampling inequalities.

This discussion of accuracy and optimality sets up the last topic in Part I, the "rectification" of the long-standing misconception about the — presumably unavoidable — trade-off between accuracy and numerical stability. This perception was seen as a major obstacle holding back the widespread use of kernel methods. Our discussion of "flat" limits, alternate bases, and the Hilbert–Schmidt SVD in Chapters 10–13 is intended to show the reader that — with a little more sophisticated approach — kernel methods are a serious contender in the field of numerical methods and may indeed be a step closer to becoming "big technology" number four, as predicted in [Fasshauer (2007, Preface)].

1.4 Other Applications

In Part II of this book we discuss a variety of applications that both build upon and extend the basic concepts from the scattered data interpolation problem.

1.4.1 *Statistical data fitting*

In Chapters 5, 14 and 17 we first set up and then use kernel-based approximation methods viewed in the context of *Gaussian random fields*. Specific applications include a detailed discussion of the solution of the geostatistical data fitting example already shown in Figure 1.1, fitting of 7 and 8-dimensional data obtained by sampling standard test functions from the literature on experimental design, and fitting of Richard Franke's classical data sets of water temperature measurements given along tracks off the coast at Big Sur, California, and of elevation measurements provided along isolines of a glacier. Two other case studies involve a smoothing spline fit of environmental data collected near the Arimas River in Colorado, and a regression example based on the car data provided in MATLAB's Statistics and Machine Learning Toolbox.

The implemented data fitting methods can be interpreted either as deterministic kernel-based interpolation or smoothing spline methods, or as different variants of (stochastic) kriging methods. Due to the potential presence of noise and uncertainty in the data we prefer the latter interpretation (although there is no significant difference as far as the actual formulas used for these two interpretations go).

An important component of our MATLAB implementations is the inclusion of different parametrization strategies (such as cross-validation and maximum likelihood estimation) that are used to find "optimal" shape parameters, smoothness parameters, or regularization parameters. Chapter 14 focusses on the development of these parametrization criteria.

1.4.2 *Machine learning*

Classification and regression solutions of machine learning or statistical learning problems are featured in Chapter 18. As the reader can see from the long list of recent monographs on this topic, there is much specialized information available, and our main aim is to build a bridge between different fields of applications for kernel-based methods. We do this by providing examples for RBF network regression and for classification via support vector machines.

1.4.3 *Numerical solution of PDEs*

Kernel-based methods as discussed here are a natural next step from *radial basis functions* (see Section 3.1, where this special case of kernels is discussed). However, our focus on *positive definite kernels* leaves out the many *conditionally positive definite functions* such as the popular *multiquadrics* and *polyharmonic splines* (including *thin plate splines*). The multiquadric, in particular, was singled out in the earliest papers by Ed Kansa (1986, 1990b) on the use of radial basis functions for the numerical solution of PDEs. However, even though one can find many papers in the literature that refer to the "multiquadric method," the same techniques can be applied to a much wider class of (conditionally) positive definite kernels. In Section 20.1 we will pick up this topic under the heading of *collocation methods*. Radial basis functions in general — and often also with applications to PDEs — are discussed in, e.g., [Cheney and Light (1999); Buhmann (2003); Iske (2004); Wendland (2005); Fasshauer (2007); Forrester *et al.* (2008); Sarra and Kansa (2009); Chen *et al.* (2014b)].

Other kernel-based methods surveyed in Chapter 20 for the numerical solution of PDEs include kernel-based finite differences (often referred to as RBF-FD), and the methods of fundamental and particular solutions, MFS and MPS, respectively. A kernel-based coupling method for the solution of PDEs is introduced in Chapter 19 and then applied in Chapter 21 in the context of financial option pricing.

In this book we specialize on kernel-based approximation methods, and in particular on the use of positive definite reproducing kernels, which we will be discussing in detail in the coming chapters. Such methods are by their very nature *meshfree methods* — sometimes also referred to as *meshless methods* — since they do not require an underlying computational mesh such as the finite element method does. The discretization of the computational domain Ω is accomplished via the set of points $\mathcal{X}$ alone. Since our discussion leaves out many other numerical methods that are also referred to as meshfree methods we want to briefly mention some of those methods and at least provide some references for the interested reader. Since most meshfree methods have been developed with the numerical solution of partial differential equations (PDEs) in mind, this is the best place to present this brief overview.

The list of related methods that can be found in the literature and on the Internet appears to be endless. Each of these is presented as a separate method with its own name and special acronym. Meshfree methods appear under such names as the *diffuse element method* (DEM) due to Nayroles *et al.* (1992), the *element-free Galerkin method* (EFG) due to Belytschko *et al.* (1994), *hp-clouds* due to Duarte and Oden (1996a,b), the *meshless local Petrov–Galerkin method* (MLPG) due to Atluri and Zhu (1998) (see also [Atluri and Shen (2002); Atluri (2004)]), the *reproducing kernel particle method* (RKPM) due to Liu *et al.* (1995), or *smoothed particle hydrodynamics* (SPH) due to Gingold and Monaghan (1977) and Lucy (1977) (see also [Liu and Liu (2003)]).

All of these methods are in some sense derivatives of the *moving least squares method* (MLS) which was introduced by Lancaster and Šalkauskas (1981) (see also [Fasshauer (2007)]). This connection was pointed out, e.g., in [Belytschko *et al.* (1996)]. The textbook [Chen *et al.* (2006)] provides an overview of many of these meshfree (or meshless) methods in the context of solid mechanics applications.

The *natural element method* (NEM) due to Sukumar *et al.* (1998) uses the concept of natural neighbor coordinates introduced by Sibson (1981).

The *method of fundamental solutions* (MFS) and *method of particular solutions* (MPS) due to Kupradze and Aleksidze (1964) (see also, e.g., [Fairweather and Karageorghis (1998); Chen *et al.* (2014b)]) are numerical methods that use Green's kernels to reduce the dimensionality of the PDE problem by one by turning it into a boundary only problem. As mentioned above, MFS and MPS will be discussed in Chapter 20.

The *extended finite element method* (XFEM) by Moës *et al.* (1999), the *partition of unity finite element method* (PUFEM) by Melenk and Babuška (1996), the *smoothed finite element method* (S-FEM) [Liu and Trung (2010)] and *smoothed point interpolation method* (S-PIM) [Liu and Zhang (2013)] all couple aspects of the finite element method with some features of meshfree methods — often with the help of the *partition of unity method* (PoUM).

The *radial point interpolation method* (RPIM) is discussed in [Liu (2002)]. For those readers interested in an even wider variety of methods, the Wikipedia page for "Meshfree Methods" contains a much more extensive list, and the paper by Nguyen *et al.* (2008) provides a survey of a large portion of these methods along with issues related to MATLAB implementation. For a recent general discretization framework that applies to many of these methods see [Schaback (2014)].

1.4.4 *Computational finance*

The use of radial basis functions in computational finance — in particular for option pricing problems formulated via the *Black–Scholes equation* — was pioneered by Hon and Mao (1999) and then later also promoted by Fasshauer *et al.* (2004), Khaliq *et al.* (2008), and especially by Larsson *et al.* (2008, 2013a) and Pettersson *et al.* (2008). In Chapter 21 we present examples where kernels serve a useful purpose in computing important financial quantities. We also study the relevance of some of the ideas developed earlier in this book, such as the kernel-based coupling technique of Section 19.4, to these topics in computational finance.

1.5 Topics We Do Not Cover

In addition to the meshfree methods mentioned in Section 1.4.3, which are used mostly for the numerical solution of PDEs, there are a number of topics for which we only scratch the surface, and others we do not discuss at all.

For example, while we do cover various approaches to solving partial differential equations, we do not discuss any of the recent work on *stochastic partial differential equations* as introduced in [Cialenco *et al.* (2012); Fasshauer and Ye (2013a, 2014)].

We provide examples of radial and translation-invariant kernels in Chapter 3 and also discuss their implementation in MATLAB in Chapter 4, but we do not get into any of the theory that applies specifically to radial or translation-invariant kernels (such as, e.g., *Bochner's theorem*). The reader interested in learning more about these special cases is referred to, e.g., [Buhmann (2003); Wendland (2005); Fasshauer (2007)].

Conditionally positive definite kernels are mentioned here and there throughout the book, but we never even define what they are (the closest thing to a definition in this book is (5.20)). Neither do we spend much time discussing advanced forms of kriging (such as universal or intrinsic kriging). A detailed discussion of conditionally positive definite radial functions is presented in [Wendland (2005)], and Scheuerer, Schaback and Schlather (2013) give a side-by-side comparison of (conditionally) positive definite kernels with various forms of kriging.

In Chapters 3 and 15 we spend a little time on interpolation on the unit sphere $\mathbb{S}^2$ of $\mathbb{R}^3$. Interpolation or approximation on other spheres or on other more general manifolds is not discussed in this book. We refer the reader to some of the more specialized literature mentioned in Chapters 3 and 15.

Similar as for radial kernels, we do not get into the theory behind compactly supported kernels, nor do we do much computation with them. Some examples of kernels are given in Chapter 3 and their basic MATLAB implementation is discussed in Chapter 4.

This book contains no specialized treatment of *vector-* or *matrix-valued* kernels. The reader who wants to learn more about these interesting and powerful methods is referred to some of the original literature such as, e.g., [Narcowich and Ward (1994); Fuselier (2008b,a); Fuselier and Wright (2009)] with some MATLAB code available in [Fuselier (2015)].

Iterative and *adaptive algorithms* (such as those employing spatially varying kernel shape parameters) play an important role for kernel approximation. However, except for the adaptive computation of Newton basis functions in Section 12.1.4, we have decided not to include them in this book. A number of algorithms are discussed, e.g., in the books [Wendland (2005); Fasshauer (2007)].

Kernel-based optimization is another interesting topic we did not have space for in this book. Some references include [Gutmann (2001); Jones (2001); McDonald *et al.* (2007); Oeuvray and Bierlaire (2009); Pearson (2013)].

Chapter 2

Positive Definite Kernels and Reproducing Kernel Hilbert Spaces

In this chapter we provide an overview of the fundamental theoretical concepts — mostly from functional analysis — that are needed to make sense of the more computational and algorithmic material in the later parts of this book. We introduce the notion of a positive definite kernel, provide a basic review of Hilbert–Schmidt theory including Mercer's theorem, and also discuss the Karhunen–Loève expansion theorem, which plays a central role for the stochastic framework in Chapter 5 as well as for many applications in uncertainty quantification. In the second half of the chapter some basic properties of reproducing kernel Hilbert spaces, including connections to feature maps, are stated and some background on the eigenvalue problem for integral operators is provided along with a few examples. Some basic facts from functional analysis that may support the discussion in this chapter are provided in Appendix C.

2.1 Positive Definite Kernels

In Section 1.3 we saw that the solution of the kernel-based scattered data interpolation problem amounts to solving a linear system (cf. (1.3b))

$$\mathsf{K}\boldsymbol{c} = \boldsymbol{y},$$

where $(\mathsf{K})_{i,j} = K(\boldsymbol{x}_i, \boldsymbol{x}_j)$, $i, j = 1, \ldots, N$. Linear algebra tells us that this system will have a unique solution whenever K is nonsingular. Since the determination of precise necessary and sufficient conditions to characterize this general nonsingular case still poses an open problem we will focus mostly on kernels K that generate *positive definite matrices*, which are known to be nonsingular since all of their eigenvalues are positive.

Therefore, before we delve into possibly less-familiar concepts from analysis, we begin by presenting the well-known definition of a positive definite matrix K as it can be found in just about any book on linear algebra (see, e.g., [Horn and Johnson (2013, Chapter 4)]) and then proceed by relating this idea to the concept of positive definite kernels.

Definition 2.1 (Positive definite matrix). *A real symmetric* $N \times N$ matrix K *is called* positive definite *if its associated quadratic form is positive for any nonzero coefficient vector* $\boldsymbol{c} = \begin{pmatrix} c_1 & \cdots & c_N \end{pmatrix}^T \in \mathbb{R}^N$, *i.e.,*

$$\boldsymbol{c}^T \mathsf{K} \boldsymbol{c} > 0.$$

The matrix is called positive semi-definite *if the quadratic form is allowed to be nonnegative.*

For the purposes of this book a *kernel* K is nothing but a real-valued *function of two variables*, i.e.,

$$K : \Omega \times \Omega \to \mathbb{R}, \qquad K : (\boldsymbol{x}, \boldsymbol{z}) \mapsto K(\boldsymbol{x}, \boldsymbol{z}).$$

Here Ω is usually a subset of $\mathbb{R}^d$, but it may also be a rather general set as happens frequently in machine learning applications. Other possible domains Ω include spheres or more general Riemannian manifolds (see, e.g., [Xu and Cheney (1992); Narcowich (1995); Hangelbroek *et al.* (2010); Fuselier and Wright (2012)]), or even locally compact groups as, e.g., in [Gutzmer (1996); Hangelbroek and Schmid (2011)]. More examples of such general settings for positive definite kernels are also reported in the survey paper by Stewart (1976).

A positive definite kernel K can be viewed as an infinite-dimensional positive semi-definite[1] matrix K. In fact, this can be done in two different ways. First, in the sense of Mathias (1923), where we assume that the matrix K is generated by the kernel K in the following sense.

Definition 2.2 (Positive definite kernel). *A symmetric* kernel K *is called* positive definite on $\Omega \times \Omega$ *if its associated* kernel matrix K *with entries* $(\mathsf{K})_{i,j} = K(\boldsymbol{x}_i, \boldsymbol{x}_j)$, $i, j = 1, \ldots, N$, *is positive semi-definite* for any $N \in \mathbb{N}$ *and* for any *set* $\mathcal{X} = \{\boldsymbol{x}_1, \ldots, \boldsymbol{x}_N\} \subset \Omega$ *of distinct points.*

Many generalizations of this basic definition exist. For example, the coefficient vector $\boldsymbol{c} = \begin{pmatrix} c_1 & \cdots & c_N \end{pmatrix}^T$ in Definition 2.1 as well as the kernel K in Definition 2.2 may be allowed to be complex (as already assumed in the work of Bochner (1932)), or the kernel may be matrix-valued as is desired for recent applications in fluid dynamics, where one may want to ensure that the kernel is divergence-free by construction (see, e.g., [Schräder and Wendland (2011)] and the earlier fundamental work by Narcowich and Ward (1994) as well as by Lowitzsch (2005) and by Fuselier (2008b,a); Fuselier *et al.* (2008); Fuselier and Wright (2009)).

[1] In order to align the notion of positive definiteness of kernels with that of matrices we would have to refer to a *strictly positive definite kernel* when we want to disallow the possibility of a zero eigenvalue. This confusion in terminology has its origins in the analysis literature of the early 20th century. The exposition in [Fasshauer (2007)] strictly maintains this distinction. While Definition 2.2 and the theorems to follow adhere to traditional nomenclature, we have decided to be more relaxed in our later discussions and generically refer to a kernel as positive definite — even if the strict notion applies. Whenever zero eigenvalues matter we specifically stress the fact.

The second, alternative, definition of a positive definite kernel comes essentially from the work of Mercer (1909), who was concerned with integral operators. It also represents an infinite-dimensional (and continuous) generalization of the finite-dimensional discrete quadratic form.

Definition 2.3 (Integrally positive definite kernel). *A symmetric* kernel K *is called* integrally positive definite on $\Omega \times \Omega$ *if*

$$\int_{\Omega}\int_{\Omega} K(\boldsymbol{x},\boldsymbol{z})f(\boldsymbol{x})f(\boldsymbol{z})\,\mathrm{d}\boldsymbol{x}\,\mathrm{d}\boldsymbol{z} \geq 0$$

for all $f \in L_2(\Omega)$.

In order to connect this definition with the previous one from Definition 2.2 we introduce the notion of a *positive self-adjoint operator* in analogy to the positive definiteness of a symmetric matrix. The following definition can be found, e.g., in [Hochstadt (1973, Section 3.5)].

Definition 2.4 (Positive operator). *A self-adjoint operator* $\mathcal{K}$ *acting on a Hilbert space* $\mathcal{H}$ *is called* positive *if* $\langle \mathcal{K}f, f\rangle_{\mathcal{H}} \geq 0$ *for all* $f \in \mathcal{H}$.

So if we define the operator $\mathcal{K}$ by $(\mathcal{K}f)(\boldsymbol{x}) = \int_{\Omega} K(\boldsymbol{x},\boldsymbol{z})f(\boldsymbol{z})\,\mathrm{d}\boldsymbol{z}$, $f \in L_2(\Omega)$, and use the standard L_2 inner product, i.e., $\langle f, g\rangle_{\mathcal{H}} = \langle f, g\rangle_{L_2(\Omega)} = \int_{\Omega} f(\boldsymbol{x})g(\boldsymbol{x})\,\mathrm{d}\boldsymbol{x}$, then the quadratic form in Definition 2.3 perfectly matches Definition 2.4. Therefore, an integrally positive definite kernel is nothing but the kernel of a positive integral operator.

Bochner (1933) showed that the notions of positive definiteness as stated in Definitions 2.2 and 2.3 are equivalent for continuous kernels. In addition, we remind the reader that Bochner is particularly well-known for the celebrated theorem that carries his name and which characterizes continuous positive definite translation-invariant kernels[2] as Fourier transforms of finite nonnegative Borel measures (see Bochner (1933), and also, e.g., [Wendland (2005, Chapter 6), Fasshauer (2007, Chapter 3)]).

Positive integral operators play an essential role for just about everything we do in this book, and we will see again in Section 2.2 — when we discuss the work of Mercer and Schmidt on integral operators and eigenfunction expansions of positive definite kernels — that a continuous positive definite kernel K can be characterized as one whose corresponding integral operator $\mathcal{K}$ is positive.

[2]A translation-invariant kernel is of the form $K(\boldsymbol{x},\boldsymbol{z}) = \widetilde{K}(\boldsymbol{x}-\boldsymbol{z})$ (see Section 3.2), and since $\widetilde{K}$ is no longer a kernel, i.e., a function of two variables, Bochner referred to such kernels as *positive definite functions*.

2.2 Hilbert–Schmidt, Mercer and Karhunen–Loève Series

Series expansions of positive definite kernels will play a crucial role throughout the remainder of this book. We therefore now introduce the basic ideas needed for their theoretical foundation. The notation and terminology we use is that found in modern textbooks on functional analysis such as, e.g., [Cheney (2001); Hunter and Nachtergaele (2001); Rynne and Youngson (2008); Atkinson and Han (2009); Brezis (2010)]. In these books the reader will find many more details (including proofs).

Historically, the development in this section is based largely on the seminal work of Ivar Fredholm (1903), David Hilbert (1904), Erhard Schmidt (1907) and James Mercer (1909) in the deterministic setting and on that of Kari Karhunen (1947) and Michel Loève (1977) in the stochastic setting. For those readers not able to consult the original papers written in French and German, an annotated translation of the first three papers quoted above is available [Stewart (2011)]. Interestingly, perhaps, both Schmidt and Karhunen performed this work as part of their Ph.D. theses.

2.2.1 *Hilbert–Schmidt operators*

Definition 2.5 (Hilbert–Schmidt operator). *Let $\mathcal{H}$ be a Hilbert space and $\mathcal{T} : \mathcal{H} \to \mathcal{H}$ a bounded linear operator. The operator $\mathcal{T}$ is called a* Hilbert–Schmidt operator *if there is an orthonormal basis $\{e_n\}$ in $\mathcal{H}$ such that*

$$\sum_{n=1}^{\infty} \|\mathcal{T}e_n\|_{\mathcal{H}}^2 < \infty.$$

Here $\|\cdot\|_{\mathcal{H}}$ denotes the norm in $\mathcal{H}$ induced by its inner product $\langle\cdot,\cdot\rangle_{\mathcal{H}}$.

It is common to refer to $\|\mathcal{T}\|_{HS}^2 = \sum_{n=1}^{\infty} \|\mathcal{T}e_n\|_{\mathcal{H}}^2$ as the *Hilbert–Schmidt norm* of $\mathcal{T}$. In the finite-dimensional case, i.e., when $\mathcal{T}$ is a matrix, the Hilbert–Schmidt norm turns into the Frobenius norm. Note that if $\mathcal{H}$ is separable then it has a countable orthonormal basis (and vice versa). It can be shown that every Hilbert–Schmidt operator $\mathcal{T}$ is compact. Then $\mathcal{H}$ has a basis consisting of eigenfunctions/-vectors of $\mathcal{T}$.

For later use we also define the *trace* of a bounded linear operator on $\mathcal{H}$. It is given by

$$\operatorname{trace} \mathcal{T} = \sum_{n=1}^{\infty} \langle \mathcal{T}e_n, e_n \rangle_{\mathcal{H}}. \tag{2.1}$$

Sometimes the Hilbert–Schmidt norm is expressed in terms of the trace. If $\mathcal{T}^*$ is the *adjoint* of $\mathcal{T}$ (see Appendix C) then

$$\|\mathcal{T}\|_{HS} = \sqrt{\operatorname{trace}(\mathcal{T}^*\mathcal{T})}. \tag{2.2}$$

An important property of any Hilbert–Schmidt operator is that it has finite trace. Moreover, any compact and self-adjoint $\mathcal{T}$, i.e., $\mathcal{T} = \mathcal{T}^*$, has real eigenvalues

λ_n, and

$$\operatorname{trace} \mathcal{T} = \sum_{n=1}^{\infty} \lambda_n, \tag{2.3}$$

where the eigenvalues are used according to their multiplicities. If the series of eigenvalues converges absolutely, then $\mathcal{T}$ has a finite trace, or is a *trace-class* operator. In fact, all compact self-adjoint trace-class operators are characterized by this property (see also the discussion of the spectral theory of compact self-adjoint operators below).

The following theorem (and its proof) on Hilbert–Schmidt operators and their kernels can be found with varying degrees of generality in, e.g., [Hochstadt (1973, Theorem 23), König (1986, Chapter 1), Bump (1998, Chapter 2), Brezis (2010, Chapter 6)]. Some of the concepts mentioned in the theorem are defined in Appendix C.

Theorem 2.1 (Hilbert–Schmidt integral operator). *Let $\mathcal{H} = L_2(\Omega, \mu)$ be a Hilbert space on a locally compact Hausdorff space Ω with positive Borel measure μ. Further, let the kernel $K : (\boldsymbol{x}, \boldsymbol{z}) \mapsto K(\boldsymbol{x}, \boldsymbol{z})$ be in $L_2(\Omega \times \Omega, \mu \times \mu)$, i.e., assume that*

$$\int_\Omega \int_\Omega |K(\boldsymbol{x}, \boldsymbol{z})|^2 \, \mathrm{d}\mu(\boldsymbol{x}) \, \mathrm{d}\mu(\boldsymbol{z}) < \infty.$$

Then the operator $\mathcal{K}$ defined by

$$(\mathcal{K}f)(\boldsymbol{x}) = \int_\Omega K(\boldsymbol{x}, \boldsymbol{z}) f(\boldsymbol{z}) \, \mathrm{d}\mu(\boldsymbol{z}), \qquad f \in L_2(\Omega, \mu), \tag{2.4}$$

is a Hilbert–Schmidt operator.

Conversely, every Hilbert–Schmidt operator on $L_2(\Omega, \mu)$ is of the form (2.4) for some unique kernel $K : (\boldsymbol{x}, \boldsymbol{z}) \mapsto K(\boldsymbol{x}, \boldsymbol{z})$ in $L_2(\Omega \times \Omega, \mu \times \mu)$.

Remark 2.1. The kernel K in Theorem 2.1 is called a *Hilbert–Schmidt kernel*. While this theorem is quite general, we will usually consider $\Omega \subseteq \mathbb{R}^d$ and work with a positive *weight function* ρ such that $\int_\Omega \rho(\boldsymbol{x}) \, \mathrm{d}\boldsymbol{x} = 1$, i.e., $\mathrm{d}\mu(\boldsymbol{x}) = \rho(\boldsymbol{x}) \, \mathrm{d}\boldsymbol{x}$.

In the following three examples (see [Stakgold (1979, Chapter 6)]) we consider the weight function $\rho(x) = 1$.

Example 2.1. The singular kernel

$$K(x, z) = \frac{1}{|x - z|^\beta}, \qquad x, z \in [0, 1],$$

generates a Hilbert–Schmidt operator if $0 < \beta < \frac{1}{2}$, but not for $\frac{1}{2} \leq \beta < 1$ (even though the operator is still compact).

Example 2.2. The Matérn kernel $K(x,z) = \mathrm{e}^{-|x-z|}$ on $\mathbb{R} \times \mathbb{R}$ (see Section 3.1) does not generate a Hilbert–Schmidt operator, nor is the integral operator compact. That is why we introduce an appropriate weight function in Example 2.5. The kernel $K(x,z) = \mathrm{e}^{-(x^2+z^2)}$ on $\mathbb{R}\times\mathbb{R}$ is Hilbert–Schmidt, but the Gaussian kernel $K(x,z) = \mathrm{e}^{-|x-z|^2}$ produces an unbounded integral and therefore requires a nontrivial weight function in order to become Hilbert–Schmidt (see Section 12.2.1).

Example 2.3. If we instead consider the Matérn kernel $K(x,z) = \mathrm{e}^{-|x-z|}$ on $[-1,1] \times [-1,1]$ with boundary conditions $K'(-1,z) = K(-1,z)$ and $K'(1,z) = -K(1,z)$ (for a fixed $z \in [-1,1]$) then we can obtain eigenvalues and eigenfunctions by solving the differential eigenvalue problem (see [Stakgold (1979, Chapter 6)])

$$-\varphi''(x) + \varphi(x) = \frac{1}{\lambda}\varphi(x), \qquad \varphi'(-1) = \varphi(-1) \text{ and } \varphi'(1) = -\varphi(1).$$

The eigenvalues and eigenfunctions of the slightly more general Matérn kernel $K(x,z) = \mathrm{e}^{-\varepsilon|x-z|}$, $\varepsilon > 0$, on $[-L, L]$ were derived in [Van Trees (2001, Chapter 3)] by solving the corresponding Hilbert–Schmidt integral eigenvalue problem discussed next. They are stated in Appendix A.2.2. For $\varepsilon = L = 1$ one recovers the solution to the above boundary value problem.

2.2.2 *The Hilbert–Schmidt eigenvalue problem*

The Hilbert–Schmidt integral eigenvalue problem on $L_2(\Omega, \rho)$ can be viewed as a *homogeneous Fredholm integral equation of the second kind*, i.e., for appropriate *eigenvalues* λ and *eigenfunctions* φ we have by (2.4)

$$\int_\Omega K(\boldsymbol{x},\boldsymbol{z})\varphi(\boldsymbol{z})\rho(\boldsymbol{z})\,\mathrm{d}\boldsymbol{z} = \lambda\varphi(\boldsymbol{x}) \quad \Longleftrightarrow \quad (\mathcal{K}\varphi)(\boldsymbol{x}) = \lambda\varphi(\boldsymbol{x}). \tag{2.5}$$

Using the ρ-weighted L_2 inner product

$$\langle f, g\rangle_{L_2(\Omega,\rho)} = \int_\Omega f(\boldsymbol{x})g(\boldsymbol{x})\rho(\boldsymbol{x})\,\mathrm{d}\boldsymbol{x}, \tag{2.6}$$

we can also write (2.5) as

$$\langle K(\boldsymbol{x},\cdot), \varphi\rangle_{L_2(\Omega,\rho)} = \lambda\varphi(\boldsymbol{x}),$$

which is reminiscent of the reproducing property (1.1), but of course applies only to the eigenfunctions of $\mathcal{K}$ so that one should not confuse the two.

L_2-orthonormality of the eigenfunctions plays an important role in Mercer's theorem (see Theorem 2.2). In this setting it means

$$\langle \varphi_m, \varphi_n\rangle_{L_2(\Omega,\rho)} = \int_\Omega \varphi_m(\boldsymbol{x})\varphi_n(\boldsymbol{x})\rho(\boldsymbol{x})\,\mathrm{d}\boldsymbol{x} = \delta_{mn}.$$

If we assume that K is a reproducing kernel (to be discussed in detail in Section 2.3), then we can use its reproducing property (1.1) and compute

$$\begin{aligned}\langle \mathcal{K}f, g\rangle_{\mathcal{H}_K(\Omega)} &= \left\langle \int_\Omega K(\cdot, \boldsymbol{z})f(\boldsymbol{z})\rho(\boldsymbol{z})\,\mathrm{d}\boldsymbol{z}, g\right\rangle_{\mathcal{H}_K(\Omega)} \\ &= \int_\Omega \langle K(\cdot,\boldsymbol{z}), g\rangle_{\mathcal{H}_K(\Omega)} f(\boldsymbol{z})\rho(\boldsymbol{z})\,\mathrm{d}\boldsymbol{z} \\ &= \int_\Omega g(\boldsymbol{z})f(\boldsymbol{z})\rho(\boldsymbol{z})\,\mathrm{d}\boldsymbol{z} \\ &= \langle f, g\rangle_{L_2(\Omega,\rho)}. \end{aligned} \tag{2.7}$$

The equality just derived implies that $\mathcal{K}$ can be interpreted as the adjoint of the operator that continuously embeds the reproducing kernel Hilbert space $\mathcal{H}_K(\Omega)$ into $L_2(\Omega, \rho)$ (see also [Wendland (2005, Proposition 10.28)]). Since $\mathcal{K}$ is self-adjoint we know that this embedding operator is $\mathcal{K}$ itself. In other words, the reproducing kernel Hilbert space $\mathcal{H}_K(\Omega)$ is a subspace of $L_2(\Omega, \rho)$.

If we now let $f = \varphi_m$ and $g = \varphi_n$ in (2.7) and employ the eigenvalue relation $\mathcal{K}\varphi_m = \lambda_m\varphi_m$ we see that

$$\langle \varphi_m, \varphi_n\rangle_{\mathcal{H}_K(\Omega)} = \begin{cases} 0, & m \neq n, \\ \dfrac{1}{\lambda_m}, & m = n, \end{cases} \tag{2.8}$$

i.e., the *eigenfunctions are also orthogonal in the reproducing kernel Hilbert space* $\mathcal{H}_K(\Omega)$ *of the kernel.*

Now the spectral theory is similar as in the familiar finite-dimensional case, i.e., the eigenvalues of a compact self-adjoint operator $\mathcal{K}$ are real and the eigenfunctions associated with different eigenvalues are orthogonal (see, e.g., [Stakgold (1979, Section 6.3)]). Therefore we can use the spectral theorem (see Appendix C) to represent the compact Hilbert–Schmidt operator $\mathcal{K}$ in terms of its eigenvalues and eigenfunctions, i.e.,

$$\mathcal{K}f(\boldsymbol{x}) = \sum_{n=1}^{\infty} \lambda_n \langle f, \varphi_n\rangle_{L_2(\Omega,\rho)}\varphi_n(\boldsymbol{x}), \qquad f \in L_2(\Omega,\rho).$$

Since this identity holds for arbitrary L_2 functions f, one might hope for a series representation of the kernel K itself in terms of the eigenvalues and eigenfunctions. Mercer's theorem, discussed in the next subsection, shows that such a representation indeed exists.

Remark 2.2. It will be important in Chapter 6 that there is an *inverse relation* between the integral operator $\mathcal{K}$ and a differential operator $\mathcal{L}$ which has K as its Green's kernel. This relation was already alluded to in Example 2.3. All regular ordinary differential operators have compact inverse integral operators. Moreover, if the differential operator is self-adjoint, so is the inverse integral operator (see, e.g., [Courant and Hilbert (1953, Chapter V), Cochran (1972, Appendix A), Hochstadt

(1973, Chapter 3), Akhiezer and Glazman (1993, Appendix II)]). As a special case, Sturm–Liouville eigenvalue problems are inverse to integral eigenvalue problems for compact integral operators. We will discuss this connection in more detail in Section 6.6.

As a consequence, we will see in Chapter 6 that both operators have the same eigenfunctions, and that the eigenvalues of the differential operator are reciprocals of the eigenvalues of the integral operator.

2.2.3 *Mercer's theorem*

Mercer's theorem [Mercer (1909)] provides the infinite series representation of a positive definite kernel mentioned above during our application of the spectral theorem. In the classical literature it is discussed, e.g., in [Courant and Hilbert (1953); Riesz and Sz.-Nagy (1955)]. The version below which holds on any locally compact Hausdorff space is essentially taken from [Rasmussen and Williams (2006, Chapter 4)] (see also [König (1986, Chapter 3)], which contains a proof). A modern discussion of Mercer's theorem with a number of generalizations can be found in [Sun (2005); Steinwart and Scovel (2012)].

Theorem 2.2 (Mercer's theorem). *Let (Ω, μ) be a locally compact Hausdorff space with positive Borel measure μ and $K \in L_2(\Omega \times \Omega, \mu \times \mu)$ be a kernel such that the integral operator $\mathcal{K} : L_2(\Omega, \mu) \to L_2(\Omega, \mu)$ defined by (cf. (2.4))*

$$(\mathcal{K}f)(\boldsymbol{x}) = \int_\Omega K(\boldsymbol{x}, \boldsymbol{z}) f(\boldsymbol{z})\, \mathrm{d}\mu(\boldsymbol{z})$$

is positive[3]. *Let $\varphi_n \in L_2(\Omega, \mu)$, $n = 1, 2, \ldots$, be the $L_2(\Omega, \mu)$ orthonormal eigenfunctions of $\mathcal{K}$ associated with the eigenvalues $\lambda_n > 0$. Then the following are true:*

(1) *The kernel has a* Mercer expansion

$$K(\boldsymbol{x}, \boldsymbol{z}) = \sum_{n=1}^{\infty} \lambda_n \varphi_n(\boldsymbol{x}) \varphi_n(\boldsymbol{z}) \tag{2.9}$$

which holds μ^2 almost everywhere, and converges absolutely and uniformly μ^2 almost everywhere.

(2) *The eigenvalues $\{\lambda_n\}_{n=1}^{\infty}$ are absolutely summable, and so $\mathcal{K}$ has finite trace.*

Item (2) follows easily from the Mercer series using term-by-term integration, which is permitted since the series converges absolutely and uniformly. A transparent proof of the other part of Mercer's theorem for a continuous kernel K in the case $\Omega = [0, 1]$ and $\mathrm{d}\mu(\boldsymbol{z}) = \mathrm{d}z$ can be found in [Hochstadt (1973, pg. 90)].

[3] Alternatively, we could say that the kernel K is positive definite.

Remark 2.3. The difference between Mercer's theorem and a slightly earlier version by Schmidt (1907) is that Mercer's theorem guarantees *uniform convergence* of the series (2.9) *provided $\mathcal{K}$ is a positive operator* (cf. Definition 2.4), while Schmidt's result establishes *only L_2 convergence* of the series, but *for arbitrary compact self-adjoint operators $\mathcal{K}$*. Following general conventions in the literature, we will refer to the series expansion of K interchangeably as Mercer series or *Hilbert–Schmidt series* since our kernels are generally positive definite kernels (and thus $\mathcal{K}$ is positive).

Remark 2.4. The rate of decay of the Hilbert–Schmidt eigenvalues determines the smoothness of the kernel K: *the faster the eigenvalues decay, the smoother the kernel* (and vice versa). More specifically, if the eigenvalues decay at an *algebraic* rate of $\mathcal{O}(n^{-\beta+1+\tau})$ with $\beta \in \mathbb{N}_0$ and arbitrarily small $\tau > 0$, then the kernel will be a *finite smoothness kernel* in C^β, and if the eigenvalues decay *geometrically*, then the kernel will be *infinitely smooth*, even analytic (see, e.g., [Reade (1983); Little and Reade (1984); Reade (1984); Ferreira and Menegatto (2009)]). Moreover, we show in Chapter 9 that the smoothness of the kernel determines the rate of convergence of the kernel-based interpolation or approximation method. In this case the general rule of thumb is: *the smoother the kernel, the faster the rate of convergence of the interpolant.*

Motivated by this connection between the decay of the eigenvalues, the kernel smoothness and the rate of convergence of the interpolant, we introduce two families of *designer Chebyshev kernels* in Section 3.9.2.

As we mentioned earlier, a compact self-adjoint trace-class operator is characterized by having a finite trace. So in particular, for a symmetric kernel K which is also *translation invariant*, i.e., $K(\boldsymbol{x},\boldsymbol{z}) = \widetilde{K}(\boldsymbol{x}-\boldsymbol{z})$ we have

$$\operatorname{trace}\mathcal{K} = \int_\Omega K(\boldsymbol{x},\boldsymbol{x})\rho(\boldsymbol{x})\,\mathrm{d}\boldsymbol{x} = \widetilde{K}(\boldsymbol{0}) = \sum_{n=1}^{\infty} \lambda_n < \infty \tag{2.10}$$

since $\int_\Omega \rho(\boldsymbol{x})\,\mathrm{d}\boldsymbol{x} = 1$.

2.2.4 *Examples of Hilbert–Schmidt integral eigenvalue problems and Mercer series*

In this section we present the solution of two Hilbert–Schmidt integral eigenvalue problems. The first example is quite straightforward and leads to the eigenvalues and eigenfunctions associated with the *Brownian motion kernel* $K(x,z) = \min(x,z)$. The second example reconsiders the Matérn kernel $K(x,z) = \mathrm{e}^{-\varepsilon|x-z|}$ on the nonnegative real line and therefore requires the use of a nontrivial weight function. In both cases, we form the Mercer series of the kernels from the computed eigenvalues and $L_2(\Omega,\rho)$-orthonormal eigenfunctions.

Example 2.4. (Brownian motion kernel)
We consider the domain $\Omega = [0,1]$, and let

$$K(x,z) = \min(x,z) = \begin{cases} x, & x \le z, \\ z, & x > z. \end{cases}$$

We can insert this kernel along with the weight function $\rho(x) \equiv 1$ into the generic Hilbert–Schmidt integral eigenvalue problem (2.5), i.e.,

$$\begin{aligned} (\mathcal{K}\varphi)(x) = \lambda\varphi(x) \quad &\iff \quad \int_\Omega K(x,z)\varphi(z)\rho(z)\,\mathrm{d}z = \lambda\varphi(x) \\ &\iff \quad \int_0^1 \min(x,z)\varphi(z)\,\mathrm{d}z = \lambda\varphi(x) \\ &\iff \quad \int_0^x z\varphi(z)\,\mathrm{d}z + \int_x^1 x\varphi(z)\,\mathrm{d}z = \lambda\varphi(x). \end{aligned} \tag{2.11}$$

The standard approach to solving such an integral eigenvalue problem is to convert it to a related differential eigenvalue problem (cf. Remark 2.2). To this end we apply the differential operator $\frac{\mathrm{d}}{\mathrm{d}x}$ to the integral equation (2.11) and obtain

$$\begin{aligned} &\frac{\mathrm{d}}{\mathrm{d}x}\left[\int_0^x z\varphi(z)\,\mathrm{d}z - x\int_1^x \varphi(z)\,\mathrm{d}z\right] = \frac{\mathrm{d}}{\mathrm{d}x}\left[\lambda\varphi(x)\right] \\ \implies \quad & x\varphi(x) - \int_1^x \varphi(z)\,\mathrm{d}z - x\varphi(x) = \lambda\varphi'(x) \\ \iff \quad & -\int_1^x \varphi(z)\,\mathrm{d}z = \lambda\varphi'(x). \end{aligned} \tag{2.12}$$

Now we differentiate this equation once more to end up with a simple second-order ordinary differential equation of the form

$$-\varphi(x) = \lambda\varphi''(x) \quad \iff \quad -\varphi''(x) = \frac{1}{\lambda}\varphi(x).$$

From here we can see that the eigenvalues of the integral operator $\mathcal{K}$ correspond to the reciprocals of the eigenvalues of the differential operator $\mathcal{L} = -\frac{\mathrm{d}^2}{\mathrm{d}x^2}$. Moreover, the eigenfunctions are the same for both operators. In order to be able to solve the differential eigenvalue problem we need to derive two appropriate boundary conditions. We get these conditions by looking at what happens to (2.11) and (2.12) for $x = 0$ and $x = 1$. Plugging $x = 0$ into (2.11) yields $\varphi(0) = 0$, and plugging $x = 1$ into (2.12) leads to $\varphi'(1) = 0$. Thus we need to solve

$$-\varphi''(x) = \frac{1}{\lambda}\varphi(x), \qquad \varphi(0) = \varphi'(1) = 0,$$

and it is well known that the solution to this problem is (see also Appendix A)

$$\lambda_n = \frac{4}{(2n-1)^2\pi^2}, \quad \varphi_n(x) = \sqrt{2}\sin\left((2n-1)\frac{\pi x}{2}\right), \qquad n = 1, 2, \ldots, \tag{2.13}$$

where the factor $\sqrt{2}$ normalizes the eigenfunctions.

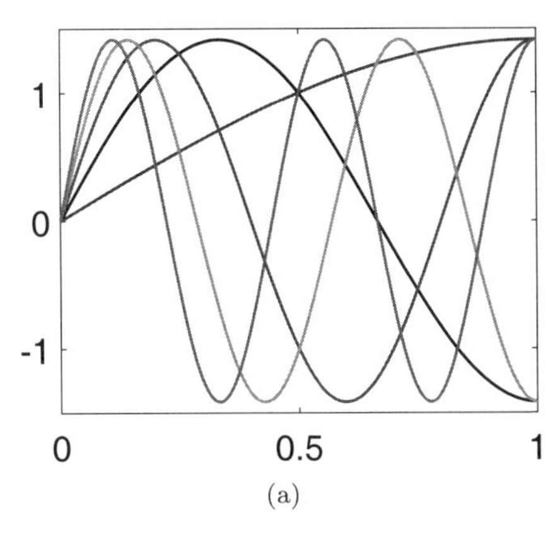

(a)

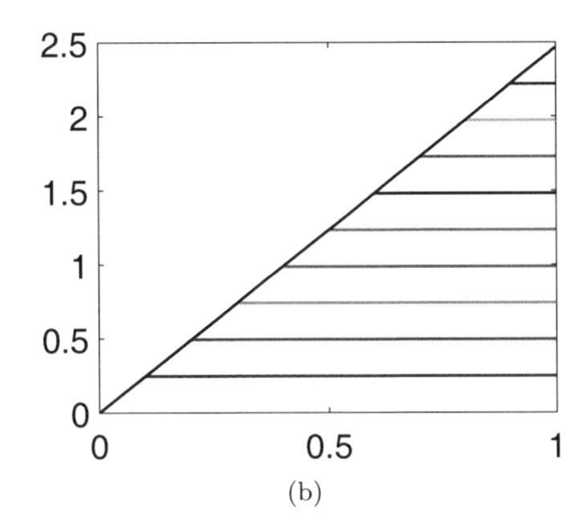

(b)

Fig. 2.1 (a) The first five eigenfunctions from (2.13). (b) Series approximation (2.14) based on 100 terms to produce ten copies of the kernel $K(x,z) = \min(x,z)$.

In Figure 2.1(a) we plot the first five eigenfunctions of the Brownian motion kernel. The Mercer series (2.9) of the Brownian motion kernel is of the form

$$\min(x,z) = \sum_{n=1}^{\infty} \frac{8}{(2n-1)^2\pi^2} \sin\left((2n-1)\frac{\pi x}{2}\right) \sin\left((2n-1)\frac{\pi z}{2}\right), \quad x,z \in [0,1]. \tag{2.14}$$

We use the first 100 terms of (2.14) to approximate the kernel and produce the plot of ten shifted copies of the kernel in Figure 2.1(b).

Example 2.5. (C^0 Matérn kernel on $[0,\infty)$)
We consider the *one-dimensional C^0 Matérn kernel* restricted to the nonnegative real line, i.e.,

$$K(x,z) = \mathrm{e}^{-\varepsilon|x-z|}, \qquad x,z \in [0,\infty), \quad \varepsilon > 0,$$

where ε is a shape parameter. Matérn kernels form an important family of positive definite kernels discussed in more detail in Chapter 3. The computations in this example are based on a paper by Juncosa (1945), who studied an equation similar to (2.15), namely $2\int_0^\infty \mathrm{e}^{-\varepsilon|x-z|-2z}\varphi(z)\,\mathrm{d}z = \lambda\varphi(x)$, in the context of random noise in radio receivers as described fully in [Kac and Siegert (1947)].

According to Example 2.2, the C^0 Matérn kernel does not give rise to a Hilbert–Schmidt operator if it is used on an unbounded interval together with the weight function $\rho(x) \equiv 1$. This implies that we must employ a nontrivial weight function. We choose $\rho(x) = \alpha\mathrm{e}^{-\alpha x}$ with the property that $\int_0^\infty \rho(x)\,\mathrm{d}x = 1$, and therefore the integral eigenvalue problem we consider is of the form

$$\alpha\int_0^\infty \mathrm{e}^{-\varepsilon|x-z|}\varphi(z)\mathrm{e}^{-\alpha z}\,\mathrm{d}z = \lambda\varphi(x), \quad x > 0, \quad \varepsilon,\alpha > 0. \tag{2.15}$$

Note that this includes Juncosa's problem as special case $\alpha = 2$. Our goal now is to find the eigenfunctions φ_n and eigenvalues λ_n of (2.15).

In order to resolve the absolute value in the kernel we break the integral at $z = x$ and get

$$\alpha\left\{\int_0^x \mathrm{e}^{-\varepsilon(x-z)}\varphi(z)\mathrm{e}^{-\alpha z}\,\mathrm{d}z + \int_x^\infty \mathrm{e}^{-\varepsilon(z-x)}\varphi(z)\mathrm{e}^{-\alpha z}\,\mathrm{d}z\right\} = \lambda\varphi(x),$$

which can be rewritten as

$$\mathrm{e}^{-\varepsilon x}\int_0^x \mathrm{e}^{(\varepsilon-\alpha)z}\varphi(z)\,\mathrm{d}z + \mathrm{e}^{\varepsilon x}\int_x^\infty \mathrm{e}^{-(\varepsilon+\alpha)z}\varphi(z)\,\mathrm{d}z = \frac{\lambda}{\alpha}\varphi(x). \tag{2.16}$$

Next we differentiate twice with respect to x and arrive at the ordinary differential equation

$$\varepsilon^2\left\{\mathrm{e}^{-\varepsilon x}\int_0^x \mathrm{e}^{(\varepsilon-\alpha)z}\varphi(z)\,\mathrm{d}z + \mathrm{e}^{\varepsilon x}\int_x^\infty \mathrm{e}^{-(\varepsilon+\alpha)z}\varphi(z)\,\mathrm{d}z\right\} - 2\varepsilon\mathrm{e}^{-\alpha x}\varphi(x) = \frac{\lambda}{\alpha}\varphi''(x).$$

This equation can be simplified by replacing the expression in the curly brackets with the right-hand side of (2.16). This gives us

$$\varphi''(x) + \left(\frac{2\varepsilon\alpha}{\lambda}\mathrm{e}^{-\alpha x} - \varepsilon^2\right)\varphi(x) = 0, \qquad x \geq 0. \tag{2.17}$$

The ODE (2.17) can be transformed into a standard *Bessel equation* using the substitution $t = \frac{2}{\alpha}\sqrt{\frac{2\varepsilon\alpha}{\lambda}}\mathrm{e}^{-\alpha x/2}$ and $\varphi(x) = \psi(t)$, i.e., $\frac{\mathrm{d}t}{\mathrm{d}x} = -\frac{\alpha}{2}t$ so that

$$\begin{aligned}
\varphi'(x) &= \frac{\mathrm{d}}{\mathrm{d}x}\psi(t) = -\frac{\alpha}{2}t\psi'(t),\\
\varphi''(x) &= \frac{\mathrm{d}^2}{\mathrm{d}x^2}\psi(t) = \frac{\alpha^2}{4}t\psi'(t) + \frac{\alpha^2}{4}t^2\psi''(t).
\end{aligned}$$

This results in

$$t^2\psi''(t) + t\psi'(t) + \left(t^2 - \frac{4\varepsilon^2}{\alpha^2}\right)\psi(t) = 0. \tag{2.18}$$

Note that we can get a "boundary condition" from the integral equation (2.15) if we consider what happens for $x \to \infty$. In this case the left-hand side of (2.15) behaves like

$$\lim_{x\to\infty}\alpha\int_0^\infty \mathrm{e}^{-\varepsilon|x-z|}\varphi(z)\mathrm{e}^{-\alpha z}\,\mathrm{d}z = 0.$$

Therefore, we should require that the eigenfunctions (of the ODE eigenvalue problem (2.17–2.18)) satisfy

$$\lim_{x\to\infty}\varphi(x) = \lim_{t\to 0}\psi(t) = 0.$$

The general solution of the Bessel equation (2.18) is given either as a linear combination of $J_{2\varepsilon/\alpha}$ and $J_{-2\varepsilon/\alpha}$, or of $J_{2\varepsilon/\alpha}$ and $Y_{2\varepsilon/\alpha}$, where J_ν and Y_ν are Bessel functions of the first and second kind of order ν, respectively. However, the boundary condition $\lim_{t\to 0}\psi(t) = 0$ rules out the use of the $J_{-2\varepsilon/\alpha}$ and $Y_{2\varepsilon/\alpha}$ components, and so the solution of the Bessel equation is given simply in terms of $J_{2\varepsilon/\alpha}$.

The work so far gives us the eigenfunctions, but the corresponding eigenvalues need to be found by plugging this solution back into the integral equation. This can be done by replacing the Bessel function by its infinite series representation and then performing term-by-term integration. The details are a bit technical, but lead to

$$\varphi_n(x) = J_{2\varepsilon/\alpha}\left(r_n \mathrm{e}^{-\alpha x/2}\right), \qquad n = 1, 2, \ldots, \tag{2.19}$$

where $J_{2\varepsilon/\alpha-1}(r_n) = 0$, i.e., the r_n are the zeros of the Bessel function $J_{2\varepsilon/\alpha-1}$ so that the corresponding eigenvalues are given by

$$\lambda_n = \frac{8\varepsilon}{\alpha} r_n^{-2}, \qquad n = 1, 2, \ldots. \tag{2.20}$$

Note that we have arrived at a parametrized representation of the eigenvalues and eigenfunctions of the C^0 Matérn kernel. We can interpret the parameter α that was introduced via the weight function as a *global scale parameter* in addition to the shape parameter ε, which we already discussed in Chapter 1.

In order to provide some more concrete eigenvalues and eigenfunctions along with some plots we now couple the choices of the parameters $\varepsilon, \alpha > 0$. In particular, we let $\alpha = 4\varepsilon$. Then we know that

$$J_{2\varepsilon/\alpha}(x) = J_{1/2}(x) = \sqrt{\frac{2}{\pi x}} \sin x,$$
$$J_{2\varepsilon/\alpha-1}(x) = J_{-1/2}(x) = \sqrt{\frac{2}{\pi x}} \cos x,$$

and r_n, the zeros of $\cos x$, are $r_n = \frac{(2n-1)\pi}{2}$.

Therefore, (2.19) implies that the eigenfunctions of the C^0 Matérn kernel $K(x,z) = \mathrm{e}^{-\varepsilon|x-z|}$ on $[0,\infty)$ under the weight function $\rho(x) = 4\varepsilon \mathrm{e}^{-4\varepsilon x}$ are given by

$$\begin{aligned}\varphi_n(x) = J_{2\varepsilon/\alpha}\left(r_n \mathrm{e}^{-\alpha x/2}\right) &= \sqrt{\frac{2}{\pi \frac{(2n-1)\pi}{2} \mathrm{e}^{-2\varepsilon x}}} \sin\left(\frac{(2n-1)\pi}{2} \mathrm{e}^{-2\varepsilon x}\right) \\ &= \frac{2\mathrm{e}^{\varepsilon x}}{\pi\sqrt{2n-1}} \sin\left(\frac{(2n-1)\pi}{2} \mathrm{e}^{-2\varepsilon x}\right).\end{aligned}$$

Since the $L_2([0,\infty),\rho)$ norm of the eigenfunctions is $\frac{2}{\sqrt{(2n-1)\pi}}$ the normalized eigenfunctions are given by

$$\varphi_n(x) = \mathrm{e}^{\varepsilon x} \sin\left(\frac{(2n-1)\pi}{2} \mathrm{e}^{-2\varepsilon x}\right). \tag{2.21}$$

The first four of these functions are plotted in Figure 2.2(a) using shape parameter $\varepsilon = 1$.

The eigenvalues are given by

$$\lambda_n = \frac{8\varepsilon}{\alpha} r_n^{-2} = 2 r_n^{-2} = \frac{8}{(2n-1)^2\pi^2}, \qquad n = 1, 2, \ldots,$$

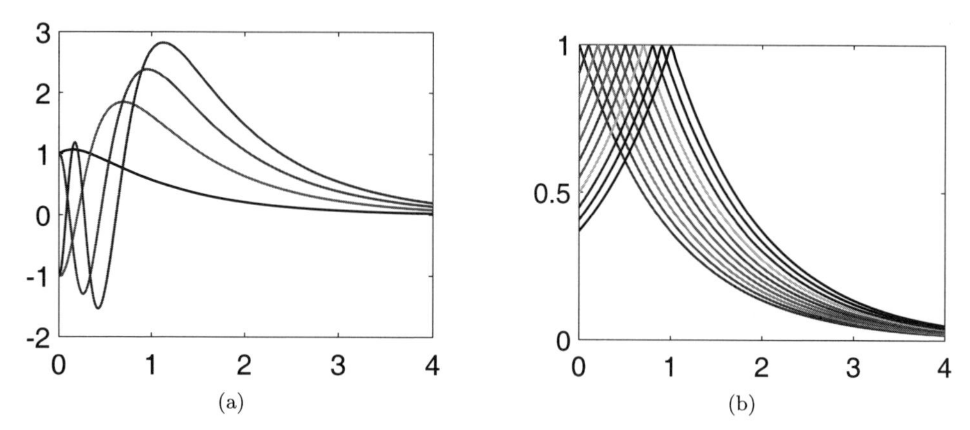

Fig. 2.2 (a) The first four eigenfunctions from (2.21) with $\varepsilon = 1$. (b) Series approximation (2.22) based on 100 terms to produce ten copies of the kernel $K(x, z) = \mathrm{e}^{-|x-z|}$.

and so we can also provide the Mercer series (see (2.9)) for the C^0 Matérn kernel, with normalized eigenfunctions, in the case $\alpha = 4\varepsilon$. It is

$$\mathrm{e}^{-\varepsilon|x-z|} = \sum_{n=1}^{\infty} \frac{8}{(2n-1)^2\pi^2} \frac{\sin\left(\frac{(2n-1)\pi}{2}\mathrm{e}^{-2\varepsilon x}\right)}{\mathrm{e}^{-\varepsilon x}} \frac{\sin\left(\frac{(2n-1)\pi}{2}\mathrm{e}^{-2\varepsilon z}\right)}{\mathrm{e}^{-\varepsilon z}}, \qquad x, z \in [0, \infty), \tag{2.22}$$

which we truncate at 100 terms and use to plot ten shifted copies of the kernel in Figure 2.2(b).

Remark 2.5. Eigenvalues and eigenfunctions of the C^0 Matérn kernel for other domains are listed in Appendix A.2.

Remark 2.6. Examples 2.4 and 2.5 are provided as demonstrations of the theory behind kernels defined through Mercer series; Section 4.5 discusses computational strategies in MATLAB.

2.2.5 *Iterated kernels*

Using the Hilbert–Schmidt integral operator (2.4) and the absolutely converging Mercer series (2.9) of a symmetric positive definite kernel K defined on $\Omega \times \Omega$ we can see that for a fixed $\boldsymbol{z} \in \Omega$ one has

$$\begin{aligned}
\mathcal{K}K(\boldsymbol{x}, \boldsymbol{z}) = \mathcal{K}\left(K(\cdot, \boldsymbol{z})\right)(\boldsymbol{x}) &= \int_\Omega K(\boldsymbol{x}, \boldsymbol{t}) K(\boldsymbol{t}, \boldsymbol{z}) \rho(\boldsymbol{t}) \,\mathrm{d}\boldsymbol{t} \\
&= \int_\Omega K(\boldsymbol{x}, \boldsymbol{t}) \sum_{n=1}^{\infty} \lambda_n \varphi_n(\boldsymbol{t}) \varphi_n(\boldsymbol{z}) \rho(\boldsymbol{t}) \,\mathrm{d}\boldsymbol{t} \\
&= \sum_{n=1}^{\infty} \lambda_n \varphi_n(\boldsymbol{z}) \int_\Omega K(\boldsymbol{x}, \boldsymbol{t}) \varphi_n(\boldsymbol{t}) \rho(\boldsymbol{t}) \,\mathrm{d}\boldsymbol{t}.
\end{aligned}$$

Now, since the φ_n are the eigenfunctions of $\mathcal{K}$ corresponding to λ_n we can further conclude that

$$\mathcal{K}K(\boldsymbol{x},\boldsymbol{z}) = \sum_{n=1}^{\infty} \lambda_n^2 \varphi_n(\boldsymbol{x})\varphi_n(\boldsymbol{z}),$$

which is a new kernel with the same eigenfunctions as K, but whose eigenvalues are the squares of those of K. We refer to such a kernel as an *iterated kernel*.

It should be noted that for a translation-invariant kernel — and therefore of course also for a radial kernel — the operation described above is a convolution. The iterated kernel is smoother than the original one, and it is clear that this iterative process can be repeated leading to a family of kernels with increasing smoothness (determined by the rate of decay of the eigenvalues) all built from a common set of eigenfunctions.

The idea of constructing smoother kernels via iteration is a classical one and already described in, e.g., [Courant and Hilbert (1953, Section III.5.3), Stakgold (1979, Section 6.3)]. In Chapter 6 we will construct iterated kernels via an iterated differential operator. We prefer that approach since there the boundary conditions for the kernel are explicitly specified.

2.2.6 *Fourier and Karhunen–Loève expansions*

Given a compact, self-adjoint operator $\mathcal{K}$ on $L_2(\Omega,\rho)$ such as a Hilbert–Schmidt operator induced by a positive definite kernel K, we can use the $L_2(\Omega,\rho)$-orthonormal eigenfunctions of $\mathcal{K}$ to represent any L_2 function f by its uniformly and absolutely convergent *generalized Fourier series* (cf. Theorem 6.1) with respect to the basis of eigenfunctions, i.e.,

$$f(\boldsymbol{x}) = \sum_{n=1}^{\infty} \langle f, \varphi_n \rangle_{L_2(\Omega,\rho)} \varphi_n(\boldsymbol{x}).$$

Now — even though we have not yet set the stage for the stochastic framework, and Mercer's theorem is not formulated in terms of stochastic processes either — we state the *Karhunen–Loève expansion* theorem since it is almost an immediate corollary to Mercer's theorem. Karhunen–Loève expansions can be considered analogous to Fourier series, but for a random field (or stochastic process) with random variables in $L_2(\mathcal{W},\mathcal{A},P)$ whose continuous covariance kernel is K (as above). Here $(\mathcal{W},\mathcal{A},P)$ denotes a probability space as in Definition 5.1. Additional helpful notions from probability theory can be found in Appendix C.

Theorem 2.3 (Karhunen–Loève expansion). *A centered mean-square continuous random field Y with continuous covariance kernel $K : \Omega \times \Omega \to \mathbb{R}$ has an orthogonal expansion of the form*

$$Y_{\boldsymbol{x}}(\omega) = \sum_{n=1}^{\infty} \sqrt{\lambda_n} Z_n(\omega)\varphi_n(\boldsymbol{x}).$$

Here λ_n are the positive eigenvalues and $\varphi_n \in \mathcal{H}_K(\Omega)$ are the associated eigenfunctions for the Hilbert–Schmidt integral eigenvalue problem associated with K. Moreover, the Z_n are "orthonormal" random variables given by

$$Z_n(\omega) = \frac{1}{\sqrt{\lambda_n}} \int_\Omega Y_{\boldsymbol{x}}(\omega)\varphi_n(\boldsymbol{x})\,\mathrm{d}\boldsymbol{x}$$

such that $\mathbb{E}[Z_m Z_n] = \delta_{mn}$.

The Karhunen–Loève theorem can be found, e.g., in [Berlinet and Thomas-Agnan (2004, Chapter 2), Xiu (2010, Chapter 4)]. As mentioned above, its proof follows almost directly from Mercer's theorem.

We will discuss the role of the Karhunen–Loève expansion in polynomial chaos approximations and uncertainty quantification in Chapter 5.

2.3 Reproducing Kernel Hilbert Spaces

As mentioned in Chapter 1, reproducing kernels are a classical concept from analysis introduced by Aronszajn and Bergman in the first half of the 20th century (see, e.g., [Aronszajn (1950); Bergman (1950)]). Detailed accounts of reproducing kernel Hilbert spaces are provided in many of the monographs listed in Chapter 1 (including [Fasshauer (2007, Chapter 13), Wendland (2005, Chapter 10)]). Therefore we give only a brief summary here and refer the reader to those other references for more information.

In Chapter 1 we already stated the reproducing property. It is repeated as item (2) below and forms the main part of the definition of a reproducing kernel Hilbert space.

Definition 2.6 (Reproducing kernel Hilbert space). *Let $\mathcal{H}_K(\Omega)$ be a Hilbert space of real-valued functions f defined on $\Omega \subseteq \mathbb{R}^d$ with inner product $\langle\cdot,\cdot\rangle_{\mathcal{H}_K(\Omega)}$. A kernel $K : \Omega \times \Omega \to \mathbb{R}$ is called* reproducing kernel for $\mathcal{H}_K(\Omega)$ *if*

(1) $K(\cdot,\boldsymbol{x}) \in \mathcal{H}_K(\Omega)$ *for all* $\boldsymbol{x} \in \Omega$,
(2) $\langle f, K(\cdot,\boldsymbol{x})\rangle_{\mathcal{H}_K(\Omega)} = f(\boldsymbol{x})$ *for all* $f \in \mathcal{H}_K(\Omega)$ *and all* $\boldsymbol{x} \in \Omega$.

The name *reproducing kernel* is motivated by the reproducing property (2) in Definition 2.6, which shows that the action of taking the inner product of any function $f \in \mathcal{H}_K(\Omega)$ with $K(\cdot,\boldsymbol{x})$ corresponds to point evaluation of f at $\boldsymbol{x}$. More formally, the Riesz representation theorem states that any continuous linear functional can be represented by an inner product with a unique element of $\mathcal{H}_K$. Since, for any $\boldsymbol{x} \in \Omega$, we have a function $K(\cdot,\boldsymbol{x})$ such that the reproducing property holds it follows that the reproducing kernel at $\boldsymbol{x}$, $K(\cdot,\boldsymbol{x})$, is the Riesz representer of function evaluation at $\boldsymbol{x}$. It is known that the reproducing kernel of a Hilbert space is unique, and so is the Hilbert space associated with a positive definite kernel.

Some useful properties of reproducing kernels and reproducing kernel Hilbert spaces are:

(1) A reproducing kernel is symmetric, i.e., for all $\boldsymbol{x}, \boldsymbol{z} \in \Omega$

$$\langle K(\cdot, \boldsymbol{x}), K(\cdot, \boldsymbol{z})\rangle_{\mathcal{H}_K(\Omega)} = K(\boldsymbol{x}, \boldsymbol{z}) = K(\boldsymbol{z}, \boldsymbol{x}).$$

This follows directly from the reproducing property (with $f = K(\cdot, \boldsymbol{z})$) and the symmetry of the inner product.

(2) Point evaluation in a reproducing kernel Hilbert space is bounded, i.e., for every $f \in \mathcal{H}_K(\Omega)$ and $x \in \Omega$ we have

$$|f(\boldsymbol{x})| \leq \sqrt{K(\boldsymbol{x}, \boldsymbol{x})}\|f\|_{\mathcal{H}_K(\Omega)}.$$

This also follows from the reproducing property and an application of the Cauchy–Schwarz inequality to the inner product along with the fact that

$$K(\boldsymbol{x}, \boldsymbol{x}) = \langle K(\cdot, \boldsymbol{x}), K(\cdot, \boldsymbol{x})\rangle_{\mathcal{H}_K(\Omega)} = \|K(\cdot, \boldsymbol{x})\|^2_{\mathcal{H}_K(\Omega)} \geq 0.$$

(3) If K_1 and K_2 are reproducing kernels of spaces $\mathcal{H}_1$ and $\mathcal{H}_2$, respectively, on the same domain Ω, then $K = K_1 + K_2$ is the reproducing kernel of the sum $\mathcal{H} = \mathcal{H}_1 \oplus \mathcal{H}_2$ with norm

$$\|f\|^2_{\mathcal{H}} = \min_{f=f_1+f_2} \left(\|f_1\|^2_{\mathcal{H}_1} + \|f_2\|^2_{\mathcal{H}_2}\right).$$

This is proved in [Berlinet and Thomas-Agnan (2004, Section 1.4.1)]. If $\mathcal{H}_1 \cap \mathcal{H}_2 = \{0\}$, then $\mathcal{H}$ is a *direct sum*, and $\mathcal{H}_2$ is the orthogonal complement of $\mathcal{H}_1$ in $\mathcal{H}$.

(4) If $K : \Omega \times \Omega \to \mathbb{R}$ is the reproducing kernel of $\mathcal{H}$ and $\Omega_0 \subset \Omega$, then K_0, the restriction of K to $\Omega_0 \times \Omega_0$, is the reproducing kernel of $\mathcal{H}_0$, a space whose elements are restrictions of elements of $\mathcal{H}$ to Ω_0. The norm in $\mathcal{H}_0$ is given by

$$\|f_0\|_{\mathcal{H}_0} = \min_{f|_{\Omega_0}=f_0} \|f\|_{\mathcal{H}}.$$

This is proved in [Berlinet and Thomas-Agnan (2004, Section 1.4.2)].

(5) If K is the reproducing kernel of the reproducing kernel Hilbert space $\mathcal{H}$ and $\mathcal{V}$ is a closed subspace of $\mathcal{H}$, then $\mathcal{V}$ is also a reproducing kernel Hilbert space whose reproducing kernel is given in terms of the orthogonal projection $P_{\mathcal{V}}$ onto $\mathcal{V}$, i.e.,

$$K_{\mathcal{V}}(\boldsymbol{x}, \boldsymbol{z}) = \left(P_{\mathcal{V}}(K(\cdot, \boldsymbol{z}))\right)(\boldsymbol{x}).$$

This is proved in [Berlinet and Thomas-Agnan (2004, Section 1.4.4)].

(6) If K_1 and K_2 are reproducing kernels of spaces $\mathcal{H}_1$ and $\mathcal{H}_2$ on domains Ω_1 and Ω_2, respectively, then the *tensor product kernel*

$$K((\boldsymbol{x}_1, \boldsymbol{x}_2), (\boldsymbol{z}_1, \boldsymbol{z}_2)) = K_1(\boldsymbol{x}_1, \boldsymbol{z}_1)K_2(\boldsymbol{x}_2, \boldsymbol{z}_2)$$

is the reproducing kernel of the *tensor product space* $\mathcal{H} = \mathcal{H}_1 \otimes \mathcal{H}_2$. This is proved in [Berlinet and Thomas-Agnan (2004, Section 1.4.6)]. Note that such tensor products are useful for "space-time" applications, where K_1 is a kernel in the spatial domain and K_2 in time (see Section 3.7), or to construct anisotropic kernels as in Section 3.4.3.

Not every Hilbert space is a reproducing kernel Hilbert space. In fact, the second property above shows that reproducing kernel Hilbert spaces are those special Hilbert spaces in which the evaluation functional is bounded, i.e., continuous. This means that they are those "smooth" Hilbert spaces in which values of functions at nearby points are closely related to each other (since this is just what the concept of continuity implies). Moreover, this interpretation also matches nicely with the analogy between reproducing kernels and covariance kernels discussed in Chapter 5. In fact, we could use the following characterization theorem (see, e.g., [Berlinet and Thomas-Agnan (2004, Theorem 1)]) as an alternative definition of a reproducing kernel Hilbert space.

Theorem 2.4. *A Hilbert space $\mathcal{H}$ of functions on Ω has a reproducing kernel if and only if the evaluation functional at any point in Ω is continuous on $\mathcal{H}$.*

Remark 2.7. While the space $L_2(\Omega)$ is a Hilbert space, it is *not* a reproducing kernel Hilbert space. First, since it is not a space of functions, but of equivalence classes of functions. Moreover, while the delta functional acts as its "reproducing kernel," i.e.,

$$\int_\Omega f(\boldsymbol{z})\delta(\boldsymbol{x}-\boldsymbol{z})\,\mathrm{d}\boldsymbol{z} = f(\boldsymbol{x}),$$

it does not belong to $L_2(\Omega)$ (see also Remark 6.1).

The third property can be generalized to nonnegative linear combinations (see, e.g., [Steinwart and Christmann (2008, Lemma 4.5)]). The fourth property concerning restrictions of kernels is actually formulated in a slightly more general setting in [Berlinet and Thomas-Agnan (2004, Section 1.4.2)]. In particular, one can show that a complex-valued kernel when restricted from a complex domain to a real subdomain not only remains a complex-valued kernel, but if the kernel is real-valued then it is also a kernel in the purely real sense (see [Steinwart and Christmann (2008, Lemmas 4.3 & 4.4)]).

If we start with a kernel instead of with the space, then [Wendland (2005, Theorem 10.10)] ensures that we can identify with each positive definite kernel a reproducing kernel Hilbert space which is its *native space*. Looking at the situation from this perspective means that one may frequently end up with a space which is not very intuitive to understand. Wendland (2005, Theorem 10.12) provides a description of native spaces associated with radial kernels in terms of Fourier transforms, but even that is not as intuitive as a Sobolev space. We list examples of reproducing kernels of many different types along with some of their reproducing kernel Hilbert spaces in Chapter 3, and in Chapter 8 we consider reproducing kernel Hilbert spaces as generalized Sobolev spaces with the aim of making them more intuitive (but also nonunique).

The definition of a reproducing kernel Hilbert space tells us that $\mathcal{H}_K(\Omega)$ contains all functions of the form

$$f = \sum_{j=1}^{N} c_j K(\cdot, \boldsymbol{x}_j)$$

provided $\boldsymbol{x}_j \in \Omega$ (and N is arbitrary). Using the properties of reproducing kernel Hilbert spaces established earlier along with this special form of f we have that

$$\begin{aligned}
\|f\|^2_{\mathcal{H}_K(\Omega)} = \langle f, f\rangle_{\mathcal{H}_K(\Omega)} &= \left\langle \sum_{i=1}^{N} c_i K(\cdot, \boldsymbol{x}_i), \sum_{j=1}^{N} c_j K(\cdot, \boldsymbol{x}_j) \right\rangle_{\mathcal{H}_K(\Omega)} \\
&= \sum_{i=1}^{N}\sum_{j=1}^{N} c_i c_j \langle K(\cdot, \boldsymbol{x}_i), K(\cdot, \boldsymbol{x}_j)\rangle_{\mathcal{H}_K(\Omega)} \\
&= \sum_{i=1}^{N}\sum_{j=1}^{N} c_i c_j K(\boldsymbol{x}_i, \boldsymbol{x}_j) = \boldsymbol{c}^T \mathsf{K} \boldsymbol{c}. \qquad (2.23)
\end{aligned}$$

Thus — for these very special types of f — we can easily calculate the Hilbert space norm of f. In particular, if $f = s$ is a kernel-based interpolant, i.e., $\boldsymbol{c} = \mathsf{K}^{-1}\boldsymbol{y}$ (cf. (1.3b)), then we also have

$$\|s\|^2_{\mathcal{H}_K(\Omega)} = \boldsymbol{y}^T \mathsf{K}^{-T} \mathsf{K} \mathsf{K}^{-1} \boldsymbol{y} = \boldsymbol{y}^T \mathsf{K}^{-1} \boldsymbol{y}. \qquad (2.24)$$

From (2.24) we see that the native space norm of a kernel-interpolant depends on the given data $\boldsymbol{y}$ — as it should. Moreover, as can be seen from (2.23), the native space norm of any function from $\mathcal{H}_K(\Omega)$ depends on the kernel K. In particular, if the kernel contains a shape or scale parameter — as many of our kernels do — then the native space norm changes as these parameters vary.

Mercer's theorem allows us to construct a reproducing kernel Hilbert space $\mathcal{H}_K(\Omega)$ for any continuous positive definite kernel K by representing the functions in $\mathcal{H}_K(\Omega)$ as infinite linear combinations of the eigenfunctions φ_n of the integral operator $\mathcal{K}$, i.e.,

$$\mathcal{H}_K(\Omega) = \left\{ f \mid f = \sum_{n=1}^{\infty} a_n \varphi_n \right\}. \qquad (2.25)$$

Thus the eigenfunctions $\{\varphi_n\}_{n=1}^{\infty}$ of $\mathcal{K}$ provide an alternative basis for $\mathcal{H}_K(\Omega)$ instead of the standard basis $\{K(\cdot, \boldsymbol{z}) \mid \boldsymbol{z} \in \Omega\}$ consisting of all possible "translates" of K. For any fixed $\boldsymbol{z}$, the corresponding *basis transformation* is given by the eigenfunction expansion

$$K(\cdot, \boldsymbol{z}) = \sum_{n=1}^{\infty} \lambda_n \varphi_n \varphi_n(\boldsymbol{z})$$

guaranteed by Mercer's theorem.

The inner product for $\mathcal{H}_K(\Omega)$ can now be written as

$$\langle f, g\rangle_{\mathcal{H}_K(\Omega)} = \langle \sum_{m=1}^{\infty} a_m \varphi_m, \sum_{n=1}^{\infty} b_n \varphi_n \rangle_{\mathcal{H}_K(\Omega)} = \sum_{n=1}^{\infty} \frac{a_n b_n}{\lambda_n},$$

where we used the fact that the eigenfunctions are not only L_2-orthonormal, but also orthogonal in $\mathcal{H}_K(\Omega)$ (cf. (2.8)), i.e.,

$$\langle \varphi_m, \varphi_n \rangle_{\mathcal{H}_K(\Omega)} = \frac{\delta_{mn}}{\sqrt{\lambda_m}\sqrt{\lambda_n}}.$$

Using the representation (2.25) of $f \in \mathcal{H}_K(\Omega)$ in terms of the eigenfunction basis, we can also verify that K is indeed the reproducing kernel of $\mathcal{H}_K(\Omega)$ since the Mercer series of K and the orthogonality of the eigenfunctions imply

$$\begin{aligned}\langle f, K(\cdot, \boldsymbol{x}) \rangle_{\mathcal{H}_K(\Omega)} &= \langle \sum_{m=1}^{\infty} a_m \varphi_m, \sum_{n=1}^{\infty} \lambda_n \varphi_n \varphi_n(\boldsymbol{x}) \rangle_{\mathcal{H}_K(\Omega)} \\ &= \sum_{n=1}^{\infty} \frac{a_n \lambda_n \varphi_n(\boldsymbol{x})}{\lambda_n} \\ &= \sum_{n=1}^{\infty} a_n \varphi_n(\boldsymbol{x}) = f(\boldsymbol{x}).\end{aligned}$$

Finally (cf. [Wendland (2005)]), we can also describe the reproducing kernel Hilbert space $\mathcal{H}_K(\Omega)$ as

$$\mathcal{H}_K(\Omega) = \left\{ f \in L_2(\Omega) \mid \sum_{n=1}^{\infty} \frac{1}{\lambda_n} |\langle f, \varphi_n \rangle_{L_2(\Omega)}|^2 < \infty \right\}$$

and write the corresponding inner product as

$$\langle f, g \rangle_{\mathcal{H}_K(\Omega)} = \sum_{n=1}^{\infty} \frac{1}{\lambda_n} \langle f, \varphi_n \rangle_{L_2(\Omega)} \langle g, \varphi_n \rangle_{L_2(\Omega)}, \qquad f, g \in \mathcal{H}_K(\Omega).$$

Remark 2.8. Since $\mathcal{H}_K(\Omega)$ is a subspace of $L_2(\Omega)$ this latter interpretation corresponds to the identification $a_n = \langle f, \varphi_n \rangle_{L_2(\Omega)}$ of the generalized Fourier coefficients in (2.25). This connection allows us to interpret the native space norm of $f \in \mathcal{H}_K(\Omega)$ in terms of a discrete ℓ_2 norm of the sequence of generalized Fourier coefficients of f, which is reminiscent of [Wendland (2005, Theorem 10.12)], where the native space norm for radial kernels is presented in terms of the Fourier transforms of f and K. We have, in fact, that

$$\|f\|_{\mathcal{H}_K(\Omega)} = \|\tilde{\boldsymbol{f}}\|_{\ell_2},$$

where $\tilde{\boldsymbol{f}} = \left(\frac{a_1}{\sqrt{\lambda_1}}, \ldots, \frac{a_n}{\sqrt{\lambda_n}}, \ldots \right)$. An analogous correspondence underlies the idea of *feature maps*, to be discussed in the next section (cf. (2.26) and (2.27)).

2.4 Feature Maps

Especially in machine learning and support vector machines one is interested in separating/classifying the given data in the input space. Ideally, one wants to do this with a hyperplane. However, using a linear separation severely limits the effectiveness and applicability of such an approach. Therefore, one looks to using a *nonlinear separation* in the input space, and the setting of reproducing kernel Hilbert spaces provides a perfect framework to accomplish this while still applying linear techniques.

In Section 2.3 we saw that the reproducing kernel K can be identified with an inner product. Namely, we had — in terms of the native space inner product — that

$$K(\boldsymbol{x},\boldsymbol{z}) = \langle K(\cdot,\boldsymbol{x}), K(\cdot,\boldsymbol{z})\rangle_{\mathcal{H}_K(\Omega)}. \tag{2.26}$$

On the other hand, we can also interpret the Mercer series representation of K in terms of an inner product — now in the sequence space ℓ_2, i.e.,

$$K(\boldsymbol{x},\boldsymbol{z}) = \sum_{n=1}^{\infty} \lambda_n \varphi_n(\boldsymbol{x})\varphi_n(\boldsymbol{z}) = \langle \Phi_{\boldsymbol{x}}, \Phi_{\boldsymbol{z}}\rangle_{\ell_2}, \tag{2.27}$$

where

$$\Phi_{\boldsymbol{x}} = \left(\sqrt{\lambda_1}\varphi_1(\boldsymbol{x}), \sqrt{\lambda_2}\varphi_2(\boldsymbol{x}), \ldots\right).$$

This decomposition of $K(\boldsymbol{x},\boldsymbol{z})$ into a *feature* that depends only on $\boldsymbol{x}$ and another that depends on $\boldsymbol{z}$ is often referred to as the *kernel trick* in the machine learning literature. Moreover, a mapping Φ from Ω to the *feature space* $\mathcal{H}$ is known as the *feature map*. Note that the feature space $\mathcal{H}$ is a Hilbert space which can be, e.g., a reproducing kernel Hilbert (function) space as in (2.26) so that we have the *canonical features* $\Phi_{\boldsymbol{x}} = K(\cdot,\boldsymbol{x})$, or it can be a sequence space as in (2.27).

In fact — as discussed in [Berlinet and Thomas-Agnan (2004, Chapter 1)] — one can find a *characterization of all possible reproducing kernels* on $\Omega\times\Omega$ in terms of inner products of (feature) maps into appropriate Hilbert spaces $\mathcal{H}$.

Theorem 2.5. *A function K defined on $\Omega\times\Omega$ is a reproducing kernel if and only if there exist a Hilbert space $\mathcal{H}$ and a mapping $\Phi:\Omega\to\mathcal{H}$, such that for all $\boldsymbol{x},\boldsymbol{z}\in\Omega$*

$$K(\boldsymbol{x},\boldsymbol{z}) = \langle \Phi_{\boldsymbol{x}}, \Phi_{\boldsymbol{z}}\rangle_{\mathcal{H}}.$$

As we saw above, in this theorem $\mathcal{H}$ can be either a Hilbert function space or a sequence space.

On the one hand, this theorem can be seen as a (nonunique) factorization theorem for the kernel K reminiscent of the spectral theorem for symmetric matrices. On the other hand, it also provides a way to construct reproducing kernels starting from a map Φ.

The following example shows that — depending on our choice of feature space and inner product — any given kernel can have many different feature maps, i.e., the feature maps associated with a given kernel are not unique.

Example 2.6. (Brownian motion kernel)
If we take $\Omega = [0,1]$, $\mathcal{H} = \ell_2$, and $\Phi_x = (\sqrt{\lambda_1}\varphi_1(x), \sqrt{\lambda_2}\varphi_2(x), \ldots)$, where

$$\lambda_n = \frac{4}{(2n-1)^2\pi^2}, \quad \varphi_n(x) = \sqrt{2}\sin\left((2n-1)\frac{\pi x}{2}\right), \qquad n = 1, 2, \ldots,$$

are eigenvalues and eigenfunctions of the *Brownian motion kernel* as computed in Example 2.4, then

$$\begin{aligned} K(x,z) &= \langle \Phi_x, \Phi_z \rangle_{\ell_2} \\ &= \sum_{n=1}^{\infty} \frac{8}{(2n-1)^2\pi^2} \sin\left((2n-1)\frac{\pi x}{2}\right) \sin\left((2n-1)\frac{\pi z}{2}\right) \qquad (2.28) \\ &= \min(x,z), \end{aligned}$$

where we can either invoke Mercer's Theorem 2.2 to obtain the closed form expression for the series, or we can derive it independently using basic trigonometric identities coupled with the fact that cosine series can be expressed in terms of Bernoulli polynomials.

Another interpretation of Theorem 2.5 would be to consider $\Phi_x = \mathbf{1}_{[0,x]}$, where $\mathbf{1}_{[a,b]}$ is the indicator function of the interval $[a,b]$, i.e.,

$$\mathbf{1}_{[a,b]}(t) = \begin{cases} 1, & t \in [a,b], \\ 0, & \text{otherwise}. \end{cases}$$

As Hilbert space we take $\mathcal{H} = L_2([0,1])$. This gives us

$$\begin{aligned} \langle \Phi_x, \Phi_z \rangle_{\mathcal{H}} &= \int_0^1 \mathbf{1}_{[0,x]}(t)\mathbf{1}_{[0,z]}(t)\,\mathrm{d}t \\ &= \int_0^{\min(x,z)} \mathrm{d}t = \min(x,z). \end{aligned}$$

And as a third possibility, we could take the canonical features $\Phi_x = K(\cdot,x)$, i.e., $\Phi_x = \min(\cdot,x)$, and the Hilbert space $\mathcal{H} = \mathcal{H}_K([0,1]) = \{f \in H^1([0,1]), f(0) = 0\}$ with inner product $\langle f,g \rangle_{\mathcal{H}} = \int_0^1 f'(t)g'(t)\,\mathrm{d}t$. Then we have

$$\begin{aligned} \langle \Phi_x, \Phi_z \rangle_{\mathcal{H}} &= \int_0^1 \frac{\mathrm{d}}{\mathrm{d}t}\min(t,x)\frac{\mathrm{d}}{\mathrm{d}t}\min(t,z)\,\mathrm{d}t \\ &= \int_0^1 \mathbf{1}_{[0,x]}(t)\mathbf{1}_{[0,z]}(t)\,\mathrm{d}t = \min(x,z). \end{aligned}$$

Note that we also know that for this example $\langle K(\cdot,x), K(\cdot,z) \rangle_{\mathcal{H}_K(\Omega)} = K(x,z)$, i.e., without having to deal with details of the integration,

$$\int_0^1 \frac{\mathrm{d}}{\mathrm{d}t}\min(t,x)\frac{\mathrm{d}}{\mathrm{d}t}\min(t,z)\mathrm{d}t = \min(x,z).$$

The canonical feature map used in Example 2.6, i.e., $\Phi : \Omega \to \mathcal{H}_K(\Omega)$ such that $\boldsymbol{x} \mapsto \Phi_{\boldsymbol{x}} = K(\cdot, \boldsymbol{x})$ transforms a given problem from Ω to $\mathcal{H}_K(\Omega)$ via the reproducing kernel K. Since Ω can be a very general set (consisting of texts, images, medical data, etc.) the feature map is at the heart of applications of kernel methods to problems in machine learning and its application is known there as the *kernel trick* (see, e.g., [Schölkopf and Smola (2002)]). In essence, the feature map allows us to compute with all sorts of quantities that are not at all numerical by turning, e.g., a Shakespeare sonnet $\boldsymbol{x}$ into a function $\Phi_{\boldsymbol{x}}$ that expresses — via the kernel K — the similarity of $\boldsymbol{x}$ to all other texts in the collection Ω. A particularly attractive feature of the kernel trick is the fact that the actual (nonlinear) feature map need not be known; simply working with the kernel K is sufficient and easy. In fact, all the techniques for determining optimal classifiers via *structural risk minimization* are therefore essentially the same as our techniques for finding optimal kernel-based approximants. We discuss this in detail in Chapter 18.

Chapter 3

Examples of Kernels

In this chapter we provide an overview of many different types of positive definite kernels. We focus on their definition and on some of their properties, while in the next chapter we add a discussion of the different kinds of implementations in MATLAB and then use the code to produce plots of some of the kernels featured below.

We start with the case of radial kernels (or radial basis functions), which may be one of the most popular types of kernels discussed in the literature. Then we move on to more general translation invariant kernels and kernels given in series form. Other types of kernels discussed below include general anisotropic kernels, dot product kernels (including zonal kernels on spheres), tensor product kernels, multiscale kernels, space-time kernels and so-called "learned" kernels. We also discuss some recent advances in compactly supported radial kernels and close the chapter with some ideas concerning designer kernels, i.e., kernels which are created either from other positive definite kernels or as infinite series.

3.1 Radial Kernels

3.1.1 *Isotropic radial kernels*

A radial kernel is of the form $K(\boldsymbol{x}, \boldsymbol{z}) = \kappa(\|\boldsymbol{x} - \boldsymbol{z}\|)$, i.e., it is *invariant under both translation* and *rotation*. In the statistics literature such a kernel is called *stationary isotropic*. We will frequently use the letter κ to denote such a function of a single scalar variable, i.e., $\kappa : \mathbb{R}_0^+ \to \mathbb{R}$, and the abbreviation $r = \|\boldsymbol{x} - \boldsymbol{z}\|$, the Euclidean distance between the points $\boldsymbol{x}$ and $\boldsymbol{z}$ in $\mathbb{R}^d$, to denote the argument of the radial kernel κ. While it is feasible to consider also other types of p-norms (see, e.g., [Fasshauer (2007, Chapter 10)]), we work exclusively with the Euclidean norm in this book[1].

Radial kernels are sometimes referred to as *radial basic functions* (or radial

[1] The geodesic distance makes a brief appearance in Section 3.4.2 but the kernels we work with on the sphere will be so-called "embedded" kernels, i.e., kernels specified in terms of the Euclidean distance in the ambient space.

Table 3.1 Some commonly used radial kernels.

Kernel	Parameters	Name
$\kappa(\varepsilon r) = \dfrac{\sin(\varepsilon r)}{\varepsilon r}$	$\varepsilon > 0$	wave kernel
$\kappa(\varepsilon r) = \dfrac{J_{d/2-1}(\varepsilon r)}{(\varepsilon r)^{d/2-1}}$	$\varepsilon > 0,\ d \in \mathbb{N}$	Poisson or Bessel kernel[a]
$\kappa(\varepsilon r) = \dfrac{K_{d/2-\beta}(\varepsilon r)}{(\varepsilon r)^{d/2-\beta}}$	$\varepsilon > 0,\ \beta > \frac{d}{2},\ d \in \mathbb{N}$	Matérn kernel[b]
$\kappa(\varepsilon r) = \mathrm{e}^{-\varepsilon^2 r^2}$	$\varepsilon > 0$	Gaussian or squared exponential kernel
$\kappa(\varepsilon r) = \mathrm{e}^{-(\varepsilon r)^\beta}$	$\varepsilon > 0,\ \beta \in (0, 2]$	powered exponential kernel
$\kappa(\varepsilon r) = \dfrac{1}{1+\varepsilon^2 r^2}$	$\varepsilon > 0$	inverse quadratic (IQ) or Cauchy kernel
$\kappa(\varepsilon r) = \dfrac{1}{\sqrt{1+\varepsilon^2 r^2}}$	$\varepsilon > 0$	inverse multiquadric (IMQ) kernel
$\kappa(\varepsilon r) = \sqrt{1+\varepsilon^2 r^2}$	$\varepsilon > 0$	multiquadric (MQ) kernel

[a]The functions $J_{d/2-1}$ are *Bessel functions of the first kind of order* $d/2-1$ [DLMF, Eq. (10.2.2)].
[b]The functions $K_{d/2-\beta}$ are *modified Bessel functions of the second kind of order* $d/2-\beta$ [DLMF, Eq. (10.31.1)].

basis functions), and most of the examples listed in Table 3.1 and discussed in the following can also be found in [Fasshauer (2007, Chapter 4)], where the reader can find additional information about radial kernels, such as their Fourier transforms (see also [Wendland (2005)]). Several different parameters appear in Table 3.1: ε is a *shape parameter* that can be thought of as an inverse length scale, d is the *space dimension* (i.e., $\boldsymbol{x}, \boldsymbol{z} \in \mathbb{R}^d$), and the parameter β affects the *smoothness of the kernel* as discussed in more detail below.

The wave kernel is a multivariate version of the univariate *sinc kernel* (see Example 3.3 and Appendix A.8). It is an infinitely differentiable kernel that is positive definite in $\mathbb{R}^d$, but only for $d \leq 3$ (see [Genton (2002)]). This kernel sometimes appears in the machine learning literature and in its univariate form plays a central role in sampling theory.

Bessel kernels are fundamental solutions, bounded at the origin, of the d-dimensional Helmholtz operator in spherical coordinates, i.e., they satisfy $\kappa''(r) + \frac{d-1}{r}\kappa'(r) + \varepsilon^2\kappa(r) = 0$ (see, e.g., [Fornberg *et al.* (2006)]). The alternative name *Poisson functions* was suggested much earlier by Schoenberg (1938). The name of the "wave kernel" is explained by the fact that it corresponds to the special case $d = 3$ of the family of Bessel kernels. Bessel kernels are infinitely smooth and positive definite on $\mathbb{R}^d$ for the corresponding value of d (and below). They also serve as a reminder that a positive definite kernel need not be positive-valued — a misconception that probably held back the discovery of compactly supported radial kernels (see Section 3.5 as well as the discussion in [Fasshauer (2011b, Section 2.4)]).

Similarly, Matérn kernels (sometimes also called *Whittle–Mátern kernels*, see, e.g., [Santner *et al.* (2003); Guttorp and Gneiting (2005)]) are fundamental solutions

Table 3.2 Some simple Matérn kernels.

k	β	Kernel	Smoothness
0	$\frac{d+1}{2}$	$\kappa(\varepsilon r) = \mathrm{e}^{-\varepsilon r}$	C^0
1	$\frac{d+3}{2}$	$\kappa(\varepsilon r) = (1+\varepsilon r)\mathrm{e}^{-\varepsilon r}$	C^2
2	$\frac{d+5}{2}$	$\kappa(\varepsilon r) = \left(1+\varepsilon r + \frac{1}{3}(\varepsilon r)^2\right)\mathrm{e}^{-\varepsilon r}$	C^4
3	$\frac{d+7}{2}$	$\kappa(\varepsilon r) = \left(1+\varepsilon r + \frac{2}{5}(\varepsilon r)^2 + \frac{1}{15}(\varepsilon r)^3\right)\mathrm{e}^{-\varepsilon r}$	C^6

of the d-dimensional iterated modified Helmholtz operator in Euclidean coordinates, i.e., $\mathcal{L} = \left(-\nabla^2 + \varepsilon^2\mathcal{I}\right)^\beta$, where $\mathcal{I}$ is the identity operator, and it is well known that they generate classical Sobolev spaces $H^\beta(\mathbb{R}^d)$ as their reproducing kernel Hilbert spaces. This fact prompted Schaback (1993) to propose the alternative name *Sobolev kernel*[2]. These kernels are positive definite on $\mathbb{R}^d$ with $\beta > \frac{d}{2}$. It should be noted that — for $\beta = \frac{d+2k+1}{2}$, $k = 0, 1, 2, \ldots$ — the expression for the Matérn kernels reduces to the product of a polynomial of degree k in r and an exponential function so that the kernel is $2k$-times continuously differentiable (see Table 3.2 for some examples). The $k = 0$ case is also known under the names *exponential kernel* or *Ornstein–Uhlenbeck kernel* — especially in the statistics and machine learning communities. We computed the eigenvalues and eigenfunctions of this special case when $\Omega = [0, \infty)$ in Example 2.5.

Remark 3.1. Whittle-Matérn kernels (along with the conditionally positive definite polyharmonic spline kernels) were recently generalized by Bozzini, Rossini and Schaback (2013), who derived Green's kernels for differential operators of the form $\prod_{j=1}^{\beta}(-\nabla^2 + \varepsilon_j^2\mathcal{I})$. We will discuss the connection between Green's kernels and positive definite kernels in Chapter 6.

While the kernels discussed so far are positive definite on $\mathbb{R}^d$ only for a certain range of d, all other kernels listed in Table 3.1, except for the multiquadric, are positive definite on $\mathbb{R}^d$ for every positive integer value of d. Multiquadrics (MQs), while not limited by any restriction on the space dimension, are only *conditionally* positive (actually negative) definite. We will occasionally mention conditionally positive definite kernels in this book, but will not discuss any of their special properties. For this we refer the reader to, e.g., [Wendland (2005); Fasshauer (2007)].

The ubiquitous *Gaussian kernel* does not need much of an introduction, although it might be curious to note that it seems to have been first mentioned in its role as a probability density function by Pierre Simon de Laplace (1778, Section XXXVII) more than 30 years before Carl Friedrich Gauß used it in [Gauß (1809, Section 177)] in the context of least squares approximation and maximum likelihood estimation

[2]We will, however, reserve the term Sobolev kernel for another family of kernels from [Berlinet and Thomas-Agnan (2004)] (see Example 8.3).

(two fundamental concepts that *are* due to Gauß). The Gaussian kernel is infinitely differentiable and positive definite on $\mathbb{R}^d$ for all $d \in \mathbb{N}$. It will be widely featured in this book. In the literature on statistics and machine learning the Gaussian kernel is sometimes referred to as the *squared exponential kernel*.

The *powered exponential kernel* is common in the statistics literature since — similar as for the Matérn family — the two parameters ε and β provide flexibility with respect to scale and smoothness. However, Stein (1999) states that an important difference between the Matérn family and the powered exponential family is that the latter kernels — except for the $\beta = 2$ case — are not mean-square differentiable, i.e., their sample paths are not smooth. Note that the C^0 Matérn kernel is a special case of the powered exponential, as is the infinitely smooth Gaussian.

The infinitely differentiable *inverse quadratic* (IQ) kernel is equivalent to the *rational quadratic kernel* (see, e.g., [Genton (2002)]) since

$$\frac{1}{1+\varepsilon^2 r^2} = 1 - \frac{r^2}{\theta + r^2}$$

with $\theta = 1/\varepsilon^2$. This kernel is sometimes recommended as a computationally cheaper alternative to the Gaussian kernel in the machine learning literature.

The *inverse multiquadric* (IMQ) kernel is quite popular in the literature on radial basis functions. It is infinitely differentiable and was introduced by Hardy and Göpfert (1975). As we will see in Section 3.4.2, it may be the "natural" kernel to use in the spherical setting.

3.1.2 *Anisotropic radial kernels*

Any isotropic radial kernel can be turned into an anisotropic radial kernel by using a weighted 2-norm instead of an unweighted one as in Section 3.1.1. For example, an *anisotropic Gaussian kernel* is of the form

$$K(\boldsymbol{x}, \boldsymbol{z}) = \mathrm{e}^{-(\boldsymbol{x}-\boldsymbol{z})^T \mathsf{E}(\boldsymbol{x}-\boldsymbol{z})}, \tag{3.1}$$

where E is a symmetric positive definite matrix. If $\mathsf{E} = \varepsilon^2 \mathsf{I}_d$, where I_d is a $d \times d$ identity matrix, then the kernel is isotropic. Use of anisotropic kernels is not a common occurrence in the RBF literature. They have been analyzed, e.g., in [Baxter (2006); Beatson *et al.* (2010)] and applied, e.g., in [Carr *et al.* (2003); Casciola *et al.* (2006, 2007, 2010)]. In the literature on information-based complexity, on the other hand, anisotropic kernels are very popular (see, e.g., [Novak and Woźniakowski (2008)]). The special case $\mathsf{E} = \mathrm{diag}(\varepsilon_1^2, \ldots, \varepsilon_d^2)$, a diagonal matrix with *dimension-dependent shape parameters*, is used, e.g., in [Fasshauer *et al.* (2012a,b)] to avoid the *curse of dimensionality* and to obtain dimension-independent error bounds (see Section 9.5 for more details on this).

Remark 3.2. Some authors have also applied a different shape parameter to each basis function in the RBF interpolation expansion (1.3a) resulting in an expansion such as

$$s(\boldsymbol{x}) = \sum_{j=1}^{N} c_j \mathrm{e}^{-\varepsilon_j^2 \|\boldsymbol{x}-\boldsymbol{x}_j\|^2}, \qquad \boldsymbol{x} \in \mathbb{R}^d,$$

where we have used Gaussian kernels to have a specific example. These kernels are still isotropic radial kernels, but with a *spatially varying shape parameter* that may be selected adaptively, e.g., depending on the local density of data sites. Now, however, the interpolant is no longer generated by a single kernel and the theoretical foundation must be reconsidered (see, e.g., [Kansa (1990b); Kansa and Carlson (1992); Bozzini *et al.* (2002); Driscoll and Heryudono (2007); Fornberg and Zuev (2007); Sarra and Sturgill (2009); Deng and Driscoll (2012)]). The most promising paper to address this approach — especially on a theoretical level — is [Bozzini *et al.* (2015)], where this problem is tackled by embedding a d-dimensional interpolation problem into $\mathbb{R}^{d+1}$ so that the additional dimension houses the locally varying shape parameter. In $\mathbb{R}^{d+1}$ one then works with a "standard" kernel with fixed global shape.

3.2 Translation Invariant Kernels

A kernel is called *translation invariant* (or *stationary* in the statistics literature) if $K(\boldsymbol{x}+\boldsymbol{h}, \boldsymbol{z}+\boldsymbol{h}) = K(\boldsymbol{x}, \boldsymbol{z})$ for any $\boldsymbol{h} \in \mathbb{R}^d$. This means that K is a function of the *difference* of $\boldsymbol{x}$ and $\boldsymbol{z}$, i.e.,

$$K(\boldsymbol{x}, \boldsymbol{z}) = \widetilde{K}(\boldsymbol{x} - \boldsymbol{z}).$$

We will frequently use the symbol $\widetilde{K}$ to denote such a function of a single vector variable. In the literature on integral equations, translation invariant kernels are often referred to as *convolution kernels*, and as such — especially in their discretized form as matrices — they find applications in image processing.

The Fourier transform is the ideal tool to analyze convolution kernels, and in fact the celebrated characterization of positive definite functions by Bochner applies precisely to functions that are translation invariant (see, e.g., [Fasshauer (2007, Theorem 3.3)]. All of the isotropic radial kernels of the previous section are, of course, also examples of translation invariant kernels.

The kernel $K(x, z) = (1-\varepsilon|x-z|)_+$, where $(x)_+ = x$ if $x > 0$ and zero otherwise, is a translation invariant kernel. It is frequently referred to as the *triangular kernel* (see, e.g., [Kailath (1966) or Stein (1999, Chapter 2)]). One can interpret it also as a radial kernel, but since it is positive definite only on the real line (and not in $\mathbb{R}^d$ for $d > 1$) it might be a bit of a stretch to call this a radially invariant kernel.

Another simple example is given by

$$K(x, z) = \cos(x - z),$$

which is a positive definite translation invariant kernel on $\mathbb{R}$. In fact, any nonnegative linear combination of kernels of the form $K_n(x,z) = \cos(n(x-z))$ is also positive definite and translation invariant on $\mathbb{R}$. We can even take infinite linear combinations, i.e., Fourier cosine series, which then also provide examples of series kernel — a form of kernel we discuss in Section 3.3.

In particular, one can use the Fourier cosine series of Bernoulli polynomials [DLMF, Eq. (24.8.1)] to represent periodic univariate splines with the kernel

$$K(x,z) = B_{2\beta}(|x-z|) = (-1)^{\beta-1}\sum_{n=1}^{\infty}\frac{2(2\beta)!}{(2n\pi)^{2\beta}}\cos\left(2n\pi(x-z)\right), \qquad \beta\in\mathbb{N}. \quad (3.2)$$

The reproducing kernel Hilbert space for these splines is $H^{\beta}_{\text{per}}(0,1)$ (see [Wahba (1990, Chapter 2)] and also Chapter 7 and Appendix A.5.2).

To get a kernel in higher dimensions we can take a *tensor product* (see also Section 3.4.3) of one-dimensional translation invariant kernels, so that

$$K(\boldsymbol{x},\boldsymbol{z}) = \prod_{\ell=1}^{d}\sum_{n=0}^{\infty} a_{n,\ell}K_n(x_\ell,z_\ell), \qquad a_{n,\ell}\ge 0,\ \boldsymbol{x} = \left(x_1 \ \cdots \ x_d\right)^T,$$

where $K_n(x,z) = \cos(n(x-z))$, is a positive definite kernel on $\mathbb{R}^d$.

Two specific examples of such kernels are given in [Steinwart and Christmann (2008, Chapter 4)]:

$$K(\boldsymbol{x},\boldsymbol{z}) = \prod_{\ell=1}^{d}\frac{1-q^2}{2-4q\cos(x_\ell-z_\ell)+2q^2}, \qquad 0<q<1,\ \boldsymbol{x},\boldsymbol{z}\in[0,2\pi]^d,$$

$$K(\boldsymbol{x},\boldsymbol{z}) = \prod_{\ell=1}^{d}\frac{\pi q\cosh(\pi q - q(x_\ell-z_\ell))}{2\sinh(\pi q)}, \qquad 1<q<\infty,\ \boldsymbol{x},\boldsymbol{z}\in[0,2\pi]^d,$$

since for all $t\in[0,2\pi]$ in the former case $\frac{1-q^2}{2-4q\cos(t)+2q^2} = \frac{1}{2}+\sum_{n=1}^{\infty}q^n\cos(nt)$, and in the latter $\frac{\pi q\cosh(\pi q-qt)}{2\sinh(\pi q)} = \frac{1}{2}+\sum_{n=1}^{\infty}\frac{\cos(nt)}{1+n^2/q^2}$.

3.3 Series Kernels

The use of closed forms for kernel evaluation is satisfying, but modern computational tools have provided the opportunity to work with kernels that have no known expression in terms of elementary functions; as we discuss in Section 7.2, even when a closed form exists it may not always be the most practical method of evaluation. One such class of kernels are kernels defined through an infinite series such as, e.g., the Mercer series introduced in Section 2.2.3.

To talk about series kernels, it behooves us to introduce the *multi-index* $\boldsymbol{\alpha} = (\alpha_1,\ldots,\alpha_d)\in\mathbb{Z}^d$, which is an array of integers that allows for concise computation

Table 3.3 Classical Taylor series kernels.

Kernel	Series	Name
$K(x,z) = \frac{1}{(1-x\overline{z})^2}$	$\sum_{n=0}^{\infty}(n+1)x^n\overline{z}^n$	Bergman kernel
$K(x,z) = \frac{1}{1-x\overline{z}}$	$\sum_{n=0}^{\infty} x^n\overline{z}^n$	Hardy or Szegő kernel
$K(x,z) = -\frac{\ln(1-x\overline{z})}{x\overline{z}}$	$\sum_{n=0}^{\infty}\frac{1}{n+1}x^n\overline{z}^n$	Dirichlet kernel

of multiple dimensions. For instance, a data point $\boldsymbol{x} = \begin{pmatrix}1 & -7 & 0.44\end{pmatrix}^T$ and multi-index $\boldsymbol{\alpha} = (2,4,9)$ could produce

$$\boldsymbol{x}^{\boldsymbol{\alpha}} = \begin{pmatrix}1\\-7\\0.44\end{pmatrix}^{(2,4,9)} = 1^2(-7)^4(0.44)^9$$

for use inside an infinite summation. Additionally, complicated quantities that appear in infinite summations can be encapsulated cleanly, such as total order $|\boldsymbol{\alpha}| = \alpha_1 + \ldots + \alpha_d$ and multinomial coefficients

$$\binom{|\boldsymbol{\alpha}|}{\boldsymbol{\alpha}} = \frac{|\alpha|!}{\prod_{\ell=1}^d \alpha_\ell!}.$$

3.3.1 *Power series and Taylor series kernels*

Power series kernels of the form

$$K(\boldsymbol{x},\boldsymbol{z}) = \sum_{\boldsymbol{\alpha}\in\mathbb{N}_0^d} w_{\boldsymbol{\alpha}} \frac{\boldsymbol{x}^{\boldsymbol{\alpha}}}{\boldsymbol{\alpha}!}\frac{\boldsymbol{z}^{\boldsymbol{\alpha}}}{\boldsymbol{\alpha}!}, \qquad \boldsymbol{x},\boldsymbol{z}\in\mathbb{R}^d,$$

where the coefficients $w_{\boldsymbol{\alpha}}$ must satisfy the summability condition $\sum_{\boldsymbol{\alpha}\in\mathbb{N}_0^d}\frac{w_{\boldsymbol{\alpha}}}{\boldsymbol{\alpha}!^2} < \infty$ were studied by Zwicknagl (2008). A typical example of such a kernel — not to be confused with the (powered) exponential kernels of Table 3.1 — is given by the *exponential kernel*

$$K(\boldsymbol{x},\boldsymbol{z}) = \mathrm{e}^{\boldsymbol{x}^T\boldsymbol{z}} = \sum_{n=0}^{\infty}\frac{1}{n!}(\boldsymbol{x}^T\boldsymbol{z})^n = \sum_{\boldsymbol{\alpha}\in\mathbb{N}_0^d}\frac{1}{|\boldsymbol{\alpha}|!}\binom{|\boldsymbol{\alpha}|}{\boldsymbol{\alpha}}\boldsymbol{x}^{\boldsymbol{\alpha}}\boldsymbol{z}^{\boldsymbol{\alpha}}.$$

Taylor series kernels were discussed in [Zwicknagl and Schaback (2013)]. They include many classical kernels such as those listed in Table 3.3, where $x, z \in \mathbb{D}$, the open complex unit disk, i.e., $\mathbb{D} = \{x \subset \mathbb{C} : |x| < 1\}$. The reproducing kernel Hilbert spaces associated with these kernels are in turn the *Bergman space* $B^2 = L^2(\mathbb{D})$, i.e., the space of analytic functions in $\mathbb{D}$ that are square summable with respect to planar Lebesgue measure; the *Hardy space* H^2, i.e., the space of analytic functions in $\mathbb{D}$ with square summable Taylor coefficients; and the *Dirichlet space* $\mathcal{D}$, i.e., the space of analytic functions in $\mathbb{D}$ whose derivatives are in B^2. Note that the Hardy space H^2 is contained in the Bergman space B^2.

All three of these kernels and spaces can be considered as *classical* examples and they are discussed, e.g., in [Aronszajn (1950); Bergman (1950); Meschkowski (1962); Hille (1972); Agler and McCarthy (2002)]. Zwicknagl and Schaback (2013) show that all three kernels can be restricted to a real (sub)-interval of $[-1, 1]$ and then yield spectral convergence orders there.

3.3.2 *Other series kernels*

Other examples of series kernels are given by kernels specified by binomial series, Fourier-type series (such as the periodic spline kernels (3.2)), or kernels specified via their Mercer or Hilbert–Schmidt series — sometimes also referred to as *Mercer kernels*. Examples of Mercer kernels that appear throughout this book include the Brownian motion kernel from Example 2.6, the *iterated Brownian bridge kernels* of Chapter 7, or the *Chebyshev kernels* introduced in Section 3.9.2. A few other Mercer kernels can be found, e.g., in [Wasilkowski and Woźniakowski (1999); Werschulz and Woźniakowski (2009); Dick *et al.* (2013b)].

In addition to some of the examples discussed in more detail in Chapter 6, we can find Green's kernels in closed form and with series expansions.

Example 3.1. (Binomial kernel)
In the machine learning literature one finds the *binomial kernel* (see, e.g., [Steinwart and Christmann (2008)]), which can be considered as a multivariate (dot product) generalization of the Bergman kernel from Table 3.3. It is defined on the open unit ball $\mathbb{B}^d$ in $\mathbb{R}^d$ as

$$\begin{aligned} K(\boldsymbol{x}, \boldsymbol{z}) &= \frac{1}{(1 - \boldsymbol{x}^T\boldsymbol{z})^{\beta}}, \qquad \beta > 0, \ \boldsymbol{x}, \boldsymbol{z} \in \mathbb{B}^d, \\ &= \sum_{n=0}^{\infty} \binom{-\beta}{n} (-1)^n \left(\boldsymbol{x}^T\boldsymbol{z}\right)^n. \end{aligned}$$

Example 3.2. (Legendre kernel)
We consider the differential operator defined by $\mathcal{L}u(x) = -\frac{\mathrm{d}}{\mathrm{d}x}\left((1 - x^2)\frac{\mathrm{d}}{\mathrm{d}x}u(x)\right)$ on $(-1, 1)$ with (regular singular) boundary condition $\lim_{|x|\to 1} u(x) < \infty$. Its eigenfunctions are the Legendre polynomials P_n of degree n (see also Example 5.3), and the corresponding eigenvalues are $n(n+1)$. In this case we have (see, e.g., [Courant and Hilbert (1953, Chapter V, Section 15)])

$$\begin{aligned} K(x, z) &= -\frac{1}{2}\log(1 - \min(x, z))(1 + \max(x, z)) + \log 2 - \frac{1}{2} \\ &= \sum_{n=1}^{\infty} \frac{(2n + 1)P_n(x)P_n(z)}{2n(n + 1)}. \end{aligned}$$

The additional factor $\frac{2n+1}{2}$ — missing from [Courant and Hilbert (1953)] — corresponds to the reciprocal of the square of the norm of the Legendre polynomials (see (6.13)). This kernel is only continuous on $(-1, 1)$.

Example 3.3. (Paley–Wiener kernel)
The reproducing kernel of the Paley–Wiener space, the space of bandlimited functions on $\mathbb{R}$, is given by the sinc function and can be expressed in series form as (see [Berlinet and Thomas-Agnan (2004); Schaback (2011a)])

$$\begin{aligned} K(x,z) &= \frac{\sin(\varepsilon(x-z))}{\varepsilon(x-z)}, \qquad x, z \in \mathbb{R}, \\ &= \sum_{n\in\mathbb{Z}} \frac{\sin(\varepsilon x - n)}{\varepsilon x - n} \frac{\sin(\varepsilon z - n)}{\varepsilon z - n}. \end{aligned}$$

Remark 3.3. More examples of series kernels can be found in Appendix A.

3.4 General Anisotropic Kernels

3.4.1 *Dot product kernels*

Dot product kernels are common in the machine learning literature (see, e.g., [Shawe-Taylor and Cristianini (2004)]) and are often mentioned together with their associated feature maps. For example, the *linear kernel* is of the form $K(\boldsymbol{x}, \boldsymbol{z}) = \boldsymbol{x}^T \boldsymbol{z}$, $\boldsymbol{x}, \boldsymbol{z} \in \mathbb{R}^d$, and can be expressed via Theorem 2.5 using the finite-dimensional features $\Phi_{\boldsymbol{x}} = (x_1, x_2, \ldots, x_d)^T = \boldsymbol{x}$, i.e.,

$$K(\boldsymbol{x}, \boldsymbol{z}) = \boldsymbol{x}^T \boldsymbol{z} = \sum_{\ell=1}^{d} x_\ell z_\ell.$$

The linear kernel is generalized by the *polynomial kernel*

$$K(\boldsymbol{x}, \boldsymbol{z}) = \left(\gamma + \boldsymbol{x}^T \boldsymbol{z}\right)^\beta, \qquad \boldsymbol{x}, \boldsymbol{z} \in \mathbb{R}^d, \quad \gamma \geq 0, \ \beta \in \mathbb{N}_0.$$

This kernel is positive definite for all $\gamma \geq 0$ and nonnegative integers β (see, e.g., [Schölkopf and Smola (2002, Section 4.6)]). As an example we may consider $\beta = d = 2$ and $\gamma = 1$ so that we get

$$\begin{aligned} K(\boldsymbol{x}, \boldsymbol{z}) &= \left(1 + \boldsymbol{x}^T \boldsymbol{z}\right)^2 = (1 + x_1 z_1 + x_2 z_2)^2 \\ &= 1 + 2x_1 z_1 + 2x_2 z_2 + 2x_1 x_2 z_1 z_2 + x_1^2 z_1^2 + x_2^2 z_2^2, \end{aligned}$$

and the corresponding features are $\Phi_{\boldsymbol{x}} = \left(1, \sqrt{2}x_1, \sqrt{2}x_2, \sqrt{2}x_1 x_2, x_1^2, x_2^2\right)$. By using a different value of γ the different powers of the coordinates x_ℓ will be weighted differently. It is well known that the dimension of the space Π_d^β of d-variate polynomials of degree β is $\binom{d+\beta}{\beta}$. This is also the dimension of the feature space associated with K (see Section 2.4 and Section 18.3.2).

Another kernel that is popular in machine learning is the *sigmoid kernel*

$$K(\boldsymbol{x}, \boldsymbol{z}) = \tanh(1 + \gamma \boldsymbol{x}^T \boldsymbol{z}),$$

even though this kernel is *not* positive definite for any choice of γ (see [Schölkopf and Smola (2002, Ex. 4.25)]). Both the sigmoid kernel and the polynomial kernels are sometimes referred to as *dot product kernels* since they depend on $\boldsymbol{x}$ and $\boldsymbol{z}$ only through their dot product. Such functions are also known as *ridge functions* (see, e.g., [Cheney and Light (1999, Chapter 22)] or [Pinkus (2013)]) and they first arose in the context of computerized tomography [Logan and Shepp (1975)].

3.4.2 *Zonal kernels*

One can analyze dot product kernels via Mercer's theorem (see Section 2.2.3) by treating them as *zonal functions* on spheres $\mathbb{S}^d$ in $\mathbb{R}^{d+1}$. Then the expansion can be written in terms of Legendre or Gegenbauer polynomials (and ultimately spherical harmonics) and the work of, e.g., Xu and Cheney (1992); Ron and Sun (1996); Menegatto (1999) applies (see also [Schölkopf and Smola (2002, Section 4.6)]). Since our examples in Chapters 4 and 15 will be for $\mathbb{S}^2$, and since the notation is a bit more transparent in this setting, we now restrict the following discussion to $\mathbb{S}^2$, the unit sphere in $\mathbb{R}^3$, even though everything we are about to say generalizes to higher dimensions.

The starting point for working with positive definite zonal kernels is a beautiful result by Schoenberg (1942) (later generalized by Chen, Menegatto and Sun (2003)) according to which any positive definite zonal kernel on $\mathbb{S}^2$ can be expressed as a series of Legendre polynomials with nonnegative coefficients, i.e.,

$$K(\boldsymbol{x},\boldsymbol{z}) = \sum_{\ell=0}^{\infty} a_\ell P_\ell(\boldsymbol{x}^T\boldsymbol{z}), \qquad a_\ell \geq 0 \text{ such that } \sum_{\ell=0}^{\infty} a_\ell < \infty.$$

Remark 3.4. The use of ℓ as the summation index contradicts our desire to use n wherever possible. This decision is made in deference to the notation for spherical harmonics and will help augment distinctions made between Mercer series (indexed with n) and spherical harmonics series that appear in Section 15.3.

The *spherical inverse multiquadric kernel* (or *multiquadric biharmonic kernel*) [Hardy and Göpfert (1975)] is given by

$$K(\boldsymbol{x},\boldsymbol{z}) = \frac{1}{\sqrt{1+\gamma^2-2\gamma\boldsymbol{x}^T\boldsymbol{z}}}, \qquad 0<\gamma<1. \tag{3.3}$$

This kernel might be the most "natural" positive definite zonal kernel since it is known to be the generating function for the *Legendre polynomials* [Szegő (1939)], i.e., the Legendre series for the spherical IMQ kernel (3.3) is given by

$$\frac{1}{\sqrt{1+\gamma^2-2\gamma\boldsymbol{x}^T\boldsymbol{z}}} = \sum_{\ell=0}^{\infty} \gamma^\ell P_\ell(\boldsymbol{x}^T\boldsymbol{z}), \qquad \boldsymbol{x},\boldsymbol{z}\in\mathbb{S}^2,\ 0<\gamma<1, \tag{3.4}$$

where γ can be interpreted as a shape parameter.

Now we can use the addition formula [Müller (1966)]

$$P_\ell(\boldsymbol{x}^T\boldsymbol{z}) = \frac{4\pi}{2\ell+1}\sum_{m=1}^{2\ell+1} Y_{\ell,m}(\boldsymbol{x})Y_{\ell,m}(\boldsymbol{z}), \qquad \boldsymbol{x},\boldsymbol{z}\in\mathbb{S}^2,$$

which expresses a Legendre polynomial P_ℓ of degree ℓ in terms of *spherical harmonics* $Y_{\ell,m}$, $m = 1,\ldots,2\ell+1$, of degree ℓ and order m. It is well known (see, e.g., [Müller (1966); Dai and Xu (2013)]) that the spherical harmonics are the eigenfunctions of the Laplace–Beltrami operator on the sphere and that they form an orthonormal basis for $L_2(\mathbb{S}^2)$. Therefore, applying this addition formula to the Legendre series (3.4) yields a form of the Mercer series

$$\frac{1}{\sqrt{1+\gamma^2-2\gamma\boldsymbol{x}^T\boldsymbol{z}}} = \sum_{\ell=0}^{\infty}\frac{4\pi\gamma^\ell}{2\ell+1}\sum_{m=1}^{2\ell+1} Y_{\ell,m}(\boldsymbol{x})Y_{\ell,m}(\boldsymbol{z}) \tag{3.5}$$

of the spherical inverse multiquadric kernel. We know that the geometric decay of the coefficients $\frac{4\pi\gamma^\ell}{2\ell+1}$, which play the role of the Hilbert-Schmidt eigenvalues λ_n in the Mercer series, implies the kernel is analytic (see Remark 2.4). More discussion about how the eigenfunctions φ_n that appear in the Mercer series relate to the spherical harmonics appears in Example 15.5.

In a sense, zonal kernels can also be interpreted as a type of radial kernel, and therefore we could just as well have discussed them in Section 3.1. On the unit sphere $\mathbb{S}^{d-1} = \{\boldsymbol{x}\in\mathbb{R}^d \mid \|\boldsymbol{x}\| = 1\}$ in any space dimension d we can relate the Euclidean distance $r = \|\boldsymbol{x}-\boldsymbol{z}\|$ between two points $\boldsymbol{x},\boldsymbol{z}\in\mathbb{S}^{d-1}$ (measured in the ambient space, i.e., cutting through the sphere), to the *geodesic distance* $g(\boldsymbol{x},\boldsymbol{z}) = \arccos\boldsymbol{x}^T\boldsymbol{z}$ (measured along a great circle on the surface of the sphere). In fact, for any $\boldsymbol{x},\boldsymbol{z}\in\mathbb{S}^{d-1}$ we have $\|\boldsymbol{x}-\boldsymbol{z}\|^2 = \|\boldsymbol{x}\|^2+\|\boldsymbol{z}\|^2-2\boldsymbol{x}^T\boldsymbol{z} = 2-2\boldsymbol{x}^T\boldsymbol{z}$. Therefore, $r = \sqrt{2-2\cos(g(\boldsymbol{x},\boldsymbol{z}))}$ or

$$r = \sqrt{2-2\boldsymbol{x}^T\boldsymbol{z}}, \qquad \boldsymbol{x},\boldsymbol{z}\in\mathbb{S}^{d-1}. \tag{3.6}$$

Since it is known that a kernel that is positive definite on a domain Ω is also positive definite on a subdomain Ω_0 of Ω (see Section 2.3), it is straightforward to obtain a large class of positive definite zonal kernels on $\Omega_0 = \mathbb{S}^{d-1}$ by restricting Euclidean radial kernels in $\mathbb{R}^d$ to the sphere via (3.6) [Fasshauer and Schumaker (1998); Hubbert and Baxter (2001); zu Castell and Filbir (2005); Fornberg and Piret (2008b); Fuselier and Wright (2009)].

Remark 3.5. An alternative spherical version of the Euclidean IMQ kernel from Table 3.1, i.e., $\kappa(\varepsilon r) = \frac{1}{\sqrt{1+(\varepsilon r)^2}}$, appears in the literature on radial — or more appropriately spherical — basis functions (see, e.g., [Hubbert and Baxter (2001); Fornberg and Piret (2008b)]). Using (3.6), and therefore $r^2 = 2-2\boldsymbol{x}^T\boldsymbol{z}$, this approach suggests that the spherical IMQ is of the form $K(\boldsymbol{x},\boldsymbol{z}) = \frac{1}{\sqrt{1+2\varepsilon^2(1-\boldsymbol{x}^T\boldsymbol{z})}}$, and subsequently the coefficients for the corresponding Legendre series provided in these references look different from those used above. We would like to point out that the two formulations of the spherical IMQ kernel are equivalent.

In fact, (3.6) also implies $2\boldsymbol{x}^T\boldsymbol{z} = 2 - r^2$. Therefore, we can rewrite the spherical IMQ (3.3) as

$$\begin{aligned} K(\boldsymbol{x}, \boldsymbol{z}) &= \frac{1}{\sqrt{1 + \gamma^2 - 2\gamma\boldsymbol{x}^T\boldsymbol{z}}} \\ &= \frac{1}{\sqrt{1 + \gamma^2 - \gamma(2 - r^2)}} \\ &= \frac{1}{\sqrt{(1-\gamma)^2 + \gamma r^2}} \\ &= \frac{1}{1-\gamma} \frac{1}{\sqrt{1 + \frac{\gamma}{(1-\gamma)^2} r^2}}, \end{aligned} \tag{3.7}$$

where we use $\gamma < 1$ in the last step. We compare this to the Euclidean form of the IMQ

$$\kappa(\varepsilon r) = \frac{1}{\sqrt{1 + \varepsilon^2 r^2}}$$

and note that we may drop the factor $\frac{1}{1-\gamma}$ from (3.7) since it is immaterial for basic interpolation (but may be relevant in the kriging setting). Then we see that setting $\frac{\gamma}{(1-\gamma)^2} = \varepsilon^2$ allows us to recover one of the IMQs given the other. Thus, both formulations are equivalent provided $\frac{\gamma}{(1-\gamma)^2}$ can take any value between $[0, \infty)$. This, however, is possible even with the restriction $0 \le \gamma < 1$ made in (3.3).

3.4.3 *Tensor product kernels*

Weighted tensor products of various univariate kernels also produce general anisotropic kernels. For example, using the notation $\boldsymbol{x} = \begin{pmatrix} x_1 & \cdots & x_d \end{pmatrix}^T \in \mathbb{R}^d$, the *product of the Brownian motion kernel* (see, e.g., [Ritter (2000, Section VI.3)])

$$K(\boldsymbol{x}, \boldsymbol{z}) = \prod_{\ell=1}^{d} (1 + \varepsilon_\ell \min(x_\ell, z_\ell)), \qquad \varepsilon_1 \ge \varepsilon_2 \ge \ldots \ge \varepsilon_d \ge 0,$$

is neither radially nor translation invariant. But it is still positive definite in $[0, 1]^d$. Such kernels play an important role in the theory of Monte-Carlo and quasi Monte-Carlo methods (see, e.g., [Novak and Woźniakowski (2008, Appendix 2.2)] or [Ritter (2000, Section VI.3)]), where they are used to avoid the *curse of dimensionality*. We have already seen the univariate Brownian motion kernel in Example 2.6. Another example is the univariate *Brownian bridge kernel* (see Section 6.4.1) and it will be featured throughout this book. Both of these kernels can be used in a tensor product formulation.

The case with uniform weights $\varepsilon_\ell = 1$, i.e.,

$$K(\boldsymbol{x}, \boldsymbol{z}) = \prod_{\ell=1}^{d} (1 + \min(x_\ell, z_\ell)),$$

Table 3.4 Some of the "original" Wendland kernels.

Kernel $\kappa_{d,k}$, where $\ell = \lfloor \frac{d}{2} + k + 1 \rfloor$	Smoothness
$\kappa_{d,0}(r) \doteq (1-r)_+^{\ell}$	C^0
$\kappa_{d,1}(r) \doteq (1-r)_+^{\ell+1}\left((\ell+1)r+1\right)$	C^2
$\kappa_{d,2}(r) \doteq (1-r)_+^{\ell+2}\left(\frac{(\ell^2+4\ell+3)}{3}r^2+(\ell+2)r+1\right)$	C^4
$\kappa_{d,3}(r) \doteq (1-r)_+^{\ell+3}\left(\frac{\ell^3+9\ell^2+23\ell+15}{15}r^3+\frac{6\ell^2+36\ell+45}{15}r^2+(\ell+3)r+1\right)$	C^6

is known as the *Brownian sheet kernel* and $K(\boldsymbol{x},\boldsymbol{z}) = \prod_{\ell=1}^{d}\min(x_\ell, z_\ell)$ is the *Wiener sheet kernel* [Ritter (2000, Ex. VI.21)]. However, the latter kernel is not positive definite, and neither of them is nearly as useful as the weighted version above is.

A rather general construction of kernels with locally varying shape parameters and smoothness was given by Stein (2005a).

Remark 3.6. Another related kernel is the kernel for *fractional Brownian motion* (see, e.g., [Berlinet and Thomas-Agnan (2004)])

$$K(\boldsymbol{x},\boldsymbol{z}) = \frac{1}{2}\left(\|\boldsymbol{x}\|^{2\beta} + \|\boldsymbol{z}\|^{2\beta} - \|\boldsymbol{x}-\boldsymbol{z}\|^{2\beta}\right), \qquad \boldsymbol{x},\boldsymbol{z} \in \mathbb{R}^d.$$

For $\beta = \frac{1}{2}$ and $d = 1$ this simplifies to the standard Brownian motion kernel. However, this kernel is not a tensor product kernel.

Remark 3.7. The *linear covariance kernel* is actually a radial kernel (see, e.g., [Ritter (2000, Section II.3.7)]) even though it is obtained by adding the kernels of two independent Brownian motion kernels, i.e.,

$$K(x,z) = \min(x,z) + \min(1-x, 1-z) = \min(x,z) + 1 - \max(x,z) = 1 - |x-z|.$$

3.5 Compactly Supported Radial Kernels

These kernels are, e.g., discussed in detail in the books [Wendland (2005); Fasshauer (2007)]. They will not play a significant role in this book, but for completeness sake we list a few typical examples. We concentrate on the Wendland family, but other families such as those suggested, e.g., by Buhmann, Gneiting or Wu (see [Fasshauer (2007, Chapter 11)]) exist in the literature. Another recent family of compactly supported piecewise polyharmonic kernels was given by Johnson (2012).

We now use the usual notation $r = \|\boldsymbol{x}-\boldsymbol{z}\|$ to indicate that we are working with radial kernels, i.e., $K(\boldsymbol{x},\boldsymbol{z}) = \kappa(\|\boldsymbol{x}-\boldsymbol{z}\|)$, and let $\doteq$ denote equality up to a constant factor. Frequently used members of the "original" family of kernels $\kappa_{d,k}$ by Wendland (1995) which are positive definite in $\mathbb{R}^d$ are listed in Table 3.4 (cf. [Fasshauer (2007, Chapter 11)]). The first subscript[3] used for the kernels $\kappa_{d,k}$

in Table 3.4 denotes the maximal space dimension d for which the functions are positive definite. The second subscript k indicates that the function is in $C^{2k}(\mathbb{R}^d)$, and $\ell = \lfloor \frac{d}{2} + k + 1 \rfloor$. The reproducing kernel Hilbert spaces associated with the "original" Wendland kernels are equivalent to the Sobolev spaces $H^{d/2+k+1/2}(\mathbb{R}^d)$.

Note that the construction of Wendland (1995) did not allow for Sobolev spaces of integer order in *even* space dimension. This gap was filled when Schaback (2011b) derived the so-called "missing" Wendland functions (see also [Hubbert (2012)]), which were subsequently generalized even further by Chernih and Hubbert (2014).

While the "original" Wendland functions were derived using a "dimension walk" involving integration and ordinary differentiation jumping down or up by two space dimensions, respectively, (see, e.g., [Wendland (2005); Fasshauer (2007)]), the analogous construction for the "missing" functions involves integration and *fractional differentiation* changing the dimension only one at a time (see [Schaback (2011b)]). Note that in contrast to the "original" Wendland functions, these new functions are no longer polynomials on their support. Typical examples are (see [Chernih *et al.* (2014)] and also [Schaback (2011b); Hubbert (2012)])

$$\begin{aligned}
\kappa_{2,\frac{1}{2}}(r) &\doteq (1+2r^2)\sqrt{1-r^2} + 3r^2 \log\left(\frac{r}{1+\sqrt{1-r^2}}\right), \\
\kappa_{3,\frac{3}{2}}(r) &\doteq \left(1 - 7r^2 - \frac{81}{4}r^4\right)\sqrt{1-r^2} - \frac{15}{4}r^4(6+r^2)\log\left(\frac{r}{1+\sqrt{1-r^2}}\right),
\end{aligned} \tag{3.8}$$

where these formulas hold for $r \in [0,1]$ and the functions are zero otherwise. Both $\kappa_{2,\frac{1}{2}}$ and $\kappa_{3,\frac{3}{2}}$ are positive definite in dimension $d \leq 2$ and their native spaces are $H^2(\mathbb{R}^2)$ and $H^3(\mathbb{R}^2)$, respectively. Here the first subscript is $\ell = \lfloor \frac{d}{2} + k + 1 \rfloor$ and it now matches the index of the Sobolev space. For the "missing" Wendland functions the second subscript k is a half-integer and is again connected to the classical smoothness.

More generally, Hubbert (2012) gives closed form representations of both the "original" and the "missing" Wendland functions in terms of associated Legendre functions (of the first and second kinds). Moreover, Chernih, Sloan and Womersley (2014) show that, as their smoothness increases, all appropriately normalized Wendland functions converge to Gaussians.

3.6 Multiscale Kernels

Introduced by Opfer (2006), a general *multiscale kernel* is of the form

$$K(\boldsymbol{x}, \boldsymbol{z}) = \sum_{j \geq 0} w_j K_j(\boldsymbol{x}, \boldsymbol{z}) = \sum_{j \geq 0} w_j \sum_{\boldsymbol{k} \in \mathbb{Z}^d} \phi(2^j \boldsymbol{x} - \boldsymbol{k}) \phi(2^j \boldsymbol{z} - \boldsymbol{k}),$$

[3]Note that the notation employed in [Chernih *et al.* (2014)] uses ℓ as the first subscript. Using that same notation, Hubbert (2012) provides an explicit closed-form representation for arbitrary values of k.

where w_j are positive weights that are used to combine the single-scale kernels K_j defined at the different scales $j = 0, 1, 2, \ldots$, and ϕ are compactly supported (possibly refinable) functions whose shifts (at scaling level j) are used to produce the single-scale kernels K_j. A typical example is given by the *multiscale piecewise linear kernel*

$$K(\boldsymbol{x}, \boldsymbol{z}) = \sum_{j=0}^{3} 2^{-2j} \sum_{\boldsymbol{k} \in \mathbb{Z}^2} \phi(2^j \boldsymbol{x} - \boldsymbol{k}) \phi(2^j \boldsymbol{z} - \boldsymbol{k})$$

with $\phi(\boldsymbol{x}) = \prod_{\ell=1}^{d} (1 - x_\ell)_+$, a tensor product of the C^0 "original" Wendland kernel $\kappa_{1,0}$ (see Table 3.4). Opfer (2006) uses these kernels in wavelet-like applications such as image compression. These kernels are just beginning to attract the interest of other researchers. For example, Ling (2006) uses them for (noisy) scattered data fitting, Jordão and Menegatto (2010) derive related integral operators, and Griebel *et al.* (2015) look at their approximation properties.

3.7 Space-Time Kernels

Many problems have both a spatial as well as a temporal component, and therefore it is quite natural to attempt to model such problems with a *space-time kernel*. Probably the most common approach is to use a kernel that factors into a spatial and a temporal component. However, some authors have identified problems for which the data does not seem to allow such separability since it contains spatio-temporal interactions which a separable model would not be able to pick up on (see, e.g., [Cressie and Huang (1999); Gneiting *et al.* (2007)]). Therefore, the use of true space-time kernels may be beneficial. While this type of kernel appears only very rarely in the RBF literature, it is much more common in the statistics literature.

Li and Mao (2011) used a space-time kernel to solve an ill-posed inverse heat conduction problem. The kernel used by these authors is an anisotropic IMQ kernel for which they augmented the one or two spatial coordinates by an additional coordinate representing time. However, since most problems live on a spatial scale that is different from the time scale the kernel contains an additional scale parameter for time, i.e.,

$$K((\boldsymbol{x}, s), (\boldsymbol{z}, t)) = \frac{1}{\sqrt{1 + \varepsilon^2 \|\boldsymbol{x} - \boldsymbol{z}\|^2 + \gamma^2 (s - t)^2}}, \qquad \boldsymbol{x}, \boldsymbol{z} \in \mathbb{R}^d, \ s, t \in \mathbb{R},$$

where $d = 1, 2$.

Stein (2005b) suggests using kernels that are translation invariant in both space and time, i.e., of the form $K((\boldsymbol{x}, s), (\boldsymbol{z}, t)) = \widetilde{K}(\boldsymbol{x} - \boldsymbol{z}, s - t)$, and then proceeds to derive appropriate generalizations of Matérn kernels, and — in [Stein (2013)] — a family of kernels that generalize polyharmonic splines (which he refers to as *power law covariance functions*).

The construction of Porcu *et al.* (2007) allows for spatial anisotropy with temporal translation invariance resulting in kernels such as, e.g.,

$$K((\boldsymbol{x},s),(\boldsymbol{z},t)) = \frac{\exp\left(-\frac{|s-t|^2}{K_{\text{space}}(\boldsymbol{x},\boldsymbol{z})}\right)}{\sqrt{K_{\text{space}}(\boldsymbol{x},\boldsymbol{z})}}, \qquad \boldsymbol{x},\boldsymbol{z} \in \mathbb{R}^d,\ s,t \in \mathbb{R},$$

where $K_{\text{space}}(\boldsymbol{x},\boldsymbol{z}) = \log\left(2 + \frac{1}{2}\left(2\varepsilon(\boldsymbol{x}+\boldsymbol{z}) - \frac{1+\varepsilon(\boldsymbol{x}+\boldsymbol{z})}{1+\varepsilon(\boldsymbol{x}-\boldsymbol{z})}\right)\right)$. This kernel is derived from the Gaussian kernel with the help of completely monotone functions.

3.8 Learned Kernels

As an extension of the idea of simply estimating (or "learning") the shape parameter of a given kernel (such as the Gaussian kernel), or more generally the shape and smoothness parameters of a given kernel family (such as the Matérn kernels), Lanckriet *et al.* (2004) suggest that the kernel matrix (instead of the actual kernel) can be learned from the given data by viewing the kernel matrix as a linear combination of given positive semi-definite kernel matrices with bounded trace and then employing semi-definite programming techniques to "learn" the coefficients (possibly constrained to be nonnegative, which improves computational efficiency and is helpful in studying the statistical properties of the kernel matrices) of this linear combination from the data. This implies that the kernel is computed only on the training and test sets, and so, it is regarded as a matrix. For example, the authors of this work maximize the margin of a binary support vector machine (SVM) trained with the kernel K (see, e.g., Chapter 18).

Micchelli and Pontil (2005) suggest learning the kernel via regularization techniques. They start with a — possibly uncountable — set $\mathbb{K}$ of kernels and then determine the optimal kernel for a given set of N pieces of data $\{(\boldsymbol{x}_i, y_i) \mid i = 1,\ldots,N\}$ as a convex combination of at most $N+2$ kernels from $\mathbb{K}$. The set $\mathbb{K}$ is assumed to be compact and convex, and then the optimal learned kernel is obtained by solving a convex optimization problem. Once the kernel K has been found, the kernel approximation is obtained by solving another finite-dimensional convex optimization problem.

3.9 Designer Kernels

In addition to the idea of learned kernels mentioned in Section 3.8, the literature contains may approaches that lead to specially designed *custom kernels* (or *designer kernels*). For example, one can use the basic properties of kernels listed in [Fasshauer (2007, Section 3.1)] or in Section 2.3, such as adding, multiplying and taking positive linear combinations of positive definite kernels to obtain another positive definite kernel which combines features of its ingredients. If one decides to work with a radial kernel then one can construct a new radial kernel by composition with multiply or completely monotone functions (see, e.g., [Fasshauer (2007)]). In Chapter 8 we discuss the use of generalized Sobolev spaces to design a kernel with desired smoothness and scale properties.

Another possibility is to essentially turn Mercer's Theorem 2.2 upside down and build a kernel by combining an appropriately decaying sequence of "eigenvalues" λ_n with a given set of orthogonal functions $\{\varphi_n\}_{n=1}^{\infty}$. As a result, the closed form of the kernel may not be known in this case. In fact, we would like to emphasize that precise knowledge of the closed form of the kernel K is not needed since we can use the *Hilbert–Schmidt SVD* (HS-SVD) of Section 13 to formulate stable and efficient algorithms for the associated kernel approximation problems. A typical example is given by the *iterated Brownian bridge kernels* of Chapter 7.

In Chapter 8 we promote the idea of not only designing our own positive definite kernels as Green's kernels of differential operators, but also pairing them with a custom inner product so that we essentially create our own *designer spaces*, which we refer to as *generalized Sobolev spaces*. That approach is motivated by the fact that the reproducing kernel Hilbert space inner product induces a norm, and it is this norm which is minimized by the kernel-based interpolant (see Chapter 9). Therefore use of such a designer space approach follows the same rationale as the use of *fairness functionals* in geometric modeling (see, e.g., [Sapidis (1987)]) or *regularization functionals* in statistical or machine learning (see, e.g., Chapter 18).

We now give details for two more examples of (Mercer) series kernels designed to have specific properties, namely periodicity in Section 3.9.1 and certain convergence rates based on the decay rate of the "eigenvalues" in Section 3.9.2.

3.9.1 *Periodic kernels*

The Hilbert–Schmidt series of a family of kernels generalizing the periodic spline kernels (3.2) is given by

$$\begin{aligned} K(x,z) &= \sum_{n=1}^{\infty} \frac{2}{(4n^2\pi^2+\varepsilon^2)^{\beta}} \left(\cos(2n\pi x)\cos(2n\pi z) + \sin(2n\pi x)\sin(2n\pi z)\right) \qquad (3.9) \\ &= \sum_{n=1}^{\infty} \frac{2}{(4n^2\pi^2+\varepsilon^2)^{\beta}} 2\cos\left(2n\pi(x-z)\right), \quad x,z\in[0,1],\ \varepsilon>0, \beta\in\mathbb{N}. \end{aligned}$$

Due to the shift in the eigenvalues these kernels are no longer piecewise polynomial splines, and we do not readily have a closed form expression available for this kernel. This situation is similar to what happens for the iterated Brownian bridge kernels in Chapter 7.

3.9.2 Chebyshev kernels

As suggested by Hickernell (2014), we can design a family of kernels by using Chebyshev polynomials as eigenfunctions and picking either a sequence of geometrically or algebraically decaying "eigenvalues." In either case we will see that it is even possible to obtain a closed form expression for the kernel.

Chebyshev polynomials of the first kind of degree n are defined as (see, e.g., [DLMF, Eq. (18.5.1)] and also Example 5.4)

$$T_n(x) = \cos(n\theta), \qquad \theta = \cos^{-1}(x), \qquad x \in [-1,1], \quad \theta \in [0,\pi].$$

They satisfy the orthogonality relation [DLMF, Eq. (18.3.1)]

$$\int_{-1}^{1} \frac{T_m(x)T_n(x)}{\sqrt{1-x^2}}\,\mathrm{d}x = \delta_{mn}\frac{\pi}{2-\delta_{m0}}, \qquad m,n = 0,1,\ldots.$$

By choosing the weight function $\rho(x) = \frac{1}{\pi\sqrt{1-x^2}}$ (so that $\int_{-1}^{1}\rho(x)\,\mathrm{d}x = 1$), and by appropriately normalizing the Chebyshev polynomials we obtain a set of "eigenfunctions" $\{\varphi_n\}_{n=0}^{\infty}$ which are orthonormal on $L_2([-1,1],\rho)$, i.e.,

$$\varphi_n(x) = \sqrt{2-\delta_{n0}}T_n(x), \qquad \int_{-1}^{1}\varphi_m(x)\varphi_n(x)\rho(x)\,\mathrm{d}x = \delta_{mn}, \quad m,n = 0,1,\ldots.$$

We will now present two different choices of "eigenvalues" $\{\lambda_n\}_{n=0}^{\infty}$ satisfying $\sum_{n=0}^{\infty}\lambda_n = 1$ such that Mercer's Theorem 2.2 gives us the positive definite, finite trace, kernel[4]

$$K(x,z) = \sum_{n=0}^{\infty}\lambda_n\varphi_n(x)\varphi_n(z).$$

Geometrically decaying eigenvalues

For fixed $a \in (0,1]$ and $b \in (0,1)$ we define the eigenvalues as

$$\lambda_0 = 1-a, \qquad \lambda_n = \frac{a(1-b)b^n}{b}, \qquad n = 1,2,\ldots, \tag{3.10}$$

so that it is clear that we have $\sum_{n=0}^{\infty}\lambda_n = 1$.

If we let $x = \cos(\theta)$ and $z = \cos(\vartheta)$, $\theta,\vartheta \in [0,\pi]$, then we can use the definition of the eigenfunctions along with some basic trigonometric identities to observe that

$$\begin{aligned}
\varphi_n(x)\varphi_n(z) &= (2-\delta_{n0})T_n(x)T_n(z)\\
&= (2-\delta_{n0})\cos(n\theta)\cos(n\vartheta)\\
&= \frac{(2-\delta_{n0})\left(\cos(n(\theta+\vartheta)) + \cos(n(\theta-\vartheta))\right)}{2} \qquad (3.11)\\
&= \frac{2-\delta_{n0}}{4}\left(\mathrm{e}^{in(\theta+\vartheta)} + \mathrm{e}^{in(\theta-\vartheta)} + \mathrm{e}^{-in(\theta-\vartheta)} + \mathrm{e}^{-in(\theta+\vartheta)}\right). \qquad (3.12)
\end{aligned}$$

[4]Note that we decide to index the eigenvalues and eigenfunctions beginning with zero in this section since it is more convenient.

Since summation of each of the four terms from (3.12) against the eigenvalues results in a geometric series, i.e.,

$$\sum_{n=0}^{\infty} \lambda_n (2-\delta_{n0}) \mathrm{e}^{\mathrm{i}n\omega} = 1-a+\frac{2a(1-b)}{b}\sum_{n=1}^{\infty} b^n \mathrm{e}^{\mathrm{i}n\omega}$$
$$= 1-a+\frac{2a(1-b)\mathrm{e}^{\mathrm{i}\omega}}{1-b\mathrm{e}^{\mathrm{i}\omega}},$$

we can use some more basic trigonometric identities along with the relationship between (x,z) and (θ,ϑ) to further simplify

$$\begin{aligned}
K_{a,b}(x,z) &= \sum_{n=0}^{\infty} \lambda_n \phi_n(x)\phi_n(z) \\
&= 1-a+\frac{a(1-b)}{2}\left(\frac{\mathrm{e}^{\mathrm{i}(\theta+\vartheta)}}{1-b\mathrm{e}^{\mathrm{i}(\theta+\vartheta)}}+\frac{\mathrm{e}^{-\mathrm{i}(\theta+\vartheta)}}{1-b\mathrm{e}^{-\mathrm{i}(\theta+\vartheta)}}\right. \\
&\qquad \left. +\frac{\mathrm{e}^{\mathrm{i}(\theta-\vartheta)}}{1-b\mathrm{e}^{\mathrm{i}(\theta-\vartheta)}}+\frac{\mathrm{e}^{-\mathrm{i}(\theta-\vartheta)}}{1-b\mathrm{e}^{-\mathrm{i}(\theta-\vartheta)}}\right) \\
&= 1-a+\frac{a(1-b)}{2}\left(\frac{\mathrm{e}^{\mathrm{i}(\theta+\vartheta)}+\mathrm{e}^{-\mathrm{i}(\theta+\vartheta)}-2b}{1+b^2-b[\mathrm{e}^{\mathrm{i}(\theta+\vartheta)}+\mathrm{e}^{-\mathrm{i}(\theta+\vartheta)}]}\right. \\
&\qquad \left. +\frac{\mathrm{e}^{\mathrm{i}(\theta-\vartheta)}+\mathrm{e}^{-\mathrm{i}(\theta-\vartheta)}-2b}{1+b^2-b[\mathrm{e}^{\mathrm{i}(\theta-\vartheta)}+\mathrm{e}^{-\mathrm{i}(\theta-\vartheta)}]}\right) \\
&= 1-a+a(1-b)\left(\frac{\cos(\theta+\vartheta)-b}{1+b^2-2b\cos(\theta+\vartheta)}+\frac{\cos(\theta-\vartheta)-b}{1+b^2-2b\cos(\theta-\vartheta)}\right) \\
&= 1-a+a(1-b)\left(\frac{xz-\sqrt{(1-x^2)(1-z^2)}-b}{1+b^2-2b[xz-\sqrt{(1-x^2)(1-z^2)}]}\right. \\
&\qquad \left. +\frac{xz+\sqrt{(1-x^2)(1-z^2)}-b}{1+b^2-2b[xz+\sqrt{(1-x^2)(1-z^2)}]}\right),
\end{aligned}$$

and obtain — after some additional simplifications — the closed form of the kernel as

$$K_{a,b}(x,z) = 1-a+2a(1-b)\frac{b(1-b^2)-2b(x^2+z^2)+(1+3b^2)xz}{(1-b^2)^2+4b\left(b(x^2+z^2)-(1+b^2)xz\right)}, \tag{3.13}$$

which we recognize as a rational function of x and z.

Algebraically decaying eigenvalues

For this case we will employ Bernoulli polynomials. They can be represented in terms of a Fourier cosine series [DLMF, Eq. (24.8.1)] (see also (3.2))

$$B_{2\beta}(|x|) = (-1)^{\beta+1}\frac{2(2\beta)!}{(2\pi)^{2\beta}}\sum_{n=1}^{\infty}\frac{\cos(2\pi n x)}{n^{2\beta}}, \qquad x\in[-1,1],$$

where $B_n(|\cdot|)$ is a polynomial of degree n.

For fixed $a \in (0,1]$ and $\beta \in \mathbb{N}$ we define the eigenvalues as

$$\lambda_0 = 1 - a, \qquad \lambda_n = \frac{a}{\zeta(2\beta) n^{2\beta}}, \qquad n = 1, 2, \ldots, \tag{3.14}$$

where we introduce $\zeta(s) = \sum_{n=1}^{\infty} \frac{1}{n^s}$, the Riemann zeta function [DLMF, Eq. (25.2.1)], so that $\sum_{n=0}^{\infty} \lambda_n = 1$. For this choice of algebraically decaying eigenvalues and for $\omega \in [-1,1]$ it follows that

$$\begin{aligned}
\sum_{n=0}^{\infty} \lambda_n (2 - \delta_{n0}) \cos(2\pi n\omega) &= 1 - a + \frac{2a}{\zeta(2\beta)} \sum_{n=1}^{\infty} \frac{\cos(2\pi n\omega)}{n^{2\beta}} \\
&= 1 - a + \frac{2a}{\zeta(2\beta)} (-1)^{\beta+1} \frac{(2\pi)^{2\beta}}{2(2\beta)!} B_{2\beta}(|\omega|) \\
&= 1 - a + \frac{a(-1)^{\beta+1}(2\pi)^{2\beta}}{(2\beta)!\zeta(2\beta)} B_{2\beta}(|\omega|).
\end{aligned}$$

And now we can get the closed form of the kernel from (3.11), i.e.,

$$\begin{aligned}
K_{a,\beta}(x,z) &= \sum_{n=0}^{\infty} \lambda_n \varphi_n(x) \varphi_n(z) \\
&= 1 - a + \frac{a(-1)^{\beta+1}(2\pi)^{2\beta}}{2(2\beta)!\zeta(2\beta)} \left(B_{2\beta}(|\theta + \vartheta|/(2\pi)) + B_{2\beta}(|\theta - \vartheta|/(2\pi))\right) \\
&= 1 - a + \frac{a(-1)^{\beta+1}(2\pi)^{2\beta}}{2(2\beta)!\zeta(2\beta)} \left(B_{2\beta} \left(\frac{|\cos^{-1}(x) + \cos^{-1}(z)|}{2\pi} \right) \right. \\
&\qquad \left. + B_{2\beta} \left(\frac{|\cos^{-1}(x) - \cos^{-1}(z)|}{2\pi} \right) \right).
\end{aligned}$$

For example, taking $\beta = 1$ we have

$$K(x,z) = 1 - a + 6a \left(B_2 \left(\frac{|\cos^{-1}(x) + \cos^{-1}(z)|}{2\pi} \right) + B_2 \left(\frac{|\cos^{-1}(x) - \cos^{-1}(z)|}{2\pi} \right) \right), \tag{3.15}$$

where $B_2(x) = x^2 - x + 1/6$.

Chapter 4

Kernels in MATLAB

Many discussions in this book focus on the theoretical underpinnings of positive definite kernels, without which this text would have no relevance. Simultaneously, however, this book is also concerned with the computational methods used to evaluate and manipulate kernels for interpolation and other mathematical problems. In this chapter, we revisit the various forms of kernels introduced in Chapter 3 and consider implementations in MATLAB that both are efficient computationally and leverage the unique functionality offered in MATLAB.

Almost all of the theoretical discussion of kernels in this book involves the general form $K(\boldsymbol{x}, \boldsymbol{z})$ which ignores any special nature enjoyed by common kernels such as the Gaussian and Matérn. While this strategy applies to the broadest variety of kernels possible, it frequently ignores special structure present in kernels such as their radial or tensor product nature; these properties provide computational opportunities during evaluation of, e.g., the interpolation matrix K and basis vector $\boldsymbol{k}$ appearing in (1.3). We take the opportunity in this section to highlight computational strategies for evaluating various kernels, while emphasizing MATLAB-specific strategies to most efficiently conduct computation.

Although we have endeavored to produce the best software we can, the development team of MATLAB is likewise constantly at work to improve their software. Consequently, the ideas described in this book are not "etched in stone" and, as is the case with any software, the library developed for this book must be considered within the context of the most current iteration of the MATLAB software. Given that, we hope that the ideas discussed here are not interpreted as the ultimate MATLAB implementation for kernels, but rather that they provide interested parties with some guidance regarding good programming strategies in MATLAB.

It is also necessary at this point to mention the `GaussQR` repository, which has been developed as supplemental content for this book. While the MATLAB programming practices described here are widely applicable, most of the code in this and other sections requires access to the `GaussQR` repository. Installation instructions are provided in Appendix D, but as a quick reminder: the command `rbfsetup` must be executed in the base `GaussQR` directory to define and access the necessary directories and data.

4.1 Radial Kernels in MATLAB

Many kernels of interest are *radial*, meaning that the kernel $K : \Omega \times \Omega \to \mathbb{R}$ can be expressed as a function of a single variable $\kappa : \mathbb{R}_0^+ \to \mathbb{R}$ of the form

$$\kappa(\|\boldsymbol{x} - \boldsymbol{z}\|) = K(\boldsymbol{x}, \boldsymbol{z}).$$

This is introduced in Section 3.1.

Radial kernels, also called RBFs, are most directly evaluated in MATLAB using its vectorized code mechanism, by which functions such as cosine or logarithm are applied to each element of the matrix. For instance, the Gaussian kernel $K(\boldsymbol{x}, \boldsymbol{z}) = \exp(-\varepsilon^2\|\boldsymbol{x} - \boldsymbol{z}\|^2)$ could be defined in MATLAB using the command

```
rbf_gaussian = @(e,r) exp(-(e*r).^2)
```

and the C^2 Wendland kernel $K(\boldsymbol{x}, \boldsymbol{z}) = (1 + 4\varepsilon\|\boldsymbol{x} - \boldsymbol{z}\|)(1 - \varepsilon\|\boldsymbol{x} - \boldsymbol{z}\|)_+^4$ (cf. Section 3.5) could be defined with

```
rbf_c2wendland = @(e,r) (1+4*e*r).*max(0,1-e*r)
```

where `r` contains the Euclidean distance between $\boldsymbol{x}$ and $\boldsymbol{z}$. Note that the use of the `.^` and `.*` notation instead of simply `^` and `*` enforces element-wise operations in lieu of matrix algebra-styled operations. This allows us to execute the same operations whether `r` is the scalar distance between just two points $\boldsymbol{x}$ and $\boldsymbol{z}$, a vector of distances between a point $\boldsymbol{x}$ and a set of points $\mathcal{Z} = \{\boldsymbol{z}_1, \ldots, \boldsymbol{z}_N\}$ (or vice versa), or even a complete matrix of all the pairwise distances between evaluation points $\mathcal{X} = \{\boldsymbol{x}_1, \ldots, \boldsymbol{x}_{N_{\text{eval}}}\}$ and *kernel centers* $\mathcal{Z}$. If the reader is unfamiliar with element-wise dot-operator notation, suitable references include [Van Loan and Fan (2010)] or [Moler (2008)].

Note that, because these kernels are radial, their evaluation in MATLAB is most naturally performed in an element-wise fashion on a matrix composed of the necessary distance values; call such a matrix R. Given data locations $\mathcal{Z} = \{\boldsymbol{z}_1, \ldots, \boldsymbol{z}_N\}$ and associated data values $\boldsymbol{y} = \begin{pmatrix} y_1 & \cdots & y_N \end{pmatrix}^T$, we learned in (1.3) that the kernel-based interpolant is $s(\boldsymbol{x}) = \boldsymbol{k}(\boldsymbol{x})^T\mathsf{K}^{-1}\boldsymbol{y}$, where $\boldsymbol{k}(\boldsymbol{x})^T = \begin{pmatrix} K(\boldsymbol{x}, \boldsymbol{z}_1) & \cdots & K(\boldsymbol{x}, \boldsymbol{z}_N) \end{pmatrix}$ and $(\mathsf{K})_{i,j} = K(\boldsymbol{z}_i, \boldsymbol{z}_j)$.

For a radial kernel, computing $(\mathsf{K})_{i,j}$ would require $(\mathsf{R})_{i,j} = \|\boldsymbol{z}_i - \boldsymbol{z}_j\|$, thus a significant computational cost in forming the K matrix occurs in forming R, which we call the *distance matrix*. The computation of this symmetric distance matrix R is discussed in Section 4.1.1. Computing $\boldsymbol{k}(\boldsymbol{x})^T$ for a radial kernel would require a vector of the distances $\begin{pmatrix} \|\boldsymbol{x} - \boldsymbol{z}_1\| & \cdots & \|\boldsymbol{x} - \boldsymbol{z}_N\| \end{pmatrix}$, which is discussed in Section 4.1.2. We consider a different norm in Section 4.1.3 with potentially different weights applied to each dimension.

4.1.1 *Symmetric distance matrices in* MATLAB

Implementing a suitable `DistanceMatrix` function in MATLAB is nontrivial, and, depending on when one was first introduced to MATLAB, different programming tendencies may surface. Our current implementation is presented in Program 4.5. There are two key mechanisms which allow for the computation of distance matrices: `repmat` (which creates a larger matrix from replicates of a smaller matrix) and `bsxfun` (which performs component-wise binary operation between two matrices). The most efficient strategy the authors have found for computing distance matrices is based on the expansion of the sum of squares: by defining $z_{i,\ell}$ to be the ℓth dimension of $\boldsymbol{z}_i$, the ith point in $\mathcal{Z}$, we know

$$\|\boldsymbol{z}_i - \boldsymbol{z}_j\|^2 = \sum_{\ell=1}^{d}(z_{i,\ell} - z_{j,\ell})^2 = \sum_{\ell=1}^{d} z_{i,\ell}^2 + \sum_{\ell=1}^{d} z_{j,\ell}^2 - \sum_{\ell=1}^{d} 2z_{i,\ell}z_{j,\ell}.$$

The i locations correspond to the rows of R and the j locations correspond to the columns of R; this is seen in the `repmat`-based code Program 4.1. This implementation is, however, not the most efficient, as the `repmat` commands can be combined into a single `bsxfun` command, as shown in Program 4.2.

Program 4.1. `DistanceMatrixRepmat.m`

```
function DM = DistanceMatrixRepmat(z)
N = size(z,1);
sz2 = sum(z.^2,2);
DM = sqrt(repmat(sz2,1,N) + repmat(sz2',N,1) - 2*z*z');
```

Program 4.2. `DistanceMatrixBsxfun.m`

```
function DM = DistanceMatrixBsxfun(z)
sz2 = sum(z.^2,2);
DM = sqrt(bsxfun(@plus,sz2,sz2') - 2*z*z');
```

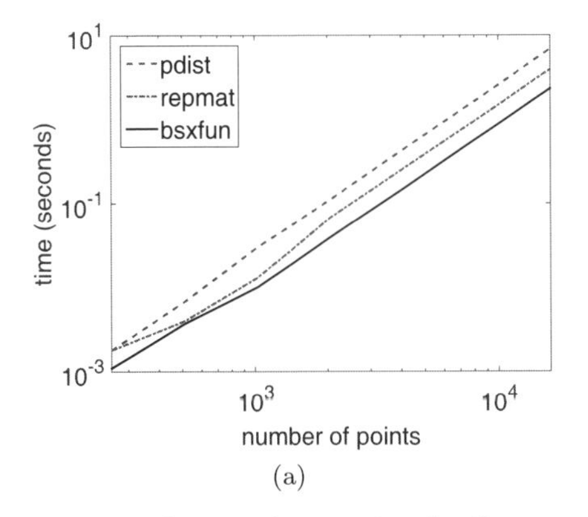

(a)

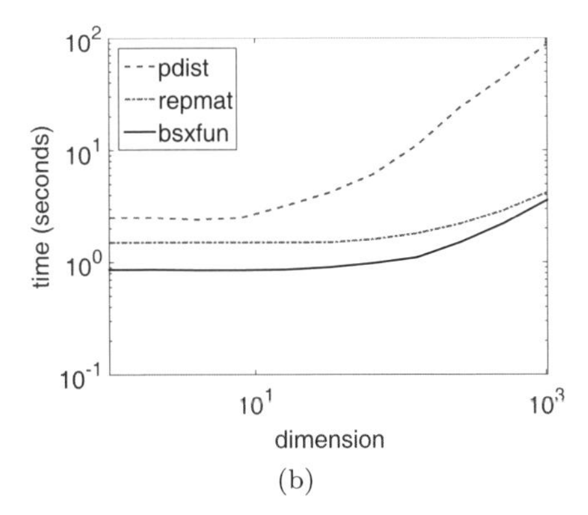

(b)

Fig. 4.1 The cost of computing the distance matrix R. These timings have been averaged over 10 runs. (a) Five-dimensional point sets of increasing size, (b) 10000 points in increasing dimensions.

A comparison between Program 4.1 and Program 4.2 when computing the R matrix of five-dimensional points is displayed in Figure 4.1; an additional comparison is made to the `pdist` function from MATLAB's Statistics and Machine Learning Toolbox. Clearly, the computational speed is highly dependent on both the strategy and the number of points, with, surprisingly to the authors, the MATLAB-supplied function performing most slowly and `bsxfun` taking the lead. These tests were run with the `KernelsDistanceTest.m` script in the library.

Results in Figure 4.1 also show the computation of R is dominated by the size of the data set N for small d: with $N = 10000$, the distance matrix computation is essentially of constant cost using `bsxfun` for $d <= 32$. It is of course possible that an improved version of the distance matrix computation could be devised to scale better for large d, and we encourage the readers to experiment with various strategies on their own. Such fundamental experiments, removed from any scientific or engineering application, are the foundation of a solid understanding of the challenges of efficient computational methods.

Remark 4.1. It should be noted that, much as fundamental experiments help develop computational intuition, fundamental aspects of numerical analysis remain unavoidable. Even just two points can demonstrate the potential problems caused by the limitations of machine precision. Running `DistanceMatrixBsxfun(z)` on `z = [0 1;1e-8 1+1e-8]` produces a matrix of all zeros, despite the fact that $\|\boldsymbol{z}_1 - \boldsymbol{z}_2\| = \sqrt{2}\times 10^{-8}$, and thus the off-diagonal values should be nonzero. In fact, errors are unavoidable with either Program 4.1 or Program 4.2 because of the subtraction of terms that have already undergone computation.

To avoid these potential issues, one would have to rearrange the distance computation to avoid subtractions of terms that have been multiplied:

$$\|\boldsymbol{z}_i - \boldsymbol{z}_j\|^2 = \sum_{\ell=1}^{d} (z_{i,\ell} - z_{j,\ell})^2, \tag{4.1}$$

as is done within the `pdist` function. Although we acknowledge the dangers of machine precision, the computational cost in circumventing it is greater than we are willing to expend, given that the values in $\mathcal{Z}$ needed to cause problems are such unrealistic data locations (locations within $\sqrt{\epsilon_{\text{mach}}}$ of each other). This remark is provided for completeness: in the `GaussQR` library our distance matrix computation always makes sure all the values in the distance matrix are positive.

4.1.2 *General distance matrices in* MATLAB

The computation of K is a special case involving the computation of $\boldsymbol{k}$ because

$$\mathsf{K} = \begin{pmatrix} \boldsymbol{k}(\boldsymbol{z}_1)^T \\ \vdots \\ \boldsymbol{k}(\boldsymbol{z}_N)^T \end{pmatrix}.$$

Therefore, the ideas developed while computing in Section 4.1.1 are relevant, but they must be adapted to the *nonsymmetric* case where $\boldsymbol{k}$ is evaluated at a location not in $\mathcal{Z}$. We can define a distance vector

$$\boldsymbol{r}(\boldsymbol{x})^T = \sqrt{\begin{pmatrix}\|\boldsymbol{x}-\boldsymbol{z}_1\| & \cdots & \|\boldsymbol{x}-\boldsymbol{z}_M\|\end{pmatrix}}$$

analogous to the distance matrix, whose computation is necessary to evaluate $\boldsymbol{k}$ in the element-wise fashion desired by MATLAB; the square root used here is applied MATLAB-style, i.e., elementwise. Performing the same expansion of squares as before, but now with $\boldsymbol{x}$ and $\mathcal{Z}$, we see

$$\|\boldsymbol{x}-\boldsymbol{z}_i\|^2 = \sum_{\ell=1}^{d} x_\ell^2 + \sum_{\ell=1}^{d} z_{i,\ell}^2 - \sum_{\ell=1}^{d} 2x_\ell z_{i,\ell}. \tag{4.2}$$

At this point, it will help to rewrite this in more of a linear algebra format. Define $\bar{x} = \sum_{\ell=1}^d x_\ell^2 = \|\boldsymbol{x}\|^2$, $\bar{z}_i = \sum_{\ell=1}^d z_{i,\ell}^2 = \|\boldsymbol{z}_i\|^2$, $\boldsymbol{e}_p$ to be the length p vector of all ones and

$$\bar{\boldsymbol{z}} = \begin{pmatrix}\bar{z}_1\\ \vdots \\ \bar{z}_N\end{pmatrix}, \qquad \mathsf{Z} = \begin{pmatrix}\boldsymbol{z}_1^T\\ \vdots \\ \boldsymbol{z}_N^T\end{pmatrix}. \tag{4.3}$$

Note that the matrix Z is actually the standard storage strategy for points returned by the function `pickpoints` from the `GaussQR` library: each row is a different point (a point $\boldsymbol{x}$ is stored as a row vector) and each column corresponds to a different dimension. Using these definitions, the distance vector $\boldsymbol{r}$ can be written as

$$\boldsymbol{r}(\boldsymbol{x})^T = \sqrt{\bar{x}\boldsymbol{e}_N^T + \bar{\boldsymbol{z}}^T - 2\boldsymbol{x}^T\mathsf{Z}^T}, \tag{4.4}$$

where, again, the square root is applied elementwise.

Admittedly, the number of transposes present in, e.g., (4.4) is undesirable as it obfuscates the actual simplicity of the computation. This is a common outcome when trying to explain computational tools using mathematical objects; for example, while the linear algebra setting demands the use of the $\boldsymbol{e}_N$ vector, such an object is easily replaced with either the `repmat` or `bsxfun` operations. The numerous transposes throughout this section occur because all vectors, including the location $\boldsymbol{x}$, are defined as column vectors mathematically but are stored and manipulated in MATLAB as row vectors. Program 4.3 provides a MATLAB computation of the distance vector.

Program 4.3. `DistanceVector.m`

```
  function DV = DistanceVector(x,z)
2 sx2 = sum(x.^2);
  sz2 = sum(z.^2,2);
4 DV = sqrt(bsxfun(@plus,sx2,sz2') - 2*x*z');
```

Extrapolating from this outcome, several distance vectors can be evaluated simultaneously as was needed for the formation of K. For a set of N kernel centers $\mathcal{Z}$ and N_{eval} evaluation points $\mathcal{X}$, the kernel matrix $(\mathsf{K}_{\text{eval}})_{i,j} = K(\boldsymbol{x}_i, \boldsymbol{z}_j)$ requires a *nonsymmetric distance matrix* $(\mathsf{R}_{\text{eval}})_{i,j} = \|\boldsymbol{x}_i - \boldsymbol{z}_j\|$. The ith row of R_{eval} is $\boldsymbol{r}(\boldsymbol{x}_i)^T$, thus the same `bsxfun` strategies as before are still applicable. By defining

$$\bar{\boldsymbol{x}} = \begin{pmatrix} \bar{x}_1 \\ \vdots \\ \bar{x}_{N_{\text{eval}}} \end{pmatrix}, \qquad \mathsf{X} = \begin{pmatrix} \boldsymbol{x}_1^T \\ \vdots \\ \boldsymbol{x}_{N_{\text{eval}}}^T \end{pmatrix},$$

the nonsymmetric distance matrix can be defined as

$$\mathsf{R}_{\text{eval}} = \sqrt{\bar{\boldsymbol{x}}\boldsymbol{e}_N^T + \boldsymbol{e}_{N_{\text{eval}}}\bar{\boldsymbol{z}}^T - 2\mathsf{X}\mathsf{Z}^T}.$$

where, again, the square root is applied elementwise. A suitable implementation is shown in Program 4.4. Because `bsxfun` is capable of creating an appropriately sized matrix with the inputs $\bar{\boldsymbol{x}}$ and $\bar{\boldsymbol{z}}$ (denoted `sx2` and `sz2` in the code), the $\boldsymbol{e}$ terms are implied and `NonsymmetricDistanceMatrix` program looks very similar to the `DistanceVector` program.

Program 4.4. `NonsymmetricDistanceMatrix.m`

```
  function NDM = NonsymmetricDistanceMatrix(x,z)
2 sx2 = sum(x.^2,2);
  sz2 = sum(z.^2,2);
4 NDM = sqrt(bsxfun(@plus,sx2,sz2') - 2*x*z');
```

4.1.3 *Anisotropic distance matrices in* MATLAB

Program 4.4 is all that is required for computing with isotropic radial kernels, e.g., Gaussians, Matérns, inverse multiquadrics, etc. Such kernels are the foundation of the RBF literature, but recent research has suggested that *anisotropic radial kernels*, where different dimensions have different scales, are necessary for certain computational situations including, e.g., dimension-independent approximation error bounds [Fasshauer *et al.* (2012a,b)]. An anisotropic radial kernel is a kernel of an anisotropic distance function (cf. (3.1)):

$$\|\boldsymbol{x} - \boldsymbol{z}\|_{\boldsymbol{\varepsilon}}^2 = \sum_{\ell=1}^{d} \varepsilon_\ell^2 (x_\ell - z_\ell)^2 = \sum_{\ell=1}^{d} \varepsilon_\ell^2 x_\ell^2 + \sum_{\ell=1}^{d} \varepsilon_\ell^2 z_\ell^2 - \sum_{\ell=1}^{d} \varepsilon_\ell^2 2 x_\ell z_\ell,$$

where $\boldsymbol{\varepsilon} = \begin{pmatrix}\varepsilon_1 & \cdots & \varepsilon_d\end{pmatrix}^T$. The ℓth dimensions has a weight ε_ℓ, and the dimensions are allowed to have different weights unlike for isotropic radial kernels, where the distance is scaled by only a single parameter ε.

The flexibility of MATLAB allows a more complicated situation like this to be handled with little change from the existing code in Program 4.4. By defining the diagonal matrix $(\mathsf{E})_{\ell,\ell} = \varepsilon_\ell$, we can perform the dimensionwise scaling by $\boldsymbol{\varepsilon}$ with

$$\mathsf{X}_{\boldsymbol{\varepsilon}} = \mathsf{X}\mathsf{E}, \qquad \mathsf{Z}_{\boldsymbol{\varepsilon}} = \mathsf{Z}\mathsf{E}.$$

This matches the notation $\|\boldsymbol{x}-\boldsymbol{z}\|_{\boldsymbol{\varepsilon}}^2 = (\boldsymbol{x}-\boldsymbol{z})^T\mathsf{E}(\boldsymbol{x}-\boldsymbol{z})$ from (3.1). Similarly, rehashing the same isotropic computations — but using an anisotropic norm — gives $\bar{x}_{\boldsymbol{\varepsilon}} = \sum_{\ell=1}^d \varepsilon_\ell^2 x_\ell^2 = \|\boldsymbol{x}\|_{\boldsymbol{\varepsilon}}^2$, $\bar{z}_{\boldsymbol{\varepsilon}} = \sum_{\ell=1}^d \varepsilon_\ell^2 z_\ell^2 = \|\boldsymbol{z}\|_{\boldsymbol{\varepsilon}}^2$, and we can call

$$\bar{\boldsymbol{x}}_{\boldsymbol{\varepsilon}} = \begin{pmatrix} \bar{x}_{1\boldsymbol{\varepsilon}} \\ \vdots \\ \bar{x}_{N_{\text{eval}}\boldsymbol{\varepsilon}} \end{pmatrix}, \qquad \bar{\boldsymbol{z}}_{\boldsymbol{\varepsilon}} = \begin{pmatrix} \bar{z}_{1\boldsymbol{\varepsilon}} \\ \vdots \\ \bar{z}_{N\boldsymbol{\varepsilon}} \end{pmatrix}.$$

Putting all of this together allows us to write the nonsymmetric anisotropic distance matrix as

$$\mathsf{R}_{\boldsymbol{\varepsilon}} = \sqrt{\bar{\boldsymbol{x}}_{\boldsymbol{\varepsilon}}\boldsymbol{e}_N^T + \boldsymbol{e}_{N_{\text{eval}}}\bar{\boldsymbol{z}}_{\boldsymbol{\varepsilon}}^T - 2\mathsf{X}_{\boldsymbol{\varepsilon}}\mathsf{Z}_{\boldsymbol{\varepsilon}}^T}$$

The symmetric distance matrix is a subset of the nonsymmetric case, so it is not discussed separately for anisotropic distances. This computation is implemented in Program 4.5, where `sxe2` and `sze2` are $\bar{\boldsymbol{x}}_{\boldsymbol{\varepsilon}}$ and $\bar{\boldsymbol{z}}_{\boldsymbol{\varepsilon}}$ respectively.

Program 4.5. `DistanceMatrix.m`

```
  function DM = DistanceMatrix(x,z,epvec)
2 if not(exist('epvec','var'))
      epvec = ones(1,size(x,2));
4 end
  xe = bsxfun(@times,x,epvec);
6 ze = bsxfun(@times,z,epvec);
  sxe2 = sum(xe.^2,2);
8 sze2 = sum(ze.^2,2);
  DM = sqrt(bsxfun(@plus,sxe2,sze2') - 2*xe*ze');
```

This anisotropic distance function essentially multiplies the ℓth dimension of $\boldsymbol{x}$ by the ℓth weight in the vector $\boldsymbol{e}$ and then performs the same distance computation as before. For the remainder of this book, we will use this `DistanceMatrix` function to perform computations involving radial kernels. Any exceptions to this are described in the remainder of this chapter, where special distance matrix computations are preferred or the kernel is not radial.

Note the use of the `exist` construct within the `if`-block on line 2: this is a standard technique for allowing optional inputs. Here, if the user chooses not to pass a vector of ε_ℓ values the function defaults to the isotropic case $\varepsilon_1 = \ldots = \varepsilon_d$. Thus, when a user wants to use an isotropic distance function, the `epvec` term can be omitted to compute `DM = DistanceMatrix(x,z)`. This function overloading strategy is common in many programming languages and can make code easier to read and implement.

Remark 4.2. The use of the notation K_{eval}, R_{eval}, $\mathsf{K}_{\boldsymbol{\varepsilon}}$ and $\mathsf{R}_{\boldsymbol{\varepsilon}}$ is cumbersome and unnecessary in most of this text as the appropriate matrix will either be defined directly or apparent in context. As a result, we use discretion when applying subscripts to kernel and distance matrices elsewhere.

4.1.4 *Evaluating radial kernels and interpolants in* MATLAB

In this section we work through some of the standard MATLAB operations used in the evaluation of kernel matrices that are based on distance matrices, and the construction and evaluation of kernel-based interpolants. We begin with Program 4.6, which evaluates three C^4 Matérn kernels (cf. Table 3.2), centered at $z = 0$, $z = 0.4$ and $z = 0.9$, and displays them all in the same plot.

Program 4.6. `KernelsRadial1.m`

```
x = pickpoints(0,1,500);
z = [0;.4;.9];
DM = DistanceMatrix(x,z);
rbf = @(e,r) (1+e*r+(e*r).^2/3).*exp(-e*r);
ep = 5;
K = rbf(ep,DM);
plot(x,K,'linewidth',3)
```

The first function to notice in this script is the `pickpoints` function which is provided in `GaussQR` and allows the user to define points in a specified interval which follow certain patterns (uniformly-spaced, Chebyshev nodes, low-discrepancy sequence, etc.). The function `pick2Dpoints` serves an analogous purpose for two-dimensional points. The distance matrix `DM` (referred to as R_{eval} in Section 4.1.2) then contains all the pairwise distances between each of the evaluation points in `x` and the centers in `z` (referred to earlier as $\mathcal{X}$ and $\mathcal{Z}$, when used as sets, and X and Z, when used as matrices).

After the evaluation of the distance matrix, the radial kernel `rbf` is defined to be the C^4 Matérn; the shape parameter is chosen to be $\varepsilon = 5$. Evaluating the Matérn kernel is as simple as evaluating `rbf` with the given shape parameter at each of the distances computed by `DistanceMatrix`. After the line `K = rbf(ep,DM)`, the matrix `K` consists of three columns, each of which containing the values of the C^4 Matérn kernel centered at the corresponding location in `z` evaluated at 500 points in `x` picked from the interval $[0, 1]$. The final line of the script plots all three kernels together in the same plot.

Similar kernel evaluations can be made in two dimensions with little change to the code, but the plotting requires more work because the `surf` command is less flexible than the `plot` command. Program 4.7 shows the computation of isotropic inverse multiquadric kernels (cf. Table 3.1),

$$K(\boldsymbol{x}, \boldsymbol{z}) = \frac{1}{\sqrt{1 + \varepsilon^2 \|\boldsymbol{x} - \boldsymbol{z}\|^2}}, \qquad \varepsilon = 4,$$

centered at five different points and how they can all be plotted on the same graph.

Program 4.7. `Excerpt from KernelsRadial2.m`

```
% Define the kernel/rbf, shape parameter and centers
rbf_iso = @(e,r) 1./sqrt(1+(e*r).^2);
```

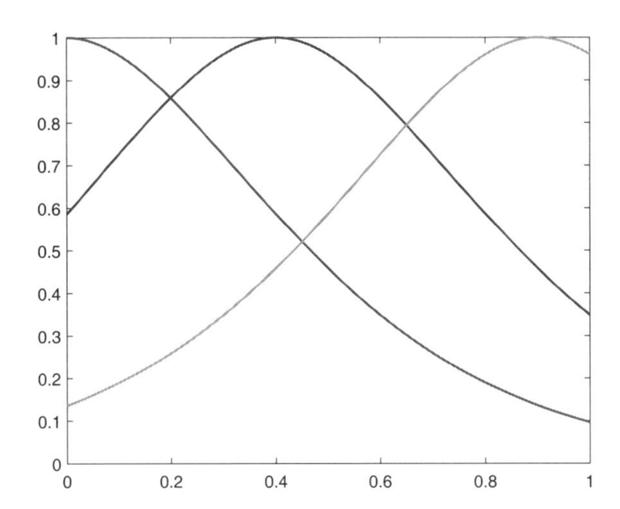

Fig. 4.2 Sample C^4 Matérn kernels with $\varepsilon = 5$ centered at $z_1 = 0$, $z_2 = 0.4$ and $z_3 = 0.9$.

```
ep = 4;
z = [.6 .6;-.6 -.6;.6 -.6;-.6 .6;0 0];
% Choose points throughout the domain at which to evaluate kernels
N = 50;
x = pick2Dpoints([-1,-1],[1,1],N);
% Reshape for surface plotting
X = reshape(x(:,1),N,N);
Y = reshape(x(:,2),N,N);
% Evaluate the distances between x & z points (R_eval)
DM = DistanceMatrix(x,z);
% Evaluate the kernels
K_iso = rbf_iso(ep,DM);
% Plot all 5 kernels on the same graph
% Each column stores one kernel evaluated at the x points
hold on
for k=1:5
    surf(X,Y,reshape(K_iso(:,k),N,N),'edgecolor','none')
end
hold off
colormap jet
view([-.2 -1 1.3])
```

Program 4.8. Anisotropic excerpt from KernelsRadial2.m

```
% Define the anisotropic kernel/rbf, shape parameter and centers
rbf_aniso = @(r) 1./sqrt(1+r.^2);
epvec = [7,2];
z = [.6 .6;-.6 -.6;.6 -.6;-.6 .6;0 0];
% Same x points and reshaping as before
N = 50;
```

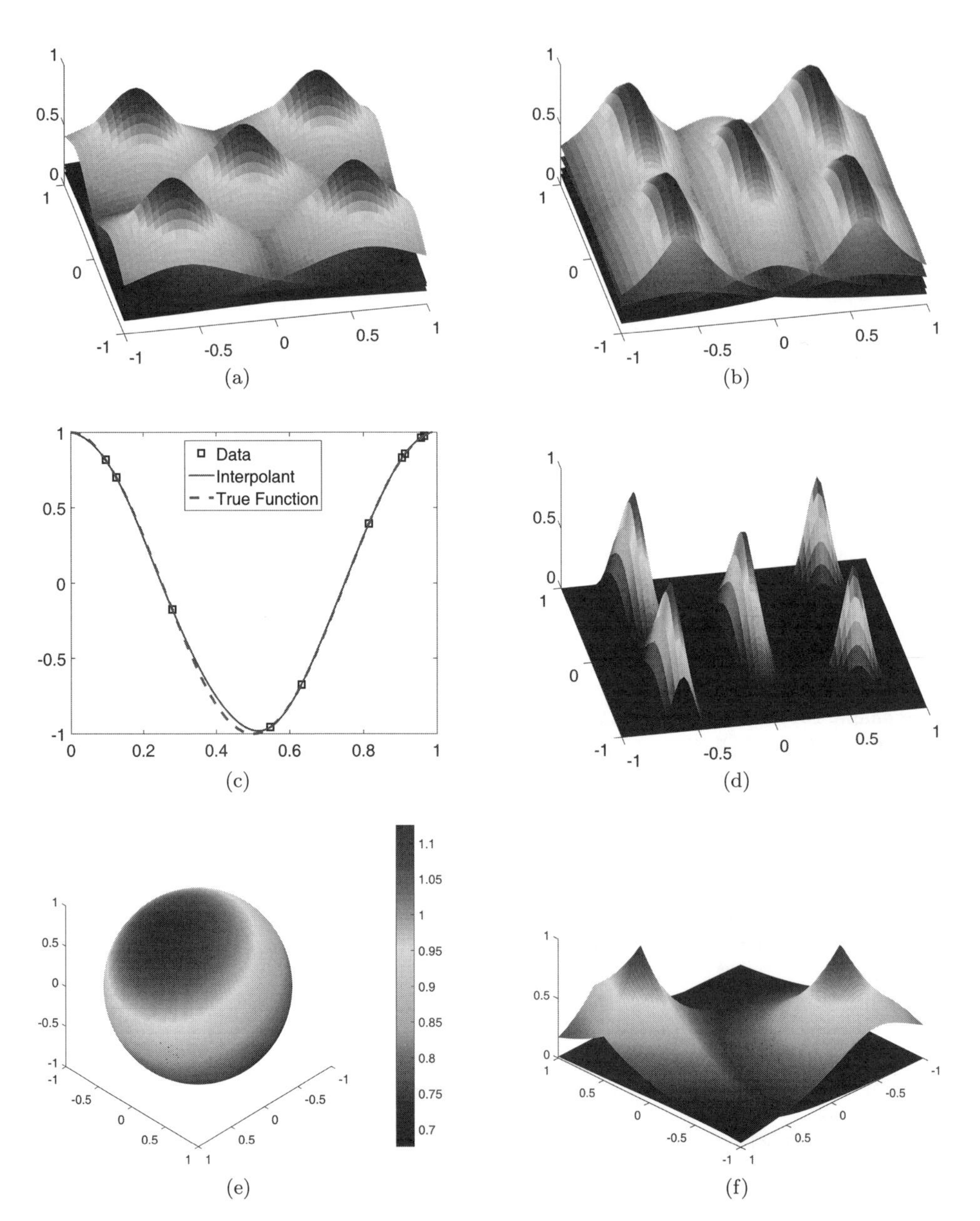

Fig. 4.3 Kernel samples and interpolants. (a) Isotropic inverse multiquadric kernels with $\varepsilon = 4$, (b) Anisotropic inverse multiquadric kernels with $\boldsymbol{\varepsilon} = (7\ 2)$, (c) Randomly sampled data from the function $f(x) = \cos(2\pi x)$ in the interval $[0, 1]$ and the associated C^2 Matérn interpolant with $\varepsilon = 4$. (d) CSRBFs from (3.8); left kernels are anisotropic with $\boldsymbol{\varepsilon} = (5\ 1)$, right kernels are isotropic with $\varepsilon = 4$. (e) Spherical inverse multiquadric with $\gamma = 0.7$. (f) Two C^0 Matérn kernels; left kernel is a tensor product of 1D kernels with $\boldsymbol{\varepsilon} = (1.5\ 2.5)$, right kernel is the standard radial kernel with $\varepsilon = 2$.

```
x = pick2Dpoints(-1,1,N);
X = reshape(x(:,1),N,N);
Y = reshape(x(:,2),N,N);
% Evaluate the anisotropic distances between x & z points
DM_aniso = DistanceMatrix(x,z,epvec);
% Evaluate the kernels
K_aniso = rbf_aniso(DM_aniso);
```

The graph in Figure 4.3(a) shows five translations of the same isotropic kernel. Program 4.7 computes all the kernel values simultaneously by using the distance matrix which is computed all at once and then reshapes the kernel values (stored in columns of the `K_iso` matrix) into the appropriate rectangular array for surface plotting. Similarly, Program 4.8 uses the anisotropic call to `DistanceMatrix` to compute anisotropic inverse multiquadrics,

$$K(\boldsymbol{x},\boldsymbol{z}) = \frac{1}{\sqrt{1+\|\boldsymbol{x}-\boldsymbol{z}\|_{\boldsymbol{\varepsilon}}^2}}, \qquad \boldsymbol{\varepsilon} = \begin{pmatrix} 7 & 2 \end{pmatrix}.$$

Notice that the definition of the radial kernel is changed from using `@(e,r)` with one shape parameter in the isotropic case (see line 2 of Program 4.7) to `@(r)` in the anisotropic case (see line 2 of Program 4.8) to account for the fact that the shape parameter vector $\boldsymbol{\varepsilon}$ is incorporated in `DistanceMatrix` rather than within the radial kernel computation. This matches the notation introduced in Section 3.1 where radial kernels are a function of only one argument.

Of course, evaluating matrices full of kernel values is simply a means to an end; the real goal is the evaluation of scattered data interpolants. As described in Section 1.3.1, the interpolant to scattered data $\mathcal{X} = \{\boldsymbol{x}_1, \dots, \boldsymbol{x}_N\}$, $\boldsymbol{y} = \begin{pmatrix} y_1 & \cdots & y_N \end{pmatrix}^T$, is

$$s(\boldsymbol{x}) = \boldsymbol{k}(\boldsymbol{x})^T \mathsf{K}^{-1} \boldsymbol{y}. \tag{1.3c}$$

Note that, in that section — as well as in much of the book — $\mathcal{X}$ plays the role of the interpolation points *and* kernel centers despite the fact that in this chapter we have introduced $\mathcal{Z}$ to represent the kernel centers. For example, in Chapters 15 and 18 we will divorce the kernel centers from the input data locations, denoting the former by $\mathcal{Z}$ and latter by $\mathcal{X}$.

Remark 4.3. In order for Program 4.9 to hew more closely to the notation in the rest of this book we now introduce the variable `x` to denote the interpolation points. However, we also use the variables `x` and `xeval` to denote, respectively, the kernel centers and evaluation locations (instead of the corresponding roles played by `Z` and `X` earlier in this chapter). This approach is simple and efficient, yet still flexible enough to ease the transition into more advanced computation later on, when we will be able to introduce an additional variable `z` to denote a separate array of kernel centers.

The elegant design of MATLAB allows for the computation of the interpolant at any desired points with surprisingly little code; see Program 4.9 for a demonstration, the results of which are on display in Figure 4.3(c). The use of the @ notation on line 5 allows for an "inline" function definition. It is worth noting that, despite the plotting of the results on $[0, 1]$, we should not expect an accurate interpolant outside the convex hull of the data locations $\mathcal{X}$ (since in that region we no longer are interpolating the data, but *extrapolating* from it).

Program 4.9. `KernelsInterpExample.m`

```
% Define data locations and evaluation locations
x = pickpoints(0,1,10,'rand');
xeval = pickpoints(0,1,500);
% Choose a function and evaluate it at those locations
yf = @(x) cos(2*pi*x);
y = yf(x);
yeval = yf(xeval);
% Choose a kernel and shape parameter
rbf = @(e,r) (1+(e*r)).*exp(-(e*r));
ep = 1;
% Form the necessary distance matrices
DM = DistanceMatrix(x,x);
DMeval = DistanceMatrix(xeval,x);
% Compute the kernel matrices from the distance matrices
K = rbf(ep,DM);
Keval = rbf(ep,DMeval);
% Evaluate the interpolant at the desired locations
seval = Keval*(K\y);
% Plot the data, interpolant, and true solution together
plot(x,y,'sk','linewidth',3)
hold on
plot(xeval,seval,'--b','linewidth',2)
plot(xeval,yeval,'r','linewidth',2)
hold off
```

4.2 Compactly Supported Kernels in MATLAB

The discussion in Section 4.1 is valid for all radial kernels, but some radial kernels have other special properties that are worth exploiting. One class of kernels which is of particular importance in a high performance setting is the class of *compactly supported kernels*. Such well-designed kernels, introduced in Section 3.5, can maintain their positive definiteness (in specific dimensions) while taking a value of zero at all points far enough beyond their centers. The benefit to this property is that the distance matrices, and therefore the kernel matrices evaluated from those distance matrices, are *sparse*. Taking advantage of this sparsity is a nontrivial task, however, and requires a different distance matrix computation scheme.

For a compactly supported kernel to be radial, it must satisfy

$$K(\boldsymbol{x}, \boldsymbol{z}) = 0, \qquad \text{for } \|\boldsymbol{x} - \boldsymbol{z}\| > r_{\text{cutoff}}$$

for some cutoff distance r_{cutoff}. The norm for $\|\boldsymbol{x}-\boldsymbol{z}\|$ is left intentionally ambiguous to allow for anisotropic kernels, i.e., $\|\boldsymbol{x} - \boldsymbol{z}\|_{\boldsymbol{\varepsilon}}$, where $r_{\text{cutoff}} = 1$ is often chosen (recall (3.8)). For isotropic CSRBFs containing the term $(1 - \varepsilon\|\boldsymbol{x} - \boldsymbol{z}\|)_+$ we use $r_{\text{cutoff}} = 1/\varepsilon$.

By calling `DistanceMatrix(x,z,[],1)`, a *sparse distance matrix* will be returned in lieu of the dense one created in Section 4.1.3. Passing a vector $\boldsymbol{\varepsilon}$ as the third argument will allow for anisotropic distance functions as before. If an isotropic kernel is being used with $r_{\text{cutoff}} = 1/\varepsilon$, one may pass `DistanceMatrix(x,z,ep,1)` with a scalar `ep` to serve as ε. It should be noted that the implementation of this sparse distance matrix computation requires Statistics and Machine Learning Toolbox in MATLAB because it calls the `rangesearch` function to compute the distances less than r_{cutoff}.

As is the case with many sparse matrix computations, memory traffic can be the performance bottleneck. To minimize the cost of forming the sparse distance matrix, some rather tricky use of MATLAB is employed within the `DistanceMatrix_SPARSE` subfunction inside `DistanceMatrix.m`. We list the implementation in Program 4.10 for the isotropic case with $r_{\text{cutoff}} = 1$ and refer readers to the `GaussQR` repository for the full implementation. The function `cellfun` is used in that program for reorganization because `rangesearch` returns cell arrays; more discussion of cell arrays takes place in Section 4.4.

Program 4.10. Simplified form of `DistanceMatrix_SPARSE.m`

```
function DM = DistanceMatrix_SPARSE(x,z)
% Accepts two sets of data points and creates a sparse
% distance matrix with zeros when ||x-z||>1
N = size(x,1); M = size(z,1);
% Use rangesearch to compute the distances between x and z less than 1
[Cidx,Cdist] = rangesearch(x,z,1);
% Reshape the results into the vectors of a sparse matrix
% ivec - rows; jvec - columns; svec - values
ivec = cell2mat(Cidx');
jvec = cell2mat(cellfun(@(x,k)k*ones(1,length(x)),Cidx',num2cell(1:M),...
                'UniformOutput',false));
svec = cell2mat(Cdist');
% Form the sparse matrix from the vectors
DM = sparse(ivec,jvec,svec,N,M);
```

The development of a sparse distance matrix is vital to taking full advantage of CSRBFs, but more changes are required to evaluate radial kernels using a sparse distance matrix. Most notably, operations such as `exp` are no longer viable for direct use because

$$\texttt{exp}\begin{pmatrix} 1 & 0 \\ 0 & 1 \end{pmatrix} = \begin{pmatrix} \mathrm{e} & 1 \\ 1 & \mathrm{e} \end{pmatrix}, \qquad \text{instead of} \quad \begin{pmatrix} \mathrm{e} & 0 \\ 0 & \mathrm{e} \end{pmatrix},$$

that is, once `exp` is applied to any matrix it is fully dense. This is because `exp` operates elementwise, even on the values of the matrix that are structurally (not just numerically) zero. To operate on only the nonzero values of a sparse matrix we call `spfun`.

Program 4.11 demonstrates the use of `spfun` to evaluate the *"missing" Wendland kernel*

$$K(\boldsymbol{x}, \boldsymbol{z}) = \kappa_{2,\frac{1}{2}}(\|\boldsymbol{x} - \boldsymbol{z}\|_{\varepsilon}),$$

where $\kappa_{2,\frac{1}{2}}$ was defined in (3.8). The kernel evaluation function `rbf` on lines 14–15 of Program 4.11 accepts a sparse distance matrix and returns a sparse kernel matrix which is evaluated using `spfun`. The auxiliary functions `sppc` and `spd` allow for pointwise scalar addition and matrix element division without introducing nonzeros. Because many standard operations such as adding a constant would change the nonzero structure of the matrix, the implementation of `rbf` is admittedly unattractive.

The complexities of working with a sparse matrix can be hidden by allowing for `DistanceMatrix` to return the kernel matrix, *not the distance matrix*. Because the formation of the sparse distance matrix, as shown in Program 4.10, creates an intermediate vector `svec` of $\|\boldsymbol{x}_i - \boldsymbol{z}_j\|_{\varepsilon}$ anywhere that $1 - \|\boldsymbol{x}_i - \boldsymbol{z}_j\|_{\varepsilon} > 0$, the radial kernel can be applied to that vector prior to forming the sparse distance matrix. The vector is treated as a dense matrix, meaning that the radial kernel can be defined in the standard way described in Section 4.1. On lines 23–24 of Program 4.11, the "dense form" of the compactly supported kernel is defined in `rbf_in_DM` and is passed as the fourth argument to `DistanceMatrix(X,Z,ε,rbf_in_DM)`.

The benefit to using the "dense form" of the compactly supported kernel is that `rbf_in_DM` looks much nicer than `rbf` and is also much easier to understand and debug. The drawback is that `DistanceMatrix` returns the kernel matrix, meaning that the full computation of the distances must be repeated to compute with other compact kernels rather than being able to reuse the $\|\boldsymbol{x} - \boldsymbol{z}\|_{\varepsilon}$ values. Users must decide which is better for their application: a small improvement in readability or a small improvement in speed when repeating computation with multiple kernels.

Program 4.11. Excerpt from `KernelsCSRBF.m`

```
   N = 70;
2  x = pick2Dpoints([-1,-1],[1,1],N);
   X = reshape(x(:,1),N,N);
4  Y = reshape(x(:,2),N,N);
   z = [.6 -.6;.6 .6];
6  % These functions are used to manipulate the sparse
   % matrices without causing them to be dense matrices
8  % sppc - sparse plus constant
   % spd  - sparse divided by sparse
10 sppc = @(const,Mat) spfun(@(x)const + x,Mat);
   spd  = @(Mat1,Mat2) Mat1.*spfun(@(x)1./x,Mat2);
```

```
12  % Definition of the compactly supported kernel to apply
    % to a sparse matrix without changing the structure
14  rbf = @(r) sppc(1,2*r.^2).*sqrt(sppc(1,-r.^2)) + 3*r.^2.* ...
          spfun(@log,sppc(eps,spd(r,sppc(1+eps,sqrt(sppc(1,-r.^2))))));
16  ep = 4;
    % Evaluate the sparse distance matrix and the kernels
18  DM_iso = DistanceMatrix(x,z,ep,1);
    K_iso = rbf(DM_iso);
20
    % Definition of the compact kernel to apply within
22  % the DistanceMatrix function as though it were dense
    rbf_in_DM = @(r) (1+2*r.^2).*sqrt(1-r.^2) + ...
24                   3*r.^2.*log(r./(1+sqrt(1-r.^2))+eps);
    epvec = [5 1];
26  z_in_DM = [-.6 .6;-.6 -.6;0 0];
    K_aniso = DistanceMatrix(x,z_in_DM,epvec,rbf_in_DM);
```

One point of note for these kernels is the use of `eps` — unit roundoff in MATLAB — to avoid divisions by 0 and `log(0)`. Although not included in Program 4.11, these kernels can be plotted using the same `surf` strategy from Program 4.7 and Program 4.8; those plots are shown in Figure 4.3(d).

Remark 4.4. Even when using builtin MATLAB functions, strange and unexpected things can occur and we as computational practitioners must be wary. Consider the following attempt to compute a missing Wendland kernel $\kappa_{3,\frac{3}{2}}$, centered at $(0,0)$, on a uniform grid:

```
rbf = @(r) (1-7*r.^2-81/4*r.^4).*sqrt(1-r.^2) - ...
           15/4*r.^4.*(6+r.^2).*log(r./(1+sqrt(1-r.^2)) + eps);
xplot = pick2Dpoints(-1,1,30);
K = DistanceMatrix(xplot,[0 0],[1 4],rbf);
norm(imag(K),'fro')
ans =
   1.0588e-22
K = DistanceMatrix(xplot,[0 0],[1 3.99999999999999],rbf);
norm(imag(K),'fro')
ans =
     0
```

In a curious result, a confluence of circumstance and minor numerical inaccuracies in the `rangesearch` computation have yielded a complex valued K matrix. Even a small deviation from the $\varepsilon = (1\ 4)$ shape parameter choices remedy the situation, but it can be very surprising to see a normally safe computation turn out with such unexpected values. Even more problematically, this is an issue on only one of the machines on which the authors have tested the code. This suggests that other code in this book could cause similar problems, but we have been unable to recognize it on our machines.

Of course, the fantastic MATLAB team is always at work trying to correct silly errors like this, so this will likely not be an issue once this work is published. Still this serves as a useful opportunity to remember the sometimes peculiar nature of computational mathematics.

4.3 Zonal Kernels in MATLAB

Although this book primarily focuses on the computation with kernels in $\mathbb{R}^d$, one of the great strengths of kernels is their ability to work in more complicated geometries. One such simple case that is discussed in this section is the interpolation of functions on a *sphere* using zonal kernels (introduced in Section 3.4.2). As discussed there, the Euclidean distance between points on a sphere of radius 1 is given by $\|\boldsymbol{x}-\boldsymbol{z}\|^2 = 2(1-\boldsymbol{x}^T\boldsymbol{z})$. The inner product term $\boldsymbol{x}^T\boldsymbol{z}$ also appeared in (4.4) in the computation of general distance matrices. Therefore computing it now will be a simple task of repurposing existing code.

By leveraging the structure of $\|\boldsymbol{x}-\boldsymbol{z}\|$, zonal kernels are often rephrased from being functions of the distance between $\boldsymbol{x}$ and $\boldsymbol{z}$ to being functions of $\boldsymbol{x}^T\boldsymbol{z}$, or dot product kernels as introduced in Section 3.4.1. Program 4.12 demonstrates the swift computation of these pairwise dot products and subsequent zonal kernel evaluation and plotting; the plot, using color to denote magnitude, is on display in Figure 4.3(e). An example of interpolating data on a sphere is provided in Section 15.3. Of course, surfaces more complicated than a sphere can also be studied with MATLAB; we refer interested readers to [Wright (2014)] for an exceptional MATLAB software package, with examples, for computing on *manifolds*.

Program 4.12. Excerpts from `KernelsZonal.m`

```
% Create evaluation locations with the builtin function sphere
% Reshape the [X,Y,Z] plotting data so each row is one point
Nplot = 105;
[X,Y,Z] = sphere(Nplot-1);
x = [X(:),Y(:),Z(:)];
% Choose a point on the sphere at which to center the kernel
z = [1 0 2]/norm([1 0 2]);
% Define the inverse multiquadric zonal kernel
% The dp term is pairwise dot product, not the distance r
zbf = @(g,dp) 1./sqrt(1+g^2-g*dp);
gam = .7;
% Evaluate the kernel
K = zbf(gam,x*z');
% Plot the kernel on the sphere, using colors for magnitude
% This vector of data must be reshaped to match the size of X,Y,Z
Kplot = reshape(K,Nplot,Nplot);
surf(X,Y,Z,Kplot,'edgecolor','none')
% Fix proportions, viewing angle, display magnitude reference
axis square
```

```
view([1 1 1])
colorbar
```

Remark 4.5. The `x*z'` computation on line 13 seems peculiar; it is referred to as a dot product but looks like an outer product. This odd notation is a compact way to compute all the inner products between the rows of X and Z (each of which is an evaluation location and kernel center similar to (4.3)) and organize them so that $(\mathsf{XZ}^T)_{i,j} = \boldsymbol{x}_i^T \boldsymbol{z}_j$.

4.4 Tensor Product Kernels in MATLAB

Although radial kernels provide a useful strategy for computing and theorizing about interpolation in higher dimensions, many kernels of relevance are *not* radial. As described in Section 3.4.3, a common strategy for creating a kernel in d dimensions is to form the tensor product of several univariate kernels. Computing tensor product kernels in MATLAB is made relatively straightforward through the manipulation of vectors and cell arrays (described below), but we will no longer be able to leverage the use of our `DistanceMatrix` framework. Program 4.14 demonstrates how kernels, both radial and nonradial, can be evaluated in the $K(x, z)$ format and proceeds to form a tensor product kernel using four 1D kernels.

Program 4.13. `KernelsTensorStandard.m`

```
% Define a series of 1D kernels which will be composed into a tensor
% kernel.  Note these are not defined in their radial form, even if some
% happen to be radial - we cannot use DistanceMatrix here.
% The C2 Matern
KM2 = @(e,x,z) exp(-e*abs(bsxfun(@minus,x,z'))).*...
                (1+e*abs(bsxfun(@minus,x,z')));
% The Inverse Multiquadric
KIM = @(e,x,z) 1./sqrt(1+(e*bsxfun(@minus,x,z')).^2);
% The Analytic Chebyshev with a = .5
% Note b is subbed for e, though they serve the same purpose
% Also note that 0<b<1
KCA = @(b,x,z) .5 + (1-b)* ...
        (b*(1-b^2) - 2*b*bsxfun(@plus,x.^2,z.^2') + (1+3*b^2)*x*z')./ ...
        ((1-b^2)^2 + 4*b*(b*bsxfun(@plus,x.^2,z.^2')-(1+b^2)*x*z'));
% The C0 Wendland kernel
KW0 = @(e,x,z) max(1-e*abs(bsxfun(@minus,x,z')),0);

% Organize these kernels into a cell array
Kcell = {KM2,KIM,KCA,KW0};

% Define the evaluation function for the kernel
% Because K is a 4D kernel, this function should accept
%         epvec - 1-by-4 vector
```

```
24  %          x - Neval-by-4 matrix
    %          z - N-by-4 matrix
26  % The columns of each of argument are passed to the 1D kernels
    % for evaluation, and the results are combined in a product
28  Kf = @(e,x,z) prod(cell2mat(reshape( ...
             cellfun(@(K,e1,x1,z1) K(e1,x1,z1), ...
30           Kcell,num2cell(e),num2cell(x,1), ...
             num2cell(z,1),'UniformOutput',0), ...
32                             [1,1,length(e)])),3);
```

Program 4.13 defines radial kernels such as the C^2 Matérn with the calling sequence `KM2 = @(e,x,z)` rather than `rbfM2 = @(e,r)` to emphasize the fact that the kernels we use here need not be radial. The computation of the tensor product kernel looks somewhat frightening, and we discuss more about cell arrays later in this section. The essential mechanism is the `cellfun` call which effectively evaluates and stores the results of each of the 1D kernels. Those kernels are then `reshaped` into a differently shaped storage unit (a multilinear algebra tensor of order 3) which MATLAB can compress into the tensor product kernel with the `prod` function.

When our data locations exhibit structure, such as a tensor product grid or points in concentric circles (see Appendix B), it may be possible to efficiently minimize the number of kernel evaluations. Using the MATLAB functions `cellfun` and `kron`, the computation of tensor kernels can be performed efficiently, albeit with more complicated code than many radial kernels. Program 4.14 demonstrates a strategy for computing 2D tensor kernels on a tensor product grid. The main tool used here is the *cell array* which allows for the organization of arbitrary data structures similar to a linked list in other programming languages. Cell arrays are declared either with the cell command or with curly braces `{}`. The plotting commands have been omitted to save space, but a sample tensor kernel is presented in Figure 4.3(f).

Program 4.14. Excerpts from `KernelsTensor.m`

```
    % Define the C0 Matern kernel for 1D only, implemented in the general
2   % x, z form (not radial) to show how general kernels can be defined
    % As always, each row is an x location, each column is a z location
4   % The shape parameter vector is defined as a cell array
    rbf_1d = @(e,x,z) exp(-e*abs(bsxfun(@minus,x,z')));
6   epcell = {1.5,2.5};
    % Choose some points in 1D to evaluate the tensor product kernel
8   % Store the data in cell arrays to allow varying N1d sizes
    % cellfun performs the same operation on each term in an array
10  N1d = {101,99};
    x1d = cellfun(@(x)pickpoints(-1,1,x),N1d,'UniformOutput',0);
12  % Choose z1d points at which to center this tensor product kernel
    % num2cell(w,1) stores the columns of w as elements in a cell array
14  z1d = num2cell(rand(11,2),1);
    % Compute and store all the 1D kernel computations in a cell array
```

```
16 K1d = cellfun(@(e,x,z) rbf_1d(e,x,z),epcell,x1d,z1d,'UniformOutput',0);
   % Form the full kernel matrix of size 101*99x11
18 % Each of the columns from the 1D kernel matrices are multiplied with
   % kron to form a full tensor product kernel in each column of K
20 K = cell2mat(cellfun(@(Kx,Ky)kron(Kx,Ky), ...
       num2cell(K1d1,1),num2cell(K1d2,1),'UniformOutput',0));
```

The kernel `rbf_1d` receives x and z values rather than the distances between them. This is necessary for nonradial kernels, and is used here as a demonstration even though the C^0 Matérn kernel is radial. To create the data locations in each dimension, `cellfun` is called on line 11 and passed a command to use `pickpoints` once for each value of `N1d`. The kernel centers are chosen randomly and then the two dimensions are split into separate arrays with the `num2cell` command on line 14. At this point, `x1d` and `z1d` are cell arrays whose k^{th} entries are the kth dimension of the x and z points, respectively. On line 16 each of these values is plugged into the kernel, along with the shape parameters stored in `epcell` using the `cellfun` function. After that happens, the kth entry of `K1d` contains the kth dimensions of univariate C^0 Matérn kernel evaluations.

To combine all of the individual dimensions into a single kernel matrix, the *Kronecker* product [Golub and Van Loan (2012)] is applied with `kron`. This has the effect of applying each value in the second dimension to every value in the first dimension. After all of these operations have taken place, having been executed with `cellfun`, the final step is to convert the output cell array into a standard kernel matrix on lines 20–21. Readers should be aware that while an extension to higher dimensions could use the same tools, the implementation becomes a good deal more complicated and would probably require either loops or additional cell arrays.

Remark 4.6. MATLAB functions such as `cellfun` and `arrayfun` will, by default return vectors. Computations in this section, however, have more complicated outputs which require the use of cell arrays to store. When this happens, the option `'UniformOutput',False` should be passed to `cellfun`. In MATLAB, as in most programming languages, the number 0 and the value `False` are treated equivalently.

4.5 Series Kernels in MATLAB

In Sections 2.2.3 and 3.3 we introduced the concept of kernels defined through a Mercer series and as a general series kernel, respectively. Although defining these kernels with an infinite series is valid mathematically, computing with an infinite series is a hopeless proposition since, if for no other reason, a computer could never perform an infinite number of operations. As a result, all the series kernels we consider in MATLAB must converge quickly enough to be acceptably accurate.

Chapter 7 discusses this in the context of a family of kernels defined through a series, but here we focus on the MATLAB implementation of such series kernels.

We begin by assuming the kernel we want to compute is defined as

$$K(\boldsymbol{x}, \boldsymbol{z}) = \sum_{n=1}^{\infty} \lambda_n \varphi_n(\boldsymbol{x}) \varphi_n(\boldsymbol{z}),$$

where $\lambda_1 \geq \lambda_2 \geq \ldots \geq \lambda_M \geq \ldots$. In the most promising of settings, φ_n and λ_n comprise a Hilbert–Schmidt eigenfunction decomposition of K as defined in Section 2.2.2 though this need not be assumed.

Because we cannot perform infinitely many computations, we choose a *truncation length* M such that

$$K(\boldsymbol{x}, \boldsymbol{z}) = \sum_{n=1}^{M} \lambda_n \varphi_n(\boldsymbol{x}) \varphi_n(\boldsymbol{z}) + \text{ small enough discrepancy},$$

where the choice of what is "small enough" may include consideration of, e.g., the quality of the data or the desired cost of computation. This issue of series truncation can be nontrivial and is studied further in Section 7.3.1.

If we assume that we want to create kernel matrices centered at locations $\mathcal{Z} = \{\boldsymbol{z}_1, \ldots, \boldsymbol{z}_N\}$ and evaluated at $\mathcal{X} = \{\boldsymbol{x}_1, \ldots, \boldsymbol{x}_{N_{\text{eval}}}\}$, we may define the objects

$$\boldsymbol{\phi}(\boldsymbol{x}) = \begin{pmatrix} \varphi_1(\boldsymbol{x}) \\ \vdots \\ \varphi_M(\boldsymbol{x}) \end{pmatrix}, \quad \Phi_{\mathcal{Z}} = \begin{pmatrix} \boldsymbol{\phi}(\boldsymbol{z}_1)^T \\ \vdots \\ \boldsymbol{\phi}(\boldsymbol{z}_N)^T \end{pmatrix}, \quad \Phi_{\mathcal{X}} = \begin{pmatrix} \boldsymbol{\phi}(\boldsymbol{x}_1)^T \\ \vdots \\ \boldsymbol{\phi}(\boldsymbol{x}_{N_{\text{eval}}})^T \end{pmatrix}, \quad \Lambda = \begin{pmatrix} \lambda_1 & & \\ & \ddots & \\ & & \lambda_M \end{pmatrix}.$$

After some manipulations, one can verify that

$$\mathsf{K} = \Phi_{\mathcal{X}} \Lambda \Phi_{\mathcal{Z}}^T + \text{ small enough discrepancy}.$$

where $(\mathsf{K})_{i,j} = K(\boldsymbol{x}_i, \boldsymbol{z}_j)$. This relatively straightforward linear algebra expression for the kernel matrix suggests that the associated implementation in MATLAB may be simple. An example applied to the Chebyshev kernels from Section 3.9.2 is provided in Program 4.15. In the code we fix the values of the kernel parameters at $a = 0.4$ and $b = 0.3$ (for (3.10)) and at $a = 0.4$ and $\beta = 1$ (for (3.14), so that $\zeta(2) = \frac{\pi^2}{6}$).

Program 4.15. Excerpts from `KernelsSeries.m`

```
% Choose an error computing scheme, explained in rbfsetup
% This is the 2-norm of a vector, and is used in errcompute
global GAUSSQR_PARAMETERS
GAUSSQR_PARAMETERS.NORM_TYPE = 2;
GAUSSQR_PARAMETERS.ERROR_STYLE = 2;
% Choose z - kernel centers and x - points to evaluate at
x = pickpoints(-1,1,500);
z = pickpoints(-1,1,10);
% Define the phi functions for the Chebyshev series kernel
% to accept a row of n values and a column of x values.
```

```
% It returns a length(x)-by-length(n) matrix
phi = @(n,x) bsxfun(@times,sqrt(2-(n==0)),cos(bsxfun(@times,n,acos(x))));
% Define the geometric (G) and algebraic (A) decay eigenvalues
% We are fixing a = .4, b = .3, beta = 1 (so zeta(2) = pi^2/6)
lamG = @(n) (n==0)*.6 + (n>0).*(.3).^(n-1)*(.4*.7);
lamA = @(n) (n==0)*.6 + (n>0).*.4./(n.^2*pi^2/6 + eps);
% Evaluate the Phix and Phiz matrices and vector of Lambda diagonal
% Consider the first M=1000 terms, starting from zero
M = 1000;
narr = 0:M-1;
Phix = phi(narr,x);
Phiz = phi(narr,z);
lamvecG = lamG(narr);lamvecA = lamA(narr);
% Define the closed form of the kernels
% It helps to define the degree 2 Bernoulli polynomial
% as well as an auxiliary term for the algebraic kernel
% Further algebraic computations are omitted; see repository
KG = @(x,z) .6 + .56*(.273 -.6*bsxfun(@plus,x.^2,z'.^2) + 1.27*x*z')./...
    (.8281 + .36*bsxfun(@plus,x.^2,z'.^2) - 1.308*x*z');
B2 = @(x) x.^2 - x + 1/6;
aux = @(op,x,z) B2(abs(bsxfun(op,acos(x),acos(z)))/(2*pi));
KA = @(x,z) .6 + 6*(.4)*(aux(@plus,x,z') + aux(@minus,x,z'));
% Evaluate the closed form kernels at the required data locations
KGcf = KG(x,z);
% Compute the M-length series approximation to the kernels
% This is the Phix*Lambda*Phiz' computation
KGser = bsxfun(@times,Phix,lamvecG)*Phiz';
% Study the convergence of the series representation with increasing terms
% Mcheck contains the series lengths for which to check the quality
% lambda values are applied to the kernel centers prior to make the code
% simpler during the sequence of evaluations for different M
% errcompute (in GaussQR) returns the error between vectors or matrices
Mcheck = [1:49,50:50:M];
PhizlamG = bsxfun(@times,lamvecG,Phiz);
errvecG = arrayfun(@(M) errcompute(Phix(:,1:M)*PhizlamG(:,1:M)',KGcf),Mcheck);
```

Remark 4.7. One common problem in writing code is the "off-by-one" error, where some index or counter is one greater or less than it should be. The line `narr = 0:M-1` is used instead of `narr = 1:M` because of this; the Chebyshev polynomials are indexed starting at 0 but MATLAB begins indexing at 1. When developing code, errors of the form `Index exceeds matrix dimensions` are sometimes attributable to an off-by-one error.

This program marks the earliest appearance in this book of the MATLAB data structure `GAUSSQR_PARAMETERS`, used in the `GaussQR` repository. `GAUSSQR_PARAMETERS` is a global object that stores parameter values relevant to computations throughout `GaussQR`. In the first few lines of this example we choose how error will be measured when using the `errcompute` function, which accepts two vectors or matrices and returns how far apart they are. The descriptions of the elements of the data structure are found in `rbfsetup`.

The definition of the `phi` function on line 12 again marks the use of `bsxfun` for operating on vectors and matrices. Because `phi` is built to return one row per `x` value and one column per `n` value, `bsxfun` is a useful tool for computing both the Chebyshev polynomials and applying the scaling which depends on n but not x. The command `n==0`, which appears both in the definition of $\varphi_n(x)$ (on line 12) and of the eigenvalues λ_n (on lines 15–16), returns a logical vector of the same size as `n` with value `1` anywhere that $n = 1$ and `0` anywhere that $n \neq 1$. This is a useful tool for performing different operations on a vector depending on what values the vector takes (e.g., case defined functions) and is applied in both the algebraic and geometric eigenvalue definitions. Again, the use of the `eps` term prevents a `0/0=NaN` computation.

The closed form definitions of the kernels on lines 28–32 are a bit dense and best left to personal study for better comprehension. We state two brief points of relevance: first, the `x*z'` computation in `KG` also appeared in Program 4.5; second, the use of auxiliary functions such as `B2` and `aux` can help make your code easier to read and easier to debug.

After computing the closed form of the kernels (only `KG` is displayed here, but `GaussQR` contains the other kernel computations as well), we perform the $\mathsf{\Phi}_{\mathcal{X}}\mathsf{\Lambda}\mathsf{\Phi}_{\mathcal{Z}}^T$ product for our series kernel approximation on line 37. While defining the matrix $\mathsf{\Lambda}$ as `Lambda=diag(lamvecG)` is correct both mathematically and computationally, it is more efficient to apply `lamvecG` through `bsxfun` rather than allocating the temporary memory required to store `Lambda` and performing a matrix-matrix product. Given that, it is (almost) always preferable to perform matrix-matrix products (Level 3 BLAS operations) rather than sequences of matrix-vector products (Level 2 BLAS operations) or inner products (Level 1 BLAS operations) when possible to minimize the cost of memory traffic [Golub and Van Loan (2012)]. This is why the computation of `KGser` is conducted at the matrix-matrix level rather than by computing each K entry individually using the definition of the Mercer series (a "Level 0" BLAS operation).

The remainder of Program 4.15 checks the quality of the series approximation as a function of M. To do so, `Mcheck` is defined as an array of truncation lengths, and `arrayfun` is used to compute the error associated with each of those lengths as measured by the Frobenius norm (recall `GAUSSQR_PARAMETERS`) and computed with `errcompute`. Series approximation quality results and sample kernels are presented in Figure 4.4. Computing `PhizlamG` separately reduces computation by computing $\mathsf{\Lambda}\mathsf{\Phi}_{\mathcal{Z}}$ once and using it several times within `arrayfun`. Incidentally, the cost of applying `lamvecG` to `Phiz` is less than to `Phix` (as we did in the `KGser` assignment) in this example because `Phiz` has fewer rows; small choices like this may save time during intense computations.

In Figure 4.4(c) we see that the series converges incredibly quickly for the ge-

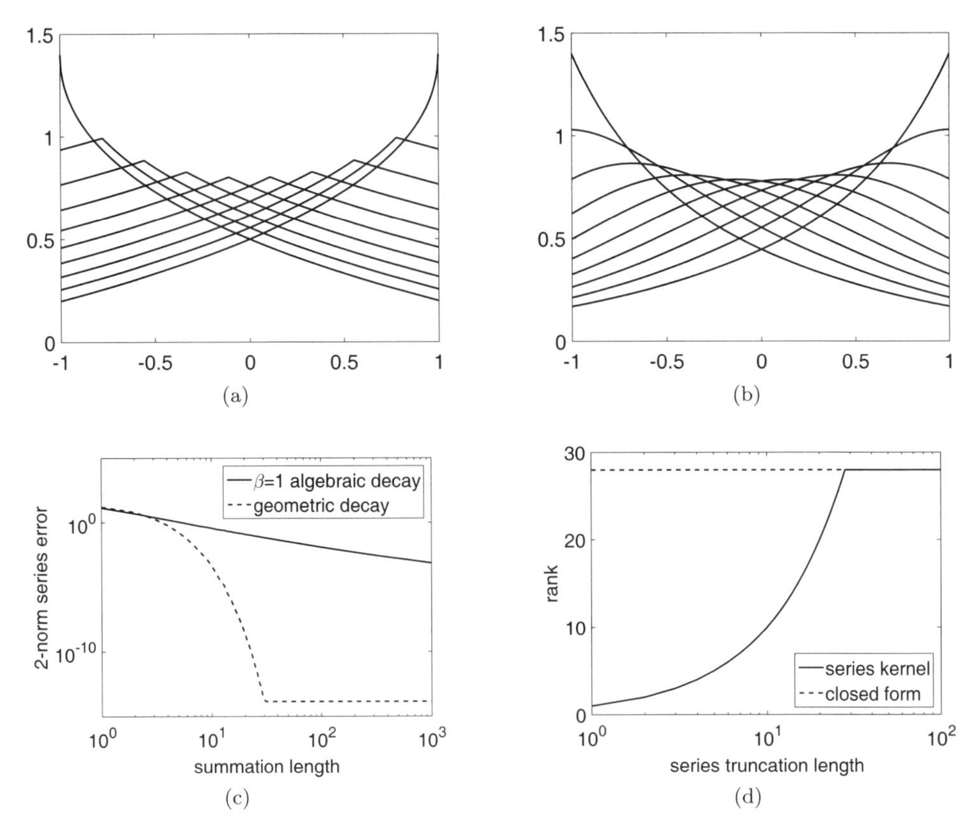

Fig. 4.4 Chebyshev series kernels from Section 3.9.2; the kernels with algebraic eigenvalue decay use $a = 0.4$ and $\beta = 1$ and those with geometric eigenvalue decay use $a = 0.4$ and $b = 0.3$. (a) 10 copies of the C^0 kernel (3.15), (b) 10 copies of the C^∞ kernel (3.13), (c) As a function of the series truncation length, polynomial convergence and spectral convergence to the true 500×10 kernel matrix are observed for the algebraic and geometric decay kernels, respectively, (d) The geometric decay Chebyshev kernel interpolation matrix of size $N = 100$ has a numerical rank of only 28.

ometrically decaying eigenvalues and rather slowly for the algebraically decaying eigenvalues. The effect of the rate of decay of the eigenvalues on series convergence rates and kernel smoothness is discussed in Section 7.3.1. The convergence rate has significant implications for the viability of inverting kernel interpolation matrices in computers with finite precision.

Figure 4.4(d) contains a graph generated by Program 4.16 which demonstrates that and includes a graph generated by Program 4.16(d) which demonstrates that an interpolation matrix with $N = 100$ Chebyshev spaced points (see Appendix B.3) using the kernel with geometrically decaying eigenvalues (which is C^∞ on $[-1, 1]$) has a *numerical rank* of only 28. Of course, any value other than rank 100 implies that this matrix — even though theoretically nonsingular — is not invertible on the computer. This means that — despite having found a nice low-rank representation for the matrix — we cannot evaluate the associated interpolant using (1.3).

A method for avoiding this low-rank issue is introduced in Chapter 13 and strategies for leveraging for computational efficiency are presented in Section 12.3 and Section 18.3.4.

Program 4.16. Excerpts from `KernelsLowRank.m`

```
% Define the points under consideration
N = 100; x = pickpoints(-1,1,N,'cheb');
% Define the closed form of the Cinf kernel
K = @(x,z) .6 + .56*(.273 -.6*bsxfun(@plus,x.^2,z'.^2) + 1.27*x*z')./...
    (.8281 + .36*bsxfun(@plus,x.^2,z'.^2) - 1.308*x*z');
% Define the phi functions and lambda values of a Chebyshev series kernel
phi = @(n,x) bsxfun(@times,sqrt(2-(n==0)),cos(bsxfun(@times,n,acos(x))));
lam = @(n) (n==0)*.6 + (n>0).*(.3).^(n-1)*(.4*.7);
% Evaluate the Phi matrix and lambda vector for K = Phi*Lam*Phi'
narr = 0:N-1;
Phi = phi(narr,x);
lamvec = lam(narr);
% Check the rank of the series interpolation matrix as a function of M
Mcheck = [1:49,50:50:N];
Philam = bsxfun(@times,lamvec,Phi);
rankvec = arrayfun(@(n) rank(Phi(:,1:n)*Philam(:,1:n)'),Mcheck);
% Plot the closed form rank of K and the series kernel rank together
semilogx(Mcheck,rankvec)
hold on
semilogx(Mcheck,ones(size(Mcheck))*rank(K(x,x)),'--')
hold off
```

Remark 4.8. Beyond the penalties, and opportunities, of working with quickly converging and potentially low-rank series kernels, issues also arise when dealing with slowly converging series kernels such as the Chebyshev series kernel with algebraically decaying eigenvalues defined in (3.15). The issue of numerical rank deficiency is rarely a problem for such kernels, but the computational cost associated with computing slowly converging kernels can be significant, and the memory cost can be prohibitive. For example, to compute the kernel interpolation matrix of $N = 100$ points with a series kernel using $M = 10^6$, 800MB of space would be required to store $\Phi_{\mathcal{X}}$. Storing such a matrix with $N = 1000$ points would be impractical on a standard 2015 computer; of course that statement is subject to change as technology improves, but there will always be some M and N combination beyond which storage is infeasible.

To combat this, a more complicated series kernel computation can be executed, one which tests the availability of memory and sums the kernel in pieces rather than all at once. One such implementation is provided in Program 4.17. This example is for more advanced users, and beginners to MATLAB may feel free to continue on without concern for this program.

Program 4.17. `KernelsMemory.m`

```
% Define the points under consideration
N = 100;x = pickpoints(-1,1,N);
% Define a necessary length for the series; large to trigger memory issues
M = 1000000;
% Define the phi functions and lambda values of a Chebyshev series kernel
phi = @(n,x) bsxfun(@times,sqrt(2-(n==0)),cos(bsxfun(@times,n,acos(x))));
lam = @(n) (n==0)*.6 + (n>0).*.4./(n.^2*pi^2/6 + eps);
% Try to allocate enough space for the kernel matrix
% If this fails, there is no hope for storing the interpolation matrix
try
    Kmat = zeros(N);
catch err
    if any(strcmp(err.identifier,...
          {'MATLAB:nomem','MATLAB:array:SizeLimitExceeded'}))
        error('Insufficient memory for interpolation matrix')
    else
        rethrow(err);
    end
end
% Set up a loop to try to allocate memory, potentially fail, and then
% allocate less memory and build the kernel in pieces
more_computing_needed = 1;
allocation_size = M;
current_narr = 1:M;
bound_by_M = @(vecstart,vecrange) unique(min(vecstart-1+(1:vecrange),M));
while more_computing_needed
    try
        Phi = phi(current_narr,x);
        lamvec = lam(current_narr);
        Philam = bsxfun(@times,lamvec,Phi);
    catch err
        if any(strcmp(err.identifier,...
              {'MATLAB:nomem','MATLAB:array:SizeLimitExceeded'}))
            allocation_size = floor(allocation_size/5);
            current_narr = bound_by_M(current_narr(1),allocation_size);
            continue
        else
            rethrow(err);
        end
    end
    Kmat = Kmat + Philam*Phi';

    if any(current_narr~=M)
        current_narr = bound_by_M(current_narr(end)+1,allocation_size);
    else
        more_computing_needed = 0;
    end
end
```

Initially, this program begins much as the others involving series kernels. The first new point is the addition of the `try/catch/rethrow` construct, which allows a MATLAB program to gracefully handle errors that would otherwise require termination. The first `try` command attempts to allocate enough space for the kernel interpolation matrix and suggests that if there is insufficient space for storing this matrix that computing it is equally a lost cause. An argument could be made that, since it is possible that the interpolation matrix is low-rank, as we saw in Program 4.16, less than `N^2` storage is required for computing with `Kmat`, but we focus only on the full-rank case now. The `rethrow` statement allows errors unrelated to memory to still terminate the program.

After successfully allocating the memory for the interpolation matrix, a loop is created to attempt to allocate the necessary space for the $\Phi_{\mathcal{X}}$ matrix. If the first pass through the `try` statement succeeds, then `Phi`$=\Phi_{\mathcal{X}}$ and `Philam`$=\Phi_{\mathcal{X}}\Lambda$ were successfully allocated and `Kmat` can be successfully computed. If an error appears in the `catch` block, then insufficient memory existed for allocating `Phi` and `Philam`. To compensate, the requested size is decreased by a factor of 5 (this factor is arbitrary) and the `continue` statement is used to restart the loop with the new `allocation_size` asking for only 1/5 of the memory.

This `continue` statement will be reached until there is sufficient space for the Φ matrices, at which point the loop begins executing the summation. Each step through the loop, the next `allocation_size` terms will be added to `Kmat`, and the current set of indices under consideration are stored in `current_narr`. The `bound_by_M` function handle increments the `current_narr` vector by `allocation_size` and prevents the indices in the summation from exceeding M, the desired truncation length.

Remark 4.9. As powerful as the MATLAB builtin functions such as `bsxfun` are, there are computations that they, by design, cannot execute. This is caused by certain assumptions which accelerate standard computations, but are incompatible with less common settings.

Consider the following situation: we want to numerically confirm the orthonormality of the Chebyshev kernels as defined in Section 3.9.2. To do this we create a matrix $(\mathsf{A})_{i,j} = \int_{-1}^{1} \varphi_{i-1}(x)\varphi_{j-1}(x)\rho(x)\,\mathrm{d}x$ which, if the φ_n functions are orthonormal, should be the identity matrix. We present Program 4.18 which demonstrates that the high performance of `bsxfun` (by evaluating multiple input values simultaneously) prevents the execution of the `integral` function, which performs as its name implies. The authors consider this a more advanced program but hope that readers challenge themselves to learn the complexity and beauty of programming in MATLAB.

Program 4.18. `KernelsOrthonormality.m`

```
  % Define the Chebyshev eigenfunctions and weight function
2 phi = @(n,x) bsxfun(@times,sqrt(2-(n==0)),cos(bsxfun(@times,n,acos(x))));
```

```
rho = @(x) 1./(pi*sqrt(1-x.^2));
% Define the orthonormality testing function
ortheval = @(n1,n2) integral(@(x)...
              bsxfun(@times,phi(n1,x).*phi(n2,x),rho(x)),-1,1);
% Define a set of eigenfunctions indices to test for orthonormality
ntest = [0,2,3:5,8];
% Causes an error
try
    orthmat = bsxfun(@(n1,n2)ortheval(n1,n2),ntest',ntest);
catch err
    if strcmp(err.identifier,'MATLAB:dimagree')
        fprintf('bsxfun calls ortheval with different sized n1 and n2\n')
    end
end
% Less elegant, but without errors
% Create a cell array with all the indices required
% Then compute orthogonality for each element of that cell array
[N1,N2] = meshgrid(ntest,ntest);
ncell = num2cell([N1(:),N2(:)],2);
orthvec = cellfun(@(nvals)ortheval(nvals(1),nvals(2)),ncell);
disp(reshape(orthvec,length(ntest),length(ntest)))
```

The final result of this program is a matrix which is roughly the identity, confirming that eigenfunctions have norm 1 and eigenfunctions of different order are orthogonal. This occurs only after encountering a dimension mismatch error brought on by an unfortunate combination of both `bsxfun` and `integral` trying to execute on vectors and not correctly speaking to each other. The program output, for MATLAB R2015a is

```
KernelsOrthonormality
bsxfun calls ortheval with different sized n1 and n2
    1.0000   -0.0000    0.0000   -0.0000    0.0000   -0.0000
   -0.0000    1.0000    0.0000   -0.0000   -0.0000   -0.0000
    0.0000    0.0000    1.0000    0.0000   -0.0000    0.0000
   -0.0000   -0.0000    0.0000    1.0000   -0.0000   -0.0000
    0.0000   -0.0000   -0.0000   -0.0000    1.0000   -0.0000
   -0.0000   -0.0000    0.0000   -0.0000   -0.0000    1.0000
```

Let this also be the first point in this book where the authors suggest that there are many ways to solve problems in MATLAB. We present our solution, but the reader should feel empowered to search for their own implementation which may be more efficient or readable. Such practice is fundamental to becoming a capable programmer, and we encourage readers to produce better code than our own.

Chapter 5

The Connection to Kriging

The kriging approach is a *stochastic prediction method* that arose in geostatistics and is named after the South African mining engineer Danie Krige (1951), who used this method in his work predicting gold ore distributions from a collection of ore samples. The method was given a solid mathematical foundation and shown to be an optimal linear prediction method by the French mathematician and geostatistician Georges Matheron (1965).

Part of our exposition in this chapter follows the discussion of Sacks, Welch, Mitchell and Wynn (1989). In that paper the kriging method was introduced into the *design of computer experiments* setting. Thus, while Krige's gold ore samples certainly would have been affected by measurement errors, the outcome of computer experiments is generally viewed as deterministic, i.e., the samples constitute exact values. Nevertheless, as we shall now see, one can apply the stochastic kriging framework in either case.

The main task in kriging is to use the values $y_1, \ldots, y_N$ sampled at locations $\boldsymbol{x}_1, \ldots, \boldsymbol{x}_N$ in order to predict the unknown value $y(\boldsymbol{x})$ at a location $\boldsymbol{x}$ (which is not among the sampling locations). Therefore — at the outset — this task is analogous to Problem 1.1, scattered data interpolation.

There are two approaches to thinking about this problem from the stochastic point of view — a regression approach (discussed in Section 5.3.1), and a Bayesian approach (discussed in Section 5.3.2), both of which proceed under the fundamental assumption that the data y_i are observed as realizations $y_{\boldsymbol{x}_i}$, $i = 1, \ldots, N$, of the random variables $Y_{\boldsymbol{x}_i}$ belonging to a random field Y — regardless of whether the data contain any random measurement error or not. Since the values $y_{\boldsymbol{x}_1}, \ldots, y_{\boldsymbol{x}_N}$ represent only one single realization of the random variables $Y_{\boldsymbol{x}_1}, \ldots, Y_{\boldsymbol{x}_N}$ our prediction models will be formulated with these random variables as input. Therefore the predictor $\overset{\triangle}{Y}_{\boldsymbol{x}}$ will also be a random variable (see (5.8)).

Following Cressie (1993), we will use language involving the notion of *prediction* when we are dealing with inference on random quantities, while in Chapter 14 we will be performing *estimation* of unknown parameters in our predictors.

In Section 5.1 we introduce the required concepts from probability theory and multivariate statistics that will allow us to present the kriging method in Section 5.3, where we also highlight the connection of the kriging method to kernel-based interpolation in the deterministic setting. A duality at the level of spaces of random variables in the stochastic setting with the reproducing kernel Hilbert spaces of Section 2.3 will be discussed in Section 5.2, and in Section 5.4 we talk briefly about the Karhunen–Loève expansion of Theorem 2.3 and its connection to (generalized) polynomial chaos.

5.1 Random Fields and Random Variables

We begin by giving the definition of a *stochastic process*, or — more appropriately in our context — a *random field*. To this end we require the notion of a *probability space* denoted by the triple $(\mathcal{W}, \mathcal{A}, P)$, where $\mathcal{W}$ is the sample space containing all possible outcomes, $\mathcal{A}$ is a σ-algebra containing the collection of all events, i.e., a set of subsets of $\mathcal{W}$, and P is a probability measure (see Appendix C for more background information from probability theory).

Definition 5.1 (Random field). *Given a probability space $(\mathcal{W}, \mathcal{A}, P)$ and an underlying parameter space Ω, a function $Y : \Omega \times \mathcal{W} \to \mathbb{R}$, i.e., $(\boldsymbol{x}, \omega) \mapsto Y(\boldsymbol{x}, \omega)$, is called a* random field *if, for every fixed $\boldsymbol{x} \in \Omega$, Y is an $\mathcal{A}$-measurable function of $\omega \in \mathcal{W}$.*

Remark 5.1.

(1) In the probability literature the sample space $\mathcal{W}$ is usually denoted by Ω, but we already use Ω to denote the domain for our spatial variables, i.e., the "parameter space" in Definition 5.1.
(2) In the numerical analysis literature one would think of the "parameter space" Ω as the (spatial) domain of the random field. Note that use of the term *stochastic process* (instead of random field) is also very common — especially when the parameter space is viewed as "time" or "space-time."
(3) In the statistics literature the parameter space is often denoted by T, and the random variables by X_t. We have chosen to use the notation Ω and $Y_{\boldsymbol{x}}$, respectively, since this agrees better with the notation used throughout the rest of this book.

As just mentioned, it is customary to omit the dependence on ω from the notation used for the random variable $Y_{\boldsymbol{x}}$. Unfortunately, random fields are commonly denoted using the same notation. We feel that this may lead to confusion and have therefore decided to deviate from this custom. Specifically, our notational conventions will be as follows:

- We will denote a *random field* by $Y = \{Y_{\boldsymbol{x}}\}_{\boldsymbol{x} \in \Omega}$.
- For a fixed $\boldsymbol{x} \in \Omega$, $Y_{\boldsymbol{x}} = Y(\boldsymbol{x}, \cdot)$ is a *random variable*. Thus, a random field is just a collection of random variables (which are actually *functions* of the

random argument $\omega \in \mathcal{W}$).

- For a fixed $\omega \in \mathcal{W}$, $y(\cdot) = Y(\cdot, \omega)$ is a *deterministic* function of $\boldsymbol{x} \in \Omega$ referred to as a *sample path* or *realization* of the random field. In this sense, a random field can also be viewed as a distribution over sample paths. The value of a sample path at $\boldsymbol{x}$ is denoted by $y(\boldsymbol{x})$.

Note that we can also look at a *realization of a random variable*, i.e., start with a random field Y and first fix $\boldsymbol{x}$ to get the random variable $Y_{\boldsymbol{x}}$ and then fix ω. Alternatively, we can *evaluate a sample path*, i.e., again start with Y but now first fix ω to obtain a deterministic function y and then fix $\boldsymbol{x}$. The former case uses the notation $y_{\boldsymbol{x}}$, while in the latter case it should be $y(\boldsymbol{x})$. These two approaches are equivalent, i.e., the numbers $y_{\boldsymbol{x}}$ and $y(\boldsymbol{x})$ are identical provided the same values of $\boldsymbol{x}$ and ω were used to determine them.

The *moments* of a random field Y provide useful information. In particular, the *expectation* or *mean* of a random field Y is a function given by

$$\mu(\boldsymbol{x}) = \mathbb{E}[Y_{\boldsymbol{x}}] = \int_{\mathcal{W}} Y_{\boldsymbol{x}}(\omega)\,\mathrm{d}P(\omega) = \int_{-\infty}^{\infty} y\,\mathrm{d}F_{Y_{\boldsymbol{x}}}(y),$$

where $F_{Y_{\boldsymbol{x}}}$ is the *cumulative distribution function (CDF)* of the random variable $Y_{\boldsymbol{x}}$ with respect to the probability measure P. If this distribution has a density $p_{Y_{\boldsymbol{x}}}$ such that

$$F_{Y_{\boldsymbol{x}}}(y) = P(-\infty < Y_{\boldsymbol{x}} < y) = \int_{-\infty}^{y} p_{Y_{\boldsymbol{x}}}(z)\,\mathrm{d}z \tag{5.1}$$

then

$$\mu(\boldsymbol{x}) = \int_{-\infty}^{\infty} y\,\mathrm{d}F_{Y_{\boldsymbol{x}}}(y) = \int_{-\infty}^{\infty} y p_{Y_{\boldsymbol{x}}}(y)\,\mathrm{d}y.$$

The second central moment of the random field Y (called its *variance*) is given by

$$\mathrm{Var}(Y_{\boldsymbol{x}}) = \mathbb{E}[Y_{\boldsymbol{x}}^2] - \mathbb{E}[Y_{\boldsymbol{x}}]^2 = \mathbb{E}[Y_{\boldsymbol{x}}^2] - \mu^2(\boldsymbol{x}),$$

and, more generally, the *covariance kernel* K of Y is defined via

$$\begin{aligned}
\sigma^2 K(\boldsymbol{x}, \boldsymbol{z}) &= \mathrm{Cov}(Y_{\boldsymbol{x}}, Y_{\boldsymbol{z}}) = \mathbb{E}\left[(Y_{\boldsymbol{x}} - \mu(\boldsymbol{x}))(Y_{\boldsymbol{z}} - \mu(\boldsymbol{z}))\right] \\
&= \mathbb{E}\left[(Y_{\boldsymbol{x}} - \mathbb{E}[Y_{\boldsymbol{x}}])(Y_{\boldsymbol{z}} - \mathbb{E}[Y_{\boldsymbol{z}}])\right] \\
&= \mathbb{E}\left[Y_{\boldsymbol{x}} Y_{\boldsymbol{z}} - Y_{\boldsymbol{x}}\mathbb{E}[Y_{\boldsymbol{z}}] - \mathbb{E}[Y_{\boldsymbol{x}}]Y_{\boldsymbol{z}} + \mathbb{E}[Y_{\boldsymbol{x}}]\mathbb{E}[Y_{\boldsymbol{z}}]\right] \\
&= \mathbb{E}[Y_{\boldsymbol{x}} Y_{\boldsymbol{z}}] - \mathbb{E}[Y_{\boldsymbol{x}}]\mathbb{E}[Y_{\boldsymbol{z}}] - \mathbb{E}[Y_{\boldsymbol{x}}]\mathbb{E}[Y_{\boldsymbol{z}}] + \mathbb{E}[Y_{\boldsymbol{x}}]\mathbb{E}[Y_{\boldsymbol{z}}] \\
&= \mathbb{E}[Y_{\boldsymbol{x}} Y_{\boldsymbol{z}}] - \mathbb{E}[Y_{\boldsymbol{x}}]\mathbb{E}[Y_{\boldsymbol{z}}] = \mathbb{E}[Y_{\boldsymbol{x}} Y_{\boldsymbol{z}}] - \mu(\boldsymbol{x})\mu(\boldsymbol{z}).
\end{aligned} \tag{5.2}$$

Remark 5.2. Here the scalar parameter σ^2 is known as the *process variance* and in the statistics literature is often included in the definition of the covariance kernel, so that the Gaussian covariance would be, e.g., $K(\boldsymbol{x}, \boldsymbol{z}) = \sigma^2 \mathrm{e}^{-\varepsilon^2\|\boldsymbol{x}-\boldsymbol{z}\|^2}$. Our kernel definitions and formulas as introduced in earlier chapters do, however, not include such an "amplification" factor. For example, we defined the Gaussian kernel as $K(\boldsymbol{x}, \boldsymbol{z}) = \mathrm{e}^{-\varepsilon^2\|\boldsymbol{x}-\boldsymbol{z}\|^2}$. The reason for this is that in the deterministic/numerical analysis setting of scattered data interpolation such a factor is irrelevant since it will be absorbed in the expansion coefficients c_j of (1.3a). However, as we will later see, in the stochastic setting the process variance plays an important role in, e.g., the correct formulation of the kriging variance and in parameter estimation.

With this notation, the variance of the random field Y corresponds to the "diagonal" of the covariance, i.e.,

$$\mathrm{Var}(Y_{\boldsymbol{x}}) = \sigma^2 K(\boldsymbol{x}, \boldsymbol{x}),$$

and it directly coincides with the process variance σ^2 if the kernel K is normalized so that $K(\boldsymbol{x}, \boldsymbol{x}) = 1$ — as is, e.g., the case for the Gaussian kernel.

There are many different kinds of stochastic processes. We will be particularly interested in *Gaussian random fields* (or *Gaussian processes*) since they model many natural phenomena and are relatively easy to work with. In particular, a Gaussian random field is completely characterized by its first two moments, namely its expectation and variance.

Definition 5.2 (Gaussian random field). *The random field* $Y = \{Y_{\boldsymbol{x}}\}_{\boldsymbol{x}\in\Omega}$ *is called a* Gaussian random field *if, for any given choice of finitely many distinct points* $\mathcal{X} = \{\boldsymbol{x}_i\}_{i=1}^N \subset \Omega$*, the vector*[1] *of random variables* $\boldsymbol{Y} = \left(Y_{\boldsymbol{x}_1} \cdots Y_{\boldsymbol{x}_N}\right)^T$ *has a multivariate normal distribution with mean vector* $\boldsymbol{\mu} = \mathbb{E}[\boldsymbol{Y}]$ *and covariance matrix* $\sigma^2\mathsf{K} = \left(\mathrm{Cov}(Y_{\boldsymbol{x}_i}, Y_{\boldsymbol{x}_j})\right)_{i,j=1}^N$*, where* σ^2 *is the* process variance.

We use the notation $\boldsymbol{Y} \sim \mathcal{N}(\boldsymbol{\mu}, \sigma^2\mathsf{K})$ to indicate that $\boldsymbol{Y}$ is a vector of Gaussian random variables, or $Y \sim \mathcal{N}(\mu, \sigma^2 K)$ to state that Y is a Gaussian random field. The *multivariate normal distribution* has the density function

$$p_{\boldsymbol{Y}}(\boldsymbol{y}) = \frac{1}{\sqrt{(2\pi\sigma^2)^N \det \mathsf{K}}} \exp\left(-\frac{1}{2\sigma^2}(\boldsymbol{y}-\boldsymbol{\mu})^T\mathsf{K}^{-1}(\boldsymbol{y}-\boldsymbol{\mu})\right), \tag{5.3}$$

where $\boldsymbol{\mu}$ and K are defined as in Definition 5.2. As long as K is a strictly positive definite kernel, K will be a positive definite matrix, and K^{-1} will exist.

Remark 5.3.

(1) The quadratic form $(\boldsymbol{y}-\boldsymbol{\mu})^T\mathsf{K}^{-1}(\boldsymbol{y}-\boldsymbol{\mu})$ in the exponent of the multivariate normal density function (5.3) is sometimes referred to as (the square of) the *Mahalanobis distance*. One can easily verify (see, e.g., [Abrahamsen (1997)]) that the computation of the Mahalanobis distance does not depend on the ordering of the points $\boldsymbol{x}_1, \ldots, \boldsymbol{x}_N$.
We may assume, for simplicity, that $\boldsymbol{\mu} = \boldsymbol{0}$. The kernel-based interpolant is defined by the linear system $\mathsf{K}\boldsymbol{c} = \boldsymbol{y}$, and this is essentially also true for the

[1] Our multivariate notation $\boldsymbol{Y}$, $\boldsymbol{\mu}$ and K does not explicitly denote the dependence on $\mathcal{X}$. However, we feel the notation $\boldsymbol{Y}_{\mathcal{X}}$, $\boldsymbol{\mu}_{\mathcal{X}}$ and $\mathsf{K}_{\mathcal{X}}$ would be excessive and unnecessary — especially by comparison to the K notation used in the numerical analysis setting. It will benefit the reader, however, to remember that $\boldsymbol{Y}$ is defined using $\mathcal{X}$.

conditional mean of the kriging predictor (see (5.8)). Then, in fact, $\boldsymbol{y}^T \mathsf{K}^{-1} \boldsymbol{y} = \boldsymbol{c}^T \mathsf{K} \boldsymbol{c}$, and this latter representation is nothing but the square of the norm of the interpolant $s = \sum_{j=1}^N c_j K(\cdot, \boldsymbol{x}_j)$ in the reproducing kernel Hilbert space $\mathcal{H}_K(\Omega)$, i.e., $\boldsymbol{y}^T \mathsf{K}^{-1} \boldsymbol{y} = \|s\|^2_{\mathcal{H}_K(\Omega)}$ (see (2.24)).

(2) The multivariate normal density function can also be defined for positive definite covariance functions which generate a positive *semi*-definite, i.e., singular, covariance matrix K. In that case the singular value decomposition of K is used (see, e.g., [Mardia *et al.* (1979, Section 2.5.4)]) and the pseudo-inverse of K replaces K^{-1}. Moreover, $\det \mathsf{K}$ is replaced by the product of the nonzero singular values of K and the vector $\boldsymbol{Y}$ is restricted to lie in a space orthogonal to the left nullspace of K.

We often encounter random fields with special properties. For example, a random field Y is called *stationary* if $\mu(\boldsymbol{x})$ is constant and $K(\boldsymbol{x}, \boldsymbol{z}) = \widetilde{K}(\boldsymbol{x} - \boldsymbol{z})$. The variance of a stationary random field is also constant and thus the correlation function of a stationary random field is just a constant multiple of its covariance function, i.e.,

$$\begin{aligned} \operatorname{Corr}(Y_{\boldsymbol{x}}, Y_{\boldsymbol{z}}) &= \frac{\operatorname{Cov}(Y_{\boldsymbol{x}}, Y_{\boldsymbol{z}})}{\sqrt{\operatorname{Var}(Y_{\boldsymbol{x}}) \operatorname{Var}(Y_{\boldsymbol{z}})}} \\ &= \frac{K(\boldsymbol{x}, \boldsymbol{z})}{\sqrt{K(\boldsymbol{x}, \boldsymbol{x}) K(\boldsymbol{z}, \boldsymbol{z})}} = \frac{\widetilde{K}(\boldsymbol{x} - \boldsymbol{z})}{|\widetilde{K}(\boldsymbol{0})|}. \end{aligned} \tag{5.4}$$

Moreover, a random field is called *isotropic* if $K(\boldsymbol{x}, \boldsymbol{z}) = \kappa(\|\boldsymbol{x} - \boldsymbol{z}\|)$ for some norm on Ω, usually the Euclidean norm. In the numerical analysis literature stationary kernels are referred to as *translation-invariant* kernels, and isotropic kernels are known as *radial* kernels. We will be flexible in our use of terminology, and we hope the reader will become comfortable as well.

Example 5.1. (Gauss–Markov kernels)
For some applications, such as in operations research (see, e.g., [Salemi (2014)]), it is important that the process be *Markovian*, i.e., without memory. In the spatial (random field) setting the Markovian property should be interpreted as the random field being determined on the domain Ω by sufficient values on the boundary, and independent of values outside of Ω. Salemi (2014) creates so-called *generalized integrated Brownian fields* via stochastic integration and shows that they are Markovian.

For one-dimensional zero-mean Gaussian processes Y one can show (see [Berlinet and Thomas-Agnan (2004, Chapter 2)] or [Kailath (1966)]) that the process Y with continuous covariance kernel K is Markovian if and only if K is of the form

$$K(x, z) = g(x) G(\min(x, z)) g(z),$$

which Kailath (1966) refers to as *Gauss–Markov kernels*. Here g is a nonvanishing continuous function defined on Ω with $g(0) = 1$ and G is a monotone increasing continuous function on Ω. Special examples — all with domain $\Omega = [0, 1]$ — are:

(1) The *Wiener process* or *Brownian motion* with $g(x) = 1$ and $G(x) = x$, so that $K(x, z) = \min\{x, z\}$.
(2) The *pinned Wiener process* or *Brownian bridge* with $g(x) = 1 - x$ and $G(x) = \frac{x}{1-x}$, so that $K(x, z) = \min\{x, z\} - xz$.
(3) The *Ornstein–Uhlenbeck process* with $g(x) = \mathrm{e}^{-\varepsilon x}$ and $G(x) = \mathrm{e}^{2\varepsilon x}$, so that $K(x, z) = \mathrm{e}^{-\varepsilon|x-z|}$. Here $\varepsilon > 0$ is a parameter. Note that this process is also stationary (depending on only $|x - z|$). In fact, it is the only Gaussian process that is both Markovian and stationary [Kailath (1966)].

Again, in the statistics literature (see, e.g., [Berlinet and Thomas-Agnan (2004)]) these kernels may be scaled by the process variance σ^2, i.e., they may appear, e.g., as $K(x, z) = \sigma^2 \min(x, z)$. The structure of the kernel is unchanged, but we prefer to consider σ^2 separately rather than absorbing it into the definition of K.

5.2 Duality Between Reproducing Kernel Hilbert Spaces and Spaces Generated by Zero-Mean Gaussian Random Fields

In discussing a general data fitting problem we can now take at least two different points of view. We can interpret our data as

(1) values $y_i = f(\boldsymbol{x}_i)$ of a deterministic function $f \in \mathcal{H}_K$, or
(2) values $y_i = y(\boldsymbol{x}_i)$ of a sample path $y(\cdot)$ of a zero-mean random field Y.

According to [Wahba (1990, Chapter 1)], the following duality goes back to the work of Parzen (1961, 1970). The restriction to zero-mean random fields in item (2) is not required but it simplifies much of the following discussion.

We discussed the reproducing kernel Hilbert space $\mathcal{H}_K(\Omega)$ in Chapter 2. Similarly, we can define a Hilbert space $\mathcal{H}_Y(\Omega)$ as the set of all linear combinations of random variables $Y_{\boldsymbol{x}}$ of the zero-mean random field $Y = \{Y_{\boldsymbol{x}}\}_{\boldsymbol{x}\in\Omega}$ together with their $L_2(\mathcal{W}, \mathcal{A}, P)$-limits.

When trying to consolidate these two view points it is important to take note of the fact that a function $f \in \mathcal{H}_K(\Omega)$ is *not* the same as a sample path $y(\cdot)$ of a stochastic process Y. In fact, reproducing kernel Hilbert space functions are generally smooth functions, while sample paths are instances of random variables which are in general *not smooth* (see, e.g., [Wahba (1990, Chapter 1)] or [Seeger (2004)]).

Brownian motion nicely illustrates this dissidence. The corresponding reproducing kernel Hilbert space is given by the standard Sobolev space $H^1(0, 1)$ of functions whose first derivative is square integrable, i.e., they must be differentiable except on a set of measure zero, but a typical sample path is nowhere differentiable.

It is possible for sample paths of a Gaussian process to belong to the space $\mathcal{H}_K(\Omega)$ with probability one; necessary and sufficient conditions are presented in [Lukić and Beder (2001)]. Those authors defined a larger reproducing kernel Hilbert space that dominates $\mathcal{H}_K(\Omega)$. This idea was employed in the context of numerical solutions of stochastic PDEs in [Cialenco *et al.* (2012)].

On the other hand, the *random field* Y — as opposed to its sample paths — does have certain smoothness properties (weak continuity and weak differentiability) that are tied to the smoothness of the kernel K (see [Berlinet and Thomas-Agnan (2004, Section 2.2.3)] for more details).

Despite these differences, there is an isometry between the reproducing kernel Hilbert space $\mathcal{H}_K(\Omega)$ and the Hilbert space $\mathcal{H}_Y(\Omega)$ generated by the zero-mean Gaussian random field Y known as *Loève's representation theorem* (see, e.g., [Wahba (1990, Chapter 1)] or [Berlinet and Thomas-Agnan (2004, Chapter 2)]). This theorem dictates that the values of the two corresponding inner products are identical and coupled by the kernel K, i.e.,

$$\langle Y_{\boldsymbol{x}}, Y_{\boldsymbol{z}} \rangle_{\mathcal{H}_Y(\Omega)} = \mathbb{E}[Y_{\boldsymbol{x}} Y_{\boldsymbol{z}}] = \sigma^2 K(\boldsymbol{x}, \boldsymbol{z}) = \sigma^2 \langle K(\cdot, \boldsymbol{x}), K(\cdot, \boldsymbol{z}) \rangle_{\mathcal{H}_K(\Omega)}.$$

According to Berlinet and Thomas-Agnan (2004), the canonical mapping from $\mathcal{H}_K(\Omega)$ to $\mathcal{H}_Y(\Omega)$ identifies the functions $k_j(\cdot) = K(\cdot, \boldsymbol{x}_j)$, where $\boldsymbol{x}_j \in \Omega$ for $j = 1, 2, \ldots$, with the eigenfunctions φ_n, $n = 1, 2, \ldots$, of the covariance kernel K of Y (see Section 2.2.6). Alternatively, Wahba (1990) suggests an identification of k_j with the random variables $Y_{\boldsymbol{x}_j}$, $j = 1, 2, \ldots$, of the random field Y, while Ghanem and Spanos (2003) suggest identifying the eigenfunctions $\varphi_n \in \mathcal{H}_K(\Omega)$ with the random variables $Z_n \in \mathcal{H}_Y(\Omega)$ used in the Karhunen–Loève expansion of Theorem 2.3 via

$$Z_n(\omega) = \frac{1}{\sqrt{\lambda_n}} \int_\Omega Y_{\boldsymbol{x}}(\omega) \varphi_n(\boldsymbol{x}) \mathrm{d}\boldsymbol{x}. \tag{5.5}$$

Using the eigenfunctions φ_n and eigenvalues λ_n, $n = 1, 2, \ldots$, of K we can represent both functions $f \in \mathcal{H}_K(\Omega)$ and realizations $y_{\boldsymbol{x}}$ of a random variable $Y_{\boldsymbol{x}} \in \mathcal{H}_Y(\Omega)$ of the random field Y, respectively, as

$$f(\boldsymbol{x}) = \langle f, K(\cdot, \boldsymbol{x}) \rangle_{\mathcal{H}_K(\Omega)} = \sum_{n=1}^{\infty} \hat{f}_n \varphi_n(\boldsymbol{x}),$$

$$\text{where } \hat{f}_n = \frac{\langle f, \varphi_n \rangle_{\mathcal{H}_K(\Omega)}}{\|\varphi_n\|^2_{\mathcal{H}_K(\Omega)}} = \lambda_n \langle f, \varphi_n \rangle_{\mathcal{H}_K(\Omega)},$$

$$y_{\boldsymbol{x}} = Y_{\boldsymbol{x}}(\omega) = \sum_{n=1}^{\infty} Z_n(\omega) \sqrt{\lambda_n} \varphi_n(\boldsymbol{x}), \qquad \text{with } Z_n(\omega) \text{ as in (5.5).}$$

Note that the latter representation follows from the Karhunen–Loève expansion. Here we have violated our notational conventions to emphasize the dependence of the KL expansion on ω. Clearly, one needs to fix a value of ω to obtain a specific realization of the "orthonormal" random variables Z_n.

The above duality can be extended in various ways. First, it is not limited to point evaluation. It can be extended to arbitrary linear functionals (see [Wahba (1990)] or [Seeger (2004)]) and therefore we have a stochastic interpretation of numerical analysis problems such as Hermite interpolation, or collocation solution of

PDEs. For example, the paper [Mardia *et al.* (1996)] discusses kriging when derivative information is available. Secondly, it can be extended to cover noncentered, non-Gaussian processes. This generalization is discussed in [Berlinet and Thomas-Agnan (2004, Chapter 2)].

5.3 Modeling and Prediction via Kriging

There are many different flavors of kriging. In *simple kriging* one assumes that the random field Y is *centered*, i.e., it has zero mean, $\mu(\boldsymbol{x}) = \mathbb{E}[Y_{\boldsymbol{x}}] = 0$. Simple kriging uses positive definite covariance kernels. In the atmospheric sciences literature simple kriging is known as *objective analysis* and due to [Gandin (1963)] (see [Cressie (1993); Stein (1999)]). If the process (and correspondingly the data) is not centered, then one can either center the data in a preprocessing step and continue to work in the simple kriging framework, or one has to modify the model to include a deterministic term that will capture the mean (also sometimes called the *trend*). This is usually done by adding polynomial basis functions[2] to the model and leads to so-called *universal kriging*.

The reader who is familiar with interpolation with conditionally positive definite kernels (such as thin plate splines or multiquadric radial basis functions) may expect some parallels here. However, this is *not* so in general since universal kriging still assumes use of a positive definite covariance kernel. A kriging variant that parallels interpolation with conditionally positive definite translation-invariant kernels augmented by polynomials is given by *intrinsic kriging*, a method that uses *intrinsic random functions* and *generalized covariance functions*, which were introduced by Matheron (1973) (see also, e.g., [Berlinet and Thomas-Agnan (2004, Chapter 2), Chiles and Delfiner (2012, Chapter 4)] or [Mardia *et al.* (1996); Scheuerer *et al.* (2013)]).

In what follows, we will focus on simple kriging and present two alternative viewpoints: a regression approach and a Bayesian approach.

5.3.1 *Kriging as best linear unbiased predictor*

In the first (standard regression) approach (see, e.g., [Sacks *et al.* (1989); Berlinet and Thomas-Agnan (2004)]) we assume that the approximate value of a realization of a zero-mean Gaussian random field is given by a linear predictor of the form

$$\hat{Y}_{\boldsymbol{x}} = \sum_{j=1}^{N} w_j(\boldsymbol{x}) Y_{\boldsymbol{x}_j} = \boldsymbol{w}(\boldsymbol{x})^T \boldsymbol{Y},$$

[2]However, it is conceivable that other types of functions are used to model the trend. For example, trigonometric functions might be used to model an oscillatory trend. The resulting method is known as *trigonometric kriging* [Seguret and Huchon (1990)].

where $\overset{\triangle}{Y}_{\boldsymbol{x}}$ and $Y_{\boldsymbol{x}_j}$ are random variables, $\boldsymbol{Y} = \left(Y_{\boldsymbol{x}_1} \cdots Y_{\boldsymbol{x}_N}\right)^T$, and $\boldsymbol{w}(\boldsymbol{x}) = \left(w_1(\boldsymbol{x}) \cdots w_N(\boldsymbol{x})\right)^T$ is a vector of *weight functions* evaluated at $\boldsymbol{x}$. Since all random variables have zero expectation this approach will automatically lead to an unbiased predictor.

This is precisely how Krige introduced the method: the unknown value to be predicted is given by a weighted average of the observed values, where the weights depend on the prediction location. Usually one assigns a smaller weight to observations further away from $\boldsymbol{x}$.

To compute "optimal" weights $\overset{*}{w}_j(\cdot)$, $j = 1, \ldots, N$, one now considers the mean-squared error (MSE) of the predictor, i.e.,

$$\mathrm{MSE}(\overset{\triangle}{Y}_{\boldsymbol{x}}) = \mathbb{E}\left[\left(Y_{\boldsymbol{x}} - \boldsymbol{w}(\boldsymbol{x})^T\boldsymbol{Y}\right)^2\right]. \;.$$

Let us rewrite the MSE in terms of the covariance kernel K by using the fact that $\sigma^2 K(\boldsymbol{x}, \boldsymbol{z}) = \mathbb{E}[Y_{\boldsymbol{x}} Y_{\boldsymbol{z}}]$ (see (5.2) and remember that we are assuming Y to be centered):

$$\begin{aligned}\mathrm{MSE}(\overset{\triangle}{Y}_{\boldsymbol{x}}) &= \mathbb{E}\left[\left(Y_{\boldsymbol{x}} - \boldsymbol{w}(\boldsymbol{x})^T\boldsymbol{Y}\right)^2\right] \\ &= \mathbb{E}[Y_{\boldsymbol{x}}Y_{\boldsymbol{x}}] - 2\mathbb{E}[Y_{\boldsymbol{x}}\boldsymbol{w}(\boldsymbol{x})^T\boldsymbol{Y}] + \mathbb{E}[\boldsymbol{w}(\boldsymbol{x})^T\boldsymbol{Y}\boldsymbol{Y}^T\boldsymbol{w}(\boldsymbol{x})] \\ &= \sigma^2 K(\boldsymbol{x}, \boldsymbol{x}) - 2\boldsymbol{w}(\boldsymbol{x})^T(\sigma^2\boldsymbol{k}(\boldsymbol{x})) + \boldsymbol{w}(\boldsymbol{x})^T(\sigma^2\mathsf{K})\boldsymbol{w}(\boldsymbol{x}),\end{aligned} \tag{5.6}$$

where the vector $\sigma^2\boldsymbol{k}(\boldsymbol{x})$ is a column vector whose j^{th} component is $\sigma^2 K(\boldsymbol{x}, \boldsymbol{x}_j) = \mathbb{E}[Y_{\boldsymbol{x}}Y_{\boldsymbol{x}_j}]$ and the matrix $\sigma^2\mathsf{K}$ has the $(i,j)^{\text{th}}$ entry $\sigma^2 K(\boldsymbol{x}_i, \boldsymbol{x}_j) = \mathbb{E}[Y_{\boldsymbol{x}}Y_{\boldsymbol{x}_j}]$.

Finding the minimum MSE is straightforward since (5.6) is a quadratic form in $\boldsymbol{w}(\boldsymbol{x})$. Differentiation yields the optimum weight vector

$$\overset{*}{\boldsymbol{w}}(\boldsymbol{x}) = \mathsf{K}^{-1}\boldsymbol{k}(\boldsymbol{x}), \tag{5.7}$$

which readers familiar with radial basis function interpolation will recognize as the vector of *cardinal basis functions* for the set $\mathcal{X} = \{\boldsymbol{x}_1, \ldots, \boldsymbol{x}_N\}$ of centers (see, e.g., [Fasshauer (2007, Chapter 14)] or (12.4)).

Thus, the (simple) *kriging predictor*

$$\overset{\triangle}{Y}_{\boldsymbol{x}} = \boldsymbol{k}(\boldsymbol{x})^T\mathsf{K}^{-1}\boldsymbol{Y} \tag{5.8}$$

is the *best linear unbiased predictor* BLUP (in the least squares sense). Recall that, despite the lack of acknowledgment in the notation, both K and $\boldsymbol{Y}$ depend on the data sites $\mathcal{X}$. Also, note that (5.8) is *independent* of the process variance σ^2. This independence is the reason why process variance does not appear in the numerical analysis setting: the interpolant is unaffected by a scaling of the kernel.

Taking a "classical frequentist stance" [Sacks *et al.* (1989)], we now look at the realizations of the random variables in (5.8), i.e., fix ω, and then get the *kriging prediction*[3] based on the given data as

$$\overset{\triangle}{y}_{\boldsymbol{x}} = \boldsymbol{k}(\boldsymbol{x})^T\mathsf{K}^{-1}\boldsymbol{y}. \tag{5.9}$$

[3]The reader should take note of the difference between a *predictor* and a *prediction*: the predictor is a random variable, and the prediction is a realization.

As we will see in Section 5.3.2 on the Bayesian approach, we can arrive at the same expression by taking conditional expectations. Notice that the expression for the kriging prediction (5.9) is formally identical to that of the kernel interpolant stated in (1.3c). Moreover, for Gaussian random fields the BLUP is also the best *nonlinear* unbiased predictor (see, e.g., [Berlinet and Thomas-Agnan (2004, Chapter 2)]).

The MSE (5.6) of the kriging predictor with the optimal weights $\mathring{\boldsymbol{w}}(\cdot)$ from (5.7),

$$\mathbb{E}\left[\left(Y_{\boldsymbol{x}} - \overset{\triangle}{Y}_{\boldsymbol{x}}\right)^2\right] = \sigma^2(K(\boldsymbol{x},\boldsymbol{x}) - \boldsymbol{k}(\boldsymbol{x})^T\mathsf{K}^{-1}\boldsymbol{k}(\boldsymbol{x})), \tag{5.10}$$

is known as the *kriging variance*. An alternative derivation using the Bayesian framework can be found in Section 5.3.3. Readers familiar with radial basis function interpolation will notice that the kriging variance is related to the square of the *power function* at $\boldsymbol{x}$ (see, e.g., [Fasshauer (2007, Chapter 14)] or (9.8) below). The significant difference between the interpolation setting and (5.10) is the presence of the process variance σ^2. The presence of the factor σ^2 explicitly distinguishes the power function from the kriging variance.

Remark 5.4.

(1) If the random field is not centered then the predictor should take the more general form

$$\overset{\triangle}{Y}_{\boldsymbol{x}} = \overset{\triangle}{Y}{}^{(0)}_{\boldsymbol{x}} + \boldsymbol{p}(\boldsymbol{x})^T\boldsymbol{\beta},$$

where $\overset{\triangle}{Y}{}^{(0)}_{\boldsymbol{x}}$ models a zero-mean random field as discussed above, and $\boldsymbol{p}(\boldsymbol{x}) = \left(p_1(\boldsymbol{x}) \cdots p_q(\boldsymbol{x})\right)^T$ is a vector of polynomial basis functions used to model the mean (or trend) of the field Y. In this more general setting the model will no longer be unbiased by default, and unbiasedness has to be enforced by adding the constraint

$$\mathbb{E}[\overset{\triangle}{Y}_{\boldsymbol{x}}] = \mathbb{E}[Y_{\boldsymbol{x}}]$$

to the quadratic MSE minimization problem. Depending on whether one performs ordinary kriging or universal kriging, the mean of the random field is then modeled, respectively, with either an additional constant parameter ($q = 1$, $p_1(\boldsymbol{x}) \equiv 1$) or a more general polynomial. The resulting universal kriging predictor is of the form

$$\overset{\triangle}{Y}_{\boldsymbol{x}} = \boldsymbol{k}(\boldsymbol{x})^T\mathsf{K}^{-1}(\boldsymbol{y} - \mathsf{P}\boldsymbol{\beta}) + \boldsymbol{p}(\boldsymbol{x})^T\boldsymbol{\beta}$$
$$\text{with} \quad \boldsymbol{\beta} = (\mathsf{P}^T\mathsf{K}^{-1}\mathsf{P})^{-1}\mathsf{P}^T\mathsf{K}^{-1}\boldsymbol{y},$$

where $\boldsymbol{k}(\cdot)$, K and $\boldsymbol{p}(\cdot)$ are as above, and the matrix P has entries $(\mathsf{P})_{i,j} = p_j(\boldsymbol{x}_i)$ (see, e.g., [Scheuerer *et al.* (2013)] or [Sacks *et al.* (1989)] for more details). Again, the process variance does not appear in this predictor.

(2) In order for the expectations above to be well-defined in general one needs to make a fundamental assumption on the underlying random field, and in particular the distribution of its random variables. Above we assumed that we were working with Gaussian random fields.

In the second (Bayesian) approach to be discussed in Section 5.3.2 (see also [Omre (1987); Handcock and Stein (1993); Sacks *et al.* (1989)]) the prediction $\hat{y}_{\boldsymbol{x}}$ is interpreted as the conditional expectation (or posterior mean), i.e.,

$$\hat{y}_{\boldsymbol{x}} = \mathbb{E}[Y_{\boldsymbol{x}}|\boldsymbol{Y} = \boldsymbol{y}],$$

where $\boldsymbol{y} = \begin{pmatrix} y_{\boldsymbol{x}_1} & \cdots & y_{\boldsymbol{x}_N} \end{pmatrix}^T$ is the vector of known realizations of random variables (the observations). Again, one needs to make assumptions on the underlying random field.

If the random field is Gaussian, then the regression approach and the Bayesian approach will lead to the same result (for general processes they may not). More specifically, the derivation leading to the best linear predictor can also be performed for data coming from a more general, non-Gaussian, random field. However, for non-Gaussian processes the Bayesian derivation that gives us the conditional mean may no longer be equivalent to the best linear predictor. As a result, we may be able to achieve a better mean-squared error using a nonlinear predictor (see, e.g., [Stein (1999, Section 1.4)]).

5.3.2 *Bayesian framework*

The use of a Bayesian perspective within a numerical analysis framework was discussed by Diaconis (1988), who suggests that such ideas go back to at least Poincaré (1896). Formulating Gaussian processes within a *Bayesian framework* is attractive because they are relatively easy to implement. Once we have accomplished this, we can apply powerful statistical methods such as *maximum likelihood estimation*, *confidence intervals*, and *Bayesian inference*.

We begin by using the notion of a *conditional probability density* to express the *joint density* of a realization, $y_{\boldsymbol{x}}$, of the random field Y at some previously unobserved location $\boldsymbol{x}$ and the vector $\boldsymbol{y}$ of observations (at locations $\mathcal{X} = \{\boldsymbol{x}_1, \ldots, \boldsymbol{x}_N\}$) in two different ways as

$$\begin{aligned} p_{Y_{\boldsymbol{x}},\boldsymbol{Y}}(y_{\boldsymbol{x}}, \boldsymbol{y}) &= p_{Y_{\boldsymbol{x}}|\boldsymbol{Y}}(y_{\boldsymbol{x}}|\boldsymbol{Y} = \boldsymbol{y})p_{\boldsymbol{Y}}(\boldsymbol{y}) \\ &= p_{\boldsymbol{Y}|Y_{\boldsymbol{x}}}(\boldsymbol{y}|Y_{\boldsymbol{x}} = y_{\boldsymbol{x}})p_{Y_{\boldsymbol{x}}}(y_{\boldsymbol{x}}). \end{aligned} \tag{5.11}$$

This immediately gives rise to *Bayes' rule*

$$p_{Y_{\boldsymbol{x}}|\boldsymbol{Y}}(y_{\boldsymbol{x}}|\boldsymbol{Y} = \boldsymbol{y}) = \frac{p_{\boldsymbol{Y}|Y_{\boldsymbol{x}}}(\boldsymbol{y}|Y_{\boldsymbol{x}} = y_{\boldsymbol{x}})p_{Y_{\boldsymbol{x}}}(y_{\boldsymbol{x}})}{p_{\boldsymbol{Y}}(\boldsymbol{y})},$$

where $p_{Y_{\boldsymbol{x}}|\boldsymbol{Y}}(y_{\boldsymbol{x}}|\boldsymbol{Y} = \boldsymbol{y})$ denotes the *posterior density* for the new value $y_{\boldsymbol{x}}$ and $p_{\boldsymbol{Y}|Y_{\boldsymbol{x}}}(\boldsymbol{y}|Y_{\boldsymbol{x}} = y_{\boldsymbol{x}})$ is the *likelihood* of this new value, also denoted by $L(y_{\boldsymbol{x}}|\boldsymbol{y})$. The likelihood tells us how compatible the new value $y_{\boldsymbol{x}}$ is with the observations $\boldsymbol{y}$. The term $p_{Y_{\boldsymbol{x}}}(y_{\boldsymbol{x}})$ denotes the *prior density* which we assume to be uninformed, and $p_{\boldsymbol{Y}}(\boldsymbol{y})$ is the *evidence* (on which we have no influence).

We hope to maximize the posterior but, without any prior beliefs on $Y_{\boldsymbol{x}}$, the uninformed (and improper) prior $p_{Y_{\boldsymbol{x}}} \equiv 1$ leaves us with only the need to search for the $y_{\boldsymbol{x}}$ to maximizes the likelihood. Using the definition of the likelihood and (5.11),

$$L(y_{\boldsymbol{x}}|\boldsymbol{y}) = p_{\boldsymbol{Y}|Y_{\boldsymbol{x}}}(\boldsymbol{y}|Y_{\boldsymbol{x}} = y_{\boldsymbol{x}}) = \frac{1}{p_{Y_{\boldsymbol{x}}}(y_{\boldsymbol{x}})} p_{Y_{\boldsymbol{x}},\boldsymbol{Y}}(y_{\boldsymbol{x}}, \boldsymbol{y}) = p_{Y_{\boldsymbol{x}},\boldsymbol{Y}}(y_{\boldsymbol{x}}, \boldsymbol{y}),$$

which shows that the likelihood is essentially given by the *joint probability density function* for the new value $y_{\boldsymbol{x}}$ and the observations $\boldsymbol{y}$. The *joint* normal density (cf. (5.3)) now is

$$p_{Y_{\boldsymbol{x}},\boldsymbol{Y}}(y_{\boldsymbol{x}}, \boldsymbol{y}) = \frac{1}{\sqrt{(2\pi\sigma^2)^{N+1} \det \widetilde{\mathsf{K}}}} \exp\left(-\frac{1}{2\sigma^2}\tilde{\boldsymbol{y}}^T \widetilde{\mathsf{K}}^{-1} \tilde{\boldsymbol{y}}\right), \tag{5.12}$$

where $\tilde{\boldsymbol{y}} = \begin{pmatrix} y_{\boldsymbol{x}} \\ \boldsymbol{y} \end{pmatrix}$, $\widetilde{\mathsf{K}} = \begin{pmatrix} K(\boldsymbol{x},\boldsymbol{x}) & \boldsymbol{k}(\boldsymbol{x})^T \\ \boldsymbol{k}(\boldsymbol{x}) & \mathsf{K} \end{pmatrix}$ and σ^2 is the process variance. Here we have assumed — as in the derivation of the simple kriging approach in Section 5.3.1 — that we are dealing with a zero-mean Gaussian process.

Now we want to choose as our prediction $\hat{y}_{\boldsymbol{x}}$ the value $y_{\boldsymbol{x}}$ which maximizes the likelihood (5.12). It is much more convenient, however, to minimize the negative log-likelihood, i.e., find $y_{\boldsymbol{x}}$ such that

$$-\log L(y_{\boldsymbol{x}}|\boldsymbol{y}) = \frac{N+1}{2}\log(2\pi\sigma^2) + \frac{1}{2}\log\det\widetilde{\mathsf{K}} + \frac{\tilde{\boldsymbol{y}}^T \widetilde{\mathsf{K}}^{-1} \tilde{\boldsymbol{y}}}{2\sigma^2} + \text{const.} \tag{5.13}$$

is minimized. The only term in (5.13) that depends on $y_{\boldsymbol{x}}$ is the quadratic form

$$Q(\tilde{\boldsymbol{y}}) = \frac{\tilde{\boldsymbol{y}}^T \widetilde{\mathsf{K}}^{-1} \tilde{\boldsymbol{y}}}{2\sigma^2}. \tag{5.14}$$

In order to compute the derivative of this term with respect to $y_{\boldsymbol{x}}$ we use the Schur complement representation of the inverse of a block matrix:

$$\widetilde{\mathsf{K}}^{-1} = \begin{pmatrix} K(\boldsymbol{x},\boldsymbol{x}) & \boldsymbol{k}(\boldsymbol{x})^T \\ \boldsymbol{k}(\boldsymbol{x}) & \mathsf{K} \end{pmatrix}^{-1} = \begin{pmatrix} \mathsf{A} & \mathsf{B}^T \\ \mathsf{B} & \mathsf{C} \end{pmatrix},$$

where

$$\begin{aligned} \mathsf{A} &= \frac{1}{K(\boldsymbol{x},\boldsymbol{x}) - \boldsymbol{k}(\boldsymbol{x})^T \mathsf{K}^{-1}\boldsymbol{k}(\boldsymbol{x})}, \\ \mathsf{B} &= -\frac{\mathsf{K}^{-1}\boldsymbol{k}(\boldsymbol{x})}{K(\boldsymbol{x},\boldsymbol{x}) - \boldsymbol{k}(\boldsymbol{x})^T \mathsf{K}^{-1}\boldsymbol{k}(\boldsymbol{x})}, \\ \mathsf{C} &= \mathsf{K}^{-1} + \frac{\mathsf{K}^{-1}\boldsymbol{k}(\boldsymbol{x})\boldsymbol{k}(\boldsymbol{x})^T\mathsf{K}^{-1}}{K(\boldsymbol{x},\boldsymbol{x}) - \boldsymbol{k}(\boldsymbol{x})^T \mathsf{K}^{-1}\boldsymbol{k}(\boldsymbol{x})}. \end{aligned} \tag{5.15}$$

All these blocks of the inverse are guaranteed to exist whenever K is a strictly positive definite covariance kernel.

We now substitute this inverse into the quadratic form (5.14) and obtain

$$Q(\tilde{\boldsymbol{y}}) = \frac{1}{2\sigma^2}\left(y_{\boldsymbol{x}}^2 \mathsf{A} + 2y_{\boldsymbol{x}}\mathsf{B}^T\boldsymbol{y} + \boldsymbol{y}^T\mathsf{C}\boldsymbol{y}\right).$$

Differentiating with respect to $y_{\boldsymbol{x}}$ and equating the result to zero gives us

$$\mathsf{A}y_{\boldsymbol{x}} + \mathsf{B}^T\boldsymbol{y} = 0$$

$$\iff \quad \frac{y_{\boldsymbol{x}}}{K(\boldsymbol{x},\boldsymbol{x}) - \boldsymbol{k}(\boldsymbol{x})^T\mathsf{K}^{-1}\boldsymbol{k}(\boldsymbol{x})} - \frac{\boldsymbol{k}(\boldsymbol{x})^T\mathsf{K}^{-1}\boldsymbol{y}}{K(\boldsymbol{x},\boldsymbol{x}) - \boldsymbol{k}(\boldsymbol{x})^T\mathsf{K}^{-1}\boldsymbol{k}(\boldsymbol{x})} = 0,$$

and therefore the optimal prediction is

$$\begin{aligned}\hat{y}_{\boldsymbol{x}} &= \operatorname*{argmax}_{y_{\boldsymbol{x}}\in\mathbb{R}} L(y_{\boldsymbol{x}}|\boldsymbol{y}) \\ &= \operatorname*{argmin}_{y_{\boldsymbol{x}}\in\mathbb{R}} (-\log L(y_{\boldsymbol{x}}|\boldsymbol{y})) \\ &= \boldsymbol{k}(\boldsymbol{x})^T\mathsf{K}^{-1}\boldsymbol{y},\end{aligned}$$

just as we derived earlier with the regression approach, cf. (5.9).

Note that the process variance dropped out when obtaining the optimal prediction. However, this process variance does play an important role in determining optimal parameters in the covariance kernel K. We will come back to this idea in Chapter 14.

The Bayesian setting in which we have performed this derivation also tells us that $\hat{y}_{\boldsymbol{x}}$ is the mean of the posterior distribution and that its variance is given by $\sigma^2(K(\boldsymbol{x},\boldsymbol{x}) - \boldsymbol{k}(\boldsymbol{x})^T\mathsf{K}^{-1}\boldsymbol{k}(\boldsymbol{x}))$, the kriging variance from above; details are presented in Section 5.3.3.

Remark 5.5.

(1) If the underlying process is not centered (but still Gaussian and with constant mean), then one obtains the following prediction (conditional mean)

$$\hat{y}_{\boldsymbol{x}} = \boldsymbol{\mu} + \boldsymbol{k}^T(\boldsymbol{x})\mathsf{K}^{-1}(\boldsymbol{y} - \boldsymbol{\mu}),$$

and the same conditional covariance as before.

(2) The Bayesian approach is common in the machine learning community. In the paper [Seeger (2004)] this framework is referred to as the *weight space view* (as opposed to the *process view*) of a Gaussian process. A "duality" of viewpoints is accomplished via the feature map representation (cf. Section 2.4), where $K(\boldsymbol{x},\boldsymbol{z}) = \langle\Phi_{\boldsymbol{x}},\Phi_{\boldsymbol{z}}\rangle_{\ell_2}$ and $\Phi_{\boldsymbol{x}} = \left(\sqrt{\lambda_1}\varphi_1(\boldsymbol{x}), \sqrt{\lambda_2}\varphi_2(\boldsymbol{x}), \ldots\right)$ is a (generally infinite) sequence — or *feature* — obtained from the Hilbert–Schmidt eigenpairs (λ_n,φ_n) of the integral operator associated with K (see Example 2.6 and also Section 5.2).

5.3.3 *Confidence intervals*

One of the benefits of studying the scattered data fitting problem in the stochastic framework based on Gaussian random fields is the ability to not only make predictions, but also to associate a level of confidence in those predictions. This acknowledgement of the uncertainty present in the predictions is an opportunity to present a more nuanced interpretation of the data by considering a range of possible results, i.e., to supplement our predictions with so-called *confidence intervals*.

In Section 5.3.2, we describe the use of the Bayesian framework as an alternate device for deriving the best linear unbiased predictor, even in the case of a nonzero-mean Gaussian random field. Here we continue the exploration of this Bayesian framework and study the posterior density, and in particular the *distribution* of $Y_{\boldsymbol{x}}|\boldsymbol{Y}$. This is motivated by the fact that — although (5.9) gives the best prediction as $\mathbb{E}(Y_{\boldsymbol{x}}|\boldsymbol{Y}=\boldsymbol{y})(\boldsymbol{x}) = \boldsymbol{k}(\boldsymbol{x})^T\mathsf{K}^{-1}\boldsymbol{y}$ — the probability of an actual realization of $Y_{\boldsymbol{x}}$ being that value is zero because $Y_{\boldsymbol{x}}$ is a continuous random variable.

Note that we focus our discussion on the zero mean case to keep the exposition transparent, but this can be extended directly to account for nonzero mean.

Rearranging (5.11), we get the posterior density as

$$p_{Y_{\boldsymbol{x}}|\boldsymbol{Y}}(y_{\boldsymbol{x}}|\boldsymbol{Y}=\boldsymbol{y}) = \frac{1}{p_{\boldsymbol{Y}}(\boldsymbol{y})} p_{Y_{\boldsymbol{x}},\boldsymbol{Y}}(y_{\boldsymbol{x}},\boldsymbol{y}), \tag{5.16}$$

where $p_{\boldsymbol{Y}}(\boldsymbol{y})$ is defined in (5.3), and the joint normal density $p_{Y_{\boldsymbol{x}},\boldsymbol{Y}}$ is given in (5.12). Because we are only interested in studying the density as a function of $y_{\boldsymbol{x}}$, we will assume that $\boldsymbol{y}$ and $\mathcal{X}$ are fixed. By defining

$$Z_1 = \left(\frac{1}{\sqrt{(2\pi\sigma^2)^N \det \mathsf{K}}} \exp\left(-\frac{1}{2\sigma^2}\boldsymbol{y}^T\mathsf{K}^{-1}\boldsymbol{y}\right) \sqrt{(2\pi\sigma^2)^{N+1}\det\widetilde{\mathsf{K}}} \right)^{-1}, \tag{5.17}$$

with $\widetilde{\mathsf{K}}$ and $\tilde{\boldsymbol{y}}$ (used below) as defined for (5.12), we can write our conditional density (5.16) as

$$p_{Y_{\boldsymbol{x}}|\boldsymbol{Y}}(y_{\boldsymbol{x}}|\boldsymbol{Y}=\boldsymbol{y}) = Z_1 \exp\left(-\frac{1}{2\sigma^2}\begin{pmatrix} y_{\boldsymbol{x}} & \boldsymbol{y}^T\end{pmatrix} \begin{pmatrix} K(\boldsymbol{x},\boldsymbol{x}) & \boldsymbol{k}(\boldsymbol{x})^T \\ \boldsymbol{k}(\boldsymbol{x}) & \mathsf{K}\end{pmatrix}^{-1} \begin{pmatrix} y_{\boldsymbol{x}} \\ \boldsymbol{y}\end{pmatrix}\right).$$

Reusing the quadratic form of (5.14) lets us identify the terms from the exponential that are independent of $y_{\boldsymbol{x}}$:

$$\begin{aligned} \tilde{\boldsymbol{y}}^T\widetilde{\mathsf{K}}^{-1}\tilde{\boldsymbol{y}} &= \begin{pmatrix} y_{\boldsymbol{x}} & \boldsymbol{y}^T\end{pmatrix} \begin{pmatrix} \mathsf{A} & \mathsf{B}^T \\ \mathsf{B} & \mathsf{C}\end{pmatrix} \begin{pmatrix} y_{\boldsymbol{x}} \\ \boldsymbol{y}\end{pmatrix} \\ &= \mathsf{A}y_{\boldsymbol{x}}^2 + 2y_{\boldsymbol{x}}\mathsf{B}^T\boldsymbol{y} + \boldsymbol{y}^T\mathsf{C}\boldsymbol{y}. \end{aligned}$$

Define now $\hat{Z}_2 = \boldsymbol{y}^T\mathsf{C}\boldsymbol{y}$, the $y_{\boldsymbol{x}}$-independent term in the exponent, and $Z_2 = Z_1\exp(-\hat{Z}_2/(2\sigma^2))$ so that

$$p_{Y_{\boldsymbol{x}}|\boldsymbol{Y}}(y_{\boldsymbol{x}}|\boldsymbol{Y}=\boldsymbol{y}) = Z_2 \exp\left(-\frac{1}{2\sigma^2}\left(\mathsf{A}y_{\boldsymbol{x}}^2 + 2y_{\boldsymbol{x}}\mathsf{B}^T\boldsymbol{y}\right)\right).$$

While not far from a Gaussian density, we must complete the square to fully expose it using the identity

$$\boldsymbol{u}^T\mathsf{Q}\boldsymbol{u} - 2\boldsymbol{u}^T\boldsymbol{v} = (\boldsymbol{u}-\mathsf{Q}^{-1}\boldsymbol{v})^T\mathsf{Q}(\boldsymbol{u}-\mathsf{Q}^{-1}\boldsymbol{v}) - \boldsymbol{v}^T\mathsf{Q}^{-1}\boldsymbol{v}, \quad \boldsymbol{u},\boldsymbol{v}\in\mathbb{R}^N,\ \mathsf{Q}\in\mathbb{R}^{N\times N}.$$

By choosing $\boldsymbol{u} = y_{\boldsymbol{x}}$, $\mathsf{Q} = \mathsf{A}$ and $\boldsymbol{v} = \mathsf{A}\boldsymbol{k}(\boldsymbol{x})^T\mathsf{K}^{-1}\boldsymbol{y}$ (because $\mathsf{B}^T = -\mathsf{A}\boldsymbol{k}(\boldsymbol{x})^T\mathsf{K}^{-1}$) we can write

$$\begin{aligned} \mathsf{A}y_{\boldsymbol{x}}^2 + 2y_{\boldsymbol{x}}\mathsf{B}^T\boldsymbol{y} =& (y_{\boldsymbol{x}} - \boldsymbol{k}(\boldsymbol{x})^T\mathsf{K}^{-1}\boldsymbol{y})^T\mathsf{A}(y_{\boldsymbol{x}} - \boldsymbol{k}(\boldsymbol{x})^T\mathsf{K}^{-1}\boldsymbol{y}) \\ &- (\mathsf{A}\boldsymbol{k}(\boldsymbol{x})^T\mathsf{K}^{-1}\boldsymbol{y})^T\mathsf{A}^{-1}(\mathsf{A}\boldsymbol{k}(\boldsymbol{x})^T\mathsf{K}^{-1}\boldsymbol{y}). \end{aligned}$$

Notice that this is written in the vector inner product form even though $y_{\boldsymbol{x}} - \boldsymbol{k}(\boldsymbol{x})^T\mathsf{K}^{-1}\boldsymbol{y}$ is a scalar, just to conform to the identity used above. More importantly, the second term is devoid of $y_{\boldsymbol{x}}$, so we define

$$\hat{Z}_3 = -\boldsymbol{y}^T\mathsf{K}^{-1}\boldsymbol{k}(\boldsymbol{x})\mathsf{A}\boldsymbol{k}(\boldsymbol{x})^T\mathsf{K}^{-1}\boldsymbol{y} = -\mathsf{A}(\boldsymbol{k}(\boldsymbol{x})^T\mathsf{K}^{-1}\boldsymbol{y})^2$$

and $Z_3 = Z_2\exp(-\hat{Z}_3/(2\sigma^2))$ so that

$$p_{Y_{\boldsymbol{x}}|\boldsymbol{Y}}(y_{\boldsymbol{x}}|\boldsymbol{Y}=\boldsymbol{y}) = Z_3\exp\left(-\frac{1}{2\sigma^2}(y_{\boldsymbol{x}} - \boldsymbol{k}(\boldsymbol{x})^T\mathsf{K}^{-1}\boldsymbol{y})^T\mathsf{A}(y_{\boldsymbol{x}} - \boldsymbol{k}(\boldsymbol{x})^T\mathsf{K}^{-1}\boldsymbol{y})\right).$$

Comparing this to the standard form of the normal density (5.3) reveals

$$Y_{\boldsymbol{x}}|\boldsymbol{Y}=\boldsymbol{y} \sim \mathcal{N}(\boldsymbol{k}(\boldsymbol{x})^T\mathsf{K}^{-1}\boldsymbol{y}, \sigma^2(K(\boldsymbol{x},\boldsymbol{x}) - \boldsymbol{k}(\boldsymbol{x})^T\mathsf{K}^{-1}\boldsymbol{k}(\boldsymbol{x})). \tag{5.18}$$

where we have substituted the value of A from (5.15).

As we already determined in Section 5.3.2, the expected value of this random variable is also the best linear unbiased predictor $\boldsymbol{k}(\boldsymbol{x})^T\mathsf{K}^{-1}\boldsymbol{y}$. The interesting result here is the variance associated with that prediction:

$$\mathrm{Var}(Y_{\boldsymbol{x}}|\boldsymbol{Y}=\boldsymbol{y}) = \sigma^2(K(\boldsymbol{x},\boldsymbol{x}) - \boldsymbol{k}(\boldsymbol{x})^T\mathsf{K}^{-1}\boldsymbol{k}(\boldsymbol{x})).$$

It is this quantity we have been searching for; with this we can produce *confidence intervals* of the standard form

$$P\left(Y_{\boldsymbol{x}} \in \boldsymbol{k}(\boldsymbol{x})^T\mathsf{K}^{-1}\boldsymbol{y} \pm z_\alpha\sigma\sqrt{K(\boldsymbol{x},\boldsymbol{x}) - \boldsymbol{k}(\boldsymbol{x})^T\mathsf{K}^{-1}\boldsymbol{k}(\boldsymbol{x})}\right) = 1-\alpha. \tag{5.19}$$

Figure 5.1 shows one such example involving a confidence interval of two standard deviations. Program 5.1 shows how this can be computed in MATLAB. Note the use of `real` in line 14 to account for numerical cancelation which may otherwise produce complex standard deviations; this cancelation issue is discussed in greater detail in Section 14.1.1. The magnitude of the standard deviation is smallest nearby data locations, and significant in gaps between data locations such as the middle of this plot where the Chebyshev points are most sparse.

Program 5.1. `KrigingConfidenceIntervals.m`

```
   % Choose a kernel, here the Gaussian, and parameters
2  rbf = @(e,r) exp(-(e*r).^2);
   ep = 3;   sigma = 1;
4  % Points at which to sample
   N = 10;          x = pickpoints(-1,1,N,'cheb');
6  Neval = 100;     xeval = unique([pickpoints(-1,1,Neval);x]);
   % Choose a function and sample it
8  yf = @(x) exp(x) - cos(2*pi*x);    y = yf(x);
   % Compute the interpolant through these points
10 DM = DistanceMatrix(x,x);   K = rbf(ep,DM);
   DMeval = DistanceMatrix(xeval,x);   Keval = rbf(ep,DMeval);
12 seval = Keval*(K\y);
   % Compute the standard deviation at all the evaluation locations
14 sd = sigma*real(sqrt(rbf(ep,0) - sum((Keval/K).*Keval,2)));
```

```
   % Plot the predictions and confidence intervals
16 plot(x,y,'or');
   hold on
18 plot(xeval,seval,'linewidth',1);
   plot(xeval,seval+2*sd,'--k','linewidth',2);
20 plot(xeval,seval-2*sd,'--k','linewidth',2);
   hold off
22 legend('Data','Prediction','+/-2 SD','location','southeast')
```

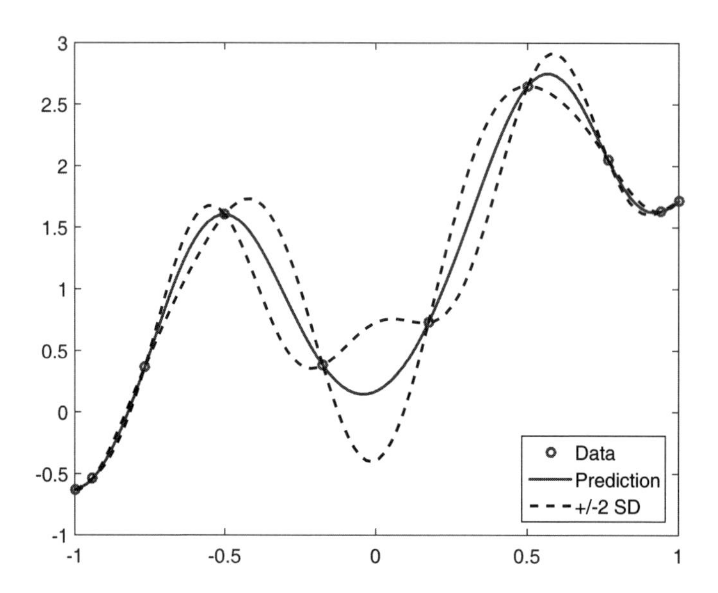

Fig. 5.1 A 95% confidence interval of the form (5.19) for predictions using a Gaussian covariance kernel with $\varepsilon = 3$ and $\sigma = 1$ given evaluations of the function $f(x) = e^x - \cos(2\pi x)$ at $N = 10$ Chebyshev points in $[-1, 1]$.

Remark 5.6. We can see that

$$\begin{aligned}\det\begin{pmatrix} K(\boldsymbol{x},\boldsymbol{x}) & \boldsymbol{k}(\boldsymbol{x})^T \\ \boldsymbol{k}(\boldsymbol{x}) & \mathsf{K}\end{pmatrix} &= \det\left(\begin{pmatrix} 1 & \boldsymbol{k}(\boldsymbol{x})^T \\ \mathbf{0}_N & \mathsf{K}\end{pmatrix}\begin{pmatrix} K(\boldsymbol{x},\boldsymbol{x}) - \boldsymbol{k}(\boldsymbol{x})^T\mathsf{K}^{-1}\boldsymbol{k}(\boldsymbol{x}) & \mathbf{0}_N^T \\ \mathsf{K}^{-1}\boldsymbol{k}(\boldsymbol{x}) & \mathsf{I}_N\end{pmatrix}\right) \\ &= \det(\mathsf{K})(K(\boldsymbol{x},\boldsymbol{x}) - \boldsymbol{k}(\boldsymbol{x})^T\mathsf{K}^{-1}\boldsymbol{k}(\boldsymbol{x})) \\ &= \det(\mathsf{K})\operatorname{Var}(Y_{\boldsymbol{x}}|\boldsymbol{Y} = \boldsymbol{y}),\end{aligned}$$

where $\mathbf{0}_N$ is a zero vector of length N. Unexpectedly, the kriging variance appears, and using this result in (5.17) gives

$$Z_1 = \frac{1}{\sqrt{2\pi}\sigma\operatorname{Var}(Y_{\boldsymbol{x}}|\boldsymbol{Y} = \boldsymbol{y})}\exp\left(-\frac{1}{2\sigma^2}\boldsymbol{y}^T\mathsf{K}^{-1}\boldsymbol{y}\right).$$

Notice that $\mathsf{K}^{-1}\boldsymbol{k}(\boldsymbol{x}_j) = \boldsymbol{e}_j$, the jth column of I_N, because the jth column of K is $\boldsymbol{k}(\boldsymbol{x}_j)$. This implies that $\operatorname{Var}(Y_{\boldsymbol{x}}|\boldsymbol{Y} = \boldsymbol{y}) = 0$ for $\boldsymbol{x} \in \mathcal{X}$, so, regardless of the value taken by $p_{Y_{\boldsymbol{x}},\boldsymbol{Y}}(y_{\boldsymbol{x}},\boldsymbol{y})$, $p_{Y_{\boldsymbol{x}}|\boldsymbol{Y}}(y_{\boldsymbol{x}}|\boldsymbol{Y} = \boldsymbol{y})$ is optimized by allowing $y_{\boldsymbol{x}}$ to replicate the data observed in $\boldsymbol{Y}$. That is, the only prediction that can be made with any confidence at $\boldsymbol{x} \in \mathcal{X}$ is the data that was observed, as we see in Figure 5.1. The reader familiar with the deterministic approach to kernel methods may recognize this as the stochastic analogue of the fact that the kernel interpolant is also the best $\mathcal{H}_K(\Omega)$-approximation of the data (see Chapter 9).

5.3.4 *Semi-variograms*

In the geostatistics literature the covariance kernel — or more precisely the correlation function (5.4) (since the processes are often stationary) — is often estimated from the data using a so-called (empirical) *semi-variogram* (see, e.g., [Hur and Baldick (2012)]).

For a stationary (i.e., translation invariant) random field the semi-variogram is given by (see [Ritter (2000, Section II.2.7)][4])

$$2\gamma(\boldsymbol{x},\boldsymbol{z}) = \sigma^2 \left(K(\boldsymbol{x},\boldsymbol{x}) - 2K(\boldsymbol{x},\boldsymbol{z}) + K(\boldsymbol{z},\boldsymbol{z})\right),$$

where K is the covariance kernel of the random field and σ^2 the process variance. From this we see that γ is a conditionally negative definite kernel[5] since

$$\begin{aligned}\sum_{i=1}^N\sum_{j=1}^N c_i c_j \gamma(\boldsymbol{x}_i,\boldsymbol{x}_j) &= \sum_{i=1}^N\sum_{j=1}^N c_i c_j \sigma^2 \left(K(\boldsymbol{x}_i,\boldsymbol{x}_i) - 2K(\boldsymbol{x}_i,\boldsymbol{x}_j) + K(\boldsymbol{x}_j,\boldsymbol{x}_j)\right) \\ &= 2\sigma^2 \left(\sum_{i=1}^N c_i K(\boldsymbol{x}_i,\boldsymbol{x}_i) \sum_{j=1}^N c_j - \sum_{i=1}^N\sum_{j=1}^N c_i c_j K(\boldsymbol{x}_i,\boldsymbol{x}_j)\right) \\ &\le 0\end{aligned} \tag{5.20}$$

provided $\sum_{j=1}^N c_j = 0$ (since K is positive definite, i.e., its quadratic form is non-negative).

Note that we can also express the semi-variogram as

$$\begin{aligned}2\gamma(\boldsymbol{x},\boldsymbol{z}) &= \sigma^2 \|K(\cdot,\boldsymbol{x}) - K(\cdot,\boldsymbol{z})\|^2_{\mathcal{H}_K(\Omega)} \\ &= \mathrm{Var}(Y_{\boldsymbol{x}} - Y_{\boldsymbol{z}}) \\ &= \mathbb{E}[(Y_{\boldsymbol{x}} - Y_{\boldsymbol{z}})^2],\end{aligned}$$

provided the mean of the random field is constant. This is how geostatisticians think about the semi-variogram: as the variance of the difference between the values of the random field at the two different points $\boldsymbol{x}$ and $\boldsymbol{z}$.

---

[4]Actually, Ritter refers to the function γ as the *variogram*, but usually the variogram in geostatistics is given by 2γ (see [Cressie (1993, Chapter 2)]). Cressie calls use of this alternative terminology "dangerous practice," and says "there is too much to lose from missing 2s."

[5]In this book we do not deal with conditionally positive (or negative) definite kernels. The interested reader is referred to, e.g., [Wendland (2005); Fasshauer (2007)] for more details.

A standard estimator for the semi-variogram is the *method of moments* or *empirical variogram* [Cressie (1993, Section 2.4)]

$$2\hat{\gamma}(\boldsymbol{h}) \approx \frac{1}{|\Omega_{\boldsymbol{h}}|} \sum_{\Omega_{\boldsymbol{h}}} (y_i - y_j)^2, \qquad \boldsymbol{h} \in \mathbb{R}^d,$$

where y_i are observations corresponding to the random variables $Y_{\boldsymbol{x}_i}$, $\Omega_{\boldsymbol{h}} = \{(\boldsymbol{x}_i, \boldsymbol{x}_j) \mid \boldsymbol{x}_i - \boldsymbol{x}_j = \boldsymbol{h},\ i, j, = 1, \ldots, N\}$ and $|\Omega_{\boldsymbol{h}}|$ denotes the number of distinct pairs in $\Omega_{\boldsymbol{h}}$. This estimator was already suggested by Matheron (1962), who also first used the name semi-variogram. According to Stein (1999), one needs to be careful when using semi-variograms to make inferences about covariance kernels. In particular, he says "Although the empirical semivariogram can be a useful tool for random fields that are not differentiable, it is much less useful and can even be seriously misleading for differentiable random fields."

5.4 Karhunen–Loève Expansions and Polynomial Chaos

The Karhunen–Loève theorem, Theorem 2.3, was presented in Chapter 2. What makes this expansion especially useful for practical applications such as the numerical solution of stochastic PDEs is the fact that the Karhunen–Loève expansion separates the random component of the process from a deterministic component. In practice, a truncated KL expansion is used. It is well known that the truncated KL expansion is optimal in the MSE sense, i.e., an M-term KL expansion is the best M-term approximation of the random field Y in the MSE sense.

However, in practical situations usually the random field Y is unknown and so the random variables Z_n must be simulated as independent and identically distributed (i.i.d.) random variables following the distribution of Y. The eigenfunctions φ_n and eigenvalues λ_n of K may also be unknown, in which case one would have to use a further numerical approximation method to solve the Hilbert–Schmidt integral eigenvalue problem.

The Karhunen–Loève expansion is frequently used to model random *coefficients* in stochastic systems. However, the *solution* to such problems is generally not modeled by KL since the covariance of the solution is assumed to be unknown. In this case one frequently uses an approach referred to as *(generalized) polynomial chaos* to approximate functionals of random variables. According to Narayan and Xiu (2012), algorithms for obtaining generalized polynomial chaos expansions for the solution of PDEs come in two flavors: Galerkin or collocation approaches. Galerkin methods are called "intrusive," which means that significant efforts are required to modify existing deterministic codes. Collocation approaches, on the other hand, are referred to as "nonintrusive" since they require a minimal amount of coding overhead. Moreover, if the dependence on the random inputs is smooth one may be able to exploit the spectral accuracy of certain collocation methods (see also [Cialenco *et al.* (2012)] for similar insights).

5.5 Generalized Polynomial Chaos

We now discuss the basic ideas behind generalized polynomial chaos (gPC). As mentioned in the previous section, we use generalized polynomial chaos if we want to approximate a function*al* f of a random variable $Z_{\boldsymbol{x}}$, i.e., we seek to approximate the random variable $Y_{\boldsymbol{x}} = f(Z_{\boldsymbol{x}})$. Many more details can be found, e.g., in [Xiu (2010)].[6]

Single random variable

We begin by explaining the case of a single random variable. We assume that $F_{Z_{\boldsymbol{x}}}$ is the cumulative distribution function (CDF) of $Z_{\boldsymbol{x}}$ and that $Z_{\boldsymbol{x}}$ has finite moments. Moreover, if $Z_{\boldsymbol{x}}$ is continuous then $F_{Z_{\boldsymbol{x}}}$ has a density $\rho_{Z_{\boldsymbol{x}}}$ such that $\mathrm{d}F_{Z_{\boldsymbol{x}}}(z) = \rho_{Z_{\boldsymbol{x}}}(z)\,\mathrm{d}z$ (see (5.1)).

If the functional dependence f is known, then $Y_{\boldsymbol{x}} = f(Z_{\boldsymbol{x}})$ also follows the distribution $F_{Z_{\boldsymbol{x}}}$ and we can perform so-called *strong approximation.* If only the distribution of $Y_{\boldsymbol{x}}$ is known, then we can perform only *weak approximation.*

We want the polynomials p_n, $n = 0, 1, 2, \ldots$, to define "orthogonal" (independent) random variables, i.e.,

$$\begin{aligned}\mathbb{E}[p_m(Z_{\boldsymbol{x}})p_n(Z_{\boldsymbol{x}})] &= \int_{-\infty}^{\infty} p_m(z)p_n(z)\rho_{Z_{\boldsymbol{x}}}(z)\,\mathrm{d}z \\ &= \delta_{mm}\mathbb{E}[p_n^2(Z_{\boldsymbol{x}})].\end{aligned}$$

Obviously, the polynomials p_n are classical orthogonal polynomials with respect to the weight function $\rho_{Z_{\boldsymbol{x}}}$. On the other hand, $\rho_{Z_{\boldsymbol{x}}}$ is the probability density function for the random variable $Z_{\boldsymbol{x}}$, and so there is an intimate connection between the distribution of the random variable and the polynomials used for the gPC expansion. In fact, different (classical) orthogonal polynomials directly correspond to (classical) probability distributions (see [Xiu (2010, Table 5.1)]).

Example 5.2. (Wiener chaos)
If our random variables are *normally distributed*, i.e., $Z_{\boldsymbol{x}} \sim \mathcal{N}(0,1)$ with $I_{Z_{\boldsymbol{x}}} = (-\infty, \infty)$ so that[7] $\rho_{Z_{\boldsymbol{x}}}(z) = \frac{1}{\sqrt{\pi}}\mathrm{e}^{-z^2}$ then $p_n = H_n$ are the classical *Hermite polynomials* (see also Example 12.1)

$$H_0(Z_{\boldsymbol{x}}) = 1, \quad H_1(Z_{\boldsymbol{x}}) = 2Z_{\boldsymbol{x}}, \quad H_2(Z_{\boldsymbol{x}}) = 4Z_{\boldsymbol{x}}^2 - 2, \quad H_3(Z_{\boldsymbol{x}}) = 8Z_{\boldsymbol{x}}^3 - 12Z_{\boldsymbol{x}}, \ldots,$$

and one obtains the "original" polynomial chaos due to Wiener (1938).

[6] The reader comparing our discussion with that in [Xiu (2010)] should note that random variables in [Xiu (2010)] are denoted without reference to the parameter space and therefore look like our random fields. We have decided to be consistent and maintain our, slightly more elaborate, notation.

[7] From the point of view of probability distributions it would be more natural to use $\rho_{Z_{\boldsymbol{x}}}(z) = \frac{1}{\sqrt{2\pi}}\mathrm{e}^{-z^2/2}$. Our choice of density function is, however, motivated by the fact that we want to be consistent with the discussion in Example 12.1 and Chapter 13.

Example 5.3. (Legendre polynomial chaos)
If our random variables are *uniformly distributed*, i.e., $Z_{\boldsymbol{x}} \sim \mathcal{U}(0,1)$ with $I_{Z_{\boldsymbol{x}}} = (-1,1)$ so that $\rho_{Z_{\boldsymbol{x}}}(z) \equiv \frac{1}{2}$ then $p_n = P_n$ are the classical *Legendre polynomials* (see also Example 3.2)

$$P_0(Z_{\boldsymbol{x}}) = 1, \quad P_1(Z_{\boldsymbol{x}}) = Z_{\boldsymbol{x}}, \quad P_2(Z_{\boldsymbol{x}}) = \frac{3}{2}Z_{\boldsymbol{x}}^2 - \frac{1}{2}, \quad P_3(Z_{\boldsymbol{x}}) = \frac{5}{2}Z_{\boldsymbol{x}}^3 - \frac{3}{2}Z_{\boldsymbol{x}}, \ldots.$$

Example 5.4. (Chebyshev polynomial chaos)
If our random variables are distributed according to a special *beta distribution*[8], i.e., $Z_{\boldsymbol{x}} \sim \text{Beta}(\frac{1}{2}, \frac{1}{2})$ and supported on $I_{Z_{\boldsymbol{x}}} = (-1,1)$ with density $\rho_{Z_{\boldsymbol{x}}}(z) = \frac{1}{\pi\sqrt{1-x^2}}$ then $p_n = T_n$ are the classical *Chebyshev polynomials of the first kind* (see also Section 3.9.2)

$$P_0(Z_{\boldsymbol{x}}) = 1, \quad P_1(Z_{\boldsymbol{x}}) = Z_{\boldsymbol{x}}, \quad P_2(Z_{\boldsymbol{x}}) = 2Z_{\boldsymbol{x}}^2 - 1, \quad P_3(Z_{\boldsymbol{x}}) = 4Z_{\boldsymbol{x}}^3 - 3Z_{\boldsymbol{x}}, \ldots.$$

Remark 5.7. As we will see in Chapter 10, there exists a deep connection between these polynomial interpolation methods and those based on radial kernels via their so-called "flat" limit (see Remark 10.4).

Definition 5.3 (Strong gPC approximation). *Let $Z_{\boldsymbol{x}}$ and $f(Z_{\boldsymbol{x}})$ be random variables with cumulative distribution function $F_{Z_{\boldsymbol{x}}}$ and support $I_{Z_{\boldsymbol{x}}}$. Then $f_N(Z_{\boldsymbol{x}}) \in \Pi^N(Z_{\boldsymbol{x}})$ forms a* strong gPC approximation *of $f(Z_{\boldsymbol{x}})$ if*

$$\|f(Z_{\boldsymbol{x}}) - f_N(Z_{\boldsymbol{x}})\| \to 0 \quad \text{as } N \to \infty.$$

Here $\Pi^N(Z_{\boldsymbol{x}})$ denotes the space of univariate polynomials up to degree N of $Z_{\boldsymbol{x}}$ (i.e., a space of random variables), and $\|\cdot\|$ is some norm on $\mathcal{W}$.

A typical norm for use in Theorem 5.3 would be the 2-norm

$$\|f(Z_{\boldsymbol{x}})\| = \sqrt{\mathbb{E}[f^2(Z_{\boldsymbol{x}})]}.$$

This allows us to get a degree-N gPC orthogonal projection

$$P_N f(Z_{\boldsymbol{x}}) = \sum_{n=0}^{N} \hat{f}_n p_n(Z_{\boldsymbol{x}}), \quad \text{with } \hat{f}_n = \frac{\mathbb{E}[f(Z_{\boldsymbol{x}})p_n(Z_{\boldsymbol{x}})]}{\mathbb{E}[p_n^2(Z_{\boldsymbol{x}})]}.$$

Classical theory implies that this projection is mean-square optimal.

When dealing with *weak gPC approximation* we again want to approximate a random variable $Y_{\boldsymbol{x}}$, but now we assume we know only the distributions $F_{Y_{\boldsymbol{x}}}$ of $Y_{\boldsymbol{x}}$ and $F_{Z_{\boldsymbol{x}}}$ of $Z_{\boldsymbol{x}}$ (not a functional dependence of $Y_{\boldsymbol{x}}$ on $Z_{\boldsymbol{x}}$).

We can perform a similar approximation, but must ensure that the expectation $\mathbb{E}[Y_{\boldsymbol{x}} p_n(Z_{\boldsymbol{x}})]$ is well-defined. This can be done with the help of an *inverse CDF*

$$Y_{\boldsymbol{x}} = F_{Y_{\boldsymbol{x}}}^{-1}(F_{Z_{\boldsymbol{x}}}(Z_{\boldsymbol{x}})).$$

[8]A linear variable transformation is needed here to match the standard domain $(0,1)$ of the beta distribution with the standard domain $(-1,1)$ of the Chebyshev polynomials.

Theorem 5.1 (Weak gPC approximation). *Let $Y_{\boldsymbol{x}} \in L_2(\mathcal{W}, \mathcal{A}, P)$ be a random variable with distribution $F_{Y_{\boldsymbol{x}}}$, and let $Z_{\boldsymbol{x}}$ be a random variable with distribution $F_{Z_{\boldsymbol{x}}}$ such that corresponding gPC basis functions exist as above. Then*

$$Y_{\boldsymbol{x},N} = \sum_{n=0}^{N} \tilde{f}_n p_n(Z_{\boldsymbol{x}}), \quad \textit{with } \tilde{f}_n = \frac{\mathbb{E}[F_{Y_{\boldsymbol{x}}}^{-1}(F_{Z_{\boldsymbol{x}}}(Z_{\boldsymbol{x}}))p_n(Z_{\boldsymbol{x}})]}{\mathbb{E}[p_n^2(Z_{\boldsymbol{x}})]}$$

forms a weak gPC approximation *of $Y_{\boldsymbol{x}}$, i.e., $Y_{\boldsymbol{x},N}$ converges to $Y_{\boldsymbol{x}}$ in probability or*

$$\forall \epsilon > 0 \quad P(|Y_{\boldsymbol{x},N} - Y_{\boldsymbol{x}}| > \epsilon) \to 0 \quad \textit{as} \quad N \to \infty.$$

For practical implementation we must use numerical methods. The inner product/expectation in Theorem 5.1 is defined as

$$\mathbb{E}[F_{Y_{\boldsymbol{x}}}^{-1}(F_{Z_{\boldsymbol{x}}}(Z_{\boldsymbol{x}}))p_n(Z_{\boldsymbol{x}})] = \int_{I_{Z_{\boldsymbol{x}}}} F_{Y_{\boldsymbol{x}}}^{-1}(F_{Z_{\boldsymbol{x}}}(z))p_n(z)\rho_{Z_{\boldsymbol{x}}}(z)\, \mathrm{d}z$$

and can be computed via standard numerical methods. Equivalently, it could be computed (linking both $Y_{\boldsymbol{x}}$ and $Z_{\boldsymbol{x}}$ to a uniform distribution on $[0,1]$) as

$$\mathbb{E}_U[F_{Y_{\boldsymbol{x}}}^{-1}(U)p_n(F_{Z_{\boldsymbol{x}}}^{-1}(U))] = \int_0^1 F_{Y_{\boldsymbol{x}}}^{-1}(u)p_n(F_{Z_{\boldsymbol{x}}}^{-1}(u))\, \mathrm{d}u.$$

Remark 5.8. Strong gPC approximation implies weak gPC approximation, which implies convergence in distribution.

Multiple independent random variables

We now consider the case when we have multiple independent random variables, i.e., we have a vector $\boldsymbol{Z}_{\boldsymbol{x}} = \begin{pmatrix} Z_{1,\boldsymbol{x}} & \cdots & Z_{d,\boldsymbol{x}} \end{pmatrix}^T$ of independent random variables, where d could be finite or infinite. If d is infinite, then we would end up with polynomials of infinite degree. Usually one truncates using a KL expansion. We now assume d to be finite.

The independence of the random variables implies the joint distribution

$$F_{\boldsymbol{Z}_{\boldsymbol{x}}} = \prod_{\ell=1}^{d} F_{Z_{\ell,\boldsymbol{x}}} \quad \text{with } I_{\boldsymbol{Z}_{\boldsymbol{x}}} = I_{Z_{1,\boldsymbol{x}}} \times \ldots \times I_{Z_{d,\boldsymbol{x}}}.$$

A multivariate gPC of total degree at most N can now be constructed as the product of univariate gPCs, i.e., using the multi-index $\boldsymbol{n} = (n_1, \ldots, n_d) \in \mathbb{N}_0^d$ with $|\boldsymbol{n}| = n_1 + \ldots + n_d$ (introduced in Section 3.3) we have the polynomials

$$p_{\boldsymbol{n}}(\boldsymbol{Z}_{\boldsymbol{x}}) = \prod_{\ell=1}^{d} p_{n_\ell}(Z_{\ell,\boldsymbol{x}}), \qquad 0 \le |\boldsymbol{n}| \le N.$$

Example 5.5. (Multivariate Wiener chaos)
Consider $d = 2$ and Hermite polynomials as before. We get

$$
\begin{aligned}
N = 0 &: H_{(0,0)}(\boldsymbol{Z}_{\boldsymbol{x}}) = 1, \\
N = 1 &: H_{(1,0)}(\boldsymbol{Z}_{\boldsymbol{x}}) = 2Z_{1,\boldsymbol{x}}, \quad H_{(0,1)}(\boldsymbol{Z}_{\boldsymbol{x}}) = 2Z_{2,\boldsymbol{x}}, \\
N = 2 &: H_{(2,0)}(\boldsymbol{Z}_{\boldsymbol{x}}) = 4Z_{1,\boldsymbol{x}}^2 - 2, \ H_{(1,1)}(\boldsymbol{Z}_{\boldsymbol{x}}) = 4Z_{1,\boldsymbol{x}}Z_{2,\boldsymbol{x}}, \ H_{(0,2)}(\boldsymbol{Z}_{\boldsymbol{x}}) = 4Z_{2,\boldsymbol{x}}^2 - 2, \\
N = 3 &: H_{(3,0)}(\boldsymbol{Z}_{\boldsymbol{x}}) = 8Z_{1,\boldsymbol{x}}^3 - 12Z_{1,\boldsymbol{x}}, \quad H_{(2,1)}(\boldsymbol{Z}_{\boldsymbol{x}}) = 8Z_{1,\boldsymbol{x}}^2 Z_{2,\boldsymbol{x}} - 4Z_{2,\boldsymbol{x}}, \\
& H_{(1,2)}(\boldsymbol{Z}_{\boldsymbol{x}}) = 8Z_{1,\boldsymbol{x}}Z_{2,\boldsymbol{x}}^2 - 4Z_{1,\boldsymbol{x}}, \quad H_{(0,3)}(\boldsymbol{Z}_{\boldsymbol{x}}) = 8Z_{2,\boldsymbol{x}}^3 - 12Z_{2,\boldsymbol{x}}.
\end{aligned}
$$

Each subset corresponds to a d-dimensional homogeneous Wiener chaos of exact degree N.

Chapter 6

The Connection to Green's Kernels

6.1 Introduction

In the first few chapters of this book we talked about positive definite kernels both from a theoretical perspective (i.e., in terms of Hilbert–Schmidt theory and the reproducing kernel Hilbert space $\mathcal{H}_K$ that is associated with each kernel K) and from a practical point of view (i.e., what kind of problems do people solve using kernels). One of the most natural questions to ask for all of these applications is *"Which kernel should I use to solve my problem?"* If we consider, e.g., the numerical solution of PDEs, then classical theory tells us to expect the solution for many of the common problems to lie in some Sobolev space; finite elements have been developed as a perfect match for this classical theory (see, e.g., [Brenner and Scott (1994)]).

As we saw in Chapters 2 and 3, the native Hilbert spaces associated with positive definite kernels in many cases do not at all look like Sobolev spaces. Especially for radial basis functions, they are usually characterized in terms of Fourier transforms, and this makes it more difficult for us to understand these spaces. It would therefore seem desirable to have a more "intuitive" interpretation for them. Can we come up with an easy way, ideally in terms of smoothness classes or derivatives, to determine what kind of functions lie in the native space of a given kernel?

Fortunately, there is a beautiful mathematical connection that can be established between the classical idea of Green's kernels and positive definite reproducing kernels. By reconsidering our kernels in this light we will be able to characterize functions in the reproducing kernel Hilbert space in terms of their *smoothness* (just as for classical Sobolev spaces). Moreover, the notion of a *length scale* associated with all functions in a reproducing kernel Hilbert space will appear naturally via the differential operator defining the Green's kernel. The concept of a length scale does not exist as part of standard Sobolev space theory, but given the huge amount of research focused on *multi-scale problems* we believe that the resulting *generalized Sobolev spaces* deserve more attention. We will discuss them separately in Chapter 8.

Connections between piecewise polynomial splines or radial basis functions and Green's functions have repeatedly been made in the literature over the past decades (see, e.g., [Atteia (1966); Parzen (1970); Schumaker (1981); Dyn *et al.* (1986); Wahba (1990); Kybic *et al.* (2002); Pesenson (2004); DeVore and Ron (2010); Mhaskar *et al.* (2010)]). We summarize some of those insights here and also present some connections that seem to go beyond the discussion in the existing literature. By clearly laying out how positive definite reproducing kernels are related to Green's kernels we can make significant progress towards better understanding the native space of a given kernel, and towards better understanding which kernels best apply to which situation.

In the literature on radial basis functions and multivariate variational splines, the first mention of a connection between splines and Green's functions was probably given by Atteia (1966), who indeed *defined* multivariate (variational) splines as Green's functions of certain differential operators and linked them to reproducing kernels of standard Sobolev spaces.

The connection between reproducing kernels and Green's kernels is also mentioned in some of the earliest work on reproducing kernels. In fact, Aronszajn (1950) begins the historical introduction of his often-cited survey article on reproducing kernel Hilbert spaces with

> *Examples of kernels of the type in which we are interested have been known for a long time, since all the Green's functions of self-adjoint ordinary differential equations (as also some Green's functions — the bounded ones — of partial differential equations) belong to this type.*

There are even papers and an entire book that specifically address this connection such as [Bergman and Schiffer (1947, 1953); Aronszajn and Smith (1957)]. Moreover, Aronszajn (1950) states that Zaremba (1907, 1909) was the first to link Green's kernels with the reproducing property in the context of harmonic and biharmonic boundary value problems.

6.2 Green's Kernels Defined

We begin by recalling some of the classical results on Green's functions and providing some simple examples before we stress the aspects of the classical theory that reveal the essential connections to kernel-based approximation methods. Some references for the classical theory are, e.g., [Courant and Hilbert (1953); Stakgold (1979); Folland (1992); Cheney (2001); Duffy (2001); Hunter and Nachtergaele (2001)] and [Ramsay and Silverman (2005, Chapters 20 and 21)].

Definition 6.1 (Green's kernel). *Given a linear (ordinary or partial) differential operator $\mathcal{L}$ on the domain $\Omega \subseteq \mathbb{R}^d$, the* Green's kernel G *of $\mathcal{L}$ is defined as the solution of*

$$\mathcal{L}G(\boldsymbol{x}, \boldsymbol{z}) = \delta(\boldsymbol{x} - \boldsymbol{z}), \qquad \boldsymbol{z} \in \Omega \textit{ fixed}. \tag{6.1}$$

Here $\delta(\boldsymbol{x}-\boldsymbol{z})$ *is the* Dirac delta functional *evaluated at* $\boldsymbol{x}-\boldsymbol{z}$, *i.e.,* $\delta(\boldsymbol{x}-\boldsymbol{z})=0$ *for* $\boldsymbol{x}\neq\boldsymbol{z}$ *and* $\int_\Omega \delta(\boldsymbol{x})\,\mathrm{d}\boldsymbol{x}=1$.

In particular, δ acts as a point evaluator for any $f\in L_2(\Omega)$, i.e.,

$$\int_\Omega f(\boldsymbol{z})\delta(\boldsymbol{x}-\boldsymbol{z})\,\mathrm{d}\boldsymbol{z}=f(\boldsymbol{x}),$$

see, e.g., [Duffy (2001, Section 1.2)].

Remark 6.1. Even though this point evaluation property of the Dirac delta functional is analogous to the reproducing property of a reproducing kernel K, δ is *not* the reproducing kernel of $L_2(\Omega)$ since $\delta\notin L_2(\Omega)$. This also implies that $L_2(\Omega)$ is *not* a reproducing kernel Hilbert space (see also Remark 2.7). Moreover, the fact that the Dirac δ functional is not a reproducing kernel should not be confused with the fact that the *Kronecker delta* symbol *is* a reproducing kernel for ℓ_2 (see, e.g., [Berlinet and Thomas-Agnan (2004, Chapter 1)]). Of course, ℓ_2 is not a Hilbert space of functions, but of (real-valued) sequences.

Since the Green's kernel of $\mathcal{L}$ is not uniquely defined via (6.1) one usually adds either some *linear homogeneous boundary conditions* or *decay conditions* (when $\Omega=\mathbb{R}^d$) to make the kernel unique, e.g., for fixed $\boldsymbol{z}$

$$G(\boldsymbol{x},\boldsymbol{z})|_{\boldsymbol{x}\in\partial\Omega}=0 \quad \text{or} \quad \lim_{\|\boldsymbol{x}\|\to\infty} G(\boldsymbol{x},\boldsymbol{z})=0.$$

Frequently, the solution of (6.1) *without* boundary conditions is referred to as a *fundamental solution of* $\mathcal{L}u=0$, or as a *full-space Green's kernel of* $\mathcal{L}$. Moreover, in the literature on signal processing one will come across the term *impulse response* used to refer to a Green's kernel, and in other engineering disciplines (such as mechanical engineering) the term *influence function* is sometimes used.

Traditionally, Green's kernels are employed in the solution of nonhomogeneous boundary value problems since the Green's kernel G of $\mathcal{L}$ defined via (6.1) satisfies $\mathcal{L}u=f$ with the appropriate boundary or decay conditions, i.e., the solution u is given in terms of f and G as

$$u(\boldsymbol{x})=\int_\Omega G(\boldsymbol{x},\boldsymbol{z})f(\boldsymbol{z})\,\mathrm{d}\boldsymbol{z}. \tag{6.2}$$

Here f is assumed to be at least piecewise continuous. This claim is easily verified by the following calculation. Since $\mathcal{L}$ acts on the variable $\boldsymbol{x}$ we have no problems interchanging the order of differentiation and integration. Thus application of $\mathcal{L}$ to both sides of (6.2) along with (6.1) and the definition of δ gives us

$$\mathcal{L}u(\boldsymbol{x})=\mathcal{L}\int_\Omega G(\boldsymbol{x},\boldsymbol{z})f(\boldsymbol{z})\,\mathrm{d}\boldsymbol{z}=\int_\Omega \mathcal{L}G(\boldsymbol{x},\boldsymbol{z})f(\boldsymbol{z})\,\mathrm{d}\boldsymbol{z}=\int_\Omega \delta(\boldsymbol{x}-\boldsymbol{z})f(\boldsymbol{z})\,\mathrm{d}\boldsymbol{z}=f(\boldsymbol{x}).$$

We can therefore *regard the integral operator*

$$\mathcal{G}f(\boldsymbol{x})=\int_\Omega G(\boldsymbol{x},\boldsymbol{z})f(\boldsymbol{z})\,\mathrm{d}\boldsymbol{z} \tag{6.3}$$

as the inverse of the differential operator $\mathcal{L}$, i.e.,

$$\mathcal{L}u = f \qquad \Longleftrightarrow \qquad u = \mathcal{G}f.$$

Here the inverse is guaranteed to exist if and only if the homogeneous equation $\mathcal{L}u = 0$ has only the trivial solution $u = 0$. Note, however, while our integral operators on $\mathcal{H}_G$ are of the Hilbert–Schmidt type and therefore compact, their inverse differential operators are unbounded whenever $\mathcal{H}_G$ has an infinite-dimensional orthonormal basis (see, e.g., [Stakgold (1979, Section 5.3)]).

6.3 Differential Eigenvalue Problems

We now consider a differential eigenvalue problem of the form

$$\mathcal{L}\varphi(\boldsymbol{x}) = \mu\rho(\boldsymbol{x})\varphi(\boldsymbol{x}), \qquad \rho(\boldsymbol{x}) > 0,\ \mu \neq 0, \tag{6.4}$$

with eigenvalues μ_n, eigenfunctions φ_n, $n = 1, 2, \ldots$, and weight function ρ, and assume that G is the Green's kernel of $\mathcal{L}$. Then (6.2) with $f(\boldsymbol{z}) = \mu\rho(\boldsymbol{z})\varphi(\boldsymbol{z})$ immediately gives us

$$\varphi(\boldsymbol{x}) = \int_\Omega G(\boldsymbol{x}, \boldsymbol{z})\mu\rho(\boldsymbol{z})\varphi(\boldsymbol{z})\,\mathrm{d}\boldsymbol{z}$$

as the Green's function-based solution of (6.4). This, however, looks precisely like our earlier Hilbert–Schmidt eigenvalue problem (2.5), albeit with eigenvalues $\lambda_n = \frac{1}{\mu_n}$ that are the *reciprocals* of those of the differential eigenvalue problem (6.4), i.e.,

$$\frac{1}{\mu}\varphi(\boldsymbol{x}) = \int_\Omega G(\boldsymbol{x}, \boldsymbol{z})\varphi(\boldsymbol{z})\rho(\boldsymbol{z})\,\mathrm{d}\boldsymbol{z} \quad \Longleftrightarrow \quad \mathcal{G}\varphi(\boldsymbol{x}) = \frac{1}{\mu}\varphi(\boldsymbol{x}).$$

Note that, since the differential eigenvalue problem (6.4) contains a weight function ρ, $\mathcal{G}$ is no longer of the form (6.3), but now corresponds to the integral operator from (2.5), i.e., it also includes the weight function ρ in its definition.

Since the Green's kernel $G \in L_2(\Omega \times \Omega, \rho \times \rho)$ we know that $\mathcal{G}$ is a compact Hilbert–Schmidt integral operator by Theorem 2.1. If we can further establish that G is positive definite, then we know that Mercer's theorem applies. For regular self-adjoint operators such as the Sturm–Liouville differential operators discussed in Section 6.6 we indeed have this positive definiteness (see, e.g., [Courant and Hilbert (1953, Chapter V), Akhiezer and Glazman (1993, Appendix II)]).

Our interest in Green's kernels will be somewhat different from their traditional use. We want to recognize them as positive definite reproducing kernels and then exploit this connection to *create new reproducing kernels*, and to *gain new insights about reproducing kernels* by drawing from known results in harmonic analysis.

6.4 Computing Green's Kernels

Being able to compute a specific Green's kernel is not an easy task. This depends on a number of things, such as

(1) the differential operator $\mathcal{L}$,
(2) the shape of the domain Ω, and the dimension d of the space it is embedded in,
(3) and the boundary or decay conditions accompanying $\mathcal{L}$.

Before we go into further details on Green's kernels we present an example that we will then continue to come back to in order to illustrate all of the theoretical concepts involved.

6.4.1 *An example: Computing the Brownian bridge kernel as Green's kernel*

We will show that the Green's kernel of the two point boundary value problem

$$-u''(x) = f(x) \quad \text{subject to} \quad u(0) = u(1) = 0$$

is given by

$$G(x,z) = \min(x,z) - xz.$$

This kernel is known as the *Brownian bridge kernel* (see, e.g., [Berlinet and Thomas-Agnan (2004); Caflisch *et al.* (1997)]) since it is also the covariance kernel for a stochastic process of this name.

Before we attempt to verify this claim we note that (6.1) implies that the Green's kernel for the boundary value problem specified above must satisfy the following properties:

- $\mathcal{L}G(x,z) = 0$ for $x \neq z$, z fixed,
- $G(x,z)|_{x\in\{0,1\}}$ satisfies homogeneous boundary conditions for all $z \in (0,1)$,
- G is continuous along the diagonal $x = z$,
- and, for any fixed $z \in (0,1)$, $\frac{\mathrm{d}G}{\mathrm{d}x}$ has a jump discontinuity at $x = z$ of the form

$$\lim_{x\to z^-} \frac{\mathrm{d}}{\mathrm{d}x} G(x,z) = \lim_{x\to z^+} \frac{\mathrm{d}}{\mathrm{d}x} G(x,z) + 1.$$

Therefore G is a piecewise defined function on $[0,1]^2$, which we can write as

$$G(x,z) = \begin{cases} G_-(x,z), & x < z, \\ G_+(x,z), & x > z. \end{cases}$$

Since $\mathcal{L} = -\frac{\mathrm{d}^2}{\mathrm{d}x^2}$ it is clear that the kernel is a piecewise linear polynomial, which we express as

$$G(x,z) = \begin{cases} a_0 + a_1 x, & x < z, \\ b_0 + b_1(x-1), & x > z. \end{cases} \tag{6.5}$$

We first look at the impact of the boundary conditions. On the "upper" triangle $x < z$ the "left" boundary condition implies

$$0 \stackrel{BC}{=} G(0, z) = a_0 \quad \text{so that} \quad a_0 = 0.$$

Similarly, for $x > z$ the "right" boundary condition yields $G(1, z) = b_0 = 0$. Thus, so far we have established that G is of the form

$$G(x, z) = \begin{cases} a_1 x, & x < z, \\ b_1(x - 1), & x > z. \end{cases}$$

In order to determine the coefficients a_1 and b_1 we use the remaining properties listed above to glue the two pieces of G together with appropriate *interface conditions at* $x = z$. The continuity of G can be formulated as

$$\lim_{x \to z^-} G(x, z) = \lim_{x \to z^+} G(x, z),$$

which implies

$$a_1 z = b_1(z - 1) \quad \text{so that} \quad a_1 = b_1 \frac{z - 1}{z}. \tag{6.6}$$

The jump condition for the first derivative states

$$\lim_{x \to z^-} \frac{\mathrm{d}}{\mathrm{d}x} G(x, z) = \lim_{x \to z^+} \frac{\mathrm{d}}{\mathrm{d}x} G(x, z) + 1,$$

which implies $a_1 = b_1 + 1$ so that, using the value of a_1 from (6.6), we get

$$b_1 \frac{z - 1}{z} = b_1 + 1 \quad \text{or} \quad b_1 = -z.$$

Putting everything together, we have $a_0 = b_0 = 0$, $b_1 = -z$ and $a_1 = 1 - z$ so that the Green's kernel (6.5) turns out to be

$$G(x, z) = \begin{cases} (1 - z)x, & x < z, \\ -z(x - 1), & x > z, \end{cases}$$

or

$$G(x, z) = \min(x, z) - xz.$$

Figure 6.1 shows copies of the Brownian bridge kernel centered at $z = \frac{j}{10}$ for $j = 1, \ldots, 9$. The piecewise linear nature of the kernel is clearly visible.

Note that $G(x, z) = \min(x, z) - xz$ is *symmetric*, and therefore the integral operator $\mathcal{G}$ is *self-adjoint*. This will be true whenever $\mathcal{L}$ is a self-adjoint differential operator.

The resulting kernel interpolant for this example will be a piecewise linear ("connect-the-dots") spline whose values at $x = 0$ and $x = 1$ will always be zero.

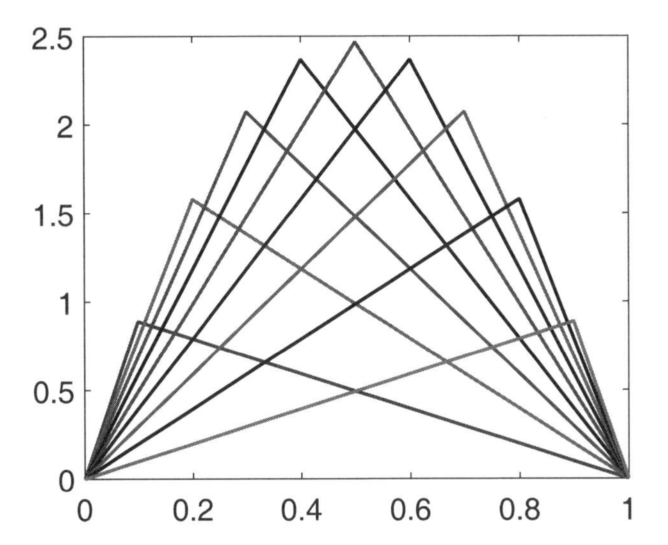

Fig. 6.1 Multiple copies of the Brownian bridge kernel centered at equally spaced points in $[0,1]$.

6.4.2 *Generalizations of the Brownian bridge kernel*

Example 6.1. (Brownian bridge kernel II)
As described in Section 3.4.3, it is straightforward to extend the one-dimensional kernel $G(x,z) = \min(x,z) - xz$ from the previous subsection to a kernel in higher dimensions using a tensor product approach. In this case, the domain will be the unit cube $[0,1]^d$. The kernel is then given by

$$K(\boldsymbol{x},\boldsymbol{z}) = \prod_{\ell=1}^{d} G(x_\ell, z_\ell) = \prod_{\ell=1}^{d} \left(\min\{x_\ell, z_\ell\} - x_\ell z_\ell\right),$$

where $\boldsymbol{x} = (x_1,\ldots,x_d)$ and $\boldsymbol{z} = (z_1,\ldots,z_d)$.

Figure 6.2 shows two instances of a two-dimensional version of this kernel. Note how the homogeneous boundary conditions are satisfied all around, and also observe that this kernel is neither radial nor translation-invariant. The multivariate version of the Brownian bridge kernel is no longer piecewise linear, i.e., it is a function of coordinate degree one instead of total degree one.

In the example in the previous subsection we dealt with a second-order differential operator. However, in Chapter 7 we will consider a higher-order operator (see also Example 6.2). In that case the properties of the Green's kernel will be analogous. If $\mathcal{L}$ is a differential operator of order β with leading coefficient 1 on $[a,b]$, then G will have to satisfy

- $\mathcal{L}G(x,z) = 0$ for $x \neq z$, z fixed,
- $G(x,z)|_{x\in\{a,b\}}$ satisfies homogeneous boundary conditions for all $z \in (a,b)$,
- for fixed z, $\frac{\mathrm{d}^\nu}{\mathrm{d}x^\nu}G$, $\nu = 0,1,\ldots,\beta-2$, is continuous along the diagonal $x = z$,

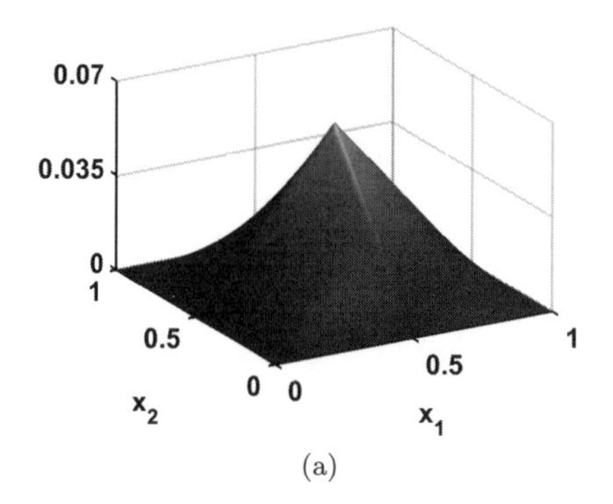

(a)

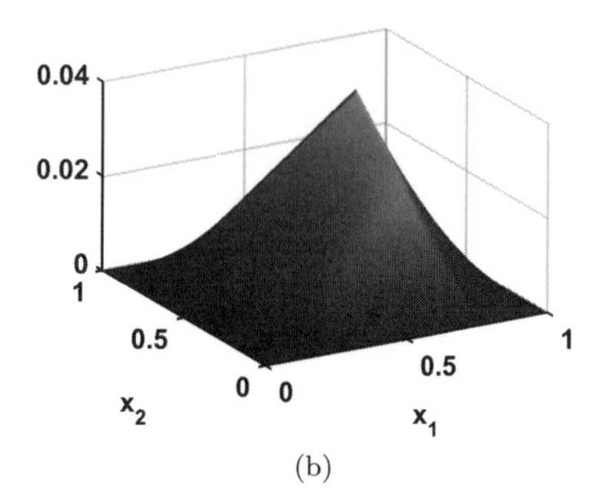

(b)

Fig. 6.2 Brownian bridge kernel centered at (a) $[\frac{1}{2}, \frac{1}{2}]$, (b) $[\frac{3}{4}, \frac{3}{4}]$.

- and, for any fixed $z \in (a,b)$, $\frac{\mathrm{d}^{\beta-1}G}{\mathrm{d}x^{\beta-1}}$ has a jump discontinuity at $x = z$ of the form

$$\lim_{x \to z^-} \frac{\mathrm{d}^{\beta-1}}{\mathrm{d}x^{\beta-1}} G(x,z) = \lim_{x \to z^+} \frac{\mathrm{d}^{\beta-1}}{\mathrm{d}x^{\beta-1}} G(x,z) + 1.$$

The nature of the interface conditions is explained by (6.1). The presence of the delta function implies the jump in the $(\beta-1)^{\mathrm{st}}$ derivative. However, all derivatives up to order $\beta - 2$ must be continuous for all x and z because otherwise the βth derivative present in the definition of $\mathcal{L}$ would involve not only the delta function but also some of its derivatives.

Example 6.2. (Iterated Brownian bridge kernels)
In Chapter 7 we will use the more general differential operator

$$\mathcal{L} = \left(-\frac{\mathrm{d}^2}{\mathrm{d}x^2} + \varepsilon^2 \mathcal{I} \right)^{\beta}, \qquad \beta \in \mathbb{N},\ \varepsilon \geq 0,$$

together with boundary conditions

$$G(0,z) = G''(0,z) = \ldots = G^{(2\beta-2)}(0,z) = 0,$$
$$G(1,z) = G''(1,z) = \ldots = G^{(2\beta-2)}(1,z) = 0,$$

$z \in (0,1)$ fixed, to introduce the so-called *iterated Brownian bridge kernels* as Green's kernels of $\mathcal{L}$.

The Brownian bridge kernel derived in Section 6.4.1 can be identified as a special case with $\beta = 1$ and $\varepsilon = 0$. We will see in Chapter 7 that one obtains the basis for *natural cubic interpolating splines* in the special case $\beta = 2$, $\varepsilon = 0$.

6.5 Classical Examples of Green's Kernels

We end this section by listing the families of full-space Green's kernels for several multivariate generalizations of the operators $\mathcal{L} = -\frac{\mathrm{d}^2}{\mathrm{d}x^2}$ and $\mathcal{L} = -\frac{\mathrm{d}^2}{\mathrm{d}x^2} + \varepsilon^2\mathcal{I}$ discussed above. While most of these kernels are *not* positive definite, and quite a few are even singular, we are interested in these kernels since we will refer to them when we discuss the *methods of fundamental and particular solutions* in Chapter 20.

Example 6.3. (Fundamental solutions of the Laplace operator)
The full-space Green's kernel (or fundamental solution) of the *Laplacian*, i.e., the solution of

$$-\nabla^2 G(\boldsymbol{x},\boldsymbol{z}) = \delta(\boldsymbol{x}-\boldsymbol{z}), \qquad \boldsymbol{x},\boldsymbol{z} \in \mathbb{R}^d, \quad d > 1,$$

is given by (see, e.g., [Folland (1992)])

$$G(\boldsymbol{x},\boldsymbol{z}) = \begin{cases} -\dfrac{1}{2\pi}\log\|\boldsymbol{x}-\boldsymbol{z}\|, & d=2,\\ \dfrac{1}{(d-2)\omega_d}\|\boldsymbol{x}-\boldsymbol{z}\|^{2-d}, & d \geq 3,\end{cases}$$

where ω_d is the volume of $\mathbb{S}^{d-1}$, the unit sphere in $\mathbb{R}^d$.

Example 6.4. (Fundamental solutions of the modified Helmholtz operator)
The full-space Green's kernel of the modified Helmholtz operator, i.e., the solution of

$$\left(-\nabla^2 + \varepsilon^2\right) G(\boldsymbol{x},\boldsymbol{z}) = \delta(\boldsymbol{x}-\boldsymbol{z}), \qquad \boldsymbol{x},\boldsymbol{z} \in \mathbb{R}^d, \quad d = 1,2,3,$$

is given by (see, e.g., [Duffy (2001)])

$$G(\boldsymbol{x},\boldsymbol{z}) = \begin{cases} \dfrac{1}{2\varepsilon}\mathrm{e}^{-\varepsilon|x-z|}, & d=1,\\ \dfrac{1}{2\pi}K_0(\varepsilon\|\boldsymbol{x}-\boldsymbol{z}\|), & d=2,\\ \dfrac{1}{4\pi\|\boldsymbol{x}-\boldsymbol{z}\|}\mathrm{e}^{-\varepsilon\|\boldsymbol{x}-\boldsymbol{z}\|}, & d=3,\end{cases}$$

where K_0 is a *modified Bessel function of the second kind* [DLMF, Eq. (10.31.1)].

Example 6.5. (Fundamental solutions of the biharmonic operator)
The full-space Green's kernel of the operator $\mathcal{L} = \nabla^4$, i.e., the solution of

$$\nabla^4 G(\boldsymbol{x},\boldsymbol{z}) = \delta(\boldsymbol{x}-\boldsymbol{z}), \qquad \boldsymbol{x},\boldsymbol{z} \in \mathbb{R}^d, \quad d = 2,3,$$

is given by (see, e.g., [Duffy (2001)], and for a more general discussion [Aronszajn *et al.* (1983)])

$$G(\boldsymbol{x},\boldsymbol{z}) = \begin{cases} \dfrac{\|\boldsymbol{x}-\boldsymbol{z}\|^2}{8\pi}\log(\|\boldsymbol{x}-\boldsymbol{z}\|-1), & d=2,\\ \dfrac{1}{8\pi}\|\boldsymbol{x}-\boldsymbol{z}\|, & d=3.\end{cases}$$

The reader familiar with *thin plate splines* may recognize these functions (see, e.g., [Fasshauer (2007, Chapter 8)]).

6.6 Sturm–Liouville Theory

The theory for Sturm–Liouville eigenvalue problems is worked out in detail in many text books (see, e.g., [Folland (1992); Haberman (2013)]). We briefly summarize the main properties and show how this theory provides insights for the positive definite reproducing kernels associated with their "inverse" Hilbert–Schmidt eigenvalue problems.

Definition 6.2 (Regular Sturm–Liouville problem). *The second-order differential equation*

$$\frac{\mathrm{d}}{\mathrm{d}x}\left(p(x)\varphi'(x)\right) + q(x)\varphi(x) + \mu\rho(x)\varphi(x) = 0, \qquad x \in (a,b), \tag{6.7}$$

with boundary conditions

$$\begin{aligned} \gamma_1\varphi(a) + \gamma_2\varphi'(a) &= 0, \\ \gamma_3\varphi(b) + \gamma_4\varphi'(b) &= 0, \end{aligned} \tag{6.8}$$

where the γ_i *are real numbers is called a* regular Sturm–Liouville problem *provided* p, q, ρ *and* p' *in* (6.7) *are real-valued and continuous on* $[a,b]$, *and if* $p(x)$ *and* $\rho(x)$ *are positive for all* x *in* $[a,b]$.

Theorem 6.1. *The following properties hold for regular Sturm–Liouville problems:*

(1) *All eigenvalues are real and there are countably many of them which can be strictly ordered:* $\mu_1 < \mu_2 < \mu_3 < \ldots$.
(2) *Every eigenvalue* μ_n *has an associated eigenfunction* φ_n *which is unique up to a constant factor. Moreover,* φ_n *has exactly* $n-1$ *zeros in the open interval* (a,b).
(3) *The set of eigenfunctions,* $\{\varphi_n\}_{n=1}^{\infty}$, *is complete, i.e., any piecewise smooth function* f *can be represented by a* generalized Fourier series

$$f(x) \sim \sum_{n=1}^{\infty} a_n\varphi_n(x) \tag{6.9}$$

with generalized Fourier coefficients

$$a_n = \frac{\int_a^b f(x)\varphi_n(x)\rho(x)\,\mathrm{d}x}{\int_a^b \varphi_n^2(x)\rho(x)\,\mathrm{d}x} = \frac{\langle f,\varphi_n\rangle_{L_2((a,b),\rho)}}{\|\varphi_n\|^2_{L_2((a,b),\rho)}}, \qquad n = 1,2,3,\ldots.$$

(4) *The eigenfunctions associated with different eigenvalues are orthogonal on* (a,b) *with respect to the weight* ρ, *i.e.,*

$$\int_a^b \varphi_n(x)\varphi_m(x)\rho(x)\,\mathrm{d}x = 0 \quad \textit{provided } \mu_n \neq \mu_m.$$

(5) *The* Rayleigh quotient *is given by*

$$\mu = \frac{-\,p(x)\varphi(x)\varphi'(x)\big|_a^b + \int_a^b \left(p(x)\left(\varphi'(x)\right)^2 - q(x)\varphi^2(x)\right)\mathrm{d}x}{\int_a^b \varphi^2(x)\rho(x)\,\mathrm{d}x}.$$

(6) *Truncating the Fourier series* (6.9) *at* M *terms yields the best* M*-term* $L_2((a,b),\rho)$ *(or weighted mean-squared) approximation:*

$$a_n = \operatorname*{argmin}_{\alpha_n} \left\| f - \sum_{n=1}^{M} \alpha_n \varphi_n \right\|_2 .$$

A proof of Theorem 6.1 can be found in many textbooks on Fourier series and boundary value problems such as [Folland (1992); Haberman (2013)].

6.7 Eigenfunction Expansions

We now review how the eigenfunctions of linear self-adjoint differential operators, such as the Sturm–Liouville operator, are related to Green's kernels.

We start from the problem

$$(\mathcal{L}G)(\boldsymbol{x},\boldsymbol{z}) = \delta(\boldsymbol{x}-\boldsymbol{z}), \qquad \boldsymbol{z} \text{ fixed},$$

where the operator $\mathcal{L}$ together with appropriate boundary conditions forms a self-adjoint *differential eigenvalue problem*

$$(\mathcal{L}\varphi)(\boldsymbol{x}) = \mu\rho(\boldsymbol{x})\varphi(\boldsymbol{x}) \tag{6.10}$$

accompanied by those same boundary conditions. Since we are primarily interested in working with positive definite kernels, we assume that zero is not an eigenvalue of (6.10) (if it is, then one needs to deal with the residues of the Green's kernel at its poles [Duffy (2001)]).

We have freedom in choosing the weight function ρ. In fact, we can use this freedom in order to come up with different families of eigenfunctions. Once ρ is chosen, we have a unique set of eigenvalues and eigenfunctions (up to their normalization constant) and we can use (6.9) to express the Green's kernel G (treated as a piecewise continuous function of $\boldsymbol{x}$ for fixed $\boldsymbol{z}$) in terms of the eigenfunction basis with expansion coefficients that are functions of $\boldsymbol{z}$, i.e.,

$$G(\boldsymbol{x},\boldsymbol{z}) = \sum_{n=1}^{\infty} a_n(\boldsymbol{z})\varphi_n(\boldsymbol{x}). \tag{6.11}$$

In order to determine the $\boldsymbol{z}$-dependent generalized Fourier coefficients $a_n(\boldsymbol{z})$ we apply the operator $\mathcal{L}$ and use linearity to get

$$(\mathcal{L}G)(\boldsymbol{x},\boldsymbol{z}) = \sum_{n-1}^{\infty} a_n(\boldsymbol{z})(\mathcal{L}\varphi_n)(\boldsymbol{x}).$$

Since the φ_n are eigenfunctions of (6.10) we have

$$(\mathcal{L}G)(\boldsymbol{x},\boldsymbol{z}) = \sum_{n=1}^{\infty} a_n(\boldsymbol{z})\mu_n\rho(\boldsymbol{x})\varphi_n(\boldsymbol{x}),$$

and the fact that G is the Green's kernel of $\mathcal{L}$ yields

$$\delta(\boldsymbol{x}-\boldsymbol{z}) = \sum_{n=1}^{\infty} a_n(\boldsymbol{z})\mu_n \rho(\boldsymbol{x})\varphi_n(\boldsymbol{x}).$$

Next we multiply both sides of this equation by $\varphi_m(\boldsymbol{x})$ and integrate over Ω so that

$$\int_\Omega \delta(\boldsymbol{x}-\boldsymbol{z})\varphi_m(\boldsymbol{x})\,\mathrm{d}\boldsymbol{x} = \sum_{n=1}^{\infty} a_n(\boldsymbol{z})\mu_n \int_\Omega \rho(\boldsymbol{x})\varphi_n(\boldsymbol{x})\varphi_m(\boldsymbol{x})\,\mathrm{d}\boldsymbol{x},$$

which becomes

$$\varphi_m(\boldsymbol{z}) = a_m(\boldsymbol{z})\mu_m \int_\Omega \rho(\boldsymbol{x})\varphi_m^2(\boldsymbol{x})\,\mathrm{d}\boldsymbol{x}$$

because of the definition of the Dirac delta functional and the orthogonality of the eigenfunctions. Therefore the generalized Fourier coefficients of $G(\cdot,\boldsymbol{z})$ end up being

$$a_n(\boldsymbol{z}) = \frac{\varphi_n(\boldsymbol{z})}{\mu_n \int_\Omega \varphi_n^2(\boldsymbol{x})\rho(\boldsymbol{x})\,\mathrm{d}\boldsymbol{x}}. \tag{6.12}$$

Putting this back into the eigenfunction expansion (6.11) for G we have

$$G(\boldsymbol{x},\boldsymbol{z}) = \sum_{n=1}^{\infty} \frac{\varphi_n(\boldsymbol{z})}{\mu_n \int_\Omega \varphi_n^2(\boldsymbol{\xi})\rho(\boldsymbol{\xi})\,\mathrm{d}\boldsymbol{\xi}}\varphi_n(\boldsymbol{x}). \tag{6.13}$$

In particular, if the eigenfunctions are normalized with respect to the $L_2(\Omega,\rho)$ inner product then

$$G(\boldsymbol{x},\boldsymbol{z}) = \sum_{n=1}^{\infty} \frac{1}{\mu_n}\varphi_n(\boldsymbol{x})\varphi_n(\boldsymbol{z}),$$

which matches the form of a Mercer series (2.9) when $\lambda_n = \frac{1}{\mu_n}$ — as we should expect.

Remark 6.2. Note that (6.12) can be read as

$$a_n(\boldsymbol{z}) = \frac{\langle G(\cdot,\boldsymbol{z}),\varphi_n\rangle_{\mathcal{H}_G(\Omega)}}{\mu_n\|\varphi_n\|^2_{L_2(\Omega,\rho)}} = \frac{\langle G(\cdot,\boldsymbol{z}),\varphi_n\rangle_{\mathcal{H}_G(\Omega)}}{\mu_n^2\|\varphi_n\|^2_{\mathcal{H}_G(\Omega)}},$$

since $\varphi_n(\boldsymbol{z}) = \langle G(\cdot,\boldsymbol{z}),\varphi_n\rangle_{\mathcal{H}_G(\Omega)}$ and the L_2 and native space norms of the eigenfunctions differ by an eigenvalue factor (see (2.8)).

Remark 6.3. This approach leads to a *series representation of* G, while the method outlined in Section 6.4 produces a *closed form representation of* G. As we will show in Chapter 13, it is not necessary to have a closed form representation of a reproducing kernel K in order to be able to use it to solve the approximation problems we are interested in. In fact, it may even be advantageous to work with its series representation, provided it is available.

Example 6.6. (Brownian bridge kernel III)
We now revisit the Brownian bridge kernel whose closed form representation, $K(x,z) = \min(x,z) - xz$, we computed earlier, and derive a series representation.

A simple exercise in standard Sturm–Liouville theory tells us that the boundary value problem

$$-\varphi''(x) = \mu\varphi(x), \qquad \varphi(0) = \varphi(1) = 0,$$

(with weight function $\rho(x) = 1$) has eigenvalues and eigenfunctions

$$\mu_n = (n\pi)^2, \quad \varphi_n(x) = \sin(n\pi x), \qquad n = 1, 2, 3, \ldots,$$

and we can verify

$$G(x,z) = \min(x,z) - xz = \sum_{n=1}^{\infty} a_n(z)\sin(n\pi x)$$

with generalized Fourier coefficients (6.12)

$$a_n(z) = \frac{1}{\mu_n}\frac{\varphi_n(z)}{\|\varphi_n\|^2} = \frac{2}{(n\pi)^2}\sin(n\pi z).$$

This could also have been computed via the L_2 projection of $G(\cdot, z)$ onto the normalized eigenfunction φ_n, i.e.,

$$\begin{aligned} a_n(z) &= \frac{1}{\|\varphi_n\|^2_{L_2((0,1))}}\langle G(\cdot,z), \varphi_n\rangle_{L_2((0,1))} \\ &= 2\int_0^1 (\min(x,z) - xz)\sin(n\pi x)\,\mathrm{d}x = \frac{2}{(n\pi)^2}\sin(n\pi z). \end{aligned}$$

Note how this calculation verifies that $\lambda_n = \frac{1}{(n\pi)^2}$ and $\varphi_n(x) = \sin(n\pi x)$ satisfy the Hilbert–Schmidt integral eigenvalue problem associated with the Brownian bridge kernel.

6.8 The Connection Between Hilbert–Schmidt and Sturm–Liouville Eigenvalue Problems

In Section 6.2 we argued that the Hilbert–Schmidt (integral) eigenvalue problem and the self-adjoint regular Sturm–Liouville (differential) eigenvalue problem are "inverse" to each other (see also [Courant and Hilbert (1953, Chapter V), Akhiezer and Glazman (1993, Appendix II)]).

In particular, if we begin with the self-adjoint differential eigenvalue problem (as we did in Section 6.6)

$$\mathcal{L}\varphi = \mu\rho\varphi$$

with an appropriate set of boundary conditions, then — assuming G is the Green's kernel for $\mathcal{L}$ — we know from (6.2) that

$$\varphi(\boldsymbol{x}) = \int_\Omega G(\boldsymbol{x},\boldsymbol{z})\mu\rho(\boldsymbol{z})\varphi(\boldsymbol{z})\,\mathrm{d}\boldsymbol{z}.$$

If we replace μ with $\frac{1}{\lambda}$ then we recognize this as a Hilbert–Schmidt eigenvalue problem (cf. (2.5)). Therefore, from now on we will always use the letter λ when we are talking about eigenvalues. However, *eigenvalues of differential operators will be denoted by* $\frac{1}{\lambda}$ *and those of integral operators with* λ.

If, conversely, we begin with a symmetric positive definite kernel K and consider its Hilbert–Schmidt eigenvalue problem

$$\mathcal{K}\varphi = \lambda\varphi,$$

then we can obtain the Sturm–Liouville eigenvalue problem via differentiation of (of sufficiently high order) of the integral equation, as already illustrated in Example 2.4 using the simple *Brownian motion kernel* $K(x,z) = \min(x,z)$. Note that the Hilbert–Schmidt eigenfunctions "inherit" their boundary behavior from the kernel K. They have to be appropriately specified for the Sturm–Liouville problem.

Since the Brownian motion kernel is also the Green's kernel of

$$-\frac{\mathrm{d}^2}{\mathrm{d}x^2}K(x,z) = \delta(x-z) \quad \text{subject to} \quad K(0,z) = \frac{\mathrm{d}}{\mathrm{d}x}K(x,z)|_{x=1} = 0$$

we can confirm that we identified the "correct" differential operator in Example 2.4, and that the boundary conditions "inherited" from the Hilbert–Schmidt integral eigenvalue problem are also appropriate for the Sturm–Liouville eigenvalue problem.

We could have also used the Brownian bridge kernel as an example. However, those calculations are very similar and can be found in [Fasshauer (2011a)].

6.9 Limitations

We close with an example showing that the approach outlined in this chapter may not always be as straightforward as it appears. In particular, if the differential operator $\mathcal{L}$ has a nontrivial null space of functions that also satisfy the boundary conditions, i.e., if $\mathcal{L}u = 0$ has a solution $u \neq 0$ such that u also satisfies the boundary conditions, then a Green's function will not exist. Take, for example,

$$-u''(x) = f(x), \qquad u'(0) = u'(1) = 0.$$

Then $u''(x) = 0$ is satisfied by any linear polynomial, and the boundary conditions are satisfied by $u(x) = c$ for any constant c. Thus, we do not have a unique solution. We will address this issue in Chapter 8.

Example 6.7. (A more complicated situation)
In [Folland (1992, Section 10.1)] one can find that the Green's kernel for the boundary value problem

$$u''(x) + \varepsilon^2 u(x) = f(x), \qquad u'(0) = u'(1) = 0,$$

with *complex* shape parameter ε is given by

$$G(x,z) = \frac{\cos(\varepsilon \min(x,z))\cos(\varepsilon(\max(x,z)-1))}{\varepsilon \sin\varepsilon}$$

provided $\varepsilon \neq n\pi$ for some integer n. Thus, if ε is real, then G will consist of trigonometric functions, while for imaginary ε we have hyperbolic functions since $\cosh(x) = \cos(\mathrm{i}x)$ and $\sinh(x) = -\mathrm{i}\sin(\mathrm{i}x)$. If ε is an integer multiple of π then the Green's kernel does not exist (see the discussion above for $\varepsilon = 0$).

Note also that here the Sturm–Liouville operator has a zero eigenvalue and the simple series formulation discussed above does not apply.

This situation is analogous to the one that arises in the solution of systems of linear equations. The linear system $\mathsf{A}\boldsymbol{x} = \boldsymbol{b}$ has a unique solution $\boldsymbol{x}$ for any given $\boldsymbol{b}$ provided $\det \mathsf{A} \neq 0$, i.e., $\mathsf{A}\boldsymbol{x} = \boldsymbol{0}$ has only the trivial solution $\boldsymbol{x} = \boldsymbol{0}$. If $\mathsf{A}\boldsymbol{x} = \boldsymbol{0}$ does have a nontrivial solution, then, depending on whether $\boldsymbol{b}$ is in the range of A, the nonhomogeneous equation $\mathsf{A}\boldsymbol{x} = \boldsymbol{b}$ may either be consistent (and have a family of solutions), or inconsistent (and have no solution at all). More generally, this is known as the *Fredholm alternative* and discussed in any textbook on functional analysis such as [Cheney (2001); Hunter and Nachtergaele (2001); Rynne and Youngson (2008); Atkinson and Han (2009); Brezis (2010)].

6.10 Summary

Summarizing the results of Chapter 2 and this chapter, we now know that we can obtain an eigenfunction expansion of the form

$$K(\boldsymbol{x}, \boldsymbol{z}) = \sum_{n=1}^{\infty} \lambda_n \varphi_n(\boldsymbol{x}) \varphi_n(\boldsymbol{z})$$

for a given positive definite kernel K. This can be done

- via Mercer's theorem using the eigenvalues and $L_2(\Omega, \rho)$-normalized eigenfunctions of the Hilbert–Schmidt integral operator $\mathcal{K}$, i.e., as solutions of

$$\mathcal{K}\varphi = \lambda\varphi, \qquad \mathcal{K}f(\boldsymbol{x}) = \int_\Omega K(\boldsymbol{x}, \boldsymbol{z}) f(\boldsymbol{z}) \rho(\boldsymbol{z}) \, \mathrm{d}\boldsymbol{z},$$

- or via a generalized Fourier series based on the eigenvalues and eigenfunctions of the corresponding Sturm–Liouville eigenvalue problem

$$\mathcal{L}\varphi = \frac{1}{\lambda}\rho\varphi, \qquad \mathcal{L}K(\boldsymbol{x}, \boldsymbol{z}) = \delta(\boldsymbol{x} - \boldsymbol{z})$$

  with appropriate boundary conditions.

In particular, we will show that such series expansions can be used to generate the Hilbert–Schmidt SVD which allows us to perform scattered data interpolation and approximation with smooth positive definite kernels such as the Gaussian in a numerically stable and highly accurate way.

Chapter 7

Iterated Brownian Bridge Kernels: A Green's Kernel Example

In this chapter we will derive series representations — and where feasible also closed-form representations — of the family of univariate anisotropic kernels we earlier referred to as *iterated Brownian bridge kernels* (cf. Example 6.2). We separate the discussion into two parts. First we discuss the case of piecewise polynomial splines (when the shape parameter $\varepsilon = 0$), and then we look at the more general case. In the second part of the chapter we will discuss various properties of these kernels in the context of examples implemented in MATLAB.

7.1 Derivation of Piecewise Polynomial Spline Kernels

7.1.1 *Recall some special Green's kernels*

In Chapter 6 we mentioned several examples of positive definite kernels that were derived as Green's kernels of the differential operator

$$\mathcal{L} = \left(-\frac{\mathrm{d}^2}{\mathrm{d}x^2} + \varepsilon^2 \mathcal{I}\right)^\beta, \qquad \beta \in \mathbb{N},\ \varepsilon \geq 0, \tag{7.1}$$

for a few specific choices of ε and β. Since we now are interested in discussing the entire family of these kernels we will use the notation $K_{\beta,\varepsilon}$ to distinguish the different cases.

For the case $\beta = 1$ and $\varepsilon = 0$ we showed earlier that the differential operator $\mathcal{L} = -\frac{\mathrm{d}^2}{\mathrm{d}x^2}$ coupled with the boundary conditions $K(0,z) = K(1,z) = 0$ gives rise to the piecewise linear Brownian bridge kernel $K_{1,0}$. The associated eigenvalue problem is

$$\mathcal{L}\varphi = \mu\varphi, \qquad \varphi(0) = \varphi(1) = 0,$$

with eigenvalues and normalized eigenfunctions

$$\mu_n = n^2\pi^2, \quad \varphi_n(x) = \sqrt{2}\sin(n\pi x), \qquad n = 1, 2, \ldots .$$

The Mercer series (2.9) of the Brownian bridge kernel is therefore given by

$$K_{1,0}(x,z) = \sum_{n=1}^{\infty} \frac{2}{n^2\pi^2} \sin(n\pi x)\sin(n\pi z), \tag{7.2a}$$

which — according to Example 6.6 — sums to

$$K_{1,0}(x,z) = \min(x,z) - xz. \tag{7.2b}$$

In Section 2.2.5 we showed that the iterated kernel $K_{2,0} = \mathcal{K}K_{1,0}$, where $\mathcal{K}$ is the usual Hilbert–Schmidt integral operator with kernel $K_{1,0}$, has the same eigenfunctions as $K_{1,0}$. Moreover, its eigenvalues are the squares of the eigenvalues of $K_{1,0}$. In the context of the differential operators we are using in this chapter, the eigenvalue problem $\mathcal{L}^2\varphi = \eta\varphi$ will have eigenvalues $\eta_n = \mu_n^2$, where we denoted the eigenvalues of $\mathcal{L}$ by μ_n above. The eigenfunctions φ_n will remain unchanged provided we use the boundary conditions

$$\varphi(0) = \varphi(1) = \varphi''(0) = \varphi''(1) = 0,$$

taken from the eigenfunctions $\varphi_n(x) = \sqrt{2}\sin(n\pi x)$. While these boundary conditions are not uniquely determined, they do follow naturally via iteration of the kernel $K_{1,0}$, which is associated with the homogeneous conditions $\varphi(0) = \varphi(1) = 0$.

Therefore, the Mercer series of the kernel $K_{2,0}$ is immediately obtained as

$$K_{2,0}(x,z) = \sum_{n=1}^{\infty} \frac{2}{n^4\pi^4}\sin(n\pi x)\sin(n\pi z).$$

On the other hand, it is a straightforward exercise to derive a closed form for this kernel via the Green's kernel approach described in Section 6.4 arriving at

$$K_{2,0}(x,z) = \begin{cases} \frac{1}{6}x(1-z)\left(1-x^2-(1-z)^2\right), & 0 \le x \le z \le 1, \\ \frac{1}{6}z(1-x)\left(1-z^2-(1-x)^2\right), & 0 \le z \le x \le 1. \end{cases} \tag{7.3}$$

For any fixed value of z this is a *natural cubic spline* that interpolates the value zero at $x = 0$ and $x = 1$ and has vanishing second-order derivatives at those end points.

As argued above, using the concept of iterated kernels from Section 2.2.5 we could also have obtained $K_{2,0}$ as the integral of $K_{1,0}$ against itself, i.e.,

$$\begin{aligned} K_{2,0}(x,z) &= \int_0^1 K_{1,0}(x,t)K_{1,0}(t,z)\mathrm{d}t \\ &= \int_0^1 (\min(x,t) - xt)\,(\min(t,z) - tz)\,\mathrm{d}t, \end{aligned}$$

which leads to the same result as in (7.3).

Remark 7.1. Both $K_{1,0}$ and $K_{2,0}$ already satisfy zero boundary conditions. Therefore, e.g., $K_{1,0}(0,z_j) = 0$ for any point z_j and as a consequence an entire row (and column because $K_{1,0}$ is symmetric) in the interpolation matrix K will be zero if $x = 0$ or $x = 1$ is one of the points at which K is evaluated; or if $z_j = 0$ or $z_j = 1$ is a center. In order to prevent this from causing any trouble we must remember to exclude $x = 0$ and $x = 1$ from the sets of kernel centers (and also from the interpolation points) when we perform numerical experiments with these kernels.

This fact imposes a restriction on the type of problems we can accurately solve with K_1 or K_2. These problems must satisfy the same homogeneous BCs as the kernel.

In order to be able to interpolate nonzero values at the endpoints $x = 0$ and $x = 1$ we could create a second (polynomial) kernel for the null space of the differential operators $\mathcal{L}$ and $\mathcal{L}^2$, respectively, i.e., a linear or cubic polynomial. This kernel can then be added to the kernel $K_{1,0}$ or $K_{2,0}$ to yield a sum kernel which can be used to handle arbitrary nonhomogeneous boundary conditions (see, e.g., [Ramsay and Silverman (2005, Chapter 21)]).

Another simple way of ensuring that we can interpolate nonzero values at the endpoints $x = 0$ and $x = 1$ is to add the linear Lagrange interpolant $p(x) = (1-x)y_1 + xy_N$, where y_1 and y_N denote the data values at $x_1 = 0$ and $x_N = 1$, respectively, to the (homogeneous) kernel interpolant. The same effect can be achieved by preprocessing the data by subtracting a corresponding linear interpolant from it.

The content of this remark can be generalized to the entire family of kernels $K_{\beta,\varepsilon}$ discussed in the following, where the null space of $\mathcal{L}$ for $\varepsilon > 0$ no longer consists of polynomials, but of appropriate exponential functions.

7.1.2 *A family of piecewise polynomial splines of arbitrary odd degree*

We now consider $K_{\beta,0}$, the Green's kernel of the differential operator

$$\mathcal{L}^\beta = (-1)^\beta \frac{\mathrm{d}^{2\beta}}{\mathrm{d}x^{2\beta}}, \qquad \beta \in \mathbb{N},$$

with boundary conditions

$$\frac{\mathrm{d}^{2\nu-2}}{\mathrm{d}x^{2\nu-2}} K_{\beta,0}(x,z)|_{x=0} = \frac{\mathrm{d}^{2\nu-2}}{\mathrm{d}x^{2\nu-2}} K_{\beta,0}(x,z)|_{x=1} = 0, \qquad \nu = 1,\ldots,\beta, \tag{7.4}$$

where z is fixed.

Our discussion of the two special cases, $K_{1,0}$ and $K_{2,0}$, in the previous subsection suggests that the kernel $K_{\beta,0}$ will be a piecewise polynomial spline. In this subsection we derive closed form representations for these piecewise polynomial spline kernels.

Remark 7.2. This idea is related to the more general concept of *L-splines*, and we refer the reader to, e.g., [Schultz and Varga (1967); Swartz and Varga (1972); Varga (1987)] for more details.

From our discussion of iterated kernels above it is clear that the Mercer series for the kernels $K_{\beta,0}$, $\beta = 1, 2, \ldots$, is given by

$$K_{\beta,0}(x,z) = \sum_{n=1}^{\infty} \frac{2}{(n\pi)^{2\beta}} \sin(n\pi x)\sin(n\pi z) \tag{7.5}$$

provided we use the boundary conditions (7.4).

Using the eigenvalues $\lambda_n = (n\pi)^{-2\beta}$ and eigenfunctions $\varphi_n(x) = \sqrt{2}\sin(n\pi x)$, $n = 1, 2, \ldots$, we now derive a closed form for the Mercer series expansion of the kernel $K_{\beta,0}$ which shows that it is indeed a piecewise polynomial of degree $2\beta - 1$.

The main ingredient in this derivation is the standard trigonometric identity $2\sin(A)\sin(B) = \cos(A-B) - \cos(A+B)$ with $A = n\pi x$ and $B = n\pi z$ along with two applications of the cosine series expansion of *Bernoulli polynomials* (see, e.g., [DLMF, Eq. (24.8.1)] and [Olver *et al.* (2010)])

$$B_{2\beta}(t) = (-1)^{\beta+1}\frac{2(2\beta)!}{(2\pi)^{2\beta}}\sum_{n=1}^{\infty} n^{-2\beta}\cos(2\pi n t), \qquad 0 \le t \le 1,\ \beta = 1, 2, \ldots, \tag{7.6}$$

setting $t = \frac{x-z}{2}$ and $t = \frac{x+z}{2}$, respectively.

Since (7.6) requires $0 \le t \le 1$ we need to treat the cases $x \ge z$ and $z \ge x$ separately. However, it is possible to combine the two resulting formulas into the desired symmetric closed form representation

$$K_{\beta,0}(x,z) = (-1)^{\beta-1}\frac{2^{2\beta-1}}{(2\beta)!}\left(B_{2\beta}\left(\frac{|x-z|}{2}\right) - B_{2\beta}\left(\frac{x+z}{2}\right)\right), \tag{7.7}$$

which is valid for any $0 \le x, z \le 1$.

Bernoulli polynomials of degree n can be expressed as (see, e.g., [DLMF, Eq. (24.6.7)] and [Olver *et al.* (2010)])

$$B_n(x) = \sum_{k=0}^{n}\frac{1}{k+1}\sum_{j=0}^{k}(-1)^j\binom{k}{j}(x+j)^n, \tag{7.8}$$

so that the first few Bernoulli polynomials are

$$B_0(x) = 1, \quad B_1(x) = x - \frac{1}{2}, \quad B_2(x) = x^2 - x + \frac{1}{6}, \quad B_3(x) = x^3 - \frac{3}{2}x^2 + \frac{1}{2}x.$$

Since it is apparent from (7.8) that the leading-order terms in x in (7.7) cancel, the kernel $K_{\beta,0}$ is indeed a piecewise polynomial of odd degree $2\beta - 1$. Knowing (7.8), we can easily verify that (7.7) for $\beta = 1, 2$ leads to piecewise linear polynomials of the form

$$K_{1,0}(x,z) = \min(x,z) - xz = \begin{cases} x - xz, & 0 \le x \le z, \\ z - xz, & z \le x \le 1, \end{cases}$$

(i.e., the Brownian bridge kernel) and piecewise cubic polynomials

$$K_{2,0}(x,z) = \begin{cases} \frac{1}{6}x(1-z)(x^2+z^2-2z), & 0 \le x \le z, \\ \frac{1}{6}(1-x)z(x^2+z^2-2x), & z \le x \le 1. \end{cases}$$

For a fixed z, $K_{3,0}(x,z)$ gives piecewise quintic polynomials in x, and so on. Thus, (7.7) allows us to compute the piecewise polynomial kernel $K_{\beta,0}$ satisfying the boundary conditions (7.4) in closed form for any β.

It is immediately apparent from (7.8) that the *derivatives* of the Bernoulli polynomials can be computed recursively via

$$B_n'(x) = nB_{n-1}(x).$$

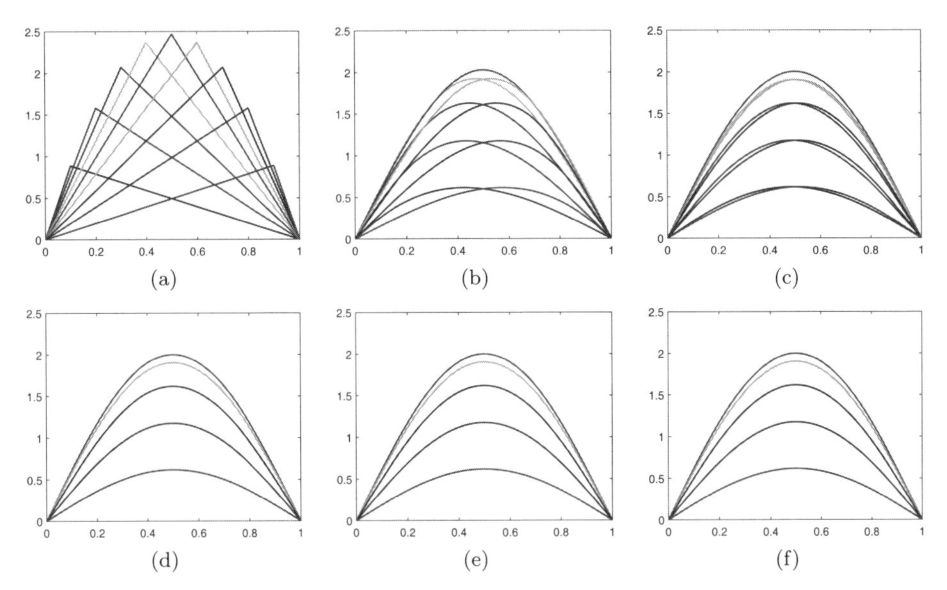

Fig. 7.1 Copies of the piecewise polynomial spline kernel $K_{\beta,0}$: (a) $\beta = 1$; (b) $\beta = 2$; (c) $\beta = 3$; (d) $\beta = 7$; (e) $\beta = 13$; (f) $\beta = 20$.

This will be useful should we later want to use kernels involving these polynomials to solve PDEs.

In Figure 7.1 we provide plots of $K_{\beta,0}(\cdot, z)$ for different values of β and $z \in \{\frac{j}{10}, j = 1, \ldots, 9\}$. Note that the kernels corresponding to $z = 0$ and $z = 1$ are identically equal to zero due to the boundary conditions we have imposed. Since the kernels get rather small as β increases we have normalized all kernel plots by multiplying the kernels by $1/\lambda_1 = \pi^{2\beta}$. Clearly, this is not a good basis for practical implementations since the set of translates of the kernel $K_{\beta,0}(\cdot, z)$ for different centers z becomes very nearly linearly dependent for larger β. In fact, the plots for $\beta \geq 7$ are nearly indistinguishable.

7.1.3 *Benefits of using a kernel representation for piecewise polynomial splines*

In spite of the poor properties of the closed form representation of $K_{\beta,0}$ just mentioned, there are several advantages of working with either the closed form representation (7.7) or its Mercer series representation (7.5). Usually, piecewise polynomial splines are represented in terms of *B-splines* (see, e.g., [de Boor (2001); Schumaker (1981)]) and that representation provides many well-known benefits such as numerical stability and efficiency. However, our representation of the kernel $K_{\beta,0}$ with a simple uniform parametrization in terms of β allows us to easily move between splines of different orders. Moreover, the Hilbert–Schmidt SVD algorithm introduced in Chapter 13 allows us to compute stably with these splines of all orders (although this may not be as efficient as doing it with the B-spline basis).

In the next section, we will generalize this family of piecewise polynomial kernels to iterated Brownian bridge kernels, and then the Mercer series representation will allow us to move effortlessly between an approximation with piecewise polynomial splines (corresponding to $\varepsilon = 0$) and one based on the more flexible iterated Brownian bridge kernels (with general ε). We will be able to accomplish all of this by simply changing the eigenvalues in the Mercer series.

7.2 Derivation of General Iterated Brownian Bridge Kernels

We now change the basic differential operator from $\mathcal{L} = -\frac{\mathrm{d}^2}{\mathrm{d}x^2}$ to the *modified Helmholtz operator* $\mathcal{L} = -\frac{\mathrm{d}^2}{\mathrm{d}x^2} + \varepsilon^2\mathcal{I}$, where the wave number ε will serve as a shape parameter for our kernels. If we consider the effects of this change in the differential operator on the eigenvalues and eigenfunctions of $\mathcal{L}$, then it is easily seen that the *eigenfunctions remain unchanged* and the *eigenvalues are shifted by* ε^2. If we then consider β-iterates of the modified Helmholtz operator, i.e.,

$$\mathcal{L} = \left(-\frac{\mathrm{d}^2}{\mathrm{d}x^2} + \varepsilon^2\mathcal{I}\right)^\beta, \qquad \beta \in \mathbb{N},\ \varepsilon > 0,$$

then the corresponding Sturm–Liouville eigenvalue problem

$$\mathcal{L}\varphi = \frac{1}{\lambda}\varphi$$

with boundary conditions

$$\varphi(0) = \varphi(1) = \varphi''(0) = \varphi''(1) = \ldots = \varphi^{(2\beta-2)}(0) = \varphi^{(2\beta-2)}(1) = 0 \tag{7.9}$$

results in the eigenpairs

$$\lambda_n = \frac{1}{(n^2\pi^2 + \varepsilon^2)^\beta}, \quad \varphi_n(x) = \sqrt{2}\sin(n\pi x), \qquad n = 1, 2, \ldots.$$

The Mercer series (2.9) of the *iterated Brownian bridge kernels* $K_{\beta,\varepsilon}$ is then a simple generalization of (7.5) for the piecewise polynomial splines of the previous section, i.e.,

$$\begin{aligned} K_{\beta,\varepsilon}(x,z) &= \sum_{n=1}^{\infty} \lambda_n \varphi_n(x)\varphi_n(z) \\ &= \sum_{n=1}^{\infty} 2\left(n^2\pi^2 + \varepsilon^2\right)^{-\beta} \sin(n\pi x)\sin(n\pi z). \end{aligned} \tag{7.10}$$

Closed form representations for the iterated Brownian bridge kernels with $\varepsilon > 0$ are more complicated than those for the piecewise polynomial spline case ($\varepsilon = 0$).

For $\beta = 1$ we can proceed in a similar fashion to what we did in Chapter 6 for the $\varepsilon = 0$ case (Brownian bridge kernel) and obtain

$$K_{1,\varepsilon}(x,z) = \frac{\sinh(\varepsilon \min(x,z)) \sinh(\varepsilon(1-\max(x,z)))}{\varepsilon \sinh(\varepsilon)} \tag{7.11}$$
$$= \begin{cases} \dfrac{\sinh(\varepsilon x) \sinh(\varepsilon(1-z))}{\varepsilon \sinh(\varepsilon)}, & 0 \le x \le z, \\ \dfrac{\sinh(\varepsilon z) \sinh(\varepsilon(1-x))}{\varepsilon \sinh(\varepsilon)}, & z \le x \le 1, \end{cases}$$

A closed form expression for $K_{2,\varepsilon}$ was found by two summer REU students [Bylund and Mayner (2012)]. It is quite a bit more complicated and turns out to be

$$\begin{aligned} K_{2,\varepsilon}(x,z) = \frac{1}{2\varepsilon^3 \sinh(\varepsilon)} \Big[& \varepsilon \min(x,z) \cosh(\varepsilon \min(x,z)) \sinh(\varepsilon(1-\max(x,z))) \\ & + \varepsilon(1-\max(x,z)) \sinh(\varepsilon \min(x,z)) \cosh(\varepsilon(\max(x,z)-1)) \\ & + (\varepsilon \coth(\varepsilon)+1) \sinh(\varepsilon \min(x,z)) \sinh(\varepsilon(\max(x,z)-1)) \Big], \end{aligned} \tag{7.12}$$

For $\beta > 2$ and $\varepsilon > 0$, closed forms of $K_{\beta,\varepsilon}$ should be derivable, but as of right now they are unknown. If the jump in complexity from $K_{1,\varepsilon}$ to $K_{2,\varepsilon}$ is indicative, writing the closed form for higher β values will quickly become impractical, even if they can be found using the same straightforward techniques.

Fortunately, as we now argue, computing the kernel values with a series rather than closed form is entirely appropriate. It might even be preferable to work with the series form of $K_{\beta,\varepsilon}$ rather than its closed form — especially for values of $\beta > 2$. In this sense, the availability of closed forms and series representations complement each other. Computing the value of an iterated Brownian bridge kernel with the series is in fact no different than evaluating the gamma function, $\Gamma(x)$, or the error function, $\mathrm{erf}(x)$, or even any common trigonometric or hyperbolic function. Even evaluating the Gaussian kernel can be done with a series expansion (see Chapter 13). However, in using an eigenfunction series directly (as in (13.9)) we ignore the special structure of this series which is not present in a standard Taylor series. Exploiting this structure allows for interpolation without directly computing the kernel values. This is accomplished by using the Hilbert–Schmidt SVD or RBF-QR technique [Fornberg and Piret (2008b); Fornberg *et al.* (2011); Fasshauer and McCourt (2012)] or Chapter 13.

Figure 7.2 should be compared with Figure 7.1 since we have used the same set of β values for both series of plots. Again we show translates of $K_{\beta,\varepsilon}(\cdot, z)$ for different values of β and $z \in \{\frac{j}{10}, j = 1, \ldots, 9\}$. As before, the kernels corresponding to $z = 0$ and $z = 1$ are identically equal to zero due to the boundary conditions. Again we normalize all kernel plots by multiplying by the first Sturm–Liouville eigenvalue, i.e., $1/\lambda_1 = \left(\pi^2 + \varepsilon^2\right)^\beta$.

Contrary to the piecewise polynomial case, where the basis of translates of $K_{\beta,0}$ becomes increasingly linearly dependent as β increases, we can counter-balance this effect by increasing the value of ε accordingly in this more general setting. In particular, Figure 7.2 shows that for large values of ε the iterated Brownian bridge kernels more and more resemble translated Gaussian kernels as $\beta \to \infty$. However, by construction, these Gaussian-like iterated Brownian bridge kernels obey zero boundary conditions. There is also potential theoretical justification for this resemblance of Gaussians since it is known that B-splines tend to Gaussians as the smoothness index β tends to infinity (see, e.g., [Unser *et al.* (1992); Allasia *et al.* (2013a,b)] and [Cavoretto (2010, Chapter 8)]).

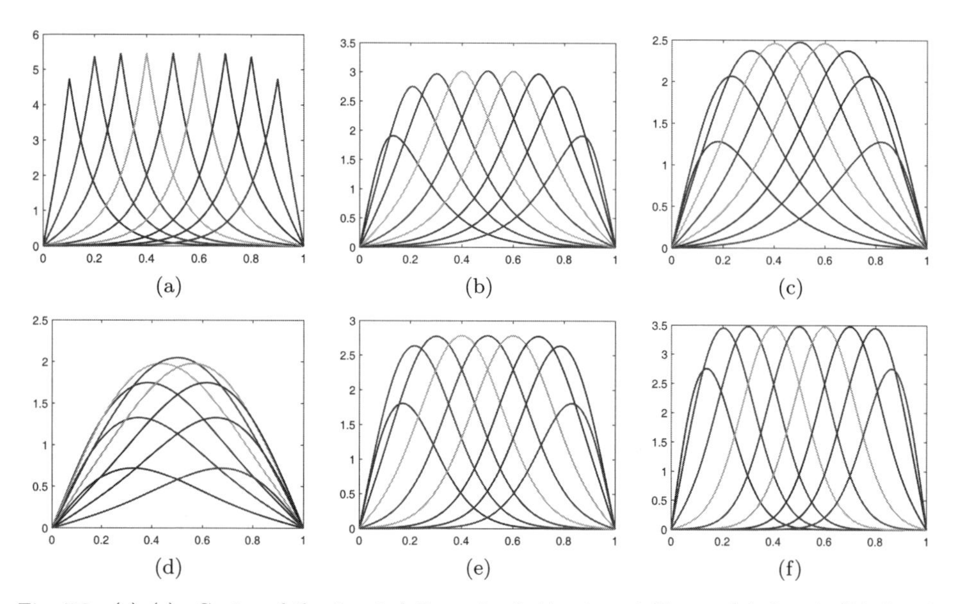

Fig. 7.2 (a)–(c): Copies of the iterated Brownian bridge kernel $K_{\beta,10}$. (a) $\beta = 1$; (b) $\beta = 2$; (c) $\beta = 3$.
(d)–(f): Copies of the iterated Brownian bridge kernel $K_{\beta,\varepsilon}$. (d) $\beta = 7$, $\varepsilon = 10$; (e) $\beta = 13$, $\varepsilon = 30$; (f) $\beta = 20$, $\varepsilon = 50$.

7.3 Properties of Iterated Brownian Bridge Kernels

We now perform and discuss a series of numerical experiments in order to aid our understanding of the behavior of the family of iterated Brownian bridge kernels. In particular, we are interested in the effects of the parameters ε and β, and on the boundary conditions. Some of the behavior observed in this section will be supported by theoretical analysis in later chapters.

7.3.1 *Truncation of the Mercer series*

The plots of the iterated Brownian bridge kernels in Figure 7.2 were obtained by evaluating their Mercer series representation (7.10) (see Program 7.1 below). Of course, in practice we cannot evaluate such an infinite series. It needs to be truncated, and we need to ensure that this truncation does not reduce accuracy. This concern is valid for any kind of infinite series in general, but the uniform convergence of the Mercer series allows us to guarantee a truncation value M exists, beyond which the remaining terms are less than some tolerance σ_{TOL}.

Because the eigenfunctions for the iterated Brownian bridge kernels are all bounded by $\sqrt{2}$ (note that this is by no means true for arbitrary positive definite kernels) we will be able to determine convergence of the Mercer series by studying the ratio of the Mth and N^{th} eigenvalues. For our examples, N will either be 1 or the number of points used for an interpolation problem. When written out explicitly, the ratio of two eigenvalues

$$\frac{\lambda_M}{\lambda_N} < \sigma_{\text{TOL}}, \qquad M > N,$$

can be solved for the necessary M yielding

$$M_{\text{TOL}} = \left\lceil \frac{1}{\pi}\sqrt{\sigma_{\text{TOL}}^{-1/\beta}(N^2\pi^2 + \varepsilon^2) - \varepsilon^2} \right\rceil, \tag{7.13}$$

where $\lceil x \rceil$ denotes the smallest integer greater than or equal to x.

Table 7.1 shows truncation values of M for various parametrizations in terms of ε and β, when $N = 1$. The precision column is the ratio of the last and first eigenvalues, λ_M/λ_1. Note that, due to rounding, the entries in the second row of the table appear identical to those in the first. In fact, we have, e.g., $M_{\text{TOL}} = 317$ for $\varepsilon = 0.1$ and $M_{\text{TOL}} = 322$ for $\varepsilon = 1$. Therefore, increases in ε require a greater M_{TOL}, whereas increases in β require a smaller M_{TOL}.

Table 7.1 Truncation values M_{TOL} required for series generated by various combinations of ε and β to reach certain precisions specified by σ_{TOL}. The precision column is the ratio of the last and first eigenvalues, λ_M/λ_1.

Parameters		Precision σ_{TOL}		
β	ε	10^{-5}	10^{-10}	10^{-15}
1	.1	3×10^2	1×10^5	3×10^7
1	1	3×10^2	1×10^5	3×10^7
1	10	1×10^3	3×10^5	1×10^8
2	.1	2×10^1	3×10^2	6×10^3
2	1	2×10^1	3×10^2	6×10^3
2	10	6×10^1	1×10^3	2×10^4
3	10	2×10^1	2×10^2	1×10^3
5	10	1×10^1	3×10^2	1×10^2
7	10	7×10^0	2×10^1	4×10^1

In Program 7.1 we provide a simple (but vectorized) MATLAB script that allows us to use the Mercer series (7.10) for iterated Brownian bridge kernels with arbitrary β and ε. Since Table 7.1 shows that the truncation criterion (7.13) may lead to very large values of M (and we should therefore switch to the closed-form representation when $\beta = 1, 2$), for simplicity we just fix $M = 1000$ in those two cases (see lines 9–13).

Program 7.1. IBBKernelEx1.m

```
% Computes and plots iterated Brownian bridge kernels via their Mercer series
x = linspace(0,1,11)'; xx = linspace(0,1,1201)';
%% iterated Brownian bridge kernel
x=x(2:end-1); N = length(x);
ep = 50; beta = 20;
phifunc = @(n,x) sqrt(2)*sin(pi*x*n);
lambdafunc = @(n) ((n*pi).^2+ep^2).^(-beta);
%% Mercer series
if beta < 3
    M = 1000;
else
    M = ceil(1/pi*sqrt(eps^(-1/beta)*(N^2*pi^2+ep^2)-ep^2));
end
Lambda = diag(lambdafunc(1:M));
Phi_interp = phifunc(1:M,x);
Phi_eval = phifunc(1:M,xx);
Kbasis = Phi_eval*Lambda*Phi_interp'/Lambda(1,1);
%% Plot kernel basis obtained via Mercer series
plot(xx,Kbasis)
```

7.3.2 *Effects of the boundary conditions*

To see what the effect of the boundary conditions built into our kernels may be, we plot the *cardinal functions*[1] for interpolation by iterated Brownian bridge kernels and by Gaussians (which are simply translated across the domain of interest). Plots (a) and (b) of Figure 7.3 show some of the cardinal functions for iterated Brownian bridge interpolation with $K_{20,50}$ at 22 equally spaced points in $(0, 1)$ on the left, and for 22 Chebyshev points on the right. Plots (c) and (d) provide the analogous plots of Gaussian cardinal functions, where the shape parameter of the Gaussian was chosen to match the shape of $K_{20,50}(\cdot, 1/2)$, i.e., $\varepsilon_{\text{Gauss}} = 5.75$.

The MATLAB script used to create these plots is listed below as Program 7.2. Gaussians cardinal functions are determined via the standard distance matrix method of Chapter 4 and are evaluated on line 29 via $\boldsymbol{k}(\boldsymbol{x})^T\mathsf{K}^{-1}$ as explained in Section 12.1.2. Iterated Brownian bridge cardinal functions are computed on line 13 using a function HSSVD_IBBSolve that solves for and evaluates the iterated Brownian bridge kernel interpolant; this function is created in Program 13.3.

Note that the cardinal functions associated with $K_{20,50}$ have to be evaluated using the Hilbert–Schmidt SVD approach since we do not know a closed form for

[1] We explain cardinal functions in detail in Section 12.1.2.

this kernel. Filling the matrix K via truncated Mercer series would be possible, but not advisable as explained in Chapter 13. On the other hand, the direct approach is fine for Gaussians due to the relatively large value of the shape parameter used for this example. If one wants to look at cardinal functions associated with "flat" Gaussians, i.e., with much smaller values of ε, then the Hilbert–Schmidt SVD of Chapter 13 can be employed.

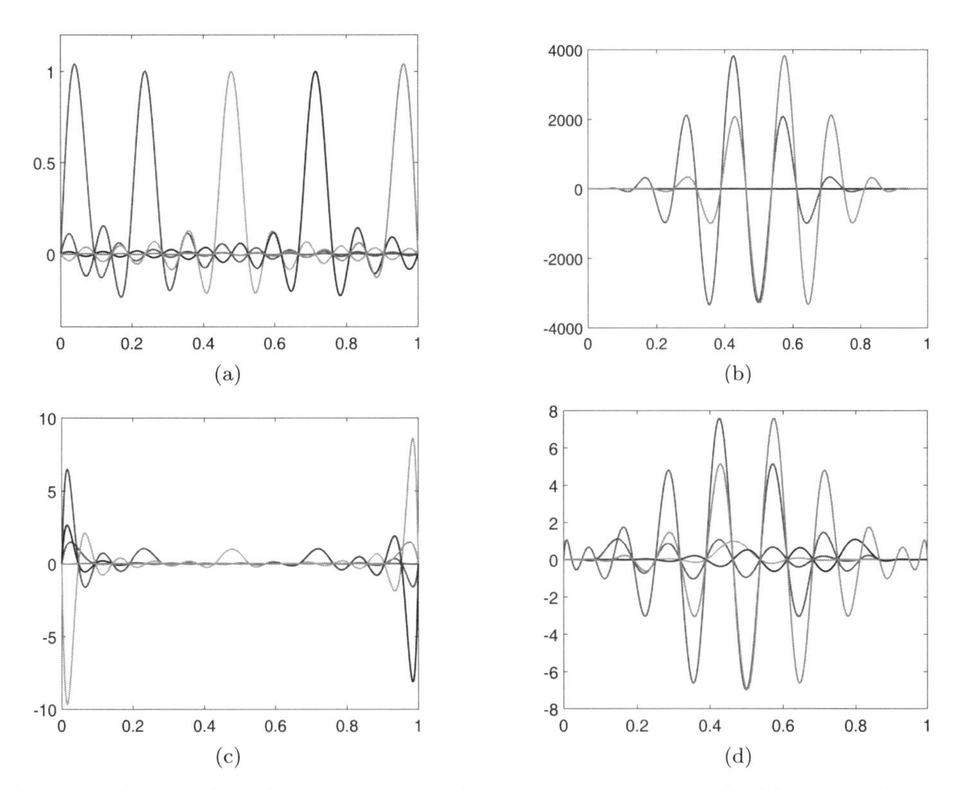

Fig. 7.3 Some cardinal functions for interpolation at 22 points in $(0, 1)$: (a) iterated Brownian bridge cardinal functions for equally spaced points; and (b) Chebyshev points; (c) Gaussian cardinal functions for equally spaced points; and (d) Chebyshev points.

It is interesting to observe that, due to the built-in boundary conditions, the iterated Brownian bridge cardinal functions indicate that uniformly distributed data is much preferred over Chebyshev data, even though one might expect something like a Runge phenomenon [Trefethen and Weideman (1991); Trefethen (2013)] to occur when interpolating on a compact interval. On the other hand, it is known that interpolation with cubic splines also favors evenly distributed data (see, e.g., [Lyche and Schumaker (1973); Marsden (1974)]), and that the optimal points for L_2 approximation with the Brownian bridge kernel are equally spaced (see, e.g., [Ritter (2000, Section II.3.7)]). In the Gaussian case, the preference is not as clear, but they will favor Chebyshev points as the shape parameter tends to zero, i.e., as one approaches polynomial interpolation (cf. the discussion of "flat" limits in Chapter 10). This justifies our use of equally spaced points for the experiments reported below.

Program 7.2. `IBBKernelCardFuncs.m`

```
% Choose kernel data size and evaluation points
N = 20;  NN = 300;  xeval = pickpoints(0,1,NN);
% Define IBB kernel
ep = 50;  beta = 20;  I = eye(N);
% Choose the cardinal functions to plot
cfuneval = [1 5 10 15 20];
% Loop through different designs
pts = {'even','cheb'};
for k=1:2
    % Create the desired points
    x = pickpoints(0,1,N+2,pts{k});  x = x(2:end-1);
    % Evaluate and plot the cardinal functions
    cardfuncs = HSSVD_IBBSolve(ep,beta,x,I,xeval);
    subplot(2,2,k),  plot(xeval,cardfuncs(:,cfuneval),'linewidth',3)
end
%% Define Gaussian kernel to resemble IBB kernel
rbf = @(e,r) exp(-(e*r).^2);
ep = 5.75;
% Gaussians nonzero at boundary so don't dump boundary values
cfuneval = cfuneval+1;
for k=1:2
    x = pickpoints(0,1,N+2,pts{k});
    % Standard interpolation objects
    DM = DistanceMatrix(x,x);
    DMeval = DistanceMatrix(xeval,x);
    K = rbf(ep,DM);
    Keval = rbf(ep,DMeval);
    % Evaluate cardinal functions instead of interpolating
    cardfuncs = Keval/K;
    subplot(2,2,k+2),  plot(xeval,cardfuncs(:,cfuneval),'linewidth',3)
end
```

Remark 7.3. The role of the identity matrix `I` $= \mathsf{I}_N$ in Program 7.2 is subtle. When the function **`HSSVD_IBBSolve(ep,beta,x,y,xeval)`** is called it accepts iterated Brownian bridge parameters **`ep`** and **`beta`**, kernel centers/data locations, `x`, function values at those points, `y`, and locations at which to evaluate the interpolant **`xeval`**. The function returns values computed with the formula $s(\boldsymbol{x}) = \boldsymbol{k}(\boldsymbol{x})^T\mathsf{K}^{-1}\boldsymbol{y}$, so when the identity matrix I_N is passed instead of $\boldsymbol{y}$, the function computes $\boldsymbol{k}(\boldsymbol{x})^T\mathsf{K}^{-1}\mathsf{I}_N = \boldsymbol{k}(\boldsymbol{x})^T\mathsf{K}^{-1}$ and returns the values of the cardinal functions at **`xeval`**. Further discussion on this function occurs in Program 13.3 and Appendix D.3.

7.3.3 *Convergence orders*

Since the differential operator $\mathcal{L} = \left(-\frac{\mathrm{d}^2}{\mathrm{d}x^2} + \varepsilon^2\mathcal{I}\right)^\beta$ from (7.1) defining our piecewise polynomial splines falls into the class of operators associated with the L-splines studied in [Schultz and Varga (1967); Swartz and Varga (1972)] — albeit coupled with different boundary conditions, it is natural to expect similar convergence behavior for interpolation using iterated Brownian bridge kernels.

For the following discussion we introduce the *meshsize* h, i.e.,

$$h = \max_{j=1,\ldots,N} (x_{j+1} - x_j), \quad x_j \in [0, 1].$$

It is known (see, e.g., [Schultz and Varga (1967)]) that an interpolant using L-splines based on the operator (7.1) provides an L_2 error of $\mathcal{O}(h^{2\beta})$ if the data comes from a function $f \in H^{2\beta}([0,1])$ and the boundary conditions are chosen such that the derivatives of f up to order β are also interpolated by the spline at the endpoints of the interval[2]. For the case of periodic boundary conditions the same order can be achieved. However, the L_2 order for natural splines[3] is in general only $\mathcal{O}(h^\beta)$.

In particular, this means that piecewise linear splines will achieve $\mathcal{O}(h^2)$ convergence order provided the spline interpolates at all data points (including at the boundary), and cubic splines will achieve $\mathcal{O}(h^4)$ provided the boundary conditions are matched, but otherwise only $\mathcal{O}(h^2)$.

Below we will use numerical experiments to show that both our piecewise polynomial splines and the more general iterated Brownian bridge kernels are able to achieve the same rates of convergence. However, we will be specifying our convergence in terms of N, which for evenly spaced data is related to h via $h = \frac{1}{N-1}$. Thus our convergence rates should be of the order $\mathcal{O}(N^{-2\beta})$ for data with homogeneous boundary conditions, and $\mathcal{O}(N^{-\beta})$ otherwise. In Section 9.4.3 we will provide theoretical proof of these rates for the kernels with $\varepsilon = 0$.

7.3.4 *Iterated Brownian bridge kernels on bounded domains*

A well-known issue that arises when global kernels (such as full-space Green's kernels) are used for interpolation problems on bounded domains is the fact that the quality of the approximation decreases significantly near the boundary (see, e.g., [Hubbert and Müller (2007); Bejancu and Hubbert (2012)], where the use of thin plate splines on an interval was studied). One strategy for dealing with this issue is to use data which has been sampled more densely near the boundary such as at

[2]Here $H^{2\beta}([0,1])$ is the usual Sobolev space of functions with 2β derivatives in $L_2([0,1])$.

[3]While our iterated Brownian bridge kernel $K_{2,0}$ does give rise to cubic natural splines that interpolate zero data on the boundary, higher-order iterated Brownian bridge kernels $K_{\beta,0}$, $\beta > 2$, which interpolate zero at the boundary and whose *even* derivatives of order $2, 4, \ldots, 2\beta - 2$ at the boundary are set to zero, do not coincide with higher-order interpolatory natural splines since natural boundary conditions correspond to setting *successive* derivatives of order $\beta, \beta+1, \ldots, 2\beta-2$ equal to zero so that the spline can be extended as a polynomial of order β (i.e., degree $\beta - 1$) beyond the endpoints.

the Chebyshev points. This latter approach was discussed in detail in [Hangelbroek (2008)] and [Rieger and Zwicknagl (2014)].

When working with an iterated Brownian bridge kernel, we can completely eliminate the boundary error if we have data that corresponds to samples of a function which satisfies the appropriate boundary conditions, i.e., all boundary conditions for data and kernel match so that the data come from our kernel's native space (see Figure 7.6 vs. Figure 7.7). In this case we should expect, as explained above, convergence orders of $\mathcal{O}(N^{-2\beta})$ on the entire interval $[0, 1]$ (see Figure 7.5) using equally spaced points (cf. Section 7.3.2).

Example 7.1. (Interpolation with iterated Brownian bridge kernels)
We consider data generated by the function

$$f_{\nu,\gamma}(x) = (1/2-\gamma)^{-2\nu}\,(x-\gamma)_+^{\nu}\,(1-\gamma-x)_+^{\nu}, \qquad x \in [0,1], \tag{7.14}$$

which is, except for a different vertical scale factor and the introduction of γ in place of the constant $\frac{1}{4}$, the function f_n from [Hubbert and Müller (2007)]. Plots of this function for a few typical values of ν are provided in Figure 7.4.

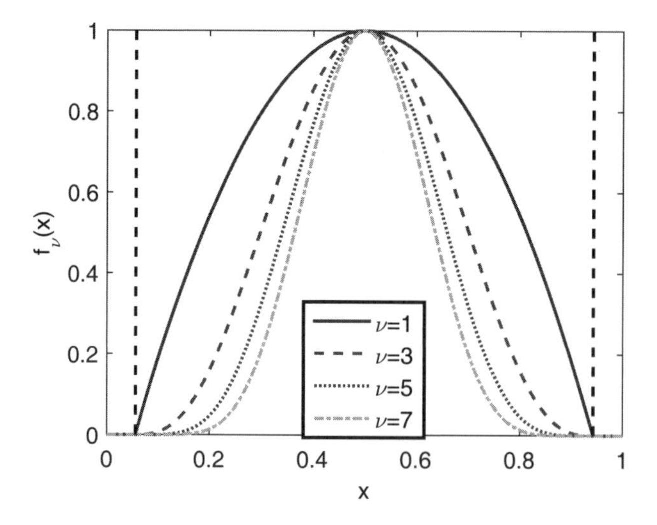

Fig. 7.4 Sample plots of the function $f_{\nu,\gamma}$ from (7.14) with $\nu = 1, 3, 5, 7$ and fixed $\gamma = 0.0567$. The vertical dashed lines are at $x = \gamma$ and $x = 1 - \gamma$, and indicate the point beyond which the function is zero.

This function was specifically designed to study the effect of error near the boundaries and used in the case $\nu = 4$ already in [Powell (1999)]. It has all its derivatives at $x = 0$ and $x = 1$ equal to 0, meaning that it definitely satisfies the boundary conditions (7.9) we imposed on our iterated Brownian bridge kernels. The parameter $\gamma \in (0, 1/2)$ causes the function to be identically zero outside of the interval $(\gamma, 1-\gamma)$, with discontinuous ν^{th} derivatives at $x = \gamma$ and $x = 1 - \gamma$. By manipulating ν we can consider functions that both satisfy or violate the necessary smoothness conditions of the underlying Green's kernel differential operator.

For this example we take $\nu = 6$ and $\gamma = 0.0567$, so that $f_{6,0.0567}$ satisfies all boundary conditions for our iterated Brownian bridge kernels $K_{\beta,\varepsilon}$ and internal smoothness conditions up to order $\beta = 3$. We also fix the shape parameter at $\varepsilon = 1$ and then interpolate data sampled from $f_{6,0.0567}$ at N evenly spaced points in $[0, 1]$, where N runs from 8 to 256.

Figure 7.5 shows that the interpolation error decreases for a fixed value of N as β increases; and the rate of improvement increases with β as well. However, once β exceeds the value for which the smoothness of the data and the kernel match, the error no longer improves.

Figure 7.6 shows how the boundary error for this example is essentially zero, and the improvement in accuracy for higher values of β proceeds unimpeded by boundary effects.

These results support our hypothesis from Section 7.3.3 that functions which satisfy the 2β homogeneous boundary conditions (7.9), and have at least 2β smooth derivatives, can be interpolated using order β iterated Brownian bridge kernels with an $O(N^{-2\beta})$ order of convergence.

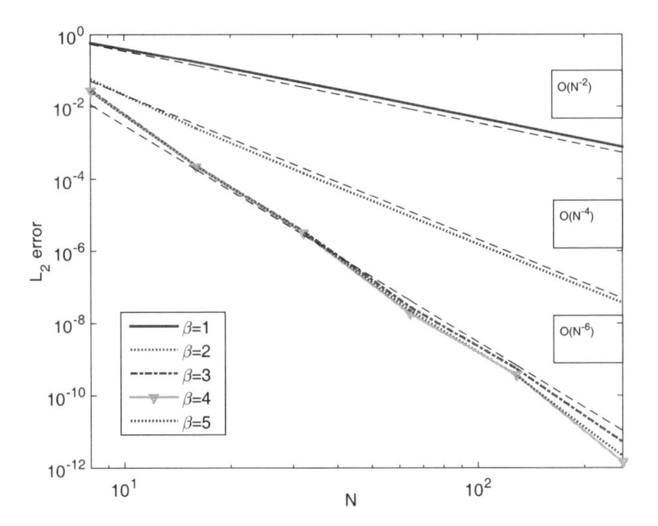

Fig. 7.5 Relative RMS errors for interpolation to the test function $f_{6,0.0567}$ with iterated Brownian bridge kernels of increasing smoothness on $[0, 1]$ based on N uniformly spaced points.

Because we have closed form expressions for $K_{1,\varepsilon}$ and $K_{2,\varepsilon}$ specified in (7.11) and (7.12), respectively, we evaluate these instead of the series. These kernels are coded in an external function `ibb.m` which is used on lines 19–24. For all other values of β the series expansion is much more efficient and we can employ the Hilbert Schmidt SVD methodology discussed in detail in Chapter 13. This happens on line 26, but is done within a function now so that we can enjoy the effect while postponing an explanation of how it takes place to Section 13.3. The plots for Figures 7.5 and 7.6 are generated (minus the labels and legends) on lines 36 and 38–43, respectively.

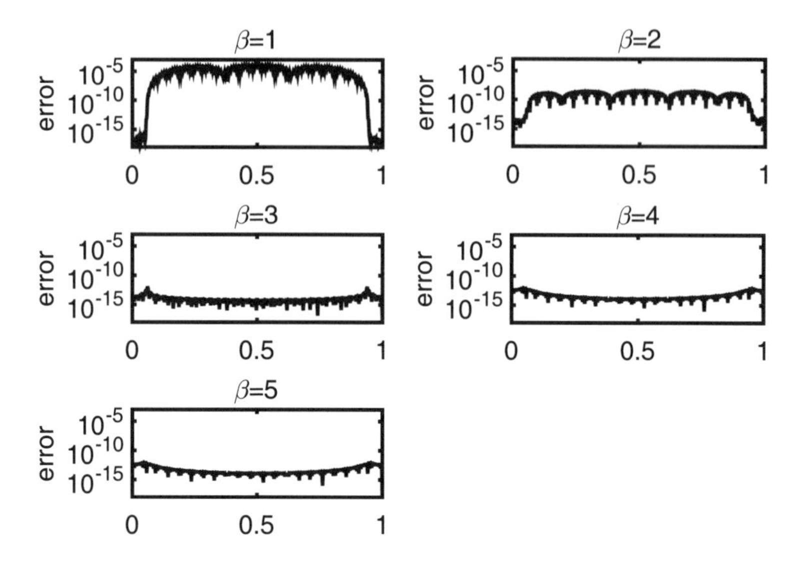

Fig. 7.6 Relative pointwise errors for nonstationary interpolation to the test function $f_{6,0.0567}$ with iterated Brownian bridge kernels of increasing smoothness on $[0,1]$ based on $N = 256$ uniformly spaced points.

Program 7.3. `IBBKernelConvergence.m`

```
% Pick the function to be studied
f = @(x) 5.0887^6.*(max(0,x-.0567)).^6.*(max(0,(1-.0567)-x)).^6;
% Define the study to be run: parameters and data
ep = 1;  betavec = 1:5;  Nvec = 2.^(3:8);
% Choose test data
Neval = 397;  xeval = linspace(0,1,Neval)';  yeval = f(xeval);
% Loop through N and beta values
errvec = zeros(length(betavec),length(Nvec));
errorForBetas = zeros(length(betavec),Neval);
k = 1; j = 1;
for N=Nvec
    % Create the data for the desired N value
    x = pickpoints(0,1,N+2);  x = x(2:end-1);  y = f(x);

    j = 1;
    for beta=betavec
        % Use the kernel closed form if stable enough
        if beta==1 || beta==2
            K = cell2mat(arrayfun(@(z) ...
                   ibb(x,z,ep,beta),x','UniformOutput',0));
            Keval = cell2mat(arrayfun(@(z) ...
                   ibb(xeval,z,ep,beta),x','UniformOutput',0));
            c = K\y;
            seval = Keval*c;
        else
```

```
26            seval = HSSVD_IBBSolve(ep,beta,x,y,xeval);
          end
28        errvec(j,k) = errcompute(seval,yeval);
          errorForBetas(beta,:) = yeval-seval;
30        j = j + 1;
      end
32    k = k + 1;
  end
34
  % Plot RMS error as beta and N vary
36 h_conv = figure;  loglog(Nvec,errvec,'linewidth',2);
  % Plot individual error
38 h_errs = figure;  title('Error as \beta varies');
  for beta = betavec;
40    errorFig = subplot(ceil(length(betavec)/2),2,beta);
      semilogy(xeval,abs(errorForBetas(beta,:)));
42    ylim([1e-18 1e-3]);  set(gca,'YTick',[1e-15 1e-10 1e-5])
  end
```

Example 7.2. (Iterated Brownian bridge kernels and boundary error)
If the data function satisfies only some boundary conditions, then using the smoothest possible iterated Brownian bridge kernel is not advisable, i.e., for a fixed shape parameter ε, there is likely to be an optimal smoothness order β.

For example, data generated by the exponential function

$$f(x) = \mathrm{e}^x + x(1 - \mathrm{e}) - 1$$

satisfies only the boundary conditions $f(0) = f(1) = 0$ required for all of our iterated Brownian bridge kernels.

We again fix $\varepsilon = 1$ and interpolate data sampled from f at N evenly spaced points in $[0, 1]$, where N runs from 10 to 100 for Figure 7.8(a), and $N = 100$ for Figure 7.7 and Figure 7.8(b).

In Figure 7.8(a) we can see that the order of convergence is *saturated* at $\mathcal{O}(h^2)$, i.e., kernels smoother than $K_{2,\varepsilon}$ do not provide a faster *rate* of convergence. However, as can be seen from both Figure 7.7 and from Figure 7.8(b), $K_{3,\varepsilon}$ is able to achieve a smaller *interior error* than $K_{2,\varepsilon}$. This happens because for smoother kernels (larger values of β), the boundary error propagates into the interior. As a result there is an *optimal value of* $\beta = 3$. Since the MATLAB code for this example is quite similar to Program 7.3 we omit it.

7.3.5 *"Flat" limits*

Earlier, we claimed that the general iterated Brownian bridge kernels associated with the differential operator from (7.1) can be used to produce piecewise polynomial splines with appropriate boundary conditions in the limiting case for $\varepsilon \to 0$. We now demonstrate this fact experimentally, and also show that there may exist

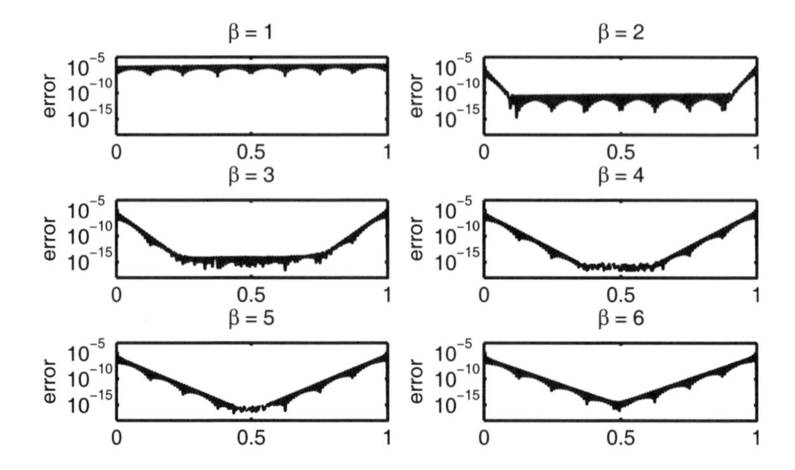

Fig. 7.7 Relative pointwise errors for nonstationary interpolation to the exponential test function with iterated Brownian bridge kernels of increasing smoothness on $[0, 1]$ based on $N = 100$ uniformly spaced points.

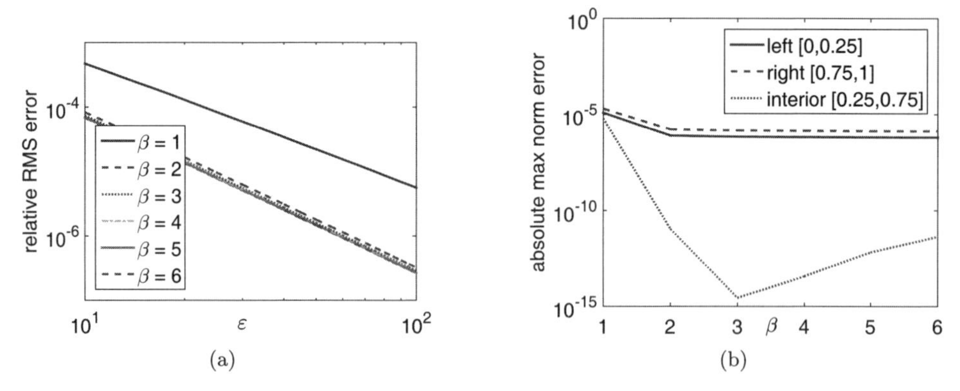

Fig. 7.8 Relative RMS errors (a) and optimal β to minimize interior error (b) for interpolation to the exponential test function with iterated Brownian bridge kernels of increasing smoothness on $[0, 1]$ based on N uniformly spaced points.

a positive value of ε for which the iterated Brownian bridge kernels are more accurate than the corresponding piecewise polynomial spline. A theoretical discussion of "flat" limits is presented in Chapter 10.

Example 7.3. (Iterated Brownian bridge kernels and their piecewise polynomial spline limits)
In Figure 7.9 we demonstrate the use of iterated Brownian bridge kernels $K_{1,\varepsilon}$ to approximate the test function $f_{1,\gamma}$ from (7.14) sampled at N evenly spaced points in $[0, 1]$. The figure displays absolute errors, computed at 400 evenly spaced points in $[0, 1]$, of the piecewise linear spline interpolants for $N = 12$, 24, and 48 (the

horizontal lines, computed with a piecewise linear interpolant using `interp1` in MATLAB), and corresponding iterated Brownian bridge interpolants for different values of ε ranging from $\varepsilon = 1$ to $\varepsilon = 100$.

The figure confirms that the iterated Brownian bridge interpolants are converging to piecewise linear spline interpolants as $\varepsilon \to 0$, and the table in the right part of Figure 7.9 displays the value ε_{opt} for which $K_{1,\varepsilon}$ produces an optimal error. The numbers in the error ratio column indicate the improvement over the piecewise polynomial spline interpolant.

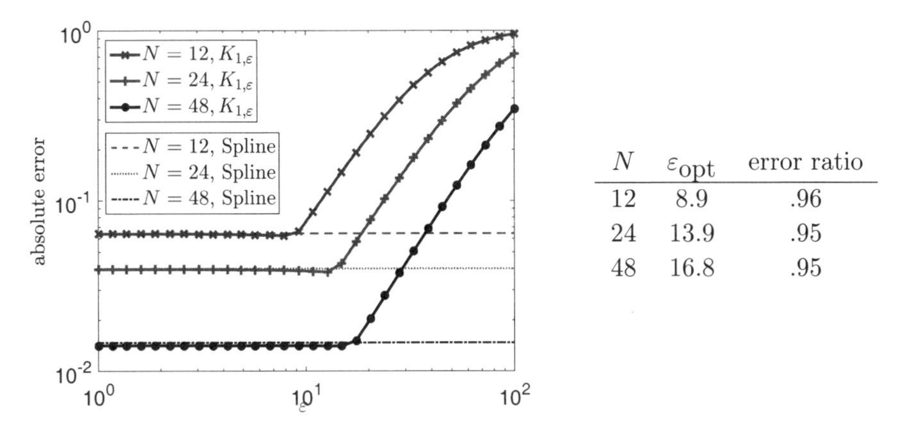

N	ε_{opt}	error ratio
12	8.9	.96
24	13.9	.95
48	16.8	.95

Fig. 7.9 Error results for iterated Brownian bridge interpolation with $K_{1,\varepsilon}$ on the function $f_{1,\gamma}$ from (7.14). Left: The effect of ε for various N values. Right: ε_{opt} value and ratio of the optimal error to the spline error.

For this example with $\beta = 1$, the optimal error for iterated Brownian bridge interpolation is only marginally better than the spline result. This is not always the case, as is demonstrated for a $\beta = 2$ interpolant in Figure 7.10.

Example 7.4. (More IBB kernels and splines)
We consider a function

$$\hat{f}_{2,\gamma}(x) = f_{2,\gamma}(x) \exp(-36(x - 0.4)^2), \tag{7.15}$$

where $f_{2,\gamma}$ was defined in (7.14), using the parameter $\gamma = 0.0567$. This function satisfies the necessary smoothness and convergence criteria, but is concentrated around $x = 0.4$ rather than $x = 0.5$. In Figure 7.10, the effect of ε is studied for $N = 12$, 24, and 48 evenly spaced points on $[0, 1]$ with errors computed at 400 evenly spaced points. We see clear benefits to using a kernel $K_{2,\varepsilon}$ with $\varepsilon > 0$ to perform the interpolation, but not for all N values.

For the smaller values of N an optimal ε, denoted ε_{opt}, produces a significantly smaller error than the cubic natural spline, which occurs in the $\varepsilon \to 0$ limit (and is computed here using `csape` from Curve Fitting Toolbox of MATLAB). In the table on the right of Figure 7.10, the ε_{opt} values and the ratios of the optimal error to the spline error are displayed. For $N = 48$, the optimal error seems to occur for $\varepsilon = 0$.

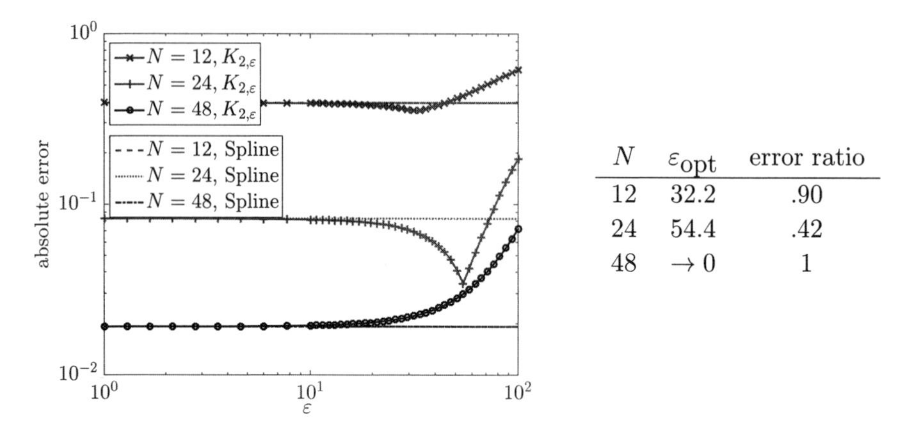

N	ε_{opt}	error ratio
12	32.2	.90
24	54.4	.42
48	$\to 0$	1

Fig. 7.10 Error results for iterated Brownian bridge interpolation with $K_{2,\varepsilon}$ on the function $\hat{f}_{2,\gamma}$ from (7.15). Left: The effect of ε for various N values. Right: ε_{opt} value and ratio of the optimal error to the spline error.

It seems that for large enough N, this function is best fit with the interpolating spline, but when the function is only partially known, the additional flexibility of $\varepsilon \neq 0$ provides a benefit.

7.3.6 *Summary for functions satisfying homogeneous boundary conditions*

The results of this section suggest two points:

- The order of the iterated Brownian bridge interpolation scheme is $O(N^{-2\beta})$ for sufficiently smooth and appropriately homogeneous functions.
- The $\varepsilon \to 0$ limit of the iterated Brownian bridge interpolant is an interpolating piecewise polynomial spline of degree $2\beta - 1$ with boundary conditions specified by (7.9).

Chapter 8

Generalized Sobolev Spaces

Our goal in this chapter is to develop "the right tool for the right job." We will show that — starting from a linear self-adjoint differential operator $\mathcal{L}$ with or without appropriate boundary conditions — we can specify an inner product and then produce the associated reproducing kernel which is optimal for our intended use. In doing so, we can rely on the norm minimization property of a kernel-based interpolant to produce, arguably, the optimal kernel interpolant by the metric given by our chosen inner product. This strategy is reminiscent of the variational approach to curve and surface design in computer-aided geometric design (see, e.g., [Sapidis (1987)]), where aesthetically looking curves or surfaces are obtained by specifying different kinds of norms to be minimized. These ideas go back to at least the *splines in tension* of Schweikert (1966).

8.1 How Native Spaces Were Viewed Until Recently

Even though piecewise polynomial splines and radial basis functions are conceptually very similar (some people do not even distinguish between the two and use the term *spline* to refer to either method), there are relatively few intersections in the literature on these two approximation methods. Perhaps the most prominent common feature of the two methods is given by the fact that they both yield *minimum norm interpolants* (see, e.g., [de Boor (2001); Schumaker (1981); Wendland (2005); Fasshauer (2007)]). In fact, it was precisely this property that led Schoenberg to refer to piecewise polynomial univariate interpolating functions as *splines* [Schoenberg (1946a,b)] since they resemble the shape of a designer's wooden spline bent under the weights of the ducks used the fix the spline. And later Atteia (1966) decided to use the same terminology for a multivariate kernel-based interpolation method most people nowadays would probably associate more readily with radial basis functions than with piecewise polynomial splines.

Let us begin with the specific example of *natural splines*. It has been known since the seminal paper of Holladay (1957)[1] that the natural spline $s_{2\beta}$ of order

[1] Holladay established the minimum norm property for cubic natural splines. The general form

2β provides the smoothest interpolant to data sampled from any function f in the Sobolev space $H^\beta([a,b])$ of functions whose β^{th} derivative is square integrable on $[a,b]$, i.e.,

$$s_{2\beta} = \operatorname*{argmin}_{f \in H^\beta([a,b])} \left\{ \int_a^b \left(f^{(\beta)}(x) \right)^2 \mathrm{d}x \mid f(x_i) = y_i,\ i = 1, \ldots, N \right\}. \tag{8.1}$$

Note that natural splines have derivatives of orders β through $2\beta - 2$ that vanish at the endpoints of the interval $[a,b]$, i.e., $f^{(\ell)}(a) = f^{(\ell)}(b) = 0,\ \ell = \beta, \ldots, 2\beta - 2$.

Now let us consider the corresponding minimum norm property as it is commonly found for radial basis functions, or more generally reproducing kernel interpolants (see, e.g., [Wendland (2005)]). The reproducing kernel interpolant s_K is optimal in the sense that it is the minimum norm interpolant to data sampled from any function f in $\mathcal{H}_K(\Omega)$, the reproducing kernel Hilbert space (or *native space*) associated with K, where the norm is that of $\mathcal{H}_K(\Omega)$. This can be stated as

$$s_K = \operatorname*{argmin}_{f \in \mathcal{H}_K(\Omega)} \left\{ \|f\|_{\mathcal{H}_K(\Omega)} \mid s_K(\boldsymbol{x}_i) = f(\boldsymbol{x}_i),\ i = 1, \ldots, N \right\}. \tag{8.2}$$

While the function space $H^\beta(a,b)$ that appears in (8.1) can be rather easily understood in terms of the smoothness conditions imposed, the native space $\mathcal{H}_K(\Omega)$ in (8.2) looks a bit more cryptic. What is this mysterious reproducing kernel Hilbert space and how is its norm defined?

For a general positive definite kernel K and domain $\Omega \subseteq \mathbb{R}^d$ the native space is commonly defined as the closure of

$$H_K(\Omega) = \operatorname{span}\left\{ K(\cdot, \boldsymbol{z}) \mid \boldsymbol{z} \in \Omega \right\}$$

with respect to the native space norm (see [Wendland (2005, Section 10.2)]), i.e., the native space is given by all linear combinations of — often infinitely many — "shifts" of the kernel K along with their limits.

This is certainly a valid definition, but what sort of functions does $\mathcal{H}_K(\Omega)$ contain? The literature is more specific for the case in which we use translation invariant kernels on $\Omega = \mathbb{R}^d$, i.e., if the kernel is really a function of only one variable, namely the difference of two points, or $K(\boldsymbol{x}, \boldsymbol{z}) = \widetilde{K}(\boldsymbol{x} - \boldsymbol{z})$ (cf. Section 3.2). In this case, if $\widetilde{K}$ continuous and absolutely integrable on $\mathbb{R}^d$, then

$$\mathcal{H}_{\widetilde{K}}(\mathbb{R}^d) = \left\{ f \in L_2(\mathbb{R}^d) \cap C(\mathbb{R}^d) \mid \frac{\mathcal{F}f}{\sqrt{\mathcal{F}\widetilde{K}}} \in L_2(\mathbb{R}^d) \right\},$$

i.e., a function f belongs to the native space $\mathcal{H}_{\widetilde{K}}(\mathbb{R}^d)$ of the kernel $\widetilde{K}$ if the decay of its Fourier transform $\mathcal{F}f$ relative to that of the Fourier transform $\mathcal{F}\widetilde{K}$ of the kernel is rapid enough (see [Wendland (2005, Theorem 10.12)]). This characterization certainly encodes some kind of smoothness information, but it is not very intuitive.

As mentioned above, we are not only interested in understanding the type of functions contained in the native space, but also the norm this space is equipped

for odd-degree natural splines stated here can be found, e.g., in the book [Ahlberg *et al.* (1967)].

with. Since both the spline and kernel spaces are Hilbert spaces it is natural to look at their inner products. In the natural spline case this is the standard Sobolev inner product whose induced (semi-)norm appears in (8.1). What does the native space inner product look like?

For a general positive definite kernel K on a general domain Ω we take functions $f, g \in \mathcal{H}_K(\Omega)$ and use the notation $N_K = \dim(\mathcal{H}_K(\Omega))$ for the dimension of the native space (note that $N_K = \infty$ is common). Then

$$\begin{aligned}\langle f, g\rangle_{\mathcal{H}_K(\Omega)} &= \langle \sum_{i=1}^{N_K} c_i K(\cdot, \boldsymbol{x}_i), \sum_{j=1}^{N_K} d_j K(\cdot, \boldsymbol{z}_j)\rangle_{\mathcal{H}_K(\Omega)} \\ &= \sum_{i=1}^{N_K}\sum_{j=1}^{N_K} c_i d_j K(\boldsymbol{x}_i, \boldsymbol{z}_j) = \boldsymbol{c}^T \mathsf{K}\boldsymbol{d},\end{aligned}$$

where $\boldsymbol{c} = \begin{pmatrix} c_1 & \cdots & c_{N_K}\end{pmatrix}^T$, $\boldsymbol{d} = \begin{pmatrix} d_1 & \cdots & d_{N_K}\end{pmatrix}^T$, and K has entries $(\mathsf{K})_{i,j} = K(\boldsymbol{x}_i, \boldsymbol{z}_j)$ (cf. (2.23)).

Once again, one might wonder how to interpret this. As before, for translation invariant kernels on $\Omega = \mathbb{R}^d$, i.e., $K(\boldsymbol{x}, \boldsymbol{z}) = \widetilde{K}(\boldsymbol{x} - \boldsymbol{z})$, we can employ Fourier transforms. Then we have

$$\langle f, g\rangle_{\mathcal{H}_{\widetilde{K}}(\mathbb{R}^d)} = \frac{1}{\sqrt{(2\pi)^d}}\langle \frac{\mathcal{F}f}{\sqrt{\mathcal{F}\widetilde{K}}}, \frac{\mathcal{F}g}{\sqrt{\mathcal{F}\widetilde{K}}}\rangle_{L_2(\mathbb{R}^d)}$$

provided $\widetilde{K} \in C(\mathbb{R}^d) \cap L_1(\mathbb{R}^d)$ and $f, g \in \mathcal{H}_{\widetilde{K}}(\mathbb{R}^d)$.

Before we bring the Green's kernels introduced in Chapter 6 into our discussion — and thereby provide an interpretation of native spaces as *generalized Sobolev spaces* — we remind the reader of a few examples of kernels whose native spaces are already known to be Sobolev spaces.

Example 8.1. (Kernels whose native spaces are Sobolev spaces)
In Section 3.1.1 on isotropic radial kernels we discussed the family of *Matérn kernels* and the fact that they are reproducing kernels for $\mathcal{H}_K(\mathbb{R}^d) = H^\beta(\mathbb{R}^d)$, the classical Sobolev spaces of functions whose β^{th} derivative is square integrable. Moreover, in Section 3.5 we discussed different families of compactly supported *Wendland kernels*. The reproducing kernel Hilbert spaces for the "original" families are norm equivalent to $H^{d/2+k+1/2}(\mathbb{R}^d)$, and those for the "missing families are equivalent to $H^{\lfloor d/2+k+1\rfloor}(\mathbb{R}^d)$.

A third example is provided by so-called *polyharmonic splines* (which we did not pay much attention to earlier since they are conditionally positive definite kernels). These are also radial kernels, i.e., $K(\boldsymbol{x}, \boldsymbol{z}) = \kappa(r)$ with $r = \|\boldsymbol{x} - \boldsymbol{z}\|$, and are of the form

$$\kappa_{d,\beta}(r) \doteq \begin{cases} r^{2\beta-d}, & d \text{ odd}, \\ r^{2\beta-d}\log r, & d \text{ even}. \end{cases}$$

The native space $\mathcal{H}_{\kappa_{d,\beta}}(\mathbb{R}^d)$ is a *Beppo–Levi space of order* β

$$\mathrm{BL}^\beta(\mathbb{R}^d) = \left\{f \in C(\mathbb{R}^d) \mid D^{\boldsymbol{\alpha}} f \in L_2(\mathbb{R}^d) \text{ for all } |\boldsymbol{\alpha}| = \beta\right\}.$$

We may consider this space as a homogeneous Sobolev space of order β (see, e.g., [Fasshauer (2007)]). Note that the natural cubic splines discussed above fit this framework for $d = 1$ and $\beta = 2$ (see also [Gutzmer and Melenk (2001)]).

Finally, the abstract kernels[2] of Atteia (1966) were conceived to be reproducing kernels of the classical Sobolev spaces $H^\beta(\Omega)$ with Ω a *bounded* domain in $\mathbb{R}^d$.

Our next example prepares us for the possibilities that arise when we equip one and the same space of functions with different norms. This will result in different reproducing kernel Hilbert spaces with different inner products which are all viewed as equivalent in the classical theory of Sobolev spaces.

Example 8.2. (Weighted Sobolev spaces)
Novak and Woźniakowski (2008, Appendix A.2.1) describe three different examples of *weighted Sobolev spaces*. In all three examples, the set of functions included in the space is the same. Namely, the authors always consider absolutely continuous real functions defined over $[0, 1]$ whose first derivatives belong to $L_2([0, 1])$, i.e., functions in the classical Sobolev space $H^1([0, 1])$. The difference between the three examples will lie in the use of *different inner products*, and therefore different norms. This will lead to the weighted Sobolev spaces $H^{1,\varepsilon}([0, 1])$ mentioned above. In other words, the three examples below feature spaces that are algebraically the same, but differ topologically since they are equipped with different norms.

The low-order smoothness was chosen since they were intended for applications of quasi-Monte Carlo methods in finance. Since all of these spaces are defined on the unit interval, high-dimensional kernels can be constructed as tensor products on the d-dimensional hypercube (see, e.g., Section 3.4.3). The first example below is due to Thomas-Agnan (1996), while the other two examples follow from [Hickernell (1998)].

The presence of the weights opens up the possibility of breaking the *curse of dimensionality* provided the (dimension-dependent) weights grow sufficiently in higher dimensions (see [Novak and Woźniakowski (2008)] for more details). Note that the reproducing kernels for these spaces are *not* radial.

(a) The norm for the first example is induced by the inner product

$$\langle f, g \rangle_{H^{1,\varepsilon}([0,1])} = \int_0^1 f'(x)g'(x)\,\mathrm{d}x + \varepsilon^2 \int_0^1 f(x)g(x)\,\mathrm{d}x,$$

i.e., it balances an integral over first derivatives (the L_2 inner product of derivatives) against an integral over function values (the standard L_2 inner product). The reproducing kernel for this example is given by (see also Appendix A.6)

$$K(x, z) = \frac{\cosh(\varepsilon \min(x, z)) \cosh(\varepsilon(1 - \max(x, z)))}{\varepsilon \sinh(\varepsilon)}.$$

[2] We refer to these kernels as *abstract kernels* since Atteia (1966) provided only a general framework, but specific kernels and their reproducing kernel Hilbert spaces depend on the domain and the boundary conditions imposed. We look at some of the required details in Section 8.3.

Remark 8.1. Note that this example is different from the generalized Brownian bridge kernel $K_{1,\varepsilon}$. While the inner products are the same for the two cases, the boundary conditions — which we see from the eigenfunctions (sines for the Brownian bridge kernel vs. cosines here, see Appendix A.6) — are different. For the Brownian bridge kernel one has the homogeneous space $H_0^{1,\varepsilon}([0,1])$, while here we are dealing with $H^{1,\varepsilon}([0,1])$.

(b) The inner product for this example is

$$\langle f, g\rangle_{H_{1,\varepsilon}([0,1])} = \int_0^1 f'(x)g'(x)\,\mathrm{d}x + \varepsilon^2 f(a)g(a),$$

where $a \in [0,1]$ is referred to as an *anchor*. The reproducing kernel for this space is given by

$$K(x,z) = 1 + \frac{\varepsilon^2}{2}\left(|x-a| + |z-a| - |x-z|\right),$$

which for $a = 0$ turns into $K(x,z) = 1 + \varepsilon^2 \min(x,z)$, and for $a = 1$ we get $K(x,z) = 1 + \varepsilon^2 \min(1-x, 1-z)$. Note that these kernels will always be piecewise linear, for any choice of ε and anchor a. However, as can be observed in Figure 8.1(b), the kernel generally has *two* points with discontinuous derivatives. Moreover, it was noted in [Novak and Woźniakowski (2001)] that, for values of $0 < a < 1$, this kernel may be less useful since multivariate integration based on a tensor product of this kernel was proven to be intractable.

(c) This time we use the inner product

$$\langle f, g\rangle_{H_{1,\varepsilon}([0,1])} = \int_0^1 f'(x)g'(x)\,\mathrm{d}x + \varepsilon^2 \int_0^1 f(x)\,\mathrm{d}x \int_0^1 g(x)\,\mathrm{d}x,$$

which uses the product of the averages of f and g over $[0,1]$ instead of their L_2 inner product as in (a). The reproducing kernel for this space is given by

$$K(x,z) = 1 + \frac{\varepsilon^2}{2}\left(B_2(|x-z|) + 2(x-\frac{1}{2})(z-\frac{1}{2})\right),$$

where B_2 is the Bernoulli polynomial of degree 2, i.e., $B_2(x) = x^2 - x + \frac{1}{6}$.

Shifted copies of reproducing kernels for the different weighted Sobolev spaces $H^{1,\varepsilon}([0,1])$ are displayed in Figure 8.1. A value of $\varepsilon = 10$ was used for the kernels from (a), while $\varepsilon = 1$ was used for the other two cases along with $a = \frac{1}{2}$ for the kernels from (b).

Similar to the weighted Sobolev spaces of Example 8.2, the generalized Sobolev spaces discussed in the following sections will all be equivalent to a common classical Sobolev space $H^\beta(\Omega)$, i.e., they all consist of the same sets of functions, but are all equipped with their own individual norms.

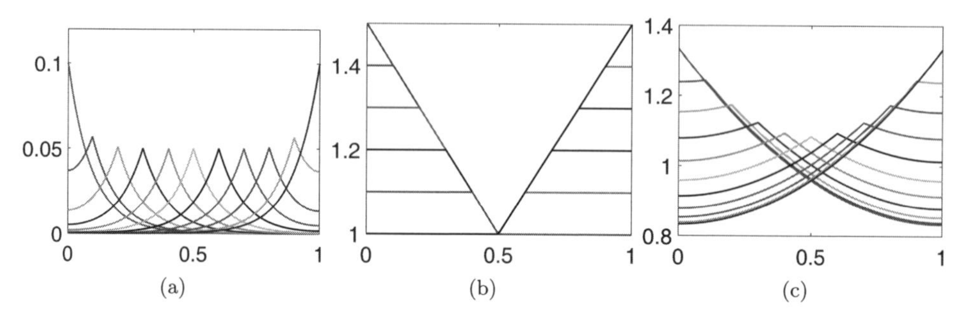

(a) (b) (c)

Fig. 8.1 Copies of weighted Sobolev kernels from Example 8.2(a)–(c), using $\varepsilon = 10$ (a), $\varepsilon = 1$ and $a = \frac{1}{2}$ (b), $\varepsilon = 1$ (c).

8.2 Generalized Sobolev Spaces on the Full Space $\mathbb{R}^d$

We use our work from Chapter 6 to interpret the reproducing kernel Hilbert space of a kernel K in terms of an associated linear self-adjoint differential operator $\mathcal{L}$. A rigorous theoretical framework supporting this interpretation for general *vector distributional operators* is provided in [Fasshauer and Ye (2011, 2013b)]. The paper [Fasshauer and Ye (2011)] contains the theory for generalized Sobolev spaces on the unbounded domain $\mathbb{R}^d$, which we discuss in this section. A theoretical framework for the more complicated case involving bounded domains is the subject of [Fasshauer and Ye (2013b)]. We will treat that setting in Section 8.3.

Our approach depends on the ability to identify a linear self-adjoint differential operator $\mathcal{L}$ that corresponds to a given Green's kernel K or vice versa. This means that we can either attempt to understand the generalized Sobolev space by starting with a known kernel and then identifying its differential operator (and subsequently the inner product and norm in the associated Hilbert space as described below), or by starting with a differential operator (which again defines an inner product and a norm in a generalized Sobolev space) and then getting the reproducing kernel as the corresponding full-space Green's kernel as discussed in Chapter 6. The latter approach is most likely the easier one, and therefore we start from a given/chosen differential operator.

Given a self-adjoint linear differential operator $\mathcal{L}$ of order 2β on $\mathbb{R}^d$, we decompose it into

$$\mathcal{L} = \mathcal{P}^*\mathcal{P},$$

with an appropriate differential operator $\mathcal{P}$ and its *formal adjoint* $\mathcal{P}^*$. It is crucial to note that this decomposition of $\mathcal{L}$ is *not unique*, especially once we also allow *vector differential operators* $\boldsymbol{\mathcal{P}} = \begin{pmatrix}\mathcal{P}_1 & \cdots & \mathcal{P}_{n_P}\end{pmatrix}^T$ with an appropriately defined adjoint $\boldsymbol{\mathcal{P}}^*$ (as proposed in [Fasshauer and Ye (2011)]), i.e.,

$$\mathcal{L} = \boldsymbol{\mathcal{P}}^{*T}\boldsymbol{\mathcal{P}} = \sum_{j=1}^{n_P} \mathcal{P}_j^*\mathcal{P}_j.$$

For example, the C^0 Matérn kernel $K(x,z) = \mathrm{e}^{-\varepsilon|x-z|}$ discussed in Chapter 3 is obtained as the full-space Green's kernel of $\mathcal{L} = -\mathcal{D}^2 + \varepsilon^2\mathcal{I}$, where $\mathcal{D} = \frac{\mathrm{d}}{\mathrm{d}x}$ denotes a univariate first-order derivative. which can be decomposed via $\mathcal{P} = \mathcal{D} + \varepsilon\mathcal{I}$ (with adjoint $\mathcal{P}^* = -\mathcal{D} + \varepsilon\mathcal{I}$), or via $\boldsymbol{\mathcal{P}} = \begin{pmatrix}\mathcal{D} & \varepsilon\mathcal{I}\end{pmatrix}^T$ (with $\boldsymbol{\mathcal{P}}^* = \begin{pmatrix}-\mathcal{D} & \varepsilon\mathcal{I}\end{pmatrix}^T$).

The Hilbert space inner products associated with $\mathcal{P}$ and $\boldsymbol{\mathcal{P}}$ are then, respectively, defined as

$$\langle f, g\rangle_{H_{\mathcal{P}}(\mathbb{R}^d)} = \int_{\mathbb{R}^d} \mathcal{P}f(\boldsymbol{x})\mathcal{P}g(\boldsymbol{x})\,\mathrm{d}\boldsymbol{x}, \tag{8.3a}$$

$$\langle f, g\rangle_{H_{\boldsymbol{\mathcal{P}}}(\mathbb{R}^d)} = \int_{\mathbb{R}^d} \left(\boldsymbol{\mathcal{P}}f(\boldsymbol{x})\right)^T \boldsymbol{\mathcal{P}}g(\boldsymbol{x})\,\mathrm{d}\boldsymbol{x} = \sum_{j=1}^{n_P}\int_{\mathbb{R}^d} \mathcal{P}_j f(\boldsymbol{x})\mathcal{P}_j g(\boldsymbol{x})\,\mathrm{d}\boldsymbol{x}. \tag{8.3b}$$

Since the reproducing kernel is obtained directly from $\mathcal{L}$, the choice of decomposition does not affect the kernel. However, it most likely will affect the inner product of the Hilbert space. For the C^0 Matérn kernel these two inner products are

$$\langle f, g\rangle_{H_{\mathcal{P}}(\mathbb{R})} = \int_{\mathbb{R}} \left(f'(x)g'(x) + \varepsilon(f(x)g'(x) + f'(x)g(x)) + \varepsilon^2 f(x)g(x)\right)\,\mathrm{d}x, \tag{8.4a}$$

$$\langle f, g\rangle_{H_{\boldsymbol{\mathcal{P}}}(\mathbb{R})} = \int_{\mathbb{R}} \left(f'(x)g'(x) + \varepsilon^2 f(x)g(x)\right)\,\mathrm{d}\boldsymbol{x}, \tag{8.4b}$$

which are clearly different since one involves terms mixing function values and derivatives, and the other does not. On the other hand, the norms induced by both of these inner products are equivalent to the norm of the classical Sobolev space $H^1(\mathbb{R})$. So, from a traditional point of view, the C^0 Matérn kernel is the reproducing kernel of the Sobolev space $H^1(\mathbb{R})$, while from our more refined point of view the kernel serves as the reproducing kernel for $H_{\mathcal{P}}(\mathbb{R})$, $\mathcal{P} = \mathcal{D} + \varepsilon\mathcal{I}$, as well as for $H_{\boldsymbol{\mathcal{P}}}(\mathbb{R})$, $\boldsymbol{\mathcal{P}} = \begin{pmatrix}\mathcal{D} & \varepsilon\mathcal{I}\end{pmatrix}^T$. Thus, the reproducing kernel Hilbert space associated with a given kernel is *not* unique if one allows for the freedom of choosing the inner product. Note that in [Wendland (2005, Section 10.2)] it was shown that the native space of a given positive definite kernel *is* unique. There, however, the native space was equipped with a very specific inner product derived from the kernel itself (cf. Section 2.3).

We can verify that the full-space Green's kernel K of $\mathcal{L}$, i.e., the function satisfying $\mathcal{L}K(\boldsymbol{x},\boldsymbol{z}) = \delta(\boldsymbol{x}-\boldsymbol{z})$, $\boldsymbol{z}$ fixed, is indeed a reproducing kernel with respect to the inner product from (8.3a):

$$\begin{aligned}
\langle f, K(\cdot,\boldsymbol{z})\rangle_{H_{\mathcal{P}}(\mathbb{R}^d)} &= \int_{\mathbb{R}^d} \mathcal{P}f(\boldsymbol{x})\mathcal{P}K(\boldsymbol{x},\boldsymbol{z})\,\mathrm{d}\boldsymbol{x} \\
&= \int_{\mathbb{R}^d} f(\boldsymbol{x})\mathcal{P}^*\mathcal{P}K(\boldsymbol{x},\boldsymbol{z})\,\mathrm{d}\boldsymbol{x} \\
&= \int_{\mathbb{R}^d} f(\boldsymbol{x})\mathcal{L}K(\boldsymbol{x},\boldsymbol{z})\,\mathrm{d}\boldsymbol{x} \\
&= \int_{\mathbb{R}^d} f(\boldsymbol{x})\delta(\boldsymbol{x}-\boldsymbol{z})\,\mathrm{d}\boldsymbol{x} = f(\boldsymbol{z}),
\end{aligned}$$

where we have used the definition of the Dirac delta functional and the fact that the adjoint is defined so that

$$\int_{\mathbb{R}^d} \left(g(\boldsymbol{x})\mathcal{P}f(\boldsymbol{x}) - f(\boldsymbol{x})\mathcal{P}^*g(\boldsymbol{x})\right) \mathrm{d}\boldsymbol{x} = 0,$$

i.e., the boundary term in integration by parts (or, more formally, the divergence theorem) vanishes due to appropriate decay conditions. The argument for the inner product from (8.3b) proceeds analogously.

For the C^0 Matérn kernel with the inner product $\langle \cdot, \cdot \rangle_{H_{\boldsymbol{\mathcal{P}}}(\mathbb{R})}$ from (8.4b) we can directly use integration by parts and properties of a Green's kernel to compute

$$\begin{aligned}
\langle f, K(\cdot, z) \rangle_{H_{\boldsymbol{\mathcal{P}}}(\mathbb{R})} &= \int_{\mathbb{R}} \left(\boldsymbol{\mathcal{P}} f(x)\right)^T \boldsymbol{\mathcal{P}} K(x, z) \, \mathrm{d}x \\
&= \int_{\mathbb{R}} \varepsilon^2 f(x) K(x, z) \, \mathrm{d}x \\
&\quad + \int_{-\infty}^{z^-} f'(x) \frac{\mathrm{d}}{\mathrm{d}x} K(x, z) \, \mathrm{d}x + \int_{z^+}^{\infty} f'(x) \frac{\mathrm{d}}{\mathrm{d}x} K(x, z) \, \mathrm{d}x \\
&= \int_{\mathbb{R}} \varepsilon^2 f(x) K(x, z) \, \mathrm{d}x \\
&\quad - \int_{-\infty}^{z^-} \left(f(x) \frac{\mathrm{d}^2}{\mathrm{d}x^2} K(x, z) \right) \mathrm{d}x + \left[f(x) \frac{\mathrm{d}}{\mathrm{d}x} K(x, z) \right]_{x=-\infty}^{z^-} \\
&\quad - \int_{z^+}^{\infty} \left(f(x) \frac{\mathrm{d}^2}{\mathrm{d}x^2} K(x, z) \right) \mathrm{d}x + \left[f(x) \frac{\mathrm{d}}{\mathrm{d}x} K(x, z) \right]_{x=z^+}^{\infty} \\
&= \int_{-\infty}^{z^-} f(x) \left(\varepsilon^2 K(x, z) - \frac{\mathrm{d}^2}{\mathrm{d}x^2} K(x, z) \right) \mathrm{d}x \\
&\quad + \int_{z^+}^{\infty} f(x) \left(\varepsilon^2 K(x, z) - \frac{\mathrm{d}^2}{\mathrm{d}x^2} K(x, z) \right) \mathrm{d}x \\
&\quad + f(z) \left(\frac{\mathrm{d}}{\mathrm{d}x} K(x, z) \Big|_{z^-} - \frac{\mathrm{d}}{\mathrm{d}x} K(x, z) \Big|_{z^+} \right) \\
&= f(z).
\end{aligned}$$

Note that we split the integral at the discontinuity of the derivative of K and that we used the fact that the derivative of the kernel decays to zero when $x \to \pm\infty$. The final simplifications are based on the differential equation $\mathcal{L}K(x, z) = 0$ for $x \neq z$ and the jump condition $\lim_{x \to z^-} \frac{\mathrm{d}}{\mathrm{d}x} K(x, z) = \lim_{x \to z^+} \frac{\mathrm{d}}{\mathrm{d}x} K(x, z) + 1$ (see Section 6.4.1).

Remark 8.2. The adjoint of $\boldsymbol{\mathcal{P}}$ (and implicitly also the scalar case $\mathcal{P}$) is defined rigorously in [Fasshauer and Ye (2011)] using a distributional framework. We leave those details to the interested reader.

Remark 8.3. The splines in tension mentioned at the beginning of this chapter were discussed in [Fasshauer and Ye (2011, Examples 2.2 and 5.2)].

8.2.1 *Two different kernels for* $H^2(\mathbb{R})$

Example 8.3. (Two kernels for $H^2(\mathbb{R})$)
We derive the generalized Sobolev spaces and reproducing kernels corresponding to two different fourth-order differential operators.

(a) We begin with the differential operator

$$\mathcal{L} = (-\mathcal{D}^2 + \varepsilon^2\mathcal{I})^2 = \mathcal{D}^4 - 2\varepsilon^2\mathcal{D}^2 + \varepsilon^4\mathcal{I}$$
$$= \left(\mathcal{D}^2 - 2\varepsilon\mathcal{D} + \varepsilon^2\mathcal{I}\right)\left(\mathcal{D}^2 + 2\varepsilon\mathcal{D} + \varepsilon^2\mathcal{I}\right) \tag{8.5a}$$
$$= \begin{pmatrix}\mathcal{D}^2 & -\sqrt{2}\varepsilon\mathcal{D} & \varepsilon^2\mathcal{I}\end{pmatrix}\begin{pmatrix}\mathcal{D}^2 \\ \sqrt{2}\varepsilon\mathcal{D} \\ \varepsilon^2\mathcal{I}\end{pmatrix}. \tag{8.5b}$$

We can view (8.5a) as a scalar decomposition $\mathcal{L} = \mathcal{P}^*\mathcal{P}$ with $\mathcal{P} = \mathcal{D}^2+2\varepsilon\mathcal{D}+\varepsilon^2\mathcal{I}$, and (8.5b) as a vector decomposition $\mathcal{L} = \boldsymbol{\mathcal{P}}^{*T}\boldsymbol{\mathcal{P}}$ with $\boldsymbol{\mathcal{P}} = \begin{pmatrix}\mathcal{D}^2 & \sqrt{2}\varepsilon\mathcal{D} & \varepsilon^2\mathcal{I}\end{pmatrix}^T$. The full-space Green's kernel of $\mathcal{L}$ on $\mathbb{R}$ (regardless of the decomposition chosen) is given by the C^2 Matérn kernel

$$K(x,z) = (1+\varepsilon|x-z|)\mathrm{e}^{-\varepsilon|x-z|}, \qquad \varepsilon > 0.$$

The usual reproducing kernel norm for the C^2 Matérn kernel is the one based on (8.5b) (see, e.g., [Fasshauer and Ye (2011)]), i.e.,

$$\|f\|_{H_{\boldsymbol{\mathcal{P}}}(\mathbb{R})} = \int_{\mathbb{R}} \left(f''(x)\right)^2 \,\mathrm{d}x + 2\varepsilon^2 \int_{\mathbb{R}} \left(f'(x)\right)^2 \,\mathrm{d}x + \varepsilon^4 \int_{\mathbb{R}} \left(f(x)\right)^2 \,\mathrm{d}x.$$

(b) Now we consider the differential operator

$$\mathcal{L} = \mathcal{D}^4 + 4\varepsilon^4\mathcal{I}$$
$$= \left(\mathcal{D}^2 - 2\varepsilon\mathcal{D} + 2\varepsilon^2\mathcal{I}\right)\left(\mathcal{D}^2 + 2\varepsilon\mathcal{D} + 2\varepsilon^2\mathcal{I}\right) \tag{8.6a}$$
$$= \begin{pmatrix}\mathcal{D}^2 & 2\varepsilon^2\mathcal{I}\end{pmatrix}\begin{pmatrix}\mathcal{D}^2 \\ 2\varepsilon^2\mathcal{I}\end{pmatrix}. \tag{8.6b}$$

Again, we have a scalar decomposition $\mathcal{L} = \mathcal{P}^*\mathcal{P}$ with $\mathcal{P} = \mathcal{D}^2 + 2\varepsilon\mathcal{D} + 2\varepsilon^2\mathcal{I}$ in (8.6a), and a vector decomposition $\mathcal{L} = \boldsymbol{\mathcal{P}}^{*T}\boldsymbol{\mathcal{P}}$ with $\boldsymbol{\mathcal{P}} = \begin{pmatrix}\mathcal{D}^2 & 2\varepsilon^2\mathcal{I}\end{pmatrix}^T$ in (8.6b).
The full-space Green's kernel of $\mathcal{L}$ on $\mathbb{R}$ (regardless of the decomposition chosen) is given by the $H^2(\mathbb{R})$ *Sobolev kernel* of Berlinet and Thomas-Agnan (2004, Chapter 7, Example 25)

$$K(x,z) = \frac{\varepsilon}{\sqrt{2}} \sin\left(\varepsilon|x-z| + \frac{\pi}{4}\right)\mathrm{e}^{-\varepsilon|x-z|}, \qquad \varepsilon > 0.$$

In [Berlinet and Thomas-Agnan (2004)] the norm is specified as

$$\|f\|_{H_{\boldsymbol{\mathcal{P}}}(\mathbb{R})} = \int_{\mathbb{R}} \left(f''(x)\right)^2 \,\mathrm{d}x + 4\varepsilon^4 \int_{\mathbb{R}} \left(f(x)\right)^2 \,\mathrm{d}x,$$

thus also coming from the vector decomposition (8.6b) of $\mathcal{L}$.

Clearly, we have come up with two different kernels and two different topologies for the classical Sobolev space $H^2(\mathbb{R})$. The two kernels are plotted in Figure 8.2(a) and two kernel-based interpolants to data sampled at 10 equally spaced points from the test function $f(x) = \cos(x) + \exp(-(x-1)^2) - \exp(-(x+1)^2)$ on $[-2, 2]$ are compared in Figure 8.2(b).

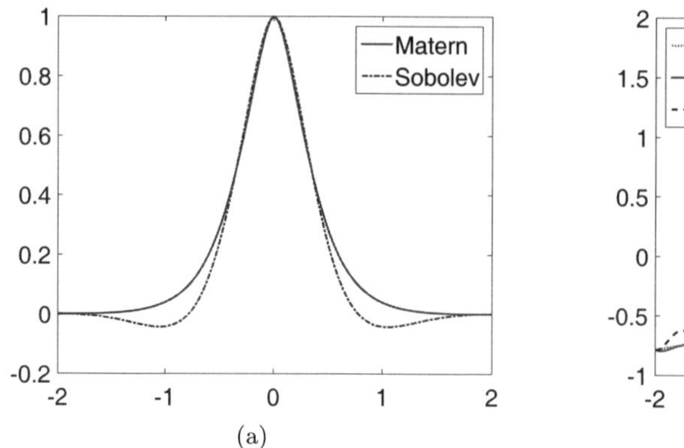

(a)

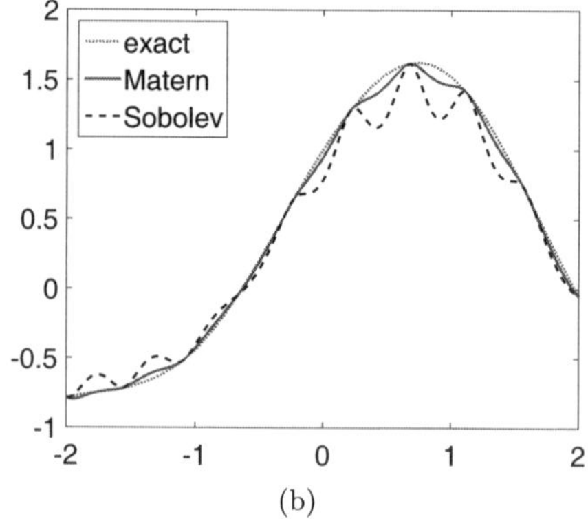

(b)

Fig. 8.2 (a) Plots of the C^2 Matérn kernel from Example 8.3(a) and H^2 Sobolev kernel from Example 8.3(b), both with $\varepsilon = 5$. (b) Interpolant for the two kernels at 10 equally spaced points sampled from $f(x) = \cos(x) + \exp(-(x-1)^2) - \exp(-(x+1)^2)$ on $[-2, 2]$.

Remark 8.4. Example 8.3 shows that it may make sense to *redefine classical Sobolev spaces* employing *different inner products in terms of scale parameters* even though the corresponding reproducing kernel Hilbert spaces are composed of functions having *identical smoothness properties* and are not distinguished under standard Hilbert space theory (i.e., considered isomorphic). The different inner products provide us with a clearer understanding of the important role of the scale parameter. We can think of ε^{-1} as the *natural length scale* dependent on the weight of various derivatives. The *choice of smoothness and scale now tell us which kernel to use for a particular application.* This choice may be performed by the user based on some *a priori* knowledge of the problem and based directly on the data.

8.2.2 *Higher-dimensional examples*

In this section we list two important examples — the full family of Matérn kernels on $\mathbb{R}^d$, and the Gaussian kernel — which are discussed in [Fasshauer and Ye (2011)]. We refer the interested reader to that paper for more details.

Example 8.4. (Generalized Sobolev spaces for multivariate Matérn kernels)
As already discussed in Chapter 3, the general multivariate Matérn kernels are of

the form

$$K(\boldsymbol{x}, \boldsymbol{z}) = \frac{K_{d/2-\beta}(\varepsilon\|\boldsymbol{x} - \boldsymbol{z}\|)}{(\varepsilon\|\boldsymbol{x} - \boldsymbol{z}\|)^{d/2-\beta}}, \qquad \beta > \frac{d}{2},$$

where $K_{d/2-\beta}$ is a modified Bessel function of the second kind. These Matérn kernels can be obtained as Green's kernels of

$$\mathcal{L} = \left(-\nabla^2 + \varepsilon^2\mathcal{I}\right)^\beta, \qquad \beta > \frac{d}{2},$$

where ∇ is the *gradient operator* and $\nabla^2 = \Delta$ is the *Laplacian operator*. A vector operator $\boldsymbol{\mathcal{P}}$ for the decomposition of $\mathcal{L}$ is given by $\boldsymbol{\mathcal{P}} = \begin{pmatrix}\boldsymbol{\mathcal{Q}}_0 & \boldsymbol{\mathcal{Q}}_1 & \cdots & \boldsymbol{\mathcal{Q}}_\beta\end{pmatrix}^T$, where

$$\boldsymbol{\mathcal{Q}}_j = \begin{cases} \sqrt{\binom{\beta}{j}}\varepsilon^{(\beta-j)}\Delta^k, & \text{when } j = 2k, \\ \sqrt{\binom{\beta}{j}}\varepsilon^{(\beta-j)}\Delta^k\nabla, & \text{when } j = 2k+1. \end{cases}$$

Here k is a nonnegative integer and $j = 0, 1, \ldots, \beta$ such that $\beta > d/2$. For example, when $\beta = 1$ we get

$$\boldsymbol{\mathcal{P}} = \begin{pmatrix}\varepsilon\mathcal{I} & \nabla\end{pmatrix}^T,$$

and for $\beta = 2$ we have

$$\boldsymbol{\mathcal{P}} = \begin{pmatrix}\varepsilon^2\mathcal{I} & \sqrt{2}\varepsilon\nabla & \Delta\end{pmatrix}^T.$$

Note, that the operators $\boldsymbol{\mathcal{Q}}_j$ are actually scalar-valued when $j = 2k$, and only in the odd case, $j = 2k+1$, do they become vector-valued operators.

The real power of the framework of [Fasshauer and Ye (2011)] is that the operator $\boldsymbol{\mathcal{P}}$ need not be a differential operator. It can be a more general *distributional operator*. Without going into any details, this allows us to describe the generalized Sobolev space of the Gaussian kernels.

Example 8.5. (Generalized Sobolev space for the Gaussian kernel)
We consider the Gaussian kernel $K(\boldsymbol{x}, \boldsymbol{z}) = \mathrm{e}^{-\varepsilon^2\|\boldsymbol{x}-\boldsymbol{z}\|^2}$. For this example, the appropriate distributional operator $\boldsymbol{\mathcal{P}}$ turns out to be an *infinite-dimensional pseudo-differential operator*, namely, $\boldsymbol{\mathcal{P}} = \begin{pmatrix}\boldsymbol{\mathcal{Q}}_0 & \boldsymbol{\mathcal{Q}}_1 & \cdots & \boldsymbol{\mathcal{Q}}_j & \cdots\end{pmatrix}^T$, where

$$\boldsymbol{\mathcal{Q}}_j = \begin{cases} \frac{1}{\sqrt{j!(2\varepsilon)^j}}\Delta^k, & \text{when } j = 2k, \\ \frac{1}{\sqrt{j!(2\varepsilon)^j}}\Delta^k\nabla, & \text{when } j = 2k+1, \end{cases}$$

with k a nonnegative integer, Δ the Laplacian, and ∇ the gradient operator as in the previous example (see also [Yuille and Grzywacz (1989)]).

8.2.3 *Summary for full-space generalized Sobolev spaces*

We now summarize the concept of generalized Sobolev spaces on $\mathbb{R}^d$ using the notation introduced above. The following theorem is based mostly on [Fasshauer and Ye (2011, Theorem 4.2)].

Theorem 8.1. *Given a linear self-adjoint differential operator $\mathcal{L} = \boldsymbol{\mathcal{P}}^{*T}\boldsymbol{\mathcal{P}}$ of order 2β on $\mathbb{R}^d$, the* full-space generalized Sobolev space *$H_{\boldsymbol{\mathcal{P}}}(\mathbb{R}^d)$ is embedded in the classical Sobolev space $H^\beta(\mathbb{R}^d)$ and is given by*

$$H_{\boldsymbol{\mathcal{P}}}(\mathbb{R}^d) = \{f \in H^\beta(\mathbb{R}^d) \mid \|f\|_{H_{\boldsymbol{\mathcal{P}}}(\mathbb{R}^d)} < \infty\},$$

equipped with the inner product

$$\langle f, g\rangle_{H_{\boldsymbol{\mathcal{P}}}(\mathbb{R}^d)} = \int_{\mathbb{R}^d} (\boldsymbol{\mathcal{P}}f(\boldsymbol{x}))^T \boldsymbol{\mathcal{P}}g(\boldsymbol{x})\, \mathrm{d}\boldsymbol{x} = \sum_{j=1}^{n_P} \int_{\mathbb{R}^d} \mathcal{P}_j f(\boldsymbol{x})\mathcal{P}_j g(\boldsymbol{x})\, \mathrm{d}\boldsymbol{x}$$

from (8.3b).

8.3 Generalized Sobolev Spaces on Bounded Domains

On a bounded domain $\Omega \subset \mathbb{R}^d$, our goal is to find the reproducing kernel K of a generalized Sobolev space $H_{\mathcal{P},\boldsymbol{\mathcal{B}}}(\Omega)$ associated with the nonhomogeneous partial (or possibly only ordinary) differential equation problem

$$\mathcal{L}u = f, \quad \text{in } \Omega, \tag{8.7a}$$

$$\boldsymbol{\mathcal{B}}u = \boldsymbol{g}, \quad \text{on } \partial\Omega. \tag{8.7b}$$

Here $\mathcal{L} = \boldsymbol{\mathcal{P}}^{*T}\boldsymbol{\mathcal{P}}$ is a self-adjoint linear vector differential operator of order 2β with vector differential operator $\boldsymbol{\mathcal{P}}$ as in Section 8.2, but now on a *bounded domain* $\Omega \subset \mathbb{R}^d$, along with a vector of boundary condition operators $\boldsymbol{\mathcal{B}} = \begin{pmatrix}\mathcal{B}_1 & \cdots & \mathcal{B}_{n_b}\end{pmatrix}^T$, where each $\mathcal{B}_j$ is of order β (or lower) chosen so that the differential equation problem is well-posed.

Our kernels now will have two parts

$$K(\boldsymbol{x}, \boldsymbol{z}) = G(\boldsymbol{x}, \boldsymbol{z}) + R(\boldsymbol{x}, \boldsymbol{z}), \tag{8.8}$$

where for fixed $\boldsymbol{z} \in \Omega$

$$\mathcal{L}G(\boldsymbol{x}, \boldsymbol{z}) = \delta(\boldsymbol{x} - \boldsymbol{z}), \quad \boldsymbol{x} \in \Omega, \tag{8.9a}$$

$$\boldsymbol{\mathcal{B}}G(\boldsymbol{x}, \boldsymbol{z}) = \mathbf{0}, \quad \boldsymbol{x} \in \partial\Omega, \tag{8.9b}$$

and

$$\mathcal{L}R(\boldsymbol{x}, \boldsymbol{z}) = 0, \qquad \boldsymbol{z} \in \Omega.$$

In contrast to the full-space Green's function (fundamental solution) of Section 8.2, the function G now is the Green's kernel of the differential operator $\mathcal{L}$ which satisfies the homogeneous boundary conditions specified in (8.9b).

The presence of the boundary correction term $R(\boldsymbol{x}, \boldsymbol{z})$ enables us to interpolate nonhomogeneous boundary data. For example, by finding an appropriate correction term for the Brownian bridge kernel $K(x, z) = \min(x, z) - xz$ we will be able to accurately interpolate arbitrary data on $[0, 1]$, not just samples of a function that is zero at $x = 0$ and $x = 1$.

To form this correction kernel, we define

$$R(\boldsymbol{x}, \boldsymbol{z}) = \sum_{i=1}^{n_a} a_i \psi_i(\boldsymbol{x}) \psi_i(\boldsymbol{z}), \qquad a_i \geq 0, \tag{8.10}$$

for n_a linearly independent functions $\psi_i \in \text{null}(\mathcal{L})$. Note that this may be an infinite-dimensional space. Since we can think of R as a designer series kernel of Mercer type as we discussed in Section 3.9, we will need to ensure that the "eigenfunctions" $\{\psi_i\}_{i=1}^{n_a}$ are orthonormal. However, we need to enforce this Fourier-like structure on the boundary $\partial\Omega$. Thus, we require that $\langle \psi_i, \psi_j \rangle_{H_{\mathcal{B}}(\partial\Omega)} = \delta_{i,j}$, where the $H_{\mathcal{B}}$-inner product is given by

$$\langle f, g \rangle_{H_{\mathcal{B}}(\partial\Omega)} = \sum_{i=1}^{n_b} (\mathcal{B}_i f, \mathcal{B}_i g)_{\partial\Omega}. \tag{8.11}$$

and the boundary inner product $(f, g)_{\partial\Omega}$ must be defined by the user. The choice of this inner product will have an impact on the generalized Sobolev space associated with K.

The series representation (8.10) of R also contains the nonnegative "eigenvalues" a_i. These values are design parameters that can be chosen by the user. Different values will yield different kernels, but — most importantly — they will impact the native space in which that kernel lives and the associated inner product, a key component of generalized Sobolev spaces. For a kernel of this composite form, its reproducing inner product is

$$\langle f, g \rangle_{H_{\mathcal{P},\mathcal{B}}(\Omega)} = \langle f, g \rangle_{H_{\mathcal{P}}(\Omega)} + \sum_{i=1}^{n_a} \hat{f}_i \hat{g}_i \frac{I_{a_i}}{a_i} - \sum_{i=1}^{n_a} \sum_{j=1}^{n_a} \hat{f}_i \hat{g}_j \langle \psi_i, \psi_j \rangle_{H_{\mathcal{P}}(\Omega)} I_{a_i a_j}, \tag{8.12}$$

where $I_x = 1$ if x is positive and 0 otherwise, and we employ the convention $0/0 \equiv 0$. The role of the indicator I_x is to drop contributions of any function ψ_i whose corresponding coefficient $a_i = 0$ from the inner product. The $\hat{f}_i$ and $\hat{g}_j$ are generalized Fourier coefficients with respect to the $H_{\mathcal{B}}$-inner product, i.e., $\hat{f}_i = \langle f, \psi_i \rangle_{H_{\mathcal{B}}(\partial\Omega)}$. A proof that K is the reproducing kernel with respect to the inner product (8.12) can be found in [Fasshauer and Ye (2013b, Theorem 3.2)].

Remark 8.5. If the functions ψ_i are restricted to be in $\text{null}(\mathcal{P})$ (instead of $\text{null}(\mathcal{L})$), then the $H_{\mathcal{P},\mathcal{B}}$-inner product simplifies and the "mixed" term that is subtracted in (8.12) is not needed (see [Fasshauer and Ye (2013b, Theorem 3.2)]). We discuss an alternative framework that makes this assumption in Section 8.3.3.

8.3.1 *Modifications of the Brownian bridge kernel: A detailed investigation*

We now give some examples in the case $d = 1$, i.e., for simple ordinary differential boundary value problems. These examples will illustrate the great flexibility in designing our own reproducing kernel Hilbert spaces permitted by the generalized Sobolev framework, but they will also demonstrate the complexity of this problem.

Example 8.6. (Generalized Sobolev space for modified Brownian bridge kernels I) The simplest example involves the domain $[0,1]$ and differential operator $\mathcal{L} = -\mathcal{D}^2$ so that $\mathcal{P} = \mathcal{D}$ and $\mathcal{B} = \mathcal{I}$ are vector differential and boundary operators, respectively, of length one.

We know from Section 6.4.1 that the Green's kernel of $\mathcal{L}$ with homogeneous boundary conditions is given by the Brownian bridge kernel $G(x,z) = \min(x,z) - xz$, and the nullspace of $\mathcal{L}$ is the space of linear polynomials so that we need only two linearly independent functions ψ_i to span it. We take these functions as

$$\psi_1(x) = x \quad \text{and} \quad \psi_2(x) = 1 - x.$$

As mentioned above, it is up to the user to pick the boundary inner product $(\cdot,\cdot)_{\partial\Omega}$ that is needed in (8.11), and we will define it as

$$(f,g)_{\partial\Omega} = f(0)g(0) + f(1)g(1). \tag{8.13}$$

Since the boundary operator $\mathcal{B}$ is the identity we get $\langle f,g\rangle_{H_{\mathcal{B}}(\partial\Omega)} = (f,g)_{\partial\Omega}$ from (8.11).

It is easy to check that $\{\psi_1,\psi_2\}$ are orthonormal with respect to the $H_{\mathcal{B}}$-inner product, i.e.,

$$\begin{aligned}
\langle\psi_1,\psi_1\rangle_{H_{\mathcal{B}}(\partial\Omega)} &= (\psi_1(0))^2 + (\psi_1(1))^2 = 0 + 1 = 1,\\
\langle\psi_1,\psi_2\rangle_{H_{\mathcal{B}}(\partial\Omega)} &= \psi_1(0)\psi_2(0) + \psi_1(1)\psi_2(1) = 0\cdot 1 + 1\cdot 0 = 0,\\
\langle\psi_2,\psi_2\rangle_{H_{\mathcal{B}}(\partial\Omega)} &= (\psi_2(0))^2 + (\psi_2(1))^2 = 1 + 0 = 1.
\end{aligned}$$

As in Section 8.2, the $H_{\mathcal{P}}$-inner product is

$$\langle f,g\rangle_{H_{\mathcal{P}}(\Omega)} = \int_0^1 \mathcal{P}f(x)\mathcal{P}g(x)\,\mathrm{d}x = \int_0^1 f'(x)g'(x)\,\mathrm{d}x. \tag{8.14}$$

According to (8.8) and (8.10), the reproducing kernel is given by $K(x,z) = G(x,z) + R(x,z)$ so that, using our choice of ψ_1 and ψ_2,

$$\begin{aligned}
K(x,z) &= \min(x,z) - xz + a_1 xz + a_2(1-x)(1-z)\\
&= \min(x,z) + (a_1 + a_2 - 1)xz + a_2(1 - x - z).
\end{aligned} \tag{8.15}$$

As mentioned above, the coefficients a_1 and a_2 can be chosen by the user, and we will look at different possibilities below.

The inner product for the generalized Sobolev space is specified in (8.12) and takes the form

$$\begin{aligned}
\langle f,g\rangle_{H_{\mathcal{P},\mathcal{B}}(\Omega)} = &\int_0^1 f'(x)g'(x)\,\mathrm{d}x + \hat{f}_1\hat{g}_1\frac{I_{a_1}}{a_1} + \hat{f}_2\hat{g}_2\frac{I_{a_2}}{a_2}\\
&- \hat{f}_1\hat{g}_1 I_{a_1} - \hat{f}_2\hat{g}_2 I_{a_2} + (\hat{f}_1\hat{g}_2 + \hat{f}_2\hat{g}_1) I_{a_1 a_2},
\end{aligned}$$

where we have used $\langle \psi_1, \psi_1 \rangle_{H_{\mathcal{P}}(\Omega)} = 1$, $\langle \psi_2, \psi_2 \rangle_{H_{\mathcal{P}}(\Omega)} = 1$ and $\langle \psi_1, \psi_2 \rangle_{H_{\mathcal{P}}(\Omega)} = -1$ as can be verified via (8.14). Moreover, we can evaluate

$$\hat{f}_1 = \langle f, \psi_1 \rangle_{H_{\mathcal{B}}(\partial\Omega)} = f(1),$$
$$\hat{f}_2 = (f, \psi_2)_{H_{\mathcal{B}}(\partial\Omega)} = f(0),$$

with similar values for $\hat{g}_1$ and $\hat{g}_2$, so that

$$\langle f, g \rangle_{H_{\mathcal{P},\mathcal{B}}(\Omega)} = \int_0^1 f'(x)g'(x)\,\mathrm{d}x + f(1)g(1)I_{a_1}\frac{1-a_1}{a_1} + f(0)g(0)I_{a_2}\frac{1-a_2}{a_2} + (f(1)g(0) + f(0)g(1))I_{a_1a_2}. \tag{8.16}$$

We now consider the implications of different choices of the parameters a_1 and a_2 on the nature of the reproducing kernel K and its associated reproducing kernel Hilbert space.

(1) If we take $a_1 = 0$ and $a_2 = 1$ in (8.15) we get the kernel

$$K(x, z) = \min(x, z) + 1 - x - z,$$

and (8.16) implies that the inner product would be

$$\langle f, g \rangle_{H_{\mathcal{P},\mathcal{B}}(\Omega)} = \int_0^1 f'(x)g'(x)\,\mathrm{d}x.$$

Now we check whether the reproducing property holds true without any restrictions on the function space. We see that

$$\langle f, K(\cdot, z) \rangle_{H_{\mathcal{P},\mathcal{B}}(\Omega)} = \int_0^1 f'(x)\left(H(z-x) - 1\right)\,\mathrm{d}x = f(z) - f(1),$$

where $H(x)$ is the Heaviside step function. This suggests that only functions with $f(1) = 0$ are in the generalized Sobolev space of this kernel.

(2) Similarly, $a_1 = 1$ and $a_2 = 0$ produces the Brownian motion kernel $K(x, z) = \min(x, z)$, which has the same inner product but instead requires $f(0) = 0$.

(3) Working this situation through using arbitrary (nonzero) a_i values requires the $H_{\mathcal{P}}$-inner product of f and $K(\cdot, z)$ as part of the full inner product. It can be computed as

$$\langle f, K(\cdot, z) \rangle_{H_{\mathcal{P}}(\Omega)} = f(z) + f(0)(z - a_1z - a_2z - 1 - a_2) + f(1)(a_2 + a_1z + a_2z - z).$$

Using this in conjunction with $K(0, z) = a_2(1 - z)$ and $K(1, z) = a_1z$ gives

$$\begin{aligned}\langle f, K(\cdot, z) \rangle_{H_{\mathcal{P},\mathcal{B}}(\Omega)} =& \langle f, K(\cdot, z) \rangle_{H_{\mathcal{P}}(\Omega)} + f(1)a_1z\frac{1-a_1}{a_1} + f(0)a_2(1-z)\frac{1-a_2}{a_2} \\ &+ f(1)a_2(1-z) + f(0)a_1z \\ =& f(z) + 2a_2\left(f(1) - f(0)\right).\end{aligned}$$

This implies that — for our choice of functions $\psi_1(x) = x$ and $\psi_2(x) = 1 - x$ — the native space of K in the general case when $a_i > 0$ consists of *periodic functions* f, i.e., $f(0) = f(1)$.

We just concluded that only periodic functions can be handled with kernels of the form (8.15). This is true, unless $a_i = 0$ is allowed in an attempt to "break" the symmetry. As seen in items (1) and (2) in Example 8.6, fixing $a_2 = 0$ requires $f(0) = 0$ and fixing $a_1 = 0$ requires $f(1) = 0$.

If we wish to avoid this kind of boundary restriction, we need to consider using a different set of "eigenfunctions" ψ_i to represent the boundary correction kernel R.

Example 8.7. (Generalized Sobolev space for modified Brownian bridge kernels II)
A generic "eigenfunction" basis for the space of linear polynomials (i.e., null$(-\mathcal{D}^2)$) for computing with the boundary inner product (8.13) is given by

$$\psi_1(x) = c_1 x + c_2(1-x), \qquad \psi_2(x) = c_3 x + c_4(1-x),$$

which are just linear combinations of the earlier basis functions. Enforcing orthonormality requires

$$\langle \psi_1, \psi_1 \rangle_{H_{\mathcal{B}}(\partial\Omega)} = c_1^2 + c_2^2 = 1, \tag{8.17a}$$
$$\langle \psi_2, \psi_2 \rangle_{H_{\mathcal{B}}(\partial\Omega)} = c_3^2 + c_4^2 = 1, \tag{8.17b}$$
$$\langle \psi_1, \psi_2 \rangle_{H_{\mathcal{B}}(\partial\Omega)} = c_1 c_3 + c_2 c_4 = 0. \tag{8.17c}$$

There is no unique solution to the nonlinear system (8.17), but for a given c_4, all other values can be determined up to a few signs:

$$c_1 = \pm c_4, \quad c_2^2 = 1 - c_1^2, \quad c_3 = \mp c_2, \tag{8.18}$$

where the $\pm, \mp$ signs must remain consistent throughout. We assume that none of these coefficients are zero since we otherwise revert to the basis from Example 8.6. Clearly, there is also a restriction that $c_k \in (0,1)$ to avoid complex values.

Using this basis of null$(\mathcal{L})$, and again employing the Brownian bridge kernel $G(x,z) = \min(x,z) - xz$ we can show that the reproducing kernel K is of the general form

$$\begin{aligned} K(x,z) = \min(x,z) &+ xz(-1 + a_1 + a_2 - 2c_1c_2(a_1 - a_2)) + a_1 - c_1^2(a_1 - a_2) \\ &+ (x+z)(c_1c_2(a_1 - a_2) - a_1 + c_1^2(a_1 - a_2)). \end{aligned} \tag{8.19}$$

Thus, the four values c_1^2, c_1c_2, a_1, a_2 uniquely define the kernel. We now derive a nonlinear system that can be used to obtain these values from a user-specified inner product.

To find the general form of the inner product, we will need to determine the generalized Fourier coefficients

$$\hat{f}_1 = \langle f, \psi_1 \rangle_{H_{\mathcal{B}}(\partial\Omega)} = c_2 f(0) + c_1 f(1),$$
$$\hat{f}_2 = \langle f, \psi_2 \rangle_{H_{\mathcal{B}}(\partial\Omega)} = c_4 f(0) + c_3 f(1),$$

and the $H_{\mathcal{P}}$-inner products

$$\langle \psi_1, \psi_1 \rangle_{H_{\mathcal{P}}(\Omega)} = \int_0^1 (c_1 - c_2)(c_1 - c_2)\, \mathrm{d}x = 1 - 2c_1c_2, \tag{8.20}$$

$$\langle \psi_2, \psi_2 \rangle_{H_{\mathcal{P}}(\Omega)} = \int_0^1 (c_3 - c_4)(c_3 - c_4)\, \mathrm{d}x = 1 - 2c_3c_4, \tag{8.21}$$

$$\langle \psi_1, \psi_2 \rangle_{H_{\mathcal{P}}(\Omega)} = \int_0^1 (c_1 - c_2)(c_3 - c_4)\, \mathrm{d}x = -c_1c_4 - c_2c_3 = \pm(1 - 2c_1^2). \tag{8.22}$$

Again, simplifications have been made exploiting orthonormality and the $\pm$ matches the choice made in defining c_1.

Substituting these into (8.12) — and performing some tedious, but elementary, simplifications — leads to

$$\begin{aligned}\langle f, g \rangle_{H_{\mathcal{P},\mathcal{B}}(\Omega)} = \langle f, g \rangle_{H_{\mathcal{P}}(\Omega)} &+ f(0)g(0) \left((1 - c_1^2)/a_1 + c_1^2/a_2 - 1\right) \\ &+ \left(f(0)g(1) + f(1)g(0)\right)\left(c_1c_2(1/a_1 - 1/a_2) + 1\right) \\ &+ f(1)g(1) \left(c_1^2/a_1 + (1 - c_1^2)/a_2 - 1\right).\end{aligned}$$

At this point, we can notice an interesting result: the $f(0)g(1)$ and $f(1)g(0)$ contributions to the inner product must be equal.

We are now ready to specify the nonlinear system alluded to above, which will allow us to find the reproducing kernel associated with some inner product of interest to us.

Any inner product we may wish to consider[3] must be in the form

$$\langle f, g \rangle_{H_{\mathcal{P},\mathcal{B}}(\Omega)} = \langle f, g \rangle_{H_{\mathcal{P}}(\Omega)} + \alpha f(0)g(0) + \beta(f(0)g(1) + f(1)g(0)) + \gamma f(1)g(1), \tag{8.23}$$

assuming we use the boundary inner product defined in (8.13). Here we can view α, β, γ as design parameters and we can obtain the values c_1, c_2, a_1, a_2 with $a_1, a_2 > 0$ that define the kernel (8.19) by solving the nonlinear equations

$$\begin{aligned}(1 - c_1^2)/a_1 + c_1^2/a_2 - 1 &= \alpha, \\ c_1c_2(1/a_1 - 1/a_2) + 1 &= \beta, \\ c_1^2/a_1 + (1 - c_1^2)/a_2 - 1 &= \gamma. \\ c_1^2 + c_2^2 &= 1\end{aligned}$$

If these equations have a solution for a given α, β, γ then we have successfully defined the kernel for a given inner product.

In Remark 8.6, we discuss some restrictions on these values to guarantee a nonnegative value for $\|f\|^2_{H_{\mathcal{P},\mathcal{B}}(\Omega)} = \langle f, f \rangle_{H_{\mathcal{P},\mathcal{B}}(\Omega)}$.

As a demonstration of this, examine the MATLAB snippet in Program 8.1 which is pulled from `GSSEx1.m` in the library and shows the required function evaluation within the execution of the nonlinear solver `fsolve`.

[3]To yield a different structure would require a different boundary inner product $(\cdot,\cdot)_{\partial\Omega}$, which we do not discuss.

Program 8.1. Part of `GSSEx1.m`, converted for readability

```
fx = @(c1,c2,a1,a2) [(1-c1^2)/a1 + c1^2/a2 - 1
                     c1*c2*(1/a1 - 1/a2) + 1
                     c1^2/a1 + (1-c1^2)/a2 - 1
                     c1^2 + c2^2];
fx_full = @(c1,c2,a1,a2,p) fx(c1,c2,a1,a2) - [p;1];
```

Using pre-selected values $c_1 = 1/4, c_2 = \sqrt{15}/4, a_1 = 1/4, a_2 = 1/2$ to create inner product coefficients $\alpha = 2.875, \beta = 1.484, \gamma = 1.125$, we can run the MATLAB code in Program 8.2 to recover those coefficients.

Program 8.2. Solving for coefficients

```
ipcoef = [2.875 1.484 1.125];
[gss_opts,x0] = gssfunc; % Get the optimization options and initial guess
gss_coef = fsolve(@(x)gssfunc(x,ipcoef),x0,gss_opts);
```

After some iterations, the solution $c_1 = 0.9683$, $c_2 = -0.2499$, $a_1 = 0.5$ and $a_2 = 0.25$ is reached. While this does not match the original input, the resulting inner product matches, as does the associated reproducing kernel,

$$\begin{aligned} \langle f, g \rangle_{H_{\mathcal{P},\mathcal{B}}(\Omega)} &= \langle f, g \rangle_{H_{\mathcal{P}}(\Omega)} \\ &\quad + 2.875 f(0)g(0) + 1.484(f(0)g(1) + f(1)g(0)) + 1.125 f(1)g(1), \\ K(x,z) &= \min(x,z) - 0.129xz - 0.3261(x+z) + 0.2656. \end{aligned} \tag{8.24}$$

Despite the multiple solutions for this nonlinear system, all the real-valued solutions we can find using different initial guesses produce the same kernel. Plots of this kernel for different z values are shown in Figure 8.3.

In a more realistic setting, a user would specify inner product values such as $\alpha = 101$, $\beta = 100$, $\gamma = 102$ and use the same procedure described in Program 8.2 to find the solution $c_1 = 0.7089$, $c_2 = 0.7053$, $a_1 = 0.0050$ and $a_2 = 0.2858$. The inner product and associated reproducing kernel for these coefficients is

$$\begin{aligned} \langle f, g \rangle_{H_{\mathcal{P},\mathcal{B}}(\Omega)} &= \langle f, g \rangle_{H_{\mathcal{P}}(\Omega)} + 101 f(0)g(0) + 100(f(0)g(1) + f(1)g(0)) + 102 f(1)g(1), \\ K(x,z) &= \min(x,z) - 0.4284xz - 0.2865(x+z) + 0.1461. \end{aligned} \tag{8.25}$$

Other examples include

$$\begin{aligned} \langle f, g \rangle_{H_{\mathcal{P},\mathcal{B}}(\Omega)} &= \langle f, g \rangle_{H_{\mathcal{P}}(\Omega)} + 8 f(0)g(0) + 20(f(0)g(1) + f(1)g(0)) + 77 f(1)g(1), \\ K(x,z) &= \min(x,z) - 0.6334xz - 0.2845(x+z) + 0.2287, \end{aligned} \tag{8.26}$$

and

$$\begin{aligned} \langle f, g \rangle_{H_{\mathcal{P},\mathcal{B}}(\Omega)} &= \langle f, g \rangle_{H_{\mathcal{P}}(\Omega)} + f(0)g(0), \\ K(x,z) &= \min(x,z) + 1. \end{aligned} \tag{8.27}$$

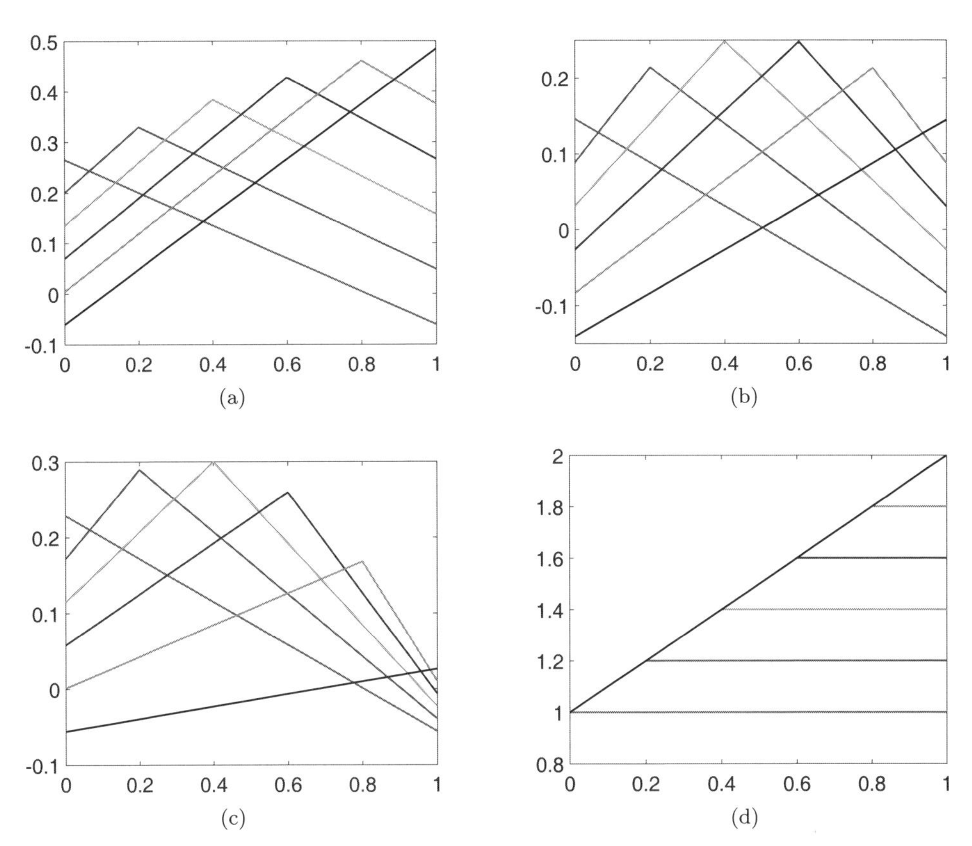

Fig. 8.3 Sample kernels designed to minimize different norms, plotted with centers at $z = 0, 0.2, 0.4, 0.6, 0.8, 1$. (a) (8.24), (b) (8.25), (c) (8.26), (d) (8.27).

Remark 8.6. The inner product from (8.23) can be rewritten as

$$\langle f, g\rangle_{H_{\mathcal{P},\mathcal{B}}(\Omega)} = \langle f, g\rangle_{H_{\mathcal{P}}(\Omega)} + \big(f(0)\ f(1)\big)\begin{pmatrix}\alpha & \beta\\ \beta & \gamma\end{pmatrix}\begin{pmatrix}g(0)\\ g(1)\end{pmatrix},$$

so, for this inner product to satisfy $\langle f, f\rangle_{H_{\mathcal{P},\mathcal{B}}(\Omega)} > 0$, the eigenvalues

$$\lambda = \frac{\alpha+\gamma}{2} \pm \sqrt{\left(\frac{\alpha+\gamma}{2}\right)^2 + \beta^2 - \alpha\gamma}$$

of the matrix

$$\begin{pmatrix}\alpha & \beta\\ \beta & \gamma\end{pmatrix}$$

must be positive. This will occur when both $\alpha + \gamma > 0$ and $\beta^2 - \alpha\gamma < 0$ are true, which is a necessary condition on choosing valid inner product components for a reproducing kernel of this form. If these inequalities are not strict, then a semi-norm will still exist; this was the case for (8.27) above.

Remark 8.7. The discussion in Example 8.7 assumed nonzero a_k values, but if $a_1 = 0$ the boundary contribution is

$$f(0)g(0)\left(c_1^2(1/a_2 - 1 - 2c_1c_2)\right) + f(1)g(1)\left((1-c_1^2)(1/a_2 - 1 - 2c_1c_2)\right) \\ +(f(0)g(1) + f(1)g(0))\left(-c_1c_2(1/a_2 - 1 - 2c_1c_2)\right),$$

and if $a_2 = 0$ we have

$$f(0)g(0)\left((1-c_1^2)(1/a_1 - 1 + 2c_1c_2)\right) + f(1)g(1)\left(c_1^2(1/a_1 - 1 + 2c_1c_2)\right) \\ +(f(0)g(1) + f(1)g(0))\left(c_1c_2(1/a_1 - 1 + 2c_1c_2)\right).$$

At this point, we note that setting one of the $a_i = 0$ decreases the number of free parameters in the inner product to three. This is not a problem because the coefficients of the boundary terms are not independent.

As an example, were $a_1 = 0$ fixed then the user could specify any two of α, β, γ — but not all three — from (8.23) as before. Once two of those coefficients are specified, c_1, c_2, a_2 can be determined and the third coefficient is fixed accordingly. The system of equations to be solved, were α and β chosen, would be

$$\begin{aligned} c_1^2(1/a_2 - 1 - 2c_1c_2) &= \alpha, \\ -c_1c_2(1/a_2 - 1 - 2c_1c_2) &= \beta, \\ c_1^2 + c_2^2 &= 1. \end{aligned}$$

A similar function to Program 8.1 would need to be coded for each of these situations to determine the appropriate coefficients.

Example 8.8. (Generalized Sobolev space for modified Brownian bridge kernels III)
By exploiting the known structure of the kernel and the inner product, we can actually form a *linear* system to solve for the kernel from Example 8.7. As described in (8.19), the kernel for $\mathcal{P} = \mathcal{D}$ must take the form

$$K(x,z) = \min(x,z) + r_1xz + r_2(x+z) + r_3. \tag{8.28}$$

The boundary inner product $(\cdot,\cdot)_{\partial\Omega}$ from (8.13) fixes the form of the $H_{\mathcal{P},\mathcal{B}}$-inner product to be (8.23). We can define a relationship between r_1, r_2, r_3 and α, β, γ by enforcing the reproducing property.

To do so, we find $\mathcal{P}K(\cdot,z)|_x = H(z-x) + r_1z + r_2$, and thus

$$\langle f, K(\cdot,z)\rangle_{H_\mathcal{P}(\Omega)} = f(z) - f(0) + f(1) + (f(1) - f(0))(r_1z + r_2).$$

Also using $K(0,z) = r_2z + r_3$ and $K(1,z) = r_2 + r_3 + z(1 + r_1 + r_2)$, we know that

$$\begin{aligned} \langle f, K(\cdot,z)\rangle_{H_{\mathcal{P},\mathcal{B}}(\Omega)} = {} & f(z) - f(0) + f(1) + (f(1) - f(0))(r_1z + r_2) \\ & + \alpha f(0)(r_2z + r_3) + \gamma f(1)(z(1 + r_1 + r_2) + r_2 + r_3) \\ & + \beta(f(0)(z(1 + r_1 + r_2) + r_2 + r_3) + f(1)(r_2z + r_3)). \end{aligned} \tag{8.29}$$

For us to ensure that $\langle f, K(\cdot,z)\rangle_{H_{\mathcal{P},\mathcal{B}}(\Omega)} = f(z)$ we need that the coefficients of the $f(0)$, $f(1)$, $zf(0)$ and $zf(1)$ terms in (8.29) are zero. This produces the linear system

$$\begin{pmatrix} 0 & \beta-1 & \beta+\alpha \\ 0 & \gamma+1 & \beta+\gamma \\ \beta-1 & \beta+\alpha & 0 \\ \gamma+1 & \beta+\gamma & 0 \end{pmatrix} \begin{pmatrix} r_1 \\ r_2 \\ r_3 \end{pmatrix} = \begin{pmatrix} 1 \\ -1 \\ -\beta \\ -\gamma \end{pmatrix}. \tag{8.30}$$

This is exploited in the program `GSSEx1.m` for comparison to the solution using the nonlinear solver. Note that solving this system is much faster and safer than solving a direct system, but in general, such a linear system may not exist.

8.3.2 *Summary for generalized Sobolev spaces on bounded domains*

Using the notation introduced earlier, we can summarize the theoretical foundation provided in this section in the following theorem which is based on [Fasshauer and Ye (2013b, Theorem 3.2)].

Theorem 8.2. *Given a linear self-adjoint differential operator* $\mathcal{L} = \boldsymbol{\mathcal{P}}^{*T}\boldsymbol{\mathcal{P}}$ *of order* 2β *and boundary operators* $\boldsymbol{\mathcal{B}}$ *on a bounded domain* $\Omega \subset \mathbb{R}^d$*, and given "eigenvalues"* $a_i \geq 0$ *and "eigenfunctions"* $\psi_i \in \mathrm{null}(\mathcal{L})$, $i = 1, \ldots, n_a$, *such that* $\langle \psi_i, \psi_j \rangle_{H_{\boldsymbol{\mathcal{B}}}(\partial\Omega)} = \delta_{i,j}$, *the* generalized Sobolev space $H_{\boldsymbol{\mathcal{P}},\boldsymbol{\mathcal{B}}}(\Omega)$ *is embedded in the classical Sobolev space* $H^\beta(\Omega)$ *and is given by*

$$H_{\boldsymbol{\mathcal{P}},\boldsymbol{\mathcal{B}}}(\Omega) = \{f \in H^\beta(\Omega) \mid \|f\|_{H_{\boldsymbol{\mathcal{P}},\boldsymbol{\mathcal{B}}}(\Omega)} < \infty\},$$

equipped with the inner product

$$\langle f, g\rangle_{H_{\boldsymbol{\mathcal{P}},\boldsymbol{\mathcal{B}}}(\Omega)} = \langle f, g\rangle_{H_{\boldsymbol{\mathcal{P}}}(\Omega)} + \sum_{i=1}^{n_a} \hat{f}_i \hat{g}_i \frac{I_{a_i}}{a_i} - \sum_{i=1}^{n_a}\sum_{j=1}^{n_a} \hat{f}_i \hat{g}_j \langle \psi_i, \psi_j\rangle_{H_{\boldsymbol{\mathcal{P}}}(\Omega)} I_{a_i a_j}$$

from (8.12).

8.3.3 *An alternative framework for boundary value problems on* $[a,b]$

In [Ramsay and Silverman (2005, Chapter 20)] one can find a description of how to construct a reproducing kernel Hilbert space based on a nonhomogeneous ordinary differential equation analogous to (8.7), i.e., they consider a univariate domain $\Omega = [a,b]$. Moreover, they focus mainly on *initial value problems*, i.e., $\Omega = [0,T]$. This point of view reflects the treatment of splines in [Wahba (1990)] (see also [Berlinet and Thomas-Agnan (2004, Chapter 6, Section 1.6.2)]). In [Dalzell and Ramsay (1993)] the constraints for the ordinary differential equation are generalized to arbitrary "boundary" conditions specified by an operator $\mathcal{B}$ such that $\mathrm{null}(\mathcal{L}) \cap \mathrm{null}(\mathcal{B}) = \{0\}$.

This latter assumption ensures that the reproducing kernel K again is of the form $K(x, z) = G(x, z) + R(x, z)$, and that, additionally, the reproducing kernel Hilbert space $\mathcal{H}_K(\Omega)$ is given as the direct sum of the spaces $\mathcal{H}_G(\Omega)$ and $\mathcal{H}_R(\Omega)$, according to Property 3 in Section 2.3. In this setting $\mathcal{H}_G(\Omega) = \text{null}(\mathcal{B})$ and $\mathcal{H}_R(\Omega) = \text{null}(\mathcal{L})$, and the inner product does not contain any subtractive "mixed" terms due to the direct sum decomposition of $\mathcal{H}_K(\Omega)$ (cf. Remark 8.5).

8.4 Conclusions

It follows directly from the definition of reproducing kernel Hilbert spaces in Definition 2.6 that — given a Hilbert function space (i.e., a space of (smooth) functions together with an inner product) and its reproducing kernel — the reproducing kernel is *unique*.

On the other hand, the discussion in this chapter has shown that we can design not only a positive definite kernel as in Section 3.9, but also a reproducing kernel Hilbert *space* (which implies that we pair the kernel with an inner product of our choice). The approach to doing so proposed here starts with a self-adjoint linear differential operator $\mathcal{L} = \boldsymbol{\mathcal{P}}^{*T}\boldsymbol{\mathcal{P}}$, where $\boldsymbol{\mathcal{P}}$ may be a scalar or a vector differential operator with appropriately defined adjoint $\boldsymbol{\mathcal{P}}^*$. The differential operator $\boldsymbol{\mathcal{P}}$ (together with potential boundary operators) defines an inner product for the Hilbert space associated with the kernel K, which is either a full-space Green's kernel of $\mathcal{L}$ (in the case of unbounded domains), or a homogeneous Green's kernel appropriately corrected for boundary effects. The resulting reproducing kernel Hilbert space is called a generalized Sobolev space.

The generalized Sobolev space approach also shows that a given positive definite kernel (e.g., a fundamental solution or full-space Green's kernel of $\mathcal{L}$) can be the reproducing kernel for *different* Hilbert spaces since it can be paired with different inner products arising from different operators $\boldsymbol{\mathcal{P}}$ representing different decompositions of $\mathcal{L}$.

This fact needs to be reconciled with the claim made in Chapter 2 that the native space associated with a given positive definite kernel is unique. That claim is proved in detail in [Wendland (2005, Section 10.2)], and one can observe there that this conclusion is arrived at after pairing the kernel with a very specific native space inner product which is assumed to come from a specific bilinear form involving the kernel.

The ability to change the correction to the Green's kernel by using different boundary operators adds a whole other layer of flexibility (and complexity) to our ability to design reproducing kernels and their associated native Hilbert spaces as explored in Section 8.3.1.

A potential future application of this *designer space* approach is motivated by the fact that the interpolant based on the reproducing kernel of a Hilbert function space has minimal Hilbert space norm among all possible interpolants of the given data. If one associates the norm with the shape of the interpolant (such as, e.g., the bending energy associated with cubic splines), then such a norm — specified via $\mathcal{P}$ — will give rise to the "optimal" kernel (via $\mathcal{L}$) and then produce the desired interpolant. Of course, this application is not limited to the interpolation setting. It can easily be extended to the least squares setting, where different Hilbert space norms (induced by different inner products) will lead to different regularization functionals (see, e.g., Section 15.1 or Chapter 18).

Slightly less complicated is the situation when the choice of the ideal differential operator $\mathcal{P}$ is not made between fundamentally different operators (such as in Example 8.3 or Section 8.3.1), but only within a parametrized family of operators such as, e.g., the operators $\mathcal{L} = (-\nabla^2 + \varepsilon^2\mathcal{I})^\beta$ which give rise to the family of Matérn kernels. We will fix the smoothness class β and study the effects of just the shape parameter ε in later chapters.

Chapter 9

Accuracy and Optimality of Reproducing Kernel Hilbert Space Methods

9.1 Optimality

A common problem in applications is the desire to have an "optimal" numerical estimate of the value of a linear functional Lf of some function f, and a very convenient approach to accomplishing this is to use as an *Ansatz* a linear combination of certain function values. This leads to the theory of *optimal recovery* (see, e.g., Golomb and Weinberger (1959); Larkin (1970) and also [Fasshauer (2007, Chapter 18)] and the references mentioned there). In the following we will briefly mention the most relevant results for the case when L corresponds to function evaluation at a point $\boldsymbol{x}$, i.e., $Lf = f(\boldsymbol{x})$, and f is from some reproducing kernel Hilbert space.

It is well known (see, e.g., [Fasshauer (2007, Chapter 18)]) that the kernel-based interpolant s to a function f in a reproducing kernel Hilbert space $\mathcal{H}_K(\Omega)$ is actually a *minimum norm interpolant*. In order to see this we first prove that $f - s$ is orthogonal to s in the Hilbert-space inner product:

$$\begin{aligned}
\langle f - s, s\rangle_{\mathcal{H}_K(\Omega)} &= \langle f - s, \boldsymbol{k}(\cdot)^T \mathsf{K}^{-1}\boldsymbol{y}\rangle_{\mathcal{H}_K(\Omega)} \\
&= \langle f - s, \boldsymbol{k}(\cdot)^T\rangle_{\mathcal{H}_K(\Omega)} \mathsf{K}^{-1}\boldsymbol{y} \\
&= \left(\langle f - s, K(\cdot,\boldsymbol{x}_1)\rangle_{\mathcal{H}_K(\Omega)} \cdots \langle f - s, K(\cdot,\boldsymbol{x}_N)\rangle_{\mathcal{H}_K(\Omega)}\right) \mathsf{K}^{-1}\boldsymbol{y} \\
&= \left(f(\boldsymbol{x}_1) - s(\boldsymbol{x}_1) \cdots f(\boldsymbol{x}_N) - s(\boldsymbol{x}_N)\right) \mathsf{K}^{-1}\boldsymbol{y} \\
&= \left(0 \cdots 0\right) \mathsf{K}^{-1}\boldsymbol{y} = 0 \qquad (9.1)
\end{aligned}$$

because we know that $f(\boldsymbol{x}_i) = s(\boldsymbol{x}_i)$, $i = 1, \ldots, N$, if s interpolates f at those points. Note that we employed the reproducing property (1.1) of K in going from the third to the forth line.

Using this, we can prove the *Pythagorean theorem* in reproducing kernel Hilbert spaces, i.e., if s is a kernel-based interpolant of $f \in \mathcal{H}_K(\Omega)$, then

$$\begin{aligned}
\|f\|^2_{\mathcal{H}_K(\Omega)} &= \|f - s + s\|^2_{\mathcal{H}_K(\Omega)} = \|f - s\|^2_{\mathcal{H}_K(\Omega)} + 2\langle f - s, s\rangle_{\mathcal{H}_K(\Omega)} + \|s\|^2_{\mathcal{H}_K(\Omega)} \\
&= \|f - s\|^2_{\mathcal{H}_K(\Omega)} + \|s\|^2_{\mathcal{H}_K(\Omega)}. \qquad (9.2)
\end{aligned}$$

From this it immediately follows that

$$\|s\|_{\mathcal{H}_K(\Omega)} \le \|f\|_{\mathcal{H}_K(\Omega)}, \tag{9.3a}$$

$$\|f - s\|_{\mathcal{H}_K(\Omega)} \le \|f\|_{\mathcal{H}_K(\Omega)}. \tag{9.3b}$$

We see from (9.3a) that the Hilbert space norm of the interpolant s never exceeds the norm of f, i.e., s has the minimum Hilbert space norm among all functions interpolating the given data, and (9.3b) will be useful below to improve several error bounds.

9.2 Different Types of Error

In this section we want to briefly summarize a number of different situations in which *error* and *uncertainty* play a role in the solution of the kernel interpolation problem. A discussion of some of these topics can also be found in [Fasshauer (2007, Chapters 15 and 17)].

We have already seen some experimental evidence of how various parameters affect the accuracy when using positive definite kernels for interpolation. For example, in Example 7.1 we saw how the meshsize h, i.e., how fine we make the discretization, or the number of samples N affects the error. Note that N and h are related to each other in a dimension-dependent way, i.e., for quasi-uniformly distributed data we have $h = \mathcal{O}(N^{-1/d})$. We will discuss this along with *dimension-independent* error bounds in Section 9.5. Example 7.1 also showed the effects of the smoothness of the kernel, i.e., the role of β within the family of iterated Brownian bridge kernels. Moreover, Example 7.2 revealed that the domain boundary and the conditions imposed there play a role in the error distribution. The role of the shape parameter ε, i.e., how localized the kernel is chosen, was explored for iterated Brownian bridge kernels in Example 7.3. And finally, the specific choice of basis, distribution of data or the *design* (see Section 16.2 and Appendix B.1), error in the data (which we deal with in Chapter 17), and data generated by a random process are other sources of error and uncertainty.

If nothing else, this brief summary provides an indication of the flexibility of kernel-based approximation methods.

9.3 The "Standard" Error Bound

There is a very elegant error bound that can be easily derived in the reproducing kernel Hilbert space setting since it follows almost immediately from the reproducing property.

As we will see below, this error bound provides an indication that the error for kernel-based interpolation depends on the choice of kernel along with its parameters, the design (both where the data is located and how many samples we use), and on the function that generated the data (in particular its smoothness relative to the kernel).

For this derivation, we need to recall the reproducing property (1.1), i.e.,

$$\langle f, K(\cdot,\boldsymbol{x})\rangle_{\mathcal{H}_K(\Omega)} = f(\boldsymbol{x}), \qquad f \in \mathcal{H}_K(\Omega),$$

where $\mathcal{H}_K(\Omega)$ is the reproducing kernel Hilbert space, or native space, generated by K. It is also important to remember from Section 2.3 that $K(\cdot,x) \in \mathcal{H}_K(\Omega)$ for $x \in \Omega$, which implies that

$$\langle K(\cdot,\boldsymbol{x}), K(\cdot,\boldsymbol{z})\rangle_{\mathcal{H}_K(\Omega)} = K(\boldsymbol{x},\boldsymbol{z}), \qquad \boldsymbol{x},\boldsymbol{z} \in \Omega.$$

Note that because $s(\boldsymbol{x})$ is a number we have $s(\boldsymbol{x}) = s(\boldsymbol{x})^T$, and therefore we can also write (cf. (1.3c))

$$s(\boldsymbol{x}) = \boldsymbol{y}^T \mathsf{K}^{-1}\boldsymbol{k}(\boldsymbol{x}). \tag{9.4}$$

The vector $\boldsymbol{y}$ is full of function values ($y_i = f(\boldsymbol{x}_i)$) so we can write

$$\begin{aligned}\boldsymbol{y}^T = \big(f(\boldsymbol{x}_1) \cdots f(\boldsymbol{x}_N)\big) &= \big(\langle f, K(\cdot,\boldsymbol{x}_1)\rangle_{\mathcal{H}_K(\Omega)} \cdots \langle f, K(\cdot,\boldsymbol{x}_N)\rangle_{\mathcal{H}_K(\Omega)}\big)\\ &= \big\langle f, \big(K(\cdot,\boldsymbol{x}_1) \cdots K(\cdot,\boldsymbol{x}_N)\big)\big\rangle_{\mathcal{H}_K(\Omega)}\\ &= \langle f, \boldsymbol{k}(\cdot)^T\rangle_{\mathcal{H}_K(\Omega)}.\end{aligned} \tag{9.5}$$

Inserting (9.5) into (9.4) allows us to write

$$s(\boldsymbol{x}) = \langle f, \boldsymbol{k}(\cdot)^T\rangle_{\mathcal{H}_K(\Omega)}\mathsf{K}^{-1}\boldsymbol{k}(\boldsymbol{x}) = \big\langle f, \boldsymbol{k}(\cdot)^T\mathsf{K}^{-1}\boldsymbol{k}(\boldsymbol{x})\big\rangle_{\mathcal{H}_K(\Omega)} \tag{9.6}$$

because $\mathsf{K}^{-1}\boldsymbol{k}(\boldsymbol{x})$ is not a function and is unaffected by the inner product. Now, using the reproducing property for $f(\boldsymbol{x})$ and (9.6) for $s(\boldsymbol{x})$, we can derive the pointwise error expression

$$\begin{aligned}|f(\boldsymbol{x}) - s(\boldsymbol{x})| &= \Big|\langle f, K(\cdot,\boldsymbol{x})\rangle_{\mathcal{H}_K(\Omega)} - \big\langle f, \boldsymbol{k}(\cdot)^T\mathsf{K}^{-1}\boldsymbol{k}(\boldsymbol{x})\big\rangle_{\mathcal{H}_K(\Omega)}\Big|\\ &= \Big|\big\langle f, K(\cdot,\boldsymbol{x}) - \boldsymbol{k}(\cdot)^T\mathsf{K}^{-1}\boldsymbol{k}(\boldsymbol{x})\big\rangle_{\mathcal{H}_K(\Omega)}\Big|\\ &\le \|f\|_{\mathcal{H}_K(\Omega)}\Big\|K(\cdot,\boldsymbol{x}) - \boldsymbol{k}^T(\cdot)\mathsf{K}^{-1}\boldsymbol{k}(\boldsymbol{x})\Big\|_{\mathcal{H}_K(\Omega)}\\ &= \|f\|_{\mathcal{H}_K(\Omega)} P_{K,\mathcal{X}}(\boldsymbol{x}),\end{aligned} \tag{9.7}$$

using the Cauchy–Schwarz inequality, and the *power function*

$$P_{K,\mathcal{X}}(\boldsymbol{x}) = \sqrt{K(\boldsymbol{x},\boldsymbol{x}) - \boldsymbol{k}^T(\boldsymbol{x})\mathsf{K}^{-1}\boldsymbol{k}(\boldsymbol{x})}. \tag{9.8}$$

This expression for the power function is obtained by again using the reproducing property on the norm present in the inequality so that

$$\begin{aligned}\Big\|K(\cdot,\boldsymbol{x}) - \boldsymbol{k}^T(\cdot)\mathsf{K}^{-1}\boldsymbol{k}(\boldsymbol{x})\Big\|^2_{\mathcal{H}_K(\Omega)} &= \langle K(\cdot,\boldsymbol{x}), K(\cdot,\boldsymbol{x})\rangle_{\mathcal{H}_K(\Omega)}\\ &\quad - 2\langle K(\cdot,\boldsymbol{x}), \boldsymbol{k}^T(\cdot)\mathsf{K}^{-1}\boldsymbol{k}(\boldsymbol{x})\rangle_{\mathcal{H}_K(\Omega)}\\ &\quad + \langle \boldsymbol{k}^T(\boldsymbol{x})\mathsf{K}^{-1}\boldsymbol{k}(\cdot), \boldsymbol{k}^T(\cdot)\mathsf{K}^{-1}\boldsymbol{k}(\boldsymbol{x})\rangle_{\mathcal{H}_K(\Omega)}\\ &= K(\boldsymbol{x},\boldsymbol{x}) - 2\boldsymbol{k}^T(\boldsymbol{x})\mathsf{K}^{-1}\boldsymbol{k}(\boldsymbol{x}) + \boldsymbol{k}^T(\boldsymbol{x})\mathsf{K}^{-1}\boldsymbol{k}(\boldsymbol{x})\\ &= K(\boldsymbol{x},\boldsymbol{x}) - \boldsymbol{k}^T(\boldsymbol{x})\mathsf{K}^{-1}\boldsymbol{k}(\boldsymbol{x}).\end{aligned}$$

Remark 9.1. Note that earlier we derived (5.10), providing a similar estimate for the *mean-squared error* and *kriging variance* in the stochastic setting. However, in (5.10) we also included the *process variance* σ^2. We continue this digression to the stochastic setting in Remark 9.2.

The standard error bound (9.7) derived above can be improved (see, e.g., [Golomb and Weinberger (1959)]) to

$$|f(\boldsymbol{x}) - s(\boldsymbol{x})| \leq \|f - s\|_{\mathcal{H}_K(\Omega)} P_{K,\mathcal{X}}(\boldsymbol{x}), \tag{9.9}$$

and (9.3b) shows that this is indeed an improvement. Even so, this tighter error bound does not seem to play a significant role in the kernel literature since most standard error bounds (see [Wendland (2005); Fasshauer (2007)]) serve mostly for theoretical purposes, and then the term involving $f - s$ on the right-hand side is of no use. The books just mentioned discuss how the power function can be bounded by powers of the fill-distance h, thus resulting in error bounds of the form

$$|f(\boldsymbol{x}) - s(\boldsymbol{x})| \leq Ch^{\beta} \|f\|_{\mathcal{H}_K(\Omega)}$$

for some appropriate exponent β depending on the smoothness of the kernel K.

One of the main drawbacks of this power function approach is that error bounds apply only to functions $f \in \mathcal{H}_K(\Omega)$, but not to functions that are possibly rougher, i.e., to larger classes of functions. The ability to address this deficiency is one of the main reasons for the use of *sampling inequalities* discussed in the next subsection.

If one is interested in *computable* error bounds, then none of the bounds above are useful since $\|f\|_{\mathcal{H}_K(\Omega)}$ usually is not computable (remember, we do not even know f, but want to reconstruct it from the data). On the other hand, if we assume that our approximation s is good, i.e.,

$$\|f - s\|_{\mathcal{H}_K(\Omega)} \leq \delta \|s\|_{\mathcal{H}_K(\Omega)}$$

for some not too large constant δ (which may depend on the kernel parameters, the data locations, and on the function f), then the Golomb–Weinberger improved error bound (9.9) yields a mostly computable error bound

$$|f(\boldsymbol{x}) - s(\boldsymbol{x})| \leq \delta \|s\|_{\mathcal{H}_K(\Omega)} P_{K,\mathcal{X}}(\boldsymbol{x}).$$

This is indeed computable since $\|s\|_{\mathcal{H}_K(\Omega)} = \sqrt{\boldsymbol{y}^T \mathsf{K}^{-1} \boldsymbol{y}}$:

$$\begin{aligned} \|s\|^2_{\mathcal{H}_K(\Omega)} = \langle s, s \rangle_{\mathcal{H}_K(\Omega)} &= \langle \boldsymbol{y}^T \mathsf{K}^{-1} \boldsymbol{k}(\cdot), \boldsymbol{k}(\cdot)^T \mathsf{K}^{-1} \boldsymbol{y} \rangle_{\mathcal{H}_K(\Omega)} \\ &= \boldsymbol{y}^T \mathsf{K}^{-1} \langle \boldsymbol{k}(\cdot), \boldsymbol{k}(\cdot)^T \rangle_{\mathcal{H}_K(\Omega)} \mathsf{K}^{-1} \boldsymbol{y} \\ &= \boldsymbol{y}^T \mathsf{K}^{-1} \mathsf{K} \mathsf{K}^{-1} \boldsymbol{y} = \boldsymbol{y}^T \mathsf{K}^{-1} \boldsymbol{y}. \end{aligned} \tag{9.10}$$

The term $\boldsymbol{y}^T \mathsf{K}^{-1} \boldsymbol{y}$ plays an important role in the context of zero-mean Gaussian fields as well, where it is known as the *Mahalanobis distance*.

Remark 9.2. In Chapter 14 we compute an optimal process variance $\sigma^2_{\text{opt}} = \frac{1}{N} \boldsymbol{y}^T \mathsf{K}^{-1} \boldsymbol{y}$ (see (14.19)) by maximizing the so-called *profile likelihood.* Note that with this value the mean-squared error (5.10) of the kriging predictor becomes

$$\mathbb{E}\left[\left(Y_{\boldsymbol{x}} - \hat{Y}_{\boldsymbol{x}}\right)^2\right] = \frac{1}{N} \boldsymbol{y}^T \mathsf{K}^{-1} \boldsymbol{y} (K(\boldsymbol{x}, \boldsymbol{x}) - \boldsymbol{k}(\boldsymbol{x})^T \mathsf{K}^{-1} \boldsymbol{k}(\boldsymbol{x})),$$

whose right-hand side is identical to the square of the right-hand side of our computable error bound (9.10) obtained in the deterministic setting, provided $\delta = 1/\sqrt{N}$.

9.4 Error Bounds via Sampling Inequalities

In this section we use the "modern" technique known as *sampling inequalities* to obtain error bounds for kernel-based interpolation methods. In some of the papers listed below this technique is referred to as a *zeros lemma*. These methods can also be extended to other settings, such as (least squares) approximation or PDE error estimates — both in strong form and weak form. However, we focus only on the interpolation case and leave it up to the reader to consult the original literature listed below for more advanced versions of error bounds based on sampling inequalities.

The first occurrence of a sampling inequality in the literature, used to obtain an error estimate for multivariate interpolation on $\mathbb{R}^d$, was probably in the seminal paper [Duchon (1978)] whose author, however, gives credit to Rémi Arcangéli for having used the same proof technique in his Ph.D. thesis [Arcangéli (1974)] under Marc Atteia a few years earlier. A few years later, [Madych and Potter (1985)] followed as another isolated paper. Then, however, it took 20 years until a recent flurry of activity in the RBF community began with the paper [Narcowich *et al.* (2005)]. Some other recent references dealing with sampling inequalities are, e.g., [Wendland and Rieger (2005); Madych (2006); Narcowich *et al.* (2006); Arcangéli *et al.* (2007); Rieger (2008); Rieger and Zwicknagl (2008); Arcangéli *et al.* (2009); Hangelbroek *et al.* (2010); Rieger *et al.* (2010); Hangelbroek *et al.* (2011); Arcangéli *et al.* (2012); Zwicknagl and Schaback (2013); Rieger and Zwicknagl (2014); Griebel *et al.* (2015)]. The unpublished lecture notes [Schaback (2011a)] also provide valuable insight and context to the foundation and implications of sampling inequalities.

9.4.1 *How sampling inequalities lead to error bounds*

A sampling inequality formalizes in a rigorous way the following very general idea: It is not possible for a smooth function u to become uncontrollably large on its domain Ω provided

(1) samples of u obtained on a sufficiently dense discrete subset $\mathcal{X}$ of Ω are small enough (or zero), and
(2) the derivatives of u are bounded on Ω.

A typical sampling inequality looks like this:

$$|u|_{W_q^m(\Omega)} \leq C \left(h^{\beta - m - d\left(\frac{1}{p} - \frac{1}{q}\right)_+} |u|_{W_p^\beta(\Omega)} + h^{-m} \|\boldsymbol{u}\|_\infty \right), \tag{9.11}$$

where $m \geq 0$ and $\beta > m + \frac{d}{p}$ define the *strong* semi-norm $|\cdot|_{W_p^\beta(\Omega)}$ which — together with the discrete norm of the values $\boldsymbol{u} = \left(u(\boldsymbol{x}_1) \ldots u(\boldsymbol{x}_N)\right)^T$ of u on the set $\mathcal{X} = \{\boldsymbol{x}_1, \ldots, \boldsymbol{x}_N\}$ with fill distance h — bounds the *weak* semi-norm $|u|_{W_q^m(\Omega)}$. Here the W_p^β-norm of u is multiplied by a positive power of h (i.e., it gets small as h gets small), while $\|\boldsymbol{u}\|_{\ell_\infty}$ is multiplied by a nonpositive power of h (i.e., it grows as h gets small).

In order for such a sampling inequality to give rise to a useful error bound we will first want to ensure that the function u is interpreted as the residual $f - s$ between the function f and its approximation s. Then, for any successful approximation method, the samples of the residual will be small and will (hopefully) outweigh the h^{-m} factor. If, in particular, s is obtained by *interpolation on $\mathcal{X}$*, then the residual vector $\boldsymbol{u}$ will have norm zero, and one obtains a bound of the form

$$|f - s|_{W_q^m(\Omega)} \leq C h^{\beta - m - d\left(\frac{1}{p} - \frac{1}{q}\right)_+} |f - s|_{W_p^\beta(\Omega)}.$$

Note that this bound is not quite in the form one would like it to be for it to serve as an error bound since it involves the residual $f - s$ also on the right-hand side. Therefore, an important assumption needs to be made: we assume that the approximation s satisfies the optimality property (9.3b) (as is certainly the case for a kernel-based interpolation method whose native space is equivalent to the Sobolev space $W_p^\beta(\Omega)$). Then we end up with

$$|f - s|_{W_q^m(\Omega)} \leq C h^{\beta - m - d\left(\frac{1}{p} - \frac{1}{q}\right)_+} |f|_{W_p^\beta(\Omega)}. \tag{9.12}$$

To make this bound look a bit more familiar to those with only a cursory exposure to approximation theory, we consider $\Omega = [a, b]$, set $m = 0$, let $p = q = 2$ and convert the Sobolev norm on the right-hand side to an L_2-norm of derivatives, i.e.,

$$\|f - s\|_{L_2([a,b])} \leq C h^\beta \|f^{(\beta)}\|_{L_2([a,b])},$$

provided $f, s \in C^\beta([a, b])$. This is, e.g., the kind of error bound that is typical for piecewise polynomial spline interpolation.

In the next two subsections we look more closely at error bounds for kernel-based interpolation methods obtained via sampling inequalities. First we consider the relatively transparent univariate case in detail, and then we outline the important steps for the much more complicated multivariate case.

9.4.2 *Univariate sampling inequalities and error bounds*

In one space dimension, sampling inequalities are relatively easy to understand and to obtain, so we begin with a discussion of that setting. Moreover, some of the kernels discussed in this book — such as the iterated Brownian bridge kernels of Chapter 7 — are defined on an interval, and therefore the 1D framework is all that is needed.

Following the outline of the previous subsection, to obtain an error formula for a univariate kernel-based interpolation to data $\{x_i, f(x_i)\}_{i=1}^N$ at N distinct points $\mathcal{X} = \{x_1, \ldots, x_N\} \subset \Omega = [a,b]$, we must first obtain a sampling inequality for a generic function u of the same smoothness class as the kernel of interest. We assume that the point set $\mathcal{X}$ has *fill distance* h in the domain Ω, i.e.,

$$h = \sup_{\boldsymbol{x} \in \Omega} \min_{\boldsymbol{x}_j \in \mathcal{X}} \|\boldsymbol{x} - \boldsymbol{x}_j\|, \tag{9.13}$$

and that $u \in C^\beta(\Omega)$ for some nonnegative integer β. Note that one can think of this fill distance as the radius of the largest empty ball that can be placed among the data sites $\boldsymbol{x}_i$.

We begin by constructing a *local* polynomial interpolant of order β (i.e., degree $\beta-1$) to the function $u \in C^\beta(\Omega)$ at the points[1] $\mathcal{X}_x = \{t_1, \ldots, t_\beta\} \subset \mathcal{X}$. Here we pick the local interpolation points in dependence on the point of evaluation x, i.e., $\mathcal{X}_x$ consists of the β nearest neighbors of x from the set of global interpolation points $\mathcal{X}$; this implies that we need $N \geq \beta$. This also ensures that the local interpolation points are chosen as "symmetrical" as possible around x. Furthermore, we define the interval I_x as the smallest closed interval which contains the points $\mathcal{X}_x$.

The uniqueness of univariate polynomial interpolation ensures that there exists a constant C such that for any polynomial p of degree at most $\beta - 1$ we have

$$\sup_{x \in I_x} |p(x)| \leq C \sup_{t_j \in \mathcal{X}_x} |p(t_j)|, \tag{9.14}$$

i.e., the polynomial p is *determined*[2] by its values on the set $\mathcal{X}_x$.

A useful exact error formula for polynomial interpolation known as *Kowalewski's exact remainder formula* (see [Davis (1963, Eq. (3.7.10))], [Madych (2006)]) can be obtained starting from a Taylor expansion of u. We include the derivation of this formula since it sheds some light on the procedure recommended for the multivariate setting. Given $u \in C^\beta(\Omega)$ and an arbitrary point $t_j \in \mathcal{X}_x$, we begin with [Madych (2006, Eq. (2.2.1))]

$$u(x) = u(t_j) - \sum_{i=1}^{\beta-1} \frac{u^{(i)}(x)}{i!}(t_j - x)^i + \int_{t_j}^{x} \frac{(t_j - t)^{\beta-1}}{(\beta-1)!} u^{(\beta)}(t)\, \mathrm{d}t, \tag{9.15}$$

which is a straightforward rearrangement of Taylor's theorem with remainder.

Now we consider the βth order Lagrange polynomials for interpolation on $\mathcal{X}_x$ which satisfy the cardinality conditions $L_j(t_i) = \delta_{i,j}$, $i, j = 1, \ldots, \beta$ and also

$$\sum_{j=1}^{\beta} p(t_j) L_j(x) = p(x), \qquad \text{for } \textit{any} \text{ polynomial } p \text{ of order } \beta \text{ or less and } x \in I_x, \tag{9.16}$$

[1]For convenience we denote these points with t_j, $j = 1, \ldots, \beta$, instead of picking appropriate subsets of indices for the original points in $\mathcal{X}$.

[2]A consequence of this property is that if $p(t_j) = 0$ for all $t_j \in \mathcal{X}_x$ then $p \equiv 0$ on I_x, i.e., the points in $\mathcal{X}_x$ determine p. In the univariate setting this is straightforward, but in the multivariate setting this will present the first hurdle, calling for the notion of *norming sets* (9.25) (see also [Madych (2006, Section 2.8)]).

so that in particular $\sum_{j=1}^{\beta} L_j(x) \equiv 1$. Using this partition of unity property of the Lagrange polynomials we can obtain a weighted average of the Taylor formula (9.15) for $j = 1, \ldots, \beta$, i.e., we multiply both sides of (9.15) by $L_j(x)$ and then sum over j from 1 to β resulting in

$$u(x) = \sum_{j=1}^{\beta} L_j(x) u(t_j) - \sum_{j=1}^{\beta} L_j(x) \sum_{i=1}^{\beta-1} \frac{u^{(i)}(x)}{i!}(t_j - x)^i$$
$$+ \sum_{j=1}^{\beta} L_j(x) \int_{t_j}^{x} \frac{(t_j - t)^{\beta-1}}{(\beta-1)!} u^{(\beta)}(t)\, \mathrm{d}t. \tag{9.17}$$

The first sum on the right-hand side in (9.17) is nothing but $p_\beta(x)$, the unique β^{th} order (i.e., degree $\beta - 1$) polynomial interpolating u on $\mathcal{X}_x$, since $u(t_j) = p_\beta(t_j)$, $j = 1, \ldots, \beta$, and p_β is defined by its Lagrange form $p_\beta(x) = \sum_{j=1}^{\beta} L_j(x) p_\beta(t_j)$. The reproduction of polynomials up to degree $\beta - 1$ (9.16) implies that the double sum in (9.17) vanishes since $\sum_{j=1}^{\beta} L_j(x)(t_j - x)^i = 0$, $i = 1, \ldots, \beta - 1$. The result is Kowalewski's exact remainder formula as stated in [Madych (2006)]:

$$u(x) = p_\beta(x) + \sum_{j=1}^{\beta} L_j(x) \int_{t_j}^{x} \frac{(t_j - t)^{\beta-1}}{(\beta-1)!} u^{(\beta)}(t)\, \mathrm{d}t. \tag{9.18}$$

By replacing p_β by its Lagrange form as in (9.17), and then applying the triangle inequality to (9.18) and bounding the values of u in the first sum by their maximum we get

$$|u(x)| \le \Lambda_\beta \max_{t_j \in \mathcal{X}_x} |u(t_j)| + \sum_{j=1}^{\beta} |L_j(x)| \left| \int_{t_j}^{x} \frac{(t_j - t)^{\beta-1}}{(\beta-1)!} u^{(\beta)}(t)\, \mathrm{d}t \right|, \tag{9.19}$$

where we have introduced the abbreviation $\Lambda_\beta = \max_{x \in I_x} \sum_{j=1}^{\beta} |L_j(x)|$ to denote the *Lebesgue constant*[3] for polynomial interpolation at the points $t_1, \ldots, t_\beta$.

In order to remove the Lagrange polynomials from the second term in (9.19) we make use of (9.14) so that $|L_j(x)| \le C_j \max_{t_i \in \mathcal{X}_x} |L_j(t_i)| \le C_j$, by the cardinality of the Lagrange polynomials. This leaves us with

$$|u(x)| \le \Lambda_\beta \max_{t_j \in \mathcal{X}_x} |u(t_j)| + C \sum_{j=1}^{\beta} \left| \int_{t_j}^{x} \frac{(t_j - t)^{\beta-1}}{(\beta-1)!} u^{(\beta)}(t)\, \mathrm{d}t \right|,$$

where $C = \max_{j=1,\ldots,\beta} C_j$. Now we bound the integral using the Cauchy–Schwarz inequality, i.e.,

$$\left| \int_{t_j}^{x} \frac{(t_j - t)^{\beta-1}}{(\beta-1)!} u^{(\beta)}(t)\, \mathrm{d}t \right| \le \left| \int_{t_j}^{x} \frac{(t_j - t)^{2\beta-2}}{((\beta-1)!)^2}\, \mathrm{d}t \right|^{1/2} \left| \int_{t_j}^{x} \left(u^{(\beta)}(t)\right)^2 \mathrm{d}t \right|^{1/2}$$
$$= \frac{|t_j - x|^{\beta-1/2}}{\sqrt{2\beta-1}(\beta-1)!} \left| \int_{t_j}^{x} \left(u^{(\beta)}(t)\right)^2 \mathrm{d}t \right|^{1/2}.$$

[3]Note that the Lebesgue constant depends on the distribution of interpolation points. It grows logarithmically in β for Chebyshev points (which is the minimal rate of growth, but not with optimal constant, see, e.g. [Trefethen (2013, Chapter 15)]), while it grows exponentially for equally spaced points. Note, however, that this growth is not important for us since β is fixed here.

This results in

$$|u(x)| \leq \Lambda_\beta \max_{t_j \in \mathcal{X}_x} |u(t_j)| + C \sum_{j=1}^{\beta} \left\{ \frac{|t_j - x|^{\beta-1/2}}{\sqrt{2\beta-1}(\beta-1)!} \left| \int_{t_j}^{x} \left(u^{(\beta)}(t)\right)^2 \mathrm{d}t \right|^{1/2} \right\}.$$

We will aim for a sampling inequality in the L_2-norm. Therefore we next square both sides of this inequality and apply the Cauchy–Schwarz estimates $(A+B)^2 \leq 2A^2 + 2B^2$ and $\left(\sum_{j=1}^{\beta} A_j\right)^2 \leq \beta \sum_{j=1}^{\beta} (A_j)^2$ to the right-hand side. This yields

$$|u(x)|^2 \leq 2\Lambda_\beta^2 \left(\max_{t_j \in \mathcal{X}_x} |u(t_j)| \right)^2 + \frac{2kC^2}{(2\beta-1)((\beta-1)!)^2} \sum_{j=1}^{\beta} |t_j - x|^{2\beta-1} \left| \int_{t_j}^{x} \left(u^{(\beta)}(t)\right)^2 \mathrm{d}t \right|. \tag{9.20}$$

To obtain an L_2-norm estimate on the local interval I_x we will have to integrate both sides over $I_x = [a_x, b_x]$ with appropriately chosen endpoints a_x and b_x. We observe what happens if we integrate one of the summands on the right-hand side of (9.20) considering that $t_j \in [a_x, b_x]$:

$$\begin{aligned}
&\int_{a_x}^{b_x} |t_j - x|^{2\beta-1} \left| \int_{t_j}^{x} \left(u^{(\beta)}(t)\right)^2 \mathrm{d}t \right| \mathrm{d}x \\
&\quad = \int_{a_x}^{t_j} \left(u^{(\beta)}(t)\right)^2 \int_{t}^{a_x} |t_j - x|^{2\beta-1} \mathrm{d}x \, \mathrm{d}t + \int_{t_j}^{b_x} \left(u^{(\beta)}(t)\right)^2 \int_{t}^{b_x} |t_j - x|^{2\beta-1} \mathrm{d}x \, \mathrm{d}t \\
&\quad \leq \frac{|t_j - a_x|^{2\beta}}{2\beta} \int_{a_x}^{t_j} \left(u^{(\beta)}(t)\right)^2 \mathrm{d}t + \frac{|t_j - b_x|^{2\beta}}{2\beta} \int_{t_j}^{b_x} \left(u^{(\beta)}(t)\right)^2 \mathrm{d}t.
\end{aligned}$$

Since both $|t_j - a_x|$ and $|t_j - b_x|$ are at most βh (because of the definition of fill distance) we get

$$\int_{a_x}^{b_x} |t_j - x|^{2\beta-1} \left| \int_{t_j}^{x} \left(u^{(\beta)}(t)\right)^2 \mathrm{d}t \right| \mathrm{d}x \leq \frac{(\beta h)^{2\beta}}{2\beta} \int_{a_x}^{b_x} \left(u^{(\beta)}(t)\right)^2 \mathrm{d}t,$$

and since $\int_{a_x}^{b_x} \left(u^{(\beta)}(t)\right)^2 \mathrm{d}t = \|u^{(\beta)}\|_{L_2(I_x)}^2$ we now plug this into the integrated version of (9.20) to obtain

$$\begin{aligned}
\|u\|_{L_2(I_x)}^2 &\leq 2\beta h \Lambda_\beta^2 \left(\max_{t_j \in \mathcal{X}_x} |u(t_j)| \right)^2 + \frac{2kC^2}{(2\beta-1)\left((\beta-1)!\right)^2} \sum_{j=1}^{\beta} \frac{(\beta h)^{2\beta}}{2\beta} \|u^{(\beta)}\|_{L_2(I_x)}^2 \\
&= 2\beta h \Lambda_\beta^2 \left(\max_{t_j \in \mathcal{X}_x} |u(t_j)| \right)^2 + \frac{kC^2 (\beta h)^{2\beta}}{(2\beta-1)\left((\beta-1)!\right)^2} \|u^{(\beta)}\|_{L_2(I_x)}^2,
\end{aligned}$$

which is of the form

$$\|u\|_{L_2(I_x)}^2 \leq C_1^2 h \left(\max_{t_j \in \mathcal{X}_x} |u(t_j)| \right)^2 + C_2^2 h^{2\beta} \|u^{(\beta)}\|_{L_2(I_x)}^2$$

provided the constants C_1 and C_2 (which depend on β and on the points in $\mathcal{X}_x$, but not on h or u) are defined accordingly. Applying the inequality $\sqrt{A^2+B^2} \leq A+B$ (for $A, B > 0$) we have the final form of the *local* sampling inequality

$$\|u\|_{L_2(I_x)} \leq C_1 \sqrt{h} \max_{t_j \in \mathcal{X}_x} |u(t_j)| + C_2 h^\beta \|u^{(\beta)}\|_{L_2(I_x)}.$$

To finally obtain a *global* sampling inequality we use a technique introduced by Duchon (1978). For his *covering argument* to apply we note that we can cover the interval $\Omega = [a, b]$ with a family of subintervals I_x as discussed above. In fact, each of these subintervals has length at most βh and each point $x \in \Omega$ is covered by at most $\beta \leq N$ subintervals. This allows us to sum up all the local estimates to arrive at

$$\|u\|_{L_2(\Omega)} \leq C_3\sqrt{b-a} \max_{x_j \in \mathcal{X}} |u(x_j)| + C_4 h^\beta \|u^{(\beta)}\|_{L_2(\Omega)}, \tag{9.21}$$

where the constants C_3 and C_4 depend on β, N, and the distribution of the points in $\mathcal{X}$, but not on h and u. Note that this looks like (9.11) with $\Omega = [a, b]$, $C = \max\{C_3\sqrt{b-a}, C_4\}$, $m = 0$, and $p = q = 2$. Estimates in other norms can be obtained similarly using Hölder's inequality instead of Cauchy–Schwarz.

The sampling inequality (9.21) says that the continuous L_2-norm of u is bounded by the βth power of the fill distance times the stronger H^β-norm of u and a discrete maximum-norm of the values of u on $\mathcal{X}$. This estimate holds for any function $u \in C^\beta(\Omega)$. So, in particular, it provides an error bound for any method that generates an interpolant $s \in C^\beta(\Omega)$ to a function $f \in C^\beta(\Omega)$ at the points in $\mathcal{X}$. In this case we have

$$\|f - s\|_{L_2(\Omega)} \leq C h^\beta \|(f - s)^{(\beta)}\|_{L_2(\Omega)}$$

as outlined in Section 9.4.1. In order to get the upper bound to depend only on the input data, i.e., f and $\mathcal{X}$ (or its fill distance h) we use the triangle inequality so that

$$\|f - s\|_{L_2(\Omega)} \leq C h^\beta \left(\|f^{(\beta)}\|_{L_2(\Omega)} + \|s^{(\beta)}\|_{L_2(\Omega)} \right).$$

If we are working in a setting in which the interpolant is also a *minimum norm interpolant* (such as in a reproducing kernel Hilbert space $\mathcal{H}_K(\Omega)$, see (9.3b)) then $\|s^{(\beta)}\|_{L_2(\Omega)} \leq \|f^{(\beta)}\|_{L_2(\Omega)}$, provided $\mathcal{H}_K(\Omega)$ is equivalent to the Sobolev space $W_2^\beta(\Omega)$, and we obtain the error bound

$$\|f - s\|_{L_2(\Omega)} \leq 2C h^\beta |f|_{W_2^\beta(\Omega)}.$$

These orders in h, however, are only "half" of what is known as the optimal bounds in the literature. In order to "double" the orders one needs to take into account additional properties of the native Hilbert space. In particular, incorporating the boundary behavior of f and s is important to achieve those higher orders. We demonstrated this experimentally for one example of iterated Brownian bridge kernels in Section 7.3.3 and will provide a proof in Section 9.4.3.

Remark 9.3. Schaback (2011a) uses a simpler approach to obtain error bounds for univariate splines. He derives his sampling inequalities only for the $\beta = 1$ case (which is considerably simpler than what we did above for general smoothness β), and then obtains an error bound for smoother functions by iterating the first-order sampling inequality in conjunction with Rolle's theorem (see [Schaback

(2011a, Theorem 8.35)]). However, since Rolle's theorem does not apply in higher dimensions, other proof techniques are needed in that more general setting. The ideas used above can be generalized to higher dimensions, and this is outlined briefly in Section 9.4.4.

9.4.3 *Application to iterated Brownian bridge kernels*

We just showed that for any C^β-smooth interpolant of samples from a C^β-smooth function f on $\Omega = [0,1]$ we get a (sub-optimal) error bound of the form

$$\|f - s\|_{L_2[0,1]} \le Ch^\beta |f|_{W_2^\beta[0,1]}. \tag{9.22}$$

This is true, in particular, for the piecewise polynomial iterated Brownian bridge kernel $K_{\beta,0}$ of Chapter 7.

If we also demand that the even-derivative boundary conditions of our iterated Brownian bridge kernels (in this $\varepsilon = 0$ case) are satisfied, then we can prove

Theorem 9.1. *Let $f \in C^{2\beta}[0,1]$ be interpolated by a $K_{\beta,0}$ iterated Brownian bridge spline s on the set $\mathcal{X} = \{x_1, \ldots, x_N\}$ and also assume that f satisfies the same boundary conditions as the kernel $K_{\beta,0}$, i.e., $f^{(2\nu)}(0) = f^{(2\nu)}(1) = 0$, $\nu = 0, \ldots, \beta - 1$, then*

$$\|f - s\|_{L_2[0,1]} \le C^2 h^{2\beta} |f|_{W_2^{2\beta}[0,1]}. \tag{9.23}$$

Proof. The proof follows [Schaback (2011a)], but since the boundary conditions for our splines are different from his case we need a slightly modified argument. Following [Schultz and Varga (1967); Swartz and Varga (1972)], we introduce the *bilinear concomitant* $P_\beta(f,g)$ as

$$P_\beta(f,g) = \sum_{\nu=0}^{\beta-1} (-1)^\nu \mathcal{D}^{\beta-\nu-1} f(x)\, \mathcal{D}^{\beta+\nu} g(x),$$

where $\mathcal{D} = \frac{\mathrm{d}}{\mathrm{d}x}$ denotes a standard first-order derivative. The bilinear concomitant arises as the boundary contribution in repeated integration by parts, i.e.,

$$\int_0^1 \mathcal{D}^\beta f(x) \mathcal{D}^\beta g(x)\,\mathrm{d}x = P_\beta(f,g)\big|_0^1 + (-1)^\beta \int_0^1 f(x) \mathcal{D}^{2\beta} g(x)\,\mathrm{d}x.$$

Now, using (9.1), i.e., orthogonality in the reproducing kernel Hilbert space of our kernel $K_{\beta,0}$, we have

$$\|f - s\|^2_{\mathcal{H}_\mathcal{P}[0,1]} = \langle f - s, f - s\rangle_{\mathcal{H}_\mathcal{P}[0,1]} = \langle f - s, f\rangle_{\mathcal{H}_\mathcal{P}[0,1]}, \tag{9.24}$$

where $\|\cdot\|_{\mathcal{H}_\mathcal{P}[0,1]}$ is the norm induced by

$$\langle f, g\rangle_{\mathcal{H}_\mathcal{P}[0,1]} = \int_0^1 \mathcal{D}^\beta f(x) \mathcal{D}^\beta g(x)\,\mathrm{d}x$$

as in Section 8.3.

Using (9.24) together with the definition of the bilinear concomitant above we get

$$
\begin{aligned}
\|f-s\|^2_{\mathcal{H}_{\mathcal{P}}[0,1]} = \langle f-s, f\rangle_{\mathcal{H}_{\mathcal{P}}[0,1]} &= \int_0^1 \mathcal{D}^\beta(f-s)(x)\mathcal{D}^\beta f(x)\,\mathrm{d}x \\
&= P_\beta(f-s,f)\big|_0^1 + (-1)^\beta \int_0^1 (f-s)(x)\mathcal{D}^{2\beta}f(x)\,\mathrm{d}x.
\end{aligned}
$$

Inspecting the bilinear concomitant we note that

$$
\begin{aligned}
&P_1(f-s,f) = (f-s)(x)\mathcal{D}f(x) \\
&P_2(f-s,f) = \mathcal{D}(f-s)(x)\mathcal{D}^2 f(x) - (f-s)(x)\mathcal{D}^3 f(x) \\
&P_3(f-s,f) = \mathcal{D}^2(f-s)(x)\mathcal{D}^3 f(x) - \mathcal{D}(f-s)(x)\mathcal{D}^4 f(x) + (f-s)(x)\mathcal{D}^5 f(x) \\
&\qquad\vdots
\end{aligned}
$$

so that $P_\beta(f-s,f)$ is zero on the boundary since each product contains an even derivative of order up to $2\beta-2$ which is zero since the corresponding derivative of f as well as that of s (and therefore their difference) is zero on the boundary.

This leaves us with

$$
\begin{aligned}
\|f-s\|^2_{\mathcal{H}_{\mathcal{P}}[0,1]} &= (-1)^\beta \int_0^1 (f-s)(x)\mathcal{D}^{2\beta}f(x)\,\mathrm{d}x \\
&\le \|f-s\|_{L_2[0,1]}|f|_{W_2^{2\beta}[0,1]}.
\end{aligned}
$$

Our earlier (sub-optimal) estimate (9.22) together with the inequality just derived now gives us

$$
\begin{aligned}
\|f-s\|^2_{L_2[0,1]} &\le C^2 h^{2\beta}\|f-s\|^2_{\mathcal{H}_{\mathcal{P}}[0,1]} \\
&\le C^2 h^{2\beta}\|f-s\|_{L_2[0,1]}|f|_{W_2^{2\beta}[0,1]}
\end{aligned}
$$

or

$$
\|f-s\|_{L_2[0,1]} \le C^2 h^{2\beta}|f|_{W_2^{2\beta}[0,1]}.
$$

□

Unfortunately, the case $\varepsilon > 0$ seems to be considerably more difficult to analyze since the inner product of the reproducing kernel Hilbert space of $K_{\beta,\varepsilon}$ is related to the differential operator $\mathcal{L}_{\beta,\varepsilon}$ as well as our specific boundary conditions. Thus, bounds analogous to (9.23) cannot be established since the corresponding bilinear concomitant no longer seems to vanish on the boundary. Nevertheless, the numerical experiments in Chapter 7 showed that the same rate of convergence as in Theorem 9.1 can be achieved also in the non-polynomial case when $\varepsilon > 0$.

Remark 9.4. Our choice of boundary conditions for the iterated Brownian bridge kernels in Chapter 7 was motivated by the simplicity of the resulting eigenexpansion and therefore a particularly transparent implementation of the Hilbert–Schmidt SVD. One could, of course, work with the same differential operator and impose different boundary conditions, such as periodic boundary conditions, prescribed successive derivatives at the boundary, or setting odd derivatives to zero instead of even ones. All of these choices have been studied in the L-spline literature. In the literature on reproducing kernels the case of periodic boundary conditions was discussed in [Wahba (1990, Chapter 2)], resulting in the reproducing kernel (3.2). For this setup it is also rather straightforward to establish error bounds using either L-spline techniques or RKHS techniques combined with sampling inequalities as done above since the associated periodic boundary conditions work out nicely with the bilinear concomitant. In fact, error bounds for the case $\varepsilon > 0$ with periodic boundary conditions can already be found in the L-spline literature [Schultz and Varga (1967)].

9.4.4 *Sampling inequalities in higher dimensions*

The main problem in higher dimensions is the misbehavior of polynomial interpolation: depending on the points in $\mathcal{X}$, the interpolating polynomial may not be unique, or it may not even exist (see, e.g., [Gasca and Sauer (2000)]). Therefore, one usually resorts to *oversampling* and an approximation in terms of local (moving) least squares polynomials. This approach then also calls for so-called *norming sets* as defined in [Jetter *et al.* (1999)] (or as *determining sets* in [Madych (2006)]). A norming set Λ of a normed linear space $\mathcal{F}$ is a finite set of linear functionals (such as the point evaluation functionals $\lambda_i(f) = f(\boldsymbol{x}_i)$, $i = 1, \ldots, N$) defined on the dual of $\mathcal{F}$. It provides a bound of the continuous $\mathcal{F}$-norm of f by only discrete samples, i.e.,

$$\|f\|_{\mathcal{F}} \le C \sup_{\lambda_i \in \Lambda, \|\lambda_i\|=1} |\lambda_i(f)| \qquad \text{for all } f \in \mathcal{F}, \tag{9.25}$$

where C is some positive constant that depends on $\mathcal{F}$ and Λ (cf. (9.14) for the case of univariate polynomials and point evaluation functionals).

A standard way to prove a typical sampling inequality of the form

$$\|\mathcal{D}^{\boldsymbol{\alpha}} u\|_{L_q(\Omega)} \le C \left(h^{\beta - |\boldsymbol{\alpha}| - d\left(\frac{1}{p} - \frac{1}{q}\right)_+} |u|_{W_p^\beta(\Omega)} + h^{-|\boldsymbol{\alpha}|} \|S_{\mathcal{X}} u\|_{\ell_\infty} \right) \tag{9.26}$$

can be sketched as follows (see [Rieger *et al.* (2010)]). Here $S_{\mathcal{X}}$ is a *sampling operator* on $\mathcal{X}$ and $\boldsymbol{\alpha}$ is a multi-index (introduced in Section 3.3) so that $\mathcal{D}^{\boldsymbol{\alpha}}$ denotes a standard multivariate derivative operator of order $|\boldsymbol{\alpha}|$.

For some domain D, star-shaped with respect to a ball[4], let $\{a_j^{(\boldsymbol{\alpha})}, j = 1, \ldots, N\}$ be a *local* polynomial reproduction of degree β with respect to a discrete set $\mathcal{X} = \{\boldsymbol{x}_1, \ldots, \boldsymbol{x}_N\} \subset D$, i.e.,

$$\mathcal{D}^{\boldsymbol{\alpha}} q(\boldsymbol{x}) = \sum_{j=1}^{N} a_j^{(\boldsymbol{\alpha})}(\boldsymbol{x}) q(\boldsymbol{x}_j)$$

[4] D is called *star-shaped with respect to a ball* B if, for all $\boldsymbol{x} \in D$, the closed convex hull of $\{\boldsymbol{x}\} \cup B$ is a subset of D (see, e.g., [Brenner and Scott (1994, (4.2.2) Definition)]).

holds for every multi-index $\boldsymbol{\alpha}$ with $|\boldsymbol{\alpha}| \leq \beta$, all $\boldsymbol{x} \in D$ and all $q \in \Pi^d_\beta(D)$, where Π^β_d denotes the space of all d-variate polynomials of degree at most β. Then one obtains

$$|\mathcal{D}^{\boldsymbol{\alpha}} u(\boldsymbol{x})| \leq \|\mathcal{D}^{\boldsymbol{\alpha}}(u-p)\|_{L_\infty(D)} + \sum_{j=1}^N \left| a_j^{(\boldsymbol{\alpha})}(\boldsymbol{x}) \right| \left(\|u-p\|_{L_\infty(D)} + \|S_{\mathcal{X}} u\|_{\ell_\infty} \right)$$

for arbitrary $u \in W_p^\beta(D)$ and any polynomial $p \in \Pi^\beta_d(D)$. Using a polynomial reproduction argument based on norming sets, the Lebesgue constant can be bounded by $\sum_{j=1}^N \left| a_j^{(\boldsymbol{\alpha})}(\boldsymbol{x}) \right| \leq 2$, if some moderate oversampling is allowed which is controlled via a Markov inequality. As a local polynomial approximation one usually chooses the averaged Taylor polynomials of degree β (see [Brenner and Scott (1994, Section 4.1)][5], where a Bramble–Hilbert lemma [Brenner and Scott (1994, Lemma 4.3.8)] is given as an analogue to our estimate based on the Kowalewski remainder formula (9.18). In contrast to the averaging process we used in the one-dimensional setting, the multivariate average needs to be constructed with an appropriately defined *measure* which reproduces polynomials.

Doing all of this leads to a *local* sampling inequality of the form

$$\|\mathcal{D}^{\boldsymbol{\alpha}} u\|_{L_\infty(D)} \leq \frac{C}{(\beta - |\boldsymbol{\alpha}|)!} \delta_D^{\beta - d/p} \left(\delta_D^{-|\boldsymbol{\alpha}|} + h^{-|\boldsymbol{\alpha}|} \right) |u|_{W_p^\beta(D)} + 2h^{-|\boldsymbol{\alpha}|} \|S_{\mathcal{X}} u\|_{\ell_\infty},$$

where δ_D denotes the diameter of D.

To derive sampling inequalities on a *global* Lipschitz domain Ω satisfying an interior cone condition, we cover Ω by domains D which are star-shaped with respect to a ball, satisfying $\delta_D \approx h$ (see [Duchon (1978)] for details on such coverings). Global estimates are obtained by summation or maximization over the local estimates (see, e.g., [Wendland (2005, Theorem 11.32)]) similar to what was done in the one-dimensional setting.

9.5 Dimension-independent error bounds

In the last section of this chapter we mention some results of Fasshauer *et al.* (2012b) on the rates of convergence for kernel-based approximation, and in particular using Gaussian kernels. To be more specific, we will discuss weighted L_2 approximation when the data is specified either by function values of an unknown function f (from the native space of the kernel) or with the help of arbitrary, unspecified, linear functionals. Our convergence results pay special attention to the dependence of the estimates on the space dimension d. We will see that the use of *anisotropic Gaussian kernels* instead of isotropic ones provides *dimension-independent convergence rates*. The related paper [Fasshauer *et al.* (2012a)] discusses the related problem in the average case setting.

[5][Brenner and Scott (1994)] give credit to [Sobolev (1963)] and suggest the name *Sobolev polynomials*.

9.5.1 *Traditional dimension-dependent error bounds*

A good resource for standard scattered data approximation results based on the use of radial basis functions up to the year 2005 is [Wendland (2005)]. There we can find two different $L_\infty(\Omega)$, $\Omega \subseteq \mathbb{R}^d$, error bounds for isotropic Gaussian interpolation to data sampled from a function f in the native space $\mathcal{H}_K(\Omega)$ of the Gaussian. Both of these results are formulated in terms of the fill distance h defined in (9.13). Since the results we mention below are in terms of N, the number of data sites, we will restate the error bounds from [Wendland (2005)] also in terms of N using the fact that for quasi-uniformly distributed data sites we have $h = \mathcal{O}(N^{-1/d})$.

If f has derivatives up to total order β and s is the interpolant based on the Gaussian kernel $K(\boldsymbol{x}, \boldsymbol{z}) = \mathrm{e}^{-\varepsilon^2\|\boldsymbol{x}-\boldsymbol{z}\|^2}$, i.e.,

$$s(\boldsymbol{x}) = \sum_{j=1}^{N} c_j K(\boldsymbol{x}, \boldsymbol{x}_j) \quad \text{such that} \quad s(\boldsymbol{x}_i) = f(\boldsymbol{x}_i), \quad i = 1, \ldots, N,$$

then the first error bound is of the form

$$\|f - s\|_{L_\infty(\Omega)} \leq C_d N^{-\beta/d} \|f\|_{\mathcal{H}_K(\Omega)}$$

with some possibly dimension-dependent constant C_d. Therefore, infinitely smooth functions can be approximated with order $\beta = \infty$. With some extra effort one can also obtain the spectral estimate

$$\|f - s\|_{L_\infty(\Omega)} \leq \mathrm{e}^{-\frac{c}{d} N^{1/d} \log N} \|f\|_{\mathcal{H}_K(\Omega)}.$$

It is apparent from both of these bounds that the rate of convergence deteriorates as d increases. Moreover, the dependence of the constants on d is not clear. Therefore, these kinds of error bounds — and in fact almost all error bounds in the literature on radial basis functions — suffer from the *curse of dimensionality*. We will now present some results from [Fasshauer *et al.* (2012b)] on *dimension-independent* convergence rates for Gaussian kernel approximation.

9.5.2 *Worst-case weighted L_2 error bounds*

As already indicated above, we will make several assumptions in order to be able to obtain dimension-independent error bounds.

We define the worst-case weighted $L_2(\mathbb{R}^d, \rho)$ error as

$$\mathrm{err}^{wc}_{L_2(\mathbb{R}^d,\rho)} = \sup_{\|f\|_{\mathcal{H}_K(\mathbb{R}^d)} \leq 1} \|f - s\|_{L_2(\mathbb{R}^d,\rho)},$$

where ρ is the weight function for the Hilbert–Schmidt integral operator $\mathcal{K}$ associated with the Gaussian kernel as in Remark 2.1 or (12.19), the ρ-weighted L_2 norm is induced by the corresponding inner product (2.6), and s is our (minimum norm) kernel approximation calculated in the usual way. Therefore

$$\|f - s\|_{L_2(\mathbb{R}^d,\rho)} \leq \mathrm{err}^{wc}_{L_2(\mathbb{R}^d,\rho)} \|f\|_{\mathcal{H}_K(\mathbb{R}^d)} \quad \text{for all } f \in \mathcal{H}_K(\mathbb{R}^d).$$

The Nth *minimal worst case error* $\mathrm{err}^{wc}_{L_2(\mathbb{R}^d,\rho)}(N)$ refers to the worst case error that can be achieved with an optimal design, i.e., data generated by N optimally chosen linear functionals. For function approximation this means that the data sites have to be chosen in an optimal way. The results in [Fasshauer *et al.* (2012b)] are nonconstructive, i.e., no such optimal design is specified. However, a Smolyak or sparse grid algorithm is a natural candidate for such a design [Garcke and Griebel (2013)], as may be some of the other designs discussed in Appendix B.1.

If we are allowed to choose arbitrary linear functionals (so that derivatives or integrals/averages are allowed), then the optimal choice for weighted L_2 approximation is known. In this case we use *generalized Fourier coefficients*, i.e., the optimal linear functionals are $\mathcal{L}_j = \langle \cdot, \varphi_j \rangle_{\mathcal{H}_K(\mathbb{R}^d)}$ and we obtain the truncated *generalized Fourier series* approximation

$$s(\boldsymbol{x}) = \sum_{n=1}^{N} \langle f, \varphi_n \rangle_{\mathcal{H}_K(\mathbb{R}^d)} \varphi_n(\boldsymbol{x}) \qquad \text{for all } f \in \mathcal{H}_K(\mathbb{R}^d),$$

where we have used the Mercer series representation of K (cf. Section 2.2.3 and Section 2.2.6), i.e.,

$$K(\boldsymbol{x}, \boldsymbol{z}) = \sum_{n=1}^{\infty} \lambda_n \varphi_n(\boldsymbol{x}) \varphi_n(\boldsymbol{z}), \qquad \int_{\Omega} K(\boldsymbol{x}, \boldsymbol{z}) \varphi_n(\boldsymbol{z}) \rho(\boldsymbol{z}) \, \mathrm{d}\boldsymbol{z} = \lambda_n \varphi_n(\boldsymbol{x}).$$

It is then known (see, e.g., [Novak and Woźniakowski (2008)]) that

$$\mathrm{err}^{wc}_{L_2(\mathbb{R}^d,\rho)}(N) = \sqrt{\lambda_{N+1}},$$

the $(N+1)^{\mathrm{st}}$ largest eigenvalue of $\mathcal{K}$, which is easy to identify in the univariate case, but takes some care to specify in the multivariate setting (see the discussion in Section 12.2.1).

In [Fasshauer *et al.* (2012b)] it is then proved that in the isotropic case, i.e., with a truly radial Gaussian kernel of the form

$$K(\boldsymbol{x}, \boldsymbol{z}) = \mathrm{e}^{-\varepsilon^2 \|\boldsymbol{x} - \boldsymbol{z}\|^2}, \qquad \boldsymbol{x}, \boldsymbol{z} \in \mathbb{R}^d,$$

one can approximate function data with an Nth minimal error of the order $\mathcal{O}(N^{-1/4+\delta})$, where the constant in the $\mathcal{O}$-notation does not depend on the dimension d, and $\delta > 0$ is arbitrarily small. Similarly, for Fourier data (i.e., arbitrary linear functional data) an Nth minimal error of the order $\mathcal{O}(N^{-1/2+\delta})$ can be achieved.

Except for the paper [Beatson *et al.* (2010)], which provides error bounds for interpolation with anisotropic radial kernels, anisotropic kernels have not received much attention in the literature on approximation theory (in particular, radial basis functions). They do play an important role in the theory of high-dimensional integration (see, e.g., [Dick *et al.* (2013a)]) or in information-based complexity (see, e.g., [Novak and Woźniakowski (2008)]), and due to the additional flexibility they provide they have the potential to outperform isotropic kernels (see, e.g., our numerical experiments in Example 17.4 vs. Example 17.5).

This is also true for function approximation using Gaussian kernels. With anisotropic kernels, i.e.,

$$K(\boldsymbol{x},\boldsymbol{z}) = \mathrm{e}^{-\varepsilon_1^2(x_1-z_1)^2-\ldots-\varepsilon_d^2(x_d-z_d)^2}, \qquad \boldsymbol{x},\boldsymbol{z} \in \mathbb{R}^d,\ \boldsymbol{x} = \begin{pmatrix} x_1 & \cdots & x_d \end{pmatrix}^T,$$

one can do much better than what was mentioned above. In this case, if the shape parameters decay like $\varepsilon_\ell = \ell^{-\alpha}$ for some $\alpha \geq 0$, then one can approximate function data with an Nth minimal error of the order $\mathcal{O}(N^{-\max(\alpha^2/(2+\alpha),1/4)+\delta})$, and Fourier data (i.e., arbitrary linear functional data) with an Nth minimal error of the order $\mathcal{O}(N^{-\max(\alpha,1/2)+\delta})$. Again, the constants in the $\mathcal{O}$-notation do not depend on the dimension d.

In order to prove these results it was essential to have explicit expressions for the eigenvalues and eigenfunctions associated with the Gaussian kernel as given in Section 12.2.1.

Even if we do not have an eigenfunction expansion of a specific kernel available, the work of [Fasshauer *et al.* (2012b)] shows that for any radial (isotropic) kernel one has a dimension-independent Monte–Carlo type convergence rate of $\mathcal{O}(N^{-1/2+\delta})$ provided arbitrary linear functionals are allowed to generate the data. For translation-invariant (stationary) kernels the situation is similar. However, the constant in the $\mathcal{O}$-notation depends — in any case — on the trace of $\mathcal{K}$, i.e., the sum of the eigenvalues of the kernel (cf. (2.3)). For the radial case[6] this sum is simply $\kappa(0)$ (independent of d), while for general translation invariant kernels it is $\widetilde{K}(\mathbf{0})$, which may depend on d.

These results show that — even though radial basis function methods are often advertised as being "dimension-blind" — their rates of convergence are only excellent (i.e., spectral for infinitely smooth kernels) if the dimension d is small. For large dimensions, the constants in the $\mathcal{O}$-notation take over. If one, however, permits an anisotropic scaling of the kernel (i.e., elliptical symmetry instead of strict radial symmetry) and if those scale parameters decay rapidly with increasing dimension, then excellent convergence rates for approximation of smooth functions can be maintained independent of d.

[6]Here we use the notation $\kappa(\|\boldsymbol{x}-\boldsymbol{z}\|) = K(\boldsymbol{x},\boldsymbol{z})$ for radial kernels, and $\widetilde{K}(\boldsymbol{x}-\boldsymbol{z}) = K(\boldsymbol{x},\boldsymbol{z})$ for translation invariant kernels as introduced in Chapter 3.

Chapter 10

"Flat" Limits

In this chapter we take a closer look at the effect of the shape parameter ε present in the definition of many of our kernels. In particular, we are interested in understanding the behavior of *radial* kernel interpolants (cf. Section 3.1) in the limiting case as $\varepsilon \to 0$, i.e., the increasingly "flat" limit.

The results in this chapter specifically address the scattered data interpolation problem. In other words, we are given data sites $\mathcal{X} = \{\boldsymbol{x}_1, \ldots, \boldsymbol{x}_N\} \subset \mathbb{R}^d$ with associated data values $\{y_1, \ldots, y_N\}$ sampled from some function f and wish to reconstruct f by a function of the form

$$s_\varepsilon(\boldsymbol{x}) = \sum_{j=1}^{N} c_j \kappa(\varepsilon \|\boldsymbol{x} - \boldsymbol{x}_j\|), \qquad \boldsymbol{x} \in \mathbb{R}^d,$$

where $K(\boldsymbol{x}, \boldsymbol{z}) = \kappa(\|\boldsymbol{x} - \boldsymbol{z}\|)$ and the coefficients c_j are determined by satisfying the interpolation conditions

$$s_\varepsilon(\boldsymbol{x}_i) = y_i, \qquad i = 1, \ldots, N.$$

10.1 Introduction

Let us begin with a simple example.

Example 10.1. ("Flat" limits of C^2 Wendland kernels)
We take the C^2 compactly supported Wendland kernel (see Section 3.5) given by $\kappa(\varepsilon r) = (1 - \varepsilon r)_+^4 (1 + 4\varepsilon r)$ and look at its graph for different values of ε in Figure 10.1(a). We can see that a larger value of ε makes the kernel more peaked, and a smaller value produces a wider or "flatter" kernel. In fact, κ is supported on $[-\frac{1}{\varepsilon}, \frac{1}{\varepsilon}]$, i.e., the shape parameter — applied as a factor multiplying the argument r — acts as an inverse length scale.

As described above, we now use translates of this particular kernel to interpolate samples of the sinc function, $\operatorname{sinc}(x) = \frac{\sin(\pi x)}{\pi x}$, obtained at N evenly spaced points in $[0, 1]$. The sinc function is included in MATLAB's Signal Processing Toolbox. If that is not available, then the following function can be used instead.

Program 10.1. `sinc.m`

```
function f = sinc(x)
f = ones(size(x));
f(x~=0) = sin(pi*x(x~=0))./(pi*x(x~=0));
```

Figure 10.1(b) shows the maximum errors for different values of N ranging from 3 to 65 and shape parameter ε ranging from 0.001 to 100. From the graphs of the error (which we compute on a fine grid of 200 evenly spaced points) it is apparent that the error curve becomes "flat" both for sufficiently small and large values of ε. Since large values of ε give rise to very peaked kernels it is clear that the limiting error for large ε reflects the fact that the interpolant matches the data function only at the data points and is essentially zero everywhere else. The resulting errors are the same for all values of N and equal to the maximum absolute value of the data function on the interval (i.e., equal to $\operatorname{sinc}(0) = 1$ for our example).

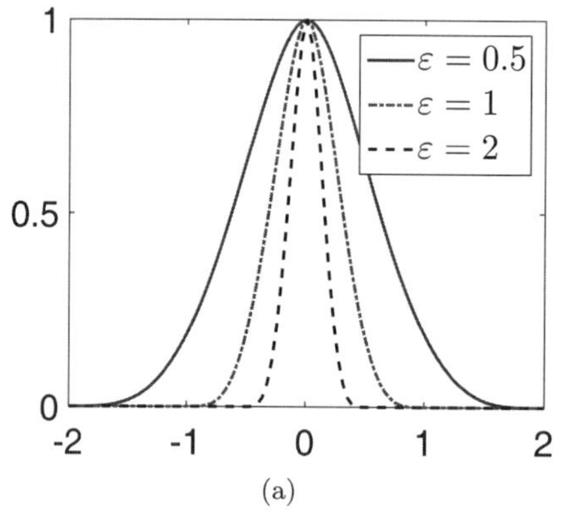

(a)

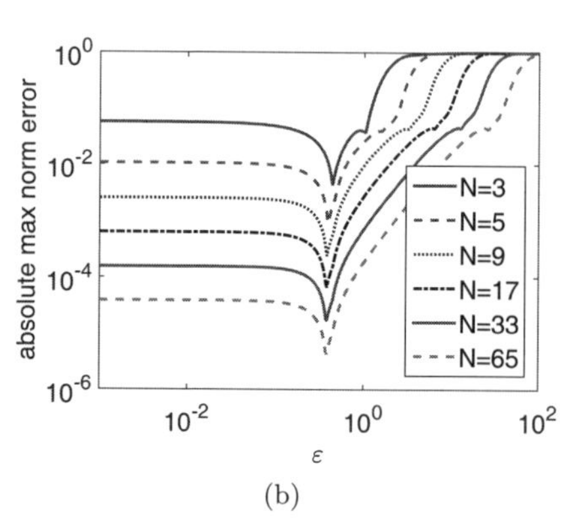

(b)

Fig. 10.1 (a) Wendland C^2 kernel with different shape parameters. (b) Maximum errors for interpolation of the sinc function at N evenly spaced points in $[0, 1]$ with C^2 Wendland kernels scaled by $\varepsilon \in [10^{-3}, 10^2]$.

But what is happening for small values of ε? Each one of the error curves levels out, but the actual value of the limit seems to decrease as N increases. Is there an explanation for this behavior? Does this happen for other radial kernels? Does it happen for all radial kernels? And what is this limit?

It is our goal to answer all of these questions below.

An additional interesting feature apparent in the error graphs is the fact that each curve seems to have a distinct minimum at a value of ε that is significantly different from zero, and the corresponding error is clearly smaller than in the "flat" limit. At this point, all we want to say is that it would seem to make sense to try to determine this "optimal" shape parameter (which we are able to locate here only because we know the function f that generated the data, and therefore can compute the error of the interpolant). This question of how to find a (near) optimal shape parameter — even if we don't know f — will be discussed in Chapter 14.

Finally, as we will see in Chapter 11, for increasingly small ε the error curves do not always look as "clean" as those in Figure 10.1 since the use of "flatter" kernels will lead to more and more similar entries in the system matrix K, and therefore to potential numerical problems due to ill-conditioning of K. In fact, this happens even for our example above if we let ε get an order or two smaller.

We now investigate the surprising answer to the "flat" limit mystery for different types of kernels — infinitely smooth kernels in Section 10.2, and finitely smooth kernels in Section 10.3.

10.2 Kernels with Infinite Smoothness

A little over 10 years ago (see, e.g., [Driscoll and Fornberg (2002); Larsson and Fornberg (2003); Fornberg and Wright (2004); Fornberg and Flyer (2005); Larsson and Fornberg (2005)]) an interesting connection was discovered between interpolants based on *infinitely smooth* radial basis functions such as Gaussians, inverse quadratics, (inverse) multiquadrics, and the Bessel/Poisson kernels (see Section 3.1): In many cases the limiting ("flat") radial basis function interpolants were identical to polynomial interpolants — especially in univariate experiments.

Thus, in recent years so-called "flat" RBFs have received much attention (see the references listed above and also, e.g., [Schaback (2005); Lee *et al.* (2007); Schaback (2008); Lee and Micchelli (2013)]). We begin by summarizing the essential insight gained in these papers, and then present some recent results from [Song *et al.* (2012)] that deal with radial kernels of *finite* smoothness in the next section.

Theorem 10.1. *Assume the positive definite radial kernel κ has an expansion of the form*

$$\kappa(r) = \sum_{n=0}^{\infty} a_n r^{2n}$$

into even powers of r (i.e., κ is infinitely smooth), and that the data locations $\mathcal{X}$ are unisolvent with respect to any set of N linearly independent d-variate polynomials of degree at most m. Then

$$\lim_{\varepsilon \to 0} s_\varepsilon(\boldsymbol{x}) = p_m(\boldsymbol{x}), \qquad \boldsymbol{x} \in \mathbb{R}^d,$$

where p_m is determined as follows:

- *If interpolation with polynomials of degree at most m is unique, then p_m is that unique polynomial interpolant.*
- *If interpolation with polynomials of degree at most m is not unique, then p_m is a polynomial interpolant whose form depends on the specific choice of kernel.*

Theorem 10.1 applies to kernels such as those listed in Table 10.1. The implications of this theorem are quite deep since it essentially establishes radial basis

Table 10.1 Some infinitely smooth globally supported radial kernels.

Name	Kernel	Series
IQ or Cauchy	$\frac{1}{1+r^2}$	$1 - r^2 + r^4 - r^6 + r^8 + \cdots$
Gaussian	e^{-r^2}	$1 - r^2 + \frac{r^4}{2} - \frac{r^6}{6} + \frac{r^8}{24} + \cdots$
IMQ	$\frac{1}{\sqrt{1+r^2}}$	$1 - \frac{r^2}{2} + \frac{3r^4}{8} - \frac{5r^6}{16} + \frac{35r^8}{128} + \cdots$
Poisson	$J_0(r)$	$1 - \frac{r^2}{4} + \frac{r^4}{64} - \frac{r^6}{2304} + \frac{r^8}{147456} + \cdots$

functions as generalizations of polynomial spectral methods. As a consequence, this opens the door to the design of algorithms for function approximation (see Chapter 15) as well as the numerical solution of partial differential equations that are more accurate than the standard polynomial spectral methods (see Chapter 20). Moreover, the scattered data setting in which radial basis functions are meant to be used allows for more flexibility with respect to geometry and adaptivity than standard polynomial approaches (for a discussion of the recent use of polynomials in the multivariate scattered data setting see Remark 10.4).

Example 10.2. (Kernels as more flexible tools than polynomials)
The Runge phenomenon [Trefethen and Weideman (1991); Trefethen (2013)] may arise when computing interpolants on evenly spaced points. Essentially, the "desire" of the interpolant to behave appropriately in the interior of the domain causes wild oscillations near the boundaries. The common example involves the so-called Runge function

$$f(x) = \frac{1}{1+25x^2}, \qquad x \in [-1,1],$$

which exposes the problematic Lebesgue constant associated with the cardinal functions of polynomial interpolants on equally spaced points. The common mechanism for fighting this behavior is to cluster points near the boundaries, often at the Chebyshev nodes (see Appendix B.3); see Figure 10.2.

Figure 10.2 contains a graph showing that the wild oscillations of polynomial interpolation on uniformly spaced points can be recovered by an increasingly flat analytic kernel, as predicted by Theorem 10.1. The kernel studied here is the Gaussian, which can achieve an accuracy that is superior to that obtained with for a uniformly spaced polynomial interpolant when more locality is enforced by choosing $\varepsilon > 0$. The small ε results for the Gaussian were computed using the Hilbert–Schmidt SVD which is discussed in Chapter 13.

A more advanced strategy for dealing with the Runge phenomenon would be to allow the use of *spatially varying* shape parameters, as suggested by Fornberg and Zuev (2007).

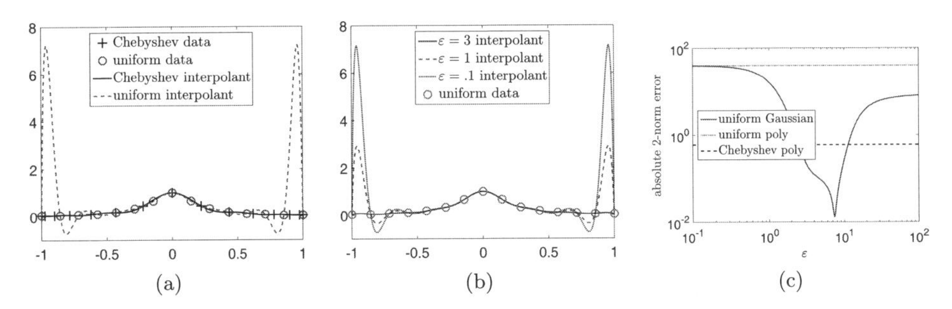

Fig. 10.2 The Runge phenomenon appears for polynomial interpolation on equally spaced points. (a) One solution is to consider interpolation on points clustered near the boundary, (b) Kernels can provide a strategy to enforce locality and prevent wild oscillations, (c) This ε profile demonstrates an optimal behavior for Gaussian kernels away from the $\varepsilon = 0$ polynomial limit.

10.3 Kernels with Finite Smoothness

To our knowledge, the flat limit of radial basis functions with finite smoothness was not studied until the recent paper [Song *et al.* (2012)] in which interpolation on $\mathbb{R}^d$ was investigated. The authors of [Lee *et al.* (2014)] use a variational approach, and obtain "flat" limit results also for translation invariant kernels and anisotropic radial kernels. As we will see below, there remain a number of unanswered questions in this area.

Before we explain the results obtained in [Song *et al.* (2012)], we remind the reader of our discussion in Chapters 6 and 7, where we focused on the connection to Green's kernels. We saw in those chapters that many of the popular kernels discussed in Chapter 3 can be obtained as Green's kernels for certain differential operators, either on all of $\mathbb{R}^d$, or on a bounded domain with appropriately defined boundary conditions.

Example 10.3. ("Convergence" of differential operators)

(1) The univariate C^0 *Matérn kernel* $K(x,z) = \mathrm{e}^{-\varepsilon|x-z|}$ is the full-space Green's function for the differential operator

$$\mathcal{L} = -\frac{\mathrm{d}^2}{\mathrm{d}x^2} + \varepsilon^2 \mathcal{I}.$$

On the other hand, it is well-known that univariate C^0 *piecewise linear splines* may be expressed in terms of kernels of the form $K(x,z) = |x-z|$. The corresponding differential operator in this case is

$$\mathcal{L} = -\frac{\mathrm{d}^2}{\mathrm{d}x^2}.$$

Note that the differential operator associated with the Matérn kernel "converges" to that of the piecewise linear splines as $\varepsilon \to 0$.

(2) As we saw in Chapter 7, on the interval $[0, 1]$ — with homogeneous boundary conditions added — the same differential operator that gives rise to the C^0 Matérn kernel produces the C^0 *iterated Brownian bridge kernel*

$$K_{1,\varepsilon}(x, z) = \frac{\sinh(\varepsilon \min(x, z)) \sinh(\varepsilon(1 - \max(x, z)))}{\varepsilon \sinh(\varepsilon)},$$

and the limiting kernel (for the differential operator $\mathcal{L} = -\frac{\mathrm{d}^2}{\mathrm{d}x^2}$ with homogeneous boundary conditions) is given by the piecewise linear *Brownian bridge kernel* $K_{1,0}(x, z) = \min(x, z) - xz$.

(3) The univariate C^2 *tension spline kernel* [Renka (1987)] $K(x, z) = \mathrm{e}^{-\varepsilon|x-z|} + \varepsilon|x - z|$ is the Green's kernel of

$$\mathcal{L} = -\frac{\mathrm{d}^4}{\mathrm{d}x^4} + \varepsilon^2 \frac{\mathrm{d}^2}{\mathrm{d}x^2},$$

while the univariate C^2 *cubic spline kernel* $K(x, z) = |x - z|^3$ corresponds to

$$\mathcal{L} = -\frac{\mathrm{d}^4}{\mathrm{d}x^4}.$$

Again, the differential operator associated with the tension spline "converges" to that of the cubic spline as $\varepsilon \to 0$.

(4) In Example 8.3 we discussed the so-called univariate *Sobolev kernel* from [Berlinet and Thomas-Agnan (2004, Chapter 7)] which is of the form $K(x, z) = \mathrm{e}^{-\varepsilon|x-z|} \sin\left(\varepsilon|x - z| + \frac{\pi}{4}\right)$ and associated with

$$\mathcal{L} = -\frac{\mathrm{d}^4}{\mathrm{d}x^4} - 4\varepsilon^4 \mathcal{I}.$$

The operator for this kernel also "converges" to that of the *cubic spline kernel*, but the effect of the scale parameter is analogous to that of the relaxation spline of Appendix A.3.2.

(5) If we instead consider the interval $[0, 1]$ with homogeneous boundary conditions for the function as well as its second derivative then we have a *cubic natural spline* limit if we consider the differential operator

$$\mathcal{L} = \frac{\mathrm{d}^4}{\mathrm{d}x^4} - 2\varepsilon^2 \frac{\mathrm{d}^2}{\mathrm{d}x^2} + \varepsilon^4 \mathcal{I}$$

and its $\varepsilon \to 0$ limit. The corresponding kernels are $K_{2,\varepsilon}$ and $K_{2,0}$ of Chapter 7.

(6) The general *multivariate Matérn kernels* (see Section 3.1.1) are of the form

$$K(\boldsymbol{x}, \boldsymbol{z}) = \frac{K_{d/2-\beta}(\varepsilon\|\boldsymbol{x} - \boldsymbol{z}\|)}{\varepsilon\|\boldsymbol{x} - \boldsymbol{z}\|^{d/2-\beta}}, \qquad \boldsymbol{x}, \boldsymbol{z} \in \mathbb{R}^d, \; \beta > \frac{d}{2},$$

and can be obtained as Green's kernels of (see [Fasshauer and Ye (2011)])

$$\mathcal{L} = \left(-\nabla^2 + \varepsilon^2 \mathcal{I}\right)^\beta, \qquad \beta > \frac{d}{2},$$

where ∇ is the standard gradient operator in $\mathbb{R}^d$. We contrast this with the *polyharmonic spline kernels*

$$K(\boldsymbol{x},\boldsymbol{z}) = \begin{cases} \|\boldsymbol{x}-\boldsymbol{z}\|^{2\beta-d}, & d \text{ odd}, \\ \|\boldsymbol{x}-\boldsymbol{z}\|^{2\beta-d}\log\|\boldsymbol{x}-\boldsymbol{z}\|, & d \text{ even}, \end{cases}$$

and

$$\mathcal{L} = (-1)^\beta \nabla^{2\beta}, \qquad \beta > \frac{d}{2}.$$

In summary, all of these examples show that the differential operators associated with finitely smooth radial kernels "converge" to those of a piecewise polynomial or polyharmonic spline kernel as $\varepsilon \to 0$. This motivates us to ask whether interpolants based on finitely smooth radial kernels converge to (polyharmonic) spline interpolants for $\varepsilon \to 0$ mimicking the relation between infinitely smooth radial kernels and polynomials. As Theorem 10.2 shows, this is indeed true.

Remark 10.1. However, we will see below (compare Table 10.2 with Remark 10.3) that — while we do get convergence of the interpolants — we may or may not get convergence of the kernels themselves.

As mentioned in Theorem 10.1, infinitely smooth radial kernels can be expanded into an infinite series of even powers of r. Finitely smooth radial kernels can also be expanded into an infinite series of powers of r. However, in this case there always exists some minimal term with nonzero coefficient which is not an even power of r (often this is an odd power, but sometimes it might be of a different form such as the logarithmic terms in Table 10.3). This term is an indicator of the smoothness of the kernel.

Remark 10.2. Use of a Taylor series expansion to determine the smoothness of the radial kernel at $r = 0$ is already suggested in [Stein (1999, Theorem 2)], where the name *principal irregular term* is used to denote the term corresponding to the index $2\nu + 1$ in Theorem 10.2.

Theorem 10.2 (Song *et al.* (2012)). *Suppose κ is a conditionally positive definite radial kernel*[1] *of order $\beta \le \nu$ with an expansion of the form*

$$\kappa(r) = a_0 + a_2 r^2 + \ldots + a_{2\nu} r^{2\nu} + a_{2\nu+1} r^{2\nu+1} + a_{2\nu+2} r^{2\nu+2} + \ldots,$$

where $2\nu+1$ denotes the smallest odd power of r present in the expansion (i.e., κ is finitely smooth). Also assume that the data locations $\mathcal{X}$ contain a unisolvent set with respect to the space $\Pi_d^{2\nu}$ of d-variate polynomials of degree less than 2ν. Then

$$\lim_{\varepsilon\to 0} s_\varepsilon(\boldsymbol{x}) = \sum_{j=1}^{N} c_j \|\boldsymbol{x}-\boldsymbol{x}_j\|^{2\nu+1} + \sum_{k=1}^{Q} d_k p_k(\boldsymbol{x}), \quad \boldsymbol{x} \in \mathbb{R}^d,$$

where $\{p_k,\ k = 1,\ldots,Q\}$ denotes a basis of Π_d^ν.

[1] The precise definition of a conditionally positive definite kernel can be found, e.g., in [Wendland (2005); Fasshauer (2007)]. However, the discussion surrounding (5.20) in Section 5.3.4 also has some details.

In other words, the "flat" limit of a piecewise smooth RBF interpolant is nothing but a polyharmonic spline interpolant. Therefore, just as infinitely smooth RBFs can be interpreted as generalizations of polynomials, we can view finitely smooth RBFs as generalizations of piecewise polynomial (or more generally polyharmonic) splines.

We list some specific Matérn kernels, Sobolev kernels from [Berlinet and Thomas-Agnan (2004)], Wendland functions, and Wu functions as examples in Tables 10.2 and 10.3.

Table 10.2 Series expansions for some globally supported Matérn and Sobolev kernels for spaces with different orders of smoothness.

Name	Kernel	Series
C^0 Matérn	e^{-r}	$1 - r + \frac{1}{2}r^2 - \frac{1}{6}r^3 + \cdots$
C^2 Matérn	$(1+r)\mathrm{e}^{-r}$	$1 - \frac{1}{2}r^2 + \frac{1}{3}r^3 - \frac{1}{8}r^4 + \cdots$
C^4 Matérn	$(1 + r + \frac{1}{3}r^2)\mathrm{e}^{-r}$	$1 - \frac{1}{6}r^2 + \frac{1}{24}r^4 - \frac{1}{45}r^5 + \frac{1}{144}r^6 + \cdots$
$H^2(\mathbb{R})$ kernel[a]	$\sqrt{2}\mathrm{e}^{-r}\sin\left(r + \frac{\pi}{4}\right)$	$1 - r^2 + \frac{2}{3}r^3 - \frac{1}{6}r^4 + \cdots$
$H^3(\mathbb{R})$ kernel[a]	$\frac{1}{2}\mathrm{e}^{-r} + \mathrm{e}^{-\frac{r}{2}}\sin\left(\frac{\sqrt{3}}{2}r + \frac{\pi}{6}\right)$	$1 - \frac{1}{4}r^2 + \frac{1}{24}r^4 - \frac{1}{80}r^5 + \cdots$

see [a][Berlinet and Thomas-Agnan (2004, Chapter 7)]

Table 10.3 Series expansions for some compactly supported Wendland (both "original" and "missing") and Wu kernels for spaces with different orders of smoothness. All kernels are evaluated for $0 \le r \le 1$ only and defined to be zero otherwise.

Name	Kernel	Series
C^0 Wendland ($d=3$)	$(1-r)^2$	$1 - 2r + r^2$
C^2 Wendland ($d=3$)	$(1-r)^4(4r+1)$	$1 - 10r^2 + 20r^3 - 15r^4 + 4r^5$
C^0 Wu kernel ($d \le 7$)	$(1-r)^4(1 + \frac{29}{16}r + \frac{5}{4}r^2 + \frac{5}{16}r^3)$	$1 - \frac{35}{16}r + \frac{35}{16}r^3 - \frac{21}{16}r^5 + \frac{5}{16}r^7$
C^2 Wu kernel ($d \le 5$)	$(1-r)^5(1 + 5r + 6r^2 + \frac{25}{8}r^3 + \frac{5}{8}r^4)$	$1 - 9r^2 + \frac{105}{8}r^3 - \frac{63}{8}r^5 + \frac{27}{8}r^7 - \frac{5}{8}r^9$
Wendland $H^2(\mathbb{R}^2)$	$(1+2r^2)\sqrt{1-r^2} + 3r^2\log\left(\frac{r}{1+\sqrt{1-r^2}}\right)$	$1 - \left(3\log(2) - \frac{3}{2}\right)r^2 + 3r^2\log(r)$ $-\frac{3}{8}r^4 - \frac{1}{32}r^6 - \cdots$
Wendland $H^3(\mathbb{R}^2)$	$(1 - 7r^2 - \frac{81}{4}r^4)\sqrt{1-r^2}$ $-\frac{15}{4}r^4(6+r^2)\log\left(\frac{r}{1+\sqrt{1-r^2}}\right)$	$1 - \frac{15}{2}r^2 - \left(\frac{135}{8} - \frac{45}{2}\log(2)\right)r^4$ $+\frac{45}{2}r^4\log(r) + \left(\frac{85}{16} + \frac{15}{4}\log(2)\right)r^6$ $-\frac{15}{4}r^6\log(r) - \frac{15}{128}r^8 - \cdots$

We point out that Theorem 10.2 does not cover Matérn kernels with odd-order smoothness, or the "missing" Wendland functions of Table 10.3. However, all other examples listed above are covered by the theorem.

Figure 10.3 illustrates the convergence of univariate C^0 and C^2 Matérn interpolants to piecewise linear and piecewise cubic spline interpolants, respectively.

Remark 10.3. While the iterated Brownian bridge kernels are not radial kernels,

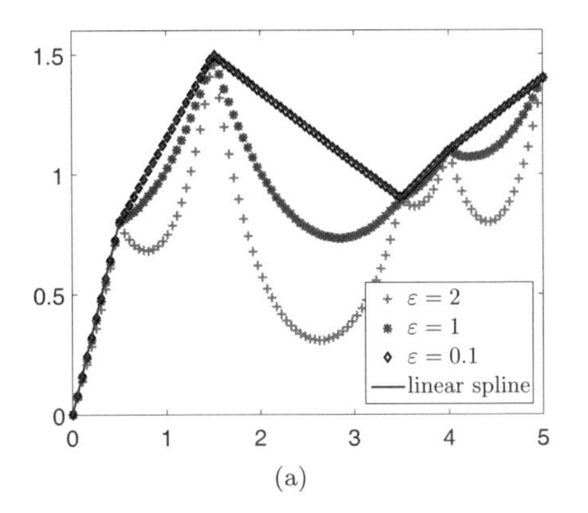

(a)

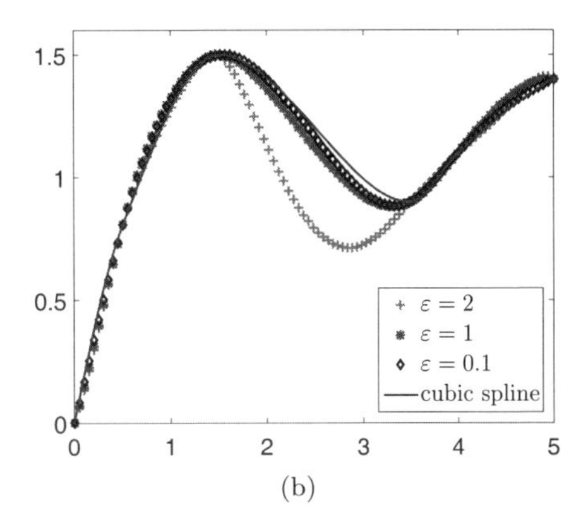

(b)

Fig. 10.3 Convergence of C^0 (a) and C^2 (b) Matérn interpolants to piecewise linear (a) and cubic (b) spline interpolants.

we can still expand them in terms of the shape parameter ε. For example, for the C^0 kernel (7.11) we get

$$\begin{aligned}
K_{1,\varepsilon}(x,z) &= \frac{\sinh(\varepsilon \min(x,z)) \sinh(\varepsilon(1-\max(x,z)))}{\varepsilon \sinh(\varepsilon)} \\
&= \min(x,z)(1-\max(x,z)) \\
&\quad + \frac{\varepsilon^2}{6}\min(x,z)(1-\max(x,z))\left(\min(x,z)^2+\max(x,z)^2-2\max(x,z)\right) \\
&\quad + \frac{\varepsilon^4}{360}\min(x,z)(1-\max(x,z))\left(3\min(x,z)^4+\cdots\right)+\cdots
\end{aligned}$$

and thus we see that the "flat" limit is indeed the Brownian bridge kernel $K_{1,0}(x,z) = \min(x,z) - xz$ (cf. (7.2a)). On the other hand, if we consider for simplicity only the case $0 \le x \le z \le 1$ of the C^2 kernel (7.12) then we get

$$\begin{aligned}
K_{2,\varepsilon}(x,z) &= \frac{x(1-z)}{6}\left(x^2+z^2-2z\right) \\
&\quad + \frac{\varepsilon^2 x(1-z)}{180}\left(3x^4+10x^2(z^2-2z)+3z^4-12z^3+8z^2+8z\right) \\
&\quad + \frac{\varepsilon^4 x(1-z)}{5040}\left(3x^6+\cdots\right)+\cdots,
\end{aligned}$$

which identifies the cubic spline kernel (7.3) as its "flat" limit, as expected.

10.4 Summary and Outlook

For infinitely smooth kernels, Lee and Micchelli (2013) show that in the univariate setting not only "flat" smooth *radial* kernel interpolants converge to polynomial interpolants, but that the same holds for interpolants based on "flat" smooth translation invariant kernels, and even for general smooth kernels. In the multivariate setting the authors consider interpolation points that are unisolvent for d-variate polynomials of some fixed total degree. In that case they also obtain a unique polynomial limiting interpolant for a given (not necessarily radial) kernel, provided it is analytic.

For finitely smooth kernels, the work of Lee *et al.* (2014) extends the results from [Song *et al.* (2012)] to anisotropic radial kernels and more general translation invariant kernels.

Remark 10.4. A connection between interpolation by increasingly flat isotropic Gaussian kernels and the concept of *least polynomial interpolation* due to de Boor and Ron (1990, 1992a,b) is provided by de Boor (2006) and Schaback (2005). The main message contained in the work of de Boor and Ron is that for every finite set of points in $\mathbb{R}^d$ one can find a space of multivariate polynomials, the so-called *least polynomial space*, and that this space gives rise to a unique polynomial interpolant. Moreover, among all polynomials permitting unique interpolation at these points, the least polynomials are of minimal degree.

The work of Fasshauer and McCourt (2012) on stable Gaussian computation and by Narayan and Xiu (2012) on *least orthogonal polynomial interpolation*, on the other hand, establishes the same connection via the Hermite polynomials which are present in the Gaussian eigenfunctions (see Section 12.2.1). Moreover, this connection suggests that similar links might exist for other radial basis function interpolants whose "flat" limit would then be given by the least polynomial interpolant associated with other classical orthogonal polynomials (cf. Section 5.5).

Chapter 11

The Uncertainty Principle – An Unfortunate Misconception

In the previous chapter we learned that there is a theoretical "flat" limit for kernel-based interpolation, and we also said that this limit might be difficult to achieve computationally. We now provide a typical plot for a very simple univariate interpolation problem with increasingly wider Gaussian kernels, i.e., taking ε to zero.

11.1 Accuracy vs. Stability

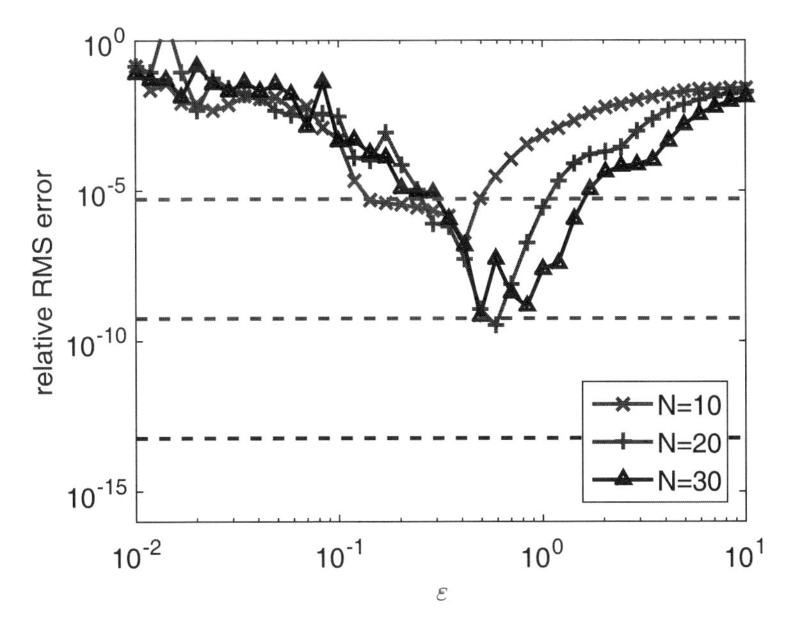

Fig. 11.1 Kernel-based interpolation to the function $f(x) = \frac{\sinh(x)}{1+\cosh(x)}$ with translates of Gaussian kernels for $\varepsilon \in [0.01, 10]$ and with polynomials of degree $N-1$ (constant dashed horizontal lines) at N Chebyshev points in $[-3, 3]$.

For the results plotted in Figure 11.1 we obtain data by sampling the function $f(x) = \frac{\sinh(x)}{1+\cosh(x)}$ at $N = 10, 20$ and 30 Chebyshev points mapped to the interval $[-3, 3]$. The curves denote the average errors for Gaussian kernel interpolation using the standard (or direct) method described in (1.3) for values of $\varepsilon \in [0.01, 10]$ and for corresponding polynomial interpolants. As we know from Theorem 10.1, the increasingly "flatter" Gaussian kernel interpolants should be converging to the polynomial interpolants for the same data. However, as the error curves for the two methods demonstrate, this is not at all happening.

For $N = 10$ points, the Gaussian error decreases as the value of the shape parameter decreases from $\varepsilon = 10$ to about $\varepsilon = 0.4$, where it even drops a bit below the polynomial error; but as ε decreases further the Gaussian error increases again, and it looks as if the Gaussian error is starting to approach that for the polynomial interpolant. However, at $\varepsilon \approx 0.1$ the Gaussian error crosses the polynomial one and continues to grow further and in fact the graph becomes rather erratic from that point on.

For $N = 20$ and $N = 30$, the erratic behavior is even more pronounced. In fact, so much so that the Gaussian error for $N = 30$ comes nowhere near that of the corresponding polynomial interpolant.

Figure 11.1 is the kind of error curve that any practitioner studying the effects of the shape parameter ε encountered for many years. Therefore it is not surprising that this led to a long-held — but mistaken — belief that

> "One can't simultaneously achieve high accuracy and numerical stability."

In addition to the countless numerical experiments that essentially sent the same message as Figure 11.1, researchers then also looked for a theoretical justification of this belief. They found it — or so they thought — in the so-called *uncertainty principle* due to Schaback (1995a,b) (see also [Fasshauer (2007, Section 16.2)], where it was referred to as a *trade-off principle*). These papers provide a rigorous mathematical foundation and show that a small theoretical error will automatically force a large condition number of the interpolation matrix K and therefore eventually lead to numerical instability so that the theoretically small error can never be reached in practice. There is, however, a fundamental assumption behind Schaback's uncertainty principle: it applies when the standard basis, i.e., $\{K(\cdot, \boldsymbol{x}_1), \ldots, K(\cdot, \boldsymbol{x}_N)\}$ is used. Therefore, the above paraphrase of the uncertainty principle should be corrected to something like

> "*Using the standard basis*, one can't simultaneously achieve high accuracy and stability."

In summary, we believe that Schaback's uncertainty principle has unfortunately led to a widespread misconception which has had a significant (negative) impact on the development and acceptance of kernel-based approximation methods by a greater portion of the scientific community.

11.2 Accuracy *and* Stability

It is one of the main goals of this book to demonstrate that it is possible to compute stably with kernels such as Gaussians — even when they become "flat." Figure 11.2 shows how the error curves in Figure 11.1 are dramatically improved when one uses a different basis for the space of Gaussian kernels. The "flat" limits predicted by Theorem 10.1 are clearly visible, and we can also see that it is possible to achieve greater accuracy with appropriately scaled Gaussians than with polynomials (which lack this scalability feature).

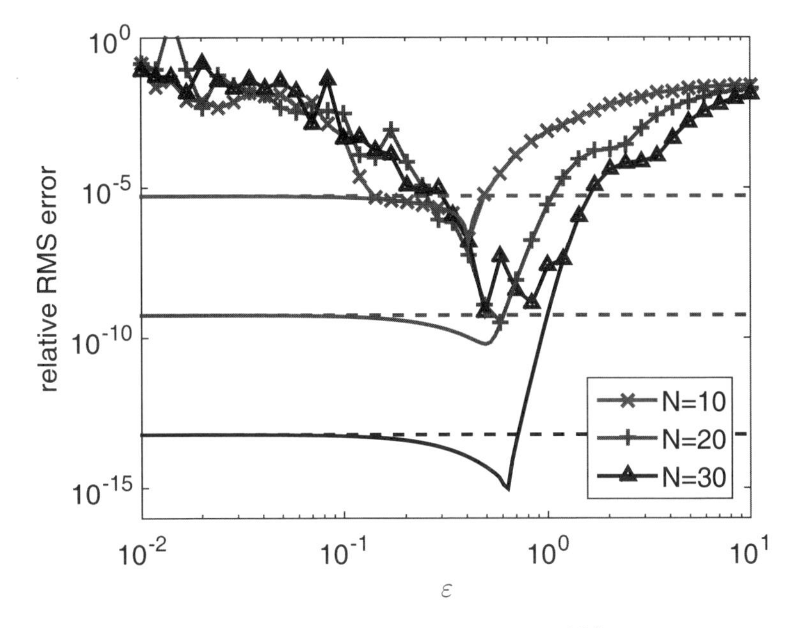

Fig. 11.2 Kernel-based interpolation to the function $f(x) = \frac{\sinh(x)}{1+\cosh(x)}$ with the stable Gaussian interpolant of Chapter 13 for $\varepsilon \in [0.01, 10]$ and with polynomials of degree $N-1$ (constant horizontal lines) at N Chebyshev points in $[-3, 3]$.

We introduce various alternatives to the standard kernel basis in Chapter 12, and then discuss the Hilbert–Schmidt SVD algorithm which was used to produce the stable and accurate results for Figure 11.2 in Chapter 13.

Figure 11.2 clearly demonstrates that we should be motivated to search for the value of ε for which kernels outperform polynomials (or at least provide their optimal accuracy), and not simply try to balance the trade-off between accuracy and numerical stability as was (and unfortunately still is being) argued by so many practitioners. This search leads to our investigation of optimal kernel parametrization strategies in Chapter 14.

Chapter 12

Alternate Bases

The main focus of this chapter is to start thinking about different ways to evaluate the kernel interpolant (see (1.3))

$$\begin{aligned} s(\boldsymbol{x}) &= \boldsymbol{k}(\boldsymbol{x})^T\boldsymbol{c}, \qquad \mathsf{K}\boldsymbol{c} = \boldsymbol{y} \\ &= \boldsymbol{k}(\boldsymbol{x})^T\mathsf{K}^{-1}\boldsymbol{y}. \end{aligned} \tag{12.1}$$

Looking at (12.1), we see that there are at least two ways to approach this task:

(1) We can compute the coefficients $\boldsymbol{c}$ by solving the linear system $\mathsf{K}\boldsymbol{c} = \boldsymbol{y}$ and then evaluate s via the dot product calculation (summation) against the basis functions $\boldsymbol{k}(\cdot)^T$.

(2) We can directly compute the *cardinal functions* $\boldsymbol{k}(\cdot)^T\mathsf{K}^{-1}$ and then evaluate s via a dot product against the data $\boldsymbol{y}$.

The first approach can be considered to be the "standard" (or "traditional") approach to this problem, i.e., we work with a basis generated by the kernel "shifts" $K(\cdot,\boldsymbol{x}_1),\ldots,K(\cdot,\boldsymbol{x}_N)$, where the points $\mathcal{X} = \{\boldsymbol{x}_1,\ldots,\boldsymbol{x}_N\} \subset \mathbb{R}^d$ are the given data sites for our interpolation (or approximation) problem. Thus, the corresponding function space is a finite-dimensional subspace, $\mathcal{H}_K(\mathcal{X})$, of the reproducing kernel Hilbert space $\mathcal{H}_K(\Omega)$. This subspace is given by

$$\mathcal{H}_K(\mathcal{X}) = \operatorname{span}\{K(\cdot,\boldsymbol{x}_1),\ldots,K(\cdot,\boldsymbol{x}_N)\}, \tag{12.2}$$

which is clearly *data-dependent*, as suggested by the Haar–Mairhuber–Curtis Theorem 1.2 and outlined in Chapters 1 and 2.

Following this strategy often involves working with an ill-conditioned system matrix K since the *numerical rank* of K is often much lower than N (see, e.g., Figure 4.4(d)). This usually means that the entries in $\boldsymbol{k}(\cdot)^T$ (i.e., the columns of K) are all very similar to each other. Therefore, the process of computing the coefficient vector $\boldsymbol{c}$ is numerically unstable. Moreover, even if we somehow manage to accurately compute $\boldsymbol{c}$, then we need to form the dot product with $\boldsymbol{k}(\boldsymbol{x})^T$ to evaluate s at $\boldsymbol{x}$. This process is again numerically unstable because there are huge oscillations in the entries of $\boldsymbol{c}$, but only very tiny variations in the entries of $\boldsymbol{k}(\boldsymbol{x})^T$. Thus, the first approach has recently been criticized by a number of researchers (see, e.g., [Fornberg and Piret (2008b); Boyd (2010)]) and has prompted research into more stable evaluation methods.

If one instead considers the second approach, then one is led to a quest for alternate *data-dependent* bases. In Section 12.1 we discuss different bases for $\mathcal{H}_K(\mathcal{X})$ such as the standard basis, cardinal (or Lagrange) basis, a Newton-type basis, and a (weighted) SVD basis. These approaches are all obtained via different factorizations of the kernel matrix K. A related idea is to precondition the standard kernel basis, and some recent developments are mentioned in Section 12.4.

Our interpretation of the second approach leads to the *Hilbert–Schmidt SVD*, which we discuss at length in Chapter 13. The Hilbert–Schmidt SVD produces a data-dependent basis[1] for $\mathcal{H}_K(\mathcal{X})$ which is obtained directly from the eigenfunctions of the Hilbert–Schmidt integral operator $\mathcal{K}$ associated with the kernel K. This means, in particular, that the Hilbert–Schmidt basis functions are obtained *without ever working with the matrix* K. Other recent stable alternate bases obtained without factoring K include the *Contour-Padé algorithm* of Fornberg and Wright (2004) (see also [Fasshauer (2007, Chapter 17)], the *RBF-QR algorithm* of Fornberg and Piret (2008b) (and also [Fornberg *et al.* (2011)]), and the *RBF-GA algorithm* of Fornberg, Lehto and Powell (2013).

In contrast to all of these data-dependent bases, we showed in Section 2.3 that the Hilbert–Schmidt eigenfunctions form a *data-independent* basis for the (infinite-dimensional) reproducing kernel Hilbert space $\mathcal{H}_K(\Omega)$ (see (2.25)).

In Section 12.2 we discuss analytical and numerical representations of the Hilbert–Schmidt eigenfunctions. This extends the discussion we started in Chapter 2. In particular, we study the eigenfunctions for the Gaussian kernel. Once we have explored these different ways to obtain eigenfunctions, we take a first look at using them for approximation in Section 12.3. That approach will not evaluate the kernel interpolant s from (12.1). Instead it will provide a low-rank approximation. We see that this is a reasonable alternative approach to (12.1) since — as mentioned above — the *numerical rank* of K is often much lower than N.

12.1 Data-dependent Basis Functions

12.1.1 *Standard basis functions*

The most commonly used approach for the solution of the scattered data interpolation problem is to employ the standard basis $\{K(\cdot,\boldsymbol{x}_1),\ldots,K(\cdot,\boldsymbol{x}_N)\}$ of the finite-dimensional function space $\mathcal{H}_K(\mathcal{X})$ (12.2). We have followed this approach in most of our discussions up to now, and many different kinds of kernels were featured in Chapter 3. As we have mentioned before, some of the most widely used kernels are the radial basis functions of the form $K(\|\cdot-\boldsymbol{x}_j\|)$.

As we saw in the previous chapter, one of the most frequently criticized aspects

[1] Up to a (small) truncation error.

of radial basis functions (and more general kernel-based methods) is rooted in the facts that

(1) the kernel matrix K is often *ill-conditioned*, along with
(2) this matrix often being a *dense matrix* and thus computationally inefficient.

We briefly focus on item (2) in Section 15.4.2.

An obvious idea for tackling the ill-conditioning of the matrix K is to use an appropriate *preconditioning* technique. Many such techniques have been proposed (and some were discussed in [Fasshauer (2007, Chapter 34)]). However, such techniques usually are successful only for moderately ill-conditioned matrices. Instead, in this chapter we want to look at a few options for dealing with item (1) by selecting a different basis for the approximation space. The recent paper [Pazouki and Schaback (2011)] — as well as its extension to conditionally positive definite kernels [Pazouki and Schaback (2013)] — are excellent references that provides a very nice discussion of this topic. In [Pazouki and Schaback (2011)] alternate bases were obtained by factoring the matrix K. In Chapter 13 we introduce the Hilbert–Schmidt SVD of the matrix K which also produces a new and numerically stable basis. However, this is accomplished *without ever forming and factoring* K.

Remark 12.1. Throughout this chapter we assume that the space $\mathcal{H}_K(\mathcal{X})$ is fixed, i.e., that the data sites have been chosen for us. This assumption is by no means the only possible scenario, nor is it necessarily a natural one since there exist plenty applications such as the *design of experiments* (see Chapter 16) for which an important part of the challenge usually lies in determining a "good" *design*, i.e., a good choice of data sites. The error bounds for kernel interpolation given in Chapter 9 also depend on the specific choice of data sites, and therefore — from that perspective — a choice that aims to minimize the error bound is certainly desirable.

If one, however, includes the data sites as variables in the data fitting problem, then one ends up with a *nonlinear* problem and we want to avoid that discussion here. In fact, to our knowledge, a satisfactory theoretical approach to this problem does not yet exist. This is in contrast to the multivariate integration problem, where so-called *low discrepancy* point sets have been studied for a long time (see, e.g., our brief summary in Appendix B.1 or the recent book [Dick and Pillichshammer (2010)]). In the literature on kernel-based methods, the papers [Iske (2000); De Marchi *et al.* (2005); Baldi Antognini and Zagoraiou (2010); Auffray *et al.* (2012)] have provided some initial progress in this direction for RBF interpolation and kriging, respectively. One special case presented in this chapter is the adaptive Newton basis computation in Section 12.1.4, which iteratively selects the "best" data sites as centers for the Newton basis functions.

12.1.2 *Cardinal basis functions*

In many ways, the *ideal* basis for $\mathcal{H}_K(\mathcal{X})$ is given by a *Lagrange* or *cardinal basis* $\{\ell_1, \ldots, \ell_N\}$. Such functions satisfy the *Lagrange property*

$$\ell_j(\boldsymbol{x}_i) = \delta_{i,j}, \qquad i, j = 1, \ldots, N, \tag{12.3}$$

so that we can find them (in a pointwise sense) by solving the linear system

$$\mathsf{K}\boldsymbol{\ell}(\boldsymbol{x}) = \boldsymbol{k}(\boldsymbol{x}), \tag{12.4}$$

for any given evaluation point $\boldsymbol{x} \in \mathbb{R}^d$. Here $\boldsymbol{\ell}(\cdot) = \left(\ell_1(\cdot) \cdots \ell_N(\cdot)\right)^T$ and we use the standard kernel matrix K with entries $(\mathsf{K})_{i,j} = K(\boldsymbol{x}_i, \boldsymbol{x}_j)$. The right-hand side vector is formed using the standard basis functions, i.e., $\boldsymbol{k}(\cdot) = \left(K(\cdot, \boldsymbol{x}_1) \cdots K(\cdot, \boldsymbol{x}_N)\right)^T$.

Once the cardinal functions are known, the interpolation problem becomes trivial since the interpolant can then immediately be written in the form

$$s(\boldsymbol{x}) = \boldsymbol{\ell}(\boldsymbol{x})^T \boldsymbol{y}, \qquad \boldsymbol{x} \in \Omega \subseteq \mathbb{R}^d, \tag{12.5}$$

i.e., the interpolation matrix for this basis is an identity matrix by virtue of (12.3).

Figure 12.1 shows three Gaussian cardinal basis functions obtained by solving (12.4) on a fine grid of evaluation points. The three functions plotted are identified with a corner, center and edge midpoint of the unit square, respectively. It is clear that these basis functions are no longer shifted copies of one single basic function. In fact, cardinal basis functions are inherently tied to the set $\mathcal{X}$ of data sites and the domain Ω. Nevertheless, if many interpolation problems with the same set $\mathcal{X}$ need to be solved, then one might consider pre-computing the cardinal basis. A value of $\varepsilon = 1$ was used for these plots.

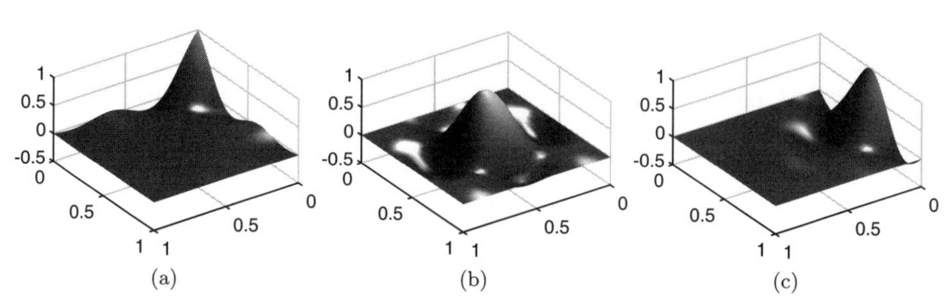

Fig. 12.1 Gaussian cardinal basis functions centered at different points in the unit square.

Remark 12.2. Historically, the term "cardinal" takes on at least two different meanings in the literature on interpolation and approximation. On the one hand, in the literature on classical piecewise polynomial splines, *cardinal interpolation* refers to the setting when the knots of the spline are restricted to lie on an infinite grid $h\mathbb{Z}^d$ (see, e.g., [Schoenberg (1969, 1973)], or [Kounchev (2001)] for a multivariate variant). Schoenberg developed this theory as a less smooth extension of the much older *cardinal series* of Whittaker (1915) (see also the survey [McNamee *et al.* (1971)] or [Fasshauer (2007, Chapter 13)]), i.e.,

$$C_{f,h}(x) = \sum_{j=-\infty}^{\infty} f(jh)\mathrm{sinc}\left(\frac{x-jh}{h}\right), \qquad x \in \mathbb{R},\ h>0,$$

which exactly recovers any band-limited function f. Note that this is also an interpolation formula based on the *sinc kernel* $K(x,z) = \mathrm{sinc}(x-z) = \frac{\sin \pi(x-z)}{\pi(x-z)}$ (see also Example 3.3 and Appendix A.8). Here the interpolation points $x_j = jh$ lie on an infinite grid. In addition, the sinc basis functions also satisfy the cardinality property defined above. It is this latter aspect that we intend to highlight whenever we use the term "cardinal". In fact, our cardinal functions will not only be defined on grids, but also on sets of scattered points as is the usual practice, e.g., in standard polynomial interpolation. The reader should be aware of this as many papers — especially in the RBF literature — refer only to the infinite grid setting when speaking of "cardinal" interpolation.

While it is straightforward to obtain values of the cardinal basis functions associated with any kernel K and any given set of data points $\mathcal{X}$ by solving a linear system of the form (12.4) for every desired evaluation point $\boldsymbol{x}$, this is also rather inefficient. Of course, for multiple evaluation points this can be done by viewing the problem as a matrix equation $\mathsf{K}\mathsf{L}_{\text{eval}} = \mathsf{K}_{\text{eval}}$ and thus having to factor the matrix K only once. Here the matrices L_{eval} and K_{eval} consist of columns of the form $\boldsymbol{\ell}(\boldsymbol{x})$ and $\boldsymbol{k}(\boldsymbol{x})$, respectively, each one determined by a desired evaluation point $\boldsymbol{x}$. Due to the cost that comes with determining the cardinal functions via such a linear system, it would be much more desirable to have closed-form expressions for the cardinal basis functions.

As we saw above, the sinc kernel is cardinal on an infinite grid. In that setting it is in fact possible to use Fourier transform techniques such as the *Poisson summation formula* to obtain closed-from expressions also for other kernels. This was done, e.g., in [Buhmann (1990); Madych and Nelson (1990)] for multiquadrics and polyharmonic splines. For the Gaussian kernel, an infinite cardinal basis was found in [Hangelbroek *et al.* (2012)], but there is also earlier work in, e.g., [Baxter and Sivakumar (1996)]. Moreover, in [Maz'ya and Schmidt (2007); Boyd and Wang (2009)] one can find so-called approximate cardinal functions for the Gaussian. In the more general — scattered — setting we are aware of only one result for univariate Gaussian cardinal functions on $[-1,1]$ from [Platte and Driscoll (2005)]. These functions are given by

$$\ell_j(\boldsymbol{x}) = \mathrm{e}^{-\varepsilon^2\left((x+1)^2-(x_j+1)^2\right)} \prod_{\substack{i=0\\ i\neq j}}^{N} \frac{\mathrm{e}^{\gamma x} - \mathrm{e}^{\gamma x_i}}{\mathrm{e}^{\gamma x_j} - \mathrm{e}^{\gamma x_i}}, \qquad j=0,1,\ldots,N.$$

Here $\gamma = \frac{4\varepsilon^2}{N}$ and the notation is based on the use of $N+1$ data points $x_0,\ldots,x_N$.

An algorithm using iterated quasi-interpolation for the family of Laguerre-Gaussian kernels (see [Fasshauer (2007, Chapter 4)]) which converges to the interpolant in cardinal form was presented in [Fasshauer and Zhang (2007a, 2009)] and similar ideas are used in [Kang and Joseph (2014)].

Cardinal functions are also very useful for theoretical purposes such as establishing error bounds for kernel-based scattered data interpolation, as was done in the seminal paper [Wu and Schaback (1993)].

If one has an explicit formula for the cardinal functions, then one immediately obtains information about the *Lebesgue constant* (cf. Section 9.4.2 or [Trefethen (2013, Chapter 15)])

$$\Lambda_{K,\mathcal{X}} = \max_{\boldsymbol{x}\in\Omega} \sum_{j=1}^{N} |\ell_j(\boldsymbol{x})|,$$

which in turn provides information about *accuracy* and *stability* of a kernel interpolant via

$$\|f - s\| \leq (1 + \Lambda_{K,\mathcal{X}})\, \|f - \mathring{s}\|,$$

where s is the kernel interpolant to the data $\boldsymbol{y}$ sampled from f on $\mathcal{X}$ and $\mathring{s}$ is the L_∞-best approximation of f from the finite-dimensional space $\mathcal{H}_K(\mathcal{X})$, i.e., the accuracy of the kernel interpolant s is not much worse than that of the best approximation of f. The stability is ensured by

$$\|s\|_{L_\infty(\Omega)} \leq \Lambda_{K,\mathcal{X}} \|\boldsymbol{y}\|_\infty,$$

which bounds the continuous norm of the interpolant on the left by a discrete norm of the data on the right and states that the interpolant behaves no worse than the data.

Unfortunately, the cardinal functions and associated Lebesgue constants of kernel-based interpolants are known only for a few special cases. Some results on the behavior of the Lebesgue constant can be found, e.g., in the papers [Platte and Driscoll (2005); De Marchi and Schaback (2010)].

Remark 12.3. One can use cardinal basis functions to obtain a *barycentric form* of the interpolant as discussed in [Berrut and Trefethen (2004)] or [Trefethen (2013, Chapter 5)], In the context of kernel-based methods this is employed for 1D Gaussians in [Platte and Driscoll (2005)], or for the sinc kernel in [Gautschi (2001)].

12.1.3 *Alternate bases via matrix factorization*

A general framework for finding alternate bases for the finite-dimensional kernel space $\mathcal{H}_K(\mathcal{X})$ by employing appropriate factorizations of the kernel matrix K was given by Pazouki and Schaback (2011). We briefly introduce this general framework in the following theorem (see [Pazouki and Schaback (2011, Theorem 3.1)]), and relate the theorem back to the cardinal basis discussed above. In the following subsections we then specialize to the factorizations that yield Newton and SVD bases.

Theorem 12.1. *Any data-dependent basis* $\{v_1, \ldots, v_N\}$ *for* $\mathcal{H}_K(\mathcal{X})$ *arises from a factorization*

$$\mathsf{K} = \mathsf{V}\mathsf{T}^{-1}$$

of the kernel matrix K *into the Vandermonde-like matrix* V *with values* $(\mathsf{V})_{i,j} = v_j(\boldsymbol{x}_i)$ *and the inverse of the* basis transformation matrix[2] T, *which relates the values of the new basis functions at a point* $\boldsymbol{x}$ *to those of the standard basis, i.e.,*

$$\boldsymbol{v}(\boldsymbol{x})^T = \boldsymbol{k}(\boldsymbol{x})^T \mathsf{T}.$$

Here $\boldsymbol{v}(\boldsymbol{x}) = \big(v_1(\boldsymbol{x}) \cdots v_N(\boldsymbol{x})\big)^T$ *and* $\boldsymbol{k}(\boldsymbol{x}) = \big(K(\boldsymbol{x}, \boldsymbol{x}_1) \cdots K(\boldsymbol{x}, \boldsymbol{x}_N)\big)^T$ *are — as usual — column vectors of the basis functions evaluated at* $\boldsymbol{x}$.

The validity of this theorem is immediately seen from the fact that the matrices K and V are obtained by stacking row vectors of the type $\boldsymbol{v}(\boldsymbol{x}_i)^T$ and $\boldsymbol{k}(\boldsymbol{x}_i)^T$, $i = 1, \ldots, N$, on top of each other.

Clearly, we can also express the kernel interpolant using any alternate basis $\{v_1, \ldots, v_N\}$. As we have already seen, in the standard basis we solve the linear system $\mathsf{K}\boldsymbol{c} = \boldsymbol{y}$ so that we have

$$s(\boldsymbol{x}) = \sum_{j=1}^{N} c_j K(\boldsymbol{x}, \boldsymbol{x}_j) = \boldsymbol{k}(\boldsymbol{x})^T \boldsymbol{c} = \boldsymbol{k}(\boldsymbol{x})^T \mathsf{K}^{-1} \boldsymbol{y}.$$

Analogously, in the alternate basis the linear system $\mathsf{V}\boldsymbol{b} = \boldsymbol{y}$ leads to

$$s(\boldsymbol{x}) = \sum_{j=1}^{N} b_j v_j(\boldsymbol{x}) = \boldsymbol{v}(\boldsymbol{x})^T \boldsymbol{b} = \boldsymbol{v}(\boldsymbol{x})^T \mathsf{V}^{-1} \boldsymbol{y}. \tag{12.6}$$

In light of Theorem 12.1, the cardinal basis (12.4) can be seen as

$$\mathsf{K}\boldsymbol{\ell}(\boldsymbol{x}) = \boldsymbol{k}(\boldsymbol{x}) \quad \Longleftrightarrow \quad \boldsymbol{\ell}(\boldsymbol{x})^T = \boldsymbol{k}(\boldsymbol{x})^T \mathsf{K}^{-1} \quad \Longrightarrow \quad \mathsf{K} = \mathsf{L}\mathsf{K}, \tag{12.7}$$

i.e., the basis transformation matrix is given by $\mathsf{T} = \mathsf{K}^{-1}$ and $\mathsf{V} = \mathsf{L} = \mathsf{I}$, as was already mentioned earlier. Equation (12.6) shows that — relative to an alternate basis, and therefore essentially *any* basis — the cardinal functions can be expressed as

$$\boldsymbol{\ell}(\boldsymbol{x})^T = \boldsymbol{v}(\boldsymbol{x})^T \mathsf{V}^{-1}, \tag{12.8}$$

where $\boldsymbol{v}$ is the vector of values of the basis functions at $\boldsymbol{x}$ and V is the Vandermonde-like matrix containing the values of the basis functions on the set $\mathcal{X}$ of data sites.

Another nice consequence of Theorem 12.1 is a representation of the *power function* in terms of the alternate basis $\{v_1, \ldots, v_N\}$. Starting from the definition

[2]Pazouki and Schaback (2011) call this matrix the *construction matrix*.

of the power function (9.8) and using (12.7), it is easy to express the power function in terms of the cardinal basis functions, i.e., we have

$$\begin{aligned} P_{K,\mathcal{X}}^2(\boldsymbol{x}) &= K(\boldsymbol{x},\boldsymbol{x}) - \boldsymbol{k}(\boldsymbol{x})^T \mathsf{K}^{-1}\boldsymbol{k}(\boldsymbol{x}) \\ &= K(\boldsymbol{x},\boldsymbol{x}) - \boldsymbol{\ell}(\boldsymbol{x})^T \mathsf{K}\boldsymbol{\ell}(\boldsymbol{x}). \end{aligned}$$

However, using Theorem 12.1 we also have

$$\begin{aligned} P_{K,\mathcal{X}}^2(\boldsymbol{x}) &= K(\boldsymbol{x},\boldsymbol{x}) - \boldsymbol{v}(\boldsymbol{x})^T \mathsf{T}^{-1}\mathsf{K}^{-1}\mathsf{T}^{-T}\boldsymbol{v}(\boldsymbol{x}) \\ &= K(\boldsymbol{x},\boldsymbol{x}) - \boldsymbol{v}(\boldsymbol{x})^T \mathsf{G}_v^{-1}\boldsymbol{v}(\boldsymbol{x}), \end{aligned} \tag{12.9}$$

where the matrix $\mathsf{G}_v = \mathsf{T}^T\mathsf{K}\mathsf{T}$ is the *Gram matrix* for the v-basis with respect to the native space inner product, i.e.,

$$\begin{aligned} (\mathsf{G}_v)_{i,j} &= (\mathsf{T}^T\mathsf{K}\mathsf{T})_{i,j} \\ &= \sum_{k=1}^N \sum_{\ell=1}^N K(\boldsymbol{x}_k,\boldsymbol{x}_\ell)\mathsf{T}_{ik}\mathsf{T}_{j\ell} \\ &= \sum_{k=1}^N \sum_{\ell=1}^N \langle K(\cdot,\boldsymbol{x}_k), K(\cdot,\boldsymbol{x}_\ell)\rangle_{\mathcal{H}_K(\Omega)}\mathsf{T}_{ik}\mathsf{T}_{j\ell} \\ &= \langle \sum_{k=1}^N K(\cdot,\boldsymbol{x}_k)\mathsf{T}_{ik}, \sum_{\ell=1}^N K(\cdot,\boldsymbol{x}_\ell)\mathsf{T}_{j\ell}\rangle_{\mathcal{H}_K(\Omega)} \\ &= \langle v_i, v_j\rangle_{\mathcal{H}_K(\Omega)}. \end{aligned}$$

Note that the Gram matrix for the cardinal basis is $\mathsf{G}_\ell = \mathsf{K}^{-T}\mathsf{K}\mathsf{K}^{-1} = \mathsf{K}^{-1}$. On the other hand, a basis whose Gram matrix is the identity — i.e., a basis consisting of functions that are $\mathcal{H}_K(\Omega)$-*orthonormal* — would be desirable, especially since one can show (see [Pazouki and Schaback (2011, Theorem 5.2)]) that

$$|s(\boldsymbol{x})|^2 \leq K(\boldsymbol{x},\boldsymbol{x})\|f\|_{\mathcal{H}_K(\Omega)}^2 \mathrm{cond}(\mathsf{G}_v).$$

Since $\mathsf{G}_v = \mathsf{I}$ minimizes this bound, one obtains optimum stability of the interpolation process in the sense that the magnitude of the interpolant is controlled as tightly as possible by the norm of the function $f \in \mathcal{H}_K(\Omega)$ to be interpolated. For more on the stability of kernel interpolants we refer the reader to [De Marchi and Schaback (2010)].

In the next two subsections we look at matrix factorizations of K that lead to $\mathcal{H}_K(\Omega)$-orthonormal bases.

12.1.4 *Newton-type basis functions*

In the setting of polynomial interpolation it is well known that the *Newton basis* is "in-between" the Lagrange basis and the Vandermonde-like monomial basis (which corresponds to our standard kernel basis) in the sense that the interpolation matrix associated with the Newton basis is triangular. In the context of Theorem 12.1 this means that we want to decompose K such that V is triangular, i.e., we want our Newton-type basis functions to satisfy a *Newton property*, i.e.,

$$v_j(\boldsymbol{x}_i) = 0, \qquad 1 \leq i < j \leq N.$$

This idea was first presented in the Ph.D. thesis of Stefan Müller (2009) (see also [Müller and Schaback (2009)]). From the point of view of Theorem 12.1 it can be deduced that a Newton-type basis can be computed via a pivoted *Cholesky decomposition* of the kernel matrix K. This fact was recognized in [Müller (2009); Müller and Schaback (2009); Bos *et al.* (2011c); Pazouki and Schaback (2011)]. Moreover, we show shortly that the Newton basis is *orthonormal with respect to native space inner product* $\langle\cdot,\cdot\rangle_{\mathcal{H}_K(\Omega)}$.

Following Pazouki and Schaback (2011), we obtain a Newton basis for our kernel space $\mathcal{H}_K(\mathcal{X})$ by applying the Cholesky factorization to K, i.e.,

$$\mathsf{K} = \mathsf{N}\mathsf{N}^T, \tag{12.10}$$

where N is lower triangular and nonsingular (since K is nonsingular). The matrix N is composed of the values $(\mathsf{N})_{i,j} = N_j(\boldsymbol{x}_i)$ of the Newton basis functions N_j at the data sites. The Cholesky factorization and Theorem 12.1 also imply that

$$\boldsymbol{n}(\boldsymbol{x})^T = \boldsymbol{k}(\boldsymbol{x})^T\mathsf{N}^{-T},$$

i.e., $\mathsf{V} = \mathsf{N}$ and the transformation matrix is given by $\mathsf{T} = \mathsf{N}^{-T}$. Here, the vector $\boldsymbol{n}(\cdot) = \big(N_1(\cdot) \cdots N_N(\cdot)\big)^T$ collects the Newton basis functions.

As claimed earlier, the Newton basis computed via the Cholesky factorization of K is $\mathcal{H}_K(\Omega)$-orthonormal. This is easily seen from the definition $\mathsf{G}_v = \mathsf{T}^T\mathsf{K}\mathsf{T}$ of the Gram matrix introduced above. For the Newton basis with transformation matrix $\mathsf{T} = \mathsf{N}^{-T}$ and factorization (12.10) we have

$$\mathsf{G}_n = \mathsf{N}^{-1}\mathsf{K}\mathsf{N}^{-T} = \mathsf{N}^{-1}\mathsf{N}\mathsf{N}^T\mathsf{N}^{-T} = \mathsf{I}.$$

In fact, [Pazouki and Schaback (2011, Theorem 6.1)] provides the more general result that *any* $\mathcal{H}_K(\Omega)$-orthonormal basis is obtained from K via a factorization of the form $\mathsf{K} = \mathsf{B}^T\mathsf{B}$ with an appropriate matrix B such that $\mathsf{B} = \mathsf{T}^{-1}$.

As mentioned above, the most straightforward way to obtain a Newton basis is to form the Cholesky factorization of K. However, that is probably also the most inefficient approach. Nevertheless, we list MATLAB code in Program 12.1 to produce the Newton basis functions of Figure 12.2.

Program 12.1. `NewtonBasisDirect.m`

```
% Define the Gaussian RBF
rbf = @(e,r) exp(-(e*r).^2);   ep = 3;
% Create some kernel centers and evaluation points
Ncenters = 5;  x = pick2Dpoints(0,1,Ncenters);
Neval = 40;  xeval = pick2Dpoints(0,1,Neval);
% Choose which Newton basis functions to plot
Nplot = [25 7 10];
```

```
% Evaluate the kernel matrices
K = rbf(ep,DistanceMatrix(x,x));
Keval = rbf(ep,DistanceMatrix(xeval,x));
% Compute the Newton basis functions
N = chol(K,'lower');
newtonbasis = Keval/N';
% Reshape the data for surface plotting
X = reshape(xeval(:,1),Neval,Neval);  Y = reshape(xeval(:,2),Neval,Neval);
Nmat = cellfun(@(nvec) reshape(nvec,Neval,Neval), ...
                        num2cell(newtonbasis,1),'UniformOutput',0);
% Create surface plots of the desired newton basis functions
for k=Nplot
    figure,  surf(X,Y,Nmat{k},'FaceColor','interp','EdgeColor','none')
    colormap autumn; camlight; lighting gouraud
end
```

The three Newton basis functions are for the Gaussian kernel with $\varepsilon = 3$ and a uniform grid of 5×5 points in the unit square. The displayed functions are once again identified with a corner, center and edge midpoint of the unit square, respectively. Clearly, these basis functions also cannot be obtained as shifted copies of a single basic function.

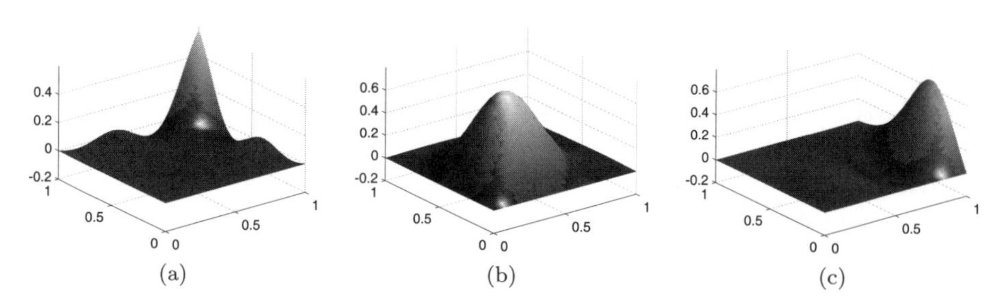

Fig. 12.2 Newton-type basis functions for the Gaussian kernel centered at different points in the unit square.

In [Müller and Schaback (2009); Pazouki and Schaback (2011)] several recursive algorithms were given for computing the Newton basis stably and efficiently. We now discuss two adaptive algorithms from [Pazouki and Schaback (2011, Section 10)]; the first selects the points independent of the data function f, while the second employs a data-dependent point selection. The most important insight for both these algorithms is provided by the expression that we now have for the power function in terms of the Newton basis. Using (12.9) and the fact that the Gram matrix for the Newton basis is the identity we have

$$P^2_{K,\mathcal{X}}(\boldsymbol{x}) = K(\boldsymbol{x},\boldsymbol{x}) - \boldsymbol{n}(\boldsymbol{x})^T\boldsymbol{n}(\boldsymbol{x}).$$

Here $\boldsymbol{n}(\cdot)$ is a vector containing the values of all N Newton basis functions $N_1(\cdot),\ldots,N_N(\cdot)$. Using a subset consisting of only n functions we have analogously

$$P^2_{K,\mathcal{X},n}(\boldsymbol{x}) = K(\boldsymbol{x},\boldsymbol{x}) - \sum_{j=1}^{n} N_j^2(\boldsymbol{x}), \tag{12.11}$$

where $P_{K,\mathcal{X},n}$ denotes the power function based on the points $\{\boldsymbol{x}_1,\ldots,\boldsymbol{x}_n\} \subseteq \mathcal{X}$. A simple telescoping argument shows that

$$N_n^2(\boldsymbol{x}) = P_{K,\mathcal{X},n-1}^2(\boldsymbol{x}) - P_{K,\mathcal{X},n}^2(\boldsymbol{x}) \le P_{K,\mathcal{X},n-1}^2(\boldsymbol{x}),$$

so that if $\boldsymbol{x}_n$ is chosen at the location that maximizes $P_{K,\mathcal{X},n-1}^2$, then $N_n^2(\boldsymbol{x}) \le N_n^2(\boldsymbol{x}_n)$ for all $\boldsymbol{x} \in \Omega$ (see [Pazouki and Schaback (2011, Theorem 9.1)]). Thus, an algorithm that selects successive new points at the maxima of the power function should perform very well and stably.

We have implemented such an algorithm in Program 12.2. In the initial step we find the point that gives rise to the maximal diagonal entry for the kernel matrix K (i.e., we use (12.11) for the case when none of the Newton basis functions are available). If we refer to this point as $\boldsymbol{\xi}_1$, then the first Newton basis is defined as

$$N_1(\boldsymbol{x}) = \frac{K(\boldsymbol{x},\boldsymbol{\xi}_1)}{\sqrt{K(\boldsymbol{\xi}_1,\boldsymbol{\xi}_1)}}. \tag{12.12}$$

After that, we use (12.11) to find the next point $\boldsymbol{\xi}_2$ (among the set $\mathcal{X}$ of data sites) for which (the square of) the power function $P_{K,\mathcal{X},1}$ has a maximum. This corresponds to finding the maximum of $\boldsymbol{z} - \boldsymbol{w}_1$, where these two vectors are given by

$$\begin{aligned} \boldsymbol{z} &= \operatorname{diag}(\mathsf{K}), \\ \boldsymbol{w}_1 &= \big(w_1(\boldsymbol{x}_1) \cdots w_1(\boldsymbol{x}_N)\big)^T, \quad \text{with } w_1(\boldsymbol{x}) = N_1^2(\boldsymbol{x}), \quad \boldsymbol{x} \in \mathcal{X}. \end{aligned}$$

The second Newton basis function is then calculated as

$$N_2(\boldsymbol{x}) = \frac{K(\boldsymbol{x},\boldsymbol{\xi}_2) - N_1(\boldsymbol{x})N_1(\boldsymbol{\xi}_2)}{\sqrt{K(\boldsymbol{\xi}_2,\boldsymbol{\xi}_2) - w_1(\boldsymbol{\xi}_2)}}. \tag{12.13}$$

The nth Newton basis function, $n = 3,\ldots,N$, requires

$$\boldsymbol{w}_n = \big(w_n(\boldsymbol{x}_1) \cdots w_n(\boldsymbol{x}_N)\big)^T, \quad \text{with } w_n(\boldsymbol{x}) = \sum_{j=1}^{n} N_j^2(\boldsymbol{x}), \quad \boldsymbol{x} \in \mathcal{X},$$

and

$$N_n(\boldsymbol{x}) = \frac{K(\boldsymbol{x},\boldsymbol{\xi}_n) - \sum_{j=1}^{n-1} N_j(\boldsymbol{x})N_j(\boldsymbol{\xi}_n)}{\sqrt{K(\boldsymbol{\xi}_n,\boldsymbol{\xi}_n) - w_n(\boldsymbol{\xi}_n)}}, \tag{12.14}$$

where $\boldsymbol{\xi}_n$ is the location at which $P_{K,\mathcal{X},n}^2 = \boldsymbol{z} - \boldsymbol{w}_n$ is maximal.

Note that the subtraction of the previous Newton basis function in (12.14) ensures the native space orthogonality of the basis since — by construction — we already have $\langle N_i, N_j\rangle_{\mathcal{H}_K(\Omega)} = \delta_{i,j}$ for $1 \le i,j \le n-1$. Thus, using (12.14) without the normalization factor and the reproducing property of K, we see that N_n is orthogonal to all the previous basis functions, i.e., for any $i = 1,\ldots,n-1$ we have

$$\langle N_i, K(\cdot,\boldsymbol{\xi}_n) - \sum_{j=1}^{n} N_j(\cdot)N_j(\boldsymbol{\xi}_n)\rangle_{\mathcal{H}_K(\Omega)} = N_i(\boldsymbol{\xi}_n) - \sum_{j=1}^{n} N_j(\boldsymbol{\xi}_n)\langle N_i, N_j\rangle_{\mathcal{H}_K(\Omega)} = 0.$$

Since the power function is a good error indicator it makes sense to potentially stop the iteration once the maximum of $P^2_{K,\mathcal{X},n} = \boldsymbol{z} - \boldsymbol{w}_n$ drops below a certain acceptable tolerance $(\mathrm{tol})^2$. The standard error estimate (9.7) then guarantees that the error of the interpolant will satisfy

$$\|f - s\|_\infty \leq \mathrm{tol}\|f\|_{\mathcal{H}_K(\Omega)}.$$

Program 12.2. Excerpts from NewtonBasisAdaptive.m

```
% Call the direct Newton code to set up the problem
NewtonBasisDirect
% Zero out the newton basis matrices
newtonbasis = zeros(Ncenters^2,Ncenters^2);
newtonbasisplot = zeros(Neval^2,Ncenters^2);
% Initialize with the first Newton basis function
z = diag(K);
[~,zind] = max(abs(z));
newtonbasis(:,1) = K(:,zind)/sqrt(z(zind));
newtonbasisplot(:,1) = Keval(:,zind)/sqrt(z(zind));
% Iterate through and compute one basis function at a time
w = zeros(Ncenters^2,1);
for iter=2:Ncenters
    % Update the power function and find the next kernel center
    w = w + newtonbasis(:,iter-1).^2;
    [~,zind] = max(z-w);
    % Form the auxiliary terms for the Newton basis computation
    y = newtonbasis(zind,1:iter-1)';
    powerfuniter = sqrt(z(zind)-w(zind));

    % Evaluate the Newton basis on the kernel centers
    u = K(:,zind) - newtonbasis(:,1:iter-1)*y;
    newtonbasis(:,iter) = u/powerfuniter;
end
```

Program 12.2 can be made more efficient by not forming and storing the full interpolation matrix K and the full evaluation matrix Keval. Instead, the required columns of these matrices can be formed "on demand" to minimize storage requirements. Even so, this program stands in stark contrast to much of the programming in this book in that a loop is fundamental to its operation. In this case, it seems to be a necessary evil, because the nth Newton basis function is dependent on all the earlier Newton basis functions. Still, one should remember to always search for an alternate to creating loops in MATLAB because using loops negates some of the significant computational prowess available in MATLAB.

Modifications to Program 12.2 could be made to produce a *function-dependent* Newton basis. In this case one would center the first basis function at that data site $\boldsymbol{\xi}_1$ whose corresponding value $y = f(\boldsymbol{\xi}_1)$ is maximal, and also base future selections on the size of the residual $f_n = f_{n-1} - \langle f, N_n\rangle_{\mathcal{H}_K(\Omega)} N_n$ rather than the power function. The plotting commands of Program 12.2 have been omitted to avoid redundancy with Program 12.1. The interested reader can consult [Pazouki and Schaback (2011)] for more details and also the MATLAB code provided by the authors of that paper. The resulting algorithm can be thought of as *orthogonal matching pursuit in* $\mathcal{H}_K(\Omega)$ as well as a version of the *greedy algorithm* of [Schaback and Wendland (2000)] which was also discussed in [Fasshauer (2007, Chapter 33)].

12.1.5 *SVD and weighted SVD bases*

In principle, an *SVD basis* can be obtained by performing an SVD of the (symmetric) kernel matrix K, i.e., $\mathsf{K} = \mathsf{Q}\Sigma^2\mathsf{Q}^T$ with the usual orthogonal matrix Q and diagonal matrix Σ^2 containing the squares of the eigenvalues of K so that — according to Theorem 12.1 — the new basis is given by $\boldsymbol{v}(\boldsymbol{x})^T = \boldsymbol{k}(\boldsymbol{x})^T\mathsf{Q}\Sigma^{-1}$ (implying $\mathsf{T} = \mathsf{Q}\Sigma^{-1}$) and the Vandermonde-like matrix containing the values of the new basis at the data sites is given by $\mathsf{V} = \mathsf{Q}\Sigma$ (see [Pazouki and Schaback (2011, Section 8)]).

A more sophisticated approach employing a *weighted SVD* is described in [De Marchi and Santin (2013)]. Essentially, it corresponds to a discretization of the integral eigenvalue problem for the Hilbert–Schmidt operator $\mathcal{K}$. As a matrix factorization of K we then end up with

$$\mathsf{K} = \mathsf{W}^{-1/2}\mathsf{Q}\Sigma^2\mathsf{Q}^T\mathsf{W}^{-1/2},$$

so that now $\mathsf{Q}\Sigma^2\mathsf{Q}^T$ is the SVD of $\mathsf{W}^{1/2}\mathsf{K}\mathsf{W}^{1/2}$ instead of K, as above. This means that

$$\boldsymbol{v}(\boldsymbol{x})^T = \boldsymbol{k}(\boldsymbol{x})^T\mathsf{W}^{1/2}\mathsf{Q}\Sigma^{-1} \tag{12.15}$$

and $\mathsf{V} = \mathsf{W}^{-1/2}\mathsf{Q}\Sigma$. Here the matrix W is a diagonal matrix of positive weights that arise in the discretization of the Hilbert–Schmidt eigenvalue problem (cf. (2.5))

$$\int_\Omega K(\boldsymbol{x},\boldsymbol{z})\varphi(\boldsymbol{z})\rho(\boldsymbol{z})\,\mathrm{d}\boldsymbol{z} = \lambda\varphi(\boldsymbol{x})$$

by employing a quadrature rule of the form

$$\int_\Omega f(\boldsymbol{z})\,\mathrm{d}\boldsymbol{z} \approx \sum_{j=1}^N w_j f(\boldsymbol{x}_j),$$

where $\mathcal{X} = \{\boldsymbol{x}_1,\ldots,\boldsymbol{x}_N\}$ are the data sites and $w_1,\ldots,w_N$ are appropriately chosen positive scalar weights. In [De Marchi and Santin (2013)] it is recommended that $\sum_{j=1}^N w_j = \mathrm{meas}(\Omega)$, which ensures that the quadrature rule is exact for constant f. A standard *Nyström method* (see, e.g., [Atkinson (1997)]) then consists of

(1) discretizing the integral via the quadrature method, and
(2) collocating the resulting equations on $\mathcal{X}$.

This results in a standard linear algebra eigenvalue problem of the form

$$\sum_{j=1}^N w_j K(\boldsymbol{x}_i,\boldsymbol{x}_j)\varphi(\boldsymbol{x}_j) = \mu\varphi(\boldsymbol{x}_i) \quad\Longleftrightarrow\quad \mathsf{K}\mathsf{W}\boldsymbol{\phi} = \mu\boldsymbol{\phi} \tag{12.16}$$

for the scaled kernel matrix KW, where $\mathsf{W} = \mathrm{diag}(w_1, \ldots, w_N)$. Here we have used μ to express the eigenvalues of this discrete eigenvalue problem. Since the matrix KW is no longer symmetric it is advantageous to rewrite (12.16) by multiplying both sides of the equation by $\sqrt{w_i}$, i.e.,

$$\sum_{j=1}^{N} \sqrt{w_i} K(\boldsymbol{x}_i, \boldsymbol{x}_j) \sqrt{w_j} \sqrt{w_j} \varphi(\boldsymbol{x}_j) = \mu \sqrt{w_i} \varphi(\boldsymbol{x}_i).$$

This latter system can now be viewed as

$$\mathsf{W}^{1/2} \mathsf{K} \mathsf{W}^{1/2} \tilde{\boldsymbol{\phi}} = \mu \tilde{\boldsymbol{\phi}}$$

with symmetric system matrix $\mathsf{W}^{1/2}\mathsf{K}\mathsf{W}^{1/2}$ and the eigenvectors $\tilde{\boldsymbol{\phi}} = \boldsymbol{\phi}\mathsf{W}^{1/2}$ representing a scaled version of $\boldsymbol{\phi}$, the discretization of the eigenfunction φ on $\mathcal{X}$.

De Marchi and Santin (2013) show that the weighted SVD basis (12.15) is $\mathcal{H}_K(\Omega)$-orthonormal and orthogonal with respect to the weighted discrete inner product in $\ell_{2,w}(\mathcal{X})$. This orthonormality leads to a formula for the power function that is similar to that for the Newton basis (which is also $\mathcal{H}_K(\Omega)$-orthonormal). For the SVD basis we get

$$P_{K,\mathcal{X}}^2(\boldsymbol{x}) = K(\boldsymbol{x}, \boldsymbol{x}) - \boldsymbol{v}(\boldsymbol{x})^T \boldsymbol{v}(\boldsymbol{x}).$$

The connection between the eigenvalues of the matrix $\mathsf{W}^{1/2}\mathsf{K}\mathsf{W}^{1/2} = \mathsf{Q}\Sigma^2\mathsf{Q}^T$, i.e., $\mu_j = \sigma_j$, and those of the integral operator $\mathcal{K}$ (denoted by λ_j above) are also discussed in [De Marchi and Santin (2013, Theorem 10)]. Namely,

$$\lambda_j \approx \sigma_j^2, \qquad j = 1, \ldots, N.$$

This follows from the fact that the SVD basis can be written as

$$\boldsymbol{v}(\boldsymbol{x})^T = \boldsymbol{k}(\boldsymbol{x})^T \mathsf{W}^{1/2} \mathsf{Q} \Sigma^{-1} = \boldsymbol{k}(\boldsymbol{x})^T \mathsf{W} \mathsf{V} \Sigma^{-2}$$

(since $\mathsf{V} = \mathsf{W}^{-1/2}\mathsf{Q}\Sigma$), and

$$\begin{aligned} \boldsymbol{k}(\boldsymbol{x})^T \mathsf{W} \mathsf{V} &= \sum_{j=1}^{N} K(\boldsymbol{x}, \boldsymbol{x}_j) w_j \boldsymbol{v}(\boldsymbol{x}_j)^T \\ &\approx \int_{\Omega} K(\boldsymbol{x}, \boldsymbol{z}) \boldsymbol{v}(\boldsymbol{z})^T \, \mathrm{d}\boldsymbol{z} = \mathcal{K} \boldsymbol{v}(\boldsymbol{x})^T. \end{aligned}$$

Since the trace of the matrix $\mathsf{W}^{1/2}\mathsf{K}\mathsf{W}^{1/2}$ equals the sum of its eigenvalues we have

$$\begin{aligned} \sum_{j=1}^{N} \sigma_j^2 &= \mathrm{trace}(\mathsf{W}^{1/2}\mathsf{K}\mathsf{W}^{1/2}) = \sum_{j=1}^{N} K(\boldsymbol{x}_j, \boldsymbol{x}_j) w_j \\ &\approx \int_{\Omega} K(\boldsymbol{z}, \boldsymbol{z}) \, \mathrm{d}\boldsymbol{z} = \mathrm{trace}\, \mathcal{K} = \sum_{n=1}^{\infty} \lambda_n, \end{aligned}$$

so that the (finite) sum of the eigenvalues of the weighted K matrix approximates the (infinite) sum of the eigenvalues of the integral operator $\mathcal{K}$ (cf. (2.3) and (2.10)).

In the case of equal weights, i.e., $w_j = \frac{1}{N}$, $j = 1, \ldots, N$, this result has been stated in various papers in the machine learning literature, either in the trace formulation (see, e.g., [Williams and Seeger (2000); Shawe-Taylor *et al.* (2002)]) or by comparing μ_j with $\frac{\lambda_j}{N}$ (see, e.g., [Williams and Seeger (2000); Bach and Jordan (2003)]).

Finally, Hansen (1988) goes into quite a bit more detail — but also uses a different matrix K with entries $(\mathsf{K})_{i,j} = \int_\Omega \int_\Omega K(\boldsymbol{x}, \boldsymbol{z}) h_i(\boldsymbol{x}) h_j(\boldsymbol{z})\, \mathrm{d}\boldsymbol{x}\, \mathrm{d}\boldsymbol{z}$, where $h_1, \ldots, h_N$ are an orthonormal basis (for the eigenfunctions) in Ω — and, e.g., bounds each individual kernel eigenvalue λ_j in terms of the corresponding matrix eigenvalue σ_j as

$$\sigma_j^2 \le \lambda_j \le \sqrt{\sigma_j^4 + \delta_N^2},$$

where $\delta_N^2 = \sum_{n=1}^\infty \lambda_n^2 - \sum_{j=1}^N \sigma_j^4$.

12.2 Analytical and Numerical Eigenfunctions

The search for data-dependent alternate bases has led primarily to strategies for approximate cardinal functions and matrix factorization schemes. As we will discuss in Section 12.3, the Hilbert–Schmidt eigenfunctions of a given kernel can also produce an *approximate alternate basis*. Before we can reach that point, we use this section to explain how Hilbert–Schmidt eigenfunctions can be found.

Some brief points on notation and conventions will help explain how these eigenfunctions are organized in more than one dimension. The first point we make is to not worry too much: when actually implementing solutions then technical details are significant, but when reading this book the details are more distracting than informative. The second point is that, in many cases, the order of eigenfunctions in higher dimensions is flexible and our strategy is not the only one.

The third point is that the ordering of series in higher dimensions is often dependent on a tensor product, with one major exception being zonal kernels on a sphere (see Section 15.3). We generally adhere to graded lexicographic ordering (which is discussed very effectively in Xiu (2010, Table 5.2)). The relationship between Mercer series ordered with a single index n and multi-index $\boldsymbol{n}$ is described in Table 12.1 for the three-dimensional setting (which generalizes in a straightforward way to arbitrary dimensions). The graded lexicographic ordering employed there ensures that

$$\lambda_n \varphi_n(\boldsymbol{x}) \varphi_n(\boldsymbol{z}) = \lambda_{\boldsymbol{n}} \varphi_{\boldsymbol{n}}(\boldsymbol{x}) \varphi_{\boldsymbol{n}}(\boldsymbol{z})$$

is satisfied for every term in the series.

The main focus of Table 12.1 is to show that eigenfunctions of the same order $|\boldsymbol{n}|$ are grouped together. The first eigenfunction has $|\boldsymbol{n}| = 3$ and all subsequent eigenfunctions are of greater order than that; the next three eigenfunctions have $|\boldsymbol{n}| = 4$, the next six after that have $|\boldsymbol{n}| = 5$ and so on. Such an arrangement with nondecreasing order is necessary in, e.g., Chapter 13 to ensure stability of the

Table 12.1 Relationship between single index notation and multi-index notation for a three-dimensional Mercer series.

Single index	Multi-index			Order
n	n_1	n_2	n_3	$\lvert\boldsymbol{n}\rvert$
1	1	1	1	3
2	2	1	1	4
3	1	2	1	4
4	1	1	2	4
5	3	1	1	5
6	2	2	1	5
7	2	1	1	5
8	1	3	1	5
9	1	2	2	5
10	1	1	3	5
11	4	1	1	6
12	3	2	1	6
		⋮		

HS-SVD basis. What is not necessary is that all the eigenfunctions with a fixed order, such as $|\boldsymbol{n}| = 5$, maintain the ordering prescribed in Table 12.1.

At times in this book, notably Section 13.4.2 and the end of Section 19.4.3, we will need to remove certain eigenfunctions from our Mercer series. These are very special circumstances though and, unless otherwise specified, we will primarily follow the ordering described above.

12.2.1 *Eigenfunctions given analytically*

Example 12.1. (Gaussian eigenfunctions)
Eigenfunctions for the Gaussian kernel $K(\boldsymbol{x},\boldsymbol{z}) = \mathrm{e}^{-\varepsilon^2\|\boldsymbol{x}-\boldsymbol{z}\|^2}$, $\boldsymbol{x},\boldsymbol{z} \in \mathbb{R}^d$, were first discussed for the case $d = 1$ in [Zhu *et al.* (1998)] (and later [Rasmussen and Williams (2006)], including the online errata). We follow here the formulation used in [McCourt (2013b)]. Since the general d-dimensional eigenfunctions and eigenvalues follow immediately from the univariate ones via the tensor product form of the Gaussian kernel, i.e.,

$$K(\boldsymbol{x},\boldsymbol{z}) = \mathrm{e}^{-\varepsilon^2\|\boldsymbol{x}-\boldsymbol{z}\|^2} = \mathrm{e}^{-\sum\limits_{\ell=1}^{d}\varepsilon^2(x_\ell-z_\ell)^2} = \prod_{\ell=1}^{d}\mathrm{e}^{-\varepsilon^2(x_\ell-z_\ell)^2},$$

where $\boldsymbol{x} = \begin{pmatrix}x_1 & \cdots & x_d\end{pmatrix}^T$, $\boldsymbol{z} = \begin{pmatrix}z_1 & \cdots & z_d\end{pmatrix}^T \in \mathbb{R}^d$, we have the Mercer series of the multivariate Gaussian kernel as

$$K(\boldsymbol{x},\boldsymbol{z}) = \sum_{\boldsymbol{n}\in\mathbb{N}^d} \lambda_{\boldsymbol{n}}\varphi_{\boldsymbol{n}}(\boldsymbol{x})\varphi_{\boldsymbol{n}}(\boldsymbol{z}),$$

where

$$\lambda_{\boldsymbol{n}} = \prod_{\ell=1}^{d} \lambda_{n_\ell}, \qquad \varphi_{\boldsymbol{n}}(\boldsymbol{x}) = \prod_{\ell=1}^{d} \varphi_{n_\ell}(x_\ell).$$

Therefore, the rest of this example will be concerned only with the univariate eigenfunctions and eigenvalues which we now conveniently index by $n = 1, 2, \ldots$. They are given by[3]

$$\varphi_n(x) = \gamma_n \mathrm{e}^{-\delta^2 x^2} H_{n-1}(\alpha\beta x), \tag{12.17}$$

$$\lambda_n = \sqrt{\frac{\alpha^2}{\alpha^2+\delta^2+\varepsilon^2}} \left(\frac{\varepsilon^2}{\alpha^2+\delta^2+\varepsilon^2}\right)^{n-1}, \tag{12.18}$$

where the H_n are *Hermite polynomials* of degree n (cf. Example 5.2), and

$$\beta = \left(1 + \left(\frac{2\varepsilon}{\alpha}\right)^2\right)^{\frac{1}{4}}, \quad \gamma_n = \sqrt{\frac{\beta}{2^{n-1}\Gamma(n)}}, \quad \delta^2 = \frac{\alpha^2}{2}\left(\beta^2 - 1\right)$$

are constants defined in terms of ε and α. The latter serves as a parametrization of the *weight function*

$$\rho(x) = \frac{\alpha}{\sqrt{\pi}} \mathrm{e}^{-\alpha^2 x^2} \tag{12.19}$$

which defines the Hilbert–Schmidt integral operator (2.4) and the associated inner product.

Remark 12.4. The eigenvalues and eigenfunctions of the Gaussian kernel are quite frightening by comparison to, say, those of the iterated Brownian bridge kernel. Their computation is complicated by numerical issues, e.g., the computation of δ^2 in the $\varepsilon \to 0$ limit is affected by numerical cancelation. The `GaussQR` library provides our best implementation of these functions using `gqr_phi`.

Instead of attempting to derive the Gaussian eigenfunctions and eigenvalues from the Hilbert–Schmidt integral operator eigenvalue problem we simply verify that the eigenfunctions are indeed orthonormal with respect to the ρ-weighted L_2 inner product, and that the Hilbert–Schmidt series sums to the Gaussian kernel.

In order to show the orthonormality of our eigenfunctions we recall the orthogonality relation for Hermite polynomials (see, e.g., [Abramowitz and Stegun (1965, Eq. (22.2.14))]), i.e.,

$$\int_{-\infty}^{\infty} H_m(x) H_n(x) \mathrm{e}^{-x^2}\, \mathrm{d}x = \sqrt{\pi} 2^n \Gamma(n+1) \delta_{m,n}. \tag{12.20}$$

[3]It should be noted that the eigenfunctions φ_n are *not* the same as the well-known classical Hermite functions, even though there is some similarity. The relative scaling of the arguments of the exponential function and the Hermite polynomials are different in the two cases.

Now, using the definition of the eigenfunctions (12.17) and of the weight function (12.19), a simple substitution $t = \alpha\beta x$ gives us

$$\int_{-\infty}^{\infty} \varphi_m(x)\varphi_n(x)\rho(x)\,\mathrm{d}x = \gamma_m\gamma_n \int_{-\infty}^{\infty} H_{m-1}(\alpha\beta x)H_{n-1}(\alpha\beta x)\mathrm{e}^{-2\delta^2x^2}\frac{\alpha}{\sqrt{\pi}}\mathrm{e}^{-\alpha^2x^2}\,\mathrm{d}x$$
$$= \frac{\beta}{\sqrt{2^{m-1}\Gamma(m)2^{n-1}\Gamma(n)}}\int_{-\infty}^{\infty} H_{m-1}(\alpha\beta x)H_{n-1}(\alpha\beta x)\mathrm{e}^{\alpha^2(1-\beta^2)x^2}\frac{\alpha}{\sqrt{\pi}}\mathrm{e}^{-\alpha^2x^2}\,\mathrm{d}x$$
$$= \frac{1}{\sqrt{\pi}\sqrt{2^{m-1}\Gamma(m)2^{n-1}\Gamma(n)}}\int_{-\infty}^{\infty} H_{m-1}(\alpha\beta x)H_{n-1}(\alpha\beta x)\mathrm{e}^{-\alpha^2\beta^2x^2}\alpha\beta\,\mathrm{d}x$$
$$= \frac{1}{\sqrt{\pi}\sqrt{2^{m-1}\Gamma(m)2^{n-1}\Gamma(n)}}\int_{-\infty}^{\infty} H_{m-1}(t)H_{n-1}(t)\mathrm{e}^{-t^2}\,\mathrm{d}t = \delta_{m,n},$$

where the last step invokes the orthogonality of the Hermite polynomials (12.20).

Verification of the sum of the Mercer series is a bit more involved. The classical result employed here is *Mehler's formula* (see [DLMF, Eq. (18.18.28)])

$$\sum_{n=0}^{\infty}\frac{H_n(x)H_n(z)}{2^n\Gamma(n+1)}t^n = \frac{1}{\sqrt{1-t^2}}\mathrm{e}^{\frac{2xzt-(x^2+z^2)t^2}{1-t^2}}, \quad |t|<1. \tag{12.21}$$

We now look at the Mercer series based on the Gaussian eigenfunctions (12.17) and eigenvalues (12.18) as defined above.

$$\sum_{n=1}^{\infty}\lambda_n\varphi_n(x)\varphi_n(z) = \sum_{n=0}^{\infty}\lambda_{n+1}\varphi_{n+1}(x)\varphi_{n+1}(z)$$
$$= \sum_{n=0}^{\infty}\lambda_{n+1}\gamma_{n+1}^2\mathrm{e}^{-\delta^2(x^2+z^2)}H_n(\alpha\beta x)H_n(\alpha\beta z)$$
$$= \mathrm{e}^{-\delta^2(x^2+z^2)}\sum_{n=0}^{\infty}\sqrt{\frac{\alpha^2}{\alpha^2+\delta^2+\varepsilon^2}}\left(\frac{\varepsilon^2}{\alpha^2+\delta^2+\varepsilon^2}\right)^n\frac{\beta}{2^n\Gamma(n+1)}H_n(\alpha\beta x)H_n(\alpha\beta z)$$
$$= \mathrm{e}^{-\delta^2(x^2+z^2)}\frac{\alpha\beta}{\varepsilon}\sqrt{\frac{\varepsilon^2}{\alpha^2+\delta^2+\varepsilon^2}}\sum_{n=0}^{\infty}\frac{H_n(\alpha\beta x)H_n(\alpha\beta z)}{2^n\Gamma(n+1)}\left(\frac{\varepsilon^2}{\alpha^2+\delta^2+\varepsilon^2}\right)^n.$$

Now we let $t = \frac{\varepsilon^2}{\alpha^2+\delta^2+\varepsilon^2}$ and apply (12.21) with $x \to \alpha\beta x$ and $z \to \alpha\beta z$ to get

$$\sum_{n=1}^{\infty}\lambda_n\varphi_n(x)\varphi_n(z) = \mathrm{e}^{-\delta^2(x^2+z^2)}\frac{\alpha\beta}{\varepsilon}\sqrt{\frac{t}{1-t^2}}\mathrm{e}^{\frac{2\alpha^2\beta^2xzt-\alpha^2\beta^2(x^2+z^2)t^2}{1-t^2}}$$
$$= \frac{\alpha\beta}{\varepsilon}\sqrt{\frac{t}{1-t^2}}\mathrm{e}^{-\frac{\alpha^2\beta^2t}{1-t^2}(x-z)^2}\mathrm{e}^{\frac{\alpha^2\beta^2t-\alpha^2\beta^2t^2-\delta^2(1-t^2)}{1-t^2}(x^2+z^2)}$$
$$= \mathrm{e}^{-\varepsilon^2(x-z)^2},$$

where we have combined all of the exponential functions, replaced $2\alpha^2\beta^2xzt$ by $-\alpha^2\beta^2(x-z)^2t+\alpha^2\beta^2(x^2+z^2)t$, and then separated into two exponential functions in terms of $(x-z)^2$ and x^2+z^2, respectively.

The remaining details for the last step can be verified by the reader (if necessary with a computer algebra system). They are

$$\frac{\alpha\beta}{\varepsilon}\sqrt{\frac{t}{1-t^2}} = 1,$$

which takes care of both the factor multiplying the exponential functions as well as the exponent of the first exponential function. The other identity is

$$\alpha^2\beta^2 t - \alpha^2\beta^2 t^2 - \delta^2(1-t^2) = 0.$$

After having performed these calculations, i.e., after having established that $K(x,z) = \sum_{n=1}^{\infty}\lambda_n\varphi_n(x)\varphi_n(z)$ and $\int_{-\infty}^{\infty}\varphi_m(x)\varphi_n(x)\rho(x)\,\mathrm{d}x = \delta_{m,n}$, it is also straightforward to verify that the Hilbert–Schmidt eigenvalue problem is satisfied, i.e.,

$$\begin{aligned}
(\mathcal{K}\varphi_n)(z) &= \int_{-\infty}^{\infty} K(x,z)\varphi_n(x)\rho(x)\,\mathrm{d}x \\
&= \int_{-\infty}^{\infty}\sum_{m=1}^{\infty}\lambda_m\varphi_m(x)\varphi_m(z)\varphi_n(x)\rho(x)\,\mathrm{d}x \\
&= \sum_{m=1}^{\infty}\lambda_m\varphi_m(z)\int_{-\infty}^{\infty}\varphi_m(x)\varphi_n(x)\rho(x)\,\mathrm{d}x \\
&= \sum_{m=1}^{\infty}\lambda_m\varphi_m(z)\delta_{m,n} \\
&= \lambda_n\varphi_n(z).
\end{aligned}$$

In fact, this argument holds as soon as we have the Hilbert–Schmidt series of an arbitrary kernel and know that the eigenfunctions are L_2-orthonormal with respect to some weight function ρ. Moreover, the argument carries over to the multivariate case with arbitrary domains $\Omega \subseteq \mathbb{R}^d$.

12.2.2 *Eigenfunctions obtained computationally*

In Section 12.1.5 we saw one approach to approximating the eigenfunctions of the Hilbert–Schmidt operator $\mathcal{K}$ based on the use of a Nyström method. We now present two alternative approaches. The first is a general *collocation method* and the second a *Galerkin method*. The Nyström method of Section 12.1.5 can be seen as a special case of the collocation method we now describe.

Collocation Method

Making the basic assumption that the eigenfunctions can be represented by the finite-dimensional basis

$$\mathbb{H} = \{h_1, h_2, \ldots, h_N\},$$

we know the eigenfunctions can be approximated as

$$\varphi(\boldsymbol{x}) \approx \sum_{j=1}^{N} c_j h_j(\boldsymbol{x}). \tag{12.22}$$

Plugging this into the Hilbert–Schmidt integral eigenvalue problem gives

$$\sum_{j=1}^{N} c_j \int_\Omega K(\boldsymbol{x}, \boldsymbol{z}) h_j(\boldsymbol{z}) \,\mathrm{d}\boldsymbol{z} = \lambda \sum_{j=1}^{N} c_j h_j(\boldsymbol{x}).$$

We must choose collocation points $\boldsymbol{x}_i \in \Omega$, $i = 1, \ldots, M$, to turn this into a system of equations

$$\sum_{j=1}^{N} c_j \int_\Omega K(\boldsymbol{x}_i, \boldsymbol{z}) h_j(\boldsymbol{z}) \,\mathrm{d}\boldsymbol{z} = \lambda \sum_{j=1}^{N} c_j h_j(\boldsymbol{x}_i), \qquad i = 1, \ldots, M.$$

By stacking all of these equations on top of each other and treating the summations as inner products of vectors we can express this system in matrix-vector form as

$$\begin{pmatrix} \int_\Omega K(\boldsymbol{x}_1, \boldsymbol{z}) h_1(\boldsymbol{z}) \,\mathrm{d}\boldsymbol{z} & \cdots & \int_\Omega K(\boldsymbol{x}_1, \boldsymbol{z}) h_N(\boldsymbol{z}) \,\mathrm{d}\boldsymbol{z} \\ & \vdots & \\ \int_\Omega K(\boldsymbol{x}_M, \boldsymbol{z}) h_1(\boldsymbol{z}) \,\mathrm{d}\boldsymbol{z} & \cdots & \int_\Omega K(\boldsymbol{x}_M, \boldsymbol{z}) h_N(\boldsymbol{z}) \,\mathrm{d}\boldsymbol{z} \end{pmatrix} \boldsymbol{c} = \lambda \begin{pmatrix} h_1(\boldsymbol{x}_1) & \cdots & h_N(\boldsymbol{x}_1) \\ & \vdots & \\ h_1(\boldsymbol{x}_M) & \cdots & h_N(\boldsymbol{x}_M) \end{pmatrix} \boldsymbol{c}, \tag{12.23}$$

where $\boldsymbol{c} = \begin{pmatrix} c_1 & \cdots & c_N \end{pmatrix}^T$. We can write this more compactly by defining the right-hand matrix H such that $(\mathsf{H})_{i,j} = h_j(\boldsymbol{x}_i)$ and the left hand side matrix A such that $(\mathsf{A})_{i,j} = \int_\Omega K(\boldsymbol{x}_i, \boldsymbol{z}) h_j(\boldsymbol{z}) \,\mathrm{d}\boldsymbol{z}$. Then this system is just

$$\mathsf{A}\boldsymbol{c} = \lambda \mathsf{H}\boldsymbol{c},$$

which is a *rectangular generalized eigenvalue problem*.

The solution of this problem in its full generality is not yet a standard numerical linear algebra problem (although some work exists such as [Stewart (1994); Boutry *et al.* (2005); Chu and Golub (2006); Das and Neumaier (2013)]). Therefore we make a simplifying assumption and set $N = M$ (i.e., the number of collocation points and basis functions are equal) so that (12.23) becomes a square system which can be treated with basic MATLAB routines.

Independent of the specific value of M and N, as long as we can analytically compute the necessary integrals (which may not be trivial), we can find a way to solve this problem and determine the coefficients c_j in the expansion of the eigenfunctions. For certain kernels, domains and bases this is indeed possible. For example, the *Brownian bridge kernel* on $[0, 1]$ can be discussed with either *Chebyshev polynomials* or *piecewise linear splines* as bases.

In the event that the integrals of concern cannot be evaluated analytically, we need to discretize them, which can be done using any sort of *quadrature* method such that

$$\int_\Omega f(\boldsymbol{z}) \,\mathrm{d}\boldsymbol{z} = \sum_{\ell=1}^{\nu} w_\ell f(\boldsymbol{z}_\ell).$$

The choice of ν and appropriate w_ℓ and $\boldsymbol{z}_\ell$ values for $\ell = 1, \ldots, \nu$ will determine the order of accuracy of this approximate integral. Substituting this discretization into (12.23) gives

$$\begin{pmatrix} \sum_{\ell=1}^{\nu} w_\ell K(\boldsymbol{x}_1, \boldsymbol{z}_\ell) h_1(\boldsymbol{z}_\ell) & \cdots & \sum_{\ell=1}^{\nu} w_\ell K(\boldsymbol{x}_1, \boldsymbol{z}_\ell) h_N(\boldsymbol{z}_\ell) \\ & \vdots & \\ \sum_{\ell=1}^{\nu} w_\ell K(\boldsymbol{x}_M, \boldsymbol{z}_\ell) h_1(\boldsymbol{z}_\ell) & \cdots & \sum_{\ell=1}^{\nu} w_\ell K(\boldsymbol{x}_M, \boldsymbol{z}_\ell) h_N(\boldsymbol{z}_\ell) \end{pmatrix} \boldsymbol{c} = \lambda \mathsf{H} \boldsymbol{c}.$$

This matrix looks complicated, but it can actually be written as the product of three matrices

$$\begin{pmatrix} K(\boldsymbol{x}_1, \boldsymbol{z}_1) & \cdots & K(\boldsymbol{x}_1, \boldsymbol{z}_\nu) \\ & \vdots & \\ K(\boldsymbol{x}_M, \boldsymbol{z}_1) & \cdots & K(\boldsymbol{x}_M, \boldsymbol{z}_\nu) \end{pmatrix} \begin{pmatrix} w_1 & & \\ & \ddots & \\ & & w_\nu \end{pmatrix} \begin{pmatrix} h_1(\boldsymbol{z}_1) & \cdots & h_N(\boldsymbol{z}_1) \\ & \vdots & \\ h_1(\boldsymbol{z}_\nu) & \cdots & h_N(\boldsymbol{z}_\nu) \end{pmatrix} \boldsymbol{c} = \lambda \mathsf{H} \boldsymbol{c}.$$

As before, we can assume $M = N$ to make this a square generalized eigenvalue problem. Supplementing this $M = N$ restriction is a much more serious restriction — that the collocation points and the integral discretization points coincide, i.e., $N = \nu$ and $\boldsymbol{x}_i = \boldsymbol{z}_i$ for $i = 1, \ldots, N$. This choice of collocation points may be valuable, or it may be better to perform a more/less accurate integral independent of the collocation points. Making this restriction comes with many benefits. For instance, the left hand side matrix with all the kernel values automatically becomes symmetric and positive definite, assuming K is a positive definite kernel.

Remark 12.5. Choosing $\nu < N$ is viable, and may be significantly more computationally efficient because of the reduced number of points used to approximate the integral. This does, however, limit the eigenvalues that may be studied because the resulting matrix product is, at most, rank ν. This has not impeded its effectiveness in, e.g., active subspace detection [Constantine *et al.* (2014a,b); Constantine and Gleich (2014)], and may be worthwhile in this context as well.

A second benefit to our $N = \nu$ choice, which may actually have bigger implications in this eigenvalue problem setting, is gained through the matrix full of basis function values $h_j(\boldsymbol{z}_i)$. The matching points restriction converts our previous system to (now using $N = M = \nu$ and $\boldsymbol{x}_i = \boldsymbol{z}_i$ for $i = 1, \ldots, N$)

$$\begin{pmatrix} K(\boldsymbol{x}_1, \boldsymbol{x}_1) & \cdots & K(\boldsymbol{x}_1, \boldsymbol{x}_N) \\ & \vdots & \\ K(\boldsymbol{x}_N, \boldsymbol{x}_1) & \cdots & K(\boldsymbol{x}_N, \boldsymbol{x}_N) \end{pmatrix} \begin{pmatrix} w_1 & & \\ & \ddots & \\ & & w_N \end{pmatrix} \begin{pmatrix} h_1(\boldsymbol{x}_1) & \cdots & h_N(\boldsymbol{x}_1) \\ & \vdots & \\ h_1(\boldsymbol{x}_N) & \cdots & h_N(\boldsymbol{x}_N) \end{pmatrix} \boldsymbol{c} = \lambda \mathsf{H} \boldsymbol{c},$$

or, in more concise notation,

$$\mathsf{KWH}\boldsymbol{c} = \lambda \mathsf{H} \boldsymbol{c}. \tag{12.24}$$

This system has the benefit of allowing us to solve a standard eigenvalue problem in place of our original generalized eigenvalue problem, assuming that H^{-1} exists (which is a standard assumption). We can first solve the eigenvalue problem

$$\mathsf{KW}\tilde{\boldsymbol{c}} = \lambda \tilde{\boldsymbol{c}}, \tag{12.25}$$

and then solve the auxiliary linear system

$$\mathsf{H}\boldsymbol{c} = \tilde{\boldsymbol{c}}.$$

Replacing (12.24) with (12.25) is nice, because it is easier to solve. Furthermore, because K and W are positive definite, the eigenvalues λ will also all be positive.

If, however, the matrix H were symmetric positive definite (which is possible depending on the choice of basis functions) then it may be preferable to solve (12.24). One way to do this is to compute the Cholesky factorization $\mathsf{H} = \mathsf{L}\mathsf{L}^T$ which converts the generalized problem into a different eigenvalue problem

$$\begin{aligned} & \mathsf{KWLL}^T\boldsymbol{c} = \lambda\mathsf{LL}^T\boldsymbol{c}, \\ \Longleftrightarrow \quad & \mathsf{L}^{-1}\mathsf{KWLL}^T\boldsymbol{c} = \lambda\mathsf{L}^T\boldsymbol{c}, \\ \Longleftrightarrow \quad & \mathsf{L}^{-1}\mathsf{KWL}\hat{\boldsymbol{c}} = \lambda\hat{\boldsymbol{c}}, \end{aligned} \tag{12.26}$$

and auxiliary problem

$$\mathsf{L}^T\boldsymbol{c} = \hat{\boldsymbol{c}}.$$

The eigenvalue system matrix $\mathsf{L}^{-1}\mathsf{KWL}$ was obtained by a similarity transformation to the matrix KW.

Yet another variation on this theme would be to make the following manipulations,

$$\begin{aligned} & \mathsf{KWH}\boldsymbol{c} = \lambda\mathsf{H}\boldsymbol{c}, \\ \Longleftrightarrow \quad & \mathsf{KW}^{1/2}\mathsf{W}^{1/2}\mathsf{H}\boldsymbol{c} = \lambda\mathsf{H}\boldsymbol{c}, \\ \Longleftrightarrow \quad & \mathsf{W}^{1/2}\mathsf{KW}^{1/2}\mathsf{W}^{1/2}\mathsf{H}\boldsymbol{c} = \lambda\mathsf{W}^{1/2}\mathsf{H}\boldsymbol{c}, \\ \Longleftrightarrow \quad & \mathsf{W}^{1/2}\mathsf{KW}^{1/2}\bar{\boldsymbol{c}} = \lambda\bar{\boldsymbol{c}}, \end{aligned} \tag{12.27}$$

where the auxiliary system would be

$$\mathsf{W}^{1/2}\mathsf{H}\boldsymbol{c} = \bar{\boldsymbol{c}}.$$

The matrix $\mathsf{W}^{1/2}$ is a diagonal matrix with the values $\sqrt{w_i}$, $i = 1, \dots, N$, on the diagonal. If we set the system in this manner, the eigenvalue system matrix $\mathsf{W}^{1/2}\mathsf{KW}^{1/2}$ is symmetric, positive definite, but the auxilliary system is not symmetric (though H likely is). This approach is near that described in Section 12.1.5.

Galerkin Method

We begin by picking any basis $\{h_n\}$ for our Hilbert space $\mathcal{H}_K(\Omega)$. Shifts of the kernel K should probably work, but Ghanem and Spanos (2003) suggest using piecewise polynomials. Using an appropriately large N to give us a finite-dimensional basis we represent a (numerical) eigenfunction as

$$v(\boldsymbol{x}) = \sum_{j=1}^{N} c_j h_j(\boldsymbol{x}), \qquad \boldsymbol{x} \in \Omega. \tag{12.28}$$

If we insert this approximate representation into the Hilbert–Schmidt integral eigenvalue problem we obtain a residual

$$r_N(\boldsymbol{x}) = \sum_{j=1}^{N} c_j \left(\int_\Omega K(\boldsymbol{x}, \boldsymbol{z}) h_j(\boldsymbol{z}) \, \mathrm{d}\boldsymbol{z} - \lambda h_j(\boldsymbol{x}) \right).$$

As is usually done for Galerkin methods, we now require the residual to be orthogonal to our finite-dimensional approximation space, i.e.,

$$\langle r_N, h_i \rangle = 0, \qquad i = 1, \ldots, N.$$

Using the standard L_2 inner product for this orthogonality we have a system of linear equations

$$\int_\Omega r_N(\boldsymbol{x}) h_i(\boldsymbol{x}) \, \mathrm{d}\boldsymbol{x} = 0, \qquad i = 1, \ldots, N.$$

To see this in more detail, we insert the expression for the residual from above and get

$$\sum_{j=1}^{N} c_j \int_\Omega \int_\Omega K(\boldsymbol{x}, \boldsymbol{z}) h_i(\boldsymbol{x}) h_j(\boldsymbol{z}) \, \mathrm{d}\boldsymbol{x} \, \mathrm{d}\boldsymbol{z} = \lambda \int_\Omega h_i(\boldsymbol{x}) h_j(\boldsymbol{x}) \, \mathrm{d}\boldsymbol{x}, \qquad i = 1, \ldots, N.$$

We can now define matrices B and H as follows:

$$(\mathsf{B})_{i,j} = \int_\Omega \int_\Omega K(\boldsymbol{x}, \boldsymbol{z}) h_i(\boldsymbol{x}) h_j(\boldsymbol{z}) \, \mathrm{d}\boldsymbol{x} \, \mathrm{d}\boldsymbol{z}, \qquad (\mathsf{H})_{i,j} = \int_\Omega h_i(\boldsymbol{x}) h_j(\boldsymbol{x}) \, \mathrm{d}\boldsymbol{x}.$$

This results in a generalized matrix eigenvalue problem of the form

$$\mathsf{B}\boldsymbol{c} = \lambda \mathsf{H} \boldsymbol{c}.$$

Once the coefficients $\boldsymbol{c} = \begin{pmatrix} c_1 & \cdots & c_N \end{pmatrix}^T$ are found, they can be inserted into (12.28) to yield an approximate eigenfunction.

The process needs to be repeated for all eigenpairs. This could, in principle, be done using matrices T and $\mathsf{\Lambda}$ that capture all coefficients and eigenvalues "in parallel."

As for the collocation method discussed above, we need to compute the integrals representing entries of B and H. If this is not possible analytically, then we have to use quadrature to evaluate these integrals.

Remark 12.6. Williams and Seeger (2001) suggest using random sampling of a small number M of numerically computed eigenfunctions and eigenvalues to reconstruct an accurate approximation to the $N \times N$ kernel matrix K in $\mathcal{O}(M^2 N)$ operations. Similar ideas are presented in a more general setting and with more detail in [Frieze *et al.* (2004)].

12.3 Approximation Using Eigenfunctions

In Section 2.2.3 we introduced the Mercer series as a mechanism for defining and interpreting positive definite kernels as

$$K(\boldsymbol{x}, \boldsymbol{z}) = \sum_{n=1}^{\infty} \lambda_n \varphi_n(\boldsymbol{x}) \varphi_n(\boldsymbol{z}), \tag{2.9}$$

where λ_n and φ_n are the Hilbert–Schmidt eigenvalues and eigenfunctions, respectively, as defined in Section 2.2.2.

As we learned in Section 4.5, the provably positive definite nature of kernels is of little solace in a computational setting, where the matrix K — which must be inverted to evaluate the interpolant (12.1) — is rank deficient to machine precision. In Figure 4.4 we interpreted this as a crisis because one of the defining features of a reproducing kernel was lost in the translation between theory and implementation. One strategy for dealing with this loss of accuracy is to rephrase the interpolant in an alternate basis which produces a better-conditioned — hopefully (numerically) full-rank — interpolation matrix, as is studied in Section 12.1 and Chapter 13. It is also possible to view this issue as an opportunity to redefine the solution to our scattered data approximation problem in such a way that the low-rank appearance of K is an advantage.

We begin by revisiting our discussion in Section 4.5 regarding low-rank computation. The notation introduced there can be repurposed to state that, for $(\mathsf{K})_{i,j} = K(\boldsymbol{x}_i, \boldsymbol{x}_j)$, we may consider $M < N$ (unlike in Section 4.5) and define

$$\boldsymbol{\phi}(\boldsymbol{x}) = \begin{pmatrix} \varphi_1(\boldsymbol{x}) \\ \vdots \\ \varphi_M(\boldsymbol{x}) \end{pmatrix}, \quad \mathsf{\Phi} = \begin{pmatrix} \boldsymbol{\phi}(\boldsymbol{x}_1)^T \\ \vdots \\ \boldsymbol{\phi}(\boldsymbol{x}_N)^T \end{pmatrix}, \quad \mathsf{\Lambda} = \begin{pmatrix} \lambda_1 & & \\ & \ddots & \\ & & \lambda_M \end{pmatrix},$$

so as to compute the quantities

$$\mathsf{K} \approx \mathsf{\Phi} \mathsf{\Lambda} \mathsf{\Phi}^T, \tag{12.29a}$$

$$\boldsymbol{k}(\boldsymbol{x})^T \approx \boldsymbol{\phi}(\boldsymbol{x})^T \mathsf{\Lambda} \mathsf{\Phi}^T. \tag{12.29b}$$

These are only approximately correct because this computation involves only M terms, while the actual series requires infinitely many terms. Fortunately, for many kernels we can get by with a relatively small M to produce an accurate K. As a result, we omit the "$\approx$" in favor of "$=$" within this section.

It is well known (see, e.g., [Novak and Woźniakowski (2008)]) that the best M-term ρ-weighted least squares approximation (or mean-squared approximation) of a function $f \in L_2(\Omega, \rho)$ is given by the generalized Fourier series of f, truncated at M terms. We now show that the *truncated Mercer series* provides the *best M-term approximation* of the kernel K.

Theorem 12.2 (Truncated Mercer series). *Let $K : \Omega \times \Omega \to \mathbb{R}$ be a positive definite kernel with Mercer series*

$$K(\boldsymbol{x}, \boldsymbol{z}) = \sum_{n=1}^{\infty} \lambda_n \varphi_n(\boldsymbol{x}) \varphi_n(\boldsymbol{z})$$

and M-term truncation

$$K_M(\boldsymbol{x}, \boldsymbol{z}) = \sum_{n=1}^{M} \lambda_n \varphi_n(\boldsymbol{x}) \varphi_n(\boldsymbol{z}).$$

$K_M(\cdot, \boldsymbol{z})$, $\boldsymbol{z} \in \Omega$ fixed, provides the best M-term ρ-weighted least squares approximation of $K(\cdot, \boldsymbol{z})$ from $L_2(\Omega, \rho)$.

Proof. We know that the eigenfunctions $\{\varphi_n\}_{n=1}^{\infty}$ of the Hilbert–Schmidt eigenvalue problem $\mathcal{K}\varphi_n = \lambda_n \varphi_n$ form an orthonormal basis of $L_2(\Omega, \rho)$ (see Section 2.2.2). We now fix $\boldsymbol{z} \in \Omega$ and use this eigenfunction basis to represent the kernel $K(\cdot, \boldsymbol{z})$ by its generalized Fourier series (cf. (6.11))

$$K(\cdot, \boldsymbol{z}) = \sum_{n=1}^{\infty} a_n(\boldsymbol{z}) \varphi_n(\cdot). \tag{12.30}$$

We now show that the generalized Fourier coefficients of $K(\cdot, \boldsymbol{z})$ are given by $a_n(\boldsymbol{z}) = \lambda_n \varphi_n(\boldsymbol{z})$ and thus the generalized Fourier series of K is identical to its Mercer series. The claim of the theorem then follows from the fact that the truncated Fourier series provides the best M-term weighted least squares approximation of $K(\cdot, \boldsymbol{z})$ (see, e.g., Theorem 6.1 or [Novak and Woźniakowski (2008)]).

To compute the generalized Fourier coefficients $a_n(\boldsymbol{z})$ we apply the Hilbert–Schmidt integral operator $\mathcal{K}$ associated with K (cf. (2.4)) to both sides of (12.30), resulting in

$$\mathcal{K}K(\cdot, \boldsymbol{z}) = \sum_{n=1}^{\infty} a_n(\boldsymbol{z}) \mathcal{K}\varphi_n(\cdot) \tag{12.31a}$$

$$\iff \quad \mathcal{K}K(\cdot, \boldsymbol{z}) = \sum_{n=1}^{\infty} a_n(\boldsymbol{z}) \lambda_n \varphi_n(\cdot) \tag{12.31b}$$

$$\iff \quad \sum_{n=1}^{\infty} \lambda_n^2 \varphi_n(\cdot) \varphi_n(\boldsymbol{z}) = \sum_{n=1}^{\infty} a_n(\boldsymbol{z}) \lambda_n \varphi_n(\cdot). \tag{12.31c}$$

Here we first used the fact the λ_n and φ_n, respectively, are the eigenvalues and eigenfunctions of $\mathcal{K}$, and then recognized $\mathcal{K}K(\cdot, \boldsymbol{z})$ as an iterated kernel with its corresponding Mercer series (see Section 2.2.5).

The form of the generalized Fourier coefficients $a_n(\boldsymbol{z})$ follows by comparing the coefficients of $\varphi_n(\cdot)$ on both sides of the identity (12.31c). The nature of the generalized Fourier coefficients ensures uniform convergence of the Fourier series, and thus justified the interchange of the integral operator with the infinite summation in (12.31a). □

Remark 12.7. With a slight reorganization, the product $\Phi\Lambda\Phi^T$ in (12.29a) can be rewritten to emphasize the construction of K as a sum of rank-1 matrices — there should exist rank(K) such matrices. Earlier we used $\boldsymbol{\phi}(\boldsymbol{x})$ to collect the values of all eigenfunctions, $\varphi_1, \ldots, \varphi_M$ at $\boldsymbol{x}$. Now we introduce $\boldsymbol{\varphi}_n$ to collect the values of the nth eigenfunction at all the data locations in $\mathcal{X} = \{\boldsymbol{x}_1, \ldots, \boldsymbol{x}_N\}$, i.e.,

$$\boldsymbol{\varphi}_n = \begin{pmatrix} \varphi_n(\boldsymbol{x}_1) \\ \vdots \\ \varphi_n(\boldsymbol{x}_N) \end{pmatrix}.$$

This allows us to easily write this sum as

$$\mathsf{K} = \Phi\Lambda\Phi^T = \sum_{n=1}^{M} \lambda_n \boldsymbol{\varphi}_n \boldsymbol{\varphi}_n^T. \tag{12.32}$$

According to Theorem 12.2, the summation (12.32) yields the best M-term approximation of each matrix entry in the $L_2(\Omega, \rho)$ sense, but it does not provide the best low-rank approximation in the matrix 2-norm sense. That honor is reserved for the singular value decomposition [Golub and Van Loan (2012)] $\mathsf{K} = \mathsf{U}\Sigma\mathsf{U}^T$, where $\mathsf{U} \in \mathbb{R}^{N\times M}$ with $\mathsf{U}^T\mathsf{U} = \mathsf{I}_M$ and $\Sigma \in \mathbb{R}^{M\times M}$ is a diagonal matrix with the nonincreasing singular values down the diagonal. While we did consider the use of the SVD to obtain a (weighted) SVD basis in Section 12.1.5, computing the SVD of K — even of just the first M singular values — is unfortunately a costly proposition and is prone to instability (recall the fact that the rank deficiency of K is numerical and that K is full-rank in "exact arithmetic"). Because of this, we only study the low-rank Mercer series here, but one should also be aware that other low-rank representations may exist and may prove useful.

Because the matrix K is numerically low-rank, our system

$$\mathsf{K}\boldsymbol{c} = \boldsymbol{y} \qquad \Longrightarrow \qquad \Phi\Lambda\Phi^T\boldsymbol{c} = \boldsymbol{y} \tag{12.33}$$

no longer has a unique solution and permits the existence of a solution if $\boldsymbol{y} \in$ range(Φ). Therefore, we choose to solve (12.33) in the *least squares sense*,

$$\boldsymbol{c} \text{ is the solution which minimizes } \|\Phi\Lambda\Phi^T\boldsymbol{c} - \boldsymbol{y}\|_2^2.$$

If we define the auxiliary term $\boldsymbol{a} = \Lambda\Phi^T\boldsymbol{c}$, the new quantity to be minimized is

$$\boldsymbol{a} \text{ is the solution which minimizes } \|\Phi\boldsymbol{a} - \boldsymbol{y}\|_2^2,$$

which is a standard overdetermined problem that can be solved with the pseudoinverse $\Phi^\dagger$. The vector $\boldsymbol{a}$ represents the linear combination of the first M eigenfunctions, evaluated at $\mathcal{X}$, which best fit the data $\boldsymbol{y}$ that has been sampled at $\mathcal{X}$. In this sense, the eigenfunctions form an *alternate basis* for the interpolant, i.e.,

$$s(\boldsymbol{x}) = \boldsymbol{k}(\boldsymbol{x})^T\mathsf{K}^{-1}\boldsymbol{y} \quad \text{is replaced by} \quad s(\boldsymbol{x}) = \boldsymbol{\phi}(\boldsymbol{x})^T\Phi^\dagger\boldsymbol{y}. \tag{12.34}$$

We refer to this result as the *low-rank approximate interpolant*.

Remark 12.8. It is reasonable to doubt the argument in support of replacing $\boldsymbol{k}$ with $\boldsymbol{\phi}$ because the loss of $\boldsymbol{c}$ seems more detrimental than it is. The standard approach to obtaining the $\boldsymbol{a}$ that minimizes $\|\Phi\boldsymbol{a}-\boldsymbol{y}\|_2^2$ uses the normal equations and results in the solution $\boldsymbol{a} = (\Phi^T\Phi)^{-1}\Phi\boldsymbol{y}$. The matrix $(\Phi^T\Phi)^{-1}\Phi$ is the *pseudoinverse* of the matrix Φ [Golub and Van Loan (2012)], it is often denoted as $\Phi^\dagger$ and is usually computed using the QR factorization or SVD of Φ. More than one definition of pseudoinverse exists (a detailed discussion is available in, e.g., [Campbell and Meyer (2009)]) but the most common definition can be computed in MATLAB with the function `pinv`.

Without going into the details of the QR factorization, for a matrix Φ with full column rank we may write $\Phi = \mathsf{QR}$, where $\mathsf{R} \in \mathbb{R}^{M\times M}$ is triangular and invertible and $\mathsf{Q} \in \mathbb{R}^{N\times M}$ satisfies $\mathsf{Q}^T\mathsf{Q} = \mathsf{I}_M$. This allows us to write the overdetermined least squares solution as $\boldsymbol{a} = \mathsf{R}^{-1}\mathsf{Q}^T\boldsymbol{y}$ or, more succinctly, as $\boldsymbol{a} = \Phi^\dagger\boldsymbol{y}$ because

$$\Phi = \mathsf{QR} \qquad \text{implies} \qquad \Phi^\dagger = \mathsf{R}^{-1}\mathsf{Q}^T.$$

Substituting this into our definition of $\boldsymbol{a} = \Lambda\Phi^T\boldsymbol{c}$ gives

$$\begin{aligned} &\Lambda\Phi^T\boldsymbol{c} = \mathsf{R}^{-1}\mathsf{Q}^T\boldsymbol{y} \\ \iff \quad &\Phi^T\boldsymbol{c} = \Lambda^{-1}\mathsf{R}^{-1}\mathsf{Q}^T\boldsymbol{y}, \end{aligned} \tag{12.35}$$

where the invertibility of Λ is guaranteed because the kernel K is positive definite (see Section 2.2). Equation (12.35) is an *underdetermined* system, which is to say that, if it has a solution, it has infinitely many solutions because the M constraints imposed by Φ^T are fewer than the N degrees of freedom afforded by $\boldsymbol{c}$.

One common strategy for choosing one of the infinitely many solutions is to minimize $\|\boldsymbol{c}\|_0$, that is, choose the solution with the fewest nonzeros. Such a strategy is common in the field of compressed sensing [Candès *et al.* (2006)]. The solution we consider here is instead the solution that minimizes $\|\boldsymbol{c}\|_2^2$, which can be found with the introduction of a Lagrange multiplier $\boldsymbol{\nu} \in \mathbb{R}^M$. We try to minimize the quantity

$$\|\boldsymbol{c}\|_2^2 + \boldsymbol{\nu}^T(\Lambda^{-1}\mathsf{R}^{-1}\mathsf{Q}^T\boldsymbol{y} - \Phi^T\boldsymbol{c}),$$

i.e., we seek to find the minimum norm solution of (12.35).

Differentiating with respect to $\boldsymbol{c}$ and setting equal to zero allows us to find $\boldsymbol{\nu}$ by solving the equation

$$2\boldsymbol{c} - \Phi\boldsymbol{\nu} = 0.$$

Multiplying through by Φ^T and enforcing (12.35) gives

$$2\Lambda^{-1}\mathsf{R}^{-1}\mathsf{Q}^T\boldsymbol{y} - \Phi^T\Phi\boldsymbol{\nu} = 0 \qquad \iff \qquad \boldsymbol{\nu} = 2(\Phi^T\Phi)^{-1}\Lambda^{-1}\mathsf{R}^{-1}\mathsf{Q}^T\boldsymbol{y},$$

which implies that

$$\begin{aligned} \boldsymbol{c} = \frac{1}{2}\Phi\boldsymbol{\nu} &= \Phi(\Phi^T\Phi)^{-1}\Lambda^{-1}\mathsf{R}^{-1}\mathsf{Q}^T\boldsymbol{y} \\ &= \mathsf{QR}((\mathsf{QR})^T\mathsf{QR})^{-1}\Lambda^{-1}\mathsf{R}^{-1}\mathsf{Q}^T\boldsymbol{y} \\ &= \mathsf{QR}^{-T}\Lambda^{-1}\mathsf{R}^{-1}\mathsf{Q}^T\boldsymbol{y} \end{aligned} \tag{12.36}$$

after cleaning up with $\Phi = \mathsf{QR}$. While this solution may seem substantially different than any involving $\boldsymbol{a}$, combining (12.36) and (12.29b) within the interpolant evaluation (12.1) reveals

$$\begin{aligned} s(\boldsymbol{x}) &= \boldsymbol{k}(\boldsymbol{x})^T \boldsymbol{c} \\ &= \boldsymbol{\phi}(\boldsymbol{x})^T \Lambda \Phi^T \mathsf{Q} \mathsf{R}^{-T} \Lambda^{-1} \mathsf{R}^{-1} \mathsf{Q}^T \boldsymbol{y} \\ &= \boldsymbol{\phi}(\boldsymbol{x})^T \Lambda \mathsf{R}^T \mathsf{Q}^T \mathsf{Q} \mathsf{R}^{-T} \Lambda^{-1} \Phi^\dagger \boldsymbol{y} \\ &= \boldsymbol{\phi}(\boldsymbol{x})^T \Phi^\dagger \boldsymbol{y} \\ &= \boldsymbol{\phi}(\boldsymbol{x})^T \boldsymbol{a} \end{aligned} \tag{12.37}$$

This result implies that, at least within the context of forming and evaluating a low-rank approximation to an interpolant, only the overdetermined system $\Phi \boldsymbol{a} = \boldsymbol{y}$ must be solved.

The implications of this alternate eigenfunction approach are interesting, as it provides a strategy for computing a low-rank approximation to the interpolant s — even for ranks that cannot be achieved to computer precision. This can occur because the matrix Λ of eigenvalues, whose swift decay is the main source of ill-conditioning and rank deficiency in interpolation matrices, is not involved in the computation. Such a theme also appears as part of our discussion of the Hilbert–Schmidt SVD in Chapter 13.

In Example 15.2 we apply the low-rank approximate interpolant and demonstrate its potential to produce superior quality approximations for a numerically singular kernel matrix K.

12.4 Other Recent Preconditioning and Alternate Basis Techniques

Fuselier *et al.* (2013) derive *localized Lagrange basis functions* for *polyharmonic splines* restricted to the unit sphere in $\mathbb{R}^3$. The construction of the local Lagrange functions is similar to what was used in [Beatson *et al.* (1999); Faul and Powell (1999, 2000)] (and also discussed in [Fasshauer (2007, Chapters 33 and 34)]). However, now a theoretical foundation is provided for the size of the local point set involved in computing the local Lagrange bases. Fuselier *et al.* (2013) show that $M \log(N)^2$ points are needed in the local systems. Here N is the total number of data sites and the constant M scales proportional to the square of the smoothness of the kernel so that smoother kernels, which have faster decaying Lagrange functions, require fewer points than less smooth kernels. In the earlier papers this number had been determined ad hoc. The paper [Fuselier *et al.* (2013)] also includes numerical experiments with data sets exceeding $N = 160,000$ points, and the preprint [Hangelbroek *et al.* (2014)] extends these ideas to bounded domains in $\mathbb{R}^d$ and on compact Riemannian manifolds. Error bounds for certain *nodal bases* (such as these localized Lagrange bases) can be found in [Schaback (2014)].

Chapter 13

Stable Computation via the Hilbert–Schmidt SVD

In Chapter 10 we established the fact that the kernel-based interpolation problem is well-posed in the flat limit, and then contrasted this in Chapter 11 with the historical belief — and misconception — that increasingly flat kernels are able to provide more accurate interpolants only in exact arithmetic, because the interpolation problem simultaneously becomes more and more ill-conditioned. As explained in Chapter 11, this misconception stems from the fact that the standard basis $\{K(\cdot, \boldsymbol{x}_1), \ldots, K(\cdot, \boldsymbol{x}_N)\}$ is often insufficient for computing a stable solution in finite precision. Therefore, we were led to investigate alternate bases for stably solving the scattered data interpolation problem in Chapter 12. Alternate bases provide a useful framework for avoiding the ill-conditioning of the flat limit, but the strategy of deriving an alternate basis by decomposing the K matrix likely suffers from the same instabilities present in interpolation. This suggests the need for techniques other than those employed in Chapter 12.

The *Hilbert–Schmidt SVD* (HS-SVD) developed in this chapter will give us an alternate basis for the approximation space $\mathcal{H}_K(\mathcal{X}) = \text{span}\{K(\cdot, \boldsymbol{x}_1), \ldots, K(\cdot, \boldsymbol{x}_N)\}$ *without* having to form — and much less decompose — the matrix K.

In this chapter we study the use of Mercer's series, introduced in Chapter 2, as an avenue to develop a general framework that allows us to stably compute interpolants for flat kernels. We show that, although the standard basis for many positive definite kernels is ill-conditioned in the flat limit, the HS-SVD provides a linear transformation which can be applied *analytically* (as opposed to the numerical basis transformation T of Theorem 12.1) and therefore is able to remove a significant portion of the ill-conditioning.

This technique is demonstrated on several examples involving iterated Brownian bridge kernels, which were introduced in Chapter 7, the Chebyshev kernel of Sections 3.9.2 and 4.5, and the Gaussian kernel, whose eigenvalues and eigenfunctions were discussed in Example 12.1.

13.1 A Formal Matrix Decomposition of K

In this section, we take advantage of the Mercer series (13.1) to obtain a formal matrix decomposition of the kernel matrix K. We refer to this decomposition as *formal* since it is not obtained by actually factoring the kernel matrix. The matrix K is, in fact, never even formed. Nevertheless, we still end up with a factorization of the kernel matrix, i.e., we can obtain K — if needed — from our factorization, but not vice versa (as is done with matrix factorizations in linear algebra, such as in Section 12.1).

In the next section we use techniques from linear algebra to develop this matrix factorization and to isolate and remove the major source of ill-conditioning in the interpolation problem. That process leads to the Hilbert–Schmidt SVD.

Referring to the discussion from Chapter 2, our starting point for the Hilbert–Schmidt SVD is the *Mercer series* representation of a positive definite kernel K given by

$$K(\boldsymbol{x}, \boldsymbol{z}) = \sum_{n=1}^{\infty} \lambda_n \varphi_n(\boldsymbol{x}) \varphi_n(\boldsymbol{z}), \tag{13.1}$$

where λ_n and φ_n are the eigenvalues and eigenfunctions from the Hilbert–Schmidt eigenvalue problem

$$\mathcal{K}\varphi_n(\boldsymbol{x}) = \lambda_n \varphi_n(\boldsymbol{x}), \qquad n = 1, 2, \ldots, \tag{13.2}$$

associated with the integral operator $\mathcal{K}$ defined as

$$\mathcal{K}f = \int_\Omega K(\cdot, \boldsymbol{z}) f(\boldsymbol{z}) \rho(\boldsymbol{z}) \,\mathrm{d}\boldsymbol{z}, \qquad f \in L_2(\Omega, \rho).$$

Here $\rho : \Omega \to \mathbb{R}$ is a positive weight function.

The eigenfunctions φ_n are $L_2(\Omega, \rho)$-orthonormal, i.e., they satisfy the orthogonality relations

$$\int_\Omega \varphi_m(\boldsymbol{x}) \varphi_n(\boldsymbol{x}) \rho(\boldsymbol{x}) \,\mathrm{d}\boldsymbol{x} = \begin{cases} 1, & m = n, \\ 0, & \text{else}, \end{cases} \tag{13.3}$$

and, because K is strictly positive definite, all the eigenvalues λ_n are positive. The eigenfunctions are also orthogonal in $\mathcal{H}_K(\Omega)$ (see (2.8)).

Examples of kernels and their Mercer series, when available, are given in various places throughout this book, such as Sections 3.3, 3.9 and Appendix A.

Remark 13.1. We may occasionally refer to the series (13.1) as a *Hilbert–Schmidt series*, because of its connection to the Hilbert–Schmidt operator $\mathcal{K}$. We may also refer to λ_n and φ_n as being the eigenvalues and eigenfunctions of the kernel K rather than the Hilbert–Schmidt operator $\mathcal{K}$. We do this only for simplicity, and this should be unambiguous since the kernel itself has no eigenvalues or eigenfunctions in the absence of the Hilbert–Schmidt operator.

Our approach can be applied to any positive definite kernel K once it is written as a Hilbert–Schmidt series. Therefore, the discussion in this and the next section involves arbitrary positive definite kernels and applies in full generality, in particular on arbitrary domains in arbitrary space dimensions d.

Recall that we assume the solution to our scattered data interpolation problem to be a linear combination of kernels, i.e.,

$$s(\boldsymbol{x}) = \sum_{j=1}^{N} c_j K(\boldsymbol{x}, \boldsymbol{x}_j). \tag{13.4}$$

In Section 12.2.1 we discussed the potential for an eigenfunction basis to approximate a kernel interpolant, but that method did not actually solve the interpolation system

$$\mathsf{K}\boldsymbol{c} = \boldsymbol{y}.$$

Instead, it solved only a nearby, least-squares system.

The approach we take now will also involve the eigenfunctions φ_n, but it will solve an interpolation system which is based on an accurate elementwise reconstruction of the kernel matrix K, which we recall to be of the form

$$\mathsf{K} = \begin{pmatrix} K(\boldsymbol{x}_1, \boldsymbol{x}_1) & \cdots & K(\boldsymbol{x}_1, \boldsymbol{x}_N) \\ & \vdots & \\ K(\boldsymbol{x}_N, \boldsymbol{x}_1) & \cdots & K(\boldsymbol{x}_N, \boldsymbol{x}_N) \end{pmatrix}.$$

We start our analysis by invoking the series representation of the kernel, noting that a single kernel value can be written as a quadratic form of the diagonal matrix containing all the eigenvalues, i.e.,

$$K(\boldsymbol{x}, \boldsymbol{z}) = \sum_{n=1}^{\infty} \varphi_n(\boldsymbol{x}) \lambda_n \varphi_n(\boldsymbol{z}) = \begin{pmatrix} \varphi_1(\boldsymbol{x}) & \cdots & \varphi_N(\boldsymbol{x}) & \cdots \end{pmatrix} \begin{pmatrix} \lambda_1 & & & \\ & \ddots & & \\ & & \lambda_N & \\ & & & \ddots \end{pmatrix} \begin{pmatrix} \varphi_1(\boldsymbol{z}) \\ \vdots \\ \varphi_N(\boldsymbol{z}) \\ \vdots \end{pmatrix}. \tag{13.5}$$

Note that the vectors and matrices involved in this representation are of infinite size[1] and therefore need to be truncated at some finite length M in order for us to be able to implement our method on a computer. From Theorem 12.2 we know that the truncated Mercer series provides the best M-term approximation of the kernel. Therefore we can now safely perform and use such a truncation with *some* value of M that makes the difference between the full (infinite) Mercer series and its M-term truncation as small as desired. We will postpone the discussion of appropriate

[1] In [Cavoretto *et al.* (2015)] the authors decided to work with this infinite formulation as long as possible, and truncate only at the very end. Here, we have instead decided to immediately truncate the series so that the entire discussion below involves traditional (finite-dimensional) vectors and matrices. Either approach can be justified, but we believe that many readers may be more comfortable with the approach taken here.

choices for M until Section 13.4.1, but will always assume that $M > N$ in this section.

We define the following size M vector $\boldsymbol{\phi}(\cdot)$ and diagonal matrix $\mathsf{\Lambda}$ as

$$\boldsymbol{\phi}(\boldsymbol{x}) = \begin{pmatrix} \varphi_1(\boldsymbol{x}) \\ \vdots \\ \varphi_M(\boldsymbol{x}) \end{pmatrix}, \qquad \mathsf{\Lambda} = \begin{pmatrix} \lambda_1 & & \\ & \ddots & \\ & & \lambda_M \end{pmatrix},$$

so that we can replace the infinite quadratic form (13.5) with

$$K(\boldsymbol{x}, \boldsymbol{z}) = \sum_{n=1}^{M} \varphi_n(\boldsymbol{x}) \lambda_n \varphi_n(\boldsymbol{z}) = \boldsymbol{\phi}(\boldsymbol{x})^T \mathsf{\Lambda} \boldsymbol{\phi}(\boldsymbol{z}). \tag{13.6}$$

We justify the use of the first "=" in (13.6) (instead of "≈") by our discussion regarding the truncation length M above.

This truncated series evaluation of K was already used in Chapter 7 to evaluate the iterated Brownian bridge kernels $K_{\beta,\varepsilon}$ in those cases when the closed form of the kernel was unknown to us.

At this point, we want to use (13.6) to rewrite the kernel matrix K. Let us use the vector $\boldsymbol{k}$ to define a generic row of the matrix K:

$$\boldsymbol{k}(\boldsymbol{x})^T = \left(K(\boldsymbol{x}, \boldsymbol{x}_1) \cdots K(\boldsymbol{x}, \boldsymbol{x}_N) \right). \tag{13.7}$$

Applying (13.6) to each element of $\boldsymbol{k}(\boldsymbol{x})^T$ allows us to write a generic row of the interpolation matrix in terms of the eigenvalues and eigenfunctions as

$$\begin{aligned} \boldsymbol{k}(\boldsymbol{x})^T &= \left(\boldsymbol{\phi}(\boldsymbol{x})^T \mathsf{\Lambda} \boldsymbol{\phi}(\boldsymbol{x}_1) \cdots \boldsymbol{\phi}(\boldsymbol{x})^T \mathsf{\Lambda} \boldsymbol{\phi}(\boldsymbol{x}_N) \right) \\ &= \boldsymbol{\phi}(\boldsymbol{x})^T \mathsf{\Lambda} \left(\boldsymbol{\phi}(\boldsymbol{x}_1) \cdots \boldsymbol{\phi}(\boldsymbol{x}_N) \right). \end{aligned} \tag{13.8}$$

The entire kernel matrix K can be formed by stacking rows of this form, i.e.,

$$\begin{aligned} \mathsf{K} &= \begin{pmatrix} \boldsymbol{k}(\boldsymbol{x}_1)^T \\ \vdots \\ \boldsymbol{k}(\boldsymbol{x}_N)^T \end{pmatrix} \\ &= \begin{pmatrix} \boldsymbol{\phi}(\boldsymbol{x}_1)^T \mathsf{\Lambda} \left(\boldsymbol{\phi}(\boldsymbol{x}_1) \cdots \boldsymbol{\phi}(\boldsymbol{x}_N) \right) \\ \vdots \\ \boldsymbol{\phi}(\boldsymbol{x}_N)^T \mathsf{\Lambda} \left(\boldsymbol{\phi}(\boldsymbol{x}_1) \cdots \boldsymbol{\phi}(\boldsymbol{x}_N) \right) \end{pmatrix} \\ &= \begin{pmatrix} \boldsymbol{\phi}(\boldsymbol{x}_1)^T \\ \vdots \\ \boldsymbol{\phi}(\boldsymbol{x}_N)^T \end{pmatrix} \mathsf{\Lambda} \left(\boldsymbol{\phi}(\boldsymbol{x}_1) \cdots \boldsymbol{\phi}(\boldsymbol{x}_N) \right). \end{aligned}$$

This final expression can be simplified further by defining the matrix

$$\mathsf{\Phi} = \begin{pmatrix} \boldsymbol{\phi}(\boldsymbol{x}_1)^T \\ \vdots \\ \boldsymbol{\phi}(\boldsymbol{x}_N)^T \end{pmatrix} = \begin{pmatrix} \varphi_1(\boldsymbol{x}_1) & \cdots & \varphi_M(\boldsymbol{x}_1) \\ & \vdots & \\ \varphi_1(\boldsymbol{x}_N) & \cdots & \varphi_M(\boldsymbol{x}_N) \end{pmatrix},$$

whose columns contain samples of the first M eigenfunctions at the interpolation locations $\boldsymbol{x}_1, \ldots, \boldsymbol{x}_N$. This allows us to write the *Hilbert–Schmidt eigen-decomposition of* K as

$$\mathsf{K} = \Phi \Lambda \Phi^T. \tag{13.9}$$

The eigen-decomposition (13.9) represents a first *formal* decomposition of K, i.e., we can obtain K from the matrices Φ (of samples of eigenfunctions) and Λ (of eigenvalues), but we do this directly from the Mercer series — without ever working with the (ill-conditioned) matrix K.

The matrix $(\Phi)_{i,j} = \varphi_j(x_i)$ in (13.9) is similar to the eigenfunction regression matrix of the same name from Section 12.2.1, but here the matrix Φ has more columns than rows, because we have assumed that the truncation length M is greater than the number of input points N.

Remark 13.2. At this point, all we have done is rewrite the kernel interpolation matrix as the product of three larger matrices. Although this helps us evaluate the kernel function using only the series expansion, the result of this process is no different than simply finding each kernel element individually according to

$$\mathsf{K} = \begin{pmatrix} \sum_{n=1}^{M} \lambda_n \varphi_n(\boldsymbol{x}_1)\varphi_n(\boldsymbol{x}_1) & \cdots & \sum_{n=1}^{M} \lambda_n \varphi_n(\boldsymbol{x}_1)\varphi_n(\boldsymbol{x}_N) \\ & \vdots & \\ \sum_{n=1}^{M} \lambda_n \varphi_n(\boldsymbol{x}_N)\varphi_n(\boldsymbol{x}_1) & \cdots & \sum_{n=1}^{M} \lambda_n \varphi_n(\boldsymbol{x}_N)\varphi_n(\boldsymbol{x}_N) \end{pmatrix}.$$

From a performance standpoint, computing the interpolation matrix with (13.9) is likely more efficient because it involves Level 3 BLAS operations (matrix-matrix products) rather than Level 1 BLAS operations (vector inner products $\boldsymbol{\phi}(x)^T \Lambda \boldsymbol{\phi}(z)$). See [Golub and Van Loan (2012)] for a review of this idea. This does not, however, address the stability concerns raised in Chapter 11 because we have not yet altered the structure of the linear system $\mathsf{K}\boldsymbol{c} = \boldsymbol{y}$ we use for interpolation, and we would still be computing with the standard basis.

13.2 Obtaining a Stable Alternate Basis via the Hilbert–Schmidt SVD

The next step in creating a stable basis with *analytically removed ill-conditioning* was an ingenious leap forward by Bengt Fornberg and his colleagues in [Fornberg and Piret (2008b); Fornberg *et al.* (2011)]. It was inspired by studying the structure of the system

$$\mathsf{K}\boldsymbol{c} = \boldsymbol{y} \quad \Longleftrightarrow \quad \Phi \Lambda \Phi^T \boldsymbol{c} = \boldsymbol{y},$$

and asking the seemingly profound question *"Why is this system ill-conditioned?"*

It would seem that — because the eigen*functions* φ_n are orthogonal in $L_2(\Omega, \rho)$ — the *matrix* Φ should be relatively well behaved[2], although we should not expect

[2] Example 13.1 provides particularly convincing support for this belief.

its columns to be orthogonal to each other since the nth column consists only of samples of the eigenfunction φ_n at the points $\boldsymbol{x}_1, \ldots, \boldsymbol{x}_N$ (and, of course, there are more columns than rows since $M > N$).

Therefore, the ill-conditioning appears to instead originate in the diagonal matrix $\mathsf{\Lambda}$, whose values are the eigenvalues of K, and whose 2-norm condition number is

$$\kappa(\mathsf{\Lambda}) = \frac{\lambda_1}{\lambda_M}.$$

This condition number is not directly relevant, since the entire $\mathsf{\Lambda}$ matrix is never inverted, but it serves to give an idea of the delicate nature of the K matrix.

We now take a moment to consider the severity of this condition number of $\mathsf{\Lambda}$. Let us compare two kernels with associated eigenfunction expansions introduced earlier: the iterated Brownian bridge kernels from Chapter 7 and the Gaussian kernel from Section 12.2.1. With both of these kernels, the flat limit is reached as $\varepsilon \to 0$. The eigenvalues of the iterated Brownian bridge kernels decay with an order of $\varepsilon^{-2\beta}$, where β defines — among other things — the smoothness of the kernel; a larger value of β yielding a smoother kernel (see Remark 2.4). For the infinitely smooth Gaussians, on the other hand, the eigenvalues decay exponentially quickly, which illustrates that, the smoother the kernel is, the more we should be concerned about ill-conditioning of the $\mathsf{\Lambda}$ matrices.

This connection between ill-conditioning of the interpolation system and the smoothness of the kernel has been studied in the literature for a long time (see, e.g., [Narcowich and Ward (1991, 1992); Narcowich *et al.* (1994)] or also [Fasshauer (2007, Chapter 16)]). It would appear then that the presence of the $\mathsf{\Lambda}$ matrix is the main source of ill-conditioning in (13.9). The rest of this section will attempt to eliminate this issue from the interpolation problem, though it cannot be removed from K.

The key step in removing the ill-conditioning is to write the component matrices that appear in the eigen-decomposition of K (13.9) using blocks, i.e.,

$$\mathsf{\Phi} = \begin{pmatrix} \mathsf{\Phi}_1 & \mathsf{\Phi}_2 \end{pmatrix}, \qquad \mathsf{\Lambda} = \begin{pmatrix} \mathsf{\Lambda}_1 & \\ & \mathsf{\Lambda}_2 \end{pmatrix},$$

where $\mathsf{\Phi}_1, \mathsf{\Lambda}_1 \in \mathbb{R}^{N\times N}$, $\mathsf{\Phi}_2 \in \mathbb{R}^{N\times(M-N)}$, $\mathsf{\Lambda}_2 \in \mathbb{R}^{(M-N)\times(M-N)}$.

Using this, we can write our vector of kernel values (13.8) as

$$\boldsymbol{k}(\boldsymbol{x})^T = \boldsymbol{\phi}(x)^T \mathsf{\Lambda} \mathsf{\Phi}^T = \boldsymbol{\phi}(\boldsymbol{x})^T \begin{pmatrix} \mathsf{\Lambda}_1 & \\ & \mathsf{\Lambda}_2 \end{pmatrix} \begin{pmatrix} \mathsf{\Phi}_1^T \\ \mathsf{\Phi}_2^T \end{pmatrix}.$$

Now we make the somewhat precarious assumption that $\mathsf{\Phi}_1^T$ is invertible, which will be discussed further in Section 13.4. We know that $\mathsf{\Lambda}_1$ is invertible because all the eigenvalues of a positive definite kernel K are positive. This allows us to write

$$\boldsymbol{k}(\boldsymbol{x})^T = \boldsymbol{\phi}(\boldsymbol{x})^T \begin{pmatrix} \mathsf{I}_N \\ \mathsf{\Lambda}_2 \mathsf{\Phi}_2^T \mathsf{\Phi}_1^{-T} \mathsf{\Lambda}_1^{-1} \end{pmatrix} \mathsf{\Lambda}_1 \mathsf{\Phi}_1^T. \tag{13.10}$$

In order to accomplish an analogous representation for the kernel matrix K we can observe that we did not touch the vector $\boldsymbol{\phi}(\boldsymbol{x})^T$ in the representation of $\boldsymbol{k}(\boldsymbol{x})^T$. Therefore, we can apply the same manipulations as above to the eigen-decomposition of K and obtain

$$\begin{aligned}\mathsf{K} &= \Phi\Lambda\Phi^T \\ &= \Phi \begin{pmatrix} \mathsf{I}_N \\ \Lambda_2\Phi_2^T\Phi_1^{-T}\Lambda_1^{-1}\end{pmatrix} \Lambda_1\Phi_1^T. \end{aligned} \tag{13.11}$$

We are now ready to look at the interpolation system $\mathsf{K}\boldsymbol{c} = \boldsymbol{y}$. Substituting (13.11) for K produces

$$\Phi \begin{pmatrix} \mathsf{I}_N \\ \Lambda_2\Phi_2^T\Phi_1^{-T}\Lambda_1^{-1}\end{pmatrix} \Lambda_1\Phi_1^T\boldsymbol{c} = \boldsymbol{y}.$$

At this point, we still have not made any structural changes to the interpolation problem. We have only decomposed K using our knowledge of the eigenvalues and eigenfunctions of K. Because we have already assumed that $\Lambda_1\Phi_1^T$ is invertible we can redefine the interpolation problem using the full-rank linear transformation

$$\boldsymbol{b} = \Lambda_1\Phi_1^T\boldsymbol{c}, \tag{13.12}$$

where the matrix $\left(\Lambda_1\Phi_1^T\right)^{-1}$ formally[3] acts as a *basis transformation* akin to the matrix T in Theorem 12.1. This allows us to solve

$$\Phi \begin{pmatrix} \mathsf{I}_N \\ \Lambda_2\Phi_2^T\Phi_1^{-T}\Lambda_1^{-1}\end{pmatrix} \boldsymbol{b} = \boldsymbol{y}$$

instead of $\mathsf{K}\boldsymbol{c} = \boldsymbol{y}$. We can think of this transformation as producing an alternate basis in which to perform our kernel interpolation, and this new basis produces a linear system

$$\Psi\boldsymbol{b} = \boldsymbol{y}, \tag{13.13}$$

where

$$\Psi = \Phi \begin{pmatrix} \mathsf{I}_N \\ \Lambda_2\Phi_2^T\Phi_1^{-T}\Lambda_1^{-1}\end{pmatrix}. \tag{13.14}$$

The new system matrix Ψ has the same structure as the original kernel matrix K, but without the $\Lambda_1\Phi_1^T$ term. In fact, we can write

$$\mathsf{K} = \Psi\Lambda_1\Phi_1^T, \tag{13.15}$$

where the equality is as valid as the approximation of $K(\boldsymbol{x},\boldsymbol{z})$ by the M-term truncated Hilbert–Schmidt expansion.

From now on we will refer to (13.15) as the *Hilbert–Schmidt SVD* (or HS-SVD) of the matrix K, motivated by its structural similarity with the traditional SVD (see,

[3]We again use the term *formally*, since we make this argument only to point out certain parallels with the discussion in Chapter 12. However, in practice we never perform this basis transformation as we most likely would not be able to safely compute with the matrix $\mathsf{T} = \Phi_1^{-T}\Lambda_1^{-1}$.

e.g., [Golub and Van Loan (2012)]). In the HS-SVD, the diagonal matrix $\mathsf{\Lambda}_1$ has positive, decreasing values, and the matrices $\mathsf{\Psi}$ and $\mathsf{\Phi}_1$ are generated by orthogonal eigenfunctions, rather than actually being orthogonal matrices.

After we solve the system $\mathsf{\Psi}\boldsymbol{b} = \boldsymbol{y}$ from (13.13), we must also be able to evaluate our interpolant, which can be done by studying the original problem. Were we to solve the standard basis interpolation system $\mathsf{K}\boldsymbol{c} = \boldsymbol{y}$, we would be able to evaluate the interpolant (see (13.4)) using $\boldsymbol{k}(\boldsymbol{x})^T$, i.e.,

$$\begin{aligned} s(\boldsymbol{x}) &= c_1 K(\boldsymbol{x}, \boldsymbol{x}_1) + \ldots + c_N K(\boldsymbol{x}, \boldsymbol{x}_N) \\ &= \boldsymbol{k}(\boldsymbol{x})^T \boldsymbol{c} \\ &= \boldsymbol{\phi}(\boldsymbol{x})^T \begin{pmatrix} \mathsf{I}_N \\ \mathsf{\Lambda}_2 \mathsf{\Phi}_2^T \mathsf{\Phi}_1^{-T} \mathsf{\Lambda}_1^{-1} \end{pmatrix} \mathsf{\Lambda}_1 \mathsf{\Phi}_1^T \boldsymbol{c}, \end{aligned}$$

where the last line invoked (13.10). Using (13.12), we can write

$$s(\boldsymbol{x}) = \boldsymbol{\phi}(\boldsymbol{x})^T \begin{pmatrix} \mathsf{I}_N \\ \mathsf{\Lambda}_2 \mathsf{\Phi}_2^T \mathsf{\Phi}_1^{-T} \mathsf{\Lambda}_1^{-1} \end{pmatrix} \boldsymbol{b},$$

where $\boldsymbol{b}$ was found from solving (13.13). Analogously to our definition of $\boldsymbol{k}(\boldsymbol{x})^T$, we can define our *stable Hilbert–Schmidt SVD basis* as

$$\boldsymbol{\psi}(\boldsymbol{x})^T = \boldsymbol{\phi}(\boldsymbol{x})^T \begin{pmatrix} \mathsf{I}_N \\ \mathsf{\Lambda}_2 \mathsf{\Phi}_2^T \mathsf{\Phi}_1^{-T} \mathsf{\Lambda}_1^{-1} \end{pmatrix} \tag{13.16}$$

and evaluate our stable interpolant with

$$s(\boldsymbol{x}) = \boldsymbol{\psi}(\boldsymbol{x})^T \boldsymbol{b}.$$

Remark 13.3. We can interpret (13.16) as providing us an *alternate basis* for the finite-dimensional space $\mathcal{H}_K(\mathcal{X})$ we use for approximation, i.e., up to the accuracy governed by the choice of our truncation length M we have

$$\text{span}\{K(\cdot, \boldsymbol{x}_1), \ldots, K(\cdot, \boldsymbol{x}_N)\} = \text{span}\{\psi_1, \ldots, \psi_N\}.$$

We can see from (13.16) that these *Hilbert–Schmidt basis functions* consist of the *data-independent* eigenfunctions $\boldsymbol{\phi}_1 = \begin{pmatrix} \varphi_1 & \cdots & \varphi_N \end{pmatrix}^T$ "corrected" by *data-dependent* linear combinations of the higher-order eigenfunctions $\boldsymbol{\phi}_2 = \begin{pmatrix} \varphi_{N+1} & \cdots & \varphi_M \end{pmatrix}^T$, i.e., the entire vector of HS-SVD basis functions is given by

$$\boldsymbol{\psi}(\boldsymbol{x})^T = \boldsymbol{\phi}_1(\boldsymbol{x})^T + \boldsymbol{\phi}_2(\boldsymbol{x})^T \mathsf{\Lambda}_2 \mathsf{\Phi}_2^T \mathsf{\Phi}_1^{-T} \mathsf{\Lambda}_1^{-1}. \tag{13.17}$$

We will occasionally refer to the matrix

$$\bar{\mathsf{C}} = \mathsf{\Lambda}_1^{-1} \mathsf{\Phi}_1^{-1} \mathsf{\Phi}_2 \mathsf{\Lambda}_2 \quad \Longleftrightarrow \quad \bar{\mathsf{C}}^T = \mathsf{\Lambda}_2 \mathsf{\Phi}_2^T \mathsf{\Phi}_1^{-T} \mathsf{\Lambda}_1^{-1} \tag{13.18}$$

as the data-dependent *corrector matrix*, and the term $\boldsymbol{\phi}_2(\cdot)^T \bar{\mathsf{C}}^T$ in (13.17) as the *correction* to the first N eigenfunctions $\boldsymbol{\phi}_1(\cdot)^T$.

The use of $\bar{\mathsf{C}}^T$ rather than simply $\bar{\mathsf{C}}$ is to maintain consistency with the matrix products for K. Because we use

$$\mathsf{K} = \mathsf{\Phi} \mathsf{\Lambda} \mathsf{\Phi}^T = \mathsf{\Phi}_1 \mathsf{\Lambda}_1 \mathsf{\Phi}_1^T + \mathsf{\Phi}_2 \mathsf{\Lambda}_2 \mathsf{\Phi}_2^T,$$

we choose to define the corrector so that

$$\mathsf{\Psi} = \mathsf{\Phi} \begin{pmatrix} \mathsf{I}_N & \bar{\mathsf{C}} \end{pmatrix}^T = \mathsf{\Phi}_1 + \mathsf{\Phi}_2 \bar{\mathsf{C}}^T.$$

Example 13.1. (Numerical rank of the HS-SVD matrix Ψ)
In Program 4.16 we described a situation where the C^∞ Chebyshev kernel

$$K_{a,b}(x,z) = 1 - a + 2a(1-b)\frac{b(1-b^2) - 2b(x^2+z^2) + (1+3b^2)xz}{(1-b^2)^2 + 4b\left(b(x^2+z^2) - (1+b^2)xz\right)}, \qquad (3.13)$$

with $a \in (0,1]$, $b \in (0,1)$, and $x, z \in [-1,1]$, produced a numerically rank-deficient, and thus numerically singular, interpolation matrix K for $N = 100$ Chebyshev nodes. The present example demonstrates that the matrix Ψ generated from the Hilbert–Schmidt SVD is full-rank and thus viable for use in interpolation. We consider $M = 120$ eigenfunctions and use Program 13.1 to both compute `Psi` and plot the HS-SVD basis as it is being constructed.

Program 13.1. Excerpts from `HSSVDPsivsK.m`

```
   % Define the points under consideration
2  N = 100;  x = pickpoints(-1,1,N,'cheb');
   % Define the closed form of the Cinf kernel
4  K = @(x,z) .6 + .56*(.273 -.6*bsxfun(@plus,x.^2,z'.^2) + 1.27*x*z')./...
       (.8281 + .36*bsxfun(@plus,x.^2,z'.^2) - 1.308*x*z');
6  % Define the eigenfunctions and eigenvalues
   phi = @(n,x) bsxfun(@times,sqrt(2-(n==0)),cos(bsxfun(@times,n,acos(x))));
8  lam = @(n) (n==0)*.6 + (n>0).*(.3).^(n-1)*(.4*.7);
   % Choose the number of eigenfunctions required for the HS-SVD
10 M = N + 20;
   % Evaluate the Phi matrix and diagonal of Lambda (as a vector)
12 narr = 0:M-1;
   Phi = phi(narr,x);
14 lamvec = lam(narr);
   % Compute the closed form kernel matrix and its rank
16 Kcf = K(x,x);  rankK = rank(Kcf);
   % Separate the components of the HS-SVD
18 Phi1 = Phi(:,1:N);
   Phi2 = Phi(:,N+1:end);
20 lamvec1 = lamvec(1:N);
   lamvec2 = lamvec(N+1:end);
22 % Compute the CbarT matrix and the correction term
   % Remember that Lambda2 and inv(Lambda1) must occur simultaneously
24 CbarT = bsxfun(@rdivide,lamvec2',lamvec1).*(Phi2'/Phi1');
   % Form the full Psi matrix; eye(N) is the N sized identity
26 Psi = Phi*[eye(N);CbarT];  rankPsi = rank(Psi);
   % Display some diagnostic comments about the HS-SVD
28 Khssvd = bsxfun(@times,Psi,lamvec1)*Phi1';
   err = norm(Kcf - Khssvd);
30 fprintf('rank(K) = %d, rank(Psi) = %d, error in HS-SVD = %g\n',rankK,rankPsi,err)
   % Choose some points at which to evaluate the components of the HS-SVD
32 xeval = pickpoints(-1,1,1000);
   Phieval = phi(narr,xeval);
34 Phieval1 = Phieval(:,1:N);
   Phieval2 = Phieval(:,N+1:end);
```

```
36  Correction = Phieval2*CbarT;
    Psieval = Phieval1 + Correction;
```

The output of the `fprintf` statement is

```
rank(K) = 28, rank(Psi) = 100, error in HS-SVD = 1.35995e-14
```

which indicates that, as expected, the matrix Ψ is invertible and thus $\boldsymbol{\psi}(\cdot)$ contains a suitable basis for interpolation. The error statement displays the value of $\|\mathsf{K} - \Psi\Lambda_1\Phi_1^T\|_2$, which confirms that the Hilbert–Schmidt SVD does accurately reproduce the kernel interpolation matrix. The two largest correction terms are plotted in Figure 13.1.

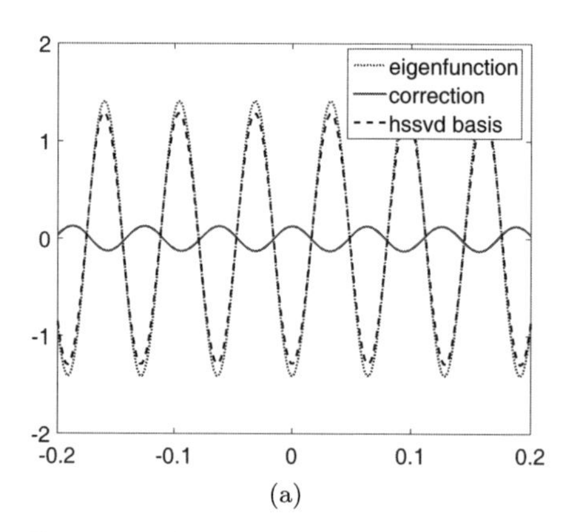

(a)

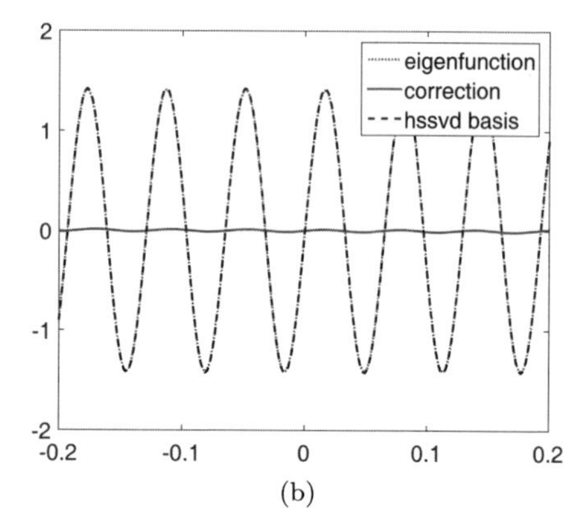

(b)

Fig. 13.1 The correction which transforms the eigenfunctions to the HS-SVD basis is plotted, and for this example only a few correction terms are significant. (a) The 99th HS-SVD basis function undergoes a minor correction, (b) The 98th HS-SVD basis function is almost identical to the 98th eigenfunction and therefore requires almost no correction.

The evaluation of the HS-SVD basis, $\boldsymbol{\psi}(\boldsymbol{x})$, can be broken down into its data-dependent and independent components for efficiency. Within MATLAB, once the data-dependent corrector matrix $\overline{\mathsf{C}}^T$ has been computed using the kernel centers, the vector $\boldsymbol{\phi}(\boldsymbol{x})^T$ is evaluated independently of the centers and is used to form $\boldsymbol{\psi}(\boldsymbol{x})^T$ as the product of data-independent and data-dependent terms: `Psi = Psi*[eye(N);CbarT]` (see lines 24–26).

As is the custom in MATLAB, the computation of all the $\boldsymbol{\psi}(\boldsymbol{x})^T$ vectors for the evaluation points are conducted simultaneously: each row of `Phieval` and `Psieval` represents one $\boldsymbol{\phi}(\boldsymbol{x})^T$ or $\boldsymbol{\psi}(\boldsymbol{x})^T$ vector. The difference in the computations of `Psi` and `Psieval` are cosmetic and meant to showcase the flexibility of computing in MATLAB: we could also have written `Psi = Phi1 + Phi2*CbarT` and `Psieval=Phieval*[eye(N);CbarT]`.

Remark 13.4. The "flat" limit behavior of the eigenfunction correction demonstrated in Example 13.1 seems to apply in similar fashion to all kernels we have encountered in practice. Namely, once we approach the flat limit, the corrector matrix $\bar{\mathsf{C}}^T$ approaches the zero matrix, and the matrix $\Psi \to \Phi_1$. We will discuss consequences of this behavior in Section 14.1.

Remark 13.5. As illustrated on lines 23–24 of Program 13.1, we always make sure to *simultaneously* apply the matrices Λ_2 and Λ_1^{-1} required in the computation of the corrector matrix $\bar{\mathsf{C}}^T$ needed for the stable HS-SVD basis (see (13.16)). This minimizes the chances that — especially as the kernel becomes flat — we would produce an overflow or underflow error caused, respectively, by dividing by the small eigenvalues at the end of Λ_1 (which, e.g., are on the order of ε^{2N-2} for Gaussian kernels) or multiplying by the even tinier ones in Λ_2 (which are on the order of ε^{2M-2} for Gaussian kernels).

Remark 13.6. The reader may wonder how the "QR" enters the RBF-QR terminology championed by Fornberg and Piret (2008b) because, thus far, we have not used any terms with the letters Q or R. The term RBF-QR was introduced because of the use of the *QR decomposition* to compute the corrector matrix $\bar{\mathsf{C}}^T = \Lambda_2 \Phi_2^T \Phi_1^{-T} \Lambda_1^{-1}$ in (13.14) or (13.16), specifically the component $\Phi_2^T \Phi_1^{-T}$.

Starting with the original matrix $\Phi = \begin{pmatrix}\Phi_1 & \Phi_2\end{pmatrix}$, the QR decomposition gives us

$$\begin{pmatrix}\Phi_1 & \Phi_2\end{pmatrix} = \mathsf{Q}\begin{pmatrix}\mathsf{R}_1 & \mathsf{R}_2\end{pmatrix},$$

where Q is a unitary matrix, and R_1 is upper triangular. Using this factorization, we can compute

$$\Phi_2^T \Phi_1^{-T} = \mathsf{R}_2^T \mathsf{Q}^T \mathsf{Q} \mathsf{R}_1^{-T} = \mathsf{R}_2^T \mathsf{R}_1^{-T}. \tag{13.19}$$

This idea was developed in [Fornberg and Piret (2008b)] and is sometimes preferable to directly inverting Φ_1^{-T} because it may be more stable.

13.2.1 *Summary: How to use the Hilbert–Schmidt SVD*

Remark 13.7. The reader who has been closely following our exposition in this book may at this point be wondering how the HS-SVD basis is related to our discussion of alternate bases in Chapter 12, in particular Theorem 12.1. From that point of view we have

$$\mathsf{K} = \Psi \Lambda_1 \Phi_1^T \quad \Longrightarrow \quad \mathsf{T} = \Phi^{-T} \Lambda_1^{-1},$$

so that

$$\boldsymbol{\psi}(\boldsymbol{x})^T = \boldsymbol{k}(\boldsymbol{x})^T \mathsf{T} = \boldsymbol{k}(\boldsymbol{x})^T \Phi^{-T} \Lambda_1^{-1}.$$

While this is certainly true — and we have made this statement above — it would be totally counterproductive to think of the stable Hilbert–Schmidt SVD basis in this way. Using a basis transformation matrix $\mathsf{T} = \Phi^{\ T} \Lambda_1^{-1}$ is infeasible since the whole point of our derivation was to *not work with the ill-conditioned matrix* K and to isolate — and remove — as much as possible of the ill-conditioning that comes from the eigenvalue matrix $\Lambda = \begin{pmatrix}\Lambda_1 & \\ & \Lambda_2\end{pmatrix}$.

Therefore, we must instead think of the stable HS-SVD basis in the form

$$\boldsymbol{\psi}(\boldsymbol{x})^T = \boldsymbol{\phi}(\boldsymbol{x})^T \begin{pmatrix} \mathsf{I}_N \\ \Lambda_2 \Phi_2^T \Phi_1^{-T} \Lambda_1^{-1} \end{pmatrix} \quad \text{or} \quad \boldsymbol{\psi}(\boldsymbol{x})^T = \boldsymbol{\phi}_1(\boldsymbol{x})^T + \boldsymbol{\phi}_2(\boldsymbol{x})^T \bar{\mathsf{C}}^T,$$

as was stated in (13.16). This representation comes directly from the Hilbert–Schmidt expansion (Mercer series) of K. While the correction still contains the eigenvalues in Λ_1^{-1} and Λ_2, we explained in Remark 13.5 that these matrices can be applied simultaneously so that the detrimental effects caused by these two matrices are balanced out as much as possible.

Correspondingly, while the matrix K does equal

$$\mathsf{K} = \Psi \Lambda_1 \Phi_1^T, \quad \text{with} \quad \Psi = \Phi \begin{pmatrix} \mathsf{I}_N \\ \Lambda_2 \Phi_2^T \Phi_1^{-T} \Lambda_1^{-1} \end{pmatrix} = \Phi_1 + \Phi_2 \bar{\mathsf{C}}^T,$$

we do not decompose K to obtain these factors, but instead "recompose" K from the matrices Ψ, Λ_1 and Φ_1, which are obtained directly from the eigenvalues and eigenfunctions.

The discussion in the previous two subsections implies that — instead of solving the standard interpolation system

$$\mathsf{K}\boldsymbol{c} = \boldsymbol{y},$$

which has the potential of yielding inaccurate expansion coefficients $\boldsymbol{c}$ that are subsequently multiplied against poorly conditioned basis functions in the vector $\boldsymbol{k}(\cdot)^T$ — we now solve the transformed system

$$\Psi \boldsymbol{b} = \boldsymbol{y}$$

with the Hilbert–Schmidt basis collected in the vector $\boldsymbol{\psi}(\cdot)^T$. This results in a new set of coefficients $\boldsymbol{b} = \Psi^{-1}\boldsymbol{y}$ with which we evaluate the interpolant

$$\begin{aligned} s(\boldsymbol{x}) &= \boldsymbol{k}(\boldsymbol{x})^T \mathsf{K}^{-1} \boldsymbol{y} \\ &= \boldsymbol{\psi}(\boldsymbol{x})^T \Lambda_1 \Phi_1^T \Phi_1^{-T} \Lambda_1^{-1} \Psi^{-1} \boldsymbol{y} \\ &= \boldsymbol{\psi}(\boldsymbol{x})^T \Psi^{-1} \boldsymbol{y}. \end{aligned} \tag{13.20}$$

Note that all the ill-conditioning from Λ_1 has been eliminated analytically.

Recall, in this Hilbert–Schmidt SVD framework, the system matrix K is *never used*. Instead, the new system matrix Ψ and the new basis function vector $\boldsymbol{\psi}(\cdot)^T$ are obtained directly from the eigenvalues and eigenfunctions of K via (13.14) and (13.16), respectively.

Remark 13.8. In Section 12.1.2 we discussed the *cardinal functions* $\boldsymbol{\ell}(\cdot)^T = \boldsymbol{k}(\cdot)^T \mathsf{K}^{-1}$. The Hilbert–Schmidt SVD provides a fresh look at these cardinal functions and shows that we can compute them stably via

$$\boldsymbol{\ell}(\cdot)^T = \boldsymbol{\psi}(\cdot)^T \Psi^{-1}. \tag{13.21}$$

We used this representation much earlier to create plots of the cardinal functions for the iterated Brownian bridge kernels in Figure 7.3.

Remark 13.9. Because the $\mathsf{\Lambda}_1$ matrix is a main cause of ill-conditioning, removing it from K *analytically* will produce a $\mathsf{\Psi}$ matrix which is better conditioned. We may draw some parallels to our discussion of *preconditioning* in Section 12.4, but emphasize two important differences:

(1) A traditional preconditioning scheme for $\mathsf{K}\boldsymbol{c} = \boldsymbol{y}$ would take the form $(\mathsf{K}\mathsf{M}^{-1})\mathsf{M}\boldsymbol{c} = \boldsymbol{y}$, where $\mathsf{K}\mathsf{M}^{-1}$ is well conditioned, and M is inverted directly, with the eventual goal of recovering $\boldsymbol{c}$. For the Hilbert–Schmidt SVD, $\mathsf{M} = \mathsf{\Lambda}_1\mathsf{\Phi}_1^T$, and is so ill-conditioned that applying M^{-1} is unsafe, even if it can be done directly. Therefore we solve a *different* linear system (13.13) which will produce the *same* interpolant. In this preconditioning context, we are choosing to work with $\mathsf{M}\boldsymbol{c}$ as our solution, rather than actually finding $\boldsymbol{c}$, thus the term preconditioning is not entirely appropriate.
(2) Our "preconditioner" $\mathsf{\Lambda}_1\mathsf{\Phi}_1^T$ must be applied analytically because once K is formed, the system is already irrevocably ill-conditioned. This idea was discussed in Chapter 11, and means that rather than forming K and right multiplying by $(\mathsf{\Lambda}_1\mathsf{\Phi}_1^T)^{-1}$, we will directly form $\mathsf{\Psi}$ using (13.14).

13.3 Iterated Brownian Bridge Kernels via the Hilbert–Schmidt SVD

To demonstrate the stable nature of the Hilbert–Schmidt SVD

$$\mathsf{K} = \mathsf{\Psi}\mathsf{\Lambda}_1\mathsf{\Phi}_1^T,$$

we will consider its application on the iterated Brownian bridge kernels developed in Chapter 7. As a reminder, these kernels $K_{\beta,\varepsilon}$ have the eigenpairs

$$\lambda_n = \left(n^2\pi^2 + \varepsilon^2\right)^{-\beta}, \qquad \varphi_n(x) = \sqrt{2}\sin\left(n\pi x\right),$$

where $\varepsilon > 0$ is a shape parameter which localizes around the kernel center for very large values, and $\beta \in \{1, 2, \ldots\}$ is a smoothness parameter which yields $2\beta - 2$ continuous derivatives for the kernel.

As shown in Sections 7.1 and 7.3.5, the "flat" piecewise polynomial limit of these kernels is obtained as $\varepsilon \to 0$. Chapter 11 discussed the ill-conditioning which occurs for iterated Brownian bridge interpolants as β increases or ε decreases. To alleviate that ill-conditioning, we will use the Mercer series of the kernel to create a stable basis and find the flat limit of the interpolant.

For our first example, we fix $\beta = 7$ on the interval $[0, 1]$; notice that these choices also determine the boundary conditions (7.9) satisfied by the kernel. Our scattered data interpolation problem will consist of N data locations evenly spaced in $[0, 1]$, although the end points $x = 0$ and $x = 1$ will be omitted because they are already fixed by the boundary conditions. The N data values are taken from the function $f_{\nu,\gamma}$ (cf. (7.14))

$$f_{\nu,\gamma}(x) = (1/2 - \gamma)^{-2\nu}\,(x - \gamma)_+^{\nu}\,(1 - \gamma - x)_+^{\nu}, \qquad x \in [0, 1],$$

with $\gamma = 0.25$ and $\nu = 14$. This function was first introduced in Section 7.3, and $f_{\nu,\gamma}$ satisfies the boundary conditions (7.9) when $2\nu \geq \beta$.

Example 13.2. (HS-SVD for IBB kernels)
Program 13.2 demonstrates the use of the Hilbert–Schmidt SVD associated with the iterated Brownian bridge kernel $K_{7,1}$ on the function $f_{14,\gamma}$ for $N = 19$ evenly spaced points in $[0, 1]$, excluding the boundary (recall Remark 7.1).

Program 13.2. Excerpts from `HSSVDInterpExample.m`

```
% Pick a function that satisfies the IBB boundary conditions
yf = @(x) .25^(-28)*max(x-.25,0).^14.*max(.75-x,0).^14;
% Set up some data points at which to sample
% Remember to dump the boundaries for the IBB kernel
N = 19;
x = pickpoints(0,1,N+2);  x = x(2:end-1);
y = yf(x);
% Set up evaluation points
Neval = 300;
xeval = pickpoints(0,1,Neval);
yeval = yf(xeval);
% Choose IBB parameters for computation
ep = 1;  beta = 7;
% Define the eigenfunctions and eigenvalues
phifunc = @(n,x) sqrt(2)*sin(pi*x*n);
lamfunc = @(b,e,n) ((pi*n).^2+e^2).^(-b);
% Determine how many eigenfunctions will be needed for accuracy
M = ceil(1/pi*sqrt(eps^(-1/beta)*(N^2*pi^2+ep^2)-ep^2));
narr = 1:M;
% Evaluate phi for the necessary eigenfunctions
Phi = phifunc(narr,x);
Phi1 = Phi(:,1:N);  Phi2 = Phi(:,N+1:end);
Phieval1 = phifunc(narr(1:N),xeval);  Phieval2 = phifunc(narr(N+1:end),xeval);
% Evaluate lamvec for the necessary eigenvalues
lamvec = lamfunc(beta,ep,narr);
lamvec1 = lamvec(1:N);  lamvec2 = lamvec(N+1:end);
% Form the Cbar^T corrector matrix for the HS-SVD basis
CbarT = bsxfun(@rdivide,lamvec2',lamvec1).*(Phi2'/Phi1');
% Compute the interpolant in the HS-SVD basis
Psi = Phi*[eye(N);CbarT];
Psieval = Phieval1 + Phieval2*CbarT;
seval = Psieval*(Psi\y);
% Create standard basis kernel matrices and evaluate their interpolant
K = bsxfun(@times,lamvec,Phi)*Phi';
Keval = bsxfun(@times,lamvec1,Phieval1)*Phi1' + ...
        bsxfun(@times,lamvec2,Phieval2)*Phi2';
sevalstandard = Keval*(K\y);
```

The interpolants from Program 13.2 are displayed in Figure 13.2. For these chosen data locations the interpolation matrix K for the standard basis is ill-

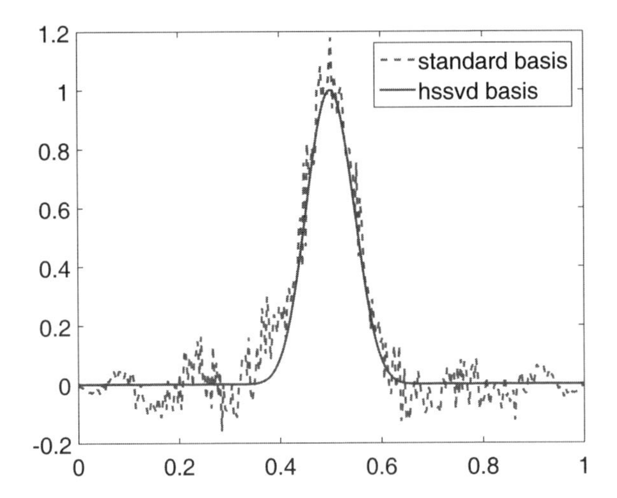

Fig. 13.2 Ill-conditioning in the standard basis yields an erratic interpolant while the more stable HS-SVD basis nearly matches the original function.

conditioned, so much so that the solution exhibits wildly erratic behavior. Because K is numerically singular, the \ operator in MATLAB "does its best" to solve the system but alerts the user that the answer may not be trustworthy. In contrast, the stable basis produced using the Hilbert–Schmidt SVD yields a solution that avoids the ill-conditioning and matches the original function to within visual accuracy.

MATLAB reports that

```
rank(K) = 11, standard basis error = 1.3e+00
rank(Psi) = 19, HS-SVD basis error = 2.9e-03
```

which implies the value in using the HS-SVD basis for the iterated Brownian bridge kernels in the $\varepsilon \to 0$ limit for larger β and N.

Note that on line 28 of Program 13.2 we use the corrector matrix $\bar{\mathsf{C}}^T$ as discussed earlier (see, e.g., (13.18) and line 24 of Program 13.1). The corrector matrix $\bar{\mathsf{C}}^T$ is required for both computing the interpolation matrix Ψ, and evaluating the interpolant using $\psi(\cdot)$, so it is in our best interests to store the matrix $\bar{\mathsf{C}}^T$ when the interpolation problem (13.13) is solved. In the example above, we simply reuse it, but for more advanced examples, when the solve and evaluation stages are conducted separately, we will save computational time by saving $\bar{\mathsf{C}}^T$.

One other point of interest is line 18, which defines the length of the eigenfunction expansion needed to approximate the iterated Brownian bridge kernel accurately. This value comes from (7.13), where we studied the ratio of the eigenvalues of the series to determine at which point machine precision dominated.

The process of solving for and evaluating the IBB interpolant with the HS-SVD basis can be encapsulated neatly into a `function`, as presented in Program 13.3. A more complicated version of this function, complete with discussion about MATLAB programming strategies, is available in Appendix D; to save space and not repeat earlier comments, the code below is presented in a simple form.

Program 13.3. `HSSVD_IBBSolve.m`

```
function seval = HSSVD_IBBSolve(ep,beta,x,y,xeval)
% Evaluate the HS-SVD basis IBB interpolant
% function seval = HSSVD_IBBSolve_Full(ep,beta,x,y,xeval)
%   Inputs:  ep    - shape parameter
%            beta  - smoothness parameter
%            x     - data locations
%            y     - data values
%            xeval - locations at which to evaluate the interpolant
%   Outputs: seval - interpolant values at xeval

phifunc = @(n,x) sqrt(2)*sin(pi*x*n);
lamfunc = @(b,e,n) ((pi*n).^2+e^2).^(-b);

N = size(x,1);

M = ceil(1/pi*sqrt(eps^(-1/beta)*(N^2*pi^2+ep^2)-ep^2));
narr = 1:M;

Phi1 = phifunc(narr(1:N),x);
Phi2 = phifunc(narr(N+1:end),x);
lamvec1 = lamfunc(beta,ep,narr(1:N));
lamvec2 = lamfunc(beta,ep,narr(N+1:end));

CbarT = bsxfun(@rdivide,lamvec2',lamvec1).*(Phi2'/Phi1');

Psi = Phi1 + Phi2*CbarT;
b = Psi\y;

Phieval1 = phifunc(narr(1:N),xeval);
Phieval2 = phifunc(narr(N+1:end),xeval);
seval = Phieval1*b + Phieval2*(CbarT*b);
```

Remark 13.10. Recall that Program 7.2 uses the function `HSSVD_IBBSolve` to create the cardinal function plots in Figure 7.3. In that setting it passed a full matrix rather than a single vector as the data values argument `y`. Although not immediately the goal of this function, evaluating the interpolant for multiple pieces of data simultaneously is an efficient use of memory (level 3 BLAS rather than level 2 BLAS evaluating each `y` vector separately). This flexibility in inputs is not by our design but rather a serendipitous result of the exceptional design of MATLAB.

Example 13.3. (Computing "flat" limits via the HS-SVD)
In Section 7.3.5, we introduced the idea that as $\varepsilon \to 0$, the kernels approach a limiting case which is severely ill-conditioned. This ill-conditioning exists only in the standard basis, and is not fundamental to the interpolant, which approaches

a piecewise polynomial spline that can be computed stably. We developed this idea further in Chapter 10, and now we will produce an example for the iterated Brownian bridge kernels which requires the use of the stable basis (as opposed to Example 7.3, where the kernels were not smooth enough to trigger any instabilities; we used only $\beta = 1, 2$ there).

Program 13.4 uses the kernel $K_{7,\varepsilon}$ and tests accuracy and stability of the interpolation scheme for $N = \{15, 30, 60\}$ as $\varepsilon \to 0$ for both the standard and Hilbert–Schmidt SVD methods. The data is sampled from the test function $f_{14,0.25}$, as before.

Program 13.4. `HSSVDFlatLimit.m`

```
yf = @(x) .25^(-28)*max(x-.25,0).^14.*max(.75-x,0).^14;
% Choose parameters to study
epvec = logspace(0,3,15);  beta = 7;
% Various input data locations to test
Nvec = [15,30,60];
% Choose points at which to test the error
Neval = 100;  xeval = pickpoints(0,1,Neval);  yeval = yf(xeval);
% Initialize error storage and loop through N values
errs = zeros(length(Nvec),length(epvec));
k = 1;
for N=Nvec
    % Create the data used to run tests
    x = pickpoints(0,1,N+2);
    x = x(2:end-1);
    y = yf(x);
    % Compute all the errors at once with array fun
    % Additional arguments are still passed to HSSVD_IBBSolve
    errs(k,:) = arrayfun(@(ep) ...
                    errcompute(HSSVD_IBBSolve(ep,beta,x,y,xeval),yeval), ...
                epvec);
    k = k + 1;
end
```

Note that, to save space, the standard basis computations and plotting commands can only be seen in the script provided in the `GaussQR` library.

Remark 13.11. Often times, MATLAB functions such as `arrayfun` or `ode45` will require the user to choose or pass a function to it. That function is supposed to have a specific calling sequence, but it is possible to pass a function with arbitrary arguments and have MATLAB think that it has the correct calling sequence. The tool we often use for this is the *function handle*, which as MATLAB describes it, is a tool used to "Pass a function to another function."

In lines 18–19 we create an *anonymous function* with the `@` symbol. This function is supposed to take in one value, `ep`, and return a scalar (since `'UniformOutput,0'` was not passed to `arrayfun`). After using the function handle, any function can be

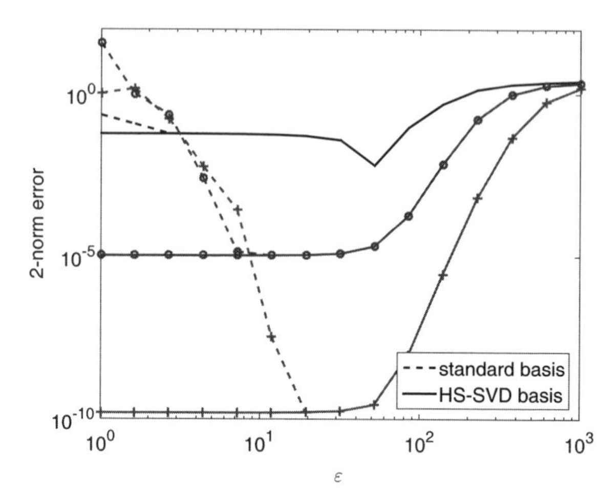

Fig. 13.3 For small ε, the standard iterated Brownian bridge interpolants (dashed lines) of $f_{14,\gamma}$ with $\beta = 7$ are unstable. When the HS-SVD is used (solid lines), the full limit can be found stably. Point sets of size $N = 15$, $N = 30$ and $N = 60$ generated the top (-), middle (o) and bottom (+) curves respectively.

defined, and the necessary additional data is stored in the handle to be called later. See Program 15.3 for more discussion on this.

13.4 Issues with the Hilbert–Schmidt SVD

We have now demonstrated the value of the Hilbert–Schmidt SVD approach when applied to the iterated Brownian bridge kernels, but the examples from Section 13.3 are only the first and simplest of the possible uses of the Hilbert–Schmidt SVD. The situation becomes more complicated when the eigenfunctions are not simply sines, but rather difficult to evaluate functions such as those of the Gaussian from Section 12.2.1.

13.4.1 *Truncation of the Hilbert–Schmidt series*

In Section 13.2 we used the truncation criterion (7.13) to determine an appropriate M-value for our Hilbert–Schmidt approximation of the iterated Brownian bridge kernel to be accurate at some tolerance. This criterion considered only the magnitude of the eigenvalues, because — for that family of kernels — the magnitude of the eigenfunctions is bounded. For the Gaussian eigenexpansion in one dimension (see Section 12.2.1 for details),

$$\lambda_n = \frac{\alpha\varepsilon^{2n}}{(\alpha^2+\delta^2+\varepsilon^2)^{2n+1/2}}, \qquad \varphi_n(x) = \gamma_n \exp(-\delta^2 x^2) H_{n-1}(\alpha\beta x),$$

the magnitude of the eigenfunctions is bounded by a nontrivial interplay between the polynomial growth of $H_{n-1}(\alpha\beta x)$ and the exponential decay of $\exp(-\delta^2 x^2)$, and it is therefore not straightforward to follow a similar truncation argument.

To use the Hilbert–Schmidt SVD on the Gaussians, we instead choose a different truncation criterion based on [Fornberg and Piret (2008b)]. If we consider the Φ matrix in its standard components Φ_1 and Φ_2 and choose a tolerance σ_{TOL}, we will choose M so that the last column of Φ_2 is a factor of σ_{TOL} less than the last column of Φ_1. In matrix notation, this might be written as

$$\|(\Phi_2)_{:,M-N}\| \le \sigma_{\mathrm{TOL}} \|(\Phi_1)_{:,N}\|.$$

The choice of norm is immaterial, but we often use the 2-norm.

This truncation design is somewhat unsatisfying, because it requires the actual construction of Φ to make a decision, and thus the entire construction process cannot be planned in advance. In a practical setting, this is not much of a bother because Φ would likely be constructed one column at a time naturally (by evaluating φ_n at all the $\boldsymbol{x}_j$ locations), but conceptually it is a bit displeasing.

Remark 13.12. A careful analysis of truncation lengths for general kernels given in series form (which includes kernels given in terms of truncated Hilbert–Schmidt series) is presented in [Griebel *et al.* (2015)]. There it is shown that the truncation length should be chosen in dependence on N and the smallest eigenvalue of the kernel matrix K. In fact, one should have

$$\sum_{n=M+1}^{\infty} \lambda_n \lesssim \frac{\lambda_{\min}(\mathsf{K})}{N},$$

where the symbol "$\lesssim$" encodes a dependence on the size of the eigenfunctions made more specific in [Griebel *et al.* (2015)]. If the truncation length is chosen according to this criterion then interpolation with the truncated kernel is shown to have the same approximation order as with the full kernel.

This criterion appears to have only limited practical applicability — especially if we are dealing with a very ill-conditioned matrix K (which is the very reason for us wanting to compute its Hilbert–Schmidt SVD). In that case it is impossible to accurately compute the smallest eigenvalue of K.

However, according to [Bach and Jordan (2003, Appendix C)] the matrix eigenvalues are roughly given by $N\lambda_n$, where λ_n are the kernel eigenvalues (this is essentially based on a Monte–Carlo approximation of the integral operator). Therefore we would have $\frac{\lambda_{\min}(\mathsf{K})}{N} \approx \lambda_N$, and we would only need an estimate for the sum of the kernel eigenvalues. And for those we know that $\sum_{n=1}^{\infty} \lambda_n = \operatorname{trace} \mathcal{K}$ (see (2.3)), which for a translation invariant kernel is $\widetilde{K}(\mathbf{0})$. Using that $\widetilde{K}(\mathbf{0}) = 1$ for Gaussian kernel, this remark implies that we want to pick M for a Gaussian kernel so that

$$1 - \sum_{n=1}^{M} \lambda_n \lesssim \lambda_N.$$

Of course, the additional bound on the eigenfunctions encoded in the "$\lesssim$" symbol is another matter.

13.4.2 *Invertibility of* Φ_1

In Section 13.2 we assumed invertibility of Φ_1. This assumption allowed the derivation of the Hilbert–Schmidt SVD to proceed smoothly, but it is conceivable that there exist point configurations and certain eigenfunction orderings — especially in higher dimensions — which lead to a singular Φ if one does not apply additional safeguards to prevent this from happening.

Having an invertible Φ_1 is fundamental to the success of the HS-SVD. In order for us to be able to compute the stable basis using (13.16), we need to compute $\Phi_1^{-1}\Phi_2$. Even if this is done using the QR factorization as in (13.19), we still need Φ_1^{-1} to exist to prevent R_1 from having zeros on its diagonal. How then can we ensure the nonsingularity of Φ_1?

Example 13.4. (Invertibility of Φ_1 for 1D Gaussians)
For the Gaussians in 1D, it is simple to show that Φ_1 is invertible: Φ_1 can be written as

$$\Phi_1 = \begin{pmatrix} e^{-\delta^2 x_1^2} & & \\ & \ddots & \\ & & e^{-\delta^2 x_N^2} \end{pmatrix} \begin{pmatrix} H_0(\alpha\beta x_1) & \cdots & H_{n-1}(\alpha\beta x_1) \\ & \vdots & \\ H_0(\alpha\beta x_N) & \cdots & H_{n-1}(\alpha\beta x_N) \end{pmatrix} \begin{pmatrix} \gamma_1 & & \\ & \ddots & \\ & & \gamma_N \end{pmatrix}.$$

Both diagonal matrices are clearly invertible. So, as long as the matrix populated by Hermite polynomials has an inverse, so does Φ_1. This Hermite polynomial matrix is a polynomial interpolation matrix which is known to be invertible since 1D polynomial interpolation at distinct points is known to have a unique solution (see, e.g., [Kincaid and Cheney (2002)]). Therefore we are assured that Φ_1^{-1} exists.

In general, the first N Sturm–Liouville eigenfunctions are linearly independent on their interval of definition (since they are even orthogonal). Moreover, due to the nestedness of their zeros (cf. Theorem 6.1), the first N eigenfunctions form a Haar system (see Definition 1.1) and therefore the matrix Φ_1 is invertible[4].

Of course, our goal for the stable basis is to work not only for 1D problems, but also for higher-dimensional problems. Unfortunately, the move to higher dimensions brings with it some additional issues as well. The first issue is that the invertibility of Φ_1 is no longer guaranteed.

Example 13.5. (Ordering of eigenfunctions)
Suppose we are given $\mathcal{X} = \{(0,0),(1,0),(0,1),(1,1)\}$ and want to interpolate with the stable basis for Gaussians using $\varepsilon = 10^{-8}$ and $\alpha = 1$. In Section 12.2.1 we discussed the topic of multiple eigenvalues existing at the same order. Because there is no unique polynomial interpolant on 4 points in 2D, there are multiple possible eigenfunction orderings which can be used to form Φ. For example, if we consider the first 6 available eigenfunctions and choose our expansion ordering as

$$[x,y] \text{ indices} \quad \underbrace{[1,1]}_{n=1} \quad \underbrace{[2,1]}_{n=2} \quad \underbrace{[1,2]}_{n=3} \quad \underbrace{[3,1]}_{n=4} \quad \underbrace{[2,2]}_{n=5} \quad \underbrace{[1,3]}_{n=6}$$

[4][Karlin (1968, Theorem 6.2)] establishes this nonsingularity for any 1D "oscillating kernel."

then the matrix Φ_1 is low-rank:

$$\Phi_1 = \begin{pmatrix} 1 & 0 & 0 & -1/\sqrt{2} \\ 1 & \sqrt{2} & 0 & 1/\sqrt{2} \\ 1 & 0 & \sqrt{2} & -1/\sqrt{2} \\ 1 & \sqrt{2} & \sqrt{2} & 1/\sqrt{2} \end{pmatrix}, \qquad \operatorname{rank}(\Phi_1) = 3.$$

If instead we had chosen the ordering

$$[x,y] \text{ indices} \qquad \underbrace{[1,1]}_{n=1} \quad \underbrace{[2,1]}_{n=2} \quad \underbrace{[1,2]}_{n=3} \quad \underbrace{[3,1]}_{n=5} \quad \underbrace{[2,2]}_{n=4} \quad \underbrace{[1,3]}_{n=6},$$

i.e., we swap the eigenvalues corresponding to indices [3,1] and [2,2], then the matrix Φ_1 is invertible:

$$\Phi_1 = \begin{pmatrix} 1 & 0 & 0 & 0 \\ 1 & \sqrt{2} & 0 & 0 \\ 1 & 0 & \sqrt{2} & 0 \\ 1 & \sqrt{2} & \sqrt{2} & 2 \end{pmatrix}, \qquad \operatorname{rank}(\Phi_1) = 4.$$

When all the eigenfunctions are of the same eigenvalue order (e.g., [3,1] and [2,2] are the same order), the order in which they appear in Φ does in principle not matter. It is, however, required that all eigenvalues of the same order appear before eigenvalues of a greater order. This is caused by the structure of the "correction" term $\phi_2(\cdot)^T \Lambda_2 \Phi_2^T \Phi^{-T} \Lambda_1^{-1} = \phi_2(\cdot)^T \bar{C}^T$ which comprises the stable basis $\psi(\cdot)^T = \phi_1(\cdot)^T + \phi_2(\cdot)^T \bar{C}^T$ (see Remark 13.3).

To form the stable basis, we need to compute $\bar{C}^T$. This term must be bounded as $\varepsilon \to 0$ in order for the Ψ matrix to stably reach an $\varepsilon \to 0$ limit. Because the eigenvalues decay, we know that the largest value in Λ_2 is less than or equal to the smallest value in Λ_1; the example above shows an instance where they would be equal. Because of these bounds we can place on the relative magnitude of the eigenvalues, we can guarantee that the cumulative effect of $\Lambda_2 X \Lambda_1^{-1}$ on any matrix X is a scaling down of the values.

If we allowed the eigenvalues to be ordered such that smaller eigenvalues appeared before larger ones, we could no longer be assured that all values in Λ_2 are less than or equal to Λ_1. This may cause the correction term to grow as $\varepsilon \to 0$, which would hinder our proposed stable basis.

This restriction means that we can only swap eigenfunctions between Φ_1 and Φ_2 if the associated eigenvalues are of the same order. In the event that an eigenfunction in Φ_1 is not helping the stable basis (i.e., provides no increase in rank), and is of very low order, we cannot move it to Φ_2 stably. We choose to remove that eigenfunction from the Hilbert–Schmidt series for that interpolation problem, since on that configuration of points it is not linearly independent from other basis functions. The use of reordering of the eigenfunctions (via a pivoted QR factorization) is also discussed in [Larsson *et al.* (2013b)].

As should be expected, the cost of implementing the stable basis ψ is significantly greater than using the standard basis; after all, there is no such thing as a free lunch. Because this section deals with theoretical aspects of the Hilbert–Schmidt SVD method, we have ignored the computational aspects. Section 15.4 will address these topics in the context of more advanced examples.

13.5 Comparison of Alternate Bases for Gaussian Kernels

Now that we have several different mechanisms for stably computing or approximating with ill-conditioned kernels, we will test them on the Gaussian kernel. Program 13.5 tests the standard basis against the Newton basis, eigenvalue basis and HS-SVD basis for a simple interpolation problem of $\cos(3\pi x)$ on $N = 20$ evenly spaced points in $[-1, 1]$. The associated errors are plotted in Figure 13.4.

Program 13.5. Excerpts from `HSSVDBasisComparison.m`

```
  % Define the Gaussian kernel
2 rbf = @(e,r) exp(-(e*r).^2);
  % Choose kernel centers: uniform in [-1 1]
4 N = 20;  x = pickpoints(-1,1,20);
  Neval = 100;  xeval = pickpoints(-1,1,Neval);
6 % Choose an analytic function to study
  yf = @(x) cos(3*pi*x);  y = yf(x);  yeval = yf(xeval);
8 % Choose a range of shape parameters
  epvec = logspace(-1,1,20);
10 % Loop through and compute the bases for different shape parameters
  k = 1;
12 for ep=epvec
      % Form the standard basis
14    K = rbf(ep,DistanceMatrix(x,x));    Keval = rbf(ep,DistanceMatrix(xeval,x));
      errvec(k) = errcompute(Keval*(K\y),yeval);
16
      % Form the Newton basis
18    Kpos = K+1e-14*eye(N);
      chol_transform = chol(Kpos,'lower');
20    Nmat = chol_transform;    Neval = Keval/chol_transform';
      errvecN(k) = errcompute(Neval*(Nmat\y),yeval);
22
      % Form the eigenfunction basis with all N eigenfunctions
24    gqr_alpha = 3;
      errvecE(k) = errcompute(gqr_eval(gqr_rsolve(x,y,ep,gqr_alpha,N),xeval),yeval);
26
      % Solve with the HS-SVD basis
28    errvecH(k) = errcompute(gqr_eval(gqr_solve(x,y,ep,gqr_alpha),xeval),yeval);

30    k = k + 1;
  end
32 errpoly = errcompute(polyval(polyfit(x,y,N-1),xeval),yeval);
```

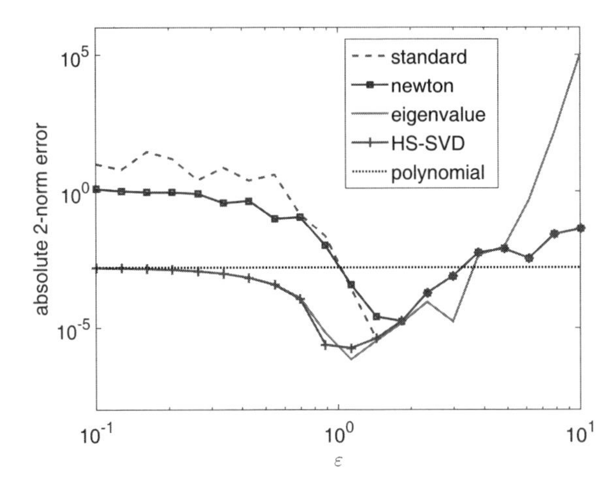

Fig. 13.4 Only the HS-SVD basis maintains stability for all values of ε, but at significant computational cost.

When ε is large, the eigenfunction basis becomes wildly inaccurate whereas the other bases maintain stability. As $\varepsilon \to 0$, the eigenfunction approximate interpolant more closely resembles the true Gaussian interpolant as the standard basis loses accuracy. The Newton basis outperforms the standard basis, but the eigenfunction and HS-SVD basis manage to approach the polynomial limit. Note that, if a low-rank eigenfunction basis were used (recall Section 12.3), the limit would not necessarily approach the error of the polynomial interpolant.

Remark 13.13. Just because alternate bases may produce more stable results than the standard basis, not all computations involving them are stable. The HS-SVD basis and (potentially low-rank) eigenfunction approximate interpolant may be troubled for inappropriate values of the Gaussian eigenfunction parameter α. For very small values of ε, the standard basis is so ill-conditioned that MATLAB's Cholesky factorization function `chol` *does not* recognize that `K` is a positive definite matrix and throws an error. To manage around this problem, line 18 adds a small amount to the diagonal of the matrix.

Perhaps this, after everything else discussed in this chapter, should be remembered when computing with alternate bases: potential exists but dangers abound. Those dangers can be new stability issues, severe computational costs, additional free parameters or some other problem that the community has yet to uncover.

Chapter 14

Parameter Optimization

So far, we have introduced two ways in which we can approach the scattered data interpolation task formalized in Problem 1.1: the deterministic approach outlined briefly in Section 1.3 resulting in an interpolant of the form $s(\boldsymbol{x}) = \boldsymbol{k}(\boldsymbol{x})^T \mathsf{K}^{-1}\boldsymbol{y}$, and the stochastic approach discussed in Chapter 5 which leads to the (simple) kriging prediction $\hat{y}_{\boldsymbol{x}} = \boldsymbol{k}(\boldsymbol{x})^T \mathsf{K}^{-1}\boldsymbol{y}$. Since the right-hand sides for these two approaches are identical we can see that under certain circumstances they yield the same result.

As we discussed in Section 10.1, for a positive definite kernel with a single shape parameter ε (such as the C^2 Wendland kernel used there), the predictive capacity of the interpolant is greatly dependent on the value of ε, as is indicated in Figure 10.1. Similarly, in Example 10.2 we showed that, for a Gaussian kernel, as $\varepsilon \to 0$ the kernel interpolant approaches the polynomial interpolant (or some polynomial approximation in higher dimensions) meaning that kernels have the potential to exceed polynomial accuracy for $\varepsilon > 0$. This is, at least in part, a result of the Runge phenomenon: polynomials are global basis functions, so because Gaussians have the ability to localize — by increasing the value of ε — they have the ability to minimize the Runge phenomenon. This concept was illustrated graphically in Figure 10.2.

Figure 10.1(b) demonstrates the opportunity for both success and failure when choosing ε. There is a minimum error attainable for a value of ε that varies with N (the number of samples used), but for this example is slightly less than 1. If ε is chosen slightly smaller or larger than this optimal value, the error could be orders of magnitude worse.

Of course, we were only able to create that graph — and make a judgement about which value of ε to consider "optimal" — because we knew the function that generated the data and therefore knew the "true" solution f. In most situations that function will not be available to us — and if it were, we would have little need to construct an interpolant[1]. This motivates the search for parametrization schemes that may allow us to find good ε values using just the available data.

As we work towards trying to identify a suitable shape parameter, we recall ex-

[1]Unless, perhaps, we are interested in some low-rank or low cost surrogate for the function f.

isting techniques in the literature which have had variable levels of success. Each of these existing parametrization techniques has some objective function which needs to be minimized, and that is what we will introduce now. Some discussion of existing parametrization schemes was already provided in [Fasshauer (2007, Chapter 17)], including various heuristic approaches — and even naive trial and error, or "educated guessing."

14.1 Modified Golomb–Weinberger Bound and Kriging Variance

As we saw in (9.7), the scattered data interpolation error is bounded by the power function times the Hilbert space norm of f. Our choice of parameter ε will have an effect on $\|f\|_{\mathcal{H}_K(\Omega)}$, but it is not one that we can immediately understand.

Fortunately, we also have the modified Golomb–Weinberger error bound

$$|f(\boldsymbol{x}) - s(\boldsymbol{x})| \leq \delta_\varepsilon \|s\|_{\mathcal{H}_K(\Omega)} P_{K,\mathcal{X}}(\boldsymbol{x}),$$

where $\|s\|_{\mathcal{H}_K(\Omega)} = \sqrt{\boldsymbol{y}^T \mathsf{K}^{-1} \boldsymbol{y}}$ according to (9.10). Alternatively, we had — via the kriging variance and profile likelihood — that

$$\mathbb{E}\left[\left(Y_{\boldsymbol{x}} - \overset{\vartriangle}{Y}_{\boldsymbol{x}}\right)^2\right] = \frac{1}{N} \boldsymbol{y}^T \mathsf{K}^{-1} \boldsymbol{y} (K(\boldsymbol{x},\boldsymbol{x}) - \boldsymbol{k}(\boldsymbol{x})^T \mathsf{K}^{-1} \boldsymbol{k}(\boldsymbol{x}))$$

(see Remark 9.2). Thus, if we want to use the criterion

$$\mathrm{C}_{\mathrm{GW}}(\varepsilon; k) = \sqrt{\boldsymbol{y}^T \mathsf{K}^{-1} \boldsymbol{y}}\, \|P_{K,\mathcal{X}}\|_k. \tag{14.1}$$

to determine the "optimal" value of the kernel parameter ε, then it is important for us to be able to safely compute both the power function

$$P_{K,\mathcal{X}}(\boldsymbol{x}) = \sqrt{K(\boldsymbol{x},\boldsymbol{x}) - \boldsymbol{k}(\boldsymbol{x})^T \mathsf{K}^{-1} \boldsymbol{k}(\boldsymbol{x})}, \tag{14.2}$$

where the k-norm can be chosen in one of several ways (2-norm, ∞-norm, $\mathcal{H}_K(\Omega)$-norm), and the native space norm of the interpolant (frequently referred to as the Mahalanobis distance in the statistics literature)

$$\|s\|_{\mathcal{H}_K(\Omega)} = \sqrt{\boldsymbol{y}^T \mathsf{K}^{-1} \boldsymbol{y}}.$$

Since both of these terms involve the notoriously ill-conditioned inverse of K we will employ the Hilbert–Schmidt SVD from Chapter 13 to help us.

The HS-SVD allows us to express the cardinal functions as $\boldsymbol{k}(\boldsymbol{x})^T \mathsf{K}^{-1} = \boldsymbol{\psi}(\boldsymbol{x})^T \Psi^{-1}$ so that

$$P_{K,\mathcal{X}}(\boldsymbol{x}) = \sqrt{K(\boldsymbol{x},\boldsymbol{x}) - \boldsymbol{\psi}(\boldsymbol{x})^T \Psi^{-1} \boldsymbol{k}(\boldsymbol{x})}. \tag{14.3}$$

In Figure 14.1 we show two examples measuring the 2-norm of the power function compared to the error for Gaussian interpolation of the functions $f(x) = \cos(\pi x)$ and $f(x) = J_0(4|x|)$, respectively, on $[-1,1]$. In both cases we see that the graph for the power function — stably computed via the HS-SVD — flattens out for small value of ε.

In fact, the 2-norm of the power function seems to decrease until reaching a value of roughly 10^{-8}. This unfortunately seems to be the result of numerical cancelation which occurs as $\boldsymbol{\psi}(\boldsymbol{x})^T \Psi^{-1} \boldsymbol{k}(\boldsymbol{x}) \to 1$ for $\varepsilon \to 0$.

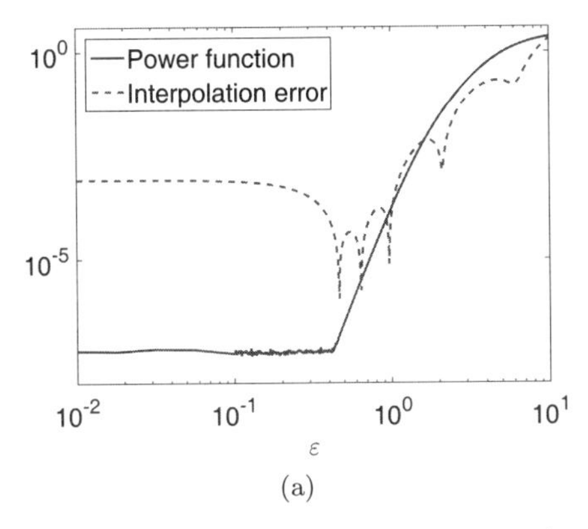

(a)

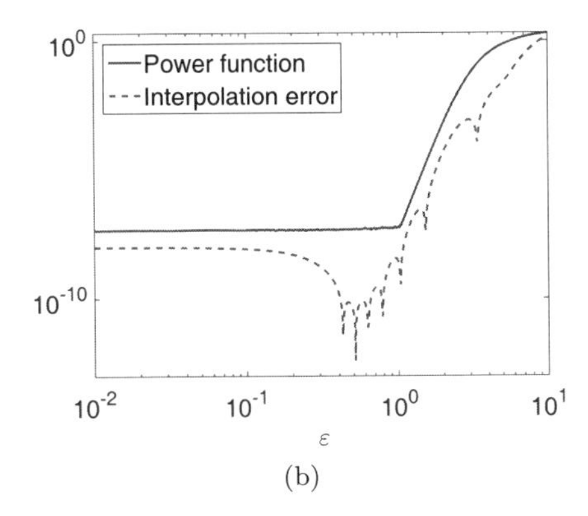

(b)

Fig. 14.1 In this example, N points in $(-1,1)$ are sampled and Gaussian interpolation is conducted with a range of ε values. These results are compared to the 2-norm of the power function, which seems to flatten out for small values of ε. The error is averaged over 100 evenly spaced points. (a) $f(x) = \cos(\pi x)$, $N = 10$ uniform points, (b) $f(x) = J_0(4|x|)$, $N = 15$ Chebyshev points.

14.1.1 *How to avoid cancelation while computing the power function (kriging variance)*

We can obtain an alternative representation for the power function which is not susceptible to numerical cancelation by considering the matrix

$$\widetilde{\mathsf{K}} = \begin{pmatrix} K(\boldsymbol{x},\boldsymbol{x}) & \boldsymbol{k}(\boldsymbol{x})^T \\ \boldsymbol{k}(\boldsymbol{x}) & \mathsf{K} \end{pmatrix}$$

from (5.12). Using a standard formula for the determinant of a block matrix [Bernstein (2009, Eqn. (2.8.13))] we get (see also Remark 5.6)

$$\det \widetilde{\mathsf{K}} = \det \begin{pmatrix} K(\boldsymbol{x},\boldsymbol{x}) & \boldsymbol{k}(\boldsymbol{x})^T \\ \boldsymbol{k}(\boldsymbol{x}) & \mathsf{K} \end{pmatrix} = \det \mathsf{K} \left(K(\boldsymbol{x},\boldsymbol{x}) - \boldsymbol{k}(\boldsymbol{x})^T \mathsf{K}^{-1} \boldsymbol{k}(\boldsymbol{x}) \right)$$

so that the power function can be computed via[2]

$$P_{K,\mathcal{X}}(\boldsymbol{x}) = \sqrt{\frac{\det \widetilde{\mathsf{K}}}{\det \mathsf{K}}}. \tag{14.4}$$

We compute these determinants using logarithms since they are likely to otherwise cause an underflow error for small enough values of the shape parameter. From a purely mathematical point of view, computing $\log \det \mathsf{K}$ is straightforward using the Hilbert–Schmidt SVD $\mathsf{K} = \Psi \Lambda_1 \Phi_1^T$, i.e.,

$$\log \det \mathsf{K} = \log \det \Psi + \log \det \Lambda_1 + \log \det \Phi_1^T. \tag{14.5}$$

[2]This formula appeared already in [De Marchi (2003); Schaback (2005)], but was derived there using a much more circuitous argument involving a determinant representation of the cardinal functions $\overset{*}{\boldsymbol{w}}(\cdot)$ which can be obtained by applying Cramer's rule to solve the linear system $\mathsf{K}\overset{*}{\boldsymbol{w}}(\boldsymbol{x}) = \boldsymbol{k}(\boldsymbol{x})$ as in [Fornberg *et al.* (2004)].

From a computational point of view, the very small eigenvalues can be handled safely by taking their logarithms, which is easy because $\mathsf{\Lambda}_1$ is diagonal. Furthermore, because $\mathsf{\Phi}_1^T$ was inverted (and therefore factored) while forming the HS-SVD basis $\boldsymbol{\psi}(\cdot)$ in (13.10) and — assuming we want to compute an interpolant — $\mathsf{\Psi}$ was inverted while computing (13.20), the cost of performing (14.5) is negligible. Program 14.1 demonstrates the computation of $\det \mathsf{K}$, and the full version of the code in the `GaussQR` library shows how to compute the power function stably using (14.4) with results plotted in Figure 14.2.

Program 14.1. Excerpts from `ParameterPower.m`

```
% Define the interpolation points
N = 11;   x = pickpoints(-1,1,N,'cheb');
% Define the eigenfunctions/eigenvalues of the a=.5 Chebyshev kernel
phifunc = @(n,x) sqrt(2)*cos(acos(x)*n);                        b = 1e-3;
lambdafunc = @(b,n) (n==0)*.5 + (n>0).*(.5*(1-b)/b*b.^n);
% Add additional eigenfunctions to form the HS-SVD basis
M = N + ceil(log(eps)/log(b));   n = 0:M-1;
% Evaluate the Chebyshev eigenfunctions
Phi = phifunc(n,x);
Phi1 = Phi(:,1:N);   Phi2 = Phi(:,N+1:end);
% Evaluate the Chebyshev eigenvalues
lamvec = lambdafunc(b,n);
lamvec1 = lamvec(1:N);   lamvec2 = lamvec(N+1:end);
% Form the corrector matrix
CbarT = bsxfun(@rdivide,lamvec2',lamvec1).*(Phi2'/Phi1');
Psi = Phi1 + Phi2*CbarT;
% Evaluate the log-determinant
logdetK = log(abs(det(Phi1))) + sum(log(lamvec1)) + log(abs(det(Psi)));
```

Everything in Program 14.1, except line 18, looks similar to Program 13.1 where we used the HS-SVD to compute an interpolant. Here, we compute (14.5), albeit carefully because $\log(\det(\mathsf{\Lambda}_1)) =$`sum(log(lamvec1))`$\approx -255$ for only this $N = 10$ sized problem. When this value reaches ≈ -708, which it will for some $b > 0$, $\det(\mathsf{\Lambda}_1)$ suffers from underflow and can only be stored in its log form. The use of `abs` on $\det(\mathsf{\Phi}_1)$ and $\det(\mathsf{\Psi})$ exist because, despite the fact that $\det(\mathsf{K})$ is positive, $\det(\mathsf{\Phi}_1)$ and $\det(\mathsf{\Psi})$ could still both be negative and we want to avoid having a $-2\pi\mathrm{i}$ meaninglessly sitting in the $\log(\det(\mathsf{K}))$ result.

14.1.2 *How to stably compute the native space norm of the interpolant (Mahalanobis distance)*

A similar strategy will allow us to approximate $\boldsymbol{y}^T\mathsf{K}^{-1}\boldsymbol{y}$. In the kernel interpolation setting, the system $\mathsf{K}\boldsymbol{c} = \boldsymbol{y}$ gives rise to the minimum norm interpolant (or best linear unbiased prediction) $s(\boldsymbol{x}) = \boldsymbol{k}(\boldsymbol{x})^T\boldsymbol{c} = \boldsymbol{k}(\boldsymbol{x})^T\mathsf{K}^{-1}\boldsymbol{y}$ mentioned earlier. As demonstrated in (13.20), using the HS-SVD basis $\boldsymbol{\psi}(\cdot)$ instead of the standard basis

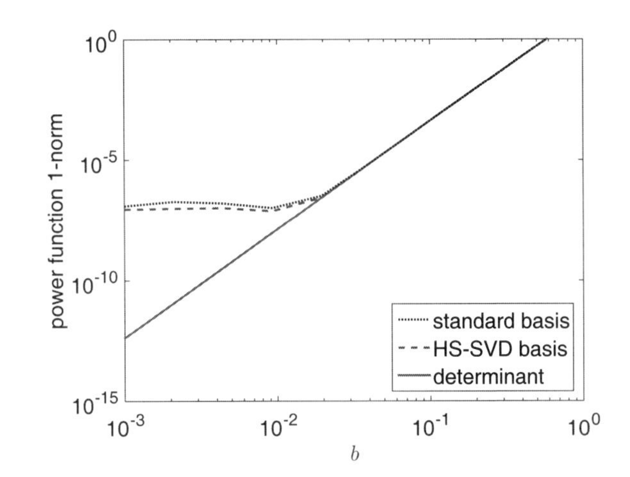

Fig. 14.2 Computation of the power function with the determinant strategy (14.4) is not subject to the cancelation that stymies the standard basis (14.2) and HS-SVD basis (14.3). This plot displays the sum of the power function evaluated at $N_{\text{eval}} = 13$ scattered points throughout a domain of $N = 10$ Chebyshev nodes for different b parameters in the Chebyshev kernel.

$\boldsymbol{k}(\cdot)$, the interpolant becomes $s(\boldsymbol{x}) = \boldsymbol{\psi}(\boldsymbol{x})^T\boldsymbol{b} = \boldsymbol{\psi}(\boldsymbol{x})^T\Psi^{-1}\boldsymbol{y}$, which corresponds to solving the system $\Psi\boldsymbol{b} = \boldsymbol{y}$. Thus we define $\boldsymbol{b} \in \mathbb{R}^N$ via

$$\Psi\boldsymbol{b} = \boldsymbol{y} \quad \Longrightarrow \quad \boldsymbol{b} = \Psi^{-1}\boldsymbol{y} \tag{14.6}$$

and note that $\boldsymbol{b}$ also would be available from the interpolant (13.20).

Applying (14.6) and the Hilbert–Schmidt SVD of K (13.15) to $\boldsymbol{y}^T\mathsf{K}^{-1}\boldsymbol{y}$ gives

$$\boldsymbol{y}^T\mathsf{K}^{-1}\boldsymbol{y} = (\Psi\boldsymbol{b})^T(\Psi\Lambda_1\Phi_1^T)^{-1}\Psi\boldsymbol{b} = \boldsymbol{b}^T\Psi^T\Phi_1^{-T}\Lambda_1^{-1}\boldsymbol{b}. \tag{14.7}$$

If we recall the formula for Ψ from (13.14), then we see that

$$\begin{aligned}\Psi^T &= \Phi_1^T + \Lambda_1^{-1}\Phi_1^{-1}\Phi_2\Lambda_2\Phi_2^T \\ &= \Phi_1^T + \bar{\mathsf{C}}\Phi_2^T\end{aligned}$$

with corrector matrix $\bar{\mathsf{C}} = \Lambda_1^{-1}\Phi_1^{-1}\Phi_2\Lambda_2$ as defined in (13.18). Inserting this into (14.7) will give us our final *exact representation of the squared native space norm of the interpolant* (or Mahalanobis distance), i.e.,

$$\boldsymbol{y}^T\mathsf{K}^{-1}\boldsymbol{y} = \boldsymbol{b}^T\Lambda_1^{-1}\boldsymbol{b} + \boldsymbol{b}^T\bar{\mathsf{C}}\Lambda_2^{-1}\bar{\mathsf{C}}^T\boldsymbol{b}. \tag{14.8}$$

The second term of (14.8) may be most efficiently computed with

$$\boldsymbol{b}^T\bar{\mathsf{C}}\Lambda_2^{-1}\bar{\mathsf{C}}^T\boldsymbol{b} = \left\|\Lambda_2^{-1/2}\bar{\mathsf{C}}^T\boldsymbol{b}\right\|_2^2. \tag{14.9}$$

At this point, we recall our observation from Remark 13.4 which stated that $\Psi \to \Phi_1$ in the "flat" limit (which for many of our kernels corresponds to small ε). Therefore, looking at (14.7), we should also expect that $\boldsymbol{b}^T\Psi^T\Phi_1^{-T}\Lambda_1^{-1}\boldsymbol{b} \to \boldsymbol{b}^T\Lambda_1^{-1}\boldsymbol{b}$. This was proved in [McCourt and Fasshauer (2014)] for Gaussian kernels.

As a consequence, $\boldsymbol{b}^T \mathsf{\Lambda}_1^{-1} \boldsymbol{b}$ is a useful approximation of the Mahalanobis distance for Gaussian kernels (and possibly also for others) in the $\varepsilon \to 0$ limit. This approximation is straightforward to compute since it is a symmetric weighted inner product with diagonal matrix $\mathsf{\Lambda}_1^{-1}$, i.e.,

$$\boldsymbol{b}^T \mathsf{\Lambda}_1^{-1} \boldsymbol{b} = \sum_{n=1}^{N} \frac{b_n^2}{\lambda_n}.$$

This ensures that cancelation will not occur. Nevertheless, this expression grows unboundedly as ε shrinks because of the growth in the eigenvalues. This observation prompts

Remark 14.1. We would like to point out that these limiting arguments are rather delicate since all four terms involved in our calculations — namely, $\boldsymbol{y}^T \mathsf{K}^{-1} \boldsymbol{y}$ itself, as well as $\boldsymbol{b}^T \mathsf{\Psi}^T \mathsf{\Phi}_1^{-T} \mathsf{\Lambda}_1^{-1} \boldsymbol{b}$, $\boldsymbol{b}^T \mathsf{\Lambda}_1^{-1} \boldsymbol{b}$ and $\boldsymbol{b}^T \bar{\mathsf{C}} \mathsf{\Lambda}_2^{-1} \bar{\mathsf{C}}^T \boldsymbol{b}$ — grow unboundedly in the "flat" limit.

14.2 Cross-Validation

Cross validation is a popular technique in statistics which uses the given data (instead of the — usually unknown — solution) to predict optimal values of model parameters for data fitting. The main idea is to split the data into a training set $\mathcal{T}$ and a validation set $\mathcal{V}$ and to then use some form of error norm obtained by gauging the accuracy of the fit built from information on the training set at points in the validation set.

Leave-one-out cross validation (LOOCV) is an especially popular version of cross-validation and corresponds to using a training set consisting of all but one of the data points, which in turn is the sole member of the validation set. In the context of kernel (or radial basis function) methods LOOCV appeared in papers such as [Hickernell and Hon (1999); Rippa (1999); Fasshauer and Zhang (2007b); Scheuerer (2011)]. The discussion in this section is based in large parts on [Hickernell (2009)].

The idea of cross-validation is described essentially by the following algorithm:

- Given data values $\boldsymbol{y}$ on a set of scattered data sites (or design) $\mathcal{X}$, partition the design into disjoint, nonempty groups $\mathcal{T}$ (called the training set) and $\mathcal{V}$ (the validation set) so that $\mathcal{T} \cup \mathcal{V} = \mathcal{X}$. The set $\mathcal{T}$ is used to create an auxiliary approximation (or partial fit) s_t, i.e., to *train* the model. Once that has been accomplished, the set $\mathcal{V}$ is used to judge the accuracy of s_t, i.e., to *validate* it.
  - The sets $\mathcal{T}$ and $\mathcal{V}$ block up the interpolation matrix K, its inverse A, and the vectors $\boldsymbol{y}$ and $\boldsymbol{c}$ as

$$\mathsf{K} = \begin{pmatrix} \mathsf{K}_{tt} & \mathsf{K}_{tv} \\ \mathsf{K}_{vt} & \mathsf{K}_{vv} \end{pmatrix}, \quad \mathsf{K}^{-1} = \mathsf{A} = \begin{pmatrix} \mathsf{A}_{tt} & \mathsf{A}_{tv} \\ \mathsf{A}_{vt} & \mathsf{A}_{vv} \end{pmatrix}, \quad \boldsymbol{y} = \begin{pmatrix} \boldsymbol{y}_t \\ \boldsymbol{y}_v \end{pmatrix}, \quad \boldsymbol{c} = \begin{pmatrix} \boldsymbol{c}_t \\ \boldsymbol{c}_v \end{pmatrix}.$$

For example, the block K_{tv} is generated using training points to evaluate and validation points as centers of the kernel. Recall that $\mathsf{K}\boldsymbol{c} = \boldsymbol{y}$, and thus $\boldsymbol{c} = \mathsf{A}\boldsymbol{y}$. Because K and A are symmetric, we know that $\mathsf{K}_{tv} = \mathsf{K}_{vt}^T$ and $\mathsf{A}_{tv} = \mathsf{A}_{vt}^T$. Also recall that the kernel K and kernel matrix K depend on the shape parameter ε (which may be a vector).

- The available data at $\mathcal{X}_v$ is $\boldsymbol{y}_v$, and the prediction at the points in $\mathcal{X}_v$ using the data $(\mathcal{X}_t, \boldsymbol{y}_t)$ is $\mathsf{K}_{vt}\mathsf{K}_{tt}^{-1}\boldsymbol{y}_t$ which can be determined by applying the structure of the interpolant $s(\boldsymbol{x}) = \boldsymbol{k}(\boldsymbol{x})^T\mathsf{K}^{-1}\boldsymbol{y}$ at $\boldsymbol{x}$ locations in $\mathcal{X}_v$. Therefore, studying

$$\left|\boldsymbol{y}_v - \mathsf{K}_{vt}\mathsf{K}_{tt}^{-1}\boldsymbol{y}_t\right|$$

tells us something about how good our approximation is. The matrix $\mathsf{K}_{vt}\mathsf{K}_{tt}^{-1}$ can be thought of as an operator which takes in values on $\mathcal{X}_t$ and interpolates them to $\mathcal{X}_v$ [McCourt (2013b)].
- By considering a set $\mathcal{V}$ partitioned into p partitions $\mathcal{V} = \{\mathcal{V}^{(1)}, \ldots, \mathcal{V}^{(p)}\}$ such that

$$\mathcal{V}^{(i)} \cap \mathcal{V}^{(j)} = \emptyset,\ i \neq j, \quad \text{and} \quad \bigcup_{i=1}^{p} \mathcal{V}^{(i)} = \mathcal{X}$$

with $\mathcal{T}^{(i)} = \mathcal{X} \setminus \mathcal{V}^{(i)}$, we could consider instances with certain pieces of data omitted for certain i values. Then we could consider the residual left over by our interpolants evaluated at those points:

$$C_{\mathrm{CV}}(\varepsilon; \mathcal{V}, k) = \sum_{\mathcal{V}^{(i)} \in \mathcal{V}} \left\|\boldsymbol{y}_v - \mathsf{K}_{vt}\mathsf{K}_{tt}^{-1}\boldsymbol{y}_t\right\|_k, \tag{14.10}$$

where k is probably 1, 2 or ∞.

The criterion to be minimized for a good ε value is (14.10). Of course, the presence of K_{tt}^{-1} suggests ill-conditioning may be a problem. If we choose to implement the Hilbert–Schmidt SVD here, we should block up the interpolation matrix

$$\mathsf{K} = \begin{pmatrix} \mathsf{K}_{tt} & \mathsf{K}_{tv} \\ \mathsf{K}_{vt} & \mathsf{K}_{vv} \end{pmatrix} = \begin{pmatrix} \Psi_{tt}\Lambda_t\Phi_t^T & \Psi_{tv}\Lambda_v\Phi_v^T \\ \Psi_{vt}\Lambda_t\Phi_t^T & \Psi_{vv}\Lambda_v\Phi_v^T \end{pmatrix} = \begin{pmatrix} \Psi_{tt} & \Psi_{tv} \\ \Psi_{vt} & \Psi_{vv} \end{pmatrix} \begin{pmatrix} \Lambda_t\Phi_t^T & \\ & \Lambda_v\Phi_v^T \end{pmatrix}. \tag{14.11}$$

Using this block structure, we see that $\mathsf{K}_{tt} = \Psi_{tt}\Lambda_t\Phi_t^T$ and $\mathsf{K}_{vt} = \Psi_{vt}\Lambda_t\Phi_t^T$, so therefore we can write the criterion as

$$C_{\mathrm{CV}}(\varepsilon; \mathcal{V}, k) = \sum_{\mathcal{V}^{(i)} \in \mathcal{V}} \left\|\boldsymbol{y}_v - \Psi_{vt}\Psi_{tt}^{-1}\boldsymbol{y}_v\right\|_k, \tag{14.12}$$

which is presumably safer to compute.

The expression in (14.10) can be rewritten in a slightly different way by exploiting the block structure of the relevant vectors and matrices. Note that $\boldsymbol{c} = \mathsf{A}\boldsymbol{y}$ implies

$$\boldsymbol{c}_v = \mathsf{A}_{vt}\boldsymbol{y}_t + \mathsf{A}_{vv}\boldsymbol{y}_v.$$

We can also invoke the fact that $\mathsf{AK} = \mathsf{I}_N$ (where I_N is the $N \times N$ identity) to note that

$$\mathsf{A}_{vt}\mathsf{K}_{tt} + \mathsf{A}_{vv}\mathsf{K}_{tv}^T = 0 \quad \Longrightarrow \quad \mathsf{A}_{vt} = -\mathsf{A}_{vv}\mathsf{K}_{tv}^T\mathsf{K}_{tt}^{-1}.$$

Plugging this in above gives

$$\begin{aligned} \boldsymbol{c}_v &= -\mathsf{A}_{vv}\mathsf{K}_{vt}\mathsf{K}_{tt}^{-1}\boldsymbol{y}_t + \mathsf{A}_{vv}\boldsymbol{y}_v \\ \mathsf{A}_{vv}^{-1}\boldsymbol{c}_v &= -\mathsf{K}_{vt}\mathsf{K}_{tt}^{-1}\boldsymbol{y}_t + \boldsymbol{y}_v \end{aligned}$$

which allows us to rewrite our optimization criterion (14.10) as

$$\mathrm{C}_{\mathrm{CV}}(\varepsilon; \mathcal{V}, k) = \sum_{\mathcal{X}_v \in \mathcal{V}} \left\| \mathsf{A}_{vv}^{-1}\boldsymbol{c}_v \right\|_k . \tag{14.13}$$

To involve the Hilbert–Schmidt SVD, we need to study (14.11) and define

$$\begin{pmatrix} \Psi_{tt} & \Psi_{tv} \\ \Psi_{vt} & \Psi_{vv} \end{pmatrix} \begin{pmatrix} \mathsf{B}_{tt} & \mathsf{B}_{tv} \\ \mathsf{B}_{vt} & \mathsf{B}_{vv} \end{pmatrix} = \begin{pmatrix} \mathsf{I}_{N_t} & \\ & \mathsf{I}_{N_v} \end{pmatrix},$$

where $\mathcal{X}_t$ and $\mathcal{X}_v$ contain N_t and N_v points, respectively. Manipulations similar to those above show that

$$\mathsf{B}_{vv}^{-1} = \Psi_{vv} - \Psi_{vt}\Psi_{tt}^{-1}\Psi_{tv},$$

which we will use shortly. We also need to define

$$\begin{pmatrix} \Psi_{tt} & \Psi_{tv} \\ \Psi_{vt} & \Psi_{vv} \end{pmatrix} \begin{pmatrix} \boldsymbol{b}_t \\ \boldsymbol{b}_v \end{pmatrix} = \begin{pmatrix} \boldsymbol{y}_t \\ \boldsymbol{y}_v \end{pmatrix} \quad \Longrightarrow \quad \begin{pmatrix} \boldsymbol{b}_t \\ \boldsymbol{b}_v \end{pmatrix} = \begin{pmatrix} \mathsf{B}_{tt} & \mathsf{B}_{tv} \\ \mathsf{B}_{vt} & \mathsf{B}_{vv} \end{pmatrix} \begin{pmatrix} \boldsymbol{y}_t \\ \boldsymbol{y}_v \end{pmatrix},$$

$$\begin{pmatrix} \boldsymbol{b}_t \\ \boldsymbol{b}_v \end{pmatrix} = \begin{pmatrix} \Lambda_t\Phi_t^T & \\ & \Lambda_v\Phi_v^T \end{pmatrix} \begin{pmatrix} \boldsymbol{c}_t \\ \boldsymbol{c}_v \end{pmatrix}$$

This implies that, in terms of the HS-SVD system, $\boldsymbol{c}_v = \Phi_v^{-T}\Lambda_v^{-1}\boldsymbol{b}_v$, and sufficient study of (14.11) knowing that $\mathsf{AK} = \mathsf{I}_N$ gives us $\mathsf{A}_{vv} = \Phi_v^{-T}\Lambda_v^{-1}B_{vv}$. Combining these in (14.13) gives us

$$\mathrm{C}_{\mathrm{CV}}(\varepsilon; \mathcal{V}, k) = \sum_{\mathcal{V}^{(i)} \in \mathcal{V}} \left\| \mathsf{B}_{vv}^{-1}\boldsymbol{b}_v \right\|_k = \sum_{\mathcal{V}^{(i)} \in \mathcal{V}} \left\| (\Psi_{vv} - \Psi_{vt}\Psi_{tt}^{-1}\Psi_{tv})^{-1}\boldsymbol{b}_v \right\|_k, \tag{14.14}$$

which is almost certainly more stable to compute.

Often times, cross-validation is conducted in one of two ways:

Leave-one-out cross-validation: All the data except a single point is used to compute the interpolant, and the residual is judged at that point. In this setting, $\mathcal{V} = \{\mathcal{V}^{(1)}, \mathcal{V}^{(2)}, \ldots, \mathcal{V}^{(N)}\} = \{\boldsymbol{x}_1, \boldsymbol{x}_2, \ldots, \boldsymbol{x}_N\}$ and the errors at each of those points are added up to find $\mathrm{C}_{\mathrm{CV}}(\varepsilon; \mathcal{V}, k)$. As explained in [Fasshauer (2007, Chapter 17)], (14.13) is most likely the preferred form to compute in this case because A_{vv} is just a number and corresponds to a diagonal entry in K^{-1}. Thus the entire LOOCV computation can be performed with little overhead compared to the computation of the interpolant/predictor.

Leave-half-out cross-validation: Half of the data is omitted to create an interpolant and the residual is judged on the other half; then the process is flipped and both results are combined to compute $CV(\varepsilon; \mathcal{V}, k)$. In this setting, $\mathcal{V} = \{\mathcal{V}^{(1)}, \mathcal{V}^{(2)}\}$ and $|\mathcal{V}^{(1)}| = |\mathcal{V}^{(2)}|$, or as close as possible. (14.10) is almost surely the preferred computation in this case because A_{vv} is roughly size $N/2 \times N/2$ and computing its inverse is costly.

Remark 14.2. As is often the case, the Hilbert–Schmidt SVD computation (14.14) is more costly to compute than the standard basis computation (14.13). What is not apparent at first glance is exactly how costly it is. In fact, the celebrated "fast" LOOCV computation described in [Rippa (1999)] is not nearly so fast for the HS-SVD.

For LOOCV in the standard basis, all of the A_{vv} terms can be determined by inverting the matrix K. For the HS-SVD basis, computing the $B_{vv} = \Psi_{vv} - \Psi_{vt}\Psi_{tt}^{-1}\Psi_{tv}$ term requires 2 HS-SVD corrector matrices: a $\bar{\mathsf{C}}_v$ to evaluate the Ψ_{vv} and Ψ_{tv} term for kernels centered at $\mathcal{X}_v$ and a $\bar{\mathsf{C}}_t$ to evaluate Ψ_{vt} and Ψ_{tt}. Then, both of those must be computed for each $\mathcal{X}_v \in \mathcal{V}$. This suggests that, if the HS-SVD is necessary, (14.12) is the preferable method.

Remark 14.3. One concern with these cross-validation computations may be the process variance (introduced in Section 5.1). It plays no role here, though, because such a term gets canceled out in the computation of $\mathsf{K}_{tv}^T\mathsf{K}_{tt}^{-1}\boldsymbol{y}_t$, in the same way that it is canceled out in $\overset{\Delta}{Y}_{\boldsymbol{x}} = \boldsymbol{k}(\boldsymbol{x})^T\mathsf{K}^{-1}\boldsymbol{y}$ (cf. (5.8)).

Remark 14.4. Although often described in a statistical setting, this cross-validation concept is equally well (or equally poorly) supported from the numerical analysis side. We are not aware of the existence of any rigorous framework that allows us to explain why this is a useful analytical technique, as opposed to just an idea that heuristically makes sense.

14.3 Maximum Likelihood Estimation

This approach leans primarily on the Gaussian random field framework, though we will show later how it can also be derived using the deterministic Hilbert space framework.

We begin by recalling Definition 5.2, i.e., that $\boldsymbol{Y} \sim \mathcal{N}(\boldsymbol{\mu}, \sigma^2\mathsf{K})$, and, as before, restricting the Gaussian random field Y to have zero mean, i.e., $\boldsymbol{\mu} \equiv \mathbf{0}$. Note that this means that we are now including the process variance in the definition of the covariance kernel $\sigma^2 K$. To keep the first part of this discussion transparent we will assume that the process variance is fixed at $\sigma^2 \equiv 1$, i.e., that it is independent from the kernel shape parameter ε. In Section 14.3.2 we will use the so-called profile likelihood approach to generalize our discussion to the case of a process variance that depends on ε.

14.3.1 *MLE independent of process variance*

We are going to treat ε as *drawn from a random variable* $\mathcal{E}$ now, with some unknown distribution. In turn we also will need to define the joint random variable $Z = (\mathcal{E}, \boldsymbol{Y})$ with density p_Z. We want to study the conditional density function $p_{\mathcal{E}|\boldsymbol{Y}}(\varepsilon|\boldsymbol{Y} = \boldsymbol{y})$, that is, how likely it is that the Gaussian random field had covariance K, parametrized by ε, when realizing the data $\boldsymbol{y}$.

As in Chapter 5, this function is often called the *likelihood* function, and it allows us to compare the relative likelihood of values of ε given the existing data. Maximizing this function yields (in some sense) the ε which most likely parametrized the covariance kernel that generated the data $\boldsymbol{y}$. The function $p_{\mathcal{E}|\boldsymbol{Y}}(\varepsilon|\boldsymbol{Y} = \boldsymbol{y})$ is a function of ε for any fixed value of $\boldsymbol{y}$.

A technical note: the notation from Chapter 5 and this section will not perfectly align. In that chapter we considered ε fixed and the joint distribution $(\boldsymbol{Y}, Y_{\boldsymbol{x}})$, but here we are studying the joint distribution $(\mathcal{E}, \boldsymbol{Y})$. Where we previously had written $p_{\boldsymbol{Y}}(\boldsymbol{y})$ in (5.3), the appropriate notation in this section is

$$p_{\boldsymbol{Y}|\mathcal{E}}(\boldsymbol{y}|\mathcal{E} = \varepsilon) = \frac{1}{\sqrt{(2\pi)^N \det(\mathsf{K})}} \exp\left(-\frac{1}{2}\boldsymbol{y}^T \mathsf{K}^{-1}\boldsymbol{y}\right), \tag{14.15}$$

where although ε does not explicitly appear on the right hand side, it appears within K.

We want to study $p_{\mathcal{E}|\boldsymbol{Y}}(\varepsilon|\boldsymbol{Y} = \boldsymbol{y})$, but we do not know that density. Using an analogous version of the joint density relations (5.11) allows us to write

$$\begin{aligned} p_{\mathcal{E}|\boldsymbol{Y}}(\varepsilon|\boldsymbol{Y} = \boldsymbol{y}) &= \frac{p_Z(\varepsilon, \boldsymbol{y})}{p_{\boldsymbol{Y}}(\boldsymbol{Y} = \boldsymbol{y})} \\ &= \frac{p_{\boldsymbol{Y}|\mathcal{E}}(\boldsymbol{y}|\mathcal{E} = \varepsilon) p_{\mathcal{E}}(\varepsilon)}{p_{\boldsymbol{Y}}(\boldsymbol{y})}. \end{aligned} \tag{14.16}$$

We can clean up our likelihood function expression somewhat by ignoring things we have no knowledge of. First off, we do not know what the marginal distribution of ε is — if we did we would just study that to determine an optimal ε parametrization. We also do not know what the marginal distribution of $\boldsymbol{Y}$ is: we do not know how $\boldsymbol{Y}$ varies independently of ε. To determine this, we would need to compute

$$p_{\boldsymbol{Y}}(\boldsymbol{y}) = \int_0^\infty p_Z(\boldsymbol{y}, \varepsilon)\, \mathrm{d}\varepsilon,$$

but since the joint density p_Z is unknown, we are not able to do that. More importantly, $p_{\boldsymbol{Y}}(\boldsymbol{y})$ is independent of ε, and therefore changing ε will have no effect on that value.

By abandoning $p_{\mathcal{E}}(\varepsilon)$ and $p_{\boldsymbol{Y}}(\boldsymbol{y})$, we can suggest that

$$p_{\mathcal{E}|\boldsymbol{Y}}(\varepsilon|\boldsymbol{Y} = \boldsymbol{y}) \propto p_{\boldsymbol{Y}|\mathcal{E}}(\boldsymbol{y}|\mathcal{E} = \varepsilon), \tag{14.17}$$

and we know this function (from (14.15)). Thus the concept of maximizing the likelihood requires maximizing $p_{\boldsymbol{Y}|\mathcal{E}}(\boldsymbol{y}|\mathcal{E}=\varepsilon)$. By itself, this function is subject to overflow and underflow, so it is common to instead work with its logarithm (cf. (5.13)):

$$\log\left(p_{\boldsymbol{Y}|\mathcal{E}}(\boldsymbol{y}|\mathcal{E}=\varepsilon)\right) = -\frac{1}{2}\log\det\mathsf{K} - \frac{1}{2}\boldsymbol{y}^T\mathsf{K}^{-1}\boldsymbol{y} - \frac{N}{2}\log 2\pi.$$

Because we have been in the practice of minimizing functions to find optimal ε parametrizations, we will multiply by -2 and ignore the constant $\log 2\pi$ to create our maximum likelihood criterion

$$\begin{aligned} \mathrm{C}_{\mathrm{MLE}}(\varepsilon) &= -2\log\left(p_{\boldsymbol{Y}|\mathcal{E}}(\boldsymbol{y}|\mathcal{E}=\varepsilon)\right) - N\log 2\pi \\ &= \log\det\mathsf{K} + \boldsymbol{y}^T\mathsf{K}^{-1}\boldsymbol{y} \end{aligned} \tag{14.18}$$

14.3.2 *MLE with process variance*

We now return to the setting that we have a zero-mean Gaussian random field with covariance kernel $\sigma^2 K$, thus generalizing our discussion from the previous subsection.

As we have pointed out in other places in this book, the process variance does not affect the kernel interpolant/kriging predictor. However, it did affect the kriging variance, and it will also affect the maximum likelihood estimate for the optimal value of the shape parameter ε.

By introducing the process variance, we are suggesting that σ^2 is a draw from a random variable Σ with unknown distribution. We would need to study the joint distribution $(\Sigma, \mathcal{E}, \boldsymbol{Y})$, and our kernel parametrization would require optimizing for both σ^2 and ε by maximizing $p_{\Sigma,\mathcal{E}|\boldsymbol{Y}}(\sigma,\varepsilon|\boldsymbol{Y}=\boldsymbol{y})$, which we will achieve by maximizing $p_{\boldsymbol{Y}|\Sigma,\mathcal{E}}(\boldsymbol{y}|\Sigma=\sigma,\mathcal{E}=\varepsilon)$ similarly to (14.17).

We could treat this as a two-dimensional optimization problem, but instead we will invoke the technique of *profile likelihood*, where σ^2 will be defined as a function of ε, i.e., $\sigma^2 = \sigma^2(\varepsilon)$. Our goal now is to choose an optimal process variance σ^2_{opt} which we will do by maximizing $p_{\Sigma|\mathcal{E},\boldsymbol{Y}}(\sigma^2|\mathcal{E}=\varepsilon,\boldsymbol{Y}=\boldsymbol{y})$. The term profile likelihood is a bit of a misnomer, because $p_{\Sigma,\mathcal{E}|\boldsymbol{Y}}(\sigma^2(\varepsilon),\varepsilon|\boldsymbol{Y}=\boldsymbol{y})$ is not derived from a cumulative distribution function and thus it is not a true density and loses some desirable properties. Even so, this is a common technique.

Using the same proportionality logic as in (14.17) we can write that

$$\begin{aligned} p_{\Sigma|\mathcal{E},\boldsymbol{Y}}(\sigma^2|\mathcal{E}=\varepsilon,\boldsymbol{Y}=\boldsymbol{y}) &\propto p_{\mathcal{E},\boldsymbol{Y}|\Sigma}(\varepsilon,\boldsymbol{y}|\Sigma=\sigma^2) \\ &= p_{\boldsymbol{Y}|\Sigma,\mathcal{E}}(\boldsymbol{y}|\Sigma=\sigma^2,\mathcal{E}=\varepsilon)p_{\mathcal{E}|\Sigma}(\varepsilon|\Sigma=\sigma^2) \\ &\propto p_{\boldsymbol{Y}|\Sigma,\mathcal{E}}(\boldsymbol{y}|\Sigma=\sigma^2,\mathcal{E}=\varepsilon). \end{aligned}$$

Therefore, our optimal σ^2 can be found by maximizing $p_{\boldsymbol{Y}|\Sigma,\mathcal{E}}(\boldsymbol{y}|\Sigma=\sigma^2,\mathcal{E}=\varepsilon)$. This function is the same as (14.15), except with K replaced by $\sigma^2\mathsf{K}$. As before, instead of maximizing, we will try to minimize the negative log of this function:

$$\begin{aligned} -2\log\left(p_{\boldsymbol{Y}|\Sigma,\mathcal{E}}(\boldsymbol{y}|\Sigma=\sigma^2,\mathcal{E}=\varepsilon)\right) - N\log 2\pi \\ = \log\det(\sigma^2\mathsf{K}) + \boldsymbol{y}^T(\sigma^2\mathsf{K})^{-1}\boldsymbol{y} \\ = N\log\sigma^2 + \log\det\mathsf{K} + \frac{1}{\sigma^2}\boldsymbol{y}^T\mathsf{K}^{-1}\boldsymbol{y} \end{aligned}$$

Differentiating this with respect to σ^2, setting it equal to 0, and solving for σ^2 gives the optimal profile variance

$$\sigma^2_{\text{opt}} = \frac{1}{N}\boldsymbol{y}^T\mathsf{K}^{-1}\boldsymbol{y}. \tag{14.19}$$

Using the profile likelihood strategy, we maximize $p_{\Sigma,\mathcal{E}|\boldsymbol{Y}}(\sigma^2,\varepsilon|\boldsymbol{Y}=\boldsymbol{y})$ by minimizing

$$\begin{aligned} &-2\log\left(p_{\boldsymbol{Y}|\Sigma,\mathcal{E}}(\boldsymbol{y}|\Sigma=\sigma^2_{\text{opt}},\mathcal{E}=\varepsilon)\right) - N\log 2\pi \\ &\quad = N\log\left(\frac{1}{N}\boldsymbol{y}^T\mathsf{K}^{-1}\boldsymbol{y}\right) + \log\det\mathsf{K} + \left(\frac{1}{N}\boldsymbol{y}^T\mathsf{K}^{-1}\boldsymbol{y}\right)^{-1}\boldsymbol{y}^T\mathsf{K}^{-1}\boldsymbol{y} \\ &\quad = N\log\left(\boldsymbol{y}^T\mathsf{K}^{-1}\boldsymbol{y}\right) + \log\det\mathsf{K} - N\log N + N \end{aligned}$$

thus defining our profile likelihood ε parametrization criterion as

$$\mathrm{C}_{\mathrm{MPLE}}(\varepsilon) = N\log\left(\boldsymbol{y}^T\mathsf{K}^{-1}\boldsymbol{y}\right) + \log\det\mathsf{K} \tag{14.20}$$

after omitting the constant $-N\log N + N$.

14.3.3 *A deterministic derivation of MLE*

The title of this subsection is a bit misleading, because of course likelihoods cannot be discussed outside of a probabilistic setting. However, the criterion $\mathrm{C}_{\mathrm{MPLE}}$ is equivalent to a criterion which can be derived deterministically (see [Hickernell (2009)]).

We now assume that the function which produced the data $\boldsymbol{y}$ is $f \in \mathcal{H}_K(\Omega)$. Let us expand the notation for our interpolant s to include the data $\boldsymbol{z}$ which generated the interpolant: $s(\cdot;\boldsymbol{z}) = \boldsymbol{k}(\cdot)^T\mathsf{K}^{-1}\boldsymbol{z}$. Also, recall that the Hilbert-space norm of an interpolant (cf. Section 14.1) is given by

$$\|s(\cdot;\boldsymbol{z})\|_{\mathcal{H}_K(\Omega)} = \sqrt{\boldsymbol{z}^T\mathsf{K}^{-1}\boldsymbol{z}}. \tag{14.21}$$

Let $V(\varepsilon)$ denote the volume of the ellipsoid in $\mathbb{R}^N$ which contains all $\boldsymbol{z}$ such that $\|s(\cdot;\boldsymbol{z})\|_{\mathcal{H}_K(\Omega)} \le \|s(\cdot;\boldsymbol{y})\|_{\mathcal{H}_K(\Omega)}$:

$$\begin{aligned} V(\varepsilon) &= \text{ volume of ellipsoid } \left\{z \in \mathbb{R}^N : \|s(\cdot;\boldsymbol{z})\|^2_{\mathcal{H}_K(\Omega)} \le \|s(\cdot;\boldsymbol{y})\|^2_{\mathcal{H}_K(\Omega)}\right\} \\ &= \text{ volume of ellipsoid } \left\{z \in \mathbb{R}^N : \boldsymbol{z}^T\mathsf{K}^{-1}\boldsymbol{z} \le \boldsymbol{y}^T\mathsf{K}^{-1}\boldsymbol{y}\right\} \\ &= \omega_N\frac{(\boldsymbol{y}^T\mathsf{K}^{-1}\boldsymbol{y})^N}{\det\mathsf{K}^{-1}} \\ &= \omega_N(\boldsymbol{y}^T\mathsf{K}^{-1}\boldsymbol{y})^N\det\mathsf{K} \\ &= \omega_N\exp(\mathrm{C}_{\mathrm{MPLE}}(\varepsilon)), \end{aligned}$$

where ω_N is the volume of the unit sphere in $\mathbb{R}^N$ as in Section 6.5. Thus, choosing ε to minimize the volume of the ellipsoid containing function data which would produce "smaller interpolants" (in the $\mathcal{H}_K(\Omega)$ norm) than the observed data $\boldsymbol{y}$ produces is equivalent to maximizing the profile likelihood that ε parametrized the Gaussian field from which $\boldsymbol{y}$ was realized.

This concept of ellipsoid volume is basically employing Occam's Razor: the interpolant that best fits the data should be the simplest, which in this case is measured using the $\mathcal{H}_K(\Omega)$ norm. The ellipsoid described in $V(\varepsilon)$ contains data that would produce a simpler interpolant. By choosing ε to minimize $V(\varepsilon)$ we are minimizing the region from which simpler interpolants could be produced, thus making it less likely that our interpolant is not the simplest.

14.4 Other Approaches to the Selection of Good Kernel Parameters

Over the years, many heuristic approaches have been proposed for determining optimal shape parameters for radial basis function interpolation and collocation methods. Most are focused on global methods, but some also address the RBF finite difference approach discussed in Chapter 19. Most strategies address the situation that involves a single shape parameter that is chosen uniformly across the entire domain, but some also talk about spatially varying shape parameter. Some of the earlier methods, such as those proposed in [Hardy (1971); Franke (1982); Carlson and Foley (1992); Kansa and Carlson (1992); Carlson and Natarajan (1994); Foley (1994); Golberg *et al.* (1996); Schaback and Wendland (2000); Bozzini *et al.* (2002); Fornberg and Zuev (2007)], were already mentioned in [Fasshauer (2007, Chapter 17)].

Early papers on cross-validation and MLE in the statistics literature include [Allen (1974); Craven and Wahba (1979); Golub *et al.* (1979)] and also the book by Wahba (1990). In the literature one can find various evaluations of the advantages and disadvantages of cross validation versus maximum likelihood estimation. For example, while Wahba (1990) states that the generalized cross-validation criterion is an "amazingly good estimate" of the minimum expected mean-squared error, Stone (1977) warns that cross-validation may sometimes be far off the mark. As another example we can learn in [Wahba (1990)] that the GMLE estimate may not be as robust as GCV, while Neumaier (1998) proclaims the GMLE criterion to be "the clear winner." In a more careful study, assuming that the stochastic model is correctly specified, Stein (1990) showed that GCV has twice the asymptotic variance of GMLE for piecewise linear smoothing splines (and worse for higher-order smoothing splines).

We now briefly mention some of the more recent contributions to this growing section of the literature on kernel-based approximation methods.

Recent modifications of cross-validation include the work of Trahan and Wyatt

(2003) who used LOOCV in a time-dependent setting, and [Fasshauer and Zhang (2007b)] where LOOCV was interpreted in the context of PDEs. In [Fasshauer (2011b, Section 5.2)] one can find various extensions of cross-validation and MLE based on Hölder means of eigenvalues of the kernel matrix K. Mongillo (2011) compared LOOCV, and MLE with some additional criteria based on theoretical error bounds. Scheuerer (2011) also proposes several generalizations of cross-validation and MLE.

For multiquadrics and Gaussians in 1D and 2D, Bayona *et al.* (2010, 2012a) used Taylor expansions to derive local truncation errors for first and second derivatives required in RBF-FD approximations. These error estimates were then coupled with (polynomial) finite difference approximations to provide the derivative "data" for the Taylor-based error formulas resulting in an error criterion which was then minimized to yield the optimal value of a constant uniform multiquadric shape parameter in [Bayona *et al.* (2011)], and for a spatially varying multiquadric parameter in [Bayona *et al.* (2012b)].

In the context of solving a Poisson equation in 2D with Gaussian-based finite differences, Davydov and Oanh (2011) proposed a multilevel algorithm to determine the optimal shape parameter. This idea was motivated by their observation that the optimal shape parameter has a strong dependence on the PDE, but much less so on the local density of the stencil points.

Huang *et al.* (2007) used the symbolic computation capabilities of Mathematica to demonstrate that frequently there is an optimal shape parameter for several different radial kernels — both for interpolation and for PDE collocation problems, which is different from the polynomial "flat" limit, and Cheng (2012) used the same approach coupled with some of the work of Luh (see, e.g., [Luh (2012)]) to come up with asymptotic formulas for the optimal inverse multiquadric or Gaussian shape parameters.

Finally, certain authors (such as, e.g., [Kansa and Hon (2000); Driscoll and Heryudono (2007); Flyer and Lehto (2010)]) have championed powerful adaptive schemes with spatially varying shape parameters.

A novel approach to computing the maximum likelihood estimate within a stochastic programming framework is presented in [Anitescu *et al.* (2012)]. One of the key ideas in this approach is to not try to maximize the likelihood function, but to instead consider the necessary condition for optimality, i.e., the gradient of the likelihood, and to set that equal to zero. This problem — which is a nonlinear equation — can be solved in a matrix-free formulation based on the use of a stochastic trace estimator proposed by Hutchinson (1990) and can be applied to very large data sets since it employs fast multipole techniques.

14.5 Goals for a Parametrization Judgment Tool

It is our goal to convince users who are considering the application of kernel methods to the solution of their particular application, that they should view the presence of a free parameter such as ε (or perhaps an additional smoothness parameter β or regularization parameter μ) as a benefit and not as a liability. To support this point of view, we need to have successful parametrization methods so that ε can be chosen to produce the most accurate results. The techniques described above show that existing parametrization methods may work or may not work depending on the situation. New schemes can be developed, but we believe that a tool is needed to judge the viability of existing and future schemes within a rigorous context.

Our goal in this section is to propose some guidelines for a metric by which the quality of a parametrization scheme can be measured with respect to other schemes, or a single scheme's success can be compared across a wide range of scattered data problems. This could be thought of analogously to how we measure the convergence rate of numerical algorithms such as Newton's method: when solving $f(\alpha) = 0$, we know

$$|\epsilon_{n+1}| = \frac{|f''(\xi_n)|}{2\,|f'(x_n)|}\,{\epsilon_n}^2, \qquad \epsilon_n = \alpha - x_n, \qquad \xi_n \text{ between } x_n \text{ and } \alpha.$$

This suggests that Newton's method converges quadratically, which is a property that can be compared to other root-finding schemes. On the other hand, this quadratic convergence is only valid when $f'(x) \neq 0$ near $x = \alpha$ and $f''(x) < \infty$ near $x = \alpha$, which is a comparison of this method across the set of problems to which it could be applied.

In our minds, the guiding principle driving the design of a successful parametrization judgment tool should be: how near is the guess to the "true" value as the amount of available data increases. Of course, the precise meaning of "true" value is open to interpretation. Some possible interpretations might be that, for any given set of data,

- the "true" ε is the value that minimizes (some norm of) the error. While this probably is the definition sought by most users, it might be an unstable definition because it may be subject to small changes in the data;
- the "true" ε is that value which defines the Hilbert-space $\mathcal{H}_K(\Omega)$ in which the function f that generated the data lies — or, analogously, the "true" ε is that value which defines the covariance kernel of the Gaussian process that generated the available data.

Just as we had error bounds in terms of the fill distance h, or the number of samples N, describing the accuracy of different interpolation schemes in Chapter 9, it is our hope that a similarly structured bound can exist in a probabilistic sense for the accuracy of a parametrization scheme. Essentially, the situation might play out as follows:

(1) Data is generated by a function $f \in \mathcal{H}_K(\Omega)$ or by a Gaussian process Y with covariance kernel K.

(a) The kernel K that appears in both settings is the same. It has some parameter ε that defines it, but ε is unknown to us.

(b) The data is $\{(\boldsymbol{x}_1, y_1), \ldots, (\boldsymbol{x}_N, y_N)\}$. This defines the design $\mathcal{X} = \{\boldsymbol{x}_1, \ldots, \boldsymbol{x}_N\}$ and the vector $\boldsymbol{y}$.

i. Knowing $\mathcal{X}$ defines the fill distance $h_{\mathcal{X}}$.

(2) A parametrization scheme is chosen to guess a value $\hat{\varepsilon}$ to be used while constructing the approximation. For example, this could be computed as

$$\hat{\varepsilon} = \underset{\varepsilon}{\operatorname{argmin}}\, \mathrm{C}_{\mathrm{GW}}(\varepsilon; 2).$$

(a) This scheme may involve $\boldsymbol{y}$ (e.g., cross-validation) or only $\mathcal{X}$ (if we, e.g., use only power function).

(b) While we have no real evidence supporting this, the accuracy of such a parametrization scheme may be of the general form

$$|\varepsilon - \hat{\varepsilon}| \leq h_{\mathcal{X}}^{\gamma} \left(C_1(\mathcal{X}) C_2(\boldsymbol{y}) \|f\|_{\mathcal{H}_K} \right),$$

for some $\gamma > 0$ and C_1, C_2.

(c) What is much more likely is that the bound will be of the form

$$P(|\varepsilon - \hat{\varepsilon}| < \alpha) \geq 1 - \nu(\mathcal{X}, \boldsymbol{y}, \alpha), \qquad \lim_{h_{\mathcal{X}} \to 0} \nu(\mathcal{X}, \boldsymbol{y}, \alpha) = 0,$$

because if the data is generated by a Gaussian process, then there is always a chance that it will be a really crummy, uninformative realization.

If something of this form is proved, then it would be able to judge (empirically at the least) the quality of parametrization schemes. Being able to compare different parametrization approaches would be very helpful to the users of kernel-based approximation methods.

With a tool like this, we would be able to make comments on

Consistency: Will the scheme recover the "true" ε for an infinitely dense design? This would only be true if $\lim_{h_{\mathcal{X}} \to 0} \nu(\mathcal{X}, \boldsymbol{y}) = 0$.

Convergence rate: How quickly is the parametrization scheme approaching the "true" ε, and thus how few points are needed before users can feel comfortable that they are doing a decent job? This would depend on the rate at which $\lim_{h_{\mathcal{X}} \to 0} \nu(\mathcal{X}, \boldsymbol{y}) = 0$, especially in comparison to other schemes.

Stability: Do small changes in $\mathcal{X}$ and $\boldsymbol{y}$ affect the consistency or convergence of the scheme?

Bounding ε: Most applications do not demand an optimal ε because of noise in the data. Can this convergence study be used to create a *region* in which the "true" ε lies?

Computational cost: If two schemes have similar convergence properties, is there any reason to not use the cheaper one?

Log scale: The plots we use throughout this book when illustrating the dependence of a method on the shape parameter always show ε on a log-scale, and often this is how we consider different ε values. Is it useful to instead study accuracy of the form

$$|\log \varepsilon - \log \hat{\varepsilon}|, \quad \text{or} \quad \log |\varepsilon - \hat{\varepsilon}|?$$

PART 2
Advanced Examples

Chapter 15

Scattered Data Fitting

In Chapter 9 we showed that the kernel interpolant s is the "smoothest" interpolant at $\mathcal{X} = \{\boldsymbol{x}_1, \ldots, \boldsymbol{x}_N\}$ with values $y_i = f(\boldsymbol{x}_i)$, $i = 1, \ldots, N$, in the sense that s is the interpolant that has minimum Hilbert space norm $\|s\|_{\mathcal{H}_K(\Omega)}$. In fact, one can formulate this problem as a constrained optimization problem, i.e., we want to find $s \in \mathcal{H}_K(\Omega)$ such that we

$$\begin{aligned} &\text{minimize } \|s\|_{\mathcal{H}_K(\Omega)} \\ &\text{subject to } s(\boldsymbol{x}_i) = y_i, \quad i = 1, \ldots, N. \end{aligned}$$

Based on our insights gained in Chapter 9, we know that the solution to this problem is $\boldsymbol{c} = \mathsf{K}^{-1}\boldsymbol{y}$, where

$$s(\boldsymbol{x}) = \sum_{j=1}^{N} c_j K(\boldsymbol{x}, \boldsymbol{x}_j) \tag{15.1}$$

and, as usual, $(\mathsf{K})_{ij} = K(\boldsymbol{x}_i, \boldsymbol{x}_j)$, $i, j = 1, \ldots, N$, and $\boldsymbol{c} = \begin{pmatrix} c_1 & \cdots & c_N \end{pmatrix}^T$, $\boldsymbol{y} = \begin{pmatrix} y_1 & \cdots & y_N \end{pmatrix}^T$.

There are, however, many situations in which we would benefit from a relaxation of strict *interpolation* conditions. For example, (1) we may have noisy data — so that we want to find some other "good" fit, which is relatively simple, but still faithful to the data; (2) we may be interested in decoupling the centers from the data sites (not only in number, but also in location) — so that we are led to work with nonsquare and nonsymmetric kernel matrices; or (3) there may just be too much data for us to handle — so that we want to work with fewer centers than data sites (which again will require a nonsquare kernel matrix). In the next few chapters we will propose a number of different ways to deal with situations such as these, all building upon the basic deterministic or stochastic kernel-based interpolation (kriging) framework laid out in the first part of this book.

We will now address the first situation involving noisy data with a method that is generally referred to as *smoothing splines*, *ridge regression* or *penalized least squares approximation* (see, e.g., [Wahba (1990); Green and Silverman (1993)]). In Chapter 18 we will discuss the more general second situation (which involves the third one as a special case) in the context of RBF network regression.

15.1 Approximation Using Smoothing Splines

Our discussion in Chapter 9, which led to the "nice" interpolation solution above, assumed that the given data did not contain any error. If, on the other hand, the data is contaminated by error, i.e., $y_i = f(\boldsymbol{x}_i) + \epsilon_i$, then one no longer wants to fit the data exactly (since overfitting of the erroneous data would be highly undesirable). Here one usually assumes that one has *Gaussian white noise*, i.e., $\boldsymbol{\epsilon} = \left(\epsilon_1 \cdots \epsilon_N\right)^T \sim \mathcal{N}(0, \sigma_{\boldsymbol{\epsilon}}^2 \mathsf{I})$. By going over to a regularization approach in which one balances a least square error term against the same smoothing term as above one can still have an analogous theory. In this setting one considers an approximation of the form (15.1), where — just as for scattered data interpolation — the number of basis functions equals the size of the data set, N, and the centers coincide with the data sites. Since we do not want to overfit the data it is customary to treat such a problem with a *Tikhonov regularization* approach so that one ends up with the unconstrained optimization problem

$$\begin{aligned} &\min_{\boldsymbol{c}\in\mathbb{R}^N} \sum_{i=1}^{N} \left(y_i - s(\boldsymbol{x}_i)\right)^2 + \mu \|s\|^2_{\mathcal{H}_K(\Omega)} \\ \iff \quad &\min_{\boldsymbol{c}\in\mathbb{R}^N} \left(\boldsymbol{y} - \mathsf{K}\boldsymbol{c}\right)^T \left(\boldsymbol{y} - \mathsf{K}\boldsymbol{c}\right) + \mu \boldsymbol{c}^T \mathsf{K}\boldsymbol{c}, \end{aligned} \tag{15.2}$$

where we have employed (2.24) for the Hilbert space norm of the kernel approximant.

Note that — at least for the time being — μ is treated as a given quantity in this minimization problem. A small value of the regularization parameter μ pushes the fit more closely to the data, while a large value of μ indicates preference for a smoother or more regularized approximation. If the variance $\sigma_{\boldsymbol{\epsilon}}^2$ of the noise is known, then a good choice of μ is discussed in [Wahba (1975)]. However, usually we have no such knowledge, so that the smoothness parameter must be estimated from the data using maximum likelihood estimation or cross validation, and we do so in Chapter 17.

Since (15.2) is a quadratic minimization problem, setting the gradient with respect to $\boldsymbol{c}$ equal to zero is not only necessary, but also sufficient:

$$\begin{aligned} &\nabla_{\boldsymbol{c}} \left[(\boldsymbol{y} - \mathsf{K}\boldsymbol{c})^T (\boldsymbol{y} - \mathsf{K}\boldsymbol{c}) + \mu \boldsymbol{c}^T \mathsf{K}\boldsymbol{c} \right] = 0 \\ \iff \quad &\nabla_{\boldsymbol{c}} \left[\boldsymbol{y}^T \boldsymbol{y} - \boldsymbol{c}^T \mathsf{K}^T \boldsymbol{y} - \boldsymbol{y}^T \mathsf{K}\boldsymbol{c} + \boldsymbol{c}^T \mathsf{K}^T \mathsf{K}\boldsymbol{c} + \mu \boldsymbol{c}^T \mathsf{K}\boldsymbol{c} \right] = 0 \\ \iff \quad &-2\mathsf{K}^T \boldsymbol{y} + 2\mathsf{K}^T \mathsf{K}\boldsymbol{c} + 2\mu \mathsf{K}\boldsymbol{c} = 0 \\ \iff \quad &-\boldsymbol{y} + \mathsf{K}\boldsymbol{c} + \mu \boldsymbol{c} = 0. \end{aligned}$$

Note that K is symmetric and positive definite so that the identification $\boldsymbol{y}^T \mathsf{K}\boldsymbol{c} = \boldsymbol{c}^T \mathsf{K}^T \boldsymbol{y}$ and left-multiplication by $\frac{1}{2}\mathsf{K}^{-1}$ in the last step are justified. Thus — for a fixed given μ — the coefficients of the smoothing spline, which is still of the form (15.1), can now be identified as the coefficients $\boldsymbol{c}$ determined by the solution of a regularized $N \times N$ linear system, i.e.,

$$\boldsymbol{c} = (\mathsf{K} + \mu \mathsf{I})^{-1} \boldsymbol{y}. \tag{15.3}$$

Inserting the coefficients from (15.3) back into (15.1), we can use the *push-through identity*[1] [Bernstein (2009, Fact 2.16.16)] for matrix inverses to rewrite the smoothing spline as follows:

$$\begin{aligned} s(\boldsymbol{x}) &= \boldsymbol{k}(\boldsymbol{x})^T(\mathsf{K}+\mu\mathsf{I})^{-1}\boldsymbol{y} \\ &= \boldsymbol{k}(\boldsymbol{x})^T\mathsf{K}^{-1}(\mathsf{K}+\mu\mathsf{I})^{-1}\mathsf{K}\boldsymbol{y} \\ &= \boldsymbol{\ell}(\boldsymbol{x})^T(\mathsf{K}+\mu\mathsf{I})^{-1}\mathsf{K}\boldsymbol{y} \\ &= \boldsymbol{\ell}(\boldsymbol{x})^T\tilde{\boldsymbol{y}}, \end{aligned}$$

where $\boldsymbol{\ell}(\boldsymbol{x})^T = \boldsymbol{k}(\boldsymbol{x})^T\mathsf{K}^{-1}$ are the *cardinal functions* (12.4) for standard interpolation (kriging prediction). This provides us with an *alternate interpretation of smoothing splines*. Instead of thinking of a smoothing spline as a smooth approximation to the given (noisy) data in $\boldsymbol{y}$, we can interpret the smoothing spline as a kernel (or kriging) interpolant to the set of smoothed out data values (or *filtered data*) in

$$\tilde{\boldsymbol{y}} = (\mathsf{K}+\mu\mathsf{I})^{-1}\mathsf{K}\boldsymbol{y}. \tag{15.4}$$

Clearly, if $\mu = 0$ then we are just performing straightforward interpolation.

Finally, we consider how to compute some of these quantities. The coefficients $\boldsymbol{c}$ in (15.3) may be computed using the standard SVD of K, i.e.,

$$\begin{aligned} \boldsymbol{c} &= (\mathsf{K}+\mu\mathsf{I})^{-1}\boldsymbol{y} \\ &= (\mathsf{U}\Sigma\mathsf{U}^T+\mu\mathsf{I})^{-1}\boldsymbol{y} \\ &= \mathsf{U}(\Sigma+\mu\mathsf{I})^{-1}\mathsf{U}^T\boldsymbol{y}. \end{aligned}$$

Moreover, the smoothing spline approximation may either be computed using the standard basis with these coefficients, or with the cardinal basis and the smoothed data. As we saw earlier, the Hilbert–Schmidt SVD allows us to stably compute the cardinal functions as $\boldsymbol{\ell}(\boldsymbol{x})^T = \boldsymbol{\psi}(\boldsymbol{x})^T\Psi^{-1}$.

Example 15.1. (Smoothing splines as interpolant of smoothed data)
We illustrate the two different interpretations of a smoothing spline by looking at a one-dimensional test case from [Wahba (1990, Chapter 4)]. The data consists of samples obtained at 100 evenly spaced points in $[0, 3.5]$ from the function $f(x) = 4.26\left(\mathrm{e}^{-x} - 4\mathrm{e}^{-2x} + 3\mathrm{e}^{-3x}\right)$ with a random error normally distributed with zero mean and standard deviation $\sigma_\epsilon = 0.2$. Program 15.1 demonstrates fitting with ridge regression in MATLAB with results displayed in Figure 15.1.

Program 15.1. Excerpts from `AppxFitRidge.m`

```
% Set up the data in the problem
N = 100;
yf = @(x) 4.26*(exp(-x)-4*exp(-2*x)+3*exp(-3*x));
sigma = .2;
```

[1] The push-through identity says that for any $n \times m$ matrix A and $m \times n$ matrix B with $\mathsf{I}_n + \mathsf{AB}$ invertible one has $(\mathsf{I}_n + \mathsf{AB})^{-1}\mathsf{A} = \mathsf{A}(\mathsf{I}_m + \mathsf{BA})^{-1}$.

```
x = pickpoints(0,3.5,N);
y = yf(x) + sigma^2*randn(N,1);
% Set up evaluation points at which to study the answer
Neval = 400;
xeval = pickpoints(0,3.5,Neval);   yeval = yf(xeval);
% Set up C2 Matern kernel and smoothing parameter to fit the data with
rbf = @(e,r) (1+e*r).*exp(-e*r);
ep = .4;
mu = 1e-5;
I = eye(N);
% Create the interpolation and evaluation matrix
DM = DistanceMatrix(x,x);
DMeval = DistanceMatrix(xeval,x);
K = rbf(ep,DM);
Keval = rbf(ep,DMeval);
K_mu = K + mu*I;
% Evaluate the interpolant and the smoothed fit with ridge regression
yint = Keval*(K\y);
yridge = Keval*(K_mu\y);
% Evaluate the smoothed data
ysmooth = K_mu\(K*y);
```

The plots in Figure 15.1(a)–(b) demonstrate that ridge regression can be interpreted both as an approximation to noisy data or an interpolant to smoothed data. Figure 15.1(c), generated by Program 15.2, displays the quality of the ridge regression as a function of the parameters ε and μ. The valley in this surface plot is the result of optimal accuracy occurring when an increase in smoothing factor μ is countered by an increase in locality (shape parameter) ε. Of course, every application and example is unique, and the results of this experiment are not guaranteed to apply widely. Rather, the goal of this example is to provide users with the tools to conduct their own investigations. We use this smoothing behavior again as a sort of spatial averaging technique in Section 16.3.2.

Program 15.2. Excerpts from `AppxFitRidge.m`

```
% This code is executed after Program 15.1
% This strategy for employing cellfun is introduced in Program 4.18
epvec = logspace(-3,1,30);
muvec = logspace(-15,0,20);
[E,M] = meshgrid(epvec,muvec);
emcell = num2cell([E(:),M(:)],2);
errvecem = cellfun(@(em)errcompute( ...
                       rbf(em(1),DMeval)*((rbf(em(1),DM)+em(2)*I)\y) ...
                    ,yeval),emcell);
ERR = reshape(errvecem,length(muvec),length(epvec));
% Create a surface plot and manipulate it to help the presentation
h_surf = figure;
surf(epvec,muvec,ERR)
xlabel('\epsilon'),ylabel('\mu'),zlabel('2-norm error')
```

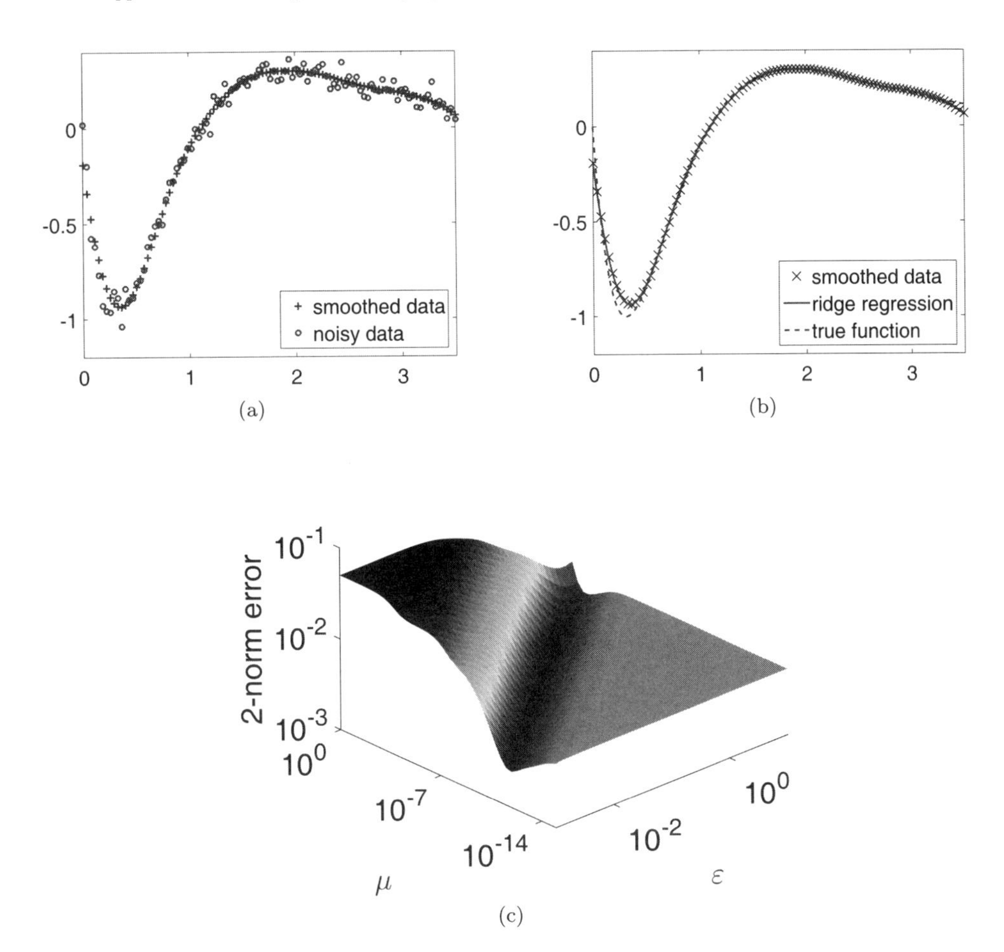

Fig. 15.1 The C^2 Matérn kernel with $\varepsilon = 0.4$ is applied to noisy data in a ridge regression with $\mu = 10^{-5}$. (a) The initial noisy data comes from an analytic function and after smoothing with (15.4) the data $\tilde{\boldsymbol{y}}$ behaves better; (b) The ridge regression fit interpolates the smoothed data and closely reconstructs the underlying function; (c) For any fixed ε value there appears to be an optimal choice of smoothing parameter μ, and vice versa.

```
set(gca,'xtick',[.01,1],'ytick',[1e-14,1e-7,1])
set(gca,'xscale','log','yscale','log','zscale','log')
colormap gray,colormap(flipud(colormap))
view([-.8,-1.2,1])
```

As a brief aside regarding MATLAB, and specifically plotting in MATLAB, manipulating plots to look “pretty” can require some heavy lifting, especially when dealing with plots in 3D or plots with multiple axes in the same figure. On one hand, functions such as `xlabel` provide a direct mechanism for labeling axes. On the other hand, some quantities such as where tick marks appear on the axes must be manipulated within the figure handle or axes handle. The current figure and axes can be accessed through `gcf` and `gca` and, when unsure of what is available, the `get` command allows viewing of all properties. Property values can be changed with the command `set`, and some more complex properties are actually managed by the `Parent` or `Children` of the `gcf`; sifting through those can often provide solutions when one is not immediately apparent.

15.2 Low-rank Approximate Interpolation

Given data sites $\mathcal{X} = \{\boldsymbol{x}_1, \ldots, \boldsymbol{x}_N\} \subset \mathbb{R}^d$ and associated values $\boldsymbol{y}$, in Section 12.3 we introduced the low-rank approximate interpolant to this data as

$$\begin{aligned} s(\boldsymbol{x}) &= \boldsymbol{\phi}(\boldsymbol{x})^T \boldsymbol{a}, \qquad \|\Phi \boldsymbol{a} - \boldsymbol{y}\|_2^2 \to \min \\ &= \boldsymbol{\phi}(\boldsymbol{x})^T \Phi^\dagger \boldsymbol{y}, \end{aligned} \tag{12.34}$$

where

$$\boldsymbol{\phi}(\boldsymbol{x})^T = \begin{pmatrix} \varphi_1(\boldsymbol{x}) & \cdots & \varphi_M(\boldsymbol{x}) \end{pmatrix}, \quad \Phi = \begin{pmatrix} \boldsymbol{\phi}(\boldsymbol{x}_1)^T \\ \vdots \\ \boldsymbol{\phi}(\boldsymbol{x}_N)^T \end{pmatrix},$$

are formed using the first M eigenfunctions of the Hilbert–Schmidt integral operator $\mathcal{K}$ associated with the kernel K.

We now investigate the use of this low-rank approximate interpolant.

Example 15.2. (Low-rank approximate interpolation)
To demonstrate the value of the low-rank replacement computation (12.34), we consider $N = 100$ Halton points in 1D and interpolation with a Gaussian kernel using $\varepsilon = 0.2$ with error computed at 500 evenly spaced points. The Gaussian eigenfunctions from Example 12.1 have global scale parameter $\alpha = 1$.

In Program 15.3 we illustrate the quality of computation using an increasing M, and thus increasingly more columns in Φ. Figure 15.2(b) shows that more columns of Φ allow for a higher-quality approximation to K, but $\operatorname{rank}(\mathsf{K})$ is bounded even while $\operatorname{rank}(\Phi)$ continues to grow because the eigenvalues in Λ decay too quickly (see Figure 15.2(a)).

The effect of this low-rank behavior in K is that the quality of the low-rank approximate interpolant using K is bounded by $\operatorname{rank}(\mathsf{K})$. Because K is numerically low-rank, $s(\boldsymbol{x}) = \boldsymbol{k}(\boldsymbol{x})^T \mathsf{K}^{-1} \boldsymbol{y}$ is replaced by $s(\boldsymbol{x}) = \boldsymbol{k}(\boldsymbol{x})^T \mathsf{K}^\dagger \boldsymbol{y}$ during computation. When increasing M no longer increases $\operatorname{rank}(\Phi\Lambda\Phi^T)$, the accuracy of s stagnates. This issue can be circumvented by using the eigenfunction basis to evaluate $s(\boldsymbol{x}) = \boldsymbol{\phi}(\boldsymbol{x})^T \Phi^\dagger \boldsymbol{y}$ because of the *analytic removal* of Λ^{-1} that took place in (12.37).

Program 15.3. Excerpts from `AppxFitEigs.m`

```
  % Define the Gaussian
2 rbf = @(e,r) exp(-(e*r).^2);ep = .2;
```

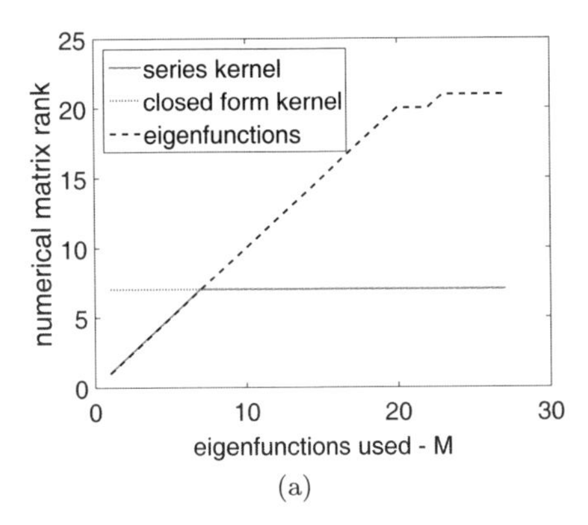

(a)

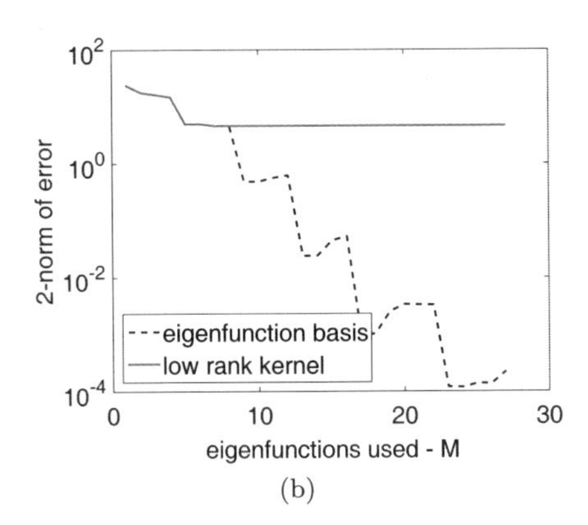

(b)

Fig. 15.2 Gaussian eigenfunctions may be more effective than the Gaussian kernel. We use the Gaussian kernel with $\varepsilon = 0.2$ to attempt to interpolate $N = 100$ points in 1D. (a) The rank of $\Phi\Lambda\Phi^T$ approaches the rank of K for increasing M, but the rank of Φ continues to grow, (b) By avoiding the formation of the K matrix and evaluating the low-rank interpolant with (12.34), a meaningful approximation can be computed.

```
   % Define the data of interest
 4 N = 100;  Neval = 500;
   x = pickpoints(-1,1,N,'halton');  xeval = pickpoints(-1,1,Neval);
 6 yf = @(x) 2*x.^3 + cos(2*pi*x.^2);  y = yf(x);  yeval = yf(xeval);
   % Form the kernel matrices and find the rank of the interpolation matrix
 8 K = rbf(ep,DistanceMatrix(x,x));  Keval = rbf(ep,DistanceMatrix(xeval,x));
   rankK = rank(K);
10 % Define the eigenfunctions and compute their matrices
   % We work with 20 more eigenfunctions than the rank of the K matrix
12 alpha = 1;  GQR = gqr_solveprep(1,x,ep,alpha,rankK+20);
   Phi = gqr_phi(GQR,x);  Phieval = gqr_phi(GQR,xeval);
14 Mvec = GQR.Marr;  lamvec = GQR.eig(Mvec);
   % Create a function to evaluate the K series approximation with M terms
16 % These functions have the data Phi, lamvec, Phieval stored in them
   MtermK = @(M) bsxfun(@times,lamvec(1:M),Phi(:,1:M))*Phi(:,1:M)';
18 MtermKeval = @(M) bsxfun(@times,lamvec(1:M),Phieval(:,1:M))*Phi(:,1:M)';
   % Study the rank of the Phi*Lambda*Phi' matrix and also the Phi matrix
20 rankvec = arrayfun(@(M)rank(MtermK(M)),Mvec);
   rankPhivec = arrayfun(@(M)rank(Phi(:,1:M)),Mvec);
22 % Compute the error in the low-rank and eigenfunction approx interpolants
   errvec = arrayfun(@(M)errcompute(MtermKeval(M)*(pinv(MtermK(M))*y),yeval),Mvec);
24 errPhivec = arrayfun(@(M)errcompute(Phieval(:,1:M)*(Phi(:,1:M)\y),yeval),Mvec);
```

We take this opportunity to further discuss function handles in MATLAB because the use of `MtermK` and `MtermKeval` in Program 15.3 allows for a cleaner implementation. In MATLAB, function handles are defined with `@` to take certain arguments, but at their definition other data can be implicitly stored in the function handle for use during computation. When `MtermK` is created, the `lamvec` and `Phi` objects are implicitly stored and frozen within that function; even if `Phi` is set to another value externally, the value within the function handle `MtermK` will persist.

This is an elegant technique for creating compact and easy to read code as evidenced above. Memory issues can arise, however, when large objects are stored implicitly in the function. The session below demonstrates this issue.

```
clear all
A = ones(1000);
addconst = @(c) c+A;
clear A
whos
  Name           Size            Bytes  Class              Attributes

  addconst       1x1                32  function_handle
info = functions(addconst);
info.workspace{1}
ans =

    A: [1000x1000 double]
whos('info')
  Name      Size             Bytes  Class     Attributes

  info      1x1            8001024  struct
```

Despite the initial `whos` call which states that the only memory in use is 32B, after further digging, the 8MB matrix stored implicitly in `addconst` is still taking up space. When a function handle is no longer needed, the `clear` command should be invoked, e.g., `clear addconst`, to guarantee any memory it may have declared has been freed. Note that, if `clear all` is invoked, `rbfsetup` should be executed again from the `GaussQR` base directory.

Remark 15.1. The error in Figure 15.2(b) does not decay monotonically, as might be expected: if the quality of the approximation were reduced by including additional basis functions then, in a perfect world, the approximation scheme should ignore the addition of that basis function. This logic is used to build models in statistics, where variables are eliminated from the model if they fail to contribute.

Unfortunately, this logic does not carry through to the evaluation of $s(\boldsymbol{x}) = \boldsymbol{\phi}(\boldsymbol{x})^T \boldsymbol{\Phi}^\dagger \boldsymbol{y}$ because $\boldsymbol{\Phi}^\dagger \boldsymbol{y}$ solves the least squares problem of fitting the data and not, necessarily, maximizing the quality of the fit to the true function f. While adding more basis functions should help asymptotically, the addition of any single basis function to the eigenfunction basis could have a negative effect as measured at a separate set of sample points.

Remark 15.2. Example 15.2 is an example where the first seven eigenfunctions summed in the series form of the kernel all contribute to the rank of the kernel. In higher dimensions this need not be the case, and some amount of pivoting/removal during the summation is necessary to distinguish between the worthwhile eigenfunctions and the eigenfunctions which make no contribution as a result of, e.g., the selection of points $\mathcal{X}$. Some implications of this are discussed in Section 13.5

and later in Section 19.4, but the authors presently have no cost-effective strategy for identifying the significant eigenfunctions in the summation.

Example 15.3. (Modeling freeform optical surfaces)
Driven by recent advances in the ability to more accurately manufacture freeform optical surfaces, there has been research [Cakmakci *et al.* (2008, 2010); Kaya and Rolland (2010, 2013)] into the use of (radial) kernel-based representations of such surfaces and how they compare with more traditional methods such as Zernicke polynomials or the orthogonal polynomials of Forbes (2010, 2012). Here we use another low-rank approximately interpolating solution obtained via an early truncation of the Mercer series of the Gaussian kernel for a spherical test surface from [Jester *et al.* (2011)].

Using $r = \|\boldsymbol{x}\| \in [0,1]$, the two-dimensional test function is given by

$$f(r) = \frac{1}{20}\left(\frac{\rho r^2}{1+\sqrt{1-(\rho^r)^2}} + 4.17\times 10^{-6}r^4 + 4.71\times 10^{-9}r^6 - 4.94\times 10^{-12}r^8 \right.$$
$$\left. - 5.42\times 10^{-15}r^{10} - 4.98\times 10^{-18}r^{12} - 1.22\times 10^{-20}r^{14}\right), \quad \rho = -3.87\times 10^{-2}. \tag{15.5}$$

We sample this surface at *weakly admissible meshes* of degree n [Bos *et al.* (2011a,b)] on the unit disk. A weakly admissible mesh $\mathcal{M}_n$ on a disk is given by the tensor product of a set of $n+1$ *Chebyshev–Lobatto points* (see Appendix B.3) in the radial direction with a set of m equally spaced angles, where $m = n+2$ for n even and $m = n+1$ for n odd, i.e.,

$$\mathcal{M}_n = \{(r_j \cos\theta_k, r_j \sin\theta_k)\}$$
$$\{(r_j,\theta_k)\}_{j,k} = \left\{\cos\frac{j\pi}{n},\ 0 \le j \le n\right\} \times \left\{\frac{k\pi}{m},\ 0 \le k \le m-1\right\}.$$

Program 15.4 computes low-rank eigenfunction approximate interpolants for various ε and M values, with results plotted in Figure 15.3. Because this program analyzes the convergence behavior, i.e., the behavior as N increases, we choose to fix the ratio M/N. This has the effect of increasing the number of eigenfunctions used to form the approximate interpolant proportionally to the amount of data sampled. These experiments use the Gaussian eigenfunctions from Example 12.1, which are available in the `GaussQR` repository via the `gqr_phi` function.

Program 15.4. Excerpts from `AppxFitOptics.m`

```
% Request access to the GaussQR management structure
global GAUSSQR_PARAMETERS
% Choose sample sizes on which to study convergence
Nvec = [8,12,16,24,32,40,64];  Neval = 20;
% Choose test values, and the eigenfunction parameter alpha
% We only show the code for the M ratio here
```

```
Mrvec = [.05,.1,.2,.3];  epvec = [.3,1,3];  alpha = 2;
% Assign the optics function that we want to approximate
yf = pickfunc('KSA2',2);
xeval = pick2Dpoints(-1,1,Neval,'wam'); yeval = yf(xeval);
% Compute the errors for the M ratio values assigned above
ep = 1;
errmat = zeros(length(Mrvec),length(Nvec));
j = 1;
for N=Nvec
    x = pick2Dpoints(-1,1,N,'wam');  y = yf(x);
    k = 1;
    for Mr=Mrvec
        GAUSSQR_PARAMETERS.DEFAULT_REGRESSION_FUNC = Mr;
        errmat(k,j) = errcompute(gqr_eval(gqr_rsolve(x,y,ep,alpha),xeval),yeval);
        k = k + 1;
    end
    j = j + 1;
end
% Plot the results of the experiment
loglog(Nvec.^2,errmat,'linewidth',3)
```

The function `pickfunc` on line 9 is available in the `GaussQR` library; it provides a set of complicated test functions which we prefer to not define inline.

The convergence behavior of the low-rank eigenfunction approximate interpolant $s(\boldsymbol{x}) = \boldsymbol{\phi}(\boldsymbol{x})\Phi^{\dagger}\boldsymbol{y}$ seems undeniable based on Figures 15.3(c) and (d); equally as relevant is their success despite the ill-conditioning of the numerically low-rank matrix K as shown with the MATLAB executions

```
x = pick2Dpoints([-1,-1],[1 1],64,'wam');  ep = 3;
rank(exp(-(ep*DistanceMatrix(x,x)).^2))
ans =

   385
```

This shows that the matrix K has numerical rank 385 — a good deal less than the 4096 required to be full rank.

We feel that results presented in this section demonstrate the potential for low-rank eigenfunction bases to serve as an alternative to the standard basis $\boldsymbol{k}(\cdot)$ when the interpolation matrix K is intolerably ill-conditioned. Many other experiments could be conducted in this optics setting to study the behavior of, for instance, the approximate interpolant as $\varepsilon \to 0$. We encourage the readers to devise experiments that they consider interesting using the computational framework we have developed.

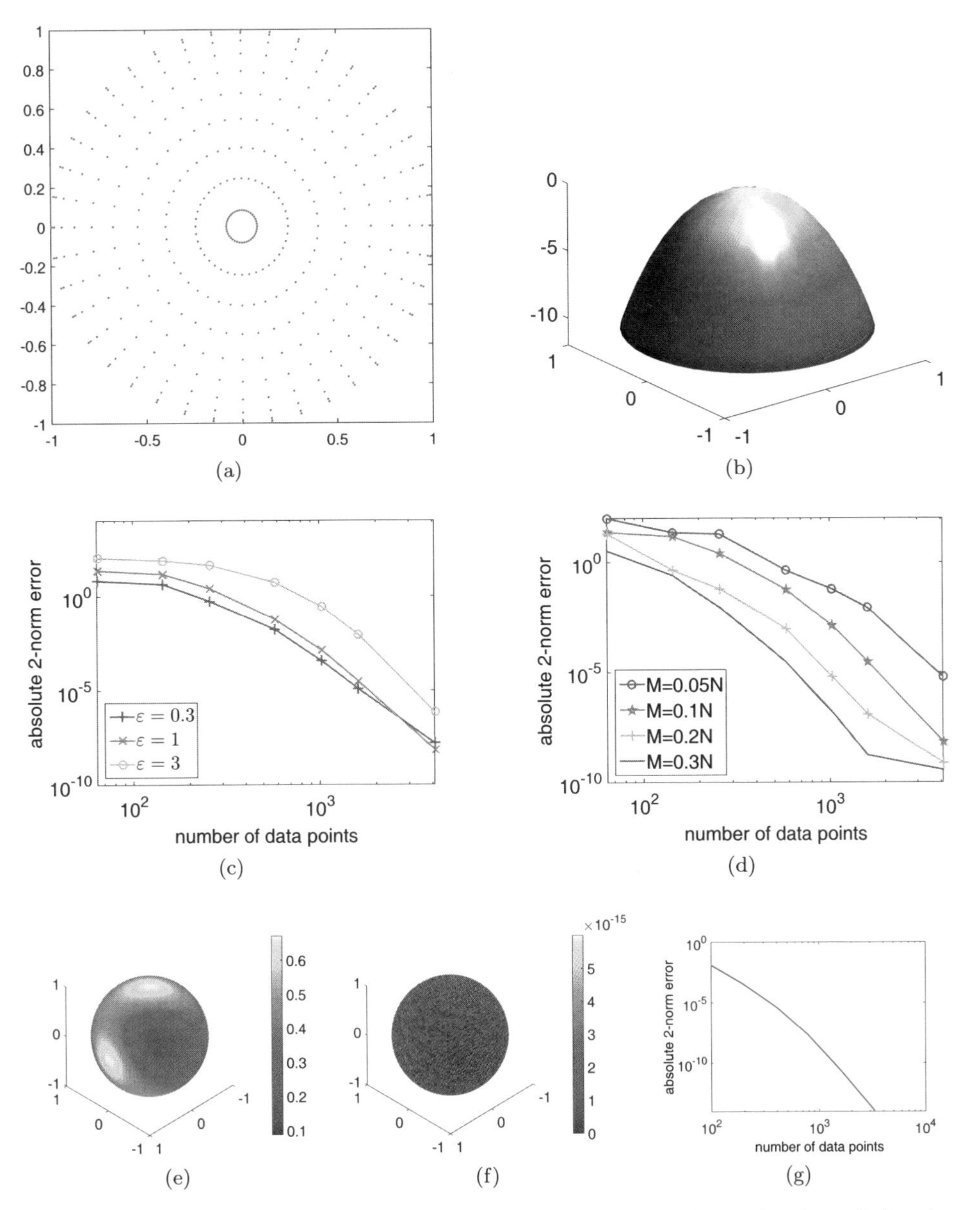

Fig. 15.3 (a)–(d): The Gaussian low-rank eigenfunction approximate interpolant is applied to the function (15.5), which is popular in optics. (a) A typical weakly admissible mesh containing 400 points; (b) Graph of the function (15.5); (c) Different ε values produce approximate interpolants of different quality, though all seem to converge with fixed M/N=0.1; (d) For fixed $\varepsilon = 1$, larger M/N ratios seem to produce better accuracy until saturation sets in for large enough N.
(e)–(g): The spherical inverse multiquadric can be used to effectively interpolate data on the sphere. (e) Graph of the test function (15.6); (f) Error for the interpolant with $N = 3249$ is evenly distributed around the sphere; (g) For fixed $\gamma = 0.569$, the rate of convergence of the interpolant appears to be better than algebraic for increasingly large N.

15.3 Interpolation on the Unit Sphere

While kernel interpolation can be performed on general compact Riemannian manifolds in $\mathbb{R}^d$ (see, e.g., [Narcowich (1995); Dyn *et al.* (1997, 1999); Levesley and Ragozin (2002, 2007); Hangelbroek *et al.* (2010); Fuselier and Wright (2012)]), we will limit ourselves to the relatively simple case of the *unit sphere* $\mathbb{S}^2$ in $\mathbb{R}^3$. We refer to the discussion of zonal kernels in Section 3.4.2 for some theoretical background, and now discuss a scattered data interpolation example.

Example 15.4. (Interpolation with the spherical inverse multiquadric)
We consider interpolation of the test function

$$f(\boldsymbol{x}) = f(x_1, x_2, x_3) = \frac{1 + 10x_1x_2x_3 + x_1^8 + \mathrm{e}^{2x_2^3} + \mathrm{e}^{2x_3^2}}{14} \tag{15.6}$$

from [Alfeld *et al.* (1996)] using the spherical inverse multiquadric kernel

$$K(\boldsymbol{x}, \boldsymbol{z}) = \frac{1}{\sqrt{1 + \gamma^2 - 2\gamma \boldsymbol{x}^T \boldsymbol{z}}}, \qquad 0 < \gamma < 1, \tag{3.3}$$

as introduced in Sections 3.4.2 and 4.3. Input data is sampled at $N = 100, 196, 400, 784, 1600, 3249$ *maximal determinant points* (see, e.g., [Womersley and Sloan (2001); An *et al.* (2010)] and convergence is studied for fixed $\gamma = 0.569$. A huge repository of spherical designs (including maximal determinant points) is available from Rob Womersley's website [Womersley (2007)]). The points used for this example are also available from the `GaussQR` library data directory in the file `sphereMDpts_data.mat`.

Program 15.5 demonstrates an implementation with accompanying graphs in Figure 15.3. Recall that line 22 computes all the inner products simultaneously, not an outer product between vectors; this was discussed in Remark 4.5.

Program 15.5. Excerpts from `AppxFitZonal.m`

```
% Download the maximal determinant points from gaussqr, if needed
2  gqr_downloaddata('sphereMDpts_data.mat')
   % Load the downloaded data into memory
4  load sphereMDpts_data
   Nvec = cellfun(@length,sphereMDpts);
6  % Define the test function (15.6)
   yf = @(x) (1 + 10*x(:,1).*x(:,2).*x(:,3) + ...
8             x(:,1).^8 + exp(2*x(:,2).^3) + exp(2*x(:,3).^2))/14;
   % Define the spherical inverse multiquadric (3.3), using the dot-product
10 gamma = .569;
   zbf = @(g,dp) 1./sqrt(1+g^2-2*g*dp);
12 % Create points at which to test the error
   Neval = 200;  xeval = zeros(Neval,3);
14 [t,testptstr] = pick2Dpoints([-pi -pi/2],[pi pi/2],sqrt(Neval),'halt');
   [xeval(:,1),xeval(:,2),xeval(:,3)] = sph2cart(t(:,1),t(:,2),1);
16 yeval = yf(xeval);
```

```
% Study the quality of the interpolation
18 errvec = zeros(size(sphereMDpts));
for k=1:length(sphereMDpts)
20     % Select data and create the interpolation, evaluation matrices
    x = sphereMDpts{k};  y = yf(x);
22     K = zbf(gamma,x*x');  Keval = zbf(gamma,xeval*x');
    % Perform the interpolation and evaluate the error at the test points
24     coef = K\y;
    errvec(k) = errcompute(Keval*coef,yeval);
26 end
```

Figure 15.3(g) shows the error of the interpolation steadily decreasing on the evaluation points as increasingly many data locations are considered. The points `xeval` that we use to test the error are chosen in a manner ill-befitting the sphere, and are not appropriate for more serious problems on manifolds because, despite the nice nature of Halton points in $\mathbb{R}^d$, their transformation to the sphere — as performed in lines 14–15 of Program 15.5 — does not yield a well-distributed sample. For this simple example they are sufficient, but much better designs exist when working on a sphere or other manifold.

Remark 15.3. In the first line of Program 15.5, we use `gqr_downloaddata` to access binary `mat` files which store data for use in MATLAB. While `gqr_downloaddata` can access any source of files on the web, it defaults to downloading from the `GaussQR` server where files needed for this book are stored. At the time of this publication, `gqr_downloaddata` uses the MATLAB builtin function `websave` which may place some limitations on the type of media that can be downloaded. Nonetheless, standard files such as `mat` files should pose no problem. The default directory for downloading can be set within `rbfsetup` as a `GAUSSQR_PARAMETERS` option.

Example 15.5. (Hilbert–Schmidt SVD in spherical harmonics)
Despite the nice convergence behavior observed in Example 15.4 with the shape parameter $\gamma = 0.569$, small values of γ can produce ill-conditioning in the interpolation matrix K. This instability, discussed in Chapter 11, can be observed by computing rank(K) for various γ values (of course, the download may be skipped in the following code snippet if Program 15.5 has been executed):

```
gqr_downloaddata('sphereMDpts_data.mat'),  load sphereMDpts_data
zbf = @(g,dp) 1./sqrt(1+g^2-2*g*dp);
x = sphereMDpts{1};
gvec = [.1,.05,.02,.01,.005];
arrayfun(@(g)rank(zbf(g,x*x')),gvec)
ans =

   100   100    64    49    36
```

The interpolation matrix — even for only $N = 100$ — becomes irrevocably ill-conditioned and numerically low-rank as $\gamma \to 0$. This rank deficiency is the result of a poor choice of basis for the interpolation: in spite of the good convergence of the inverse multiquadric basis in Figure 15.3(g), that basis performs increasingly poorly as $\gamma \to 0$. As discussed in Chapter 12, an *alternate basis* could provide a stable mechanism for, at least, approximating the interpolant. In Example 15.3 we studied the use of Hilbert–Schmidt eigenfunctions to create a low-rank eigenfunction approximate interpolant.

In Chapter 13 we introduced the Hilbert–Schmidt SVD, which describes a different use of the eigenfunctions. It takes advantage of the Mercer series expansion for the spherical inverse multiquadric derived in Section 3.4.2,

$$\frac{1}{\sqrt{1+\gamma^2-2\gamma \boldsymbol{x}^T\boldsymbol{z}}} = \sum_{\ell=0}^{\infty} \frac{4\pi\gamma^\ell}{2\ell+1} \sum_{m=1}^{2\ell+1} Y_{\ell,m}(\boldsymbol{x})Y_{\ell,m}(\boldsymbol{z}), \tag{3.5}$$

where $Y_{\ell,m}$ is the spherical harmonic of degree ℓ and order m. In this example we compute the Hilbert–Schmidt SVD for the spherical inverse multiquadric and show that it allows for stable evaluation of the interpolant s.

Program 15.6 provides such an implementation (with much of the other content omitted for space concerns and also to minimize redundant overlap with Program 15.5). The accompanying plot is displayed in Figure 15.4. Note that we use the terms "spherical harmonics" and "eigenfunctions" interchangeably in this example. The computation here mirrors that of Program 13.2 with the eigenvalues and eigenfunctions defined in Section 3.4.2. We thank Grady Wright for his spherical harmonics function, available from [Wright (2014)], which was adapted for use in the GaussQR library for this example.

Program 15.6. Excerpts from AppxFitZonalHSSVD.m

```
% Create data; !assumes yf, xeval and sphereMDpts already in memory!
x = sphereMDpts{3};  N = size(x,1);  y = yf(x);
% Identify the range and order required for the HS-SVD
Nlevels = ceil(sqrt(N)) + 3;
Neig = Nlevels^2;
lrange = 0:Nlevels-1;
larr = cell2mat(arrayfun(@(l)l*ones(1,2*l+1),lrange,'UniformOutput',0));
marr = cell2mat(arrayfun(@(l)1:2*l+1,lrange,'UniformOutput',0));
% Evaluate the spherical harmonics at the needed degree & order
Phi = cell2mat(arrayfun(@(l,m) ...
               sphHarm(l,m,x),larr,marr,'UniformOutput',0));
Phieval = cell2mat(arrayfun(@(l,m) ...
                   sphHarm(l,m,xeval),larr,marr,'UniformOutput',0));
% Break Phi and Phieval into the blocks for the HSSVD
Phi1 = Phi(:,1:N);  Phi2 = Phi(:,N+1:Neig);
Phieval1 = Phieval(:,1:N);  Phieval2 = Phieval(:,N+1:Neig);
Phi2TinvPhi1 = Phi2'/Phi1';
% Choose a small shape parameter that can cause ill-conditioning
g = .01;
% Evaluate the eigenvalues, which are a function of gamma
lamvec = 4*pi*g.^larr./(2*larr+1);
```

```
22 % Form CbarT: Apply the eigenvalues to the Phi2'/Phi1' quantity
   LamFull = bsxfun(@rdivide,lamvec(N+1:Neig)',lamvec(1:N));
24 CbarT = LamFull.*Phi2TinvPhi1;
   % Form the stable basis using the eigenfunctions and CbarT
26 Psi = Phi1 + Phi2*CbarT;
   Psieval = Phieval1 + Phieval2*CbarT;
28 % Evaluate the interpolant at the desired locations
   yevalhs = Psieval*(Psi\y);
```

The assignment `Nlevels = ceil(sqrt(N)) + 3` guarantees that more eigenfunctions will be present than interpolation points N; the choice of three additional levels is arbitrary. The use of cell arrays in the evaluation of the eigenfunction matrices provides a concise mechanism for arranging the various indices present in the double summation of (3.5). This is necessary because the Mercer series underlying the Hilbert–Schmidt SVD has a different form than the series (3.5):

$$\sum_{n=1}^{\infty} \lambda_n \varphi_n(\boldsymbol{x}) \varphi_n(\boldsymbol{z}) = \sum_{\ell=0}^{\infty} \frac{4\pi\gamma^{\ell}}{2\ell+1} \sum_{m=1}^{2\ell+1} Y_{\ell,m}(\boldsymbol{x}) Y_{\ell,m}(\boldsymbol{z})$$

This problem is organizational, not mathematical, in nature and the `arrayfun` and `cell2mat` calls can identify the appropriate eigenfunctions and eigenvalues from the spherical harmonic expansion.

Recall that, from the discussion in Remark 13.5, the diagonal matrices Λ_2 and Λ_1^{-1} should be applied simultaneously which necessitates the formation of `LamFull`. Were this not the case, the formation of `LamFull` could be omitted and the memory overhead could be reduced using

```
lamvec1 = lamvec(1:N);  lamvec2 = lamvec(N+1:Neig);
CbarT = bsxfun(@(times),lamvec2',bsxfun(@rdivide,Phi2TinvPhi1,lamvec1));
```

Also, within the actual `GaussQR` script `AppxFitZonalHSSVD.m`, this interpolation and error computation takes place within a loop for multiple γ values. The eigenfunction matrices can be evaluated once outside of the interpolation loop and reused at each iteration because they do not depend on the parameter γ, but this is generally not the case as we saw for the Gaussian eigenfunctions in Chapter 13. Looking for such optimizations can greatly accelerate code, but there is also potential for mistakes if values that are believed to be constant must actually change.

Of course, much more advanced applications on spherical domains arise in geophysics and other geosciences. The interested reader can find many more details in books such as [Freeden *et al.* (1998); Michel (2013)] or papers such as [Flyer and Wright (2007); Fornberg and Piret (2008a); Flyer and Wright (2009); Flyer and Lehto (2010); Flyer *et al.* (2012); Fuselier and Wright (2013)], where approximation/interpolation problems as well as PDEs are solved using zonal kernels.

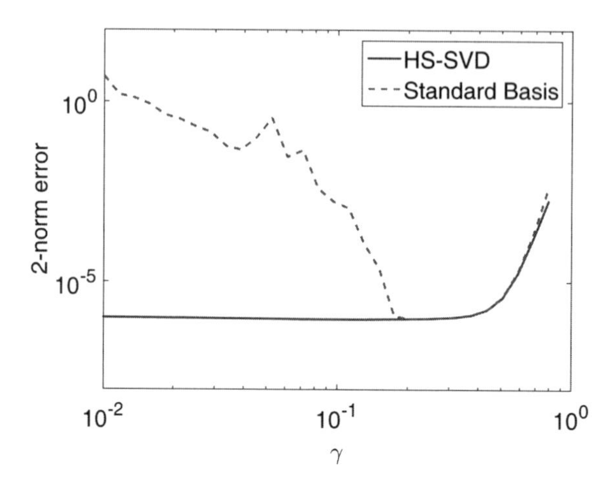

Fig. 15.4 As $\gamma \to 0$, the standard basis falters but the HS-SVD basis remains stable. These experiments used $N = 400$ maximal determinant input points, and evaluated the error on the same test points as Example 15.4.

15.4 Computational Considerations for Scattered Data Fitting

While much of this text is focused on the derivation and properties of kernel-based interpolation and approximation methods, and their subsequent development in MATLAB, the computational cost of these methods is of great significance when judging their viability in applications. We use this section to briefly consider cost-related issues for some of the methods introduced thus far as well as alternate strategies which are more cost-conscious.

The standard interpolant $s(\boldsymbol{x}) = \boldsymbol{k}(\boldsymbol{x})^T \mathsf{K}^{-1}\boldsymbol{y}$ involves three primary computational tasks: the evaluation of the objects $\boldsymbol{k}(\boldsymbol{x})$ and K, the solution of $\mathsf{K}\boldsymbol{c} = \boldsymbol{y}$ to form $\boldsymbol{c} = \mathsf{K}^{-1}\boldsymbol{y}$, and the computation of $\boldsymbol{k}(\boldsymbol{x})^T\boldsymbol{c}$ at any desired $\boldsymbol{x}$ locations. For various choices of K, the examples in Chapter 4 demonstrate our best attempts to efficiently conduct the evaluations of $\boldsymbol{k}$ and K. Furthermore, under the assumption that "K is well-conditioned and enjoys no *special structure*," the use of the "backslash" operator in MATLAB, i.e., `c=K\y`, is assumed to be as optimal a solver as is easily accessible. Similarly, without knowledge of special structure in $\boldsymbol{k}$ or $\boldsymbol{x}$, the computation of $\boldsymbol{k}(\boldsymbol{x})^T\boldsymbol{c}$ is most efficiently conducted as an inner product, or matrix-vector product, in MATLAB (see Program 4.9).

Much as the issues with the standard basis pose practical problems in computing, as discussed in Section 12.1.1, the computational picture is not so clear when one of two situations occurs:

- Evaluation of the standard interpolant is unacceptable because of ill-conditioning in the K matrix, or
- The kernel objects $\boldsymbol{k}(\boldsymbol{x})$ and K exhibit some special structure which allows for more efficient computation.

Here we briefly study these two issues and reference recent work that addresses the computational cost of these topics.

15.4.1 *The cost of computing/implementing an alternate basis*

Historically, preconditioning was the main tool available to combat ill-conditioning in K, but such a strategy is only appropriate if matrix-vector products involving K can be evaluated at less than $\mathcal{O}(N^2)$ cost, and that requires special structure. Furthermore, if the matrix K is so ill-conditioned that it appears numerically singular, then any matrix-vector products involving it will be projected into a lower-dimensional vector space, thus preventing the full space of $\mathbb{R}^N$ from being accessible to solutions. This feeling that "the damage is already done by forming K" was discussed in [Fornberg and Piret (2008b); Boyd (2010)], and thus we do not discuss preconditioning strategies here. Some references related to preconditioning were discussed in Section 12.4.

As a consequence, we restrict our cost discussion regarding the resolution of ill-conditioning to the implementation of alternate basis-type strategies as explained in Chapters 12 and 13. We study two such classes of methods that exist to replace the $\boldsymbol{k}(\boldsymbol{x})^T\mathsf{K}^{-1}$ form of the cardinal functions: those that use a full-rank basis transformation matrix as introduced in Section 12.1.3; and those that produce a low-rank approximate interpolant as introduced in Section 12.3.

The cost of forming and computing with a full-rank basis transformation matrix and a stable basis $\boldsymbol{v}(\cdot)$ varies. The Newton basis from Section 12.1.4 incurs no significant computational penalty over the standard basis since the Cholesky factorization $\mathsf{K} = \mathsf{N}\mathsf{N}^T$ to form $\boldsymbol{n}(\boldsymbol{x})^T = \boldsymbol{k}(\boldsymbol{x})^T\mathsf{N}^{-T}$ is of the same cost as computing K^{-1}. The SVD bases described in Section 12.1.5 are more costly than the simplest methods of inverting K^{-1} because the full SVD of the symmetric positive definite matrix K costs between 7 and 20 times more than a simpler decomposition such as Cholesky or LU [Golub and Van Loan (2012)].

Similarly, the HS-SVD stable basis introduced in Chapter 13 costs more to implement than the standard basis, and the cost is unknown *a priori* because the necessary number of eigenfunctions, M, depends on both the data and the kernel (see Section 13.4.1). Even once the appropriate $\Phi = \begin{pmatrix}\Phi_1 & \Phi_2\end{pmatrix}$ matrices are formed, the cost of computing $\bar{\mathsf{C}}^T = \Lambda_2\Phi_2^T\Phi_1^{-T}\Lambda_1^{-1}$ (which is often the main expense when using the HS-SVD) may be further magnified if the QR factorization is applied as is discussed in Section 13.6. Even in our best implementations available in `GaussQR`, the cost of computing with the HS-SVD requires a factor of 10 greater than working with the standard basis. As a result, we do not suggest its use when working with only moderately flat kernels, or when working with a problem where other stabilization schemes such as, e.g., the smoothing splines from Section 15.1, work well.

Unlike the high cost of computing/implementing many basis transformation matrices, low-rank approximate interpolants can, at times, require less cost than computing K^{-1}. One low-rank basis can be generated with the *adaptive Newton basis* scheme implemented in Program 12.2, and this can be formed at cost less than $\mathcal{O}(N^3)$ [Pazouki and Schaback (2011)]. Similarly, if an analytic form of the Hilbert–Schmidt eigenfunctions is known, then "forming" the eigenfunction basis matrix Φ is of no cost. Furthermore, the low-rank eigenfunction approximate interpolant coefficients $\boldsymbol{a} = \Phi^\dagger \boldsymbol{y}$ can potentially be evaluated at $\mathcal{O}(MN^2)$ complexity (the cost of computing the QR factorization of an $N \times M$ matrix) allowing for a faster solution than computing $\mathsf{K}^{-1}\boldsymbol{y}$ at complexity $\mathcal{O}(N^3)$. Thus, if the quality of the M-term low-rank approximate interpolant is acceptable and $M \ll N$, the low-rank approximate interpolant could prove to be a cost-effective approximation scheme.

15.4.2 *Exploiting structure in kernel computations*

Unlike instability issues which often cause significant increases in computational cost, the existence of special structure in the kernel K or kernel matrix K provides an opportunity to reduce the cost of evaluating, e.g., $s(\boldsymbol{x}) = \boldsymbol{k}(\boldsymbol{x})^T \mathsf{K}^{-1}\boldsymbol{y}$. Exploiting the most common forms of kernel structure, such as tensor product or radial kernels, is discussed in Section 4. That discussion focuses on the mechanisms for computing $\boldsymbol{k}$ or K as efficiently as possible in MATLAB.

A more complicated problem, and currently the topic of significant research, is the computation of $\mathsf{K}^{-1}\boldsymbol{y}$ and $\boldsymbol{k}(\boldsymbol{x})^T\boldsymbol{c}$ by exploiting matrix structure. The first strategy, and perhaps the strategy closest to popular tools such as finite elements, involves the use of *compactly supported kernels* such as those mentioned in Section 3.5. Computational tools exist for efficient computation with sparse matrices for RBF networks [McCourt *et al.* (2015)], and can even be used for kernels which traditionally produce dense matrices [Yokota *et al.* (2010)].

On the other hand, one can design more efficient methods for dealing with global kernels. Most such approaches are either *iterative* in nature, such as, e.g., the work in [Beatson *et al.* (1999); Faul and Powell (1999); Faul *et al.* (2005); Gumerov and Duraiswami (2007)], or aim at using fast summation techniques such as *fast tree codes* and *fast multipole methods* (see, e.g., [Beatson and Newsam (1992, 1998); Cherrie *et al.* (2002); Krasny and Wang (2011); Deng and Driscoll (2012); Chen *et al.* (2014a)]) or *domain decomposition* techniques [Beatson *et al.* (2001)]. Virtually every one of the iterative approaches to working with global kernels cited above employ approximations of the cardinal (or Lagrange) basis functions which are obtained by performing interpolation on only a small (local) subset of the data. Some of those ideas are also covered in [Fasshauer (2007, Chapters 34 and 35)]. In Section 12.1.2 we discussed the general idea of using cardinal basis functions for interpolation from $\mathcal{H}_K(\mathcal{X})$.

Recent work of note is the adaptation of structured/recursively low-rank matrices, alternately referred to as hierarchically semiseparable matrices or $\mathcal{H}$-matrices in the linear algebra literature [Hackbusch (1999); Vandebril *et al.* (2008); Xia *et al.* (2010); Li *et al.* (2012); Xia *et al.* (2012)], to positive definite kernels. These matrices take advantage of off-diagonal low-rank blocks on which operations can be performed at reduced cost. Because of the decay of positive definite kernels $K(\boldsymbol{x}, \boldsymbol{z})$ for values of $\boldsymbol{x}$ increasingly far from $\boldsymbol{z}$, many kernel matrices enjoy this property. Promising results from Chen (2014) suggest this may provide significant benefits to the kernel community.

Chapter 16

Computer Experiments and Surrogate Modeling

16.1 Surrogate Modeling

For the longest time, scientific research has been classified as being either theoretical or experimental in nature. More recently, however, *computer simulation* (or *computer experimentation*) has emerged as a third pillar of science. Sometimes computer experimentation is the only way to obtain an understanding of a physical process, and sometimes it is just considerably cheaper, safer, ethically more responsible or environmentally friendlier than performing an actual physical experiment. Computer simulation arises quite naturally in situations where one has a complex mathematical model (e.g., specified in terms of a system of partial differential equations) and one can use a computational approach to produce an output (or *response*) based on a given set of initial conditions and/or model parameters. A few examples of typical computer simulations are weather and climate modeling, drug design, financial risk management or the simulation of nuclear reactors. Another example arises within the context of this book, where in Remark 17.1 we suggest the possibility of using a surrogate model to reduce the cost of the search for the minimum of a complex cross-validation surface that is otherwise only accessible pointwise via relatively expensive evaluations of the cross-validation criterion.

In this chapter we are not interested in the actual computer simulation procedure itself, i.e., we treat this process as a black box that produces an accurate output which we will use as input for our *surrogate model*. If the computer simulation indeed corresponds to the solution of a system of PDEs, then many methods exist in the numerical analysis literature such as finite element, finite difference, and finite volume methods[1]. Instead, our main concern here — and the motivation for the surrogate modeling approach — is the fact that the computer simulation can be very costly, i.e., take a long time or require the use of expensive high-performance computing equipment.

The main task in surrogate modeling is to employ a scattered data fitting approach to produce a surrogate for such a computer simulation. This surrogate will

[1] In Chapter 20 we will discuss how kernels can themselves be used for the numerical solution of PDEs.

allow us to quickly and accurately predict the output of the computer simulation for a previously not used combination of input parameters. Here each data point for the scattered data fitting (surrogate modeling) problem consists of a combination of the input parameters for the computer simulation, and the corresponding data value is given by the output obtained from running the computer simulation with these inputs. In other words, one needs to run the computer simulation multiple times in order to produce a set of data, and then one can use a scattered data fitting technique to produce a *response surface* which in turn can be evaluated efficiently and accurately for any desired set of input parameters. The number of these input parameters determines the dimensionality of the data fitting problem, and it is not difficult to imagine that this dimension can be rather large. For example, Salemi (2014) constructs a surrogate model for a 75-dimensional multi-product Jackson network using 250,000 data points.

Example 16.1. (Benefits of surrogate modeling)
A computer simulation can be used to perform a "what if" analysis, i.e., it can be used to evaluate whether a process might be improved by tuning its parameters appropriately. Tongarlak *et al.* (2008) describe how they used a computer simulation of a fuel injector production line as a test bed for candidate designs without the need for their client having to invest in new equipment. For this specific example, the computer simulation of a single new scenario took approximately eight hours to run. On the one hand, this provides a vast improvement (both in time and cost) over having to configure and install an entire production line. But on the other hand, this simulation tool is still rather impractical for direct decision-making. Therefore, we could instead run the simulation for a series of carefully chosen scenarios called *design points*, and then we can use the simulation output at these design points as data to construct a surrogate model. This surrogate model can then be used in place of the computer simulation model, so that we can perform the "what if" analysis more efficiently.

Once an accurate surrogate model has been obtained, it is frequently used to optimize the response (as, e.g., with the cross-validation example discussed in Remark 17.1). We will not discuss this general optimization problem here, but instead refer the interested reader to, e.g., [McDonald *et al.* (2007)]. According to Fang *et al.* (2006), surrogate models are useful for preliminary studies simply based on visualization of the response surface, for sensitivity analysis and as a form of dimension reduction tool where one considers the functional relationship between possibly a very large set of input parameters and the response and aims to identify possibly a much smaller subset of input parameters which have the most significant impact on the response. Surrogate models also play an important role in uncertainty quantification where one is interested in the effect of input uncertainty on the variability of the response. Finally, Forrester *et al.* (2008) suggest even more applications of surrogate modeling, such as their use as a calibration and bridging mechanism for computer simulation codes of varying accuracies (e.g., to create a multi-level or multi-scale framework for solving the Navier–Stokes equations based on low-resolution Euler codes as well as high-resolution large eddy simulations), or their use as a way to deal with noisy or missing data.

Alternate names for surrogate modeling that frequently appear in the computer experiment literature are *response surface modeling*, *metamodeling* or *modeling with emulators*. Among the representations of these response surfaces one can find, e.g., (quadratic) polynomial approximations [Box and Wilson (1951)], radial basis functions [McDonald *et al.* (2007); Haaland and Qian (2011)], and kriging methods [Krige (1951); Matheron (1965); Sacks *et al.* (1989); Morris *et al.* (1993)], which are probably the most popular of the bunch. A paper that surveys several alternative approaches is [Jones (2001)], and some books on the subject are [Santner *et al.* (2003); Fang *et al.* (2006); Box and Draper (2007); Forrester *et al.* (2008)].

16.2 Experimental Design

An important issue is not only the surface fitting problem itself, but also the choice of *design*, i.e., the data locations or input parameter settings, and *model parameters* (such as, e.g., shape and smoothness parameters in the kernels used for the fit). We postpone this aspect of the surrogate modeling approach until Chapter 17, but emphasize that there is a clear-cut difference between computer experimentation and physical experimentation. Computer experiments produce a deterministic output, and so common techniques used to ensure the reliability of physical experiments such as randomization, blocking or replication are not required in the computer experimentation setting. Here, the main objectives for finding a “good” design are that it successfully deal with the *curse of dimensionality*, i.e., that it effectively fill out the high-dimensional design space, and that it ensure an accurate solution of the scattered data fitting problem.

In the examples described below — as is also true in many realistic surrogate modeling applications — we are free to decide where and how often we want to sample the computer experiment to generate the data required for our surrogate model. As mentioned above, we will not go into the details of how one might choose these samples in the best possible way. Instead, we will apply a few standard sampling techniques provided in MATLAB's Statistics and Machine Learning Toolbox, such as *Latin hypercube designs* [McKay *et al.* (1979)] or *quasi-random designs* (such as Halton (1960) or Sobol′ (1967) sets) and use them to illustrate the surrogate modeling approach. Other popular designs in the statistics literature look at so-called *D-optimality*, *A-optimality* or *E-optimality*, i.e., they aim to maximize the determinant of the Gram matrix $\mathsf{K}^T\mathsf{K}$, or minimize the trace or the smallest eigenvalue of its inverse, respectively (see, e.g., [Morris *et al.* (1993); Fang *et al.* (2006)]). In approximation theory, so-called (approximate) *Fekete points* or *maximal determinant points* are analogous to D-optimal designs, i.e., they maximize the determinant of the interpolation matrix K (see, e.g., [Womersley and Sloan (2001); De Marchi (2003); Briani *et al.* (2012)]). Since Fekete points also minimize the Lebesgue constant other points that have a similar effect are also studied in approximation theory. For kernel methods, De Marchi (2003) constructs so-called (approximate) *Leja points*. Moreover, *Padua points* [Caliari *et al.* (2008)] and *weakly admissible meshes* [Bos *et al.* (2011b)] might also be good choices for kernel methods — even though these designs are constructed with the goal of being (near-)optimal for low-dimensional polynomial interpolation methods. An additional discussion of different designs can be found in Appendix B.

16.3 Surrogate Models for Standard Test Functions

Many standard test functions used in the surrogate modeling literature can be found in the online library [Surjanovic and Bingham (2014)]. We now pick two of those test functions and apply our kernel-based data fitting techniques. Since the data for these examples are generated by known functions we will be able to judge the accuracy of our method.

One difficulty that arises in using kernels for surrogate modeling is the need to choose appropriate shape parameters. As discussed in Section 9.5 in the context of dimension-independent error bounds, higher-dimensional problems require anisotropic shape parameters to function effectively. While seven or eight dimensions can hardly be considered "high-dimensional," this already presents enough dimensions that we will only consider the use of *anisotropic kernels*, which were introduced in Section 3.4 and whose implementation based on the use of the vector $\boldsymbol{\varepsilon}$ of shape parameters we discussed in Section 4.1.3.

To ease the selection of shape parameters, we will *rescale* all problems from their physical domain into $[0, 1]^d$ with a linear transformation. This has no effect on the function being modeled but it absolves us from having to find an appropriate $\boldsymbol{\varepsilon}$ for data which varies between $[0.01, 0.1]$ in one dimension and $[10000, 220000]$ in another dimension.

16.3.1 *Piston simulation function*

The *piston simulation function*, defined via the following chain of nonlinear functions,

$$C(\boldsymbol{x}) = 2\pi\sqrt{\frac{W}{k + S^2 \frac{P_0 V_0}{T_0} \frac{T_a}{V^2}}}, \tag{16.1}$$

where

$$V = \frac{S}{2k}\left(\sqrt{A^2 + 4k\frac{P_0 V_0}{T_0} T_a} - A\right) \quad \text{and} \quad A = P_0 S + 19.62M - \frac{kV_0}{S},$$

models the circular motion of a piston within a cylinder (see, e.g., [Ben-Ari and Steinberg (2007); Kenett *et al.* (2014)]). The input variables $\boldsymbol{x} = (W, S, V_0, k, P_0, T_a, T_0) \in \mathbb{R}^7$ and their usual input ranges are:

$$
\begin{aligned}
W &\in [30, 60], && \text{piston weight (kg)},\\
S &\in [0.005, 0.020], && \text{piston surface area (m}^2\text{)},\\
V_0 &\in [0.002, 0.010], && \text{initial gas volume (m}^3\text{)},\\
k &\in [1000, 5000], && \text{spring coefficient (N/m)},\\
P_0 &\in [90000, 110000], && \text{atmospheric pressure (N/m}^2\text{)},\\
T_a &\in [290, 296], && \text{ambient temperature (K)},\\
T_0 &\in [340, 360], && \text{filling gas temperature (K)}.
\end{aligned}
$$

The response C is the time it takes to complete one cycle, in seconds.

Remark 16.1. An interesting point about this function is that it can be phrased as a function of six variables instead of seven: T_a and T_0 only appear jointly in the quantity T_a/T_0. Although we ignore this simplification in our implementation, such insights, either analytically/asymptotically or through the physical properties of the system, can improve the quality of the surrogate model, or at least help decrease the potential for significant problems.

Remark 16.2. Unlike other instances where test functions can be defined using a single line of MATLAB— think `yf = @(x) cos(2*pi*x)` from Program 4.9 — many functions of interest are too complicated to be written inline effectively. One should probably look at the piston cycle function (16.1) and think that C is complicated enough to warrant a separate M-file for its computation. Indeed, to call functions in a MATLAB script they must be defined in a separate function with the same title as the name of the function that is being called. Therefore, to define a function `DifficultFunc()` one would need a file `DifficultFunc.m` in which that function is defined.

For the piston cycle function C defined in (16.1), such a file might look as follows:

Program 16.1. Example of possible `piston.m`

```
function C = piston(W,S,V0,k,P0,Ta,T0)
% function C = piston(W,S,VO,k,PO,Ta,TO)
lb = [30,.005,.002,1000,90000 ,290,340];
ub = [60,.020,.010,5000,110000,296,360];
if any(any(bsxfun(@lt,[W,S,V0,k,P0,Ta,T0],lb) |...
           bsxfun(@gt,[W,S,V0,k,P0,Ta,T0],ub)))
    error('At least one input value outside acceptable range')
end
A = P0.*S + 19.62*W - k.*V0./S;
PVTT = P0.*V0.*Ta./T0;
```

```
V = S./(2*k).*(sqrt(A.^2 + 4*k.*PVTT) - A);
C = 2*pi*sqrt(W./(k+(S./V).^2.*PVTT));
end
```

The function accepts seven arguments and returns the function value at those arguments. Because of the use of the `.*` notation, matrices of the same size can all be passed and a matrix of that size will be returned. Repeating the calling sequence as a comment immediately below the function definition line allows the user to call `help piston` to view the calling sequence; the same is true for any other comments immediately following the function definition.

The `lb` and `ub` vectors contain the lower and upper bounds, respectively, of the acceptable input ranges for the C function. Immediately after their definition, we test to make sure that all of the values passed to this function fall in those ranges; if any are unsatisfactory we throw an `error` alerting the user to that fact. Because the `any` command operates on a matrix by testing all the values in each column, it must be called twice to check if every value passed to `piston` was acceptable.

Rather than having numerous M-files containing function definitions, we have encapsulated many of the more complicated functions into the `pickfunc` function in `GaussQR`. Explore that file to learn about the various test functions available.

Example 16.2. (A surrogate model for the 7D piston cycle function)
In this example we study the convergence behavior of surrogate models based on Matérn kernels with different smoothness and different shape parameter vectors $\boldsymbol{\varepsilon}$. Program 16.2 forms these surrogate models using samples at increasingly many *Halton points*. When working with very high-dimensional problems, the Halton points can present difficulties, as discussed briefly in Appendix B.1. Since this model, however, contains only seven dimensions there are no sampling problems.

We choose some data points for our problem and evaluate the function to be modeled. A cell array is created to allow multiple kernels (all radial in this case, for simplicity) to be tested with a single `cellfun` call.

Program 16.2. `SurrModelPiston.m`

```
% Define the function of interest, available in the GaussQR library
yf = pickfunc('piston_scaled');
% Halton points generator function
point_generator = haltonset(7,'Skip',1);
haltpts7D = @(N) net(point_generator,N);
% Define some (potentially anisotropic) kernels to test
rbfM0 = @(r) exp(-r);
rbfM2 = @(r) (1+r).*exp(-r);
rbfM4 = @(r) (1+r+r.^2/3).*exp(-r);
% Select a shape parameter vector
ep = 3.^(2:-.5:-1);
% Choose numbers of points for data and testing
convN = 10;
Nvec = floor(logspace(2,4,convN));
```

```
    Neval = 500;
16  % Organize the kernels and their names in cell arrays
    rbfarr = {rbfM0,rbfM2,rbfM4};
18  labelarr = {'C0 Matern','C2 Matern','C4 Matern'};
    % Create a waitbar to advise the user of the progress
20  h_waitbar = waitbar(0,'Initializing surrogate model','Visible','on');
    % Loop through the desired N values and test the model
22  errmat = zeros(length(rbfarr),length(Nvec));
    k = 1;
24  for N=Nvec
        waitbar((k-1)/testN,h_waitbar,sprintf('Preparing model, N=%d',N));
26      % Generate enough points; split them into data and evaluation points
        Ntot = N + Neval;
28      xtot = haltpts7D(Ntot);
        xeval = xtot(1:Neval,:);
30      x = xtot(Neval+1:end,:);
        % Evaluate the function to create the data and test values
32      y = yf(x);  yeval = yf(xeval);
        % Compute the anisotropic distance matrices
34      DM = DistanceMatrix(x,x,ep);
        DMeval = DistanceMatrix(xeval,x,ep);
36      % Fit the surrogate model for all the kernels at once
        waitbar((k-1)/testN,h_waitbar,sprintf('Computing model, N=%d',N));
38      errmat(:,k) = cellfun(@(rbf) ...
                          errcompute(rbf(DMeval)*(rbf(DM)\y),yeval),...
40                    rbfarr);
        k = k + 1;
42  end
    waitbar(1,h_waitbar,'Plotting piston convergence');
44
    h = figure;
46  loglog(Nvec,errmat,'linewidth',2)
    xlabel('number of input points')
48  ylabel('relative RMS 2-norm error')
    legend(labelarr,'location','northeast')
50
    close(h_waitbar) % Close the waitbar after the work is done
```

Convergence plots when tested with various shape parameters are provided in Figure 16.1.

Among the many difficulties when computing with seven-dimensional anisotropic kernels is the need to "correctly" choose the seven values in the shape parameter vector $\boldsymbol{\varepsilon}$. As we have seen in various ε profiles (for example, Figure 10.2), the shape parameter can greatly affect the quality of the kernel fit. The plots in Figure 16.1 seem to offer further confirmation of that fact and suggest that strategies such as cross-validation (see Section 14.2 or [Forrester *et al.* (2008, Chapter 2.1.2)]) are fundamental to the success of surrogate modeling in higher dimensions. In subplots (b)–(d), the vector $\boldsymbol{\varepsilon}$ contains the same set of values, but they have been shuffled among dimensions, leading us to the conclusion that enforcing anisotropy (the dimension-independent error strategy from Section 9.5) is most effective when the "correct" dimensions are emphasized.

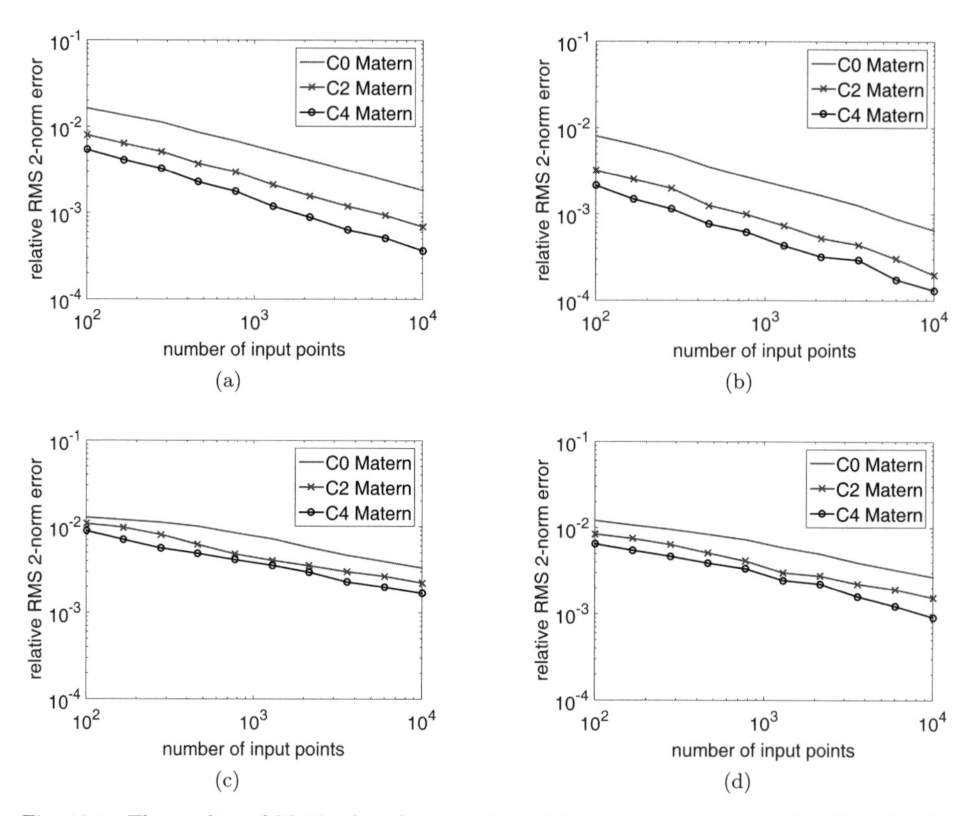

Fig. 16.1 The quality of Matérn-based surrogate model convergence, at least for $N < 10000$, is dependent on the choice of ε, though not in a predictable way. (a) `ep=4*ones(7,1)`; (b) `ep=3.^(2:-.5:-1)`; (c) `ep=[1.73 .33 9 .58 5.2 3 1]`; (d) `ep=[5.2 3 .33 1 9 1.73 .58]`.

The use of the relative RMS 2-norm for computing error is reasonable, but sometimes an analysis of which model values are most/least accurate is necessary. This is especially true when choosing the test set adaptively to ensure that a wide range of simulation operations is effectively modeled by the surrogate. Figure 16.2 displays the predictions from the C^4 Matérn surrogate with `ep = 3.^(2:-.5:-1)` (which appears in Figure 16.1(b)) and compares them to the test data.

We see that, as the number of points used to train the model increases, more of the predicted data falls on the line $y = x$, indicating that the predictions match the test values. Results that lie above the line suggest that the predicted value was

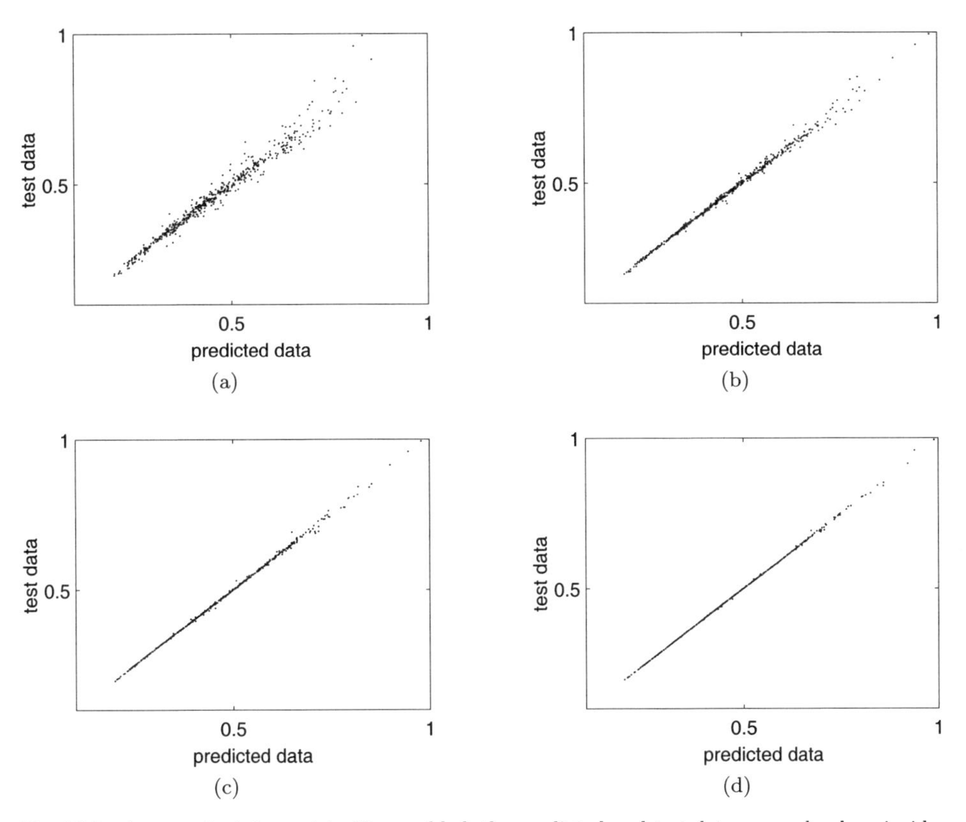

Fig. 16.2 As more training points N are added, the predicted and test data more closely coincide, indicating that the model is performing appropriately. (a) $N = 100$, (b) $N = 400$, (c) $N = 1500$, (d) $N = 5000$.

below the test value, and results below the line indicate an overestimation of the true value. Because almost all the test results fall on $y = x$ for $N = 5000$, we can conclude that little is to be gained from increasing N with this test data and that additional test points should be considered before adding more training data.

Remark 16.3. Program 16.2 marks the only appearance of the `waitbar` in the text of this book. This is a useful tool for professionally presenting a program's progress to the user if they are working in a visual environment. The option `'Visible','off'` can be passed to the instantiation of the waitbar to prevent a window from opening if a script is being run remotely and no figure window is desired. Within `GaussQR`, many example scripts take advantage of the waitbar to assuage user fears about a lack of visible progress, but we will omit their further presence from code presented in this book to save space and avoid redundancy.

16.3.2 *Borehole function*

The *borehole function*

$$f(\boldsymbol{x}) = \frac{2\pi T_u(H_u - H_\ell)}{\ln\left(\frac{r}{r_w}\right)\left(1 + \frac{2LT_u}{\ln\left(\frac{r}{r_w}\right) r_w^2 K_w} + \frac{T_u}{T_\ell}\right)}$$

models the flow of water through a borehole in an impermeable layer of rock separating an upper from a lower aquifer. This model follows from Bernoulli's law if one assumes that the flow is steady-state laminar and isothermal. The simplicity and quick evaluation of this function make it a commonly used function for testing a wide variety of methods in computer experiments (see, e.g., [Morris *et al.* (1993); An and Owen (2001); Fang *et al.* (2006)]).

The input variables $\boldsymbol{x} = (r_w, r, T_u, H_u, T_\ell, H_\ell, L, K_w) \in \mathbb{R}^8$ and their usual input ranges are:

$$\begin{aligned}
r_w &\in [0.05, 0.15], && \text{radius of borehole (m)},\\
r &\in [100, 50000], && \text{radius of influence (m)},\\
T_u &\in [63070, 115600], && \text{transmissivity of upper aquifer (m}^2\text{/yr)},\\
H_u &\in [990, 1110], && \text{potentiometric head of upper aquifer (m)},\\
T_\ell &\in [63.1, 116], && \text{transmissivity of lower aquifer (m}^2\text{/yr)},\\
H_\ell &\in [700, 820], && \text{potentiometric head of lower aquifer (m)},\\
L &\in [1120, 1680], && \text{length of borehole (m)},\\
K_w &\in [9855, 12045], && \text{hydraulic conductivity of borehole (m/yr)}.
\end{aligned}$$

The output is the water flow rate, in m^3/yr. As is the case for the piston function presented in Section 16.3.1, this flow function f could be considered a function of as few as 6 variables: H_u and H_ℓ only appear in the form $H_u - H_\ell$ and L and K_w only appear in the form L/K_w.

For this borehole function we construct a surrogate model and plot the error of the surrogate model in pairwise contour plots where each dimension is compared to the others. Program 16.3 presents the code for fitting a surrogate model with inverse quadratic kernels using 2000 data points and some of the necessary code for the plot in Figure 16.3; see the `GaussQR` library for the full details.

Program 16.3. Excerpts from `SurrModelBorehole.m`

```
% Define the function of interest, available in the GaussQR library
yf = pickfunc('borehole_scaled');
% Halton points generator, refer to Program 16.2
haltpts8D = @(N) haltonseq(N,8);
% Define the kernel for modeling, in its anisotropic form
rbfIQ = @(r) 1./(1+r.^2);
```

```
% Select an anisotropic shape parameter vector
ep = logspace(1,-1,8);
% Create the data
N = 2000;  Neval = 500;
xtot = haltpts8D(N+Neval);  xeval = xtot(1:Neval,:);  yeval = yf(xeval);
x = xtot(Neval+1:end,:);  y = yf(x);
% Train the surrogate model and evaluate the pointwise errors
K = rbfIQ(DistanceMatrix(x,x,ep));
Keval = rbfIQ(DistanceMatrix(xeval,x,ep));
seval = Keval*(K\y);
errs = (seval - yeval)/norm(yeval)*sqrt(Neval);

% Define smoothing spline to average errors
muSS = 1e-2;  epSS = 2;
rbfM4 = @(r) (1+r+r.^2/3).*exp(-r);
Npx = 34;  Npy = 35;
xinterp = pick2Dpoints([0 0],[1 1],[Npx Npy]);
% Loop through the dimensions and create contour plots of the averaged errors
for i=1:7
    for j=i:7
        dims = [i,j+1];
        Kdata = rbfM4(DistanceMatrix(xeval(:,dims),xeval(:,dims),epSS));
        Kinterp = rbfM4(DistanceMatrix(xinterp,xeval(:,dims),epSS));
        zinterp = Kinterp*((Kdata + muSS*eye(Neval))\errs);

        X = reshape(xinterp(:,1),Npx,Npy);
        Y = reshape(xinterp(:,2),Npx,Npy);
        Z = abs(reshape(zinterp,Npx,Npy));
        subplot(7,7,i+7*(7-j)+(i-1)*7)
        contourf(X,Y,Z,[0,.03,.06,.1],'linewidth',1.5);
    end
end
```

Some parameter pairs see no concentration of inaccuracy: the r_w vs. H_ℓ plot shows a flat reading of errors less than 0.03. Other plots show regions where the error is strongly clustered, such as the bottom/right-most graph which denotes inaccurate modeling with a large white region in its corner. For that region of parameters, all of the model predictions are significantly off, which, perhaps, suggests that we must sample more data with small L and large K_w values to improve the quality of the model.

Remark 16.4. In line 30 we compute the 2D projection of the errors at enough points to create a contour plot. We do this with the use of a smoothing spline from Section 15.1 to help average the behavior of the error. This prevents single values of high or low accuracy from overwhelming the projection and produces plots that more easily display the parameter regions that we have a hard time modeling well.

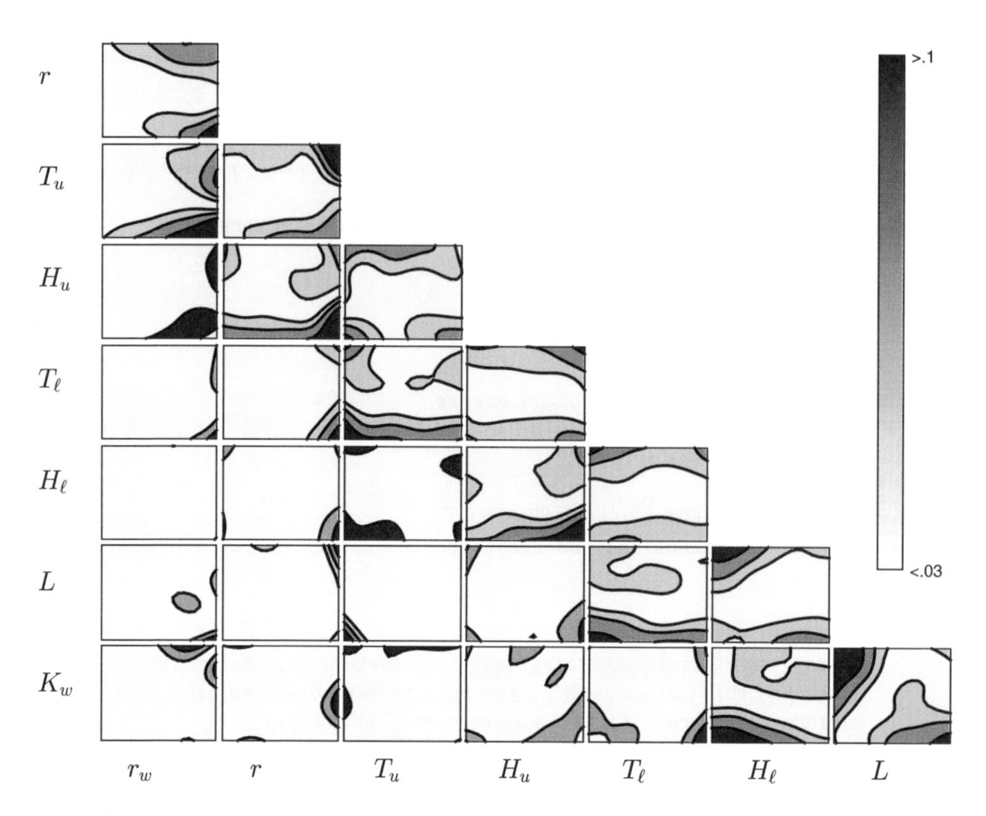

Fig. 16.3 After forming the surrogate model, the pointwise error is evaluated at the test points. A filled contour plot is created for each pair of dimensions, e.g., r_w and K_w in the bottom left plot, showing the error as influenced by those dimensions by averaging (through smoothing splines) in the other dimensions. Contours occur at $\{0.03, 0.06, 0.1\}$ relative error.

16.4 Modeling From Data

Although the ideal setting for surrogate modeling involves the ability to independently choose the input parameters for our experiments, many situations are not so accommodating. We may have to generate a surrogate model without the flexibility of choosing the experimental design, which may yield difficulties in high-dimensional spaces where only subsets of the full space are actually explored. This section considers such a surrogate model from data and how we conduct the associated fitting and parameter study.

Example 16.3. (A surrogate model for 4D car data)
The data set `carsmall` from MATLAB's Statistics and Machine Learning Toolbox contains historical data about cars from 1970, 1976, and 1982. In this example we study the impact of the attributes `Acceleration`, `Displacement`, `Horsepower` and `Weight` on a car's fuel efficiency (indicated by its `MPG` value). In other words, we will construct a surrogate model which relates the input parameters

$\boldsymbol{x} = (\texttt{Acceleration}, \texttt{Displacement}, \texttt{Horsepower}, \texttt{Weight}) \in \mathbb{R}^4$ to the response `MPG`. Although the original data in four dimensions appears in the natural physical units, e.g., pounds for weight, we use the GaussQR function `rescale_data` to convert our problem to $[-1, 1]^4$ as well as remove `NaN` values. Once we eliminate the records with incomplete data from the dataset we are left with $N = 93$ data points. The implementation of this and the remaining computations in this example are provided in `SurrModelCarSmall.m` in the `GaussQR` library, but it is omitted here to save space.

For this example, we consider a surrogate model using Gaussian kernel interpolation with $\varepsilon = 4$. Although this value was chosen in an ad hoc manner, we explore applicable parametrization strategies in Chapter 14 with examples in Chapter 17. To visualize these results, we plot cross-sections of the surrogate model in Figure 16.4(a) where two of the parameters are fixed and the other two vary over the input region. We also provide *parallel coordinates* plots of the surrogate model in Figure 16.4(b).

These plots are made possible by the presence of a four-dimensional surrogate model which can be evaluated throughout the domain. Using them we can explore the entire region of $[-1, 1]^4$, in our scaled coordinated. Unfortunately, the lack of ability to design the experiment and sample at desired locations has detrimental effects on the quality of the model.

Specifically, large regions exist in the plots of Figure 16.4(a) where the response surface predicts nearly 0 MPG despite the fact that the minimum observed MPG in the data is 9. This is simply a function of the lack of data in that region producing inaccuracy in the response surface. In the statistical setting discussed in Chapter 5, we might interpret this as the result of having very high kriging variance at these locations because of the distance from the nearest given data location. This lack of data presents itself in the plots of Figure 16.4(b) where no connections occur between, e.g., high acceleration potential and high displacement because no MPG value between 9 and 55 is observed on the response surface with those related coordinates.

16.5 Fitting Empirical Distribution Functions

The field of nonparametric density estimation has a rich history in the statistical literature [Good and Gaskins (1980); Silverman (1982, 1986); Scott (2009, 2012)]. For a d-dimensional random variable X, the *cumulative distribution function* (CDF) F_X is defined as

$$F_X(\boldsymbol{x}) = P(X_1 \le x_1, \ldots, X_d \le x_d),$$

and is needed for computing quantities of X, e.g., $\mathbb{E}[h(X)] = \int_{\mathbb{R}^d} h(\boldsymbol{x})\, \mathrm{d}F_X(\boldsymbol{x})$. For continuous random variables, the *probability density function* (PDF) is defined as

$$f_X(\boldsymbol{x}) = \mathcal{D}_{x_1} \cdots \mathcal{D}_{x_d} F_X(\boldsymbol{x}), \tag{16.2}$$

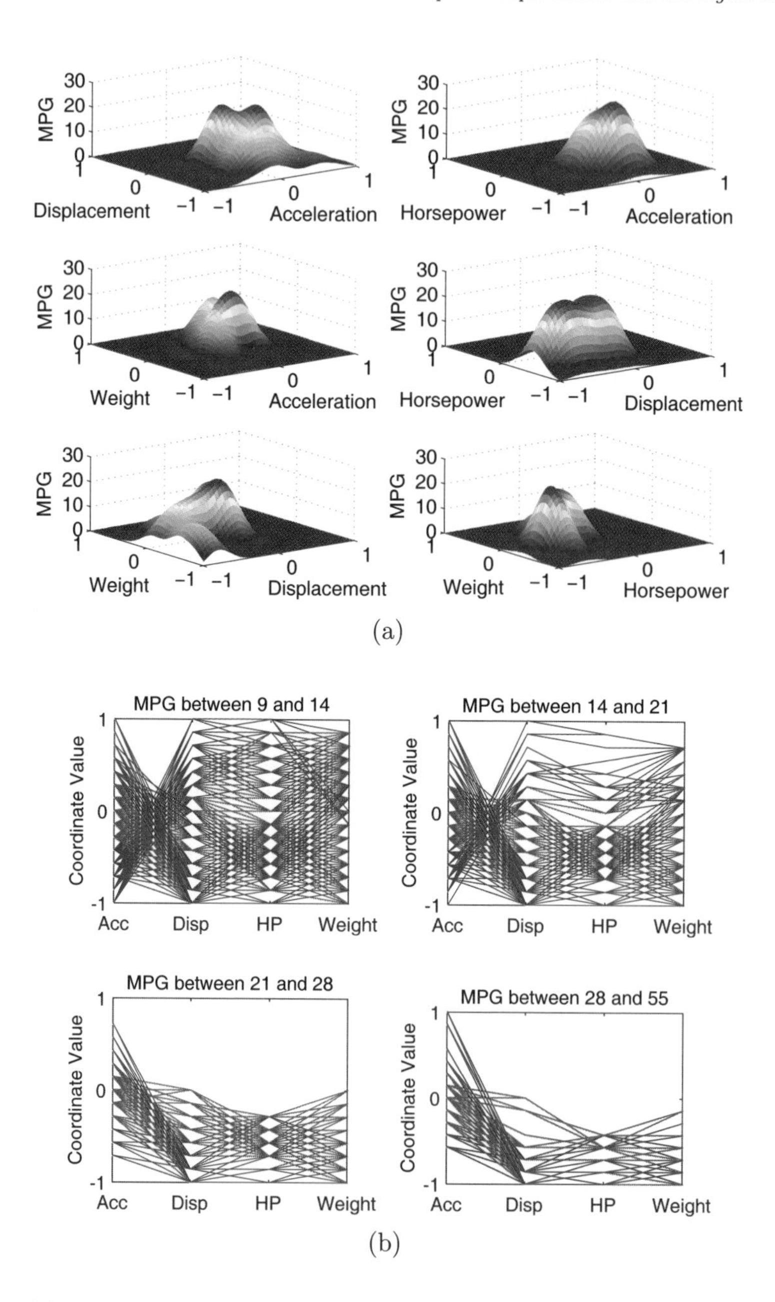

Fig. 16.4 (a) Each plot contains a cross-section of the surrogate model with two inputs allowed to vary and two fixed at the mean of the input data. The means for acceleration, displacement, horsepower and weight are -0.1336, -0.3708, -0.2922 and -0.2050 in the scaled coordinates, or 15.19, 201.4, 109.3 and 2963 in the original physical coordinates. (b) Each line in these graphs relates a single MPG value to a set of inputs. While these graphs can be presented in various ways, the key is to notice that they look different depending on the MPG range being studied. Specifically, high MPG seems to require lower displacement, horsepower and weight. The colors correspond to acceleration values less than -0.33, between -0.33 and 0.33, and greater than 0.33.

where $\mathcal{D}_{x_i}$ denotes the (partial) derivative with respect to the ith component of $\boldsymbol{x}$. We assume that any f_X, and thus F_X, of interest to us is continuous, although discontinuous densities are also important [Cline and Hart (1991)].

Our immediate focus in this section is to create a surrogate model of the cumulative distribution function. Doing so serves two purposes: we can use the CDF to evaluate the probability of certain events (because $P(a < X \leq b) = F_X(b) - F_X(a)$), and we may differentiate the CDF to produce a response surface that approximates the PDF. The idea of differentiating an interpolant is discussed in Section 19.1, but we use the computations discussed in that section here to create an approximate PDF.

We do not truly have the ability to sample from F_X in most settings, leaving us scrambling for data with which to create a surrogate model. The "data" in this setting is generally N independent samples of the random variable X, $\mathcal{X} = \{\boldsymbol{x}_1, \ldots, \boldsymbol{x}_N\}$. These could be N measurements of the time required to hail a cab, or N measurements of student heights at a school. These $\boldsymbol{x}$ values are so labeled because, despite being the only measurements of a given experiment, they play the role of the data locations. Data values $\boldsymbol{y} = \begin{pmatrix} y_1 & \cdots & y_N \end{pmatrix}$ at those locations are generated to mimic the CDF using the *empirical distribution function* (EDF) $\hat{F}_{\mathcal{X}}$. The EDF is dependent on the samples $\mathcal{X}$ and has the relatively simple form,

$$\hat{F}_{\mathcal{X}}(\boldsymbol{x}) = \frac{1}{N} \sum_{i=1}^{N} \mathbf{1}_{\boldsymbol{x}_i}(\boldsymbol{x}) \tag{16.3}$$

where $\mathbf{1}_{\boldsymbol{x}_i}$ is an indicator function checking that no component of the input vector is larger than the index $\boldsymbol{x}_i$,

$$\mathbf{1}_{\boldsymbol{z}}(\boldsymbol{x}) = \begin{cases} 1, & x_1 \leq z_1, \text{ and } x_2 \leq z_2, \text{ and ... and } x_d \leq z_d, \\ 0, & \text{else.} \end{cases}$$

Evaluating the empirical distribution function at the data locations produces the data values: $\boldsymbol{y}_i = \hat{F}_{\mathcal{X}}(\boldsymbol{x}_i)$, $i = 1, \ldots, N$.

MATLAB code for evaluating the EDF is displayed in Program 16.4, where random data is generated and the associated EDF is plotted. This program demonstrates that, through the use of MATLAB's multidimensional linear algebra structure, $\hat{F}_{\mathcal{X}}$ can be evaluated with an inline function.

Program 16.4. EDF with x input evaluated at xe

```
x = randn(200,2);
Ne = [40;40];
xe = pick2Dpoints(min(x(:))*[1 1],max(x(:))*[1 1],Ne);
Fhat = @(xe,x) reshape(sum( ...
   all(repmat(x,[1,1,size(xe,1)]) ...
   < repmat(reshape(xe',[1,size(xe,2),size(xe,1)]),[size(x,1),1,1]),2) ...
                         ,1),size(xe,1),1)/size(x,1);
surf(reshape(xe(:,1),Ne),reshape(xe(:,2),Ne),reshape(Fhat(xe,x)),Ne);
```

Remark 16.5. The code written in Program 16.4 looks especially dense, even for this text. In reality, it could be written in a less opaque fashion with the use of a separate file, a `function` definition and multiple auxiliary terms. It is, however, a testament to the number of builtin functions in MATLAB, and their ability to handle inputs of high-dimensional arrays, that this computation can be written in a single line, albeit split over four lines for readability. We doubt that our implementation of this is the cleanest "one-liner" possible, and encourage the readers to understand what we have written and improve upon it.

Because we are interested in studying the derivative of our surrogate CDF, we want it to be as smooth as feasible while still maintaining fidelity to the observed data. Section 15.1 discusses this balance in the context of smoothing splines computed with Tikhonov regularization. In Example 16.4 we study a CDF response surface in 1D and show that kernels can be used to model this and therefore generate approximate PDFs from i.i.d. realizations of a random variable.

Example 16.4. (1D cumulative distribution function)
We generate $N = 700$ i.i.d. random variables X_i with a very simple choice of *generalized Pareto distribution* [Coles (2001)],

$$f_X(x) = \begin{cases} 1 - x/2, & 0 \le x \le 2, \\ 0, & \text{else}, \end{cases} \tag{16.4a}$$

$$F_X(x) = \begin{cases} 0, & x < 0, \\ x - x^2/4, & 0 \le x < 2, \\ 1, & x \ge 2, \end{cases} \tag{16.4b}$$

and define $\mathcal{X} = \{x_1, \dots, x_N\}$. The data values are created from the empirical distribution function (16.3):

$$\boldsymbol{y} = \begin{pmatrix} \hat{F}_{\mathcal{X}}(\boldsymbol{x}_1) \\ \vdots \\ \hat{F}_{\mathcal{X}}(\boldsymbol{x}_N) \end{pmatrix} = \begin{pmatrix} 1/N \\ \vdots \\ N/N \end{pmatrix}.$$

A smoothing spline approximation (see Section 15.1) based on a C^2 Matérn kernel is created using $\varepsilon = 0.3$ and $\mu = 0.01$. The surrogate model for the CDF along with the approximate PDF computed through differentiating the kernels are plotted in Figure 16.5. The code to conduct these computations is shown in Program 16.5.

Program 16.5. Excerpts from `SurrModelEDF1D.m`

```
% Define the RBF we use for this problem
rbf = @(r) (1+r).*exp(-r);
rbfdx = @(r,dx,ep) -ep^2*exp(-r).*dx;
% Define the empirical distribution function, see Program 16.4
Fhat = EDF from Program 16.4
% Define the generalized Pareto distribution of interest
```

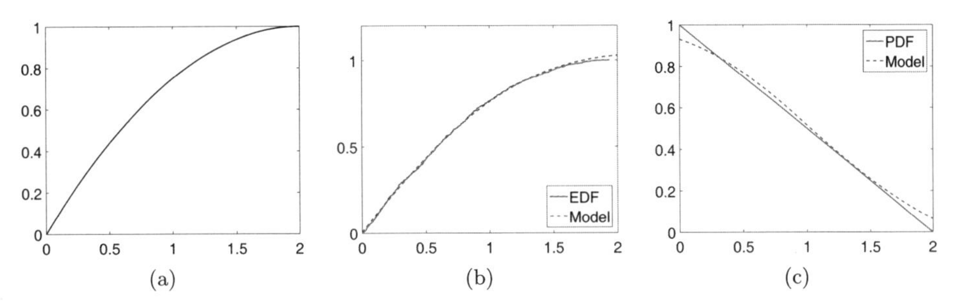

Fig. 16.5 A kernel approximation to the empirical distribution function provides an approximation to the density function, albeit with some failings. (a) The true CDF F_X that we want to model, (b) Our EDF "data" and the smoothing spline fit, (c) The modeled PDF has some similarity to the true PDF.

```
   % With these parameters, the distribution only exists on [0,2]
8  gp_k = -1/2;  gp_sigma = 1;  gp_theta = 0;
   % Create the data locations for this problem
10 N = 700;
   x = sort(icdf('gp',rand(N,1),gp_k,gp_sigma,gp_theta));
12 % Evaluate the EDF at the given data locations
   y = Fhat(x,x);
14 % Create the surrogate model
   ep = .3;  mu = 1e-2;
16 % Create the smoothing spline matrix coefficients
   cdf_coef = (rbf(DistanceMatrix(x,x,ep))+mu*eye(N))\y;
18 % Define functions to evaluate the surrogate CDF and PDF
   cdf_eval = @(xeval) rbf(DistanceMatrix(xeval,x,ep))*cdf_coef;
20 pdf_eval = @(xeval) rbfdx(DistanceMatrix(xeval,x,ep),...
                           DifferenceMatrix(xeval,x),ep)*cdf_coef;
```

The plots in Figure 16.5 suggest that a simple kernel-based approximation, even with the benefit of the smoothing spline strategy from Section 15.1, does not preserve the desired properties of a cumulative distribution function. For example, Figure 16.5(b) clearly shows the modeled CDF exceeding the value 1 which is not allowed by the axioms of probability. Enforcement of such boundedness (that $s(x) \in [0, 1]$ is required) can be performed with, e.g., MATLAB's Optimization Toolbox and solvers such as `fmincon`. The cost of using such methods can, however, be significant, especially in higher dimensions.

Although we can clearly recognize problems with these results, it is more difficult to identify exactly their cause. Example 16.5 discusses one possible cause of difficulties.

Example 16.5. (1D cumulative distribution function II)
One of the problems with creating a surrogate model of the CDF (16.4b) from Example 16.4 is that the data from the EDF used to create $\boldsymbol{y}$ may not have normally distributed independent errors. We consider Program 16.6 which constructs a random sample with a small additional set of points below the median, $2 - \sqrt{2}$. The

data from the EDF is then compared to the CDF to form a vector of "errors" in the data; this vector of differences is called `ECdiff`.

Program 16.6. Excerpts from `SurrModelEDFNonnormal.m`

```
% Create some slightly off data, with an extra 10% below the median
x = sort([rand(20,50)*(2-sqrt(2));...
          icdf('gp',rand(200,50),-.5,1,0)]);
% Form the EDF values at the data points (no need for Fhat here)
EDFs = repmat((1:220)'/220,1,50);
% Compute the difference between the EDF and CDF
% Only store the results for x>1 (outside of bad region)
ECdiff = EDFs(x>1)-cdf('gp',x(x>1),-.5,1,0);
n = length(ECdiff);
% Plot the results in a QQplot
h_fig = figure;
h = normplot(ECdiff);
legend(h([1,3]),'Data','Normal Expectation','location','northwest')
% Compute the skewness and add it to the plot
skewness = sqrt(n^2-n)/(n-2)*...
          sum((ECdiff-mean(ECdiff)).^3)/n/std(ECdiff)^3;
annotation('textbox',[.6,.2,.2,.07],'string',...
          sprintf('skewness = %3.2f',skewness'))
% Run the KS test to see if it fits a normal distribution
[h,p] = kstest((ECdiff-mean(ECdiff))/std(ECdiff))
```

We then plot these errors on a probability plot of the normal distribution, which can be viewed in Figure 16.6. Using the test for skewness described in [Doane and Seward (2011)], we can conclude that there is a nonzero skew in `ECdiff`, i.e., errors do not appear equally positive and negative. When the Kolmogorov–Smirnov test for goodness-of-fit to the standard normal distribution is run with `kstest`, the resulting p-value of

```
p =
   1.1360e-05
```

lets us safely conclude that the `ECdiff` vector does not have a normal distribution.

What does all this mean? We believe that it suggests that errors in fitting a CDF with data from an EDF are not independent, since the extra data added to these random samples was added in the region $[0, 2-\sqrt{2}]$ but the study above takes place on $[1, 2]$. Furthermore, the errors do not seem normally distributed, thus violating the Gaussian random field requirement for kriging, as discussed in Section 5.3.

If we postulate that the lack of independence between errors causes skewness which impairs the normality of the data, one strategy might be to force independence within the data. Example 16.6 proposes one possible way for achieving this through separating the random samples into groups and considering the EDFs on each group.

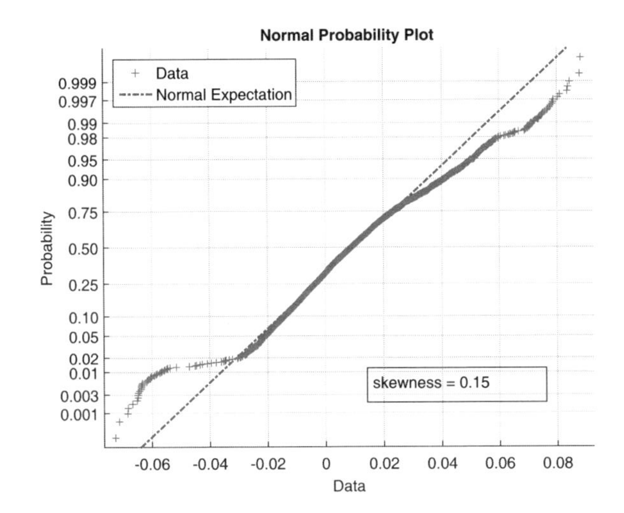

Fig. 16.6 For a sample of size 2518, a skewness of 0.15 is too large to accept that the underlying data is normally distributed. This probability plot would show data that follows the dashed line if the data were normal, but it strays significantly.

Example 16.6. (1D cumulative distribution function III)
We revisit the generalized Pareto example from before, and consider the Kolmogorov–Smirnov test on that data.

```
SurrModelEDF1D
errs = y - cdf('gp',x,gp_k,gp_sigma,gp_theta);
[h,p] = kstest((errs-mean(errs))/std(errs))

h =
     1


p =
    0.0064
```

This is a small enough p-value to conclude that the data was likely not normally distributed. To combat this, we study Program 16.7 which breaks the $N = 700$ random samples into five groups. The EDF is computed for each of these five groups and the data is recombined into a single sample on which `kstest` is run.

Program 16.7. `SurrModelEDFRand.m`

```
% Set up the same problem as before
SurrModelEDF1D
% Divide the data into groups, stored in a cell array
Ncuts = 5;
xall = icdf('gp',rand(N,1),gp_k,gp_sigma,gp_theta);
xcell = arrayfun(@(n)xall((n-1)*N/Ncuts+1:n*N/Ncuts,:),...
                  (1:Ncuts)','UniformOutput',0);
% Evaluate the EDF at the given data locations
```

```
ycell = cellfun(@(x)Fhat(x,x),xcell,'UniformOutput',0);
% Reorganize the cell arrays into the normal vectors
[x,isort] = sort(cell2mat(xcell));
yunsorted = cell2mat(ycell);
y = yunsorted(isort);
% Conduct the KS test for normality on the difference
ydiff = y - cdf('gp',x,gp_k,gp_sigma,gp_theta);
[h,p] = kstest((ydiff-mean(ydiff))/std(ydiff))
```

The result of this `kstest` call is a p-value of 0.3716, which is large enough to believe the data *might* have been normally distributed. The plots of the data for this problem, as well as the resulting fit are provided in Figure 16.7. The code for fitting the surface and plotting has been omitted from the book but is available in the `GaussQR` library. They show that this is certainly not a perfect solution to the problem (the PDF in Figure 16.7(b) arguably looks worse than in Figure 16.5(c)), but the model CDF is better constrained than before.

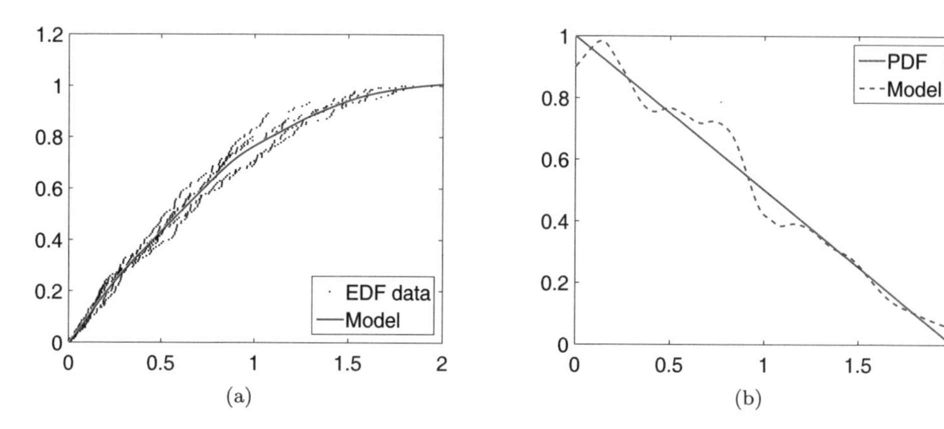

Fig. 16.7 An alternate strategy for producing data for a kernel approximation to the CDF (16.4b). (a) The model CDF seems better behaved than in Figure 16.5(a), (b) The model PDF appears more multimodal than it should be.

Clearly, the most direct methods for fitting a distribution function using kernel approximation have limited viability given the special properties (i.e., monotonically increasing, limits at boundaries 0 and 1) of distribution functions. Approximation methods specific to density estimation (most often penalized likelihood methods) have been developed in the references listed at the top of this section with more recent advances including [Holmes *et al.* (2007); Singham *et al.* (2013); Gu *et al.* (2013); Chang and Wu (2015)]. Multivariate density estimation remains a complicated and open research topic with great potential for a breakthrough to help multiple applications.

Chapter 17

Statistical Data Fitting via Gaussian Processes

In Chapter 5 we introduced the use of Gaussian processes as a statistical technique for conducting scattered data fitting. In this chapter we demonstrate the use of these kriging strategies on examples from geostatistics. These examples employ the criteria for optimal parametrizations of the covariance kernel, as introduced in Chapter 14, and some of our examples include the use of estimation to find nonzero mean functions.

It should be noted that the fitting of "real world" data can be considered both easier and more difficult than artificially generated data. On one hand, data from a real application often includes noise; this absolves the user from having to seek machine precision level accuracy because it cannot be achieved and, in some sense, makes the problem "easier." Of course, few in such applications would argue the presence of noise is a benefit, and it is often seen as a complicating factor in any estimation or prediction. It is especially troubling in situations with a limited or no definition of a true solution because studying convergence is a complicated concept. In this chapter we will work with potentially noisy data but rarely address it directly. Chapter 18 will go into more detail on handling noise.

17.1 Geostatistics

The original application behind kriging and the use of Gaussian processes in data fitting was attempting to predict gold deposits in South Africa [Krige (1951)]. In this section we demonstrate these techniques on data drawn from geostatistics and study the effectiveness of different parametrization options discussed in Chapter 14.

Example 17.1. (Kriging prediction for Animas River data)
Our first example deals with data drawn from [McCafferty *et al.* (2011)] involving the amount (in parts per million) of lanthanum in samples taken from around the Animas River in Colorado. This data is available from the `GaussQR` library data directory in the file `AnimasRiver_data.mat`, which is accessible with the `gqr_downloaddata` function introduced in Remark 15.3.

In this example, we focus primarily on the use of cross-validation to parametrize our C^0 Matérn kriging covariance, i.e., to determine an "optimal" shape parameter ε. In Section 15.1 we showed that the smoothing spline fit $s(\boldsymbol{x}) = \boldsymbol{k}(\boldsymbol{x})^T(\mathsf{K}+\mu\mathsf{I})^{-1}\boldsymbol{y}$ corresponds to an (interpolating) kriging fit $s(\boldsymbol{x}) = \boldsymbol{k}(\boldsymbol{x})^T\mathsf{K}^{-1}\tilde{\boldsymbol{y}}$ to smoothed data $\tilde{\boldsymbol{y}}$ as defined in (15.4). We follow this philosophy, and therefore will always be referring to our fits as *kriging* fits, regardless of whether we do any smoothing of the data or not.

Our code is designed to handle a general kriging fit of smoothed data (smoothing spline fit). However, in this first example we set the regularization parameter $\mu = 0$. Program 17.1 presents our study of this geologic data. In Example 17.2 we will continue this study with a nonzero smoothing parameter.

Program 17.1. Excerpts from `StatFitColorado.m`

```
  % Download the necessary data and load it into memory
2 gqr_downloaddata('AnimasRiver_data.mat')
  load AnimasRiver_data
4 % Scale the data into a predictable domain: [-1,1]^2
  [x,y,latlong_shift,latlong_scale] = rescale_data(latlong,Lappm,1);
6 N = size(x,1);
  % Choose the C0 Matern to fit the data
8 rbf = @(e,r) exp(-(e*r));
  % Choose a range of epsilon values and smoothing spline parameter
10 epvec = logspace(-1,1,20);    mu = 0;
  % Choose locations at which to make predictions
12 Neval = 60;  xeval = pick2Dpoints(-1,1,Neval);
  % Define a function which makes the kriging predictions
14 sf = @(ep,mu,x,y,xeval) rbf(ep,DistanceMatrix(xeval,x))*...
                      (rbf(ep,DistanceMatrix(x,x)+mu*eye(length(y)))\y);
16 % Define the LOOCV computation as a function of epsilon
  cv = @(ep,mu) crossval('mse',x,y,...
18                          'Predfun', @(x,y,xeval) sf(ep,mu,x,y,xeval),...
                           'leaveout',1);
20 % Evaluate the cross-validation and plot for all epsilon values
  cvepvec = arrayfun(@(ep) cv(ep,0),epvec);
22 semilogx(epvec,cvepvec)
  % Find the LOOCV minimizing epsilon and evaluate with it
24 epopt = fminbnd(@(ep) cv(ep,0),1,2);
  seval = sf(epopt,0,x,y,xeval);
```

The function `rescale_data` used here is from the `GaussQR` library (see also our discussion in Example 16.3). This is a basic function that linearly moves data from its physical domain to the domain $[-1, 1]^d$. This is a useful strategy when studying the effect of shape parameters such as ε because it helps focus the search region more predictably.

Although in Program 17.1 we fix $\mu = 0$, in line 14 we define the evaluation function as a function of both ε and μ in preparation for further tests involving both parameters. Similarly, some of the cost of evaluating the kriging prediction `sf`

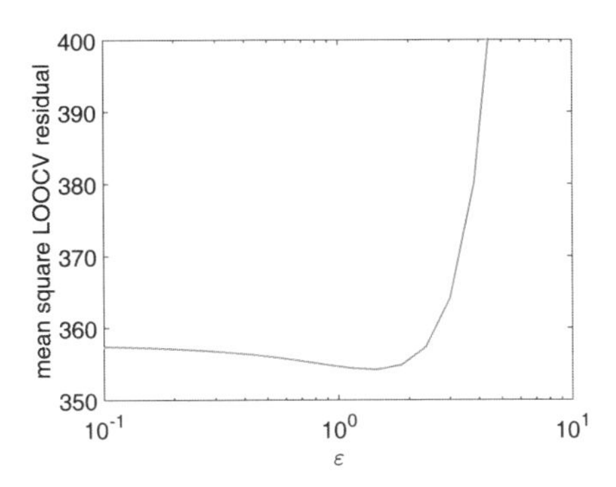

Fig. 17.1 From Example 17.1. LOOCV mean-squared residual for $\mu = 0$.

could be reduced by storing commonly computed values for reuse, but the function is defined here in its most general form for use in the `crossval` function. That function is a tool in MATLAB's Statistics and Machine Learning Toolbox that easily allows for cross validation if given a prediction function similar to `sf`.

Line 17 defines a function handle `cv` which conducts the cross-validation and, because we pass `crossval` the option `mse`, returns the mean-squared residual. The `crossval` function requires a prediction function of a specific form, which we create with an anonymous function in line 18. In line 19 we choose to use leave-one-out cross-validation, although any k-fold cross-validation choice could be made with the option `kfold`.

We use `arrayfun` in line 21 to compute the entire vector of LOOCV residuals and then plot the results which can be seen in Figure 17.1. After looking at this graph, we determine a region where the LOOCV-optimal ε exists and find that optimal ε in line 24 with the MATLAB function `fminbnd`. Figure 17.2(a) displays the corresponding solution with LOOCV-optimal $\varepsilon = 1.40$ and μ fixed at 0.

Example 17.2. (Kriging prediction for Animas River data II)
In Program 17.1 we demonstrated our computation with a fixed smoothing parameter of $\mu = 0$, but the definition of `sf` allows for computation with variable μ as well. We can use a `for`-loop to compute for a range of μ values as well as ε values.

```
muvec = logspace(-4,-1,30);  cvmat = zeros(length(muvec),length(epvec))
for k=1:length(muvec)
    cvmat(k,:) = arrayfun(@(ep) cv(ep,muvec(k)),epvec);
end
```

We could actually replace this loop with a single, more complicated, `arrayfun` call, but the use of a loop here is a simple extension of our earlier computations and allows us to update a `waitbar` during the computation, as is done in the full version of `StatFitColorado.m` in the `GaussQR` library. These cross-validation residuals are plotted in Figure 17.2(b).

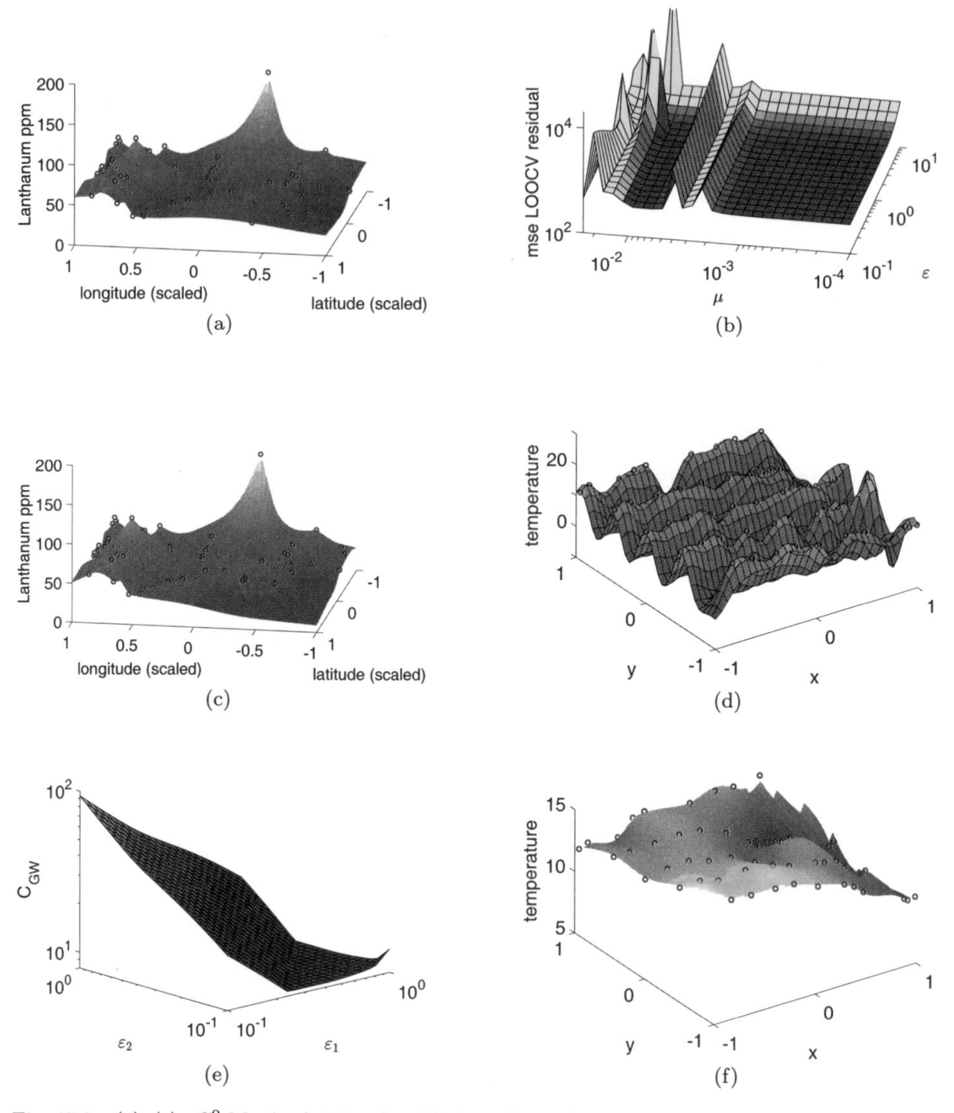

Fig. 17.2 (a)–(c): C^0 Matérn kriging fit of Animas River data; parameter optimization based on LOOCV. (a) From Example 17.1. Kriging interpolant of data using $\varepsilon = 1.40$ and $\mu = 0$. (b) From Example 17.2. LOCCV profile for variable ε and μ. Two local minima exist: at $\varepsilon = 1.40$ and $\mu \to 0$ with residual 354, and at $\varepsilon = 1.54$ and $\mu = 0.0061$ with residual 353. (c) From Example 17.2. Kriging interpolant of smoothed data using $\varepsilon = 1.54$ and $\mu = 0.0061$.
(d)–(f): Fit of Big Sur data with isotropic Gaussian and anisotropic C^2 Matérn tensor product kernels; parameter optimization based on LOOCV, $C_{\mathrm{MPLE}}(\boldsymbol{\varepsilon})$, and $C_{\mathrm{GW}}(\boldsymbol{\varepsilon}; \infty)$. (d) From Example 17.4. Isotropic Gaussian interpolant of debatable quality with $\varepsilon = 8$, chosen between the different criteria. (e) From Example 17.5. Anisotropic shape parameters for a C^2 Matérn tensor product kernel satisfying $\varepsilon_1 \approx 2.4\varepsilon_2$ seem to minimize $C_{\mathrm{GW}}(\boldsymbol{\varepsilon}; \infty)$. (f) From Example 17.5. Anisotropic C^2 Matérn kernel interpolant (kriging prediction) with optimality determined by minimizing the kriging variance $C_{\mathrm{GW}}(\boldsymbol{\varepsilon}; \infty)$.

When we look at the surface in Figure 17.2(b), there are two clear regions where the LOOCV residual encounters a local minimum. We can use `fminunc` from MATLAB's Optimization Toolbox to minimize this function of two variables (ε, μ). This presents a bit of a problem because `fminunc` performs *unconstrained* optimization, but both ε and μ must be positive. To eliminate this issue, we instead ask MATLAB to optimize a function of $(\log \varepsilon, \log \mu)$ and exponentiate within the function:

```
logepmuopt = fminunc(@(epmu) cv(exp(epmu(1)),exp(epmu(2))),log([1,1e-2]));
exp(logepmuopt)
ans =
    1.5375    0.0061
```

Note that the initial conditions are passed in log form, and the result must be exponentiated to recover the local minimum $\varepsilon = 1.5375$ and $\mu = 0.0061$. This is actually the global minimum as well, and it occurs in the small region around $\mu = 10^{-2}$. The kriging predictions are displayed in Figure 17.2(c) with a noticeable difference from Figure 17.2(a) near the point $(-0.5, 0.5)$. If we had chosen an initial guess on the other side of the ridge in the middle of Figure 17.2(b) — such as $(\varepsilon, \mu) = (1, 10^{-3})$, then `fminunc` would have tended toward $\mu \to 0$ and we would have recovered the optimum used in Figure 17.1.

Remark 17.1. Optimizing parametrization criteria — such as the cross-validation residual — can be a challenging endeavor, as demonstrated by the complicated shape of the surface in Figure 17.2(b) and the fact that its global minimum is difficult to access. Even with MATLAB's Optimization Toolbox and functions such as `fminunc` and `fmincon`, one must provide suitable initial data in order to be able to find good results with reasonable computational expense. As a result, it is conceivable that building a *surrogate model* for the LOOCV residual could help identify initial guesses for potential minima, while minimizing the required number of kriging predictions. This is especially true as the number of parameters in the kriging fit increases. We discussed surrogate modeling in Chapter 16.

Remark 17.2. There are legitimate questions as to the validity of applying kriging to data on which a smoothing spline has been applied through (15.4). The underlying assumption for kriging is that the data $\boldsymbol{y}$ has been generated by a Gaussian random field, and it is possible that the smoothed data $\tilde{\boldsymbol{y}} = (\mathsf{K} + \mu \mathsf{I})^{-1}\mathsf{K}\boldsymbol{y}$ does not share that property. While this is unlikely to be an issue for very small μ, since $\mu = 0$ applies no smoothing, larger μ could negate some of the nice predictive properties of kriging by introducing unacceptable kriging input. We ignore these concerns to focus on the use of parametrization schemes in this chapter, but this may be relevant in certain applications.

Example 17.3. (Kriging prediction for Mt. Eden data)
This example uses the `volcano` data set available in the statistical software package R [R Core Team (2014)]. The data represents 5307 elevation measurements obtained from Maunga Whau (Mt. Eden) in Auckland, NZ, sampled on a 10m×10m grid. The data file is also available at the `GaussQR` repository as `volcano_data.mat`.

Our goal here is two-fold: first, to demonstrate the use of compactly supported kernels and sparse matrices on this relatively large problem of size $N = 5307$, and second, to compute the parametrization criteria (cf. (14.1) and (14.20))

$$C_{GW}(\varepsilon;\infty) = \sqrt{\boldsymbol{y}^T\mathsf{K}^{-1}\boldsymbol{y}}\,\|P_{K,\mathcal{X}}\|_\infty,$$
$$C_{MPLE}(\varepsilon) = N\log\left(\boldsymbol{y}^T\mathsf{K}^{-1}\boldsymbol{y}\right) + \log\det\mathsf{K},$$

for a range of ε values. The necessary computation is performed in Program 17.2. Note that we call our data with the letter `h` for heights rather than $\boldsymbol{y}$ as has been our custom.

Program 17.2. Excerpts from `StatFitVolcano.m`

```
% Download the necessary data and load it into memory
gqr_downloaddata('volcano_data.mat'),  load volcano_data
x = locations;  h = heights;  N = Nx*Ny;
% Define the 2D C2 Wendland kernel (in anisotropic dense form)
rbf = @(r) (1-r).^4.*(4*r+1);
% Consider a range of epsilon values to check the GW criterion
epvec = [.05,.1,.25,.5,1:20];  k = 1;
for ep=epvec
    % Compute the sparse kernel matrix by passing rbf to DistanceMatrix
    [K,DMtimevec(k)] = DistanceMatrix(x,x,ep,rbf);
    densevec(k) = length(find(K))/N^2;
    if densevec(k)>.2
        K = full(K);   p = 1:N;
    else
        p = symamd(K);
    end

    tic
    % Factor the kernel matrix: Lp*Lp' = Kp
    Kp = K(p,p);   Lp = chol(Kp,'lower');
    % Compute the solution to the system
    hp = h(p);    ctemp = Lp'\(Lp\hp);
    c = zeros(N,1);   c(p) = ctemp;
    % Compute the Hilbert space norm
    hsnorm = sqrt(h'*c);
    % Compute the log(det(K))
    logdetK = 2*sum(log(full(diag(Lp))));
    % Compute the power function
    [Keval,evaltime] = DistanceMatrix(xeval,x,ep,rbf);
    pow = norm(sqrt(1 - sum(full(Keval(:,p)/Lp').^2,2),'inf'));
```

```
32      solvetimevec(k) = toc - evaltime;
        DMtimevec(k) = DMtimevec(k) + evaltime;
34      gwvec(k) = hsnorm*pow;
        mplevec(k) = 2*N*log(hsnorm) + logdetK;
36      k = k + 1;
    end
```

The technical detail provided in Program 17.2 is greater than in many other of our examples in order to help discuss performance analysis in MATLAB. The first point of emphasis is the use of `DistanceMatrix` to evaluate not the sparse distance matrix but the sparse *kernel matrix* by passing `rbf`. This was discussed as an option in Section 4.2. In line 10, `DistanceMatrix` also returns the amount of time required to evaluate the kernel matrix, which we store for comparison later. Lines 11–16 are also sparse matrix specific, where `symamd` is called if the matrix is sparse enough to help reorder the matrix and minimize the cost of inverting it. If the matrix is too dense (we have arbitrarily chosen 20% dense as too great) we revert to a dense matrix and execute no reordering by defining the permutation vector `p` to be the identity permutation.

We use `tic` and `toc` in lines 18 and 32 to record the time required to compute $C_{GW}(\varepsilon, \infty)$. In line 20 we compute the (potentially sparse) Cholesky factorization, $\mathsf{K} = \mathsf{L}\mathsf{L}^T$, and the next two lines solve for the interpolation coefficients $\boldsymbol{c} = \mathsf{K}^{-1}\boldsymbol{y}$ while using any reordering provided by `symamd`.

The Hilbert space norm of the interpolant (see (2.24)) is computed using these coefficients:

$$\|s\|_{\mathcal{H}_K(\Omega)} = \sqrt{\boldsymbol{y}^T\mathsf{K}^{-1}\boldsymbol{y}} = \sqrt{\boldsymbol{y}^T\boldsymbol{c}} = \texttt{sqrt(h'*c)}.$$

The computation of $\log(\det(\mathsf{K}))$ on line 27 is a bit cryptic, and leverages the Cholesky decomposition,

$$\log(\det(\mathsf{K})) = \log(\det(\mathsf{L}\mathsf{L}^T)) = 2\log(\det(\mathsf{L})).$$

Line 35 combines these two terms to form $C_{MPLE}(\varepsilon)$.

The computation of the power function on line 30 is even more cryptic. It leverages the fact that $K(\boldsymbol{x}, \boldsymbol{x}) = 1$ for the Wendland kernels and, for a given $\boldsymbol{x}$,

$$\begin{aligned} P_{K,\mathcal{X}}^2(\boldsymbol{x}) &= K(\boldsymbol{x}, \boldsymbol{x}) - \boldsymbol{k}(\boldsymbol{x})^T\mathsf{K}^{-1}\boldsymbol{k}(\boldsymbol{x}) \\ &= 1 - \boldsymbol{k}(\boldsymbol{x})^T\mathsf{L}^{-T}\mathsf{L}^{-1}\boldsymbol{k}(\boldsymbol{x}) \\ &= 1 - \|\boldsymbol{k}(\boldsymbol{x})^T\mathsf{L}^{-T}\|_2^2. \end{aligned}$$

The command `sum(full(Keval(:,p)/Lp').^2,2)` executes the 2-norm computation for all the `xeval` points at which we are testing $P_{K,\mathcal{X}}$. Here `full` is not required, but it speeds up the summation by storing `Keval(:,p)/Lp'` as a dense matrix even if those matrices were originally sparse. In line 30, the `'inf'` term appears because we are evaluating $C_{GW}(\varepsilon; \infty)$ on line 34.

Figure 17.3(a) presents the results of these computations. As we can see, there appears to be a minimum around $\varepsilon \approx 1$ for $C_{MPLE}(\varepsilon)$, but in the full plot, available in the GaussQR library, there is a limit as $\varepsilon \to 0$ which is almost indistinguishable from the minimum at $\varepsilon \approx 1$. This may be interpreted as stating that the $C_{\text{MPLE}}(\varepsilon)$ criterion suggests that all values $\varepsilon < 1$ are equally as good. Ill-conditioning does not affect this kriging fit until $\varepsilon < 0.025$, at which point `chol` throws an error because it believes K to be not positive definite. A plot similar to the kriging prediction for this example is on display in Figure 1.1.

The graphs in Figure 17.3(b)–(c) illustrate the increase in density of the kernel matrix associated with compactly supported kernels with increasingly smaller values of ε, and the effects of switching between sparse and dense matrix computations on the time it takes to compute the parametrization criteria.

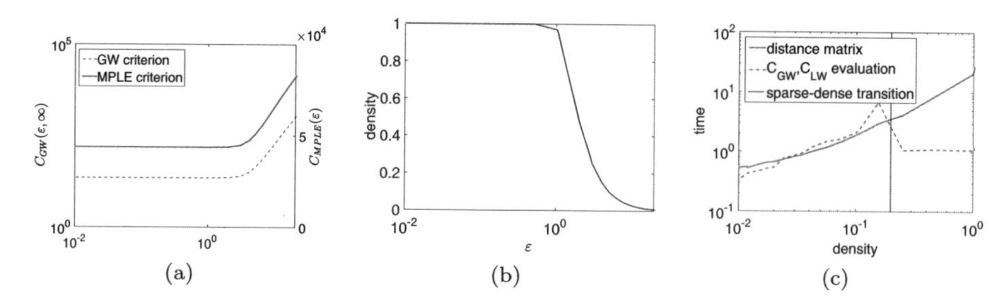

Fig. 17.3 Results from experiments involving a C^2 Wendland kernel kriging fit on the `volcano` data. (a) $C_{\text{MPLE}}(\varepsilon)$ has a global minimum near $\varepsilon \approx 1$, but it is only slightly less than all other $\varepsilon < 1$. C_{GW} seems to have its minimum as $\varepsilon \to 0$. We use two y-axes because C_{GW} is on a log scale but C_{MPLE} is on a linear scale. (b) The density of the sparse matrix increases as ε decreases. (c) The time required to compute the sparse kernel matrices and the parametrization criteria scales poorly with density. When the sparse matrices are converted to dense (after the sparse density is greater than 20%) the cost of computing C_{GW} and C_{MPLE} is roughly constant.

Remark 17.3. The commands `tic` and `toc` for timing in MATLAB are simple and effective tools. There can even be nested `tic` calls to push multiple timers, which then pop out in the order that the `toc` commands are called. MATLAB, however, provides more advanced tools for studying a program's performance which are particularly useful when developing software in MATLAB and trying to optimize performance. For more information, see the `profile` command and MATLAB's array of profiling tools.

Example 17.4. (Kriging prediction for Big Sur data)
This example demonstrates the use of the Hilbert–Schmidt SVD in computing the LOOCV, $C_{\text{GW}}(\varepsilon;\infty)$ and $C_{\text{MPLE}}(\varepsilon)$ parametrization criteria in the context of fitting the `BigSur_data.mat` data set (see, e.g., [Carlson and Foley (1992)]). This data set was provided by Richard Franke and is now hosted at Oleg Davydov's website [Davydov (2014)]). It is also available from the GaussQR library data directory in the file `BigSur_data.mat`. The data consist of 64 water temperature measurements taken from a boat traveling on tracks approximately perpendicular to the coast of Big Sur, California.

Program 17.3 contains an implementation for Gaussians and investigates their use into the ill-conditioned $\varepsilon \to 0$ limit. Associated plots can be viewed in Figure 17.4 and Figure 17.2(d).

Program 17.3. Excerpts from `StatFitBigSur.m`

```
% Load the Big Sur data set
gqr_downloaddata('BigSur_data.mat'),load BigSur_data
[x,t] = rescale_data(locations,temperatures);  N = size(x,1);
% Choose some points at which to evaluate the power function
xeval = pick2Dpoints(-1,1,4,'halt');
% Consider a range of epsilon values
epvec = logspace(0,1.1,15);   k = 1;   gqr_alpha = 1;
for ep = epvec
    % Form the GaussQR matrices and solve for the coefficients
    GQR = gqr_solveprep(0,x,ep,gqr_alpha);
    Phi1 = GQR.stored_phi1;    CbarT = GQR.CbarT;
    lamvec1 = GQR.eig(GQR.Marr(:,1:N));
    lamvec2 = GQR.eig(GQR.Marr(:,N+1:end));
    Psi = Phi1 + GQR.stored_phi2*CbarT;
    b = Psi\t;
    % Compute the log(det(K))
    logdetK = log(abs(det(Phi1))) + sum(log(lamvec1)) + log(abs(det(Psi)));
    % Compute the Hilbert space norm
    hsnorm = sqrt(norm(bsxfun(@ldivide,sqrt(lamvec1)',b)).^2 + ...
                  norm(bsxfun(@ldivide,sqrt(lamvec2)',CbarT*b)).^2);
    % Compute the maximum of the power function at the sample points
    poweval = zeros(size(xeval,1),1);  j = 1;
    for xp=xeval'
        GQRp = gqr_solveprep(0,[x;xp'],ep,gqr_alpha);
        Phi1 = GQRp.stored_phi1;
        lamvec1 = GQRp.eig(GQRp.Marr(:,1:N));
        Psi = Phi1 + GQRp.stored_phi2*GQRp.CbarT;
        logdetKp = log(abs(det(Phi1))) + sum(log(lamvec1)) + log(abs(det(Psi)));
        poweval(j) = sqrt(exp(logdetKp - logdetK));
        j = j + 1;
    end
    pow = norm(poweval,'inf');
    % Compute the various optimization criteria
    cvvec(k) = crossval('mse',x,t,...
        'Predfun',@(x,y,xeval) gqr_eval(gqr_solve(x,y,ep,gqr_alpha),xeval),...
                    'leaveout',1);
    gwvec(k) = hsnorm*pow;
    mplevec(k) = 2*N*log(hsnorm) + logdetK;
    k = k + 1;
end
```

The computations performed in Program 17.3 follow directly from the derivations in Chapter 14. Lines 19–20 compute

$$\sqrt{\boldsymbol{y}^T \mathsf{K}^{-1} \boldsymbol{y}} = \sqrt{\boldsymbol{b}^T \Lambda_1^{-1} \boldsymbol{b} + \boldsymbol{b}^T \overline{\mathsf{C}} \Lambda_2^{-1} \overline{\mathsf{C}}^T \boldsymbol{b}}$$

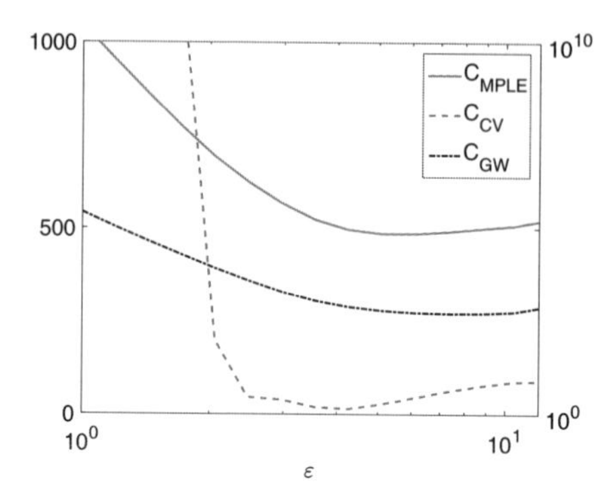

Fig. 17.4 From Example 17.4. Minima exist for all the parametrization criteria for $\varepsilon \in [1, 20]$: $C_{GW}(\varepsilon, \infty)$ is minimized near $\varepsilon = 10.5$, $C_{MPLE}(\varepsilon)$ is minimized near $\varepsilon = 5.5$ and $C_{CV}(\varepsilon; 1, \infty)$ is minimized near $\varepsilon = 4.5$. $C_{MPLE}(\varepsilon)$ and $C_{CV}(\varepsilon; 1, \infty)$ are measured by the left y-axis, $C_{GW}(\varepsilon, \infty)$ is measured by the right y-axis.

as described in Section 14.1.2. The power function is evaluated at each of the test points in `xeval` at line 29 with the strategy laid out in (14.4). We use the notation $C_{CV}(\varepsilon; 1, \infty)$ to denote the leave-one-out cross-validation performed by `crossval` in lines 34–36. Figure 17.4 displays the values of $C_{CV}(\varepsilon; 1, \infty)$, $C_{MPLE}(\varepsilon)$ and $C_{GW}(\varepsilon; \infty)$ for a range of ε values. $C_{GW}(\varepsilon, \infty)$ is minimized near $\varepsilon = 10.5$, $C_{MPLE}(\varepsilon)$ is minimized near $\varepsilon = 5.5$ and $C_{CV}(\varepsilon; 1, \infty)$ is minimized near $\varepsilon = 4.5$. We therefore — rather arbitrarily — picked $\varepsilon = 8$ as representative value predicted by all three criteria for the kriging fit displayed in Figure 17.2(d).

From looking at the surface plot in Figure 17.2(d), the predictive capabilities of this kriging fit to the Big Sur temperature data are limited. Clusters of points near $(0.5, 0.5)$ and $(0.8, -0.5)$ seem to have dominated the parametrization criteria and forced a choice of ε which is too large to make effective predictions. In Section 17.2 we reconsider this problem with anisotropic ε to allow for more flexibility in the covariance kernel.

We conclude this example by again emphasizing the significant computational cost presently associated with the HS-SVD. While this cost is, often, only a fraction of the cost needed to perform the computation in the standard basis with arbitrary precision, it is still too great for consideration on large problems. We hope to remedy this shortcoming in the near future, but believe that this example demonstrates the potential of this strategy since use of the HS-SVD allowed us to investigate the small-ε region — even if the kriging fit for the "optimal" value of $\varepsilon = 8$ can be safely (and more cheaply) computed in the standard basis.

17.2 Anisotropic Data Fitting

Data fitting in more than one dimension allows for the use of *anisotropic* kernels, i.e., kernels with different shape parameters in each dimension (recall Section 3.4). In

Section 17.1 we already discussed fitting kernels with more than one free parameter (including a smoothing parameter), but those kernels had only one shape parameter. Here we reconsider the Big Sur example from Example 17.4 using coordinate-dependent shape parameters.

Example 17.5. (Kriging prediction for Big Sur data II)
The Big Sur data set is an example of *track data*, where one or more dimension takes values in only fixed locations (essentially on a grid) and the other variables can be sampled in scattered locations. Such data may benefit from spatially varying shape parameters, as we addressed in Section 14.4, but we choose instead here to go with a fixed (but possibly different) shape parameter for each coordinate direction, and thus to employ anisotropic kernels. One reason for following this strategy is that it fits into our `DistanceMatrix` approach with radial kernels more cleanly.

At the time of this printing, the current version of the `GaussQR` library cannot compute a Hilbert–Schmidt SVD for anisotropic Gaussian kernels. The theory is relatively straightforward but the implementation is not. As a result, we study here the C^2 Matérn kernel and the $\mathrm{C_{GW}}(\varepsilon;\infty)$ parametrization criterion for the shape parameter array $\boldsymbol{\varepsilon}$.

To make the code easier to read and understand we define an auxiliary function which encapsulates the $\mathrm{C_{GW}}(\varepsilon;\infty)$ computation.

Program 17.4. `CGW_eval.m`

```
function gw = CGW_eval(x,y,rbf,eparr,xeval)
% function gw = CGW_eval(x,y,rbf,eparr,xeval)
% Inputs: x     - data sites
%         y     - data values
%         rbf   - radial kernel in anisotropic form
%         eparr - anisotropic shape parameter
%         xeval - data sites at which to test the power function
% Outputs: gw   - the C_GW criterion (kriging variance)
K = rbf(DistanceMatrix(x,x,eparr));
Keval = rbf(DistanceMatrix(xeval,x,eparr));
L = chol(K,'lower');
Linvy = L\y;
hsnorm = norm(Linvy);
pow = norm(sqrt(1 - sum((Keval/L').^2,2)),'inf');
gw = hsnorm*pow;
```

This function follows the same form as Program 17.2, but assumes that the kernel matrix is dense. Using a function overloading strategy, it would be possible to consider either case in a single function (this is discussed in Appendix D), but we consider only the dense case here.

Using this `CGW_eval` function we can quickly test a range of ε values and determine which of them produces the smallest $\mathrm{C_{GW}}(\varepsilon;\infty)$. That minimum can then be used as an initial guess for a call to `fminunc` to potentially find a global minimum. Program 17.5 demonstrates this computation and plots the results: the computed $C_{GW}(\varepsilon;\infty)$ values are plotted in Figure 17.2(e) and the $C_{GW}(\varepsilon;\infty)$-optimal kriging prediction using an anisotropic C^2 Matérn covariance kernel is plotted in Figure 17.2(f).

Program 17.5. Excerpts from `StatFitBigSurAnisotropic.m`

```
% Load the Big Sur data set
gqr_downloaddata('BigSur_data.mat'),load BigSur_data
[x,t] = rescale_data(locations,temperatures);  N = size(x,1);
% Choose points, only in the convex hull, to test the power function
cind = convhull(x(:,1),x(:,2));
xeval = pick2Dpoints(-1,1,6,'halt');
xeval = xeval(inpolygon(xeval(:,1),xeval(:,2),x(cind,1),x(cind,2)),:);
% Define the anisotropic C2 Matern kernel
rbf = @(r) (1+r).*exp(-r);
% Consider a range of epsilon values for each dimension
Nep = 25;  ep2dvec = 10.^pick2Dpoints(-1,0,Nep)';
% Evaluate the C_GW criterion for all the anisotropic epsilon choices
gwvec = cellfun(@(eparr)CGW_eval(x,t,rbf,eparr',xeval),num2cell(ep2dvec,1));
% Find the minimum of the tested epsilon values
[~,imin] = min(gwvec);  epguess = ep2dvec(:,imin)';
% Compute the minimum C_GW with fminunc, using initial guess
epopt = exp(fminunc(@(eparr)CGW_eval(x,t,rbf,exp(eparr),xeval),log(epguess)));
% Create a surface plot of the optimal result
Nplot = 55;  xplot = pick2Dpoints(-1,1,Nplot);
splot = rbf(DistanceMatrix(xplot,x,epopt))*(rbf(DistanceMatrix(x,x,epopt))\t);
X = reshape(xplot(:,1),Nplot,Nplot);  Y = reshape(xplot(:,2),Nplot,Nplot);
S = reshape(splot,Nplot,Nplot);
% Only plot points in the convex hull
S(not(inpolygon(X,Y,x(cind,1),x(cind,2)))) = NaN;
surf(X,Y,S)
```

The first new elements in this program are the `convhull` and `inpolygon` functions: `convhull` returns the indices of `x` that form the convex hull of the data locations, and `inpolygon` checks to make sure that new points fall inside the convex hull. This appears in line 7 to check if the power function evaluation points are acceptable and in line 24 to avoid making predictions at locations outside the convex hull. The use of `NaN` to prevent plotting inappropriate data also appears in Program 20.5 and `inpolygon` is used in Program 20.6 to help determine how points should be distributed.

In line 11 we create a matrix where each column is a different ε to test and proceed to test those ε values in line 13 with a quick conversion to a cell array and a call to `cellfun`. The optimal ε is found in line 17 using `fminunc` with the same log/exp trick that appeared in Example 17.2 and the optimal kriging predictions are presented in a surface plot.

Remark 17.4. The surface plot in Figure 17.2(e) is rather well-structured by comparison to other similar parametrization profiles such as Figure 17.2(b). In fact, there appears to be an entire *line* of locally optimal $\boldsymbol{\varepsilon}$ values. We can use MATLAB's `polyfit` function to fit values taken from the graph to a line satisfying the presumed relationship $\log \varepsilon_2 = a \log \varepsilon_1 + b$. Using the commands

```
ex = [.2371;1];ey = [.1;.4217];
polyfit(log(ex),log(ey),1)
ans =
    0.9999   -0.8635
```

we see that $a \approx 1$ and $b = -.8635$. A more useful interpretation might come from writing this relationship as $\varepsilon_2 \approx 2.37\varepsilon_1$, and along this line we can roughly minimize $\mathrm{C}_{\mathrm{GW}}(\boldsymbol{\varepsilon}; \infty)$ for an entire range of $\boldsymbol{\varepsilon}$ vectors. This strengthens our belief that an anisotropic kernel is the right choice for fitting this, and potentially other, track data.

It is hard to argue that the anisotropic fit in Figure 17.2(f) is not a better fit to the data than the one in Figure 17.2(d). We cannot attribute this only to the anisotropic shape parameter, because it may also be the result of the use of the C^2 Matérn kernel rather than the Gaussian kernel from Example 17.4. The question of which *family of kernels* is optimal, rather than choosing an optimal shape for a given family of kernels, could also be answered using the parametrization strategies defined in Chapter 14.

17.3 Data Fitting Using Universal Kriging and Maximum Likelihood Estimation

In all of the kriging and parametrization examples presented thus far in this chapter, we have assumed our data to have zero mean, as is required for simple kriging. We discussed briefly in Remark 5.4 a strategy for fitting a Gaussian random field with *nonzero mean* μ and covariance kernel K: the goal is to make predictions of the form

$$\overset{\triangle}{Y}_{\boldsymbol{x}} = \boldsymbol{k}(\boldsymbol{x})^T \tilde{\boldsymbol{c}} + \mu(\boldsymbol{x})$$

where $\boldsymbol{k}(\boldsymbol{x})^T \tilde{\boldsymbol{c}}$ will be developed using simple kriging *after* the nonzero mean has been removed from the data. The notation $\tilde{\boldsymbol{c}}$ is used here to emphasize that these are, likely, different coefficients than from the simple kriging system $\mathsf{K}\boldsymbol{c} = \boldsymbol{y}$.

Of course, we likely have no knowledge of what the mean μ is, so we must estimate it from the data. Doing so for arbitrary mean μ is impractical [Scheuerer *et al.* (2013)], so we make an assumption about the possible forms that μ can take. In this section we assume that the mean is a polynomial with q terms,

$$\mu(\boldsymbol{x}) = \boldsymbol{p}(\boldsymbol{x})^T \boldsymbol{\beta} = \sum_{k=1}^{q} p_q(\boldsymbol{x}) \beta_q,$$

and that we have data sampled at $\mathcal{X} = \{\boldsymbol{x}_1, \ldots, \boldsymbol{x}_N\}$ with values $\boldsymbol{y} = \begin{pmatrix} y_1 & \cdots & y_N \end{pmatrix}^T$. The coefficients $\boldsymbol{\beta}$ defining this mean function can be estimated with a *generalized least squares fitting* [Sacks *et al.* (1989)],

$$\boldsymbol{\beta} = (\mathsf{P}^T\mathsf{K}^{-1}\mathsf{P})^{-1}\mathsf{P}^T\mathsf{K}^{-1}\boldsymbol{y}. \tag{17.1}$$

The matrix $(\mathsf{K})_{i,j} = K(\boldsymbol{x}_i, \boldsymbol{x}_j)$ is our standard covariance matrix and the matrix $(\mathsf{P})_{i,j} = p_j(\boldsymbol{x}_i)$ contains entries from the components of our model for the mean function and may be referred to as a *design matrix* in the statistics literature (see, e.g., [Sacks *et al.* (1989); Everitt and Skrondal (2010)]).

Remark 17.5. One could argue that $\boldsymbol{\beta}$ should be written as $\hat{\boldsymbol{\beta}}$ since these values are only estimates of the true coefficients, but we ignore such concerns here for notational simplicity. That notation would be more important if we wanted to study the statistical properties of these coefficients.

The role of the mean function in universal kriging is, in a sense, to handle long-range trends in the data and allow the covariance kernel to manage more localized behavior. After finding these $\boldsymbol{\beta}$ coefficients, we subtract out the long-range deterministic behavior of the Gaussian random field and produce data $\boldsymbol{y} - \mathsf{P}\boldsymbol{\beta}$ that has zero mean. We fit our standard simple kriging predictor to the data by solving the system $\mathsf{K}\tilde{\mathsf{c}} = \boldsymbol{y} - \mathsf{P}\boldsymbol{\beta}$ to arrive at predictions of the form

$$\hat{Y}_{\boldsymbol{x}} = \boldsymbol{k}(\boldsymbol{x})^T\mathsf{K}^{-1}(\boldsymbol{y} - \mathsf{P}\boldsymbol{\beta}) + \boldsymbol{p}(\boldsymbol{x})^T\boldsymbol{\beta}. \tag{17.2}$$

Remark 17.6. As pointed out in the introduction to Chapter 5, the deterministic analogy to universal kriging would be to use an interpolant based on a positive definite kernel, but with a polynomial appended to it. This approach is rather uncommon in the approximation theory literature — especially the literature on radial basis functions — since the addition of such a polynomial is not "necessary" from a theoretical point of view. However, as the following study in Examples 17.6 and 17.7 shows, addition of a polynomial to fit the nonzero mean present in the data (as outlined above) can generate a more accurate fit — at least when judged according to our parametrization criteria.

In the remainder of this section we consider an example involving scattered data and attempt to apply the $\mathrm{C}_{\mathrm{MPLE}}$ criterion to fit both the mean and the covariance to make predictions.

Example 17.6. (Kriging prediction for glacier data)
This example involves the `glacier` data set provided by Richard Franke. The data consist of height measurements of a glacier sampled along 44 different level curves. This results in 8338 irregularly distributed data locations along with corresponding height values. The data is now housed at Oleg Davydov's website [Davydov (2014)], but can also be obtained via the `GaussQR` data directory in the file `glacier_data.mat`.

Since this is a relatively large data set we again turn to compactly supported kernels and choose the radial "missing" Wendland kernel $\kappa_{3,\frac{3}{2}}$ from Section 3.5 to help mitigate the computational cost.

Our goal now is to determine the effect of assuming a nonzero mean as measured by the $C_{\mathrm{MPLE}}(\varepsilon, \boldsymbol{p})$ parametrization criterion, whose notation we now augment to indicate that it is a function of the mean approximation scheme as well. Computing it in this setting requires the Mahalanobis distance for the zero mean data (cf. Remark 5.3)

$$(\boldsymbol{y} - \mathsf{P}\boldsymbol{\beta})^T \mathsf{K}^{-1} (\boldsymbol{y} - \mathsf{P}\boldsymbol{\beta}),$$

so that

$$C_{\mathrm{MPLE}}(\varepsilon, \boldsymbol{p}) = N \log((\boldsymbol{y} - \mathsf{P}\boldsymbol{\beta})^T \mathsf{K}^{-1} (\boldsymbol{y} - \mathsf{P}\boldsymbol{\beta})) + \log\det \mathsf{K}$$

Necessary computations are conducted in Program 17.6.

Program 17.6. Excerpts from `StatFitMLEPolynomial.m`

```
% Load the data from the GaussQR web server
gqr_downloaddata('glacier_data.mat'),  load glacier_data
[x,h] = rescale_data(datasites,heights,1);  N = size(x,1);
% Define an anisotropic missing Wendland kernel in dense form
rbf = @(r) (1-7*r.^2-81/4*r.^4).*sqrt(1-r.^2) - ...
           15/4*r.^4.*(6+r.^2).*log(r./(1+sqrt(1-r.^2)) + eps);
% Define a shape parameter vector
ep = [20 21];
% Prepare the kriging matrix for use in studying polynomial means
K = DistanceMatrix(x,x,ep,rbf);
[Lp,~,p] = chol(K,'lower','vector');
logdetK = 2*sum(log(full(diag(Lp))));
% Define a grid of max 2D polynomial degrees to test
pd1dvec = 0:8;  [P1,P2] = meshgrid(pd1dvec,pd1dvec);
pd2dvec = [P1(:)';P2(:)'];
% Define a function to evaluate a polynomial x(1).^a(1).*x(2).^a(2)
peval = @(x,a) prod(bsxfun(@power,x,a),2);
Pmat = @(x,arr) cell2mat(cellfun(@(a)peval(x,a),arr,'UniformOutput',0)');
% For each of the polynomials, compute the MPLE
mplevec = zeros(1,size(pd2dvec,2));  k = 1;
for pd=pd2dvec
    % Compute the design matrix, all the polynomials of degree <= pd
    alldegrees = num2cell((gqr_formMarr(pd+1)-1)',2);
    P = Pmat(x,alldegrees);
        % Compute the polynomial mean term
    LinvP_permuted = full(Lp\P(p,:));
    Linvh_permuted = Lp\h(p);
    PtKP = LinvP_permuted'*LinvP_permuted;
    beta = PtKP\(LinvP_permuted'*Linvh_permuted);
    % Compute the kernel coefficients K\(h-P*beta)
    cuk = zeros(N,1);   cuk(p) = Lp'\(Linvh_permuted - LinvP_permuted*beta);
    % Evaluate the likelihood
```

```
    mplevec(k) = N*log((h-P*beta)'*cuk) + logdetK;    k = k + 1;
34  end
    bar3(pd1dvec,reshape(mplevec,size(P1)))
```

This program reuses the same concepts of pivoting and computing with the Cholesky factorization that we introduced in Program 17.2. We create a function handle in line 17 to evaluate polynomials of a given degree and another in line 18 to evaluate the design matrix given a cell array of polynomial degrees, e.g., `{[0 0];[1 0];[0 1];[2 0]}`. With these components we can test $C_{\text{MPLE}}(\varepsilon, \boldsymbol{p})$ for a fixed ε and range of possible polynomials $\boldsymbol{p}$.

In line 23 we store all the polynomials with degree at most the current degree in a cell array using the `gqr_formMarr` function in the `GaussQR` library. Its intended purpose is in the evaluation of eigenfunctions in a Mercer series, but it can be applied here as well. After some permutations, defined during the `chol` call to minimize the computational cost of the sparse matrix factorization, we compute β in line 28, $\tilde{\boldsymbol{c}}$ in line 31 and the likelihood in line 33. Line 35 produces a bar plot, which can be seen for four different choices of ε in Figure 17.5(a)–(d).

Clearly, the polynomials can play a significant role in reducing the parametrization criterion $C_{\text{MPLE}}(\varepsilon, \boldsymbol{p})$. The effect is greatest for the very narrow kernels using $\varepsilon = (20\ 21)$, where the simple kriging value $C_{\text{MPLE}}(\varepsilon) = 1.78 \times 10^5$ is dropped to 1.06×10^5 by adding a $\boldsymbol{p}$ consisting of all 45 terms with total degree at or below 8. Even for the much wider covariance kernel generated with $\varepsilon = (5\ 4)$, we see a significant improvement from 1.16×10^5 to 0.92×10^5 by simply adding a constant mean, i.e., $\boldsymbol{p}$ is only a single term of total degree zero.

What is more interesting perhaps, is the lack of improvement when increasing the polynomial degree for covariance kernels with smaller values of ε as compared to when ε is larger. This seems to suggest that flatter kernels benefit much less from the addition of a higher-order $\boldsymbol{p}$, whereas the narrow kernels gain a great deal from including higher-order polynomials. In Example 17.7, we discuss the computational implications this has on universal kriging.

In Figure 17.5(e)–(g) we present a fit of very poor quality (Figure 17.5(f)) — obtained by using a narrow covariance kernel with assumed zero mean — in between a universal kriging fit using the same narrow covariance, but adding a polynomial mean approximation of degree $\boldsymbol{p} = (8\ 8)$ (Figure 17.5(g)) and an ordinary kriging fit (i.e., using only a constant polynomial $\boldsymbol{p}$) based on a relatively flat kernel (Figure 17.5(e)). The results argue strongly in favor of some nonzero mean given the unrecognizable state of Figure 17.5(f), but the difference between the flat kernel with constant mean and narrow kernel with high-order mean are less easily distinguished. This provides further support to our discussion in Chapter 10 regarding the relationship between "flat" limits of kernel-based interpolants and polynomials.

Remark 17.7. We have argued in Chapter 14 and in this chapter that minimizing criteria such as $C_{\text{MPLE}}(\varepsilon, \boldsymbol{p})$ is a strategy for optimally choosing a covariance

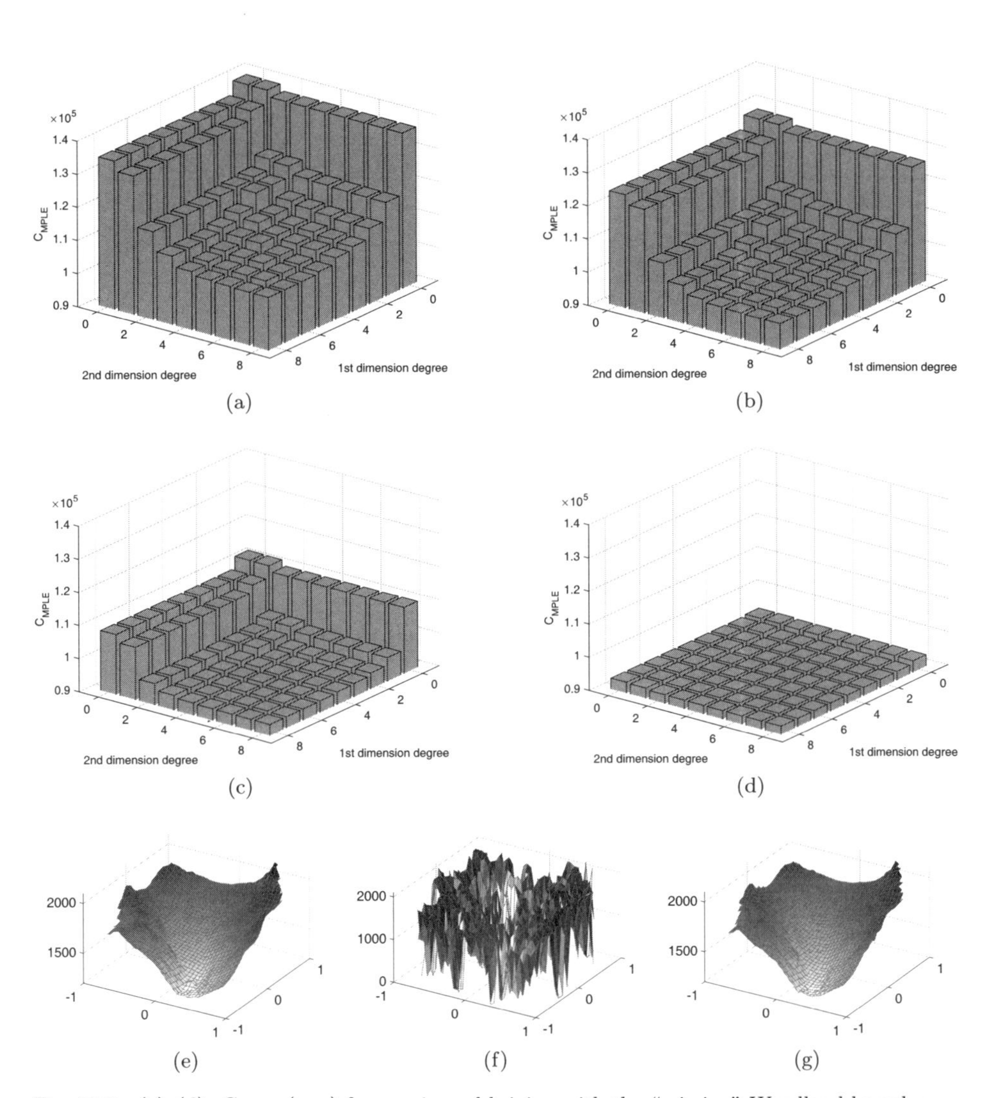

Fig. 17.5 (a)–(d): $\mathrm{C}_{\mathrm{MPLE}}(\boldsymbol{\varepsilon}, \boldsymbol{p})$ from universal kriging with the "missing" Wendland kernel $\kappa_{3,\frac{3}{2}}$ on the **glacier** data set for a fixed $\boldsymbol{\varepsilon}$ (listed below) and a range of polynomials. Also listed is the simple kriging $\mathrm{C}_{\mathrm{MPLE}}(\boldsymbol{\varepsilon})$ value, as a baseline for comparison. (a) $\boldsymbol{\varepsilon} = (20\ 21)$, with simple kriging $\mathrm{C}_{\mathrm{MPLE}}(\boldsymbol{\varepsilon}) = 1.78 \times 10^5$; (b) $\boldsymbol{\varepsilon} = (16\ 14)$, with simple kriging $\mathrm{C}_{\mathrm{MPLE}}(\boldsymbol{\varepsilon}) = 1.67 \times 10^5$; (c) $\boldsymbol{\varepsilon} = (8\ 11)$, with simple kriging $\mathrm{C}_{\mathrm{MPLE}}(\boldsymbol{\varepsilon}) = 1.50 \times 10^5$; (d) $\boldsymbol{\varepsilon} = (5\ 4)$, with simple kriging $\mathrm{C}_{\mathrm{MPLE}}(\boldsymbol{\varepsilon}) = 1.16 \times 10^5$.
(e)–(g): Universal and simple kriging predictions with different polynomials added to capture the mean. (e) Universal kriging predictions with $\boldsymbol{\varepsilon} = (5\ 4)$ and mean approximated by constant $\boldsymbol{p}$; (f) simple (i.e., zero mean) kriging predictions with $\boldsymbol{\varepsilon} = (20\ 21)$ are lousy; (g) universal kriging predictions with $\boldsymbol{\varepsilon} = (20\ 21)$ and mean approximated by polynomial of total degree 8.

kernel. Despite the fact that mixed integer nonlinear programming strategies exist for optimizing this quantity over the domain $(\boldsymbol{\varepsilon}, \boldsymbol{p}) \in \mathbb{R}^2 \times \mathbb{N}^2$, we neglect to search for this optimal value in Example 17.6.

This seeming oversight is intentional, and results from the mechanism by which we estimate $\boldsymbol{\beta}$: generalized least squares. Adding additional polynomials to the mean can never hurt the approximation because the least squares solver will ignore them if they do not contribute. As a result, $C_{\text{MPLE}}(\boldsymbol{\varepsilon}, \boldsymbol{p})$ should always decrease for polynomials of higher total degree, as evidenced by the plots in Figure 17.5(a)–(d). This presents a problem for an optimization scheme because the minimum exists in the limit as the polynomial degree grows unbounded. To counteract this, one could simply choose to terminate after the improvement between subsequent polynomial orders is below a certain threshold.

Alternatively, one could convert the problem into one with a local minimum by applying *prior beliefs* to the optimization criterion. Such a strategy is often referred to as *maximum a posteriori estimation* (rather than maximum likelihood estimation) because the combination of prior beliefs and the likelihood produce a posterior distribution. This has been applied to kriging problems in, e.g., [Berger *et al.* (2001); Nagy *et al.* (2007)].

Example 17.7. (Kriging prediction for glacier data II)
In Example 17.6 we considered the use of universal kriging with a compactly supported kernel to predict heights from glacial data. Compactly supported kernels are ideal when one can exploit them to create a sparse covariance matrix K. Recalling Program 17.6, we check the density of a sparse matrix with the command

```
nnz(K)/prod(size(K))
```

to see that the choice of $\boldsymbol{\varepsilon} = (20\ 21)$ produces K which is 0.25% dense, whereas $\boldsymbol{\varepsilon} = (5\ 4)$ produces K which is already 4.17% dense. While many of us in the kernel community may consider a 4% dense matrix to be very sparse, it is still a good deal more dense than matrices of comparable size in, e.g., the finite element community. Furthermore, the cost of computing with a sparse matrix that is "not sparse enough" can be unreasonable, as we saw in Figure 17.3(c) before the switch to dense storage.

Given that the kriging predictions for Figure 17.5(e) and Figure 17.5(g) were visually similar, we use `StatFitMLEPolynomialTime.m` to measure the time required to compute with each kernel. The results are displayed in Table 17.1.

Clearly, the most significant additional cost when dealing with the less dense (5 4)-based K matrix is Cholesky factorization, which is roughly seven times more expensive. All the operations are slower than in the (20 21) setting but that computation dominates. The introduction of the polynomial terms increases the cost of computing C_{MPLE}, primarily by increasing the cost in solving for $\boldsymbol{\beta}$. Our conclusion from these, admittedly simple, experiments is that there is a potentially large computational benefit in using polynomial terms to handle long-range trends and

Table 17.1 Computational time (in seconds) for universal kriging with narrow and flat compactly supported kernels. These have been averaged over 10 runs to minimize the effects of randomness.

ε	total polynomial degree	$\mathsf{K} = \mathsf{L}\mathsf{L}^T$	$\mathrm{C}_{\mathrm{MPLE}}(\varepsilon, \boldsymbol{p})$	Predictions
(20 21)	0 (constant)	0.59	0.03	0.32
(20 21)	8	0.59	0.21	0.33
(5 4)	0 (constant)	4.04	0.35	0.63
(5 4)	8	4.04	1.45	0.64

allowing compactly supported covariance kernels to act only locally. This benefit would be magnified as the size of the data set grows, leading to more opportunity to use polynomials and keep the K matrix as sparse as possible.

Remark 17.8. The need for the Cholesky factorization, the bottleneck computation, in Example 17.7 is not absolute. Systems involving the kernel matrix K could be solved with iterative schemes; some preconditioning strategies were discussed in Section 12.4. The main benefit to computing the Cholesky factorization $\mathsf{K} = \mathsf{L}\mathsf{L}^T$ is the accessibility of $\det \mathsf{K}$ on the diagonal of L. Alternate strategies for computing the determinant of a matrix without a decomposition can be found in [Bai *et al.* (1996); Barry and Pace (1999); Pace and LeSage (2004)] and specifically for kernel matrices in McCourt (2013a, Chapter 8.4). Other efficient strategies for spatial statistics are discussed in, e.g., [Smirnov and Anselin (2001); Cressie and Johannesson (2008)].

Chapter 18

Machine Learning

In problems we have studied thus far, we have been given data of the form $\{(\boldsymbol{x}_i, y_i)\}_{i=1}^N$, where $\boldsymbol{x}_i \in \Omega \subset \mathbb{R}^d$ and $y_i \in \mathbb{R}$, and have asked to develop an approximating function s which can be used to predict $s(\boldsymbol{x}) \approx y$ for some previously unobserved $\boldsymbol{x}$ value. In Section 1.3, we introduced the scattered data fitting approach, which chose a set of N basis functions and constructed s as a linear combination of those basis functions such that $\|\boldsymbol{s} - \boldsymbol{y}\| = 0$, where $\boldsymbol{y} = \begin{pmatrix} y_1 & \cdots & y_N \end{pmatrix}^T$ and $\boldsymbol{s} = \begin{pmatrix} s(\boldsymbol{x}_1) & \cdots & s(\boldsymbol{x}_N) \end{pmatrix}^T$. Along the same lines, in Section 12.3, we discussed a modification to this strategy, where s was a linear combination of only $M < N$ basis functions and was constructed to solve $\min_{\boldsymbol{s}} \|\boldsymbol{s} - \boldsymbol{y}\|$. Another technique to fit this data was introduced in Chapter 5 where we assumed that the data was realized by a Gaussian random field Y with a specified covariance and we constructed s such that $s(\boldsymbol{x}) = \hat{y}_{\boldsymbol{x}} = \mathbb{E}[Y_{\boldsymbol{x}} | \boldsymbol{Y} = \boldsymbol{y}]$.

If the problem is ill-posed, e.g., if we do not have enough data to capture the complexity of the model, or if we do not have enough complexity in our model to match the data (or do not want to match all the complexity of the data because some of it might be due to errors in the measurements), then a classical way to solve the data fitting problem is via a *regularization* approach. In this chapter we consider such a general regularization strategy to fitting data by first providing an overview in this introduction, and then going through the special cases of RBF network regression, support vector machine (SVM) classification and SVM regression. Each of our algorithms will involve its own particular loss function coupled with an appropriate regularization term. Such regularization is necessary because data in machine learning applications is often tainted by errors that need to be smoothed out to prevent wild oscillatory behavior in the approximation s. The discussion below is only the briefest of introductions to the topic, and more material can be found in specialized books or survey papers on machine learning or statistical learning such as, e.g., [Evgeniou *et al.* (2000); Herbrich (2002); Schölkopf and Smola (2002); Shawe-Taylor and Cristianini (2004); Rasmussen and Williams (2006); Cucker and Zhou (2007); Steinwart and Christmann (2008); Alpaydin (2009); Hastie *et al.* (2009)].

18.1 Regularization Networks

In the literature on learning one starts with some sort of *loss function* L, which depends on an input measurement $\boldsymbol{x}$, its associated value y and a value $s(\boldsymbol{x})$ predicted by the learning algorithm. The goal of the training phase in machine learning algorithms is then to determine the predictor s such that the *empirical risk*

$$R_L = \frac{1}{N}\sum_{i=1}^{N} L\left(y_i, s(\boldsymbol{x}_i)\right)$$

is minimized. Moreover, since — as mentioned above — this problem may not have a unique solution, one frequently couples the risk functional with a *regularization functional* with the help of a regularization parameter $\mu > 0$. The regularization functional frequently measures the smoothness[1] of s.

Use of the *quadratic loss* $L\left(y, s(\boldsymbol{x})\right) = \left(y - s(\boldsymbol{x})\right)^2$, or with vector inputs $L\left(\boldsymbol{y}, \boldsymbol{s}\right) = \|\boldsymbol{y} - \boldsymbol{s}\|^2$, in conjunction with a quadratic regularization functional is known in the statistics literature as *spline smoothing*, *ridge regression* (see Section 15.1), or in the approximation literature as *penalized least squares*, and this will be the approach taken in Section 18.2 for learning via RBF networks. In the literature on inverse problems this approach appears under the name *Tikhonov regularization*.

Before we get into the specific learning applications we want to point out that the regularization theory formulation is quite natural when working in a reproducing kernel Hilbert space $\mathcal{H}_K(\Omega)$. In particular, if we consider the square of the native space norm of s as the associated regularization term, then it can be shown (see, e.g., the *representer theorem* in [Schölkopf and Smola (2002, Section 4.2)], originally due to [Kimeldorf and Wahba (1971)]) that the predictor s can be expressed as a linear combination of kernel functions, i.e.,

$$s(\boldsymbol{x}) = \sum_{j=1}^{N} c_j K(\boldsymbol{x}, \boldsymbol{x}_j),$$

so that the resulting minimization problem is of the form

$$\min_{\boldsymbol{c}} L\left(\boldsymbol{y}, \mathsf{K}\boldsymbol{c}\right) + \mu \boldsymbol{c}^T \mathsf{K}\boldsymbol{c},$$

where K is our usual kernel matrix and $\boldsymbol{y} = \begin{pmatrix} y_1 & \cdots & y_N \end{pmatrix}^T$. This problem is especially easy to solve in case of the squared loss and a square system matrix K as just specified since the optimal solution is obtained by simply solving the linear system $\mathsf{K}\boldsymbol{c} = \boldsymbol{y}$, i.e., the empirical risk is zero and the native space norm of s is automatically minimized (see also [Fasshauer (2007, Chapter 19)], or the discussion following (18.2) below).

[1]The regularization term can also be interpreted as a measure of the complexity of the model. This can be seen, e.g., via the eigenfunction expansion of s since a smooth function has rapidly decaying eigenvalues and therefore the high-frequency eigenfunctions or modes contribute very little to s, i.e., s is not very complex.

18.2 Radial Basis Function Networks

Machine learning networks are a supervised learning strategy which serve to *learn* patterns that exist in a given set of inputs/outputs $\{(\boldsymbol{x}_i, y_i)\}_{i=1}^N$. In its simplest form, the pattern is approximated by a function s through a linear combination of basis functions, similar to earlier interpolants, and common choices of basis functions include the Gaussian, trigonometric functions and sigmoids, which have an "on/off" behavior and can effectively capture the behavior of neurons in a human brain (see, e.g., [Orr (1996); Hastie *et al.* (2009)]).

Additionally, some in the field [Sonnenburg *et al.* (2006); Gönen and Alpaydın (2011)] like to use linear combinations of *different types of kernels* as their basis functions (e.g., each basis function in the network might be a linear combination of a Matérn, Gaussian and Wendland kernel, where their relative weights could be predetermined or determined through optimization, cf. our discussion of *learned kernels* in Section 3.8). Although the simplest neural networks involve only one layer (that is, one linear combination of basis functions) to define s, multiple layers can be used, where the output from one layer is used as the input to the next [Cristianini and Shawe-Taylor (2000); LeCun *et al.* (2012)]. We will consider only single layer learning algorithms using shifts of the Gaussian kernel as the basis functions for s. Matías *et al.* (2004) compare RBF networks with other computational strategies, including kriging, for estimation in a geophysical setting.

Remark 18.1. In most machine learning literature, the Gaussian kernel is referred to as *the* RBF kernel and is written as

$$K(\boldsymbol{x}, \boldsymbol{z}) = \exp\left(-\frac{(\boldsymbol{x}-\boldsymbol{z})^2}{\sigma^2}\right).$$

In our notation, $\varepsilon = 1/\sigma$, and for us RBF represents any radial kernel, not just the Gaussian.

Constructing s using a linear combination of basis functions is the same strategy that we have used up until now to fit our data: if $M \leq N$ kernels are used with centers at locations $\{\boldsymbol{z}_j\}_{j=1}^M$ then

$$s(\boldsymbol{x}) = \sum_{j=1}^{M} c_j K(\boldsymbol{x}, \boldsymbol{z}_j) = \boldsymbol{k}(\boldsymbol{x})^T \boldsymbol{c}, \tag{18.1}$$

using a vector notation similar to the one defined in Section 1.3, but with an M-vector $\boldsymbol{c}$ and basis vector $\boldsymbol{k}(\cdot)^T = \big(K(\cdot, \boldsymbol{z}_1) \cdots K(\cdot, \boldsymbol{z}_M)\big)$. If $M = N$ and $\boldsymbol{z}_i = \boldsymbol{x}_i$ for $i = 1, \ldots, N$ then s is an interpolant, and otherwise s is just an approximation/regression (cf. also Section 15.1). Either way, the coefficients $\boldsymbol{c}$ are defined as the vector which minimizes $\|\mathsf{K}\boldsymbol{c} - \boldsymbol{y}\|$, where the matrix K is now a (nonsymmetric) rectangular matrix with entries $(\mathsf{K})_{i,j} = K(\boldsymbol{x}_i, \boldsymbol{z}_j)$, $i = 1, \ldots, N$, $j = 1, \ldots, M$. In the interpolation setting — since K is square and symmetric positive definite — this can be minimized to 0 with $\boldsymbol{c} = \mathsf{K}^{-1}\boldsymbol{y}$. In the context of RBF networks, $\boldsymbol{c}$ is often defined as the solution to the minimization problem with squared loss of the form

$$\begin{aligned} &\min_{\boldsymbol{c}\in\mathbb{R}^M} \sum_{i=1}^{N} \left(y_i - \sum_{j=1}^{M} c_j K(\boldsymbol{x}_i, \boldsymbol{z}_j) \right)^2 + \mu \sum_{j=1}^{M} c_j^2 \\ \iff \quad &\min_{\boldsymbol{c}\in\mathbb{R}^M} \|\boldsymbol{y} - \mathsf{K}\boldsymbol{c}\|^2 + \mu\|\boldsymbol{c}\|^2 \end{aligned} \tag{18.2}$$

after having specified the kernel K and regularization parameter μ.

Since this is a convex minimization problem, the necessary condition obtained by the standard strategy of differentiating and setting the result equal to zero is also sufficient. We see that

$$\begin{aligned} &\nabla_{\boldsymbol{c}} \left[(\boldsymbol{y} - \mathsf{K}\boldsymbol{c})^T (\boldsymbol{y} - \mathsf{K}\boldsymbol{c}) + \mu \boldsymbol{c}^T \boldsymbol{c} \right] = 0 \\ \iff \quad &\nabla_{\boldsymbol{c}} \left[\boldsymbol{y}^T\boldsymbol{y} - \boldsymbol{c}^T\mathsf{K}^T\boldsymbol{y} - \boldsymbol{y}^T\mathsf{K}\boldsymbol{c} + \boldsymbol{c}^T\mathsf{K}^T\mathsf{K}\boldsymbol{c} + \mu\boldsymbol{c}^T\boldsymbol{c} \right] = 0 \\ \iff \quad &-2\mathsf{K}^T\boldsymbol{y} + 2\mathsf{K}^T\mathsf{K}\boldsymbol{c} + 2\mu\boldsymbol{c} = 0 \end{aligned}$$

using the fact that $\boldsymbol{y}^T\mathsf{K}\boldsymbol{c} = \boldsymbol{c}^T\mathsf{K}^T\boldsymbol{y}$ since it is just a scalar. Thus we can solve the optimization problem (18.2) by solving the linear system

$$(\mathsf{K}^T\mathsf{K} + \mu\mathsf{I}_M)\boldsymbol{c} = \mathsf{K}^T\boldsymbol{y}, \tag{18.3}$$

which is guaranteed to be well-defined (i.e., the inverse exists) for any $\mu > 0$.

As a reminder, we solve the minimization problem (18.2) rather than forming an interpolant because machine learning and other applications generally deal with data where significant errors are present (perhaps on the order of 10% error). This leaves the observed data looking "rough," as though generated by a nonsmooth function. This happens despite the common assumption that the data was generated by a function with some smoothness, since the observations have been corrupted by a nonsmooth error term. By combining smooth kernels with $\mu > 0$ we can force $\boldsymbol{c}$ to not grow too large, and prevent the wild oscillations which would be needed to exactly fit data from a nonsmooth function.

The choice of the "optimal" μ is an open problem, much in the same way that choosing a shape parameter ε for the Gaussians is, or a smoothness parameter β for the iterated Brownian bridge kernels of Chapter 7, or a smoothing parameter μ for smoothing splines in Section 15.1. In practice, however, parametrization strategies such as cross-validation (discussed in Chapter 14) can be used to find "good" values, and we pursue this idea below.

It is interesting to consider the effects of the regularization parameter μ vs. those of the shape parameter ε. Although a change in μ affects the resulting approximation s, it *does not* change the space in which s lives (the reproducing kernel Hilbert space $\mathcal{H}_K(\Omega)$); rather it pushes the interpolant toward a set of functions which try to minimize $\|\boldsymbol{c}\|_2$ rather than $\|s\|_{\mathcal{H}_K(\Omega)}$. This is in contrast to changing ε, which changes K and, therefore, $\mathcal{H}_K(\Omega)$ as well as $\|\cdot\|_{\mathcal{H}_K(\Omega)}$.

Treating μ as a free parameter gives the potential to improve the quality of the approximation but also the potential to cripple it, making an appropriate choice vital as it was in Chapter 14. As in that chapter, some paramtrization tools exist to help guide our judgment. A common tool in the learning community is *generalized cross validation* [Craven and Wahba (1979); Golub *et al.* (1979); Golub and Von Matt (1997)], which is a variant on traditional cross-validation. In an RBF network as described above it can be computed as [Orr (1996); Arlot and Celisse (2010)]

$$\mathrm{C_{GCV}} = \boldsymbol{y}^T \mathsf{P}^2 \boldsymbol{y} \frac{N}{(\operatorname{trace} \mathsf{P})^2}, \qquad \mathsf{P} = \mathsf{I}_N - \mathsf{K}(\mathsf{K}^T\mathsf{K} + \mu \mathsf{I}_M)^{-1}\mathsf{K}^T. \tag{18.4}$$

Note that $\mathrm{C_{GCV}}$ is not an error in the true sense, though it is related to the residual: $\mathsf{P}\boldsymbol{y} = \boldsymbol{y} - \mathsf{K}\boldsymbol{c}$ meaning that $\boldsymbol{y}^T\mathsf{P}^2\boldsymbol{y} = \|\boldsymbol{y} - \mathsf{K}\boldsymbol{c}\|_2^2$. Even so, $\mathrm{C_{GCV}}$ is one criterion that can be used to choose a good μ value.

Remark 18.2. We have assumed that $\mu > 0$ above to guarantee invertibility of $\mathsf{K}^T\mathsf{K} + \mu\mathsf{I}_M$, but the $\mu = 0$ case may also have a unique solution: using the reduced singular value decomposition [Golub and Van Loan (2012)] $\mathsf{K} = \mathsf{U}\Sigma\mathsf{V}^T$ (such that $\Sigma, \mathsf{V} \in \mathbb{R}^{M\times M}$ and $\mathsf{U}^T\mathsf{U} = \mathsf{I}_M$) we can see that

$$\boldsymbol{c} = (\mathsf{K}^T\mathsf{K})^{-1}\mathsf{K}^T\boldsymbol{y} = (\mathsf{V}\Sigma\mathsf{U}^T\mathsf{U}\Sigma\mathsf{V}^T)^{-1}\mathsf{V}\Sigma\mathsf{U}^T\boldsymbol{y} = \mathsf{V}\Sigma^{-2}\mathsf{V}^T\mathsf{V}\Sigma\mathsf{U}^T\boldsymbol{y} = \mathsf{V}\Sigma^{-1}\mathsf{U}^T\boldsymbol{y},$$

where $\mathsf{V}\Sigma^{-1}\mathsf{U}^T$ is the pseudoinverse of K. Thus $\mu > 0$ is not *required* if K has full column rank; given that, it still serves a useful regularization purpose for noisy data.

Even for $\mu > 0$, the computations in (18.3) and (18.4) can be carried out using the reduced SVD:

$$\begin{aligned}
\boldsymbol{c} &= (\mathsf{V}\Sigma\mathsf{U}^T\mathsf{U}\Sigma\mathsf{V}^T + \mu\mathsf{I}_M)^{-1}\mathsf{V}\Sigma\mathsf{U}^T\boldsymbol{y} = \mathsf{V}(\Sigma + \mu\Sigma^{-1})^{-1}\mathsf{U}^T\boldsymbol{y}, \\
\mathsf{P} &= \mathsf{I}_N - \mathsf{U}\Sigma\mathsf{V}^T(\mathsf{V}\Sigma\mathsf{U}^T\mathsf{U}\Sigma\mathsf{V}^T + \mu\mathsf{I}_M)^{-1}\mathsf{V}\Sigma\mathsf{U}^T = \mathsf{I}_N - \mathsf{U}(\mathsf{I}_M + \mu\Sigma^{-2})^{-1}\mathsf{U}^T.
\end{aligned}$$

In the examples in the section below, these formulas are used to help minimize instabilities that arise when μ is small and $\mathsf{K}^T\mathsf{K}$ is ill-conditioned.

Remark 18.3. RBF networks can be used for classification, i.e., to predict categorical patterns rather than continuous patterns. Support vector machines (introduced in Section 18.3) are more popular for this purpose, and as such we omit a discussion of network classification. Some results on RBF networks for classification include [Nabney (2004); Fernández-Navarro *et al.* (2011)].

18.2.1 *Numerical experiments for regression with RBF networks*

We now have some intuition as to the impact of noisy data and the regularization approach which is used to combat it in machine learning. We must, however, study the role of the regularization parameter μ in the context of the other parameters which define our kernel.

Example 18.1. (Basic RBF network)
Beginning with $N = 50$ data values sampled randomly on $[-1, 1]$ from the function

$$f(x) = (1 - 4x + 32x^2)e^{-16x^2}$$

we added normally distributed noise with zero mean and standard deviation 0.2. This is displayed in Figure 18.1(a). Computations in MATLAB are presented in Program 18.1, and the randomness in the example is made consistent in line 1.

Program 18.1. Excerpts from `RBFNetworkBasic.m`

```
rng(0)
% Choose the function to study
yf = @(x) (1-4*x+32*x.^2).*exp(-16*x.^2);
% Form random data locations, but include boundaries
N = 50;  x = pickpoints(-1,1,N-2,'rand');  x = [x;-1;1];
% Create noisy data
noise = .2;  y = yf(x) + noise*randn(N,1);
% Create test points for checking the error
xeval = pickpoints(-1,1,300);  yeval = yf(xeval);
% Choose our radial kernel (the RBF)
rbf = @(e,r) exp(-(e*r).^2);    ep = 8;
% Choose points at which to center the RBFs
M = 15;  z = pickpoints(-1,1,M,'halton');
% Form the kernel matrix and SVD for use in fitting the network
K_fit = rbf(ep,DistanceMatrix(x,z));  [U,S,V] = svd(K_fit,0);  Sv = diag(S);
K_predict = rbf(ep,DistanceMatrix(xeval,z));
% Loop through a range of mu, the regularization parameters
muvec = logspace(-10,5,100);   k = 1;
for mu=muvec
    % Solve for the network weights, using the SVD
    c = V*((U'*y)./(Sv+mu./Sv));
    % Evaluate predictions and their error on the testing points
    ypred = K_predict*c;   errvec(k) = errcompute(ypred,yeval);
    % Find the projection matrix and use it to evaluate the residual
    P = eye(N) - bsxfun(@rdivide,U,(1+mu./Sv.^2)')*U';   Py = P*y;
    % Evaluate the parametrization schemes
    loovec(k) = Py'*(Py./diag(P).^2)/N;
    gcvvec(k) = N*(Py'*Py)/trace(P)^2;   k = k + 1;
end
```

To make predictions, we choose to form an RBF network using $M = 15$ Gaussian kernels with $\varepsilon = 8$ centered at evenly spaced points in the domain. To smooth out the noise, we consider regularization parameters μ in the interval $[10^{-10}, 10^5]$. The resulting root mean-squared relative error (sampled at 300 evenly spaced points in the domain) is on display in Figure 18.1(b), along with the values of err_{GCV}. We compute the SVD of the $N \times M$ kernel matrix in line 15 and follow the strategy from Remark 18.2 in line 21 to compute the network coefficients $\boldsymbol{c}$.

For this example, the "Least GCV" μ value (computed without the true solution) is close to the "Least Error" μ value (which requires the true solution), although they do not align exactly. As expected, the network designed using larger μ is less oscillatory than the network produced with the smaller μ value.

Clearly, the μ parameter can affect the quality of the network predictions in an understandable (larger μ yields less oscillation) but unpredictable (the optimal amount of oscillation is unknown *a priori*) way, much as the shape parameter ε for Gaussians works in unpredictable ways. In fact, because of the ill-conditioned nature of $\mathsf{K}^T\mathsf{K}$ as $\varepsilon \to 0$ (as discussed in Chapter 11), for small ε values, μ can also serve to improve the condition of (18.3). This further complicates the role of μ and the relationship between μ and ε because for some choice of basis functions μ may improve the accuracy through smoothing out noise and for others it may improve the quality by reducing ill-conditioning.

We have, however, already discussed other tools for computing stably with Gaussians in the flat limit: the Hilbert–Schmidt SVD Chapter 13 and the eigenfunction basis Section 12.2.1. The role of the μ parameter is studied in the following example, where the use of more stable bases than the standard Gaussian basis will affect the smoothing/stability performance of the regularization.

Example 18.2. (RBF network with HS-SVD)
We repeat the setup from Example 18.1 but now choose $\varepsilon = 0.01$, for which the matrix $\mathsf{K}^T\mathsf{K}$ is severely ill-conditioned. Three options are compared for forming an RBF network (18.1) to fit this noisy data: using the standard $K(\cdot, z_j)$ basis with z_j evenly spaced throughout the domain, using the stable $\psi_j(\cdot)$ basis with z_j evenly spaced throughout the domain, and using the eigenfunction basis $\varphi_j(\cdot)$. The error of these fits is computed across various regularization values μ in the `GaussQR` library in `RBFNetworkOtherBases`, but the code is omitted here to save space. Results are displayed in Figure 18.1(d).

For large μ values, the regularization component of (18.2) overwhelms the contribution of the network basis which causes the overlap near $\mu \approx 10^4$ for all basis choices. As μ decreases, the stable and eigenfunction bases outperform the standard basis, with little difference between the stable and eigenfunction bases.

The similarity between the stable and eigenfunction bases in Example 18.2 can be rationalized by recalling from Section 12.3 that the correction term used to form the stable basis decreases in magnitude as ε decreases. For this reason, when small ε is used in RBF networks, the eigenfunction basis is used rather than the stable basis in the remainder of this section.

This example also demonstrates that the stability provided from regularizing with $\mu > 0$ has a different effect on the resulting RBF network than the stability afforded through the Hilbert–Schmidt SVD. As is clear from the behavior in Figure 18.1(a)–(d), the effect of the μ parameter can vary greatly depending on the value of the ε parameter. In Example 18.3 we study the interaction of μ and ε in forming an RBF network.

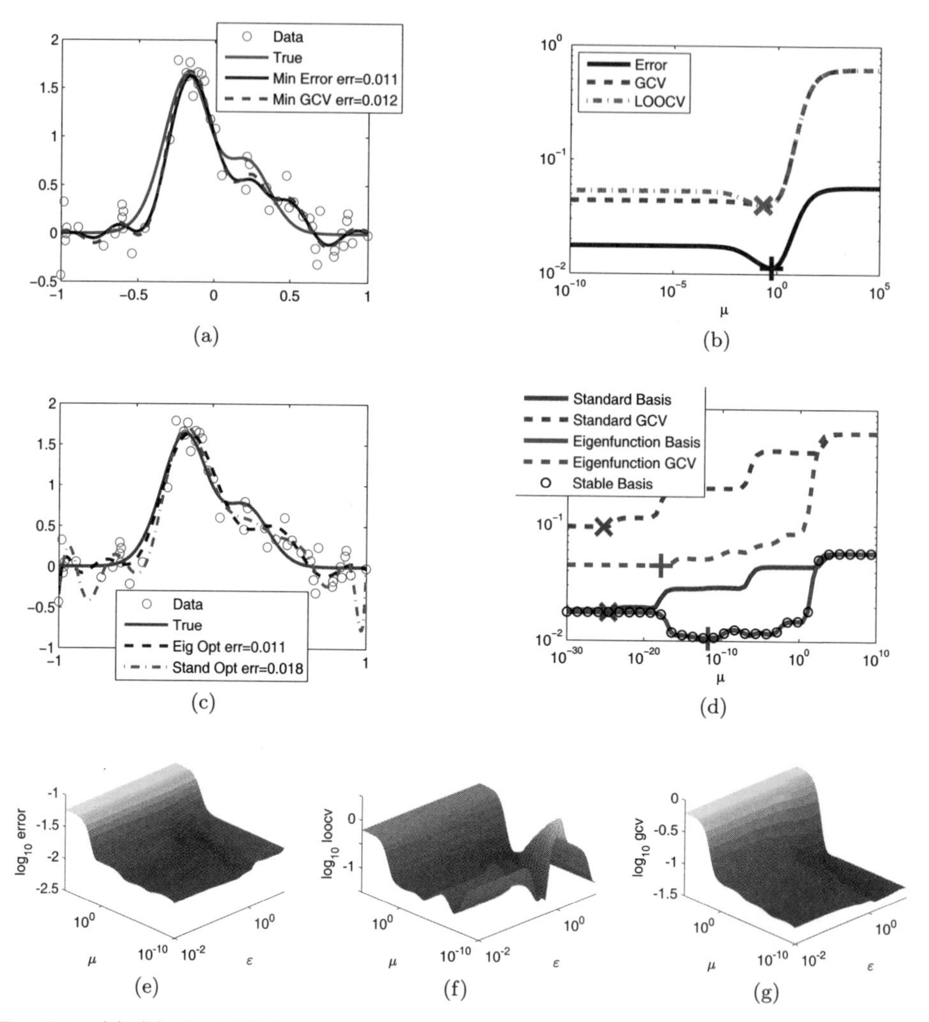

Fig. 18.1 (a)–(b): Basic RBF network fit. (a) $N = 50$ noisy data points and the original pattern are displayed along with RBF network fits generated by $M = 15$ Gaussian kernels (using $\varepsilon = 8$) with centers evenly spaced in [-1,1]. The black curve uses $\mu \approx 0.56$ (based on minimum error) and the blue one uses $\mu \approx 0.22$ (based on minimum GCV criterion (18.4)). The plot for LOOCV-minimal μ is omitted as it matches the one for GCV; (b) Error compared to true function and parameter optimization criteria.
(c)–(d): The stable basis outperforms the standard basis for most μ values. (c) $N = 50$ noisy data points and the original pattern are displayed along with RBF network fits generated with optimal μ values using the standard Gaussian kernel basis (magenta, using $\varepsilon = 0.01$) with centers evenly spaced in [-1,1], and with Gaussian eigenfunctions (black). The stable basis is not displayed as it overlaps with the eigenfunction basis; (d) The eigenfunction basis and stable basis essentially overlap with only minor differences for any μ value. Optimal values for each μ profile are denoted with $\times$ for the standard basis and $+$ for the eigenfunction basis.
(e)–(g): Effects of varying both ε and μ. (e) An optimal error for predictions exists at $\mu \approx 0.011$ and $\varepsilon \approx 2.8$, values that neither CV scheme located. The error in the $\varepsilon \to 0$ and $\mu \to 0$ limits is roughly twice the error at these optimal values; (f) LOOCV predicts optimal $\mu \approx 0.53$, $\varepsilon \approx 4.9$; (g) GCV predicts optimal $\mu \approx 0.0017$, $\varepsilon \approx 3.7$.

Example 18.3. (Optimizing parameters in an RBF network)

Again, starting from the foundation of Example 18.1 we consider the effect of allowing both ε and μ to change. Only the eigenfunction basis is considered, given its superior performance for smaller ε in Example 18.2. The errors for $\mu \in [10^{-10}, 10^5]$ and $\varepsilon \in [10^{-2}, 10^1]$, along with the LOOCV (from Section 14.2) and GCV plots, can be seen in Figure 18.1(e)–(g). The necessary computations are carried out in the `GaussQR` library in the function `RBFNetworkEpMuProfile`, but are not presented here to save space and avoid redundancy with Program 18.1.

The optimal accuracy is not attained by either the LOOCV ($\mu \approx 0.53$, $\varepsilon \approx 4.9$) or GCV ($\mu \approx 0.0017$, $\varepsilon \approx 3.7$) predictions. For small μ or even moderately small ε, the LOOCV measure performs very erratically, likely because losing the value nearest the boundary causes the predictions to miss significantly. While neither measure is perfect, they are both decent guides toward the optimal region, and they both suggest that the optimal result involves $\mu > 0$ and $\varepsilon > 0$, which is true.

In all of these examples the choice of $\boldsymbol{z}_j$ values was made arbitrarily, and the selection of optimal, or even good, kernel centers is a nonlinear problem and still an open field of research. In the numerical analysis setting, this is discussed in [Hangelbroek and Ron (2010); Schaback (2013)] among other articles. In the statistical setting, this is discussed in [Chen *et al.* (1991); Orr (1995); Mao (2002)], and more recently in [Kryżyak (2011); Deng *et al.* (2012); Du *et al.* (2012); Zhou and Yang (2013)], often in the context of subset or node selection, where a subset of the provided data points are used as centers in the fitted network.

Furthermore, the existence of multilayer networks, where the predictions of one network are used as data for the next layer (common for perceptrons but less so for RBF networks), introduces flexibility in finding and predicting a pattern but requires a nonlinear solver to determine the appropriate coefficients at most levels [Wang and Zhang (2013)]. Such a strategy can have a similar effect to freely choosing the RBF centers, but is beyond the scope of this work.

Predicting pattern behavior with RBF networks is an old strategy that is structurally similar to the regression with positive definite kernels discussed in Chapter 15: both try to fit scattered data using linear combinations of kernels dispersed throughout the domain of interest. This setting generally produces a different linear combination to account for the noise that is often present in statistical applications. By choosing $\mu > 0$ in (18.2) and demanding a "less active" RBF network, the obscuring effect of the noise is attenuated. In Section 18.3 we explore a more modern strategy to fit noisy data.

18.3 Support Vector Machines

In this section we will discuss the two main applications of *support vector machines* (SVMs) in the context of supervised machine learning: classification and regression.

Both of these applications can be formulated within the regularization framework outlined at the beginning of this chapter.

For classification we are given a set of *training data* $\{(\boldsymbol{x}_i, y_i) \mid i = 1, \ldots, N\}$ with *measurements* $\boldsymbol{x}_i \in \mathbb{R}^d$ and data values in the form of *labels* $y_i \in \{-1, +1\}$. The standard (binary) classification problem consists of finding a predictor s that will allow us to assign an appropriate label, either -1 or $+1$, to a future measurement $\boldsymbol{x}$. Such a predictor might be of the form $s(\boldsymbol{x}) = \operatorname{sign}(h(\boldsymbol{x}))$, where h denotes a hyperplane separating the given measurements. A typical loss function for this kind of problem is given by the *hinge loss* (or *soft margin loss*)

$$L(y, h(\boldsymbol{x})) = \max(1 - yh(\boldsymbol{x}), 0)$$

since $L(y, h(\boldsymbol{x})) = 0$ if and only if $yh(\boldsymbol{x}) \geq 1$, i.e., y and $h(\boldsymbol{x})$ have the same sign and $|h(\boldsymbol{x})| \geq 1$ so that we have enough *confidence* in our prediction (see, e.g., [Schölkopf and Smola (2002, Chapter 3)]). An appropriate regularization term will be given by some norm of h (see below for more details).

For regression, i.e., our estimation of continuous numeric values as already discussed in the previous section, the standard loss function is the squared loss $L(y, s(\boldsymbol{x})) = (y - s(\boldsymbol{x}))^2$ (and that is what we used for the RBF network formulation). Alternatively, the so-called *ϵ-insensitive loss*

$$L(y, s(\boldsymbol{x})) = \max(|y - s(\boldsymbol{x})| - \epsilon, 0)$$

is used for SVM regression as a symmetric analogue of the hinge loss (which is used for SVM classification). According to the ϵ-insensitive loss function, deviations of the predicted value $s(\boldsymbol{x})$ from the correct value y are only penalized if they exceed ϵ, and therefore it will be possible to obtain *sparse* representations using only a subset of the data referred to as *support vectors* (see below for details). While for RBF networks one also strives for a representation based on only a small set of RBF centers, in SVM regression the support vectors are automatically determined by the optimization algorithm.

The machine learning literature also contains other loss functions, such as the *Huber loss* which is popular among statisticians. We will, however, deal only with the three main loss functions mentioned above.

18.3.1 *Linear classification*

In the simplest case, our predictor is given by $s(\boldsymbol{x}) = \operatorname{sign}(h(\boldsymbol{x}))$, where h denotes a hyperplane — directly in input space — of the form

$$h(\boldsymbol{x}) = \boldsymbol{x}^T\boldsymbol{w} + b = 0, \qquad \boldsymbol{x} \in \mathbb{R}^d,$$

that separates the measurements with label -1 from those with label $+1$. The *weights* $\boldsymbol{w}$ (which serve as the unit normal vector to the hyperplane) and the *bias* b can be determined by maximizing the margin or gap to both sides of this hyperplane (see, e.g., [Hastie *et al.* (2009, Chapter 12)]).

Since the size of this margin is equal to $\frac{1}{\|\boldsymbol{w}\|}$, and we would ideally like to have this margin as large as possible, we are provided with a regularization functional for this problem: we want to minimize $\|\boldsymbol{w}\|$, the norm of the coefficients of h. Thus, using the hinge loss function and remembering that $h(\boldsymbol{x}) = \boldsymbol{x}^T\boldsymbol{w} + b$, we would have an unconstrained minimization problem of the form

$$\min_{\boldsymbol{w},b} \left[\frac{1}{N} \sum_{i=1}^{N} \max\left(1 - y_i h(\boldsymbol{x}_i), 0\right) + \mu \frac{1}{2} \boldsymbol{w}^T \boldsymbol{w} \right],$$

where μ is an appropriately chosen regularization parameter.

Instead of this formulation, the following constrained optimization with *slack variables* ξ_i is more common since it also allows us to deal with the case where the given measurements are not perfectly separable by h:

$$\begin{aligned} &\min_{\boldsymbol{w},b,\boldsymbol{\xi}} \left[\frac{1}{2} \boldsymbol{w}^T \boldsymbol{w} + C \sum_{i=1}^{N} \xi_i \right] \\ &\text{subject to } y_i h(\boldsymbol{x}_i) \geq 1 - \xi_i, \quad i = 1, \ldots, N, \\ &\qquad\qquad \xi_i \geq 0, \end{aligned}$$

where the regularization parameter μ is transformed into the parameter $C = \frac{1}{N\mu}$.

The formulation considered thus far is known in the SVM literature as the *primal problem* (and — ironically — as the dual problem in the optimization literature). The corresponding (SVM) *dual problem* can be derived via Lagrange multipliers α_i (see, e.g., [Hastie *et al.* (2009, Chapter 12)]) and is of the form

$$\begin{aligned} &\max_{\boldsymbol{\alpha}} \left(\sum_{i=1}^{N} \alpha_i - \frac{1}{2} \sum_{i=1}^{N} \sum_{j=1}^{N} \alpha_i \alpha_j y_i y_j \boldsymbol{x}_i^T \boldsymbol{x}_j \right) \\ &\text{subject to } \sum_{i=1}^{N} \alpha_i y_i = 0, \\ &\qquad\qquad 0 \leq \alpha_i \leq C, \end{aligned} \tag{18.5}$$

where C is known as a *box constraint* and

$$\boldsymbol{w} = \sum_{i=1}^{N} \alpha_i y_i \boldsymbol{x}_i$$

(which follows from setting the $\boldsymbol{w}$-gradient of the primal Lagrange multiplier functional equal to zero). The bias b is given by $b = y_i - \boldsymbol{x}_i^T \boldsymbol{w}$ for any i such that the optimal $\alpha_i > 0$. For stability purposes we actually consider all qualifying indices and find b using the mean. Note that the box constraint C is a free parameter which needs to be either set by the user or determined by an additional parameter optimization methods such as cross validation.

18.3.2 *Kernel classification*

In Section 2.4 we talked about feature maps, i.e., the fact that the kernel values $K(\boldsymbol{x}, \boldsymbol{z})$ can be viewed as the dot product of the transformed data in *feature space*, i.e., given $\boldsymbol{x}$ and $\boldsymbol{z}$ in *input space* and a feature map Φ we have[2]

$$K(\boldsymbol{x}, \boldsymbol{z}) = \Phi_{\boldsymbol{x}}^T \Phi_{\boldsymbol{z}}. \tag{18.6}$$

Since the objective function in (18.5) above was expressed in terms of dot products in *input space* we can now use the concept of feature maps and related kernels to talk about separating hyperplanes in *feature space*. Note that this feature space is potentially infinite-dimensional (as, e.g., in the case of the Gaussian kernel) and therefore offers much more flexibility for separating the data than the finite-dimensional input space. This fact has a theoretical foundation in the form of *Cover's theorem* [Cover (1965)] which ensures that data which can not be separated by a hyperplane in input space most likely will be linearly separable after being transformed to feature space by a suitable feature map. Thus, support vector machines — and kernel machines, in particular — are a good tool to use in order to tackle intricate data classification problems.

The algorithms for kernel classification are now more or less the same as before; simply replace the measurements $\boldsymbol{x}_i$ in input space by their features $\Phi_{\boldsymbol{x}_i}$ in feature space. The separating hyperplane can be expressed in the form

$$h(\boldsymbol{x}) = \Phi_{\boldsymbol{x}}^T \boldsymbol{w} + b = 0, \qquad \boldsymbol{x} \in \mathbb{R}^d,$$

and the dual problem for SVM classification using the transformed input data is given by

$$\begin{aligned} \max_{\boldsymbol{\alpha}} & \left(\sum_{i=1}^{N} \alpha_i - \frac{1}{2} \sum_{i=1}^{N} \sum_{j=1}^{N} \alpha_i \alpha_j y_i y_j \Phi_{\boldsymbol{x}_i}^T \Phi_{\boldsymbol{x}_j} \right) \\ \text{subject to } & \sum_{i=1}^{N} \alpha_i y_i = 0, \\ & 0 \le \alpha_i \le C. \end{aligned}$$

Moreover, since we have the kernel decomposition (18.6) we do not actually have to compute (possibly infinite) dot products in feature space, but instead just fill the

[2]Here we assume that the features lie in a sequence space such as the space ℓ_2 associated with the Mercer series representation of K. The approach discussed in this section can also be generalized to other feature spaces $\mathcal{H}$ where $K(\boldsymbol{x}, \boldsymbol{z}) = \langle \Phi_{\boldsymbol{x}}, \Phi_{\boldsymbol{z}} \rangle_{\mathcal{H}}$, but then the representation of the separating hyperplane h below would have to be of a more general convolution type, i.e., $h(\boldsymbol{x}) = \langle \Phi_{\boldsymbol{x}}, w \rangle_{\mathcal{H}} + b$, where w is an unknown weight function.

kernel matrix and solve

$$\max_{\boldsymbol{\alpha}} \left(\sum_{i=1}^{N} \alpha_i - \frac{1}{2} \sum_{i=1}^{N} \sum_{j=1}^{N} \alpha_i \alpha_j y_i y_j K(\boldsymbol{x}_i, \boldsymbol{x}_j) \right) \tag{18.7}$$

$$\text{subject to } \sum_{i=1}^{N} \alpha_i y_i = 0,$$

$$0 \le \alpha_i \le C,$$

where, as before, C is the box constraint (which can be viewed as a tuning parameter) and

$$\boldsymbol{w} = \sum_{i=1}^{N} \alpha_i y_i \Phi_{\boldsymbol{x}_i}. \tag{18.8}$$

The classifier is now given by

$$\begin{aligned} s(\boldsymbol{x}) &= \operatorname{sign}(h(\boldsymbol{x})) \\ &= \operatorname{sign}\left(\Phi_{\boldsymbol{x}}^T \sum_{j=1}^{N} \alpha_j y_j \Phi_{\boldsymbol{x}_j} + b \right) \\ &= \operatorname{sign}\left(\sum_{j=1}^{N} \alpha_j y_j K(\boldsymbol{x}, \boldsymbol{x}_j) + b \right), \end{aligned} \tag{18.9}$$

where b is obtained as before, i.e., $b = y_i - \sum_{j=1}^{N} \alpha_j y_j K(\boldsymbol{x}_i, \boldsymbol{x}_j)$ with i denoting the index of an α_i which is strictly between 0 and C. For stability purposes we can again average over all such candidates.

What does the separating hyperplane in this case look like? First, we note that the hyperplane will be linear only in feature space (which we usually have no concrete knowledge of). In the input space the data will be separated by a nonlinear manifold. Moreover, the representation of this manifold is *sparse* in the sense that not all basis functions are needed to specify it. In fact, only those centers $\boldsymbol{x}_j$ whose corresponding α_j are nonzero define meaningful basis functions. These special centers are referred to as *support vectors*.

Remark 18.4. Since the decision boundary can be expressed in terms of a limited number of support vectors, i.e., it has a sparse representation, learning is possible in very high-dimensional input spaces (see, e.g., [Steinwart and Christmann (2008)]). Moreover, SVMs are robust against several types of model violations and outliers, and they are computationally efficient, e.g., by using the sequential minimal optimization (SMO) algorithm of [Platt (1999)] to perform the quadratic optimization task required for classification as well as regression. Another way to make SVMs perform more efficiently is to consider a low-rank representation for the kernel as in [Fine and Scheinberg (2002)]. We test our own version of this latter idea below in Section 18.3.4.

Remark 18.5. For positive definite kernels it is also possible to formulate the separating hyperplane without the bias term b. In that case the equality constraint $\sum_{i=1}^{N} \alpha_i y_i = 0$ (which may be somewhat of a nuisance during the optimization process) can be omitted (see, e.g., [Poggio *et al.* (2001)]).

Remark 18.6. The primal and dual formulations each have their advantages. For example, the primal formulation (in input space) is good for large amounts of rather low-dimensional data, while the dual formulation (with kernels in feature space) is good for high-dimensional data (since only the number of support vectors matter).

Remark 18.7. In addition to the linear SVM (with dot product kernel $K(\boldsymbol{x}, \boldsymbol{z}) = \boldsymbol{x}^T \boldsymbol{z}$) and the Gaussian kernel which we use in our numerical experiments, polynomial kernels of degree β in the form $K(\boldsymbol{x}, \boldsymbol{z}) = (1 + \boldsymbol{x}^T \boldsymbol{z})^\beta$ and the sigmoid (or multilayer perceptron) kernel $K(\boldsymbol{x}, \boldsymbol{z}) = \tanh(1 + \varepsilon \boldsymbol{x}^T \boldsymbol{z})$ (see also Section 18.2) are common in the machine learning community. Moreover, kernels may be defined via the feature map (instead of in closed form), and this feature map can be picked depending on the specific application at hand (e.g., as a string kernel for text mining).

MATLAB's Statistics and Machine Learning Toolbox provides outstanding functions and documentation for training and predicting with support vector machines. We provide our own much less comprehensive and efficient training implementation in the GaussQR library function gqr_fitsvm.m. Part of it is listed here to demonstrate how the quadratic program described in this section can be written in MATLAB and for comparison to the training of RBF networks.

Program 18.2. Excerpts from gqr_fitsvm.m

```
function SVM = gqr_fitsvm(x,y,ep,bc,low_rank)
% function SVM = gqr_fitsvm(x,y,ep,bc,low_rank)
% This function fits a support vector machine to given data
%   Inputs: x - data locations
%           y - classifications
%           ep - Gaussian shape parameter
%           bc - box constraint
%           low_rank - (optional, default=0) use the eigenfunction decomp
%   Output: SVM - object to classify future data
%           SVM.eval(x_new) - classify the point x_new
%           SVM.sv_index - the indices of the support vectors
N = length(y);
% Define the RBF and GQR alpha parameter as needed
rbf = @(e,r) exp(-(e*r).^2);
% Define the kernel matrix
DM = DistanceMatrix(x,x);
K = rbf(ep,DM);
% Define the quadratic programming problem
A = (y*y').*K;
f = -ones(N,1);
```

```
Ain = [];
bin = [];
Aeq = y';
beq = 0;
lb = zeros(N,1);
ub = bc*ones(N,1);
x0 = bc/2*ones(N,1);
% Solve the quadratic programming problem
optimopt = optimset('LargeScale','off','Display','off','MaxIter',500);
sol_QP = quadprog(A,f,Ain,bin,Aeq,beq,lb,ub,x0,optimopt);
% Store the coefficients in the SVM object
SVM.coef = y.*sol_QP;
SVM.sv_index = sol_QP>0;
% Determine the bias
bias_find_coef = sol_QP>0 & sol_QP<bc;
SVM.bias = mean(y(bias_find_coef) - K(bias_find_coef,:)*SVM.coef);
% Create a function to classify points with the SVM
SVM.eval = @(x_new) sign( ...
    rbf(ep,DistanceMatrix(x_new,x(SVM.sv_index,:)))*SVM.coef(SVM.sv_index) ...
                   + SVM.bias);
```

The function definition includes an option `low_rank` to use the eigenfunction decomposition to approximate the kernel matrix; we will revisit this option in Section 18.3.4. This training algorithm returns an object we have labeled as `SVM` which contains the members

- `coef` — The $\alpha_i y_i$ values in the SVM prediction (18.9),
- `sv_index` — The indices of the support vectors, which are the only indices required for making prediction,
- `bias` — Unsurprisingly, the bias term b in (18.9), and
- `eval` — A function which evaluates the $s(\boldsymbol{x})$ prediction in (18.9).

The beginning of this function contains the same elements as many of the interpolation problems appearing earlier in the book, most noticeably, the kernel matrix K. New elements are then defined to solve the quadratic programming problem (18.7).

MATLAB's Optimization Toolbox provides a function `quadprog` which can solve such problems, but we must format our problem in the form

$$\min_{\boldsymbol{\alpha}} \boldsymbol{f}^T\boldsymbol{\alpha} + \frac{1}{2}\boldsymbol{\alpha}^T\mathsf{A}\boldsymbol{\alpha}, \quad \text{subject to} \tag{18.10a}$$

$$\mathsf{A}_{\text{ineq}}\boldsymbol{\alpha} \le \boldsymbol{b}_{\text{ineq}}, \tag{18.10b}$$

$$\mathsf{A}_{\text{eq}}\boldsymbol{\alpha} = \boldsymbol{b}_{\text{eq}}, \tag{18.10c}$$

$$\boldsymbol{b}_{\text{lower}} \le \boldsymbol{\alpha} \le \boldsymbol{b}_{\text{upper}}. \tag{18.10d}$$

Line 19 defines the A matrix which is the kernel matrix with some additional contributions from the $\boldsymbol{y}$ vector. The $\boldsymbol{f}$ vector is defined in line 20, and the minus sign is needed because `quadprog` is solving the minimization problem, not the maximization problem as (18.7) was phrased.

No inequality constraints are required, thus the `[]` in lines 21–22. The equality constraint is defined in the next two lines, and the bounds on acceptable $\boldsymbol{\alpha}$ values are defined in lines 25–26; note the presence of the box constraint `bc` imposing an upper bound. We make a rather uninformed initial guess in line 27.

In line 30 we actually call `quadprog` to solve (18.10) with some options chosen on the previous line. The solution $\boldsymbol{\alpha}$ is element-wise multiplied by the $\boldsymbol{y}$ vector to create the necessary coefficients for SVM classification using (18.9). We identify and store the indices of the support vectors in line 33 and at the end we create a classification function using the solution to the quadratic program and the bias computed just above. This classification could be written as

$$s(\boldsymbol{x}) = \operatorname{sign}\left(\boldsymbol{k}(\boldsymbol{x})^T \boldsymbol{c} + b\right), \qquad \boldsymbol{c} = \begin{pmatrix} \alpha_1 y_1 \\ \vdots \\ \alpha_N y_N \end{pmatrix},$$

but, because of the sparsity enjoyed by support vector machines, only a portion of the $\boldsymbol{k}(\boldsymbol{x})^T \boldsymbol{c}$ inner product actually contributes nonzero values. Those support vectors are the only values used in the `SVM.eval` function, and they are accessed with the `SVM.sv_index` vector of indices.

The time tests run throughout this section use this `gqr_fitsvm.m`, which is by no means as efficient or robust as the function `fitcsvm` in MATLAB's Statistics and Machine Learning Toolbox. We present timing results to provide a comparison between different parametrizations and training strategies for support vector machines and not to suggest that the actual times observed are what should be expected when running high-quality SVM algorithms such as `fitcsvm`.

18.3.3 *Numerical experiments for classification with kernel SVMs*

We look at the last example in the MATLAB documentation for SVMs, taken from [Hastie *et al.* (2009, Section 2.3)]. Two normal populations are studied, both with the identity for covariance: population 1 is centered at $(1,0)$ and population 2 is centered at $(0,1)$. A test population of 10 points from each population is selected, and 100 training points are generated from these test sets to learn each population. These points are displayed in Figure 18.2(a), where the population centers $(1,0)$ and $(0,1)$ are marked by a filled green square and filled red circle, respectively. The red $\times$ and blue ○ points belong to the $(0,1)$ population and denote the training and test points, respectively. The green $+$ and blue □ points play the analogous role for the $(1,0)$ population. The function `SVM_setup` in the `GaussQR` library creates this data set.

Example 18.4. (Decision contours for support vector machines)
This example shows the decision contours from various kernel parametrizations: Figure 18.2(b)–(c) alternately fixes C and ε values and allows the other parameter to vary to show the impact on the SVM. These tests are run with the script

`SVMContourSamples.m` in the `GaussQR` library, but it is not listed here as much of the code involves repetitive testing and intricate plotting commands.

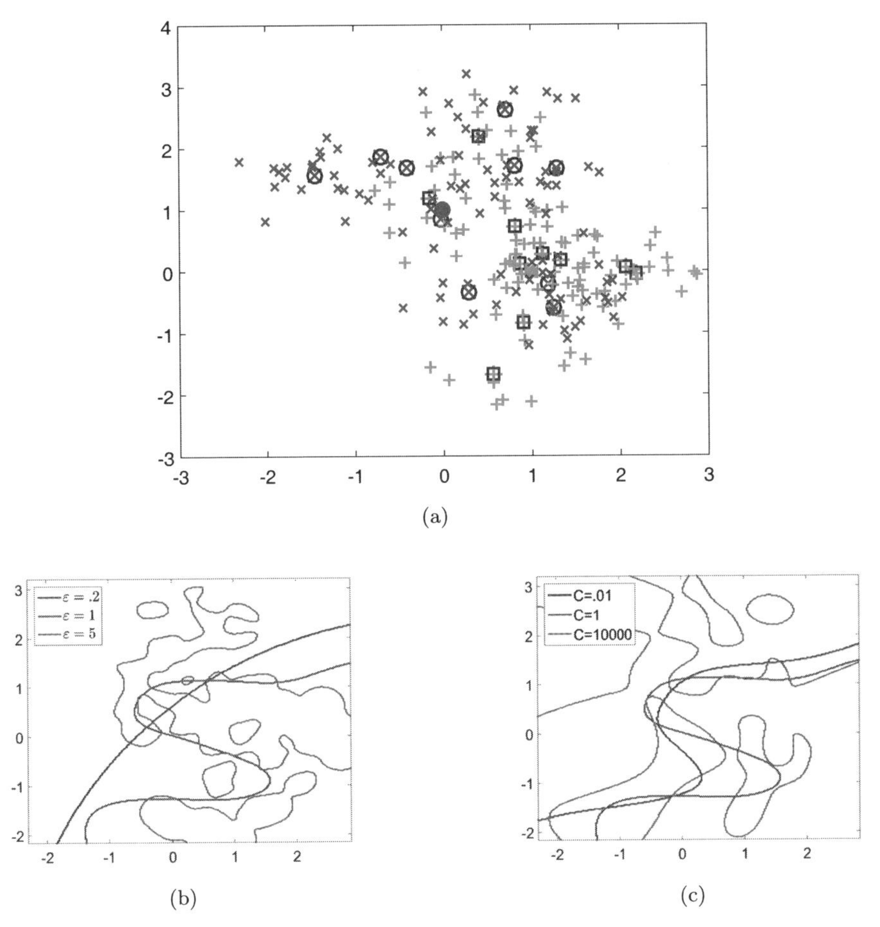

Fig. 18.2 (a) Input points and test points for the support vector machine. The $\times$ and $\bigcirc$ points belong to the $(0, 1)$ population and the $+$ and $\square$ points belong to the $(1, 0)$ population. (b)–(c): The green curves match in both plots. (b) $C = 1$, (c) $\varepsilon = 1$.

Clearly, the choice of ε plays a significant role, where larger ε encourages an SVM with more locality and smaller ε encourages less localized influence; this matches the standard localization behavior for Gaussians in an interpolation setting.

When $\varepsilon = 1$ is fixed and different C values are considered, a similar impact occurs. It seems that smaller C values produce a less active decision contour, whereas large C encourages more local fluctuations.

Example 18.5. (SVM classification)
We consider fixing $C = 0.6$ with a variety of ε values. Figure 18.3 shows the number of missed classifications (out of 20 possible) as well as the margin $1/\|\boldsymbol{w}\|$

and the required number of support vectors. The relevant experiments, along with some additional experiments not described here, are available in the script `SVMParameterProfiles` in the `GaussQR` library.

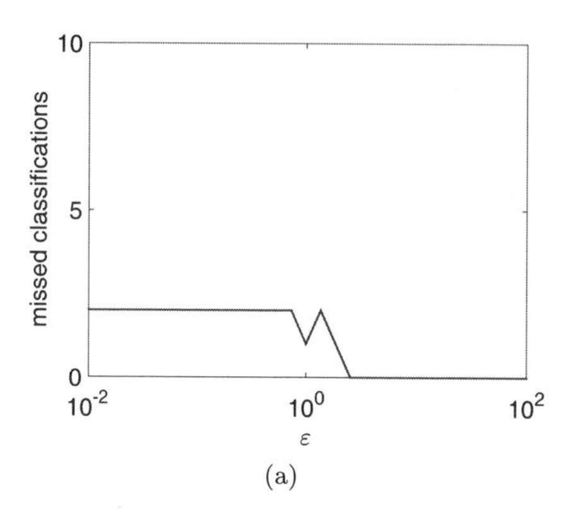

(a)

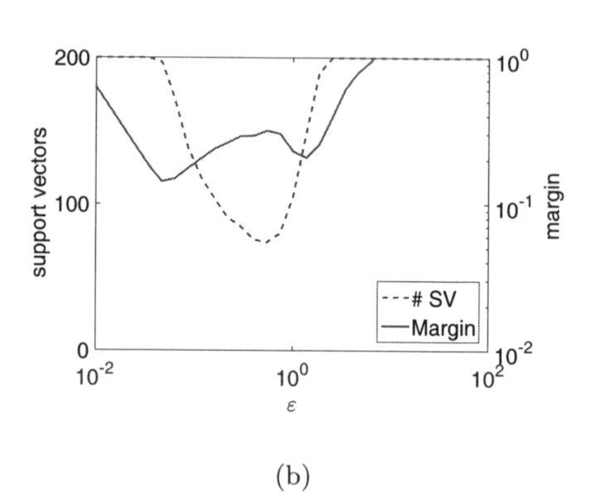

(b)

Fig. 18.3 (a) Number of missed classifications for a given ε value. (b) Computed margin and required number of support vectors. These experiments used $C = 0.6$.

The margin does not appear to be a useful guide for determining an optimal ε value, as the margin grows unboundedly as $\varepsilon \to 0$; on the other hand, for a very large ε this example is perfectly classified. Minimizing the number of support vectors is optimal from a computational standpoint, and also seems to suggest a viable region for predictions.

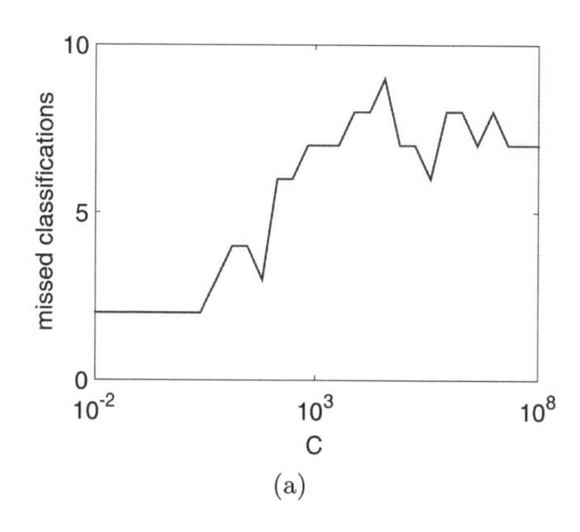

(a)

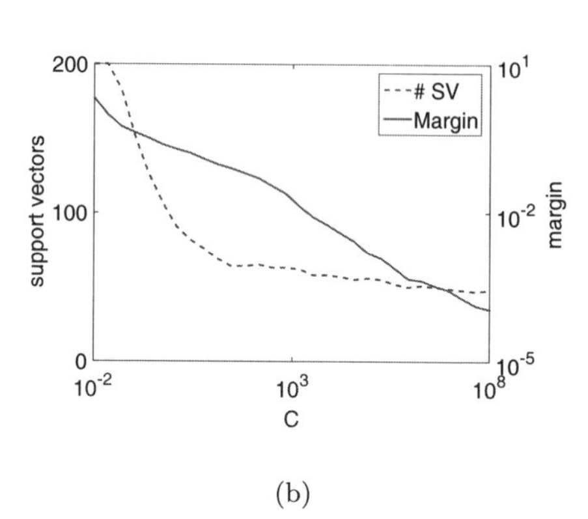

(b)

Fig. 18.4 (a) Number of missed classifications for a given C value. (b) Computed margin and required number of support vectors. These experiments used $\varepsilon = 1$.

When we fix $\varepsilon = 1$ and consider a range of C values, on display in Figure 18.4, the situation is no clearer. Good prediction results seem to occur for smaller C values, and very large values of C require few support vectors. It is worth noting that larger C values require more computing time to solve the quadratic program because a larger search space is under consideration; some discussion on this is presented in Section 18.3.4.

Remark 18.8. The margin $1/\|\boldsymbol{w}\|$ is most easily computed immediately after solving the quadratic program in `gqr_fitsvm`. In the `GaussQR` library version of that program, the `quadprog` call is

```
[sol_QP,fval,SVM.exitflag,SVM.output] = quadprog(...
```

Because we can write $\|\boldsymbol{w}\|^2 = \boldsymbol{\alpha}^T \mathsf{A} \boldsymbol{\alpha}$ (recall (18.8)), and `quadprog` returns the minimum value of $\frac{1}{2}\boldsymbol{\alpha}^T \mathsf{A} \boldsymbol{\alpha} + \boldsymbol{f}^T \boldsymbol{\alpha}$ in `fval`, the margin can be computed within `gqr_fitsvm` with the line

```
SVM.margin = sqrt(.5/(fval + sum(sol_QP)));
```

The `sum` appears because the $\boldsymbol{f}$ vector is a vector of all ones.

Example 18.5 demonstrates that finding an optimal SVM parametrization using either the margin or the number of support vectors is not always a useful strategy. A more common technique in the machine learning community is to use cross-validation, as discussed in Section 14.2 and Section 18.2. In contrast to Chapter 17, which used the `crossval` function from MATLAB's Statistics and Machine Learning Toolbox, we have implemented our own cross-validation algorithm for support vector machines in the `GaussQR` library in `gqr_svmcv.m`. While the use of MATLAB's functions is certainly recommended, we provide that function as an example of what is required to perform cross-validation.

Example 18.6. (SVM with cross-validation)
Here, a 10-fold cross-validation scheme is used to measure the effectiveness of each of the ε and C parameters. In Figure 18.5(a)–(b) we show the associated cross-validation residuals with one parameter fixed and the other varied over $[10^{-2}, 10^2]$. We also show that there is an optimal region for the 10-fold cross-validation residual, where decreases in ε are matched by increases in C. These examples can be recreated with the MATLAB script `SVMCrossValTests.m`.

Remark 18.9. The example shown in Figure 18.2(a) and discussed in Examples 18.4–18.6 happens to represent a linearly separable pattern since the population centers $(0, 1)$ and $(1, 0)$ are linearly separable. Because the $\varepsilon \to 0$ limit of Gaussians is a polynomial, it is reasonable to conclude that, with infinitely much data drawn from those populations, the optimal SVM would have $\varepsilon \to 0$ to produce a line. Example 18.7 considers a different pattern which is not linearly separable.

Example 18.7. (Nonseparable data for SVM classification)
This example uses a pattern which is not linearly separable and attempts to classify it: population 1 (denoted by blue $\square$ and green $+$) has centers at $\{(0, 1), (1, 0), (2, 1)\}$ and population 2 (denoted by blue $\bigcirc$ and red $\times$) has centers at $\{(0, 0), (1, 1), (2, 0)\}$.

Test points are chosen from a normal distribution from those populations and training data is generated from the test points. These distribution centers (filled green $\square$ and red $\bigcirc$), test points (large green $+$ and red $\times$) and training points (small green $+$ and red $\times$) are on display in Figure 18.5(c). These tests can be recreated with the script `SVMCrossValNonSep` in the `GaussQR` library.

This pattern is not linearly separable, and this fact is evident in the surface plot where increasingly small ε can no longer produce the optimal CV residual that was possible in Figure 18.5(a)–(b). Instead, smaller ε causes an increase in the CV residual, likely because the tendency towards polynomial behavior as $\varepsilon \to 0$ is not desirable when learning this pattern.

18.3.4 *Computational consideration for classification with kernel SVMs*

The popularity of support vector machines over RBF networks is in their sparsity; because SVMs may require many fewer kernel centers for evaluation (only the nonzero coefficients must be included) their predictions can come at a much lower cost. This benefit is balanced by the fact that solving the quadratic program (18.7) is more expensive than solving the linear system (18.3). This section studies that cost as a function of the SVM parameters ε and C and also presents a strategy for exploiting the low-rank eigenfunction representation for small ε to decrease cost.

Remark 18.10. The quadratic program solver used here, the MATLAB `quadprog` function, is an iterative solver, and it accepts an initial guess. As is the case with most iterative solvers, use of a good initial guess will result in quick convergence and a poor initial guess with lead to slow convergence. The initial guess used for these examples is always $C/2$ times a vector of ones; the algorithm used here is MATLAB's `interior-point-convex`.

Example 18.8. (Timing tests for SVM)
The same experimental setup from Example 18.5 is considered, except 400 training points are used rather than 100 as before. Figure 18.5(e) displays the required time to train the SVM for a range of ε and C values. Within the `GaussQR` library, the script `SVMTimeTests` runs these examples.

There is a clear dependence on ε and C in the time required to solve the quadratic program, although that dependence is not entirely clear. Very large ε and very small C seem to be solved quickly: large ε because of the RBF locality, and small C because the solution domain is very small and quickly searched. Increases in C seem to always yield longer solution times, likely because the search space is increasing.

While other problems will have a unique profile different from Figure 18.5(e), the computational cost of using certain SVM parametrizations can be significantly greater and may prove an important role in finding an acceptable prediction tool.

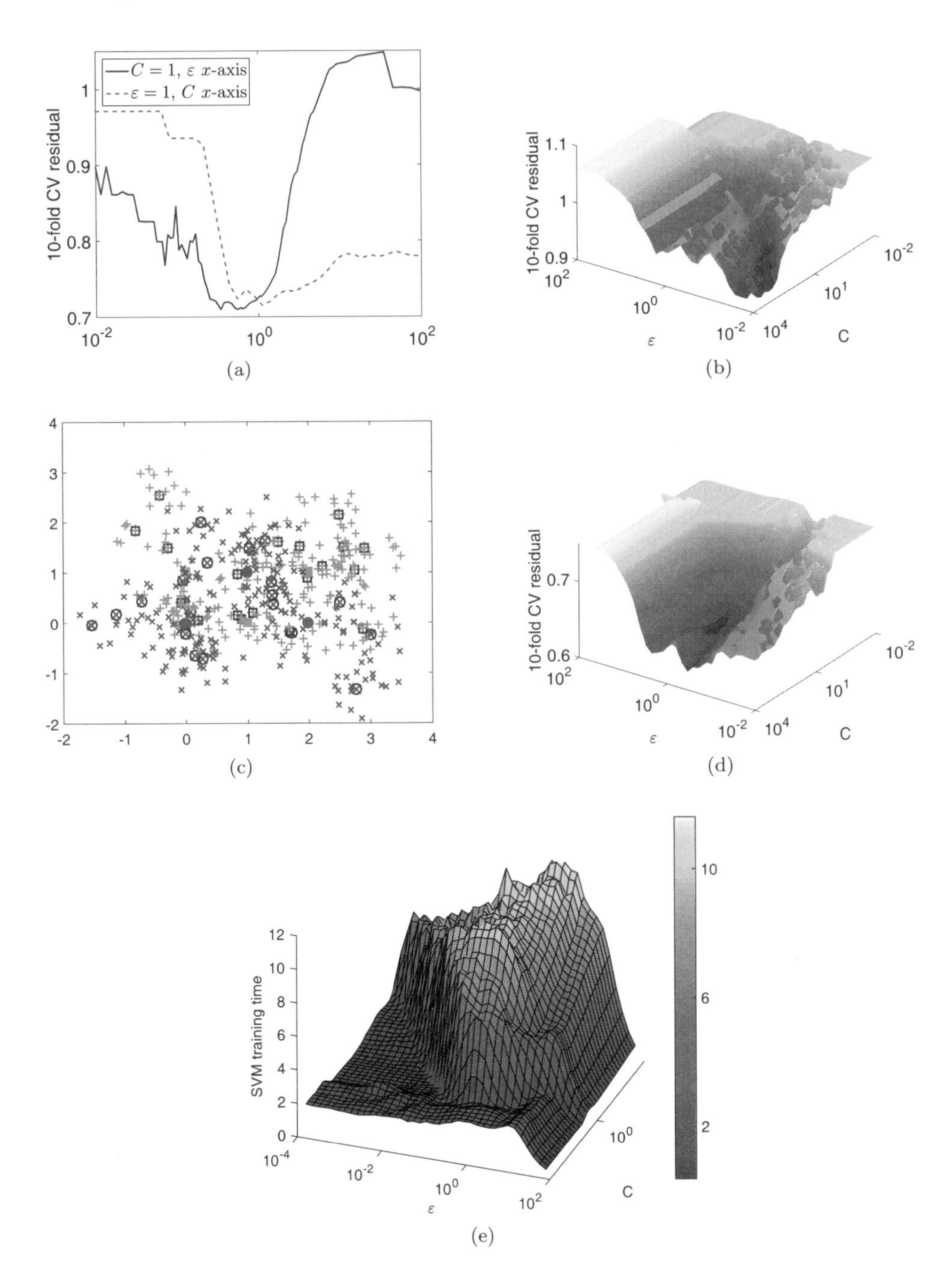

Fig. 18.5 (a)–(b): (a) One parameter fixed and the other may vary; (b) Surface plot allowing both ε and C to vary. These measure the sum residual left from a 10-fold cross-validation.
(c)–(d): (c) Scatter plot of input data and test points; (d) Surface plot of 10-fold cross-validation allowing both ε and C to vary.
(e) The cost (in seconds) of the solution is strongly dependent on ε and C.

As was discussed in Chapter 10 in the context of interpolation, some kernel parametrizations can yield ill-conditioned linear systems; for Gaussians, that ill-conditioning occurs as $\varepsilon \to 0$. We learned in Section 12.3 that this ill-conditioning — while problematic for the invertibility of K — allows for a low-rank representation in terms of the eigenfunctions (12.29a). Given N input points and a small ε, it may be possible to use $M \ll N$ eigenfunctions to accurately approximate the kernel matrix, i.e., to keep $\|\mathsf{K} - \Phi\Lambda\Phi^T\|$ small for $\Phi \in \mathbb{R}^{N\times M}$. This was exploited to produce low-rank eigenfunction approximate interpolants which were more stable that standard basis interpolants, but here we hope to use this $\mathsf{K} = \Phi\Lambda\Phi^T$ structure for efficiency.

We rephrased our dual-form quadratic program (18.7) to fit MATLAB's Optimization Toolbox function `quadprog`'s calling sequence as explained in (18.10). Now, to emphasize the role of the kernel matrix K, we rewrite it in a slightly different form,

$$\begin{aligned} \min_{\boldsymbol{\alpha}} \ & \frac{1}{2}\boldsymbol{\alpha}^T \mathsf{D}_{\boldsymbol{y}} \mathsf{K} \mathsf{D}_{\boldsymbol{y}} \boldsymbol{\alpha} - \boldsymbol{e}^T\boldsymbol{\alpha} \\ \text{subject to } & \boldsymbol{y}^T\boldsymbol{\alpha} = 0, \\ & \boldsymbol{\alpha} \in [0, C]^N, \end{aligned} \tag{18.11}$$

where $\mathsf{D}_{\boldsymbol{y}}$ is a diagonal matrix with $\boldsymbol{y}$ on the diagonal, and $\boldsymbol{e}$ is a vector of all ones. If we can state $\mathsf{K} \approx (\Lambda^{1/2}\Phi)^T(\Lambda^{1/2}\Phi)$, then this problem can be rewritten (see [Fine and Scheinberg (2002)] or [Zhang *et al.* (2008)]) as

$$\begin{aligned} \min_{\boldsymbol{\varpi},\boldsymbol{\alpha}} \ & \frac{1}{2}\begin{pmatrix}\boldsymbol{\varpi}^T & \boldsymbol{\alpha}^T\end{pmatrix}\begin{pmatrix}\mathsf{I}_M & 0\\ 0 & 0\end{pmatrix}\begin{pmatrix}\boldsymbol{\varpi}\\ \boldsymbol{\alpha}\end{pmatrix} - \begin{pmatrix}0 & \boldsymbol{e}^T\end{pmatrix}\begin{pmatrix}\boldsymbol{\varpi}\\ \boldsymbol{\alpha}\end{pmatrix} \\ \text{subject to } & \begin{pmatrix}0 & \boldsymbol{y}^T\\ -\mathsf{I}_M & \mathsf{V}^T\end{pmatrix}\begin{pmatrix}\boldsymbol{\varpi}\\ \boldsymbol{\alpha}\end{pmatrix} = 0, \\ & \boldsymbol{\alpha} \in [0, C]^N, \quad \boldsymbol{\varpi} \in \mathbb{R}^M. \end{aligned} \tag{18.12}$$

where $\mathsf{V} = \mathsf{D}_{\boldsymbol{y}}\Phi\Lambda^{1/2}$ so that $\mathsf{V}\mathsf{V}^T \approx \mathsf{D}_{\boldsymbol{y}}\mathsf{K}\mathsf{D}_{\boldsymbol{y}}$. Although this system is of size $N + M$ (and the original system was only size N), the cost of solving this system may be much lower because of the extremely simple structure of the Hessian. This sparsity, in comparison to the Hessian of the true kernel, which may be fully dense, allows for cheap matrix-vector products and decompositions, both of which may lead to a faster quadratic program solve. Note that the $\boldsymbol{\varpi}$ values are inconsequential in making predictions with the SVM.

Earlier, we mentioned that the `gqr_fitsvm` function in the `GaussQR` library had a `low_rank` option that was, at the time, ignored. That argument may be activated to train the SVM using (18.12) instead of using the K matrix as was done in Program 18.2. Program 18.3 shows the part of `gqr_fitsvm` which executes the low-rank solve. We have omitted most SVM training code that would have been duplicated from Program 18.2.

Program 18.3. Low-rank excerpts from gqr_fitsvm.m

```
% Duplicate content from gqr_fitsvm.m has been omitted from above
% Create the GQR object to evaluate the eigenfunctions
gqr_alpha = 1e6;
GQR = gqr_solveprep(1,x,ep,gqr_alpha);
lamvecsqrt = sqrt(GQR.eig(GQR.Marr));
Phi = gqr_phi(GQR,x);
V = Phi.*(y*lamvecsqrt);
% Define the low-rank quadratic programming problem
% This is done using blocks with sizes N and M
M = size(V,2);
H = sparse(1:M,1:M,ones(1,M),N+M,N+M);
f = -[zeros(M,1);ones(N,1)];
A = [];
b = [];
Aeq = [zeros(1,M),y';-eye(M),V'];
beq = [0;zeros(M,1)];
lb = [-inf*ones(M,1);zeros(N,1)];
ub = [inf*ones(M,1);bc*ones(N,1)];
x0 = [V'*(bc/2*ones(N,1));bc/2*ones(N,1)];
% Solve the quadratic programming problem
optimopt = optimset('LargeScale','off','Display','off','MaxIter',500);
sol_QP = quadprog(A,f,Ain,bin,Aeq,beq,lb,ub,x0,optimopt);
% Extract out only the relevant solution components
sol_QP = sol_QP(M+1:M+N);
% Duplicate content from gqr_fitsvm.m has been omitted from below
```

Program 18.3 opens with the formation of the `GQR` object, similarly to Program 13.5 and Section 15.2 where we earlier applied the Gaussian eigenfunctions. The call to `gqr_solveprep` in line 4 chooses an M based on the value of N choice of the global parameter `GAUSSQR_PARAMETERS.DEFAULT_REGRESSION_FUNC` defined when `rbfsetup` was called to access the `GaussQR` library. This may, of course, be set by the user to fit the needs of a specific application, but we have found the value 0.1, so that $M = 0.1N$, to be effective.

The $N \times M$ matrix Φ is evaluated in line 6, and line 7 executes the $\mathsf{V} = \mathsf{D}_{\boldsymbol{y}}\Phi\Lambda^{1/2}$ computation. Forming the components of the quadratic program and solving it in lines 11–22 mirrors the steps in Program 18.2, albeit with augmentations to account for the $N+M$ size of the problem in (18.12). After the call to `quadprog`, only the N solution components required to evaluate the SVM are retained, and the remainder of the SVM development matches that of the full-rank K.

Example 18.9. (Improving the efficiency of SVM classification with the low-rank eigenfunction basis)
Reusing again the setup from Example 18.5 we study the cost of solving the quadratic program and training the SVM. Increasingly large sets of input points are considered and the cost of solving the full-rank problem (18.11) is compared to solving the low-rank problem (18.12). The kernel is parametrized with $\varepsilon = 0.01$

and $C = 1$ and the low-rank version of the problem uses $M = 0.1N$ Gaussian eigenfunctions with global scale parameter $\alpha = 10^6$ (recall Example 12.1).

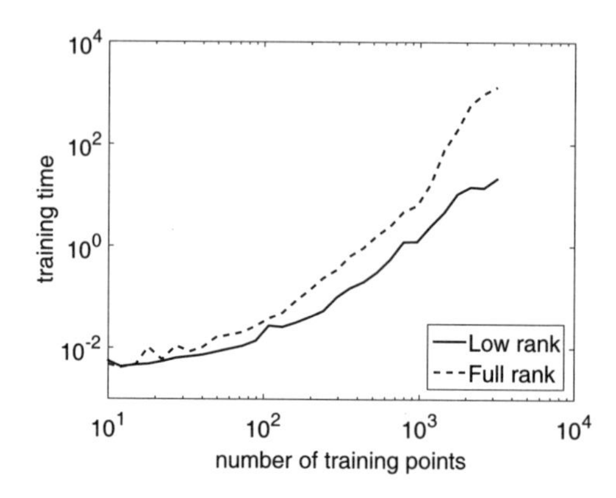

Fig. 18.6 The cost (in seconds) of the training is less for the low-rank system for sufficiently large sets of input data.

Minimizing the cost of SVMs through efficient training is an important topic in machine learning; sources such as [Flake and Lawrence (2002); Yang *et al.* (2004); Lin *et al.* (2011)] discuss various strategies.

18.3.5 *Linear support vector regression*

As for classification, we again start with a linear approximation and assume that $s(\boldsymbol{x}) = \boldsymbol{x}^T\boldsymbol{w} + b$. If we use the ϵ-insensitive loss function

$$L\left(y, s(\boldsymbol{x})\right) = \max\left(|y - s(\boldsymbol{x})| - \epsilon, 0\right).$$

Then the primal unconstrained minimization problem is given by

$$\min_{\boldsymbol{w},b} \left[\frac{1}{N}\sum_{i=1}^{N} \max\left(|y_i - s(\boldsymbol{x}_i)| - \epsilon, 0\right) + \mu\frac{1}{2}\boldsymbol{w}^T\boldsymbol{w}\right],$$

where, as before, μ is an appropriately chosen regularization parameter. Using slack variables as in the classification case we have the analogous constrained minimization problem

$$\begin{aligned} &\min_{\boldsymbol{w},b,\boldsymbol{\xi},\boldsymbol{\xi}^*} \left[\frac{1}{2}\boldsymbol{w}^T\boldsymbol{w} + C\sum_{i=1}^{N}(\xi_i + \xi_i^*)\right] \\ &\text{subject to } s(\boldsymbol{x}_i) - y_i \le \epsilon + \xi_i, \quad i = 1, \dots, N, \\ &\qquad\qquad y_i - s(\boldsymbol{x}_i) \le \epsilon + \xi_i^*, \quad i = 1, \dots, N, \\ &\qquad\qquad \xi_i, \xi_i^* \ge 0. \end{aligned}$$

In the dual formulation we need to solve the constrained quadratic programming problem

$$\min_{\boldsymbol{\alpha},\boldsymbol{\alpha}^*} \epsilon \sum_{i=1}^{N}(\alpha_i^* + \alpha_i) - \sum_{i=1}^{N} y_i(\alpha_i^* - \alpha_i) + \frac{1}{2}\sum_{i=1}^{N}\sum_{j=1}^{N}(\alpha_i^* - \alpha_i)(\alpha_j^* - \alpha_j)\boldsymbol{x}_i^T\boldsymbol{x}_j$$

$$\text{subject to } 0 \le \alpha_i, \alpha_i^* \le C,$$

$$\sum_{i=1}^{N}(\alpha_i^* - \alpha_i) = 0.$$

Once the dual variables α_i and α_i^* have been obtained by solving this problem, the SVM regression function is given by

$$\boldsymbol{w} = \sum_{i=1}^{N}(\alpha_i^* - \alpha_i)\boldsymbol{x}_i$$

$$s(\boldsymbol{x}) = \boldsymbol{x}^T\boldsymbol{w} + b = \sum_{i=1}^{N}(\alpha_i^* - \alpha_i)\boldsymbol{x}^T\boldsymbol{x}_i + b.$$

The computation of the bias term b follows from the KKT conditions (for details see [Schölkopf and Smola (2002)]) and is similar in spirit to the classification setting, i.e.,

$$b = y_i - \boldsymbol{x}_i^T\boldsymbol{w} - \epsilon \quad \text{for } \alpha_i \in (0, C),$$
$$b = y_i - \boldsymbol{x}_i^T\boldsymbol{w} + \epsilon \quad \text{for } \alpha_i^* \in (0, C).$$

As before, any one of these will theoretically suffice, but for stability reasons it is better to compute b via an average over all candidates.

As in the classification setting, $\alpha_i^* - \alpha_i \neq 0$ only for some i, and the corresponding measurements $\boldsymbol{x}_i$ are called the support vectors.

For more details see, e.g., [Schölkopf and Smola (2002, Chapter 9), Hastie *et al.* (2009, Chapter 12)].

18.3.6 *Nonlinear support vector regression*

As in the case of classification, we obtain a nonlinear "kernelized" regression fit if we map the data into feature space and then use kernels. This is straightforward and completely analogous to the classification setting. The resulting dual problem is

$$\min_{\boldsymbol{\alpha},\boldsymbol{\alpha}^*} \epsilon \sum_{i=1}^{N}(\alpha_i^* + \alpha_i) - \sum_{i=1}^{N} y_i(\alpha_i^* - \alpha_i) + \frac{1}{2}\sum_{i=1}^{N}\sum_{j=1}^{N}(\alpha_i^* - \alpha_i)(\alpha_j^* - \alpha_j)K(\boldsymbol{x}_i, \boldsymbol{x}_j)$$

$$\text{subject to } 0 \le \alpha_i, \alpha_i^* \le C,$$

$$\sum_{i=1}^{N}(\alpha_i^* - \alpha_i) = 0,$$

so that

$$\boldsymbol{w} = \sum_{i=1}^{N} (\alpha_i^* - \alpha_i)\Phi_{\boldsymbol{x}_i}$$

$$s(\boldsymbol{x}) = \Phi(\boldsymbol{x}^T)\boldsymbol{w} + b = \sum_{i=1}^{N} (\alpha_i^* - \alpha_i)K(\boldsymbol{x}, \boldsymbol{x}_i) + b,$$

and

$$b = y_i - \sum_{j=1}^{N} (\alpha_j^* - \alpha_j)K(\boldsymbol{x}_i, \boldsymbol{x}_j) - \epsilon \quad \text{for } \alpha_i \in (0, C),$$

$$b = y_i - \sum_{j=1}^{N} (\alpha_j^* - \alpha_j)K(\boldsymbol{x}_i, \boldsymbol{x}_j) + \epsilon \quad \text{for } \alpha_i^* \in (0, C).$$

Remark 18.11. Many more details on all aspects of machine learning can be found, e.g., in the books [Schölkopf and Smola (2002); Shawe-Taylor and Cristianini (2004); Rasmussen and Williams (2006); Steinwart and Christmann (2008); Alpaydin (2009); Hastie *et al.* (2009)] or the survey papers [Orr (1996); Evgeniou *et al.* (2000); Moguerza and Muñoz (2006)].

Chapter 19

Derivatives of Interpolants and Hermite Interpolation

In this chapter we consider two tasks involving derivatives:

(1) Approximating the derivative of an unknown function u from given data of the form $\{\boldsymbol{x}_i, u(\boldsymbol{x}_i)\}$, $i = 1, \ldots, N$. This task is sometimes referred to as *derivative estimation*, and will give us an opportunity to introduce the reader to *differentiation matrices*. Since this problem is approached by first computing a standard kernel-based interpolant to the given data (as discussed throughout this book) and then simply differentiating this interpolant, this process is sometimes referred to as *simultaneous approximation* in the approximation theory literature.

(2) Approximating an unknown function u from given data of the form $\{\boldsymbol{x}_i, L_i u\}$, $i = 1, \ldots, N$, where the L_i are bounded linear functionals (including point evaluation and derivative evaluation). This task is known in the numerical analysis and approximation theory literature under the names *Hermite interpolation*, *Hermite–Birkhoff interpolation*, or *generalized Hermite interpolation*; and in the statistics literature as *kriging with derivative information* or *gradient-enhanced kriging*.

We will provide details for the first task in Section 19.1, and the second in Section 19.2. Both approaches will be useful in Chapter 20, when we solve PDEs.

The first approach corresponds to various *differentiation matrix approaches*: a kind of *pseudospectral method* if we use global basis functions, a type of *finite difference method* if we use local basis functions, and a type of *partition of unity method* if we are in-between. The second approach provides the foundation for *collocation methods*, where the division is similar.

In Section 19.3 we look at an approach to Hermite interpolation based on Hilbert–Schmidt eigenfunctions, specifically for the Gaussian kernel. This can be thought of as an extension of Section 12.3.

And — finally — in Section 19.4 we add an application where the idea of Hermite interpolation is used to couple two (essentially arbitrary) PDE solvers along an interface separating two different computational domains. Here we might be using one kind of solver (or model) on one domain, and a completely different one on the other domain. Such problems arise commonly in scientific applications.

19.1 Differentiating Interpolants

In Section 16.5 we constructed an approximate cumulative distribution function (CDF) from data consisting of samples of the empirical distribution function (EDF), but then also wanted to know the corresponding probability density function (PDF). In order to obtain the latter, the interpolant to the EDF data gave us an approximate CDF, which we then differentiated to obtain the approximate PDF. This is in contrast to the Hermite interpolation problems discussed in Section 19.2, where we will have data that consists of values *and* derivatives of the unknown function we are trying to reconstruct.

19.1.1 *Cardinal function representation of derivatives*

Since our interpolants for the given data[1] $(\boldsymbol{z}_i, u(\boldsymbol{z}_i))$, $i = 1, \ldots, N$, are linear combinations of the (smooth) kernel basis functions, it is straightforward to differentiate such an interpolant. For any kernel-based interpolant

$$\hat{u}(\boldsymbol{x}) = \sum_{j=1}^{N} c_j K(\boldsymbol{x}, \boldsymbol{z}_j) = \boldsymbol{k}(\boldsymbol{x})^T \boldsymbol{c}, \qquad \boldsymbol{x} \in \Omega \subseteq \mathbb{R}^d, \tag{19.1}$$

where — as usual — $\boldsymbol{k}(\boldsymbol{x})^T = \big(K(\boldsymbol{x}, \boldsymbol{z}_1) \cdots K(\boldsymbol{x}, \boldsymbol{z}_N)\big)$, the "derivative" $\mathcal{L}\hat{u}$ of $\hat{u}$, where $\mathcal{L}$ is any linear differential operator, is simply given by

$$\mathcal{L}\hat{u}(\boldsymbol{x}) = \sum_{j=1}^{N} c_j \mathcal{L} K(\boldsymbol{x}, \boldsymbol{z}_j) = \mathcal{L}\boldsymbol{k}(\boldsymbol{x})^T \boldsymbol{c}. \tag{19.2}$$

We find the coefficients $\boldsymbol{c}$ — which are the same for both (19.1) and (19.2) — by solving the linear system $\mathsf{K}\boldsymbol{c} = \boldsymbol{u}$ associated with (19.1). Here $(\mathsf{K})_{ij} = K(\boldsymbol{z}_i, \boldsymbol{z}_j)$ and $\boldsymbol{u} = \big(u(\boldsymbol{z}_1) \cdots u(\boldsymbol{z}_N)\big)^T$. As in the earlier parts of this book, we do not actually know the function u, only its values collected in the vector $\boldsymbol{u}$.

Remark 19.1. This approach to derivative estimation is a rather natural one. The idea behind it is that — since the interpolant $\hat{u}$ from (19.1) presumably provides a good approximation to u — we expect the derivative $\mathcal{L}\hat{u}$ from (19.2) to also be a good approximation for the derivative $\mathcal{L}u$. We will see below how good this approximation actually is — both in theory (cf. Section 19.1.2) and with a matching numerical experiment provided in Example 19.1.

[1] Note that, as in Chapter 4, we denote the kernel centers with $\boldsymbol{z}_1, \ldots, \boldsymbol{z}_N$ and evaluation locations, when required later on, will be denoted by $\boldsymbol{x}_1, \ldots, \boldsymbol{x}_{N_{\text{eval}}}$. At this point we will also change our notation from f and s — denoting the function generating our data and the kernel interpolant/approximant — to u and $\hat{u}$ since we view this chapter as a transition to our discussion of PDEs in Chapter 20, and there the new notation is much more appropriate.

The discussion below will be simplified if we consider the interpolant in its cardinal form (cf. (12.8)), i.e., instead of (19.1) we use

$$\hat{u}(\boldsymbol{x}) = \boldsymbol{\ell}(\boldsymbol{x})^T \boldsymbol{u}, \qquad \boldsymbol{\ell}(\boldsymbol{x})^T = \boldsymbol{v}(\boldsymbol{x})^T \mathsf{V}^{-1}, \tag{19.3}$$

where $\boldsymbol{v}(\cdot)$ is the basis function vector for any of the alternate basis representations considered in Chapters 12 and 13, and V is the Vandermonde-like matrix containing the values of those basis functions on the set of input points $\boldsymbol{z}_1, \ldots, \boldsymbol{z}_N$. If we use, e.g., the standard basis then we have

$$\hat{u}(\boldsymbol{x}) = \boldsymbol{k}(\boldsymbol{x})^T \mathsf{K}^{-1} \boldsymbol{u},$$

while in the stable HS-SVD basis of Chapter 13 it would be

$$\hat{u}(\boldsymbol{x}) = \boldsymbol{\psi}(\boldsymbol{x})^T \Psi^{-1} \boldsymbol{u}.$$

Application of the differential operator $\mathcal{L}$ to the cardinal representation yields

$$\mathcal{L}\hat{u}(\boldsymbol{x}) = \mathcal{L}\boldsymbol{v}(\boldsymbol{x})^T \mathsf{V}^{-1} \boldsymbol{u}, \tag{19.4}$$

which we will from now on use instead of (19.2).

19.1.2 *Error bounds for simultaneous approximation*

While we did not discuss details of any interpolation error estimates that also hold for the simultaneous approximation of derivatives in Chapter 9, the specialized literature contains many such bounds. In fact, our generic multivariate sampling inequality (9.26) in the case $p = q = 2$ will lead to an interpolation error bound of the form

$$\|\mathcal{D}^{\boldsymbol{\alpha}}(u - \hat{u})\|_{L_2(\Omega)} \leq C h^{\beta - |\boldsymbol{\alpha}|} |u|_{W_2^\beta(\Omega)}, \tag{19.5}$$

where $\boldsymbol{\alpha}$ is a multi-index and $\mathcal{D}^{\boldsymbol{\alpha}}$ denotes a multivariate derivative of order $|\boldsymbol{\alpha}|$ (see Section 3.3 for more details on multi-index notation). The exponent β characterizes the smoothness of the kernel being used for the interpolation. This estimate shows that the *order of convergence drops by one for every derivative taken.*

The interested reader will find much more on the theoretical aspects of derivative approximation in the original literature mentioned in Chapter 9 such as, e.g., the survey by Rieger, Schaback and Zwicknagl (2010) on sampling inequalities.

Remark 19.2. An interesting contribution is [Fuselier and Wright (2015)], in which the authors show that one experiences no "loss of derivatives" in the computation of approximate — even higher-order — derivatives of *periodic* functions at the given data sites by interpolation with radial kernels whose Fourier coefficients satisfy a certain decay condition (see [Fuselier and Wright (2015)] for details). The approach used in that paper is a bit different from the one described above since they do not compute the β-order derivative of an interpolant, but instead apply an iterative procedure according to which one differentiates only once, and then keeps on — up to the desired order of differentiation — repeating to interpolate the (derivative) data and differentiate that new interpolant.

We now discuss two fundamentally different approaches to this sort of simultaneous derivative approximation, namely the use of kernel-based global and local differentiation matrices. These two approaches are reminiscent of global pseudospectral methods and finite difference methods, respectively (see, e.g., [Fornberg (1998); Trefethen (2000); Boyd (2001)] for the fundamentals of these traditional polynomial or Fourier-based approximation methods).

In Chapter 20 we will apply these kernel-based pseudospectral and finite difference methods to the numerical solution of partial differential equations, where they are commonly referred to as RBF collocation and RBF-FD.

19.1.3 *Global differentiation matrices*

The goal of so-called *differentiation matrices* is to obtain a discretized version of any linear differential operator $\mathcal{L}$, i.e., we want to find a matrix L such that

$$\mathsf{L}\boldsymbol{u} = \hat{\boldsymbol{u}}_{\mathcal{L}}, \tag{19.6}$$

where $\boldsymbol{u} = \left(u(\boldsymbol{z}_1) \cdots u(\boldsymbol{z}_N)\right)^T$ as above, and $\hat{\boldsymbol{u}}_{\mathcal{L}} = \left(\mathcal{L}\hat{u}(\boldsymbol{x}_1) \cdots \mathcal{L}\hat{u}(\boldsymbol{x}_{N_{\text{eval}}})\right)^T$ is an approximation of $\boldsymbol{u}_{\mathcal{L}} = \left(\mathcal{L}u(\boldsymbol{x}_1) \cdots \mathcal{L}u(\boldsymbol{x}_{N_{\text{eval}}})\right)^T$ obtained by differentiating an interpolant[2] of $\boldsymbol{u}$. This implies that L is an $N_{\text{eval}} \times N$ matrix that maps function values at the input points $\mathcal{Z} = \{\boldsymbol{z}_1, \ldots, \boldsymbol{z}_N\}$ to derivative values at the evaluation points $\mathcal{X} = \{\boldsymbol{x}_1, \ldots, \boldsymbol{x}_{N_{\text{eval}}}\}$.

From the cardinal representation (19.4) of the derivative of the interpolant we immediately see — just stack N_{eval} copies of (19.4) on top of each other — that

$$\hat{\boldsymbol{u}}_{\mathcal{L}} = \begin{pmatrix} \mathcal{L}\hat{u}(\boldsymbol{x}_1) \\ \vdots \\ \mathcal{L}\hat{u}(\boldsymbol{x}_{N_{\text{eval}}}) \end{pmatrix} = \begin{pmatrix} \mathcal{L}\boldsymbol{v}(\boldsymbol{x}_1)^T \mathsf{V}^{-1}\boldsymbol{u} \\ \vdots \\ \mathcal{L}\boldsymbol{v}(\boldsymbol{x}_{N_{\text{eval}}})^T \mathsf{V}^{-1}\boldsymbol{u} \end{pmatrix} = \begin{pmatrix} \mathcal{L}\boldsymbol{v}(\boldsymbol{x}_1)^T \\ \vdots \\ \mathcal{L}\boldsymbol{v}(\boldsymbol{x}_{N_{\text{eval}}})^T \end{pmatrix} \mathsf{V}^{-1}\boldsymbol{u}.$$

If we introduce the notation

$$\mathsf{V}_{\mathcal{L}}^{\mathcal{X}} = \begin{pmatrix} \mathcal{L}\boldsymbol{v}(\boldsymbol{x}_1)^T \\ \vdots \\ \mathcal{L}\boldsymbol{v}(\boldsymbol{x}_{N_{\text{eval}}})^T \end{pmatrix}, \tag{19.7}$$

then the differentiation matrix L can be seen as the matrix product

$$\mathsf{L} = \mathsf{V}_{\mathcal{L}}^{\mathcal{X}} \mathsf{V}^{-1}, \tag{19.8}$$

which can be interpreted in terms of *any* of our alternate basis representations.

Remark 19.3. At times below we may expand this notation for differentiation matrices to more explicitly state the input data locations $\mathcal{Z}$ and the evaluation points $\mathcal{X}$. For instance,

$\mathsf{V}_{\mathcal{L}}^{\mathcal{X}} \mathsf{V}_{\mathcal{Z}}^{-1}$ is read "$\mathcal{L}$ is applied to an interpolant on $\mathcal{Z}$ and evaluated at $\mathcal{X}$."

[2] Note that in (19.6) we could also write $\hat{\boldsymbol{u}}$ in place of $\boldsymbol{u}$ since, by our interpolation assumption, we know that $\hat{\boldsymbol{u}} = \boldsymbol{u}$.

In some contexts it may be convenient to add this additional information to L, perhaps in the form $\mathsf{L}_{\mathcal{Z}\to\mathcal{X}}$, which more explicitly states that data at $\mathcal{Z}$ is being used to evaluate at $\mathcal{X}$.

Remark 19.4. Note that, depending on the number of evaluation points, N_{eval}, the differentiation matrix L can vary in size from a $1 \times N$ row vector to a tall and skinny $N_{\text{eval}} \times N$ matrix (if $N_{\text{eval}} > N$), so that — based on simple size arguments — the rank of L can range from 1 to N. Moreover, as was discussed in [Fasshauer (2007, Chapter 42)] for the case of $N_{\text{eval}} = N$ and evaluation points coinciding with kernel centers, the matrix L can be a low-rank matrix and thus one needs to be careful when attempting to invert it (as may be desirable in the context of implicit time-stepping schemes for PDEs).

Fornberg *et al.* (2010) performed a study of the quality of derivative approximations calculated via standard pseudospectral methods vs. that obtained via radial kernels. In Example 19.1 we demonstrate how to perform the latter, but leave it to the reader to consult [Fornberg *et al.* (2010)] for some deeper insights that go beyond our simple example.

Example 19.1. (Convergence behavior of global differentiation matrices)
This example studies the use of a global differentiation matrix to approximate derivatives given data values from the function

$$u(x,y) = (x^2 + y^4)^{7/2} + xy,$$

$$\frac{\partial^2}{\partial x^2}u(x,y) + \frac{\partial^2}{\partial y^2}u(x,y) = 7(x^2+y^4)^{3/2}(6x^2y^2 + 6x^2 + 26y^6 + y^4).$$

Data locations form a 2D Chebyshev grid (see Figure B.4(a)) and we increase the sample size from $N = 6^2$ to $N = 100^2$ for our study. We use Chebyshev points for this example, but scattered data is just as viable. The values of the interpolant and approximate Laplacian are tested against the true solution on a uniform tensor grid with 24 points in each dimension. The isotropic C^4 Matérn kernels of Table 3.2 are used with $\varepsilon = 4$. Program 19.1 conducts the necessary tests, and the associated plot is located in Figure 19.1(a).

Program 19.1. Excerpts from `HermiteConvergence.m`

```
  % Define the kernel and its Laplacian
2 rbf = @(e,r) (1+(e*r)+(e*r).^2/3).*exp(-e*r);    ep = 4;
  rbfL = @(e,r) 1/3*e^2*exp(-e*r).*((e*r).^2-2*e*r-2);
4 % Define the function to be studied and its Laplacian
  u = @(x,y) (x.^2+y.^4).^(7/2) + x.*y;
6 uL = @(x,y) 7*(x.^2+y.^4).^(3/2).* ...
           (6*x.^2.*y.^2+6*x.^2+26*y.^6+y.^4);
8 % Define points for testing both values and derivatives
  xeval = pick2Dpoints(-1,1,24);
10 ueval = u(xeval(:,1),xeval(:,2));  uLeval = uL(xeval(:,1),xeval(:,2));
```

```
% Define a range of N values to test convergence over
Nvec = [6,10,17,30,50,100];
% Loop over the required N values and compute error
errvec = zeros(size(Nvec));  errvecL = zeros(size(Nvec));
hvec = zeros(size(Nvec));
k = 1;
for N = Nvec
    % Create the data to be interpolated; here we choose Chebyshev points
    x = pick2Dpoints(-1,1,N,'cheb');  uhat = u(x(:,1),x(:,2));

    % Interpolate and evaluate the interpolant
    % We store the evaluation coefficients intcoef for efficient reuse
    DM = DistanceMatrix(x,x);  DMeval = DistanceMatrix(xeval,x);
    K = rbf(ep,DM);  Keval = rbf(ep,DMeval);  KLeval = rbfL(ep,DMeval);
    intcoef = K\uhat;
    uhateval = Keval*intcoef;  uhatLeval = KLeval*intcoef;

    % Compute the errors and the fill distance
    errvec(k) = errcompute(uhateval,ueval);
    errvecL(k) = errcompute(uhatLeval,uLeval);
    hvec(k) = max(min(DM+2*eye(N^2)));
    k = k + 1;
end

% Compute the lines of best fit
p = polyfit(log10(hvec),log10(errvec),1);
pL = polyfit(log10(hvec),log10(errvecL),1);

% Plot the results, best fit lines, and legend
h = figure;
hvplot = logspace(-2,0,3);
loglog(hvec,errvec,'-^','linewidth',2)
hold on
loglog(hvec,errvecL,'linewidth',2)
loglog(hvplot,10^p(2)*hvplot.^p(1),'--^k')
loglog(hvplot,10^pL(2)*hvplot.^pL(1),'--k')
hold off
legend('Interpolant','Laplacian',...
       sprintf('O(h^%2.1f)',p(1)),sprintf('O(h^%2.1f)',pL(1)),...
       'location','southeast')
```

Of particular note in this example is the computation of the fill distance (9.13) on line 31. While this could certainly be done more efficiently were it necessary, MATLAB again flexes its muscles in performing a relatively complicated computation in a single line. The `2*eye(N^2)` term is used to prevent each point from identifying itself as its nearest neighbor which, while true, is not helpful when computing the fill distance, i.e., the largest gap between two different points.

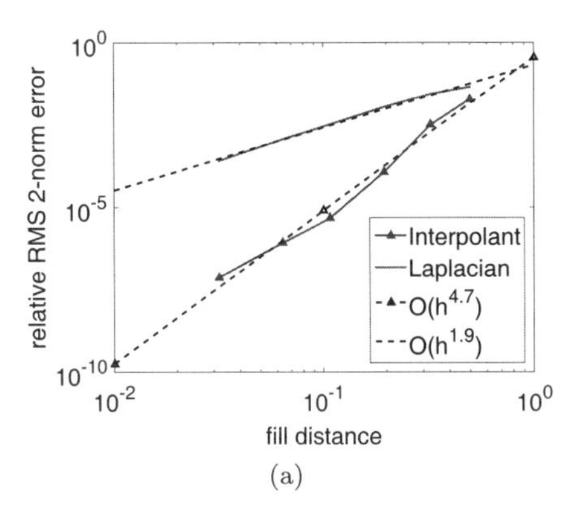

(a)

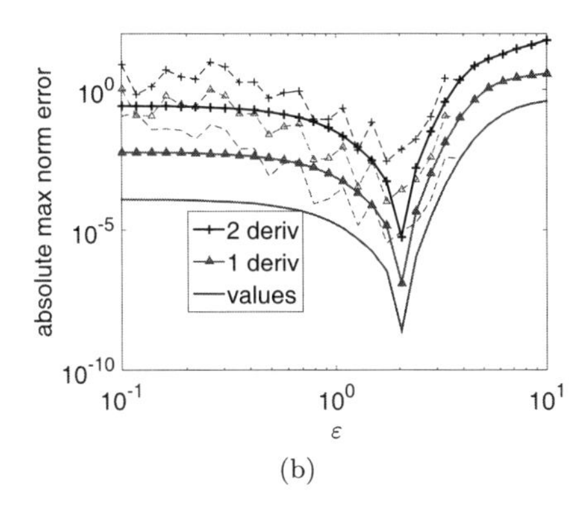

(b)

Fig. 19.1 (a) From Example 19.1. The convergence of the C^4 Matérn interpolant and its derivatives matches the order predicted by the theory. The dashed lines provide the slope of the lines of best fit for the interpolant and Laplacian error. (b) From Example 19.2. Accuracy is lost when differentiating an interpolant; here using Gaussians. The HS-SVD basis provides stability over the range of small ε values; dashed lines use the standard basis, solid lines use the HS-SVD basis.

Remark 19.5. In Figure 19.1(a) we include the lines of best fit, as computed with the `polyfit` function. We are forced to compute these on a log scale because the error, as described in (19.5), is of the form `err`$= Ch^{\nu}$. To fit C and ν we conduct a linear fit on $\log_{10}$`err`$= \log_{10} C + \nu \log_{10} h$, but to plot the results we must exponentiate the fitted parameters with base 10. This is why the plot commands in lines 45–46 look so odd.

The differentiation matrix formula (19.8) does not specify a basis for computing derivatives of an interpolant. As the reader might suspect, this allows for the evaluation of derivatives of interpolants using alternate bases instead of the standard basis computations conducted in Example 19.1. Indeed, the use of alternate bases is recommended for the stable computation of derivatives in the kernel flat-limit. In Example 19.2 we study the use of the HS-SVD basis for the Gaussians by utilizing the `GaussQR` library to compute and evaluate interpolants and derivatives with the HS-SVD basis.

Example 19.2. ("Flat" limits for derivative approximation)
This is a simple example involving a 1D function

$$u(x) = \frac{1}{1+x^2}, \quad u'(x) = -2\frac{x}{(1+x^2)^2}, \quad u''(x) = \frac{6x^2-2}{(1+x^2)^3}.$$

Data locations are provided by $N = 30$ Halton points and errors are evaluated on 50 evenly spaced points. The standard basis and HS-SVD basis interpolations to the functions are computed and the absolute max-norm of the interpolant and its first and second derivatives are evaluated. Various ε values (with Gaussian eigenfunction parameter $\alpha = 3$ as defined in (12.17)) are tested in Program 19.2, and the associated approximation errors are plotted in Figure 19.1(b).

Program 19.2. Excerpts from `HermiteEpProfile1D.m`

```
% Choose the Gaussian RBFs for HS-SVD comparison
rbf = @(e,r) exp(-(e*r).^2);
rbfx = @(e,r,dx) -2*e^2*dx.*exp(-(e*r).^2);
rbfxx = @(e,r) 2*e^2*exp(-(e*r).^2).*(2*(e*r).^2-1);
% Choose a range of shape parameters to consider
epvec = logspace(-1,1,30);
% Choose a function for testing
uf = @(x) 1./(1+x.^2);  ufx = @(x) -2*x./(1+x.^2).^2;
ufxx = @(x) (6*x.^2-2)./(1+x.^2).^3;
% Create the data for fitting and evaluation
N = 30;  Neval = 50;
x = pickpoints(-1,1,N,'halt');  xeval = pickpoints(-1,1,Neval);
u = uf(x);  ueval = uf(xeval);
uxeval = ufx(xeval);  uxxeval = ufxx(xeval);
% Pre-evaluate the distance and difference matrices, for efficiency
DM = DistanceMatrix(x,x);  DMeval = DistanceMatrix(xeval,x);
DiffMxeval = DifferenceMatrix(xeval,x);
% Loop through the epsilon values; plotting has been omitted
k = 1;
for ep=epvec
    % Solve and evaluate with the standard basis
    c = rbf(ep,DM)\u;
    uhateval = rbf(ep,DMeval)*c;
    uhatxeval = rbfx(ep,DMeval,DiffMxeval)*c;
    uhatxxeval = rbfxx(ep,DMeval)*c;

    % Solve and evaluate with the HS-SVD basis
    % Passing 1 or 2 to gqr_eval requests that many derivatives
    gqr_alpha = 3;
    GQR = gqr_solve(x,u,ep,gqr_alpha);
    uhatHSeval = gqr_eval(GQR,xeval);
    uhatHSxeval = gqr_eval(GQR,xeval,1);
    uhatHSxxeval = gqr_eval(GQR,xeval,2);
end
```

This program marks the first appearance of the function `DifferenceMatrix`, which is just a function encapsulation and some error checking for the differencing between data locations $\mathcal{X}$ and kernel centers $\mathcal{Z}$ *in a single dimension*. Line 17 has the same effect as

```
DiffMxeval = bsxfun(@minus,xeval,x)
```

but `DifferenceMatrix` handles some degenerate cases which `bsxfun` sees as errors. This difference matrix is required for odd-order derivative computations of radial kernels. More discussion on derivatives of, especially, radial kernels and their implementation in MATLAB is available in [Fasshauer (2007, Chapter 37)].

Program 19.2 also marks the first use of `gqr_eval` from the `GaussQR` library to evaluate derivatives of a Gaussian interpolant in the HS-SVD basis. Again, as discussed in Section 13.5, the quality of the HS-SVD basis for Gaussian approximation is strongly dependent on the eigenfunction parameter α. This example exists to demonstrate the dependence on the shape parameter ε of the accuracy of the derivative of the interpolant and to depict that the flat limit, which exists for interpolants, also exists for their derivatives.

19.1.4 *Local differentiation matrices*

In (19.8) we derived that a (global) differentiation matrix L is given by

$$\mathsf{L} = \mathsf{V}_{\mathcal{L}}^{\mathcal{X}}\mathsf{V}^{-1}.$$

Since the ith row of $\mathsf{V}_{\mathcal{L}}^{\mathcal{X}}$ corresponds to $\mathcal{L}\boldsymbol{v}(\boldsymbol{x}_i)^T$ (see (19.7)) we observe that the ith row of L, $(\mathsf{L})_{i,:}$, is given by

$$(\mathsf{L})_{i,:} = \mathcal{L}\boldsymbol{v}(\boldsymbol{x}_i)^T\mathsf{V}^{-1} \quad \Longleftrightarrow \quad \mathsf{V}^T(\mathsf{L}^T)_{:,i} = \mathcal{L}\boldsymbol{v}(\boldsymbol{x}_i). \tag{19.9}$$

The system of linear equations on the right side of (19.9) corresponds to [Fornberg and Flyer (2015, Eqn. (15))] — except for the additional (optional) polynomial terms in that paper — and the identity on the left side of (19.9) shows that the so-called *finite difference weights* for the evaluation point (or *stencil center*) $\boldsymbol{x}_i$ are nothing but the entries of one row of the differentiation matrix L. Moreover (as the system on the right side of (19.9) shows), these weights are chosen so that they reproduce the derivatives $\mathcal{L}\boldsymbol{v}(\boldsymbol{x}_i)$ at $\boldsymbol{x}_i$ of the basis functions collected in the vector $\boldsymbol{v}(\cdot)$.

Thus far, the *stencil* for the point $\boldsymbol{x}_i$ is global, i.e., it involves contributions from all of the input points (kernel centers) $\boldsymbol{z}_1, \ldots, \boldsymbol{z}_N$ via the Vandermonde-like interpolation matrix V. The purpose of a finite difference method is to consider only a small local neighborhood of $\boldsymbol{x}_i$ as the stencil and then perform a local interpolation problem to produce the local finite difference weights associated with *only* the point $\boldsymbol{x}_i$. This local interpolation problem will consider a stencil of points only in a local neighborhood of $\boldsymbol{x}_i$ and thus will generate a local interpolation matrix $\mathsf{V}_{\boldsymbol{x}_i}$ constructed from the local input points (e.g., the $N_{\boldsymbol{x}_i}$ nearest neighbors of $\boldsymbol{x}_i$).

The vector of local weights can be viewed as a compact representation of one row of the global sparse differentiation matrix[3] L^{FD}, and therefore needs to be mapped into the appropriate columns of L^{FD} before we can use this global differentiation matrix. While the mechanism for redistributing the local differentiation matrices into L^{FD} can be implemented in multiple efficient ways, the tool from a linear algebra perspective is simply a matrix of zeros and ones. We now walk through the process of defining and constructing a kernel-based finite difference operator L^{FD}.

Define $\mathcal{Z} = \{\boldsymbol{z}_1, \ldots, \boldsymbol{z}_N\}$ as the set of points at which we have (or may sample) data $\boldsymbol{u} = \big(u(\boldsymbol{z}_1) \cdots u(\boldsymbol{z}_N)\big)^T$. Define $\mathcal{X} = \{\boldsymbol{x}_1, \ldots, \boldsymbol{x}_{N_{\text{eval}}}\}$ as the set of points at

[3]We now introduce the notation L^{FD} to distinguish the global (sparse) differentiation matrix associated with the finite difference scheme from the global (dense) differentiation matrix L used earlier.

which we want to approximate the derivatives (we may think of these points as our stencil centers). For the ith evaluation point $\boldsymbol{x}_i$, the number of neighbors in the stencil is denoted $N_{\boldsymbol{x}_i}$ and the points in that stencil are collected in $\mathcal{Z}_i \subset \mathcal{Z}$. Using notation from Remark 19.3, the local differentiation matrix which takes in values at the sites $\mathcal{Z}_i$ and returns the approximate operation of the differential operator $\mathcal{L}$ at $\boldsymbol{x}_i$ is denoted as

$$\mathsf{L}_i = \mathsf{V}_{\mathcal{L}}^{\boldsymbol{x}_i} \mathsf{V}_{\mathcal{Z}_i}^{-1}, \qquad i = 1, \dots, N_{\text{eval}}.$$

Although we call it a local differentiation matrix because it applies only to a local stencil around $\boldsymbol{x}_i$, *on that stencil* $\mathsf{L}_{\boldsymbol{x}_i}$ acts globally in that it uses all the data of that stencil to create an interpolant from which to differentiate.

Collectively, these L_i matrices contain all the nonzeros in the sparse matrix L^{FD}, but their locations in that matrix must still be determined. We study this one row at a time: the ith row of L^{FD}, denoted $(\mathsf{L}^{\text{FD}})_{i,:}$ as is the MATLAB standard, contains nonzero values from the matrix L_i (which is really a row vector since it has only one evaluation point $\boldsymbol{x}_i$). The points in $\mathcal{Z}_i$ are used to form L_i, so the columns of L^{FD} associated with those points are the nonzero columns of row i. We can correctly place those points in the sparse row $(\mathsf{L}^{\text{FD}})_{i,:}$ by defining a sort of incidence matrix $\mathsf{P}_i \in \{0,1\}^{N_{\boldsymbol{x}_i} \times N}$ which has the elements

$$(\mathsf{P}_i)_{k,\ell} = \begin{cases} 1, & \text{if the } k^{\text{th}} \text{ point in } \mathcal{Z}_i \text{ matches the } \ell^{\text{th}} \text{ point in } \mathcal{Z}, \\ 0, & \text{else.} \end{cases}$$

Using this, the full sparse matrix is defined as

$$\mathsf{L}^{\text{FD}} = \begin{pmatrix} \mathsf{L}_1 \mathsf{P}_1 \\ \vdots \\ \mathsf{L}_{N_{\text{eval}}} \mathsf{P}_{N_{\text{eval}}} \end{pmatrix} = \begin{pmatrix} \mathsf{V}_{\mathcal{L}}^{\boldsymbol{x}_1} \mathsf{V}_{\mathcal{Z}_1}^{-1} \mathsf{P}_1 \\ \vdots \\ \mathsf{V}_{\mathcal{L}}^{\boldsymbol{x}_{N_{\text{eval}}}} \mathsf{V}_{\mathcal{Z}_{N_{\text{eval}}}}^{-1} \mathsf{P}_{N_{\text{eval}}} \end{pmatrix}. \tag{19.10}$$

In this notation, painful though it was to derive, it should be clear that the ith row accepts values on $\mathcal{Z}_i$, applies $\mathcal{L}$ to an interpolant on those values, and evaluates the result at the point $\boldsymbol{x}_i$.

Of course, in practice, actually forming all the P_i matrices is unnecessary because the movement of data can be conducted more directly. Program 19.3 demonstrates one possible approach in MATLAB which stores the effect of the P_i matrices at the time the stencil $\mathcal{Z}_i$ is determined. Lines 20–21 in that program show the process of correctly associating the nonzeros from L_i with their appropriate locations in L^{FD}.

Example 19.3. (Building a global finite difference operator from differentiation matrices on local stencils)
We present Program 19.3 to demonstrate a strategy for computing kernel-based finite differences. Associated convergence plots are provided in Figure 19.2. The function of interest here is

$$u(x,y) = \mathrm{e}^{-2xy}, \qquad \frac{\partial^2}{\partial x^2}u + \frac{\partial^2}{\partial y^2}u = 4(x^2+y^2)\mathrm{e}^{-2xy}. \tag{19.11}$$

This code again runs into the `xeval` and `x` confusion described in Remark 4.3 because the derivation above used $\boldsymbol{x}$ and $\boldsymbol{z}$ for the evaluation and kernel centers, respectively. The MATLAB code presented here follows the `xeval` and `x` standard used throughout the rest of this text.

Program 19.3. Single stencil form of `HermiteFiniteDiffConv.m`

```
% Choose a function to test
uf = @(x) exp(-2*x(:,1).*x(:,2));
% Pick some points on which to provide data
N = 100;  x = pick2Dpoints(0,1,N,'halt');  u = uf(x);
% Pick some points on which to evaluate the error
Neval = 25;  xeval = pick2Dpoints(.05,.95,Neval);
% Choose the C4 Matern kernel for computation
rbf = @(e,r) (1+(e*r)+(e*r).^2/3).*exp(-e*r);
rbfL = @(e,r) 1/3*e^2*exp(-e*r).*((e*r).^2-2*e*r-2);
ep = 1;
% Choose a stencil size
Nx = 90;
% Find the stencil of Nx nearest neighbors to the evaluation points
nearest = num2cell(knnsearch(x,xeval,'K',Nx),2);
% Find the local differentiation matrix for each evaluation point
FDcell = cellfun(@(xe,xi)rbfL(ep,DistanceMatrix(xe,x(xi,:)))/...
                      rbf(ep,DistanceMatrix(x(xi,:),x(xi,:))),...
                  num2cell(xeval,2),nearest,'UniformOutput',0);
% Form the vectors needed to construct a sparse matrix
FDvecs = cell2mat(cellfun(@(row,cols,vals)[row*ones(1,Nx);cols;vals],...
              num2cell((1:Neval^2)',2),nearest,FDcell,'UniformOutput',0)');
% Create the sparse matrix from those vectors
FDmat = sparse(FDvecs(1,:),FDvecs(2,:),FDvecs(3,:),Neval^2,N^2);
% Evaluate the finite difference Jacobian at xeval
uLappx = FDmat*u;
```

We now explain each of the steps in the construction of the finite difference Laplacian from Program 19.3 in the context of the notation leading up to (19.10).

(1) A single stencil size of $N_{\boldsymbol{x}_i} = 90$ is defined for all the evaluation points in $\mathcal{X}$ on line 12. The formulation above allows for different stencil sizes, but we consider the simpler case of a uniform stencil size only.

(2) Line 14 determines the set $\mathcal{Z}_i$, i.e., the stencil, for each $\boldsymbol{x}_i$. In typical MATLAB fashion, this occurs all at once in `knnsearch` from MATLAB's Statistics and Machine Learning Toolbox. That function returns the indices of the points from $\mathcal{Z}$ appearing in $\mathcal{Z}_i$, which makes the task of applying the P_i matrix later much easier.

(3) Lines 16–18 compute the local differentiation matrices $\mathsf{L}_i = \mathsf{K}_{\mathcal{L}}^{\boldsymbol{x}_i}\mathsf{K}_{\mathcal{Z}_i}^{-1}$ using the MATLAB `/` operator: `A/B` is equivalent to `A*inv(B)`. Notice that we explicitly state the use of the standard basis here to match the code; an alternate basis is used in Example 19.4. The use of cell arrays and `cellfun` here allows

us to compute all the local differentiation matrices simultaneously and is not fundamental to the finite difference construction.

(4) The task of forming the sparse matrix L^{FD}, defined in MATLAB as `FDmat`, requires the creation of vectors storing the matrix values along with their associated rows and columns; we presented an example of these vectors in Program 4.10. Lines 20–21 associate the correct row with each column (determined earlier with `knnsearch`) and value (stored in `FDcell`). Again the use of cell arrays simply allows all rows to be organized simultaneously. The sparse matrix is constructed from these vectors with the `sparse` command.

This L^{FD} formation can be encapsulated into a function if desired, similarly to the `DistanceMatrix` function from Chapter 4, though we do not do so here.

The error computation is omitted from the code above because of its similarity to all other error computations in this book, but it is provided in the `GaussQR` library. Program 19.3 was run with several different `ep` and `xeval` choices, described in Figure 19.2, to produce convergence plots. The errors are computed on different regions in the interior of the interval $[0, 1]$ (indicated in the legend of Figure 19.2(a)). As we extend the evaluation region closer to the boundary, the local stencil becomes increasingly nonsymmetric and the quality of the approximation is limited; in fact the error is saturated.

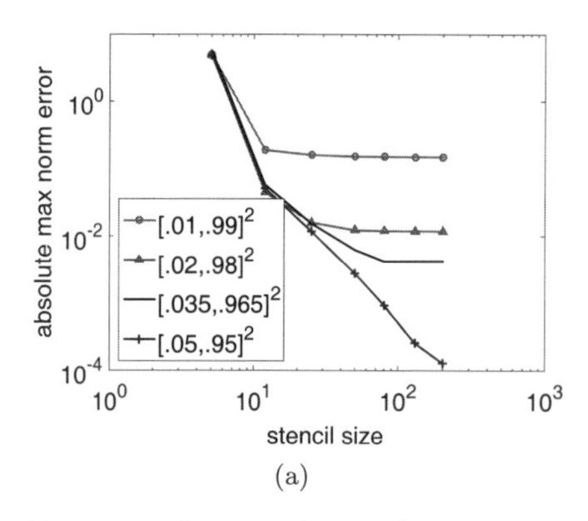

(a)

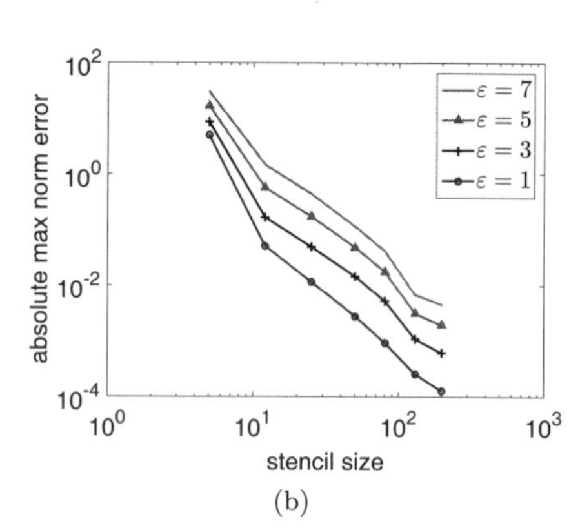

(b)

Fig. 19.2 The error for a C^4 Matérn kernel-based finite difference scheme is studied on 25^2 uniform tensor grid points. $N = 10000$ Halton points in $[0, 1]^2$ are used as data locations and an approximate Laplacian is computed to (19.11) using stencils of increasing size. (a) Error is computed on different regions (denoted in the legend) which exclude the boundary. As the points of evaluation approach the boundary, the stencil becomes increasingly nonsymmetric and the quality of the approximation is bounded. These only consider $\varepsilon = 1$. (b) When considering the error only on $[0.05, 0.95]^2$, smaller ε values produce better finite difference Laplacians. These results are limited by stability issues for smaller values of ε.

In Example 19.3 we observed that approximating derivatives near the boundaries is more difficult. Figure 19.2 shows stagnant error even as the size of the stencil increased. While there is nothing inherently difficult about working near the boundary, the asymmetry in the stencil, caused by a skewed distribution of points towards the interior of the domain, yields a loss of accuracy. The effect of the stencil shape is discussed in, e.g., [Fornberg *et al.* (2010)], and Example 19.4 explores the effect on the shape parameter.

Example 19.4. ("Flat" limits for kernel-based FD methods)
We reuse (19.11) and consider a finite difference approximation to $\mathcal{D}_{xy}u$ using Gaussian kernels on the local FD stencils. The stencil size is fixed and evaluation points are considered at the middle of the stencil and skewed to the side; Figure 19.3(a) shows these points. Program 19.4 computes the error in the mixed derivative, and the results are plotted for a variety of ε values in Figure 19.3(b).

Program 19.4. Single ε form of `HermiteFiniteDiffEpProfile.m`

```
% Choose a function to test
uf = @(x) exp(-2*x(:,1).*x(:,2));
ufxy = @(x) (4*x(:,1).*x(:,2)-2).*exp(-2*x(:,1).*x(:,2));
% Choose a shape parameter to test
ep = .1;   gqr_alpha = 1;
% Two points: one at the center and one at the edge
xeval = [0 0;.5 0];
% Choose a stencil size and possible stencil points
Nx = 80;  xtest = pick2Dpoints(-1,1,12,'halt');
% Form a stencil of nearest neighbors and evaluate the function
x = xtest(knnsearch(xtest,xeval(1,:),'K',Nx),:);
u = uf(x);  uxyeval = ufxy(xeval);
% Create the HS-SVD basis
GQR = gqr_solveprep(0,x,ep,gqr_alpha);
CbarT = GQR.CbarT;
% Form the differentiation matrix for the two points
Psi = GQR.stored_phi1 + GQR.stored_phi2*CbarT;
Phixy = gqr_phi(GQR,xeval,[1 1]);
Psixy = Phixy*[eye(Nx);CbarT];
FDmat = Psixy/Psi;
% Evaluate the kernel finite difference mixed derivative
uhatxyeval = FDmat*u;
error = abs(uhatxyeval-uxyeval);
```

The initial selection of possible stencil points in line 9 is what yields the circular stencil in Figure 19.3(a); to create Figure 19.3(c) one could change to `xtest = pick2Dpoints(-1,1,10,'halt')` which provides fewer possible stencil points. Offering fewer points forces the stencil into less of a circular shape and more of a rounded square shape. The geometry of the stencil, both in the actual distribution of stencil points and the location at which the derivative is to be approximated, is clearly a significant factor in the quality of a kernel-based finite difference method.

Thus far, essentially all the literature on kernel-based finite differences empha-

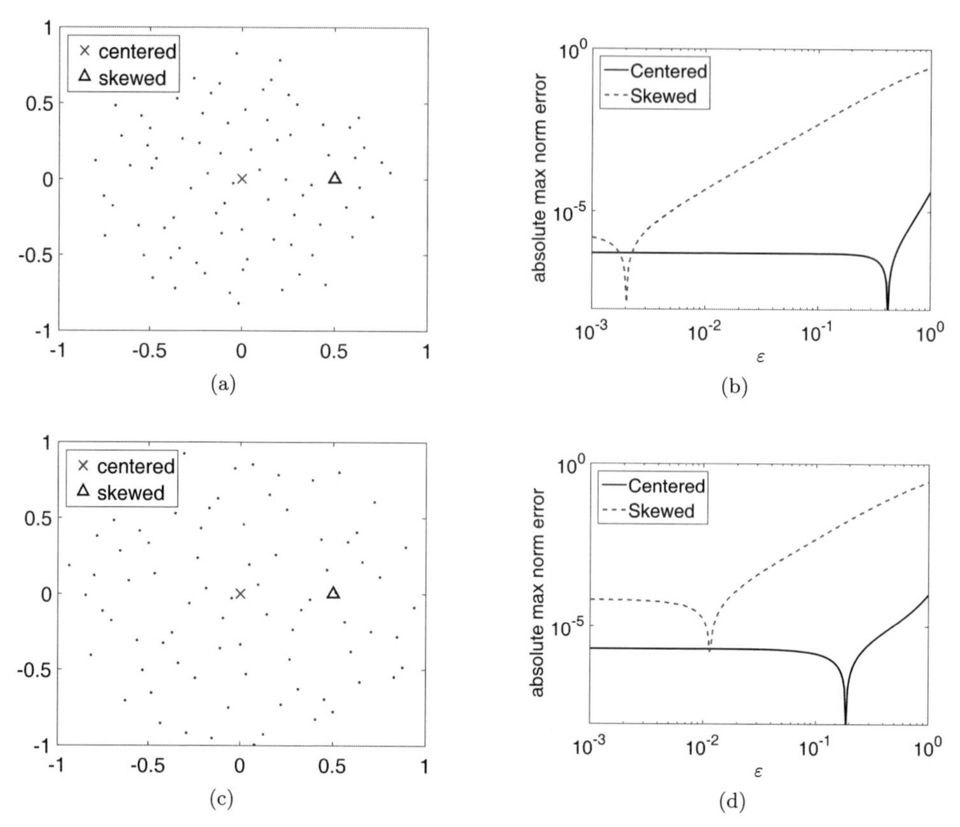

Fig. 19.3 Different stencils and evaluation points are considered for a Gaussian kernel finite difference computation. The differentiation matrices were computed with the HS-SVD basis for stability with small ε. The $N_x = 80$ points stencils are centered around $[0, 0]$ but are skewed from the perspective of evaluation at $[0, 0.5]$. (a) This stencil is radially symmetric. (b) While most ε values yield significantly worse error for the skewed finite difference mixed derivative at $[0, 0.5]$, it is possible for it to reach nearly the accuracy of the $[0, 0]$ approximation, albeit for a very small ε range. (c) This stencil is more in the shape of a square with rounded corners. (d) Evaluating the skewed finite difference computation at $[0, 0.5]$ for the rounded square stencil produces significantly worse accuracy.

sizes the use of radial kernels, but nothing we have explained thus far requires the kernels of interest to be radial. Indeed, for the purposes of differentiation, it is entirely possible that a tensor product kernel (recall Section 3.4.3 and Section 4.4) would be preferable since the action of the derivatives occurs only on one of the 1D kernels that make up the full kernel. Example 19.5 explores this possibility.

Example 19.5. (Tensor product formulation of kernel-based finite differences) This example refers in large part to Program 19.3. We explain the necessary steps for altering that program to use a *tensor product kernel* of 1D C^4 Matérn kernels instead of the radial 2D C^4 Matérn presently in use. The only significant change is the need to define the tensor product kernel. Unlike in Program 4.13, where we used cell arrays to cleanly encapsulate the tensor product structure, here we use "brute force" to define the kernel and its derivatives.

We assume that the user has a length 2 vector `epvec`, an $N_{\text{eval}} \times 2$ matrix `x` of evaluation points and an $N \times 2$ matrix `z` of kernel centers. The product kernel is defined with

```
Kf = @(e,x,z) (1+e(1)*abs(bsxfun(@minus,x(:,1),z(:,1)'))+ ...
                (e(1)*abs(bsxfun(@minus,x(:,1),z(:,1)'))).^2/3).* ...
              exp(-e(1)*(abs(bsxfun(@minus,x(:,1),z(:,1)')))).* ...
              (1+e(2)*abs(bsxfun(@minus,x(:,2),z(:,2)'))+ ...
                (e(2)*abs(bsxfun(@minus,x(:,2),z(:,2)'))).^2/3).* ...
              exp(-e(2)*(abs(bsxfun(@minus,x(:,2),z(:,2)'))));
```

Using that definition, `Kf(epvec,x,z)` returns an $N_{\text{eval}} \times N$ matrix $(\mathsf{K})_{i,j} = K_1(x_{i,1}, z_{i,1})K_2(x_{j,2}, z_{j,2})$ with K_1 and K_2 the (univariate) radial C^4 Matérn kernels from Table 3.2 with shape parameters ε_1 and ε_2 (`e(1)` and `e(2)`), respectively.

Differentiating this twice in the first dimension (physically the x direction or, in MATLAB, the first column of the arguments to `Kf`) is arguably simpler than in the radial case since only the K_1 component is affected:

```
Kfxx = @(e,x,z) (e(1)^2/3).*(-1-e(1)*abs(bsxfun(@minus,x(:,1),z(:,1)'))+ ...
                         (e(1)*abs(bsxfun(@minus,x(:,1),z(:,1)'))).^2).* ...
                         exp(-e(1)*(abs(bsxfun(@minus,x(:,1),z(:,1)')))).* ...
                (1+e(2)*abs(bsxfun(@minus,x(:,2),z(:,2)'))+ ...
                  (e(2)*abs(bsxfun(@minus,x(:,2),z(:,2)'))).^2/3).* ...
                exp(-e(2)*(abs(bsxfun(@minus,x(:,2),z(:,2)'))));
```

An analogous definition could be used to define `Kfyy`, the second derivative in the physical y direction, although the code is omitted here. The Laplacian can easily be defined as

```
KfL = @(e,x,z) Kfxx(e,x,z) + Kfyy(e,x,z);
```

With this definition, the computation of the differentiation matrices comprising the finite difference operator becomes

```
FDcell = cellfun(@(xe,xi)KfL(epvec,xe,x(xi,:))/...
                          Kf(epvec,x(xi,:),x(xi,:)),...
             num2cell(xeval,2),nearest,'UniformOutput',0);
```

in place of lines 16–18. This replacement and the subsequent finite difference experiments are available in the `GaussQR` library in `HermiteFiniteDiffTensor.m`.

One additional point of note: because tensor product kernels are essentially unused in the finite difference literature, there is no guarantee that the advice regarding radial stencils from Figure 19.3 is equally applicable. MATLAB's Statistics and Machine Learning Toolbox provides an option to find the nearest neighbors in a square (according to the ∞-norm) rather than in a circle (according to the 2-norm). To do so, the `knnsearch` on line 14 should be called as

```
nearest = num2cell(knnsearch(x,xeval,'K',Nx,'Distance','Chebychev'),2);
```

Also, recall that many kernel derivatives, including those in use here, were computed and are available in [Fasshauer (2007, Appendix D)].

Remark 19.6. At the time of writing of this chapter, the literature contains very few contributions addressing the theoretical foundation of local differentiation matrices. Exceptions are [Davydov and Schaback (2015)], where the consistency of the kernel-based differentiation matrix approach is established. Some results on the local truncation error for kernel-based finite difference stencils (in particular for Gaussians and multiquadrics) are provided in [Bayona *et al.* (2010, 2012a)].

Remark 19.7. The criterion for choosing the stencil points associated with $\boldsymbol{x}_i$ is open for discussion. The most popular choice is probably to use the *nearest neighbors* of $\boldsymbol{x}_i$. That is what we do in our implementation. However, Beatson *et al.* (2011) (mentioned in Remark 19.8 below), e.g., recommend using the *mean value coordinates* of Floater (2003) to select points in a well-balanced neighborhood. Other choices are certainly possible — especially when we are free to choose the locations of the input points. Under those circumstances, Fornberg and Flyer (2015), e.g., recommend the use of *hexagonal lattices* coupled with the nearest neighbor criterion since hexagonal lattices are optimally distributed in a square domain, as was proved in [Iske (2000)].

Remark 19.8. If we consider the case when $\mathcal{L} = \mathcal{I}$, i.e., we are in the context of standard (Lagrange) interpolation, then the "differentiation" matrix consists simply of the values of cardinal basis functions, i.e., from (19.9) we see

$$(\mathsf{L})_{i,:} = \boldsymbol{v}(\boldsymbol{x}_i)^T \mathsf{V}^{-1} = \boldsymbol{\ell}(\boldsymbol{x}_i)^T.$$

As in the finite difference context discussed above, we can interpret this either as interpolation on an isolated local stencil, or as a compact representation of a local Lagrange basis within a global "differentiation" matrix L^{FD}. This scenario was approached from a different angle in the work of Beatson, Cherrie and Mouat (1999); Faul and Powell (1999); Mouat (2001); Faul, Goodsell and Powell (2005); Beatson, Levesley and Mouat (2011), where the main role of L^{FD} was as a *preconditioner* for the global interpolation matrix K (and thus used in the context of *iterative algorithms*, see also [Fasshauer (2007, Chapters 33–34)]), and more recently by Hangelbroek, Narcowich and Ward (2010); Hangelbroek, Narcowich, Sun and Ward (2011); Fuselier, Hangelbroek, Narcowich, Ward and Wright (2013); Hangelbroek, Narcowich, Rieger and Ward (2014), where the convergence behavior of interpolation methods using these *local Lagrange bases* was carefully investigated.

Remark 19.9. Another interesting — as of yet largely unexplored — connection to other work on meshfree methods is the interpretation of the RBF-FD approach in the context of (interpolatory) *moving least squares* (MLS) approximation (see, e.g., [Fasshauer (2007, Chapters 22–24)] for a discussion of that method). The local

Lagrange bases now play the role of the MLS generating functions, and polynomial reproduction can be enforced by adding the appropriate terms to the local kernel interpolation problem. The latter idea is already quite common with practitioners in the RBF-FD community (see, e.g., [Fornberg and Flyer (2015, Eqn. (15))]) when defining the local differentiation weights referred to earlier.

Remark 19.10. Methods classified as kernel-based finite difference schemes often attempt to approximate derivatives at only a single point per stencil. This relationship of each $\boldsymbol{x}_i$ using its own coefficients $\mathsf{V}_{\mathcal{L}}^{\boldsymbol{x}_i}\mathsf{V}_{\mathcal{X}_i}^{-1}$ lends itself nicely to high performance computation on GPUs [Bollig *et al.* (2012)].

Alternatively, *partition of unity methods*, also referred to as RBF-PU, allow for the possibility of approximating the derivative at multiple locations using the same stencil [Safdari-Vaighani *et al.* (2014)]. This is advantageous because it allows for a single $\mathsf{V}_{\mathcal{X}}^{-1}$ computation for multiple $\boldsymbol{x}$ evaluation points using the same stencil $\mathcal{X}$, thus reducing the cost of approximating derivatives. The quality of such methods, however, is somewhat dependent on the location of evaluation points within a stencil, as we saw in Figure 19.3.

19.2 Hermite Interpolation

The kinds of tasks that lend themselves to interpolation of both function *and* derivative values include *sensitivity analysis* such as in model validation, *data assimilation* and other *inverse problems*, as well as *design optimization* by choosing "optimal" simulation parameters. Such design optimization is commonly treated via surrogate modeling, computer experiments or response surface modeling. Sometimes derivative information is directly provided by the computer simulation, or — in its absence — an increasingly popular approach is to use *automatic differentiation* [Griewank and Walther (2008)] to compute derivative data.

Specific examples of the use of kernel methods for such computer experiments include uncertainty propagation in nuclear reactor design (see [Lockwood and Anitescu (2011, 2012)]) or the identification of nonlinear dynamical systems from experimental data [Solak *et al.* (2003)]. Other applications can be found in modeling of implicit surfaces defined in terms of point clouds and their associated normals — information commonly provided by laser range scanners (see, e.g., [Macedo *et al.* (2009)]).

The advantage of Hermite interpolation over basic scattered data interpolation is that fewer input *locations* are required to obtain comparable predictive accuracy. If data at a single input location corresponds to running an expensive computer experiment with a specific choice of simulation parameters (and this run can generate both a function and a derivative value), then generating, e.g., gradient data is far more efficient than generating the equivalent number of $d+1$ function values, where

d is the dimension of the space of simulation parameters. If higher-order derivatives — such as Hessian information — are included, then the savings in generating Hermite data become even more pronounced.

In [Fasshauer (2007, Chapters 36–37), Wendland (2005, Chapter 16)] a *generalized Hermite interpolation* problem is described. We will align our discussion in this section with those expositions and also refer to [Mardia *et al.* (1996)] for the stochastic/kriging setting. Our goal is two-fold:

(1) We want to show how the Hilbert–Schmidt SVD can be applied to interpolation problems that include data which is more general than simple function values.
(2) We want to discuss — with references to Chapter 5 — the connection between kernel-based (generalized) Hermite interpolation (as is common in the approximation theory literature) and a generalized form of kriging, sometimes referred to as *gradient-enhanced kriging*.

We will begin with a rather general formulation and then specialize to a very specific example of a first-order Hermite interpolation problem, where function values along with values of the gradient at the points $\{\boldsymbol{x}_1, \ldots, \boldsymbol{x}_N\} \subset \mathbb{R}^2$ are specified. This example will illustrate the details required for a practical implementation which can be obscured by the elegant formulation in the general setting.

The literature on Hermite interpolation — both in its deterministic incarnation in approximation theory and in its statistics-based kriging version — emphasizes the use of a *symmetric* form of Hermite interpolation. Before we discuss that approach in Sections 19.2.2–19.2.5, we first look at a nonsymmetric version of Hermite interpolation that is extremely popular as the basis for a nonsymmetric collocation approach to solving PDEs often referred to as *Kansa's method* which we will feature in Chapter 20.

19.2.1 *Nonsymmetric kernel-based Hermite interpolation*

In Section 19.2.2 we discuss the use of symmetric Hermite collocation, for which a solid theoretical foundation exists. In this section we introduce a very simple, but potentially ill-posed, strategy for interpolating (generalized) functional data.

We consider data $\{\boldsymbol{x}_i, L_i u\}$, $i = 1, \ldots, N_H$, $\boldsymbol{x}_i \in \mathbb{R}^d$, with $\{L_i\}_{i=1}^{N_H}$ a linearly independent set of *bounded linear functionals* (usually associated with the points $\boldsymbol{x}_i$) that provide samples of a function u. For example, we can think of L_i as a point evaluation functional so that $L_i u = u(\boldsymbol{x}_i)$, or it might denote a derivative evaluation so that $L_i u = \mathcal{L}u(\boldsymbol{x}_i)$, where $\mathcal{L}$ is some linear differential operator. Other functionals are also possible, such as, e.g., a local average provided in terms of an integral operator associated with a domain around $\boldsymbol{x}_i$.

The reader should note that in this problem formulation there is no requirement that the points $\boldsymbol{x}_i$ be distinct; the only requirement is that the linear functionals L_i are linearly independent. Thus, we could have a problem where multiple values —

such as function value, x-partial and y-partial — are all specified for each $\boldsymbol{x}_i$, and thus, even though only N distinct data sites are used, the number of data values and subsequently also basis functions for this example would be $N_H = 3N$. Moreover, there is no requirement that derivative data needs to be provided in successive order, or that the same type of data be provided at all data sites. In other words, it is perfectly fine to have (only) derivative data at certain locations (such as data sites in the interior of the domain), and (only) function values at other locations (e.g., on the boundary). This is precisely the scenario we will consider for our PDE collocation approach in Chapter 20.

Even if we are given this kind of more complex data rather than basic function values, we can still assume that the interpolant takes the form $\hat{u}(\boldsymbol{x}) = \boldsymbol{k}(\boldsymbol{x})^T\boldsymbol{c}$.

In the following discussion we keep things as simple as possible and take $L_i u = \mathcal{D}u(\boldsymbol{x}_i)$ and $N_H = N$. As we explain in Example 19.6, this leads to an interesting situation where the continuous problem is ill-posed, but the kernel-based discretization seems to be well-posed.

To fit this data, we simply apply the derivative to the interpolant, evaluate at the data locations $\mathcal{X} = \{\boldsymbol{x}_1, \ldots, \boldsymbol{x}_N\}$ and match those interpolant derivatives to the derivative data $\boldsymbol{u_L} = \begin{pmatrix} L_1 u & \cdots & L_N u\end{pmatrix}^T = \begin{pmatrix}\mathcal{D}u(\boldsymbol{x}_1) & \cdots & \mathcal{D}u(\boldsymbol{x}_N)\end{pmatrix}^T$ to arrive at a square linear system of the form

$$\begin{pmatrix} \mathcal{D}\boldsymbol{k}(\boldsymbol{x}_1)^T \\ \vdots \\ \mathcal{D}\boldsymbol{k}(\boldsymbol{x}_N)^T \end{pmatrix} \boldsymbol{c} = \boldsymbol{u_L}. \tag{19.12}$$

While the system (19.12) is more complex than the standard interpolation system $\mathsf{K}\boldsymbol{c} = \boldsymbol{u}$, it is much simpler than the comparable system (19.14) in the symmetric Hermite interpolation setting.

Example 19.6. (Nonsymmetric Hermite interpolation)
We revisit the example from Program 4.9 and attempt to interpolate derivative data from the same underlying function $u(x) = \cos(2\pi x)$ at $N = 10$ random locations in $[0, 1]$. Program 19.5 provides a MATLAB implementation with the C^2 Matérn kernel for a fixed shape parameter ε.

Program 19.5. `HermiteNonsymmetric.m`

```
% Define data locations and evaluation locations
x = pickpoints(0,1,10,'rand');   xeval = pickpoints(0,1,500);
% Choose a function and evaluate it at those locations
uxf = @(x) -2*pi*sin(2*pi*x);   ux = uxf(x);
% Consider the C2 Matern kernel
rbf = @(e,r) (1+(e*r)).*exp(-(e*r));   rbfx = @(e,r,dx) -e^2*dx.*exp(-(e*r));
ep = 5;
% Form the necessary distance and difference matrices
DM = DistanceMatrix(x,x);   DiffM = DifferenceMatrix(x,x);
% Compute the Hermite interpolation matrix and solve
```

```
Kx = rbfx(ep,DM,DiffM);
c = Kx\ux;
% Create a function to evaluate the interpolant anywhere
uhatf = @(xeval) rbf(ep,DistanceMatrix(xeval,x))*c;
uhateval = uhatf(xeval);
```

The continuous version of this problem is phrased as:

Given the function $u'(x) = -2\pi \sin(2\pi x)$, what is the function u?

Of course, without specifying at least one function value, such a problem is an *ill-posed* differential equation, for which there are infinitely many answers, $u(x) = \sin(2\pi x) + C$ for any C. How, then, is it that the discretized version of this problem is well posed, as demonstrated by the unique solution in line 12?

As Figure 19.4(a) indicates, there is a unique nonsymmetric Hermite interpolant for any given ε, but seemingly each choice of ε produces a different solution. All of these solutions vary by a constant, suggesting that ε plays a role in implicitly, and uniquely, choosing a C for the solution. We can confirm this in MATLAB by redefining the function that evaluates the solution $\hat{u}$ from line 14 to

```
uhatepf = @(ep,xeval) rbf(ep,DistanceMatrix(xeval,x))*(rbfx(ep,DM,DiffM)\ux)
```

so as to allow ε to also be a variable. Using this function we can create Figure 19.4(b) with the commands

```
epvec = logspace(0,1,30);
semilogx(epvec,arrayfun(@(ep) uhatepf(ep,0),epvec));
```

where it seems that there is a specific value $\varepsilon \in [1, 10]$ which yields the effect of $C = 0$. We can determine this value approximately in MATLAB using the `fzero` function,

```
fzero(@(ep)uhatepf(ep,0)-1,1)
ans =
    3.0164
```

Of course, this value is only appropriate if $C = 0$ were the true constant, and if all we actually have is derivative data then we would have no reason to believe that is the case.

The relationship between the choice of ε and a standard interpolant, with data and not derivative data, is discussed in Chapter 9. Because each choice of ε produces a different kernel with a different reproducing kernel Hilbert space, and the kernel interpolant is the minimum norm interpolant in that space, the choice of ε has a significant effect on the resulting interpolant. A similar concept may apply to interpolation of derivative data — that the ill-posed interpolant minimizes some norm — but we are, as of yet, unaware of any such proof.

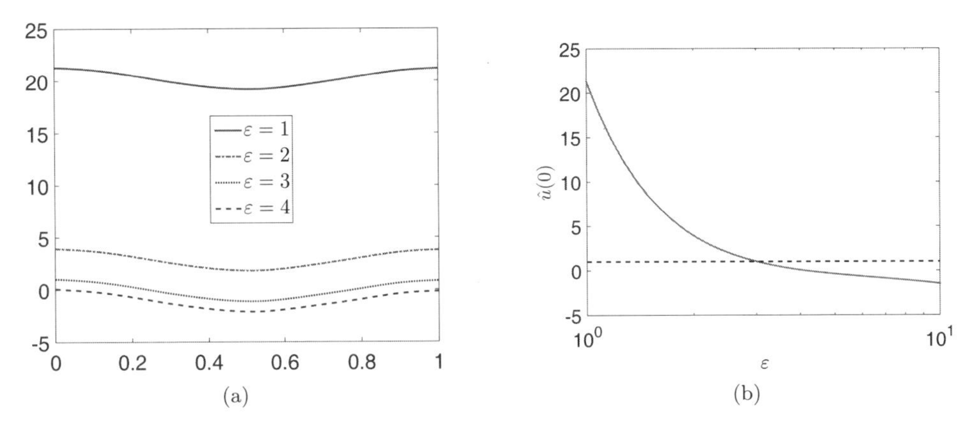

Fig. 19.4 The choice of ε affects the solution to the ill-posed problem. (a) Various ε values and their corresponding Hermite interpolants. (b) The solid curve is the value of the Hermite interpolant at $x = 0$, which passes through the dashed line $\hat{u}(0) = 1$ for $\varepsilon \approx 3$.

As such, solving ill-posed Hermite interpolation problems is not something we focus on in this text. There are, however, instances in the literature where ill-posed problems were solved by leveraging this optimality/regularization structure afforded by positive definite kernels [Hon and Wei (2004); Li and Mao (2011)]. In Section 20.4, we will use the solution to an ill-posed problem as part of a larger PDE solution, where the lack of uniqueness does not matter.

Remark 19.11. There are two points of potential ill-posedness in this nonsymmetric Hermite formulation: the ill-posedness in the continuous setting and the ill-posedness in the discretized setting. As seen in Example 19.6, ill-posedness in the continuous setting need not result in ill-posedness in the discrete setting because of the regularity imposed by the positive definite kernel basis. Even so, it is entirely possible that the discretized problem in (19.12) has a singular matrix (see the discussion in [Hon and Schaback (2001)]). This outcome can be eliminated with a symmetric formulation, as described in Section 19.2.2.

19.2.2 *Symmetric kernel-based Hermite interpolation*

We consider the same type of data as in Section 19.2.1, i.e., we have $\{\boldsymbol{x}_i, L_i u\}$, $i = 1, \ldots, N_H$, $\boldsymbol{x}_i \in \mathbb{R}^d$, with the data values generated by linearly independent bounded linear functionals $\{L_i\}_{i=1}^{N_H}$.

The main difference to the nonsymmetric Hermite formulation of Section 19.2.1 is that we now no longer assume the interpolant to have the simple form $\hat{u}(\boldsymbol{x}) = \boldsymbol{k}(\boldsymbol{x})^T \boldsymbol{c}$. The reason for using the more complex formulation we are about to explain is that we are able to support this new framework with a solid theoretical foundation.

As discussed in [Fasshauer (2007, Chapters 36–37), Wendland (2005, Chapter 16)], we assume that for $\boldsymbol{x}, \boldsymbol{z} \in \mathbb{R}^d$ the *generalized Hermite interpolant* is of the form

$$\begin{aligned}\hat{u}(\boldsymbol{x}) &= \sum_{j=1}^{N_H} c_j L_j^{\boldsymbol{z}} K(\boldsymbol{x}, \cdot) \\ &= \boldsymbol{k}_{\boldsymbol{L}^{\boldsymbol{z}}}(\boldsymbol{x})^T \boldsymbol{c}, \end{aligned} \tag{19.13}$$

where the vector of basis functions is $\boldsymbol{k}_{\boldsymbol{L}^{\boldsymbol{z}}}(\boldsymbol{x})^T = \left(L_1^{\boldsymbol{z}} K(\boldsymbol{x}, \cdot) \cdots L_{N_H}^{\boldsymbol{z}} K(\boldsymbol{x}, \cdot)\right)$ and the superscript $\boldsymbol{z}$ used on the linear functional L_j indicates that the functional acts on the second argument of the kernel K.

Remark 19.12. For example, if we for simplicity assume that L_j corresponds to point evaluation, then $L_j^{\boldsymbol{z}} u = u(\boldsymbol{z}_j)$, so that $L_j^{\boldsymbol{z}}$ not only acts on the second argument of the kernel, but also evaluates at a point $\boldsymbol{z}_j$ instead of $\boldsymbol{x}_j$. In other words, we will allow for a decoupling of the kernel centers $\boldsymbol{z}_1, \ldots, \boldsymbol{z}_{N_H}$ from the data points $\boldsymbol{x}_1, \ldots, \boldsymbol{x}_{N_H}$ similarly to what we did in Chapter 18 and Section 19.1. Creating a kernel basis with centers $\boldsymbol{z}_j$ that are different from the data points $\boldsymbol{x}_i$ can certainly produce a linearly independent set of functions (and thus a basis for an approximation space), but it is not clear how to show in general that the resulting linear system in (19.14) is uniquely solvable. If the centers and data sites do coincide, then there are various contributions in the literature that ensure a unique solution for rather general — even conditionally positive definite — kernels (see, e.g., [Wu (1992); Narcowich and Ward (1994); Sun (1994)] for generalized Hermite interpolation in $\mathbb{R}^d$, and [Narcowich (1995); Fasshauer (1999)] for problems on Riemannian manifolds and on spheres, respectively). Moreover, identifying centers with data sites provides a data-dependent basis and thus is in line with our discussion in Section 1.3.1.

If we write the *generalized interpolation conditions* for the given data as

$$L_i^{\boldsymbol{x}} \hat{u} = L_i^{\boldsymbol{x}} u, \qquad i = 1, \ldots, N_H,$$

then the coefficients $\boldsymbol{c}$ in (19.13) are found by solving the linear system

$$\mathsf{H}\boldsymbol{c} = \boldsymbol{u}_{\boldsymbol{L}^{\boldsymbol{x}}}, \tag{19.14}$$

where the matrix H has entries $(\mathsf{H})_{i,j} = L_i^{\boldsymbol{x}} L_j^{\boldsymbol{z}} K(\cdot, \cdot)$ and the right-hand side vector $\boldsymbol{u}_{\boldsymbol{L}^{\boldsymbol{x}}} = \left(L_1^{\boldsymbol{x}} u \cdots L_{N_H}^{\boldsymbol{x}} u\right)^T$. Therefore, using (19.13) and (19.14), the Hermite interpolant can be expressed in the form

$$\hat{u}(\boldsymbol{x}) = \boldsymbol{k}_{\boldsymbol{L}^{\boldsymbol{z}}}(\boldsymbol{x})^T \mathsf{H}^{-1} \boldsymbol{u}_{\boldsymbol{L}^{\boldsymbol{x}}} \tag{19.15}$$

with the vector of *Hermite cardinal functions* $\boldsymbol{\ell}_{\boldsymbol{L}^{\boldsymbol{z}}}(\boldsymbol{x}) = \boldsymbol{k}_{\boldsymbol{L}^{\boldsymbol{z}}}(\boldsymbol{x})^T \mathsf{H}^{-1}$.

Remark 19.13. Convergence orders for symmetric Hermite interpolation can be inferred from the work of Franke and Schaback (1998a,b) (see also [Wendland (2005, Chapter 16)]), and are directly addressed in Arcangéli *et al.* (2012) using the sampling inequalities introduced in Chapter 9.

19.2.3 *Generalized Hermite interpolation via the Hilbert–Schmidt SVD*

It turns out that we can apply the Hilbert–Schmidt SVD of Chapter 13 (developed for flat-limit stability but also to allow for kernels defined via their Mercer series) in a relatively straightforward manner to the generalized Hermite interpolation setting.

We begin with the truncated Mercer series of the kernel basis functions (cf. (13.10)), i.e.,

$$\begin{aligned}\boldsymbol{k}(\boldsymbol{x})^T &= \boldsymbol{\phi}(\boldsymbol{x})^T \begin{pmatrix} \mathsf{I}_{N_H} \\ \Lambda_2\Phi_2\Phi_1^{-T}\Lambda_1^{-1} \end{pmatrix} \Lambda_1\Phi_1^T \\ &= \boldsymbol{\psi}(\boldsymbol{x})^T\Lambda_1\Phi_1^T,\end{aligned}$$

and then form the vector $\boldsymbol{k}_{\boldsymbol{L}^z}(\boldsymbol{x})^T = \left(L_1^z K(\boldsymbol{x},\cdot) \cdots L_{N_H}^z K(\boldsymbol{x},\cdot)\right)$ needed for (19.15) by applying the linear functionals $\boldsymbol{L}^z = \left(L_1^z \cdots L_{N_H}^z\right)$ to the components of $\boldsymbol{k}$. If we apply the same linear functionals also to the right-hand side of the identities above, then we obtain

$$\begin{aligned}\boldsymbol{k}_{\boldsymbol{L}^z}(\boldsymbol{x})^T &= \boldsymbol{\phi}(\boldsymbol{x})^T \begin{pmatrix} \mathsf{I}_{N_H} \\ \Lambda_2\Phi_{2,\boldsymbol{L}^z}\Phi_{1,\boldsymbol{L}^z}^{-T}\Lambda_1^{-1} \end{pmatrix} \Lambda_1\Phi_{1,\boldsymbol{L}^z}^T \\ &= \boldsymbol{\psi}_{\boldsymbol{L}^z}(\boldsymbol{x})^T\Lambda_1\Phi_{1,\boldsymbol{L}^z}^T, \end{aligned} \tag{19.16}$$

since the values of the eigenfunctions in $\boldsymbol{\phi}(\boldsymbol{x})$ do not depend on $\boldsymbol{z}_1,\ldots,\boldsymbol{z}_{N_H}$, but the matrices Φ_1 and Φ_2 do. Here the vector of stable basis functions is given by

$$\boldsymbol{\psi}_{\boldsymbol{L}^z}(\boldsymbol{x})^T = \boldsymbol{\phi}(\boldsymbol{x})^T \begin{pmatrix} \mathsf{I}_{N_H} \\ \Lambda_2\Phi_{2,\boldsymbol{L}^z}\Phi_{1,\boldsymbol{L}^z}^{-T}\Lambda_1^{-1} \end{pmatrix}.$$

In order to get the Hilbert–Schmidt SVD of H we need to stack N_H copies of (19.16) and also apply the linear functionals $\boldsymbol{L}^x = \left(L_1^x \cdots L_{N_H}^x\right)$. This produces

$$\mathsf{H} = \Psi_{\boldsymbol{L}^x,\boldsymbol{L}^z}\Lambda_1\Phi_{1,\boldsymbol{L}^z}^T \tag{19.17}$$

with

$$\Psi_{\boldsymbol{L}^x,\boldsymbol{L}^z} = \Phi_{\boldsymbol{L}^x} \begin{pmatrix} \mathsf{I}_{N_H} \\ \Lambda_2\Phi_{2,\boldsymbol{L}^z}\Phi_{1,\boldsymbol{L}^z}^{-T}\Lambda_1^{-1} \end{pmatrix}.$$

The decomposition just performed is desirable because substitution of (19.16) and (19.17) into (19.15) gives us

$$\begin{aligned}\hat{u}(\boldsymbol{x}) &= \boldsymbol{\psi}_{\boldsymbol{L}^z}(\boldsymbol{x})^T\Lambda_1\Phi_{1,\boldsymbol{L}^z}^T\left(\Psi_{\boldsymbol{L}^x,\boldsymbol{L}^z}\Lambda_1\Phi_{1,\boldsymbol{L}^z}^T\right)^{-1}\boldsymbol{u}_{\boldsymbol{L}^x} \\ &= \boldsymbol{\psi}_{\boldsymbol{L}^z}(\boldsymbol{x})^T\Psi_{\boldsymbol{L}^x,\boldsymbol{L}^z}^{-1}\boldsymbol{u}_{\boldsymbol{L}^x} = \boldsymbol{\psi}_{\boldsymbol{L}^z}(\boldsymbol{x})^T\boldsymbol{b}, \end{aligned} \tag{19.18}$$

with $\boldsymbol{b} = \Psi_{\boldsymbol{L}^x,\boldsymbol{L}^z}^{-1}\boldsymbol{u}_{\boldsymbol{L}^x}$. As in the standard interpolation setting, the basis transformation matrix $\Lambda_1\Phi_{1,\boldsymbol{L}^z}^T$ — and therefore the ill-conditioning associated with Λ_1 — "cancels out" since it appears in the Hilbert–Schmidt SVD representation of both H and $\boldsymbol{k}_{\boldsymbol{L}^z}(\boldsymbol{x})^T$.

From (19.18) we obtain an HS-SVD version of the Hermite cardinal functions (cf. (19.15))

$$\begin{aligned}\boldsymbol{\ell}_{\boldsymbol{L}^z}(\boldsymbol{x}) &= \boldsymbol{k}_{\boldsymbol{L}^z}(\boldsymbol{x})^T\mathsf{H}^{-1} \\ &= \boldsymbol{\psi}_{\boldsymbol{L}^z}(\boldsymbol{x})^T\Psi_{\boldsymbol{L}^x,\boldsymbol{L}^z}^{-1}.\end{aligned}$$

19.2.4 *An example: Gradient interpolation*

We now consider data $\{\boldsymbol{x}_i, (y_i, \boldsymbol{g}_i)\}$, $i = 1, \ldots, N$, $\boldsymbol{x}_i \in \mathbb{R}^d$, and think of y_i as a sample of an unknown function u at $\boldsymbol{x}_i$. We also use the notation $\boldsymbol{g}_i$ to denote the value of the gradient ∇u at $\boldsymbol{x}_i$. This means that we have $N_H = (d+1)N$ data values, and therefore also need the same amount of basis functions. Since the linear functionals now are of two types — function evaluation and first partial derivative evaluation — we break the representation of the interpolant into two parts and assume that for $\boldsymbol{x}, \boldsymbol{z} \in \mathbb{R}^d$ it takes the form

$$\begin{aligned}\hat{u}(\boldsymbol{x}) &= \sum_{j=1}^{N} c_j^{(0)} K(\boldsymbol{x}, \boldsymbol{z}_j) + \sum_{\ell=1}^{d}\sum_{j=1}^{N} c_j^{(\ell)} \left.\frac{\partial K(\boldsymbol{x}, \boldsymbol{z})}{\partial z_\ell}\right|_{\boldsymbol{z}=\boldsymbol{z}_j} \\ &= \boldsymbol{k}(\boldsymbol{x})^T \boldsymbol{c}^{(0)} + \sum_{\ell=1}^{d} \boldsymbol{k}_\ell(\boldsymbol{x})^T \boldsymbol{c}^{(\ell)}, \end{aligned} \tag{19.19}$$

As above, we dissociate the kernel centers from the interpolation points, so that $\boldsymbol{z}_j$ and $\boldsymbol{x}_i$ need not be related; this is discussed further in Remark 19.14. We have extended our usual vector notation $\boldsymbol{k}(\boldsymbol{x})^T = \big(K(\boldsymbol{x}, \boldsymbol{z}_1) \cdots K(\boldsymbol{x}, \boldsymbol{z}_N)\big)$ to $\boldsymbol{k}_\ell(\boldsymbol{x})^T = \left(\left.\frac{\partial K(\boldsymbol{x},\boldsymbol{z})}{\partial z_\ell}\right|_{\boldsymbol{z}=\boldsymbol{z}_1} \cdots \left.\frac{\partial K(\boldsymbol{x},\boldsymbol{z})}{\partial z_\ell}\right|_{\boldsymbol{z}=\boldsymbol{z}_N}\right)$, and $\boldsymbol{c}^{(\ell)} = \left(c_1^{(\ell)}, \ldots, c_N^{(\ell)}\right)^T$, $\ell = 0, \ldots, d$, are appropriate coefficient vectors. To determine these coefficients we enforce the *gradient interpolation conditions*

$$\begin{aligned} \hat{u}(\boldsymbol{x}_i) &= y_i, \qquad i = 1, \ldots, N, \\ \nabla\hat{u}(\boldsymbol{x}_i) &= \boldsymbol{g}_i, \qquad i = 1, \ldots, N, \end{aligned} \tag{19.20}$$

where we must remember that $\nabla\hat{u}$ and $\boldsymbol{g}_i$ are vectors with d components each.

This leads to a linear system involving a $(d+1)N \times (d+1)N$ matrix H with a specific block structure

$$\mathsf{H}\boldsymbol{c} = \boldsymbol{y}, \qquad \mathsf{H} = \begin{pmatrix} \mathsf{K} & \mathsf{K}_{\nabla^{\boldsymbol{z}}} \\ \mathsf{K}_{\nabla^{\boldsymbol{x}}} & \mathsf{K}_{\nabla^{\boldsymbol{x}}\nabla^{\boldsymbol{z}}} \end{pmatrix} \tag{19.21}$$

with K the usual $N \times N$ kernel matrix with entries $(\mathsf{K})_{i,j} = K(\boldsymbol{x}_i, \boldsymbol{x}_j)$. The other blocks have themselves a certain block structure. The $N \times dN$ matrix $\mathsf{K}_{\nabla^{\boldsymbol{z}}}$ looks like

$$\mathsf{K}_{\nabla^{\boldsymbol{z}}} = \left(\mathsf{K}_{\nabla^{\boldsymbol{z}},1} \cdots \mathsf{K}_{\nabla^{\boldsymbol{z}},d}\right),$$

where $(\mathsf{K}_{\nabla^{\boldsymbol{z}},\ell})_{i,j} = \left.\frac{\partial K(\boldsymbol{x}_i,\boldsymbol{z})}{\partial z_\ell}\right|_{\boldsymbol{z}=\boldsymbol{z}_j}$. Similarly, the $dN \times N$ matrix $\mathsf{K}_{\nabla^{\boldsymbol{x}}}$ looks like

$$\mathsf{K}_{\nabla^{\boldsymbol{x}}} = \begin{pmatrix} \mathsf{K}_{\nabla^{\boldsymbol{x}},1} \\ \vdots \\ \mathsf{K}_{\nabla^{\boldsymbol{x}},d} \end{pmatrix},$$

where $(\mathsf{K}_{\nabla^{\boldsymbol{x}},\ell})_{i,j} = \left.\frac{\partial K(\boldsymbol{x},\boldsymbol{z}_j)}{\partial x_\ell}\right|_{\boldsymbol{x}=\boldsymbol{x}_i}$ and $\mathsf{K}_{\nabla^{\boldsymbol{x}}\nabla^{\boldsymbol{z}}}$ is a $d \times d$ block matrix of the form

$$\mathsf{K}_{\nabla^{\boldsymbol{x}}\nabla^{\boldsymbol{z}}} = \begin{pmatrix} \mathsf{K}_{\nabla^{\boldsymbol{x}}\nabla^{\boldsymbol{z}},1,1} & \cdots & \mathsf{K}_{\nabla^{\boldsymbol{x}}\nabla^{\boldsymbol{z}},1,d} \\ \vdots & & \vdots \\ \mathsf{K}_{\nabla^{\boldsymbol{x}}\nabla^{\boldsymbol{z}},d,1} & \cdots & \mathsf{K}_{\nabla^{\boldsymbol{x}}\nabla^{\boldsymbol{z}},d,d} \end{pmatrix}$$

with each $N \times N$ block having entries $(\mathsf{K}_{\nabla^{\boldsymbol{x}}\nabla^{\boldsymbol{z}},k,\ell})_{i,j} = \left.\frac{\partial^2 K(\boldsymbol{x},\boldsymbol{z})}{\partial x_k \partial z_\ell}\right|_{\boldsymbol{x}=\boldsymbol{x}_i,\boldsymbol{z}=\boldsymbol{z}_j}$. Moreover, $\boldsymbol{c} = \left(\boldsymbol{c}^{(0)^T},\ldots,\boldsymbol{c}^{(d)^T}\right)^T$, and the right-hand side is given by $\boldsymbol{u}_{\nabla^{\boldsymbol{x}}} = \left(y_1,\ldots,y_N,\boldsymbol{g}_{1,1},\ldots,\boldsymbol{g}_{1,N},\ldots,\boldsymbol{g}_{d,1},\ldots,\boldsymbol{g}_{d,N}\right)^T$. Note that here we have ordered the gradient interpolation conditions (19.20) so that they agree with (19.19).

Solution of the linear system (19.21) leads to the Hermite interpolant

$$\hat{u}(\boldsymbol{x}) = \boldsymbol{h}(\boldsymbol{x})^T \mathsf{H}^{-1} \boldsymbol{u}_{\nabla^{\boldsymbol{x}}}$$

as defined in (19.19). Here we have collected all of the basis functions in the vector $\boldsymbol{h}$, i.e.,

$$\boldsymbol{h}(\boldsymbol{x})^T = \left(\boldsymbol{k}(\boldsymbol{x})^T\ \boldsymbol{k}_1(\boldsymbol{x})^T\ \cdots\ \boldsymbol{k}_d(\boldsymbol{x})^T\right).$$

Remark 19.14. At first, it may appear as though the construction of the interpolant $\hat{u}$ is rather convoluted: why would the use of N kernels and dN associated derivatives be a logical choice of basis when any $(d+1)N$ $\boldsymbol{z}_j$ points could be chosen as kernel centers and the derivative components could be eliminated? Indeed such a choice is entirely acceptable in many circumstances, and may be preferable from a complexity standpoint.

Thus, if a guarantee for invertibility of H is important, using (19.19) with kernel centers and interpolation points coinciding is recommended. For many situations, the use of a single kernel and $(d+1)N$ kernel centers may be preferable because of the simpler interpolant structure.

If we want to apply the Hilbert–Schmidt SVD to this example then we will have to respect the block structure of (19.21) in all of our work. Thus, we write the gradient interpolant

$$\hat{u}(\boldsymbol{x}) = \boldsymbol{h}(\boldsymbol{x})^T \mathsf{H}^{-1} \boldsymbol{u}_{\nabla^{\boldsymbol{x}}}$$

as

$$\hat{u}(\boldsymbol{x}) = \left(\boldsymbol{\psi}(\boldsymbol{x})^T \boldsymbol{\psi}_{\nabla^{\boldsymbol{z}},1}(\boldsymbol{x})^T \cdots \boldsymbol{\psi}_{\nabla^{\boldsymbol{z}},d}(\boldsymbol{x})^T\right) \begin{pmatrix} \Psi & \Psi_{\nabla^{\boldsymbol{z}},1} & \cdots & \Psi_{\nabla^{\boldsymbol{z}},d} \\ \Psi_{\nabla^{\boldsymbol{x}},1} & \Psi_{\nabla^{\boldsymbol{x}}\nabla^{\boldsymbol{z}},1,1} & \cdots & \Psi_{\nabla^{\boldsymbol{x}}\nabla^{\boldsymbol{z}},1,d} \\ \vdots & \vdots & \ddots & \vdots \\ \Psi_{\nabla^{\boldsymbol{x}},d} & \Psi_{\nabla^{\boldsymbol{x}}\nabla^{\boldsymbol{z}},d,1} & \cdots & \Psi_{\nabla^{\boldsymbol{x}}\nabla^{\boldsymbol{z}},d,d} \end{pmatrix}^{-1} \boldsymbol{u}_{\nabla^{\boldsymbol{x}}}$$

since

$$\mathsf{H} = \begin{pmatrix} \Psi & \Psi_{\nabla^{\boldsymbol{z}},1} & \cdots & \Psi_{\nabla^{\boldsymbol{z}},d} \\ \Psi_{\nabla^{\boldsymbol{x}},1} & \Psi_{\nabla^{\boldsymbol{x}}\nabla^{\boldsymbol{z}},1,1} & \cdots & \Psi_{\nabla^{\boldsymbol{x}}\nabla^{\boldsymbol{z}},1,d} \\ \vdots & \vdots & \ddots & \vdots \\ \Psi_{\nabla^{\boldsymbol{x}},d} & \Psi_{\nabla^{\boldsymbol{x}}\nabla^{\boldsymbol{z}},d,1} & \cdots & \Psi_{\nabla^{\boldsymbol{x}}\nabla^{\boldsymbol{z}},d,d} \end{pmatrix} \begin{pmatrix} \Lambda_1 \Phi_1^T & & & \\ & \Lambda_1 \Phi_{1\,\nabla^{\boldsymbol{z}},1}^T & & \\ & & \ddots & \\ & & & \Lambda_1 \Phi_{1\,\nabla^{\boldsymbol{z}},d}^T \end{pmatrix}$$

and

$$\boldsymbol{h}(\boldsymbol{x})^T = \left(\boldsymbol{\psi}(\boldsymbol{x})^T\ \boldsymbol{\psi}_{\nabla^{\boldsymbol{z}},1}(\boldsymbol{x})^T\ \cdots\ \boldsymbol{\psi}_{\nabla^{\boldsymbol{z}},d}(\boldsymbol{x})^T\right) \begin{pmatrix} \Lambda_1 \Phi_1^T & & & \\ & \Lambda_1 \Phi_{1\,\nabla^{\boldsymbol{z}},1}^T & & \\ & & \ddots & \\ & & & \Lambda_1 \Phi_{1\,\nabla^{\boldsymbol{z}},d}^T \end{pmatrix}.$$

Here the matrix Ψ and vector $\boldsymbol{\psi}(\cdot)^T$ are defined as usual and the other pieces are given by

$$\boldsymbol{\psi}_{\nabla^{\boldsymbol{z}},\ell}(\boldsymbol{x})^T = \boldsymbol{\phi}(\boldsymbol{x})^T \begin{pmatrix} \mathsf{I}_N \\ \Lambda_2\Phi_{2,\nabla^{\boldsymbol{z}},\ell}^T\Phi_{1,\nabla^{\boldsymbol{z}},\ell}^{-T}\Lambda_1^{-1} \end{pmatrix},$$

$$\Psi_{\nabla^{\boldsymbol{z}},\ell} = \Phi \begin{pmatrix} \mathsf{I}_N \\ \Lambda_2\Phi_{2,\nabla^{\boldsymbol{z}},\ell}^T\Phi_{1,\nabla^{\boldsymbol{z}},\ell}^{-T}\Lambda_1^{-1} \end{pmatrix},$$

$$\Psi_{\nabla^{\boldsymbol{x}},\ell} = \Phi_{\nabla^{\boldsymbol{x}},\ell} \begin{pmatrix} \mathsf{I}_N \\ \Lambda_2\Phi_2^T\Phi_1^{-T}\Lambda_1^{-1} \end{pmatrix},$$

$$\Psi_{\nabla^{\boldsymbol{x}}\nabla^{\boldsymbol{z}},k,\ell} = \Phi_{\nabla^{\boldsymbol{x}},k} \begin{pmatrix} \mathsf{I}_N \\ \Lambda_2\Phi_{2,\nabla^{\boldsymbol{z}},\ell}^T\Phi_{1,\nabla^{\boldsymbol{z}}\ell}^{-T}\Lambda_1^{-1} \end{pmatrix}.$$

An example involving this is too tedious to list in the text, but is available in the `GaussQR` library in `HermiteSymmetricHSSVD.m`.

19.2.5 *Kriging interpretation*

For a Gaussian process or Gaussian random field[4] $Y = \{Y_{\boldsymbol{x}}\}_{\boldsymbol{x}\in\Omega}$ we define the derivative as (see, e.g., [Adler (2009, Section 2.2), Parzen (1962, Section 3.3)])

$$\frac{\partial Y_{\boldsymbol{x}}}{\partial x_j} = \lim_{h\to 0} \frac{Y_{(x_1,\ldots,x_j+h,\ldots,x_d)} - Y_{(x_1,\ldots,x_j,\ldots,x_d)}}{h}.$$

This derivative exists if the mean μ of the original process is differentiable and if the original covariance kernel K possesses a mixed second derivative. Then the corresponding mean and covariance functions involving the derivative process are

$$\mathbb{E}\left[\frac{\partial Y_{\boldsymbol{x}}}{\partial x_j}\right] = \frac{\partial \mu(\boldsymbol{x})}{\partial x_j} = \mu_j(\boldsymbol{x}),$$

$$\operatorname{Cov}\left(Y_{\boldsymbol{x}}, \frac{\partial Y_{\boldsymbol{z}}}{\partial z_j}\right) = \frac{\partial K(\boldsymbol{x},\boldsymbol{z})}{\partial z_j},$$

$$\operatorname{Cov}\left(\frac{\partial Y_{\boldsymbol{x}}}{\partial x_i}, Y_{\boldsymbol{z}}\right) = \frac{\partial K(\boldsymbol{x},\boldsymbol{z})}{\partial x_i},$$

$$\operatorname{Cov}\left(\frac{\partial Y_{\boldsymbol{x}}}{\partial x_i}, \frac{\partial Y_{\boldsymbol{z}}}{\partial z_j}\right) = \frac{\partial^2 K(\boldsymbol{x},\boldsymbol{z})}{\partial x_i \partial z_j}.$$

By linearity the derivative process is also a Gaussian random field.

Following the same line of argument as in the standard kriging setting of Chapter 5, the kriging predictor takes the form

$$\hat{y}_{\boldsymbol{x}} = \boldsymbol{\mu} + \boldsymbol{h}^T(\boldsymbol{x})\mathsf{H}^{-1}(\boldsymbol{y} - \boldsymbol{\mu}),$$

where $\boldsymbol{h}(\boldsymbol{x})^T$, H and $\boldsymbol{y}$ are defined in analogy to $\boldsymbol{u}_{\nabla^{\boldsymbol{x}}}$ in the previous subsection and $\boldsymbol{\mu}$ is the vector of means $\boldsymbol{\mu}^T = (\boldsymbol{\mu}^T, \boldsymbol{\mu}_1^T, \ldots, \boldsymbol{\mu}_d^T)$. Of course, while the matrix

[4]The role of the letter y, Y, $\boldsymbol{y}$ in this section corresponds to that of the letter u in the previous section.

H ends up formally being the same as in the previous subsection, in the kriging setting one *interprets* it as

$$\mathsf{H} = \begin{pmatrix} \mathsf{C} & \mathsf{C}_{\nabla^{\boldsymbol{z}}} \\ \mathsf{C}_{\nabla^{\boldsymbol{x}}} & \mathsf{C}_{\nabla^{\boldsymbol{x}}\nabla^{\boldsymbol{z}}} \end{pmatrix},$$

where $(\mathsf{C})_{i,j} = \mathrm{Cov}(Y_{\boldsymbol{x}_i}, Y_{\boldsymbol{x}_j})$, and analogous modifications to the formulas from the previous subsection need to be made for those matrices involving covariances of the derivative processes.

As in the case of basic kriging interpolation, the formulas that result from the stochastic formulation are closely related to those in the kriging interpolation setting.

This variant of *gradient-enhanced kriging* can be found, e.g., in [Mardia *et al.* (1996), Fang *et al.* (2006, Section 5.5), Rasmussen and Williams (2006, Section 9.4)].

19.3 Doing Hermite Interpolation via Derivatives of Eigenfunctions

Using the Hilbert–Schmidt SVD for the symmetric Hermite setup described in Section 19.2 may not be convenient given the complexity of the design. However, we may be able to work with an early truncated Hilbert–Schmidt series, i.e., where $M < N$, in which case the basis functions will be the data-independent eigenfunctions. As we now show, doing Hermite interpolation in this case is relatively straightforward. The idea was introduced in Section 12.3 for the standard scattered data interpolation problem, and we showed in Theorem 12.2 that an M-term truncation provides optimal accuracy.

The following discussion assumes that we work with a kernel K given in tensor product form so that we need only an expression for the one-dimensional kernel and its derivatives. If we also know the Hilbert–Schmidt series of this kernel, i.e., we know its eigenvalues and eigenfunctions, then we can discuss how to evaluate the derivative of low-rank multidimensional eigenfunction approximate interpolants produced using the Hilbert–Schmidt series.

As always, the kernel interpolant takes the form

$$\hat{u}(\boldsymbol{x}) = \sum_{j=1}^{N} c_j K(\boldsymbol{x}, \boldsymbol{z}_j), \tag{19.22}$$

where the coefficient vector $\boldsymbol{c}$ is determined by solving the interpolation system

$$\mathsf{K}\boldsymbol{c} = \boldsymbol{u}. \tag{19.23}$$

The kernel matrix has entries $(\mathsf{K})_{i,j} = K(\boldsymbol{z}_i, \boldsymbol{z}_j)$, and $(\boldsymbol{u})_i = u(\boldsymbol{z}_i)$ and $(\boldsymbol{c})_i = c_i$ as usual. We can write $\hat{u}(\boldsymbol{x}) = \boldsymbol{k}(\boldsymbol{x})^T\boldsymbol{c}$, where

$$\boldsymbol{k}(\boldsymbol{x})^T = \big(K(\boldsymbol{x}, \boldsymbol{z}_1) \cdots K(\boldsymbol{x}, \boldsymbol{z}_N)\big).$$

Replacing the kernel matrix K by the eigenfunction expansion using the truncated Hilbert–Schmidt series $K(\boldsymbol{x},\boldsymbol{z}) = \sum_{n=1}^{M} \lambda_n \varphi_n(\boldsymbol{x})\varphi_n(\boldsymbol{z})$ produces

$$\mathsf{K} \approx \Phi\Lambda\Phi^T, \tag{19.24}$$

where the approximation can be made as accurate as desired by increasing the series length M (because the Hilbert–Schmidt series is uniformly convergent). Here, $\Phi \in \mathbb{R}^{N\times M}$ has i^{th} row $(\Phi)_{i,:} = \boldsymbol{\phi}(\boldsymbol{x}_i)^T$, where

$$\boldsymbol{\phi}(\boldsymbol{x})^T = \left(\varphi_{\boldsymbol{n}_1}(\boldsymbol{x}) \cdots \varphi_{\boldsymbol{n}_M}(\boldsymbol{x})\right),$$

and $\Lambda \in \mathbb{R}^{M\times M}$ is a diagonal matrix with i^{th} diagonal value $\lambda_{\boldsymbol{n}_i}$. The ith multi-index (recall the notation from Section 12.2) $\boldsymbol{n}_i$ has d values and describes the order of the components of the ith eigenfunction,

$$\varphi_{\boldsymbol{n}_i}(\boldsymbol{x}) = \prod_{\ell=1}^{d} \varphi_{(\boldsymbol{n}_i)_\ell}(x_\ell). \tag{19.25}$$

As an example, $\varphi_{[3,5]}(x_1,x_2) = \varphi_3(x_1)\varphi_5(x_2)$.

We now consider differentiation of the eigen-decomposition in the case $M < N$ since this often produces sufficiently accurate results, and since the general Hermite formulation as discussed in Section 19.2.2 becomes too messy.

19.3.1 *Differentiation of a low-rank eigenfunction approximate interpolant*

If we choose $M < N$, which is more likely when N is large, the low-rank approximation system

$$\Phi\boldsymbol{b} = \boldsymbol{u} \tag{19.26}$$

replaces (19.23)[5]. The solution $\boldsymbol{b} \in \mathbb{R}^M$ can be thought of as $\Lambda\Phi^T\boldsymbol{c}$, although not computed that way because of the danger of computing $\boldsymbol{c} = \mathsf{K}^{-1}\boldsymbol{y}$. In this formulation, $\hat{u}(\boldsymbol{x}) = \boldsymbol{\phi}(\boldsymbol{x})^T\boldsymbol{b}$.

Differentiating $\hat{u}$ is straightforward because the only dependence on $\boldsymbol{x}$ appears in the eigenfunctions. Using $\mathcal{D}_\ell$ to indicate the derivative with respect to the ℓth dimension,

$$\begin{aligned} &\mathcal{D}_\ell\hat{u}(\boldsymbol{x}) = \mathcal{D}_\ell\boldsymbol{\phi}(\boldsymbol{x})^T\boldsymbol{b}, \\ \Longleftrightarrow\quad &\mathcal{D}_\ell\hat{u}(\boldsymbol{x}) = (\mathcal{D}_\ell\varphi_{\boldsymbol{n}_1}(\boldsymbol{x}) \quad \cdots \quad \mathcal{D}_\ell\varphi_{\boldsymbol{n}_M}(\boldsymbol{x}))\boldsymbol{b}. \end{aligned} \tag{19.27}$$

Since higher-dimensional eigenfunctions are formed by the tensor product structure in (19.25), the derivative in one dimension does not affect the others, thus

$$\mathcal{D}_\ell\varphi_{\boldsymbol{n}}(\boldsymbol{x}) = \varphi'_{(\boldsymbol{n})_\ell}(x_\ell)\prod_{\substack{j=1\\ j\neq\ell}}^{d}\varphi_{(\boldsymbol{n})_j}(x_j). \tag{19.28}$$

[5]This low-rank regression method was referred to as GaussQRr in [Fasshauer and McCourt (2012)].

For some applications, it is useful to describe this differentiation process again using the *differentiation matrix* framework from Section 19.1. If we define the matrix $(\boldsymbol{\Phi}_{\mathcal{D}_\ell})_{i,:} = \mathcal{D}_\ell \boldsymbol{\phi}(\boldsymbol{x}_i)^T$ then the relationship

$$\mathcal{D}_\ell \begin{pmatrix} \hat{u}(\boldsymbol{x}_1) \\ \vdots \\ \hat{u}(\boldsymbol{x}_N) \end{pmatrix} = \boldsymbol{\Phi}_{\mathcal{D}_\ell} \boldsymbol{\Phi}^\dagger \begin{pmatrix} u_1 \\ \vdots \\ u_N \end{pmatrix}$$

must hold, and $\boldsymbol{\Phi}_{\mathcal{D}_\ell}\boldsymbol{\Phi}^\dagger$ would be the differentiation matrix in the ℓth dimension. Because $M < N$ for a low-rank eigenfunction approximate interpolant, $\boldsymbol{\Phi}^{-1}$ does not exist, and (19.26) must be solved with the pseudoinverse [Golub and Van Loan (2012)], denoted $\boldsymbol{\Phi}^\dagger$. This was discussed in Section 12.3.

The term differentiation matrix is appropriate because $\hat{u}$ is built to approximate u, so the matrix $\boldsymbol{\Phi}_{\mathcal{D}_\ell}\boldsymbol{\Phi}^\dagger$ accepts function values and returns approximate values of the derivative of the function evaluated at those points. Note that $\boldsymbol{\Phi}_{\mathcal{D}_\ell}\boldsymbol{\Phi}^\dagger$ is an outer product, and can have column rank at most M. Combining (19.26) and (19.27) gives a similar structure for the derivative at any point $\boldsymbol{x}$ given $\boldsymbol{u}$,

$$\mathcal{D}_\ell \hat{u}(\boldsymbol{x}) = \mathcal{D}_\ell \boldsymbol{\phi}(\boldsymbol{x})^T \boldsymbol{\Phi}^\dagger \boldsymbol{u}.$$

More generally, the derivative evaluated at N_{eval} points $(\boldsymbol{x}_j)_{j=1}^{N_{\text{eval}}}$ can be written as

$$\mathcal{D}_\ell \begin{pmatrix} \hat{u}(\boldsymbol{x}_1) \\ \vdots \\ \hat{u}(\boldsymbol{x}_{N_{\text{eval}}}) \end{pmatrix} = \begin{pmatrix} \mathcal{D}_\ell \boldsymbol{\phi}(\boldsymbol{x}_1) \\ \vdots \\ \mathcal{D}_\ell \boldsymbol{\phi}(\boldsymbol{x}_{N_{\text{eval}}}) \end{pmatrix} \boldsymbol{\Phi}^\dagger \boldsymbol{u}$$

$$= \hat{\boldsymbol{\Phi}}_{\mathcal{D}_\ell} \boldsymbol{\Phi}^\dagger \boldsymbol{u}, \tag{19.29}$$

where $\hat{\boldsymbol{\Phi}}_{\mathcal{D}_\ell}\boldsymbol{\Phi}^\dagger$ is the $N_{\text{eval}} \times N$ differentiation matrix that accepts function values at $\mathcal{Z} = \{\boldsymbol{z}_j\}_{j=1}^N$ and returns derivatives at $\mathcal{X} = (\boldsymbol{x}_j)_{j=1}^{N_{\text{eval}}}$.

19.3.2 *An example: Derivatives of Gaussians eigenfunctions*

We studied the eigenfunctions for the Gaussian kernel in Section 12.2.1, and we saw there that the 1D Gaussian has the truncated Hilbert–Schmidt expansion (also called a Mercer series)

$$\mathrm{e}^{-\varepsilon^2|x-z|^2} \approx \sum_{n=1}^{M} \lambda_n \varphi_n(x) \varphi_n(z), \tag{19.30}$$

where the eigenvalues and eigenfunctions are

$$\lambda_n = \sqrt{\frac{\alpha^2}{\alpha^2+\delta^2+\varepsilon^2}} \left(\frac{\varepsilon^2}{\alpha^2+\delta^2+\varepsilon^2}\right)^{n-1},$$

$$\varphi_n(x) = \gamma_n \mathrm{e}^{-\delta^2 x^2} H_{n-1}(\alpha\beta x), \qquad n = 1, 2, \ldots. \tag{19.31}$$

Above, M is the truncation length of an otherwise infinite series, H_{n-1} is the Hermite polynomial of degree $n-1$, and α is the global scale parameter defined in Section 12.2.1. The free parameter α must be chosen to uniquely define the eigenfunctions; this is analogous to how ε must be chosen to uniquely define the reproducing kernel Hilbert space spanned by the Gaussians. The other parameters are fixed to α and ε:

$$\beta = \left(1 + \left(\frac{2\varepsilon}{\alpha}\right)^2\right)^{\frac{1}{4}}, \quad \gamma_n = \sqrt{\frac{\beta}{2^{n-1}\Gamma(n)}}, \quad \delta^2 = \frac{\alpha^2}{2}\left(\beta^2 - 1\right).$$

As discussed earlier, the expansion above can be extended to multiple dimensions by exploiting the tensor product structure of the Gaussian in multiple dimensions.

To consider the derivatives of these eigenfunctions, we can directly differentiate (19.31),

$$\frac{\mathrm{d}}{\mathrm{d}x}\varphi_n(x) = -2\gamma_n\delta^2 x \mathrm{e}^{-\delta^2x^2} H_{n-1}(\alpha\beta x) + \gamma_n \mathrm{e}^{-\delta^2x^2}\frac{\mathrm{d}}{\mathrm{d}x}H_{n-1}(\alpha\beta x).$$

From [Szegő (1939)], the derivative of a Hermite polynomial is

$$\frac{\mathrm{d}}{\mathrm{d}x}H_n(x) = 2nH_{n-1}(x), \quad n \geq 1, \qquad \frac{\mathrm{d}}{\mathrm{d}x}H_0(x) = 0.$$

Plugging this in above gives

$$\frac{\mathrm{d}}{\mathrm{d}x}\varphi_n(x) = -2\gamma_n\delta^2 x \mathrm{e}^{-\delta^2x^2} H_{n-1}(\alpha\beta x) + 2(n-1)\alpha\beta\gamma_n \mathrm{e}^{-\delta^2x^2} H_{n-2}(\alpha\beta x),$$

to which we can apply the identity $\sqrt{2(n-1)}\gamma_n = \gamma_{n-1}$, resulting in

$$\frac{\mathrm{d}}{\mathrm{d}x}\varphi_n(x) = -2\delta^2 x\gamma_n \mathrm{e}^{-\delta^2x^2} H_{n-1}(\alpha\beta x) + 2(n-1)\alpha\beta\frac{\gamma_{n-1}}{\sqrt{2(n-1)}}\mathrm{e}^{-\delta^2x^2} H_{n-2}(\alpha\beta x).$$

Finally, making the substitution of (19.31) produces

$$\frac{\mathrm{d}}{\mathrm{d}x}\varphi_n(x) = -2\delta^2\, x\, \varphi_n(x) + \sqrt{2(n-1)}\alpha\beta\, \varphi_{n-1}(x). \tag{19.32}$$

This allows us to compute the derivatives of eigenfunctions using the eigenfunctions themselves. Higher-order derivatives can also be computed via direct differentiation and substitution, producing the second derivative

$$\begin{aligned}\frac{\mathrm{d}^2}{\mathrm{d}x^2}\varphi_n(x) =& 2\delta^2\left(2\delta^2 x - 1\right)\varphi_n(x) \\ & - 4\sqrt{2}\delta^2\alpha\beta\sqrt{n-1}\, x\, \varphi_{n-1}(x) \\ & + 2(\alpha\beta)^2\sqrt{(n-1)(n-2)}\, \varphi_{n-2}(x),\end{aligned}$$

the third derivative

$$\begin{aligned}\frac{\mathrm{d}^3}{\mathrm{d}x^3}\varphi_n(x) =& -4\delta^4\, x(2\delta^2x^2 - 3)\, \varphi_n(x) \\ & + 6\sqrt{2}\delta^2\alpha\beta\sqrt{n-1}\,(2\delta^2x^2 - 1)\, \varphi_{n-1}(x) \\ & - 12\delta^2(\alpha\beta)^2\sqrt{(n-1)(n-2)}\, x\, \varphi_{n-2}(x) \\ & + 2\sqrt{2}(\alpha\beta)^3\sqrt{(n-1)(n-2)(n-3)}\, \varphi_{n-3}(x),\end{aligned}$$

and the fourth derivative

$$\begin{aligned}\frac{\mathrm{d}^4}{\mathrm{d}x^4}\varphi_n(x) =& 4\delta^4\,(4\delta^2x^4 - 12\delta^2x^2 + 3)\,\varphi_n(x)\\ &- 16\sqrt{2}\delta^4\alpha\beta\sqrt{n-1}\,x(2\delta^2x^2 - 3)\,\varphi_{n-1}(x)\\ &+ 24\delta^2(\alpha\beta)^2\sqrt{(n-1)(n-2)}\,(2\delta^2x^2 - 1)\,\varphi_{n-2}(x)\\ &- 16\sqrt{2}\delta^2(\alpha\beta)^3\sqrt{(n-1)(n-2)(n-3)}\,x\,\varphi_{n-3}(x)\\ &+ 4(\alpha\beta)^4\sqrt{(n-1)(n-2)(n-3)(n-4)}\,\varphi_{n-4}(x).\end{aligned}$$

As discussed in [McCourt (2013c)], the structure of the Hermite polynomials induces a three-term recurrence underlying the eigenfunctions:

$$\begin{aligned}\varphi_1(x) &= \sqrt{\beta}e^{-\delta^2x^2},\\ \varphi_2(x) &= \sqrt{2}\alpha\beta x\sqrt{\beta}e^{-\delta^2x^2},\\ \varphi_n(x) &= \sqrt{\frac{2}{n-1}}\alpha\beta x\varphi_{n-1}(x) - \sqrt{\frac{n-2}{n-1}}\varphi_{n-2}(x).\end{aligned} \tag{19.33}$$

Applying (19.33) multiple times produces the relations

$$\begin{aligned}\varphi_{n-2}(x) &= \frac{1}{\sqrt{n-2}}\sqrt{2}\alpha\beta\,x\,\varphi_{n-1}(x) - \sqrt{\frac{n-1}{n-2}}\,\varphi_n(x),\\ \varphi_{n-3}(x) &= \frac{2(\alpha\beta)^2x^2 - n + 2}{\sqrt{(n-2)(n-3)}}\,\varphi_{n-1}(x) - \frac{\sqrt{2}\sqrt{n-1}\alpha\beta\,x}{\sqrt{(n-2)(n-3)}}\,\varphi_n(x),\\ \varphi_{n-4}(x) &= \frac{\sqrt{2}\alpha\beta x(2(\alpha\beta)^2x^2 - 2n + 5)}{\sqrt{(n-2)(n-3)(n-4)}}\,\varphi_{n-1}(x) - \frac{\sqrt{n-1}(2(\alpha\beta)^2x^2 - n + 3)}{\sqrt{(n-2)(n-3)(n-4)}}\,\varphi_n(x).\end{aligned}$$

This allows all the derivatives of φ_n to be written only in terms of φ_n and φ_{n-1}:

$$\begin{aligned}\frac{\mathrm{d}^2}{\mathrm{d}x^2}\varphi_n(x) =& [2\delta^2(2\delta^2x^2 - 1) - 2(\alpha\beta)^2(n-1)]\,\varphi_n(x)\\ &+ 2\alpha\beta((\alpha\beta)^2 - 2\delta^2)\sqrt{2(n-1)}\,x\,\varphi_{n-1}(x),\\ \frac{\mathrm{d}^3}{\mathrm{d}x^3}\varphi_n(x) =& [-8\delta^6x^3 + (4(\alpha\beta)^2(n-1)(3\delta^2 - (\alpha\beta)^2) + 12\delta^4)x]\,\varphi_n(x)\\ &+ 2\alpha\beta\sqrt{2(n-1)}[(6\delta^4 + 2(\alpha\beta)^4 - 6(\alpha\beta)^2\delta^2)x^2\\ &\qquad - ((\alpha\beta)^2(n-2) + 3\delta^2)]\,\varphi_{n-1}(x),\\ \frac{\mathrm{d}^4}{\mathrm{d}x^4}\varphi_n(x) =& 4[4\delta^8x^4 + 2(4\delta^2(\alpha\beta)^4(n-1) - 6\delta^2(\alpha\beta)^2(n-1) - (\alpha\beta)^6(n-1) - 6\delta^6)x^2\\ &+ (\alpha\beta)^4(n-3)(n-1) + 3\delta^4 + 6\delta^2(\alpha\beta)^2(n-1)]\,\varphi_n(x)\\ &+ 4\alpha\beta\sqrt{2(n-1)}[2((\alpha\beta)^6 - 4\delta^2(\alpha\beta)^4 + 6\delta^4(\alpha\beta)^2 - 4\delta^6)x^2\\ &+ 2(\alpha\beta)^2(n-1)(2\delta^2 - (\alpha\beta)^2) + 12\delta^4 + 3(\alpha\beta)^4 - 10\delta^2(\alpha\beta)^2]\,x\,\varphi_{n-1}(x).\end{aligned}$$

We list here only the first four derivatives because those will be the only ones considered in this book. Higher derivatives can be derived following the same pattern as above.

19.4 Multiphysics Coupling

Multiphysics systems [Keyes *et al.* (2013)] have many possible forms, but for the purposes of this book we will consider only one form: two independent boundary value problems which are coupled through an interface. Moreover, we will use kernel-based Hermite interpolation — even though there are many other coupling mechanisms (such as, e.g., mortar finite elements [Bernardi *et al.* (1994)]). An example of such a system, where two 2D models are coupled through a shared 1D boundary is depicted in Figure 19.5.

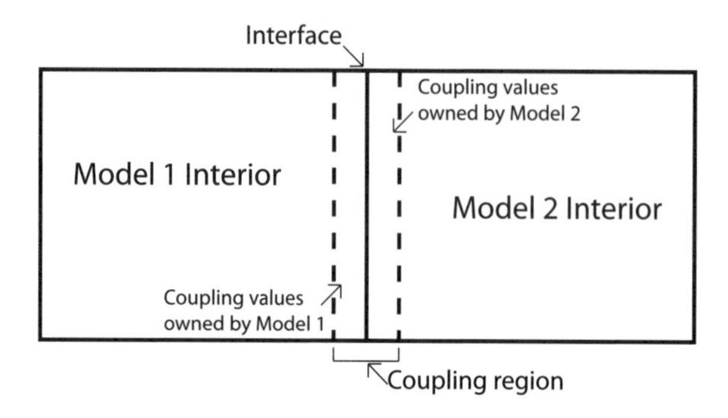

Fig. 19.5 In this image, two independent models are coupled through an interface. Each model has three distinct regions: the interior, coupling and interface regions.

For a multiphysics problem of this form, three regions are present in each model and are defined by their governing equations and their interaction with other regions.

Interior region: Nodes are governed by the PDE and its boundary conditions, and can be determined without using data from the other model. In fact, these values are also independent of the interface region, which is what differentiates them from the coupling region.

Coupling region: Nodes are governed by the PDE and its boundary conditions, and can be determined without using data from the other model. These values require data from both the local interior region and the local interface region.

Interface region: Nodes are governed by the interface conditions, and involve values from the external model as well as the local model. Data is required from the local coupling region as well as the external coupling and interface regions.

The general form of a fully coupled multiphysics problem fitting the criteria described above is $F(\boldsymbol{u}) = 0$. Note that other multiphysics problems exist which cannot be written in this form and will not be considered here; see [Keyes *et al.* (2013)] for a more thorough discussion. When written in components, the solution

Table 19.1 Regions of the coupled multiphysics system required to compute residuals. If a "×" symbol is absent, then those components can be ignored when computing that residual.

Residuals	Solution components					
	$\boldsymbol{u}_1$	$\boldsymbol{u}_2$	$\boldsymbol{u}_1^C$	$\boldsymbol{u}_1^I$	$\boldsymbol{u}_2^C$	$\boldsymbol{u}_2^I$
F_1	×		×			
F_2		×			×	
F_1^C	×		×	×		
F_1^I			×	×	×	×
F_2^C		×			×	×
F_2^I			×	×	×	×

vector and nonlinear residual function are

$$\boldsymbol{u} = \begin{pmatrix} \boldsymbol{u}_1 \\ \boldsymbol{u}_2 \\ \boldsymbol{u}_1^C \\ \boldsymbol{u}_1^I \\ \boldsymbol{u}_2^C \\ \boldsymbol{u}_2^I \end{pmatrix}, \qquad F(\boldsymbol{u}) = \begin{pmatrix} F_1(\boldsymbol{u}) \\ F_2(\boldsymbol{u}) \\ F_1^C(\boldsymbol{u}) \\ F_1^I(\boldsymbol{u}) \\ F_2^C(\boldsymbol{u}) \\ F_2^I(\boldsymbol{u}) \end{pmatrix}. \tag{19.34}$$

Here, $\boldsymbol{u}_1$, $\boldsymbol{u}_1^C$ and $\boldsymbol{u}_1^I$ are the solutions of Model 1 in the interior, coupling and interface regions respectively. $\boldsymbol{u}_2$, $\boldsymbol{u}_2^C$ and $\boldsymbol{u}_2^I$ are the corresponding solution blocks for Model 2.

Each of the function blocks has analogous notation as the solution vector, e.g., $F_1^C(\boldsymbol{u})$ is the residual of the coupling region of Model 1 and $F_2^I(\boldsymbol{u})$ is the residual of the interface region of Model 2. As described in the itemized list above, F_1 and F_1^C are defined by the PDE in Model 1, F_2 and F_2^C are defined by the PDE in Model 2, and F_1^I and F_2^I are defined by the interface conditions. These relationships are encapsulated in Table 19.1.

The interpolation approach to coupling uses data present in one model to produce approximate boundary conditions for the other model. The effects of this approximation are difficult to judge (errors become present in the definition of the problem) and there are few results concerning the *a priori* convergence order of models coupled in this fashion. More solid footing can be found in *a posteriori* error estimation of multiphysics simulations [Estep *et al.* (2008); Carey *et al.* (2009); Arbogast *et al.* (2014)], but their implementation is beyond the scope of this overview. Even with the meager error analysis, interpolation is a popular technique because of its simplicity, both in concept and in execution: take data that you have and use it to create data that you need. In some applications it is necessary to satisfy a divergence-free property [Bochev and Shashkov (2005)] or a local conservation property [Jiao and Heath (2004)]; we will, however, not consider such issues in this basic introduction. The simplicity of interpolation allows us to easily incorporate the

Hermite interpolation method described in Section 19.2.2 to improve the coupling between models.

The coupling and interior equations are defined by the PDE and boundary conditions and their discretization is not of concern here — we want to study the effect of different discretizations of the *interface conditions*. In fact, one of the benefits of this meshfree coupling tool is its flexibility in coupling other discretizations, including the example in Section 19.4.2 where interior domains are discretized with finite differences.

Thus far, little has been said about the interface conditions. To ensure some level of consistency throughout the coupled model, the actual choice of interface conditions needs to be made by the scientists running the simulation.

Because it is impractical to consider all the possible coupling strategies in use today (see [Keyes *et al.* (2013)] for a survey) we will restrict our study here to interface conditions involving matched values and derivatives along the interface. Thus, the general form of interface conditions considered here is

$$F^I(\boldsymbol{u}) = \sum_{k=0}^{r-1} a_{1,k}\mathcal{L}_k\boldsymbol{u}_1^I - \sum_{k=0}^{r-1} a_{2,k}\mathcal{L}_k\boldsymbol{u}_2^I = 0, \tag{19.35}$$

where r is the order of the PDEs, and $\mathcal{L}$ is some linear differential operator involving the derivatives, such as $\mathcal{L}_k(\boldsymbol{u}) = \boldsymbol{u}$, $\mathcal{L}_k(\boldsymbol{u}) = \frac{\partial}{\partial \boldsymbol{n}}\boldsymbol{u}$ or $\mathcal{L}_k(\boldsymbol{u}) = \nabla^2\boldsymbol{u}$. The coefficients $a_{1,k}$, and $a_{2,k}$ may depend on $\boldsymbol{x}$, and they define the relationship between the models in the interface conditions.

Interface conditions should be thought of analogously to boundary conditions, with similar terminology:

Dirichlet: $a_{1,0} = a_{2,0} = 1$, all other $a_k = 0$, and $\mathcal{L}_0(\boldsymbol{u}) = \boldsymbol{u}$ is the identity. This produces the interface condition $\boldsymbol{u}_1^I = \boldsymbol{u}_2^I$ which matches solution values at the interface.

Neumann: $a_{1,1} = a_{2,1} = 1$, all other $a_k = 0$, and $\mathcal{L}_1(\boldsymbol{u}) = \frac{\partial}{\partial \boldsymbol{n}}(\boldsymbol{u})$. This produces the interface condition $\frac{\partial}{\partial \boldsymbol{n}}\boldsymbol{u}_1^I = \frac{\partial}{\partial \boldsymbol{n}}\boldsymbol{u}_2^I$, i.e., that the normal derivatives are equal.

Robin $a_{1,1} = a_{2,1} = 1$, $a_{1,0}$ and $a_{2,0}$ arbitrary, and all other $a_k = 0$, $\mathcal{L}_0(\boldsymbol{u}) = \boldsymbol{u}$, and $\mathcal{L}_1(\boldsymbol{u}) = \frac{\partial}{\partial \boldsymbol{n}}(\boldsymbol{u})$. This produces the interface condition $a_{1,0}\boldsymbol{u}_1^I + \frac{\partial}{\partial \boldsymbol{n}}\boldsymbol{u}_1^I = a_{2,0}\boldsymbol{u}_2^I + \frac{\partial}{\partial \boldsymbol{n}}\boldsymbol{u}_2^I$, and may relate to a flux being conserved across the interface.

Laplace: $a_{1,2} = a_{2,2} = 1$, all other $a_k = 0$, and $\mathcal{L}_2(\boldsymbol{u}) = \nabla^2\boldsymbol{u}$. This produces the interface condition $\nabla^2\boldsymbol{u}_1^I = \nabla^2\boldsymbol{u}_2^I$, and is more common in biharmonic, or other fourth-order, problems.

Each model needs an interface condition, as shown in (19.34), and the interface conditions must be independent for the problem to be well-posed [Keyes *et al.* (2013)]. This issue will not be considered here, but rather mentioned to explain why both F_1^I and F_2^I are necessary. This general form of the interface conditions is made more concrete in the following examples.

The discretization of the $\mathcal{L}_k$ operators will almost certainly involve values off the interface because of the presence of derivatives. The values required to approximate $\mathcal{L}_k$ on the interface define the coupling region: choosing a specific interface discretization scheme fixes the coupling region to provide the necessary values. Conversely, a coupling region could be chosen and then a discretization could be built which best approximates $\mathcal{L}_k$ given that data. That situation is of more significance when trying to minimize the communication between models. This relationship between choosing a coupling region and the quality of the resulting interface discretization will be analyzed in the example below.

Solving the nonlinear system of the simple two-component model described in Figure 19.5, $F(\boldsymbol{u}) = 0$, with Newton's method requires the Jacobian $\mathsf{J}(F)(\boldsymbol{u})$, which has the nonzero structure

$$\begin{pmatrix} \mathsf{J}_1(F_1) & & \mathsf{J}_1^C(F_1) & & & \\ & \mathsf{J}_2(F_2) & & & \mathsf{J}_2^C(F_2) & \\ \mathsf{J}_1(F_1^C) & & \mathsf{J}_1^C(F_1^C) & \mathsf{J}_1^I(F_1^C) & & \\ & & \mathsf{J}_1^C(F_1^I) & \mathsf{J}_1^I(F_1^I) & \mathsf{J}_2^C(F_1^I) & \mathsf{J}_2^I(F_1^I) \\ & \mathsf{J}_2(F_2^C) & & & \mathsf{J}_2^C(F_2^C) & \mathsf{J}_2^I(F_2^C) \\ & & \mathsf{J}_1^C(F_2^I) & \mathsf{J}_1^I(F_2^I) & \mathsf{J}_2^C(F_2^I) & \mathsf{J}_2^I(F_2^I) \end{pmatrix}. \tag{19.36}$$

The notation used for the blocks is similar to the notation for the components of the residual, e.g., $\mathsf{J}_1^C(F_2^I)$ is the Jacobian with respect to the coupling variables of Model 1 applied to the interface equations of Model 2. Note that the location of the nonzero blocks corresponds to the dependencies described in Table 19.1. Also, all the Jacobians are evaluated with the same argument $\boldsymbol{u}$, so it is omitted from (19.36) to save space.

19.4.1 *Meshfree coupling*

Thus far we have described the interface conditions (19.35) in the continuous setting only. Actually performing the simulation requires discretizing the interface conditions. As mentioned near the end of Section 19.4, there are two choices when enforcing the interface conditions: a specific discretization can be chosen, the required nodes of which form the resulting coupling region; or a coupling region can be fixed, and the discretization is designed to best accommodate that choice. Because the interior is the dominant portion of the simulation, the discretization is generally chosen in deference to the needs of the PDE, not the interface conditions.

Typically, the discretization scheme on the interface is the same as the discretization on the interior and coupling regions. This is a logical choice because the placement of the nodes in the coupling region is designed around the interior discretization, and using the same discretization on the interface should take advantage of the placement of the coupling nodes. In spite of the common sense nature of this approach, there is nothing to guarantee that a good approximation to the interface conditions will be achieved by using the interior discretization scheme. Below we will present an example using finite differences (FD) on the interior and coupling regions, and compare the results of a finite difference interface discretization to an RBF interface discretization using Gaussian kernels.

The use of a meshfree approximation scheme in the interface region of PDEs discretized by other means was discussed by Wendland and collaborators in [Beckert and Wendland (2001); Ahrem *et al.* (2005); Wendland (2006, 2008)]. The improved approximation and stability results using an alternate basis provides an opportunity to build on the already existing meshfree coupling framework of Wendland.

19.4.2 *An example: coupled 2D heat equation*

This example is a linear parabolic boundary value problem

$$u_t - (u_{xx} + u_{yy}) = e^{-t}\left(1 + \frac{1}{2}(x^2 + y^2)\right), \qquad (x, y) \in \Omega, \tag{19.37a}$$

$$u = e^{-t}\left(1 - \frac{1}{2}(x^2 + y^2)\right), \qquad (x, y) \in \partial\Omega, \tag{19.37b}$$

$$u = 1 - \frac{1}{2}(x^2 + y^2), \qquad t = 0. \tag{19.37c}$$

Ω is the full domain of the coupled problem, such that $\Omega = \Omega_1 \cup \Omega_2$. For Model 1,

$$\Omega_1 = \{(x, y) : -1 \le x \le 0,\ 0 \le y \le 1\},$$

and for Model 2,

$$\Omega_2 = \{(x, y) : 0 \le x \le 1,\ 0 \le y \le 1\}.$$

Second-order finite differences are used to discretize in space, meaning that if N uniformly spaced points are distributed within one of these models then $\Delta y = \Delta x = 1/(\sqrt{N} - 1)$. This spatial discretization was chosen so that the order of accuracy on the interior would be superior to the weakest coupling schemes, but comparable to stronger coupling schemes. Other choices, including the kernel-based methods discussed in Chapter 20, would also work for this example.

The backward Euler method is used to discretize in time, and the vector $\boldsymbol{u}^{(k)}$ represents the computed solution at time $t = k\Delta t$. After discretizing, we are left with the "nonlinear" system

$$F(\boldsymbol{u}^{(k)}) = \frac{1}{\Delta t}(\boldsymbol{u}^{(k)} - \boldsymbol{u}^{(k-1)}) - \mathsf{L}_{FD,BC}(\boldsymbol{u}^{(k)}) - \boldsymbol{f} = 0,$$

where $\mathsf{L}_{FD,BC}$ is the Laplacian operator or identity depending on whether (19.37a) or (19.37b), respectively, is applicable. The vector $\boldsymbol{f}$ is the right hand side associated with either (19.37a) or (19.37b), and $\boldsymbol{u}^{(0)}$ is generated by (19.37c). Although the discussion earlier focused on solving nonlinear systems, the discussion is still applicable here because this linear system is just a special case of a nonlinear system.

Since the boundary condition (19.37b) is only defined on $\partial\Omega$, there must be a second-order interface condition on the shared boundary for the problem to be well-posed. Fundamentally, this is because each model needs a simulated boundary condition to allow it to be well-posed as an independent problem; once it has a "boundary condition" on the interface, it inherits the well-posedness properties associated with any single domain boundary value problem. As mentioned earlier, guaranteeing accuracy, either *a priori* or *a posteriori*, for the coupled problem is still an open problem and beyond the scope of this example.

The choice of second-order condition we make here is to require values and normal derivatives of u to be equal at the shared interface of the two models. By defining u_1 and u_2 as the solutions for Models 1 and 2, respectively, and $\Omega_I = \Omega_1 \cap \Omega_2$, we can write the interface conditions as

$$\left.\begin{aligned} u_1(x,y) &= u_2(x,y) \\ \frac{\partial}{\partial x}u_1(x,y) &= \frac{\partial}{\partial x}u_2(x,y) \end{aligned}\right\}, \qquad (x,y) \in \Omega_I.$$

It is noteworthy that these are not the only acceptable choice of interface conditions. Any choice which will produce one unique manufactured "boundary condition" per model would be acceptable. Choosing Dirichlet and Robin interface conditions would be acceptable, but it would not be acceptable to use Dirichlet to match solution values, and then have a second condition which matched twice the values. Doing so would produce a Jacobian of the form (19.36) with fourth and sixth block rows equivalent except for a factor of two, and such a system is underdetermined.

We will choose to use the Dirichlet coupling condition as F_1^I and the Neumann coupling condition as F_2^I, although this choice is arbitrary. Discretizing these interface conditions using the low-rank series approximation (19.26) will convert the general Jacobian matrix (19.36) to the matrix

$$\begin{pmatrix} \mathsf{J}_1(F_1) & & \mathsf{J}_1^C(F_1) & & & \\ & \mathsf{J}_2(F_2) & & & \mathsf{J}_2^C(F_2) & \\ \mathsf{J}_1(F_1^C) & & \mathsf{J}_1^C(F_1^C) & \mathsf{J}_1^I(F_1^C) & & \\ & & & \mathsf{I} & \multicolumn{2}{c}{[\,\Phi_{2\to 1}\Phi_2^\dagger\,]} \\ & \mathsf{J}_2(F_2^C) & & & \mathsf{J}_2^C(F_2^C) & \mathsf{J}_2^I(F_2^C) \\ & & \multicolumn{2}{c}{[\,\Phi_{1\to 2}^{\mathcal{D}_x}\Phi_1^\dagger\,]} & \multicolumn{2}{c}{[\,\Phi_2^{\mathcal{D}_x}\Phi_2^\dagger\,]} \end{pmatrix}, \qquad \boldsymbol{u} = \begin{pmatrix} \boldsymbol{u}_1 \\ \boldsymbol{u}_2 \\ \boldsymbol{u}_1^C \\ \boldsymbol{u}_1^I \\ \boldsymbol{u}_2^C \\ \boldsymbol{u}_2^I \end{pmatrix}.$$

Note the inclusion of the solution vector to the right, for reference. Two block rows have been converted in this matrix in order to satisfy the interface conditions:

- Fourth block row: This row defines F_1^I as the Dirichlet condition, and thus requires that solution values from Model 2 be mapped to the discretization of Model 1.
  - $(\,\mathsf{J}_1^C(F_1^I) \;\; \mathsf{J}_1^I(F_1^I)\,) \longrightarrow (\,0 \;\; \mathsf{I}\,)$. Because the residual F_1^I is being evaluated on the interface nodes of Model 1, $\boldsymbol{x}_1^I$, we need to produce the solution $\boldsymbol{u}_1^I$. Doing so requires only the identity I located as seen above.
  - $(\,\mathsf{J}_2^C(F_1^I) \;\; \mathsf{J}_2^I(F_1^I)\,) \longrightarrow [\,\Phi_{2\to 1}\Phi_2^\dagger\,]$. Here we are given values on $\boldsymbol{x}_2^C$ and $\boldsymbol{x}_2^I$ and asked to produce values on $\boldsymbol{x}_1^I$. The matrix $\Phi_{2\to 1}\Phi_2^\dagger$ is like the differentiation matrix defined in (19.29), where the differential operator is

replaced by the identity. Using that design, we have input points $\mathcal{Z} = \{\boldsymbol{x}_2^C, \boldsymbol{x}_2^I\}$, evaluation locations $\mathcal{X} = \boldsymbol{x}_1^I$ and $\boldsymbol{u} = \begin{pmatrix} \boldsymbol{u}_2^C \\ \boldsymbol{u}_2^I \end{pmatrix}$.

- Sixth block row: This row defines F_2^I as the Neumann condition, and thus requires that derivative values from Model 1 be mapped to the discretization of Model 2.
  - $(\, \mathsf{J}_1^C(F_2^I) \quad \mathsf{J}_1^I(F_2^I)\,) \longrightarrow [\, \Phi_{1\to 2}^{\mathcal{D}_x} \Phi_1^\dagger \,]$. This follows directly from the differentiation matrix formula (19.29) using $\mathcal{Z} = \{\boldsymbol{x}_1^C, \boldsymbol{x}_1^I\}$, $\mathcal{X} = \boldsymbol{x}_2^I$, $\boldsymbol{u} = \begin{pmatrix} \boldsymbol{u}_1^C \\ \boldsymbol{u}_1^I \end{pmatrix}$, and $\mathcal{D}_x$ is the derivative in the x direction (normal to the interface).
  - $(\, \mathsf{J}_2^C(F_2^I) \quad \mathsf{J}_2^I(F_2^I)\,) \longrightarrow [\, \Phi_2^{\mathcal{D}_x} \Phi_2^\dagger \,]$. Following again from the differentiation matrix formula with $\mathcal{Z} = \{\boldsymbol{x}_2^C, \boldsymbol{x}_2^I\}$, $\mathcal{X} = \boldsymbol{x}_1^I$, $\boldsymbol{u} = \begin{pmatrix} \boldsymbol{u}_2^C \\ \boldsymbol{u}_2^I \end{pmatrix}$, and $\mathcal{D}_x$ is the derivative in the x direction (normal to the interface).

Note that we could also have created the differentiation matrices using the full Hilbert–Schmidt SVD basis (or some other alternate basis), but the low-rank approach is less costly and produces good results for this example. Also, the code is much too long to fit comfortably in this book. The example can be found in the `GaussQR` repository under `HermiteCoupling.m`.

Figure 19.7 shows error results when the simulation is run for a single time step of size $\Delta t = .01$ and $N = 256$ uniformly distributed points in each domain. It compares the error present in the fully coupled system when the interface conditions are enforced using finite differences and low-rank Gaussian eigenfunction approximate interpolation. The standard one-sided finite difference approximations

$$\text{Couple width 1:} \qquad \frac{\partial}{\partial x} u(x) = \frac{u(x+\Delta x) - u(x)}{\Delta x} \tag{19.38a}$$

$$\text{Couple width 2:} \qquad \frac{\partial}{\partial x} u(x) = \frac{u(x+2\Delta x) - 4u(x+\Delta x) + 3u(x)}{2\Delta x} \tag{19.38b}$$

derived from the Taylor series are used here because the interface nodes of both models are aligned. A mismatch between the Model 1 and Model 2 discretizations would necessitate a more complicated finite difference approximation. The finite difference equations (19.38) define the coupling region which is used for both coupling schemes, and are described graphically in Figure 19.6, with results presented for couple width 1 in Figure 19.7(a) and couple width 2 in Figure 19.7(b).

Because of the relationship between finite difference approximation and Taylor series, the finite difference quality at each point in $\boldsymbol{u}^I$ is bounded by the accuracy which can be obtained by a polynomial in the x direction. The kernel-based interpolant has the free parameter ε, which in the limit $\varepsilon \to 0$ will reproduce a polynomial, leaving the potential to see better accuracy than the finite difference coupling for some $\varepsilon > 0$. In addition to the well-known stability issues for small values of ε, in this Hermite interpolation setting, the magnitude of the Gaussians grows with ε, which can actually cause stability issues for very large values.

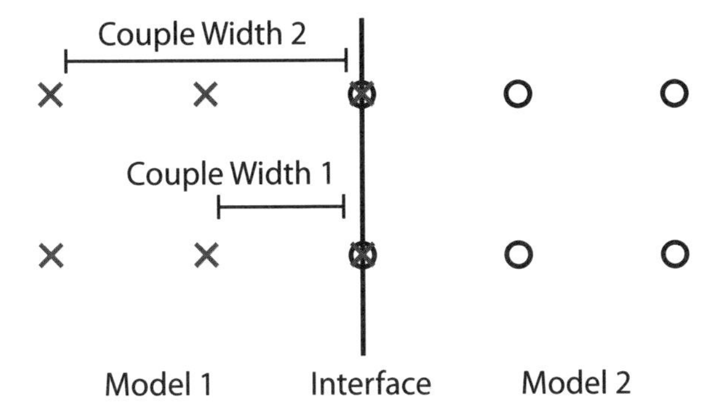

Fig. 19.6 The order of the finite difference stencil (either (19.38a) or (19.38b)) determines the coupling region. To fairly compare the basic finite difference approach to the low-rank eigenfunction approximate interpolant approach of Section 19.3, the same coupling domain is considered for both coupling schemes.

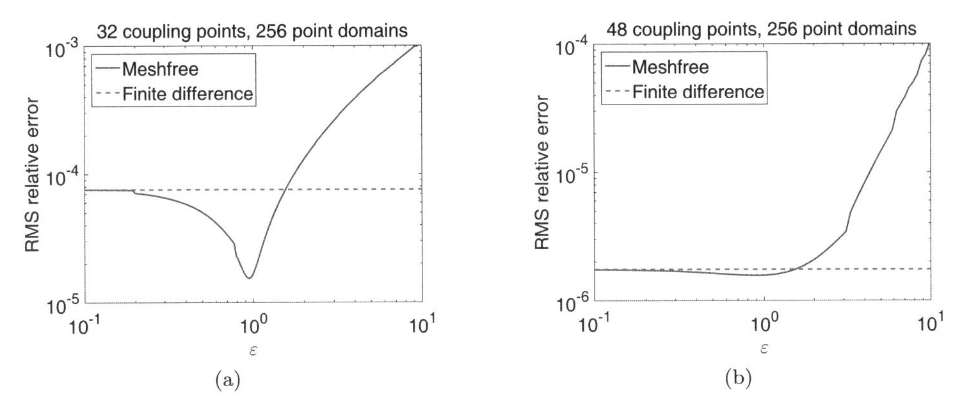

Fig. 19.7 There is a range of ε values for which the kernel-based approximation produces accuracy that can not be attained by finite differences. (a) Coupling width Δx: we see a pronounced benefit to using Gaussians; (b) Coupling width $2\Delta x$: there is little gain because the coupling error is now on level with the interior error.

Decreasing ε will produce wider Gaussians which will allow more interaction and produce more accurate interpolants, until error due to the Runge phenomenom emerges. This source of error for very large and very small ε may result in some intermediate value where accuracy is optimal (see Chapters 10–11). Such is the case in Figure 19.7(a), where — when using the same coupling region (i.e., the same available data) as the finite difference coupling method — the kernel-based Hermite interpolant produced four times more accurate results for $\varepsilon \approx 1$.

This improvement in accuracy is also related to the idea that using values in all directions around a point better approximates the derivative at that point, as discussed in [Fornberg *et al.* (2010)]. Furthermore, we see that the Gaussian interpolant is approaching the finite difference interpolant as $\varepsilon \to 0$. For the finite difference coupling to catch up to the kernel-based coupling, more points must be included in the derivative computation.

If we instead use the second order finite difference coupling (19.38b) (i.e., including any point within $2\Delta x$ of the interface), then we see in Figure 19.7(b) that the optimal coupling strategy and the finite difference approach are very close.

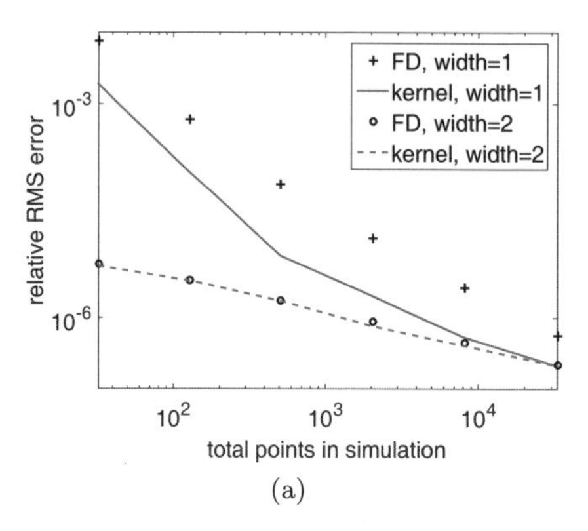

(a)

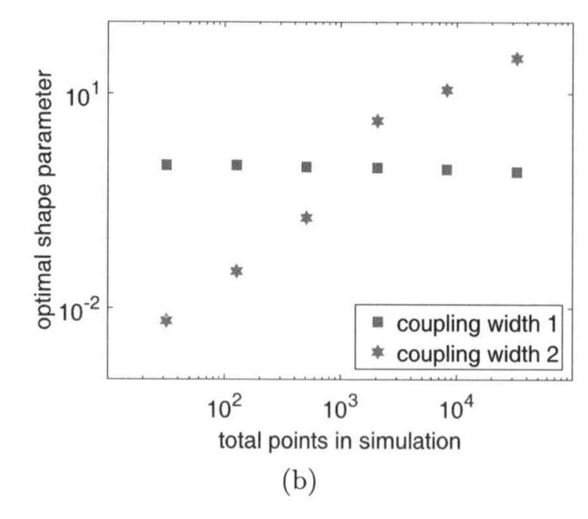

(b)

Fig. 19.8 We consider the effect of the coupling strategy given varying values of N, the number of points in the simulation. For these tests, the low-rank Gaussian eigenfunction approximation uses the parameters $\alpha = 1$, and M set to half of the number of points in the coupling region. (a) The coupling scheme may play a large role in the accuracy. With fewer points in the coupling region, the Gaussian approximation outperforms finite differences. For a large enough simulation, the Gaussian scheme can recover the couple width$=2\Delta x$ solution with only couple width$=\Delta x$ points. The ε values used here are described in the plot on the right; (b) For couple width 1, the optimal ε (best accuracy) stays relatively constant, allowing more points to be involved in computing the interpolant. With a thicker coupling region, more information is provided in the x direction, and the kernel-based interpolant prefers a larger ε to emphasize those points over points further away in the y direction.

Figure 19.8(a) shows the convergence of the coupling scheme as the number of points in the fully coupled simulation is increased. Here we can see that the accuracy of the kernel-based approach is consistently better than the basic finite difference approach for the thinner coupling region. At $N = 32768$ we see that the accuracy of the kernel-based coupling for Δx thickness is actually comparable to the $2\Delta x$ thickness case for the finite difference coupling. This means that for discretizations of that size or greater, the kernel-based coupling allows for the same accuracy while involving fewer points in the coupling.

Earlier, in Figure 19.7, we showed that there was a strong dependence on the value of ε in producing an optimal approximation for the kernel-based coupling approach. In Figure 19.8(b) we can see that the optimal value of ε changes very little for the thin coupling case. The opposite is true when the coupling width is set to $2\Delta x$, i.e., the optimal ε increases as N increases.

An ε that grows in this way limits the effect of points further away. This tells us that — when given more data in the x direction — data in the y direction becomes less valuable. When that data is unavailable, as is the case for the thin coupling strip, the optimal interpolant in this example will continue to consider all the points available to it.

19.4.3 *Computational considerations*

Another significant benefit of using kernel-based methods to perform the coupling is that these methods work equally well in arbitrary dimensions and on arbitrary node distributions. The differentiation matrix computations for kernels (see Sections 19.1 and 19.2) are carried out similarly regardless of the point distribution, whereas new finite difference relations would need to be computed at each point on scattered data. Furthermore, using the technique of producing multi-dimensional, higher-order finite difference schemes by annihilating certain polynomials will fail on many point sets in a scattered data setting. We encourage users interested in faster computation of the coupling to employ the local kernel differentiation matrices of Section 19.1.4 in lieu of standard finite difference schemes.

Jacobian-Vector Products

There are some practical concerns about the viability of kernel-based interpolation as a multiphysics coupling method. One such concern regards the use of Newton's method,

$$\boldsymbol{u}^k = \boldsymbol{u}^{k-1} - \left[\mathsf{J}(F)(\boldsymbol{u}^{k-1})\right]^{-1} F(\boldsymbol{u}^{k-1}), \qquad \boldsymbol{u}^0 \text{ given},$$

to solve the system $F(\boldsymbol{u}) = 0$, where $\boldsymbol{u}^k$ is the kth Newton iteration. Note that this notation is now overriding the previous notation where $\boldsymbol{u}^{(k)}$ was the kth time step. Newton's method suffers from several logistical barriers including evaluating the Jacobian.

To avoid directly computing the full Jacobian at each Newton iteration, Jacobian-free Newton-Krylov [Knoll and Keyes (2004)] is instead used to solve $F(\boldsymbol{u}) = 0$. This allows for matrix-vector products $\mathsf{J}(F)(\boldsymbol{u})\boldsymbol{b}$ to be approximated using the finite difference formula

$$\mathsf{J}(F)(\boldsymbol{u})\boldsymbol{b} \approx \frac{1}{h}(F(\boldsymbol{u} + h\boldsymbol{b}) - F(\boldsymbol{u})) \tag{19.39}$$

and then computes $\left[\mathsf{J}(F)(\boldsymbol{u}^{k-1})\right]^{-1} F(\boldsymbol{u}^{k-1})$ using an iterative method. The use of (19.39) is appropriate when computing the matrix-vector components involving the PDE and boundary conditions because they may be nonlinear (although not in the example above), but the discretization of the interface through GaussQRr is linear. The rows associated with F_1^I and F_2^I can be computed using only matrix-vector products involving eigenfunctions, and that should be exploited.

Beyond simply using the available matrices to compute $\mathsf{J}(F_1^I)(\boldsymbol{u})\boldsymbol{b}$ and $\mathsf{J}(F_2^I)(\boldsymbol{u})\boldsymbol{b}$ rather than (19.39), it is also useful to note the structure of the differentiation matrices. In Section 19.3.1 it was mentioned that differentiation matrices of the form $\hat{\Phi}^{\mathcal{D}_k}\Phi^\dagger$ have an outer product structure. Because of this it is much more efficient to evaluate $\hat{\Phi}^{\mathcal{D}_k}(\Phi^\dagger \boldsymbol{b})$ rather than form the differentiation matrix and then conduct the matrix-vector product $(\hat{\Phi}^{\mathcal{D}_k}\Phi^\dagger)\boldsymbol{b}$. Although the results in Figure 19.8(a) were produced for a value of M fixed in proportion to N, it may be possible to reduce M and still maintain a high order of accuracy. For example, the results for $N = 65536$ can be obtained using $M \approx 40$. Doing so would save on computational cost for Jacobian-vector products, but finding the minimum M value to acceptably approximate derivatives depends on many factors.

Eigenfunction selection and ordering

One option which allows the use of smaller M is to choose eigenfunctions which take advantage of the location of the available coupling points[6]. All the previous computations used eigenfunctions of limited order in the x direction; the first 12 eigenfunction indices are:

Couple width 1:		m_1	m_2	m_3	m_4	m_5	m_6	m_7	m_8	m_9	m_{10}	m_{11}	m_{12}
	x order	1	2	1	2	1	2	1	2	1	2	1	2
	y order	1	1	2	2	3	3	4	4	5	5	6	6

Couple width 2:		m_1	m_2	m_3	m_4	m_5	m_6	m_7	m_8	m_9	m_{10}	m_{11}	m_{12}
	x order	1	2	1	3	2	1	3	2	1	3	2	1
	y order	1	1	2	1	2	3	2	3	4	3	4	5

These eigenfunctions are chosen so that the maximum order in the x dimension is one more than the couple width: this was a logical choice because in the polynomial limit $\varepsilon \to 0$ the highest unique polynomial that could be fit to those points is of degree couple width plus 1. Adding more complexity in the x direction would just overfit the data, and add more unnecessary computation. This is one of the few times where we suggest an ordering other than that described in Table 12.1.

[6]Note that this reordering of eigenfunctions is different from the one discussed in Section 13.4.2, where we were mostly concerned with keeping the matrix Φ_1 invertible.

Chapter 20

Kernel-Based Methods for PDEs

In Chapter 19 we developed the use of differentiation matrices and similar strategies for evaluating derivatives of interpolants. We also applied these ideas in Section 19.4 to couple boundary value problems through an interface by exchanging derivative data between two different computational models. There the individual components, however, were solved using standard (polynomial) finite difference methods. In this chapter we discuss how to add to the list of traditional PDE solvers, such as those finite difference methods, by developing kernel-based methods for the numerical solution of boundary value problems.

Our first example will focus on the solution of time-independent boundary value problems using the standard kernel-based (nonsymmetric) collocation method, the Hilbert–Schmidt SVD basis and a collocation solution derived from the low-rank eigenvalue approximation introduced in Section 12.3. We will then consider the solution of time-dependent problems using the method of lines to distinguish the time and space components of the solution.

In the following two sections we will study the method of fundamental solutions and the method of particular solutions as ideal techniques for solving boundary value problems with specific differential operators whose Green's kernels are known (see, e.g., Section 6.5).

This is followed with an analysis of the use of kernel methods to produce a compact differential operator which can then be applied to the boundary value problem in the spirit of polynomial finite difference methods. In the final section we briefly investigate the use of the concept of a space-time kernel (introduced in Section 3.7) to simultaneously deal with the space and time components of the problem.

20.1 Collocation for Linear Elliptic PDEs

The original method for solving elliptic boundary value problems (BVPs) with kernel collocation was introduced by Ed Kansa (1986) (see also [Kansa (1990a,b)]). It consisted of a *nonsymmetric collocation* method based on multiquadric basis functions centered at points chosen throughout the domain. This nonsymmetric *Kansa's method* has spawned a huge body of literature — mostly by practitioners from many diverse areas of science and engineering (see, e.g., [Sarra and Kansa (2009); Chen *et al.* (2014b)] and references therein).

Since the initial work of Kansa — and in addition to the many applications of the method — a theoretical analysis of the convergence behavior of this collocation method (and various variations thereof) has been performed by Robert Schaback and collaborators [Ling *et al.* (2006); Schaback (2007); Lee *et al.* (2009); Schaback (2010, 2014)]. The huge popularity among practitioners, in addition to this theoretical support has encouraged the use of this nonsymmetric collocation method despite its potential for failure [Hon and Schaback (2001)] if applied in its most-straightforward and naive implementation.

A *symmetric collocation* technique was also developed (see, e.g., [Wu (1992); Fasshauer (1997, 2007)]), which ensured invertibility of the collocation system by using a modified set of basis functions directly motivated by the Hermite interpolation approach of Section 19.2.2. Convergence of the symmetric collocation method was analyzed by Franke and Schaback (1998a,b).

Since the nonsymmetric collocation method is both far more popular and also simpler to implement than its symmetric cousin, we will focus only on the nonsymmetric Kansa's method in this book. Let us now briefly review the basic structure of nonsymmetric kernel collocation methods.

We restrict ourselves, for this section, to *boundary value problems* of the form

$$\begin{aligned} \mathcal{L}u &= f, \qquad \text{in the interior } \Omega, \\ \mathcal{B}u &= g, \qquad \text{on the boundary } \partial\Omega, \end{aligned}$$

where $\mathcal{L}$ is a linear elliptic partial differential operator, and $\mathcal{B}$ is a linear boundary operator. However, in Section 20.2 we will also consider time-dependent problems. The domain $\Omega \subset \mathbb{R}^d$ is assumed to have a Lipschitz boundary.

20.1.1 *Nonsymmetric collocation in the standard basis*

A standard collocation method begins with the same *Ansatz* as an interpolation method, namely that the approximate solution $\hat{u}$ is a linear combination of basis functions

$$\begin{aligned} \hat{u}(\boldsymbol{x}) &= \sum_{j=1}^{M} c_j K(\boldsymbol{x}, \boldsymbol{z}_j) \\ &= \boldsymbol{k}(\boldsymbol{x})^T \boldsymbol{c}, \end{aligned} \tag{20.1}$$

where the vector notation in the second line is analogous to (1.3). Now, however, we understand $\boldsymbol{k}(\cdot)^T$ to be the vector of basis functions or *trial functions* which are formed by centering the kernel K at the *trial centers* $\mathcal{Z} = \{\boldsymbol{z}_1, \ldots, \boldsymbol{z}_M\}$. These kernel centers can, in principle, be chosen anywhere inside $\Omega \cup \partial\Omega$. It is common practice to choose these centers in a quasi-uniform fashion throughout the domain, but a specifically structured choice may yield better results (see, e.g., [Iske (1999); Fedoseyev *et al.* (2002); Larsson and Fornberg (2003); Fornberg and Zuev (2007)]).

Remark 20.1. We see here another instance where we use $\mathcal{Z}$ to denote the kernel centers (as is the case in the early chapters) rather than $\mathcal{X}$ (as is the case in many examples involving MATLAB). The formulation of these PDE problems allows for the flexibility to disassociate the kernel centers from the data locations, thus we begin the discussion using $\mathcal{Z}$ as kernel centers.

Substituting our *Ansatz* into the BVP produces

$$\begin{aligned} \mathcal{L}\boldsymbol{k}(\boldsymbol{x})^T\boldsymbol{c} &= f(\boldsymbol{x}), \qquad \boldsymbol{x} \in \Omega, \\ \mathcal{B}\boldsymbol{k}(\boldsymbol{x})^T\boldsymbol{c} &= g(\boldsymbol{x}), \qquad \boldsymbol{x} \in \partial\Omega. \end{aligned}$$

Strategies for defining and computing derivatives of kernel-based approximations are discussed in Chapter 19. At this point, we have *semi-discretized* our problem by enforcing our assumption that the solution be a linear combination of M kernels. The specific combination that approximately solves the problem is determined by the coefficient vector $\boldsymbol{c}$ and depends on the points at which we choose to collocate.

Let us define these N *collocation points* as $\boldsymbol{x}_i$, $i = 1, \dots, N$, and let us assume that $N = N_{\mathcal{L}} + N_{\mathcal{B}}$, where $N_{\mathcal{L}}$ and $N_{\mathcal{B}}$ are the number of collocation points in the interior Ω and on the boundary $\partial\Omega$, respectively. The fully discretized BVP now takes the form

$$\begin{pmatrix} \mathcal{L}\boldsymbol{k}(\boldsymbol{x}_1)^T \\ \vdots \\ \mathcal{L}\boldsymbol{k}(\boldsymbol{x}_{N_{\mathcal{L}}})^T \\ \mathcal{B}\boldsymbol{k}(\boldsymbol{x}_{N_{\mathcal{L}}+1})^T \\ \vdots \\ \mathcal{B}\boldsymbol{k}(\boldsymbol{x}_{N_{\mathcal{L}}+N_{\mathcal{B}}})^T \end{pmatrix} \boldsymbol{c} = \begin{pmatrix} f(\boldsymbol{x}_1) \\ \vdots \\ f(\boldsymbol{x}_{N_{\mathcal{L}}}) \\ g(\boldsymbol{x}_{N_{\mathcal{L}}+1}) \\ \vdots \\ g(\boldsymbol{x}_{N_{\mathcal{L}}+N_{\mathcal{B}}}) \end{pmatrix}, \tag{20.2}$$

where the collocation matrix is of size $N \times M$. Depending on the number of collocation points, this may be a square, overdetermined or underdetermined linear system. This suggests a natural question:

How do we choose our collocation points?

In the interpolation setting, we required that the kernel centers and interpolation points be the same in order to guarantee a unique solution to the scattered data interpolation problem (recall Section 1.3.1). Unfortunately, for nonsymmetric collocation, there are no such guarantees; research has even shown that some configurations of coupled collocation points/kernel centers will yield a singular collocation matrix [Hon and Schaback (2001)].

For simplicity, we ignore these potential complications and consider only the case where $N = M$ and the collocation points coincide with the kernel centers, i.e.,

$\boldsymbol{x}_i = \boldsymbol{z}_i$ for $i = 1, \ldots, N$. When this is true, the collocation matrix is square and (hopefully) has an inverse. These methods, as implemented in the standard basis, are discussed in [Fasshauer (2007)].

However, when implemented in the standard basis, these collocation schemes are subject to the same ill-conditioning issues as the interpolation problems discussed in Chapter 11. Let us consider first an example of this on a simple 1D boundary value problem.

Example 20.1. (Solving a BVP in the standard basis)
We take the boundary value problem

$$u''(x) = -\sin(x), \qquad x \in (0, \pi),$$
$$u(x) = 0, \qquad x \in \{0, \pi\},$$

for which we know the true solution is $u(x) = \sin(x)$. In Program 20.1 we use inverse multiquadric kernel collocation at $N = 25$ points, two of which appear on the boundary. The solutions for different ε values (defined in line 2) are displayed in Figure 20.1(a).

Program 20.1. `PDEUnstable2ptBVP.m`

```
   % Define the inverse multiquadric and its 2nd derivative
2  rbf = @(e,r) 1./sqrt(1+(e*r).^2);  ep = .045;
   rbfxx = @(e,r) e^2*(2*(e*r).^2-1)./(1+(e*r).^2).^(5/2);
4  % Define functions for the interior and boundary
   fint = @(x) -sin(x);
6  fbc = @(x) zeros(size(x,1),1);
   uf = @(x) sin(x);
8  % Choose a set of collocation points
   % Separate them into a interior and boundary region
10 N = 25;  xall = pickpoints(0,pi,N);
   xint = xall(xall~=0 & xall~=pi);
12 xbc = xall(xall==0 | xall==pi);
   x = [xint;xbc];
14 % Choose some evaluation points
   Neval = 200;  xeval = pickpoints(0,pi,Neval);
16 % Define the necessary distance matrices
   DMint = DistanceMatrix(xint,x);  DMbc = DistanceMatrix(xbc,x);
18 DMeval = DistanceMatrix(xeval,x);
   % Form the right hand side, in the same order as the points
20 rhsint = fint(xint);  rhsbc = fbc(xbc);
   rhs = [rhsint;rhsbc];
22 % Form the kernel matrices, both collocation and evaluation
   Kxxint = rbfxx(ep,DMint);  Kbc = rbf(ep,DMbc);  Keval = rbf(ep,DMeval);
24 A = [Kxxint;Kbc];
   % Find the solution coefficients and evaluate the solution
26 c = A\rhs;
   ucoll = Keval*c;
```

This program looks surprisingly similar to the interpolation problems that we have discussed thus far. The main difference is that the linear system is no longer uniformly defined, but rather broken into regions: the interior and the boundary. Lines 10–13 perform this dissection of the full domain into `xint` on the interior and `xbc` on the boundary. Similarly, two distance matrices are defined on line 17 with different evaluation points but always the same full set of kernel centers. The correct functions are applied to the correct domains and the matrix and right hand side are built in the same interior/boundary order in lines 21 and 24. The matrix in (20.2) is named `A` because it had not previously been named.

Despite having a shape similar to that of the sine function, only one solution in Figure 20.1(a) is reasonable while the other suffers from instability, causing its erratic and seemingly discontinuous behavior. MATLAB reports the condition number of the collocation matrix as roughly 2×10^{20}, which is so extreme that we might even expect worse behavior than we already see. Needless to say, the ill-conditioning will continue for smaller ε as we saw in the interpolation setting, so an alternate solution technique must be used to produce reasonable results as $\varepsilon \to 0$.

20.1.2 *Nonsymmetric collocation using the Hilbert–Schmidt SVD*

In collocation settings where ill-conditioning is a severe problem, and the Mercer series of the desired kernel is available, the Hilbert–Schmidt SVD (recall Chapter 13 and Section 19.1) can be used to create a more stable scheme. Instead of representing the approximate solution $\hat{u}(\boldsymbol{x}) = \boldsymbol{k}(\boldsymbol{x})^T\boldsymbol{c}$ in the standard basis as in (20.1), we now utilize the stable HS-SVD basis to write

$$\hat{u}(\boldsymbol{x}) = \boldsymbol{\psi}(\boldsymbol{x})^T\boldsymbol{b},$$

where we showed in Section 13.2 that

$$\boldsymbol{\psi}(\boldsymbol{x})^T = \boldsymbol{\phi}(\boldsymbol{x})^T \begin{pmatrix} \mathsf{I}_N \\ \Lambda_2\Phi_2^T\Phi_1^{-T}\Lambda_1^{-1} \end{pmatrix}, \tag{20.3}$$

and the coefficients $\boldsymbol{b}$ for the stable HS-SVD basis are related to the coefficients $\boldsymbol{c}$ for the standard basis by $\boldsymbol{b} = \Lambda_1\Phi_1^T\boldsymbol{c}$. This latter relation, however, is not important for practical considerations since we generally have no need to convert the solution to its standard basis representation.

Using the HS-SVD basis we can then set up a nonsymmetric collocation system in analogy to (20.2), i.e.,

$$\begin{pmatrix} \mathcal{L}\boldsymbol{\psi}(\boldsymbol{x}_1)^T \\ \vdots \\ \mathcal{L}\boldsymbol{\psi}(\boldsymbol{x}_{N_\mathcal{L}})^T \\ \mathcal{B}\boldsymbol{\psi}(\boldsymbol{x}_{N_\mathcal{L}+1})^T \\ \vdots \\ \mathcal{B}\boldsymbol{\psi}(\boldsymbol{x}_{N_\mathcal{L}+N_\mathcal{B}})^T \end{pmatrix} \boldsymbol{b} = \begin{pmatrix} f(\boldsymbol{x}_1) \\ \vdots \\ f(\boldsymbol{x}_{N_\mathcal{L}}) \\ g(\boldsymbol{x}_{N_\mathcal{L}+1}) \\ \vdots \\ g(\boldsymbol{x}_{N_\mathcal{L}+N_\mathcal{B}}) \end{pmatrix}. \tag{20.4}$$

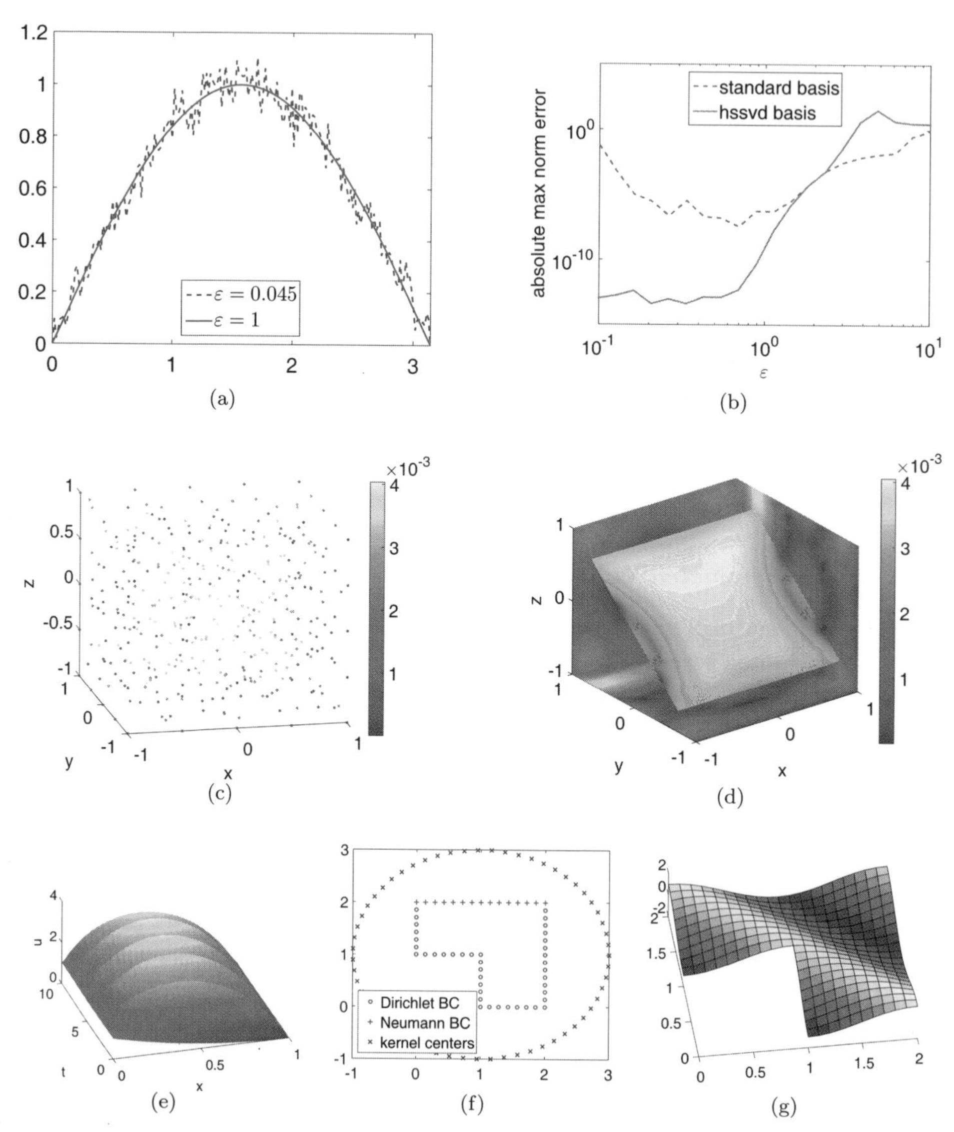

Fig. 20.1 (a) From Example 20.1. Inverse multiquadrics can perform well, but can falter for small ε. (b) From Example 20.2. The use of the HS-SVD basis allows for stable and accurate collocation with the Gaussian kernel, though only for small ε. (c) From Example 20.3. Errors sampled at the collocation points. (d) From Example 20.3. A `slice` plot in MATLAB, demonstrating some dependence on y in the solution. (e) From Example 20.4. Solution to the ODE (20.13) (f) From Example 20.5. The method of fundamental solutions uses collocation points on the boundary and kernel centers outside the domain. (g) From Example 20.5. The solution to (20.17).

As we realized in Chapter 19, we can apply differential operators to the HS-SVD basis using only derivatives of the eigenfunctions $\boldsymbol{\phi}(\cdot)^T$ (see also (20.3)). This allows us to solve the collocation problem without the ill-conditioning we encountered in Example 20.1.

Example 20.2. (Solving a BVP with HS-SVD)
We reuse the problem setting from Example 20.1 and consider now the Gaussian for collocation, rather than the inverse multiquadric, because we know the Mercer series for the Gaussian. Program 20.2 executes collocation for a range of ε values using both the standard and HS-SVD basis. A plot detailing the relationship between ε and the error is provided in Figure 20.1(b).

Program 20.2. Excerpts from `PDEStable2ptBVP.m`

```
  % Reuse the setup from the unstable experiment
2 PDEStable2ptBVP
  % Define the Gaussian and its 2nd derivative
4 rbf = @(e,r) exp(-(e*r).^2);  rbfxx = @(e,r) 2*e^2*(2*(e*r).^2-1).*exp(-(e*r).^2);
  % Evaluate the solution to test for accuracy
6 ueval = uf(xeval);
  % Define a range of epsilon values and loop through them
8 epvec = logspace(-1,1,20);
  for ep=epvec
10    % Solve with the standard basis
      ustandard = rbf(ep,DMeval)*([rbfxx(ep,DMint);rbf(ep,DMbc)]\rhs);
12
      % Use the HS-SVD basis
14    GQR = gqr_solveprep(0,x,ep,1);
      IdenCorr = [eye(N);GQR.CbarT];
16    Psixxint = gqr_phi(GQR,xint,2)*IdenCorr;
      Psibc = gqr_phi(GQR,xbc)*IdenCorr;
18    A = [Psixxint;Psibc];
      GQR.coef = A\rhs;
20    uhssvd = gqr_eval(GQR,xeval);
  end
```

The standard basis solution has been compressed to simply line 11 now that we understand its structure from Program 20.1. Computing with the HS-SVD basis first requires the construction of the corrector matrix $\overline{\mathsf{C}}^T = \Lambda_2 \Phi_2^T \Phi_1^{-T} \Lambda_1^{-1}$ which we perform using `gqr_solveprep` (recall Section D.2). In line 15 we extract $\overline{\mathsf{C}}^T$ from the `GQR` data object and construct the `[eye(N);GQR.CbarT]` matrix appearing in (20.3).

We form the necessary eigenfunction matrices for the interior (with 2 derivatives) and boundary in lines 16–17 and apply the corrector matrix to form the interior and boundary $\boldsymbol{\psi}$ terms. The next two lines form and solve the system (20.4) and the solution vector $\boldsymbol{b}$ is stored as the evaluation coefficients in the `GQR` object. That object, exactly as was done in Section D.2, can now be evaluated to compute the collocation solution wherever it is requested.

The quality of this solution is compared to the standard basis in Figure 20.1(b) and the HS-SVD basis solution performs well for $\varepsilon \to 0$ but very poorly for $\varepsilon > 1$. Fortunately, the standard basis is stable enough before the HS-SVD basis falters.

Example 20.3. (Low-rank approximation for PDEs)
We next adapt Chen *et al.* (2014b, Example 4.5) and attempt to produce a low-rank approximate solution using the eigenfunction basis. This strategy mirrors that of Section 12.3 for creating a low-rank eigenfunction approximate interpolant. The BVP in question is posed on $\Omega = (-1,1)^3$ with two different boundary conditions: a Neumann condition on $\Gamma_N = \partial\Omega \cap \{x_2 = 1\}$ and a Dirichlet condition on $\Gamma_D = \partial\Omega \setminus \Gamma_N$,

$$\Delta u(\boldsymbol{x}) = 2, \qquad \text{for } \boldsymbol{x} \in \Omega, \tag{20.5a}$$

$$\frac{\partial u}{\partial x_2}(\boldsymbol{x}) = \frac{2}{3}, \qquad \text{for } \boldsymbol{x} \in \Gamma_N, \tag{20.5b}$$

$$u(\boldsymbol{x}) = \frac{x_1^2 + x_2^2 + x_3^2}{3}, \qquad \text{for } \boldsymbol{x} \in \Gamma_D. \tag{20.5c}$$

Program 20.3 demonstrates how the Gaussian eigenfunctions can be used to create an approximate solution, with error visualizations available in Figure 20.1(c)–(d).

Program 20.3. Excerpts from `PDEEigenfunctionBasis3D.m`

```
% Define the necessary functions
fint = @(x) 2*ones(size(x,1),1);  fneu = @(x) 2/3*ones(size(x,1),1);
fdir = @(x) (x(:,1).^2+x(:,2).^2+x(:,3).^2)/3;
% Define the interior points as scattered throughout the domain
Nint = 300;  point_generator = haltonset(3,'Skip',1);
xint = 2*net(point_generator,Nint) - 1;
% Define the boundary points, Neumann and Dirichlet
Nbc = 5;  x2d = pick2Dpoints(-1,1,Nbc);
xneu = [x2d(:,1),ones(Nbc^2,1),x2d(:,2)];
xdir = [x2d(:,1),-ones(Nbc^2,1),x2d(:,2);
        -ones(Nbc^2,1),x2d(:,1),x2d(:,2);ones(Nbc^2,1),x2d(:,1),x2d(:,2);
        x2d(:,1),x2d(:,2),-ones(Nbc^2,1);x2d(:,1),x2d(:,2),ones(Nbc^2,1)];
x = [xint;xneu;xdir];
% Set up the Gaussian eigenfunctions
GAUSSQR_PARAMETERS.DEFAULT_REGRESSION_FUNC = .3;  alpha = 1;  ep = .1;
GQR = gqr_solveprep(1,x,ep,alpha);
% Evaluate the Phi components of the collocation problem
Phixx = gqr_phi(GQR,xint,[2 0 0]);
Phiyy = gqr_phi(GQR,xint,[0 2 0]);
Phizz = gqr_phi(GQR,xint,[0 0 2]);
Phiy  = gqr_phi(GQR,xneu,[0 1 0]);
Phi   = gqr_phi(GQR,xdir);
% Evaluate the right hand side function
rhsint = fint(xint);  rhsneu = fneu(xneu);  rhsdir = fdir(xdir);
rhs = [rhsint;rhsneu;rhsdir];
```

```
26  % Form the collocation matrix, same order as rhs
    A = [Phixx+Phiyy+Phizz;Phiy;Phi];
28  % Solve the system and store the coefficients
    GQR.coef = A\rhs;
30  % Evaluate the error at the collocation points
    x = [xint;xneu;xdir];
32  ucoll = gqr_eval(GQR,x);    utrue = fdir(x);
    uerr = abs(ucoll - utrue);
```

Some relatively complicated point selection takes place in lines 5–12 with the goal of choosing scattered collocation points throughout the interior of the domain and uniformly spaced points on the boundaries. In line 16, we call `gqr_solveprep` to set up the necessary number of Gaussian eigenfunctions using the chosen ε and α. In this example, the `DEFAULT_REGRESSION_FUNC=.3` value says to use $M = .3N$, where N is the total number of collocation points, to form the eigenfunction approximate solution. Exactly which eigenfunctions are used can be determined by accessing `GQR.Marr`.

`gqr_phi` is then called several times to compute the components of the collocation matrix, and those components are then combined in line 27 to form the $N \times M$ collocation matrix. As was the case in Program 20.2, the solution coefficients can be stored in the `GQR` object and then `gqr_eval` can be used to evaluate the approximate solution. Although it is omitted from the text, the code for plotting these results in three dimensions is available in the full version of this program in the `GaussQR` library.

One tricky component of this problem is that the true solution is also the Dirichlet boundary condition (20.5c). For a more complicated problem this would be unavailable. Furthermore, the PDE is defined with dimensions defined as $\boldsymbol{x} = (x_1, x_2, x_3)$ but the program denotes the dimensions using the more physical notation $\boldsymbol{x} = (x, y, z)$.

20.2 Method of Lines

Many PDEs of interest are time-dependent, incorporating both spatial and time derivatives in the BVP. To solve these time-dependent problems, we choose to use the method of lines because of its simplicity in dealing with the time component. Other time-dependent methods such as Laplace transforms [Stehfest (1970a,b); Davies and Martin (1979); Moridis and Kansa (1994)] exist, but will not be presented here; see [Chen *et al.* (2012)] for more information.

The method of lines solves time-dependent problems by separating the spatial and physical derivatives, and dealing with each separately. Suppose that the PDE of interest is

$$\frac{\partial}{\partial t} u(\boldsymbol{x}, t) = \mathcal{L} u(\boldsymbol{x}, t), \tag{20.6}$$

where $\mathcal{L}$ is some linear elliptic differential operator. At any fixed time $t_\star$, the function

$$u_\star(\boldsymbol{x}) = u(\boldsymbol{x}, t_\star)$$

has only spatial arguments, meaning that standard kernel-based methods can be used to approximate it. We will develop the method of lines by thinking of it as an interpolation problem at a fixed time, and then allowing time to vary.

If we use kernels to approximate $u_\star$, we must choose some kernel centers $\{\boldsymbol{x}_1, \ldots, \boldsymbol{x}_N\}$, which allows us to uniquely define

$$\boldsymbol{k}(\boldsymbol{x})^T = \big(K(\boldsymbol{x}, \boldsymbol{x}_1) \cdots K(\boldsymbol{x}, \boldsymbol{x}_N)\big)$$

as in Chapter 13. We can evaluate our kernel solution with

$$\hat{u}_\star(\boldsymbol{x}) = \sum_{j=1}^{N} c_{\star j} K(\boldsymbol{x}, \boldsymbol{x}_j) = \boldsymbol{k}(\boldsymbol{x})^T \boldsymbol{c}_\star,$$

where $\boldsymbol{c}_\star$ must be solved in the interpolation problem. By fixing the centers we have also defined the interpolation matrix, assuming that we choose to collocate at the kernel centers, which is the standard choice we made in Section 20.1.

Because we have not yet introduced boundary conditions (i.e., this PDE is ill-posed at the moment) we can write our "interpolation" problem at any time $t_\star$ as

$$\begin{pmatrix} \mathcal{L}\boldsymbol{k}(\boldsymbol{x}_1)^T \\ \vdots \\ \mathcal{L}\boldsymbol{k}(\boldsymbol{x}_N)^T \end{pmatrix} \boldsymbol{c}_\star = \begin{pmatrix} u_\star(\boldsymbol{x}_1) \\ \vdots \\ u_\star(\boldsymbol{x}_N) \end{pmatrix}. \tag{20.7}$$

Notice that if the kernel centers are fixed with respect to time, the interpolation matrix is agnostic to the choice of $t_\star$ — only $\boldsymbol{c}_\star$ varies with the choice of time. This is the key factor which will allow us to separate the time and spatial components.

Using the independence of the kernel centers with respect to time, we can write

$$\hat{u}(\boldsymbol{x}, t) = \boldsymbol{k}(\boldsymbol{x})^T \boldsymbol{c}(t),$$

where now we expressly denote the time-dependence of the "interpolation" coefficients. Plugging this into the original PDE (20.6) gives

$$\boldsymbol{k}(\boldsymbol{x})^T \frac{\partial}{\partial t}\boldsymbol{c}(t) = \mathcal{L}\boldsymbol{k}(\boldsymbol{x})^T \boldsymbol{c}(t).$$

By choosing to collocate at the same N centers we selected to form $\boldsymbol{k}$, we produce

$$\begin{pmatrix} \boldsymbol{k}(\boldsymbol{x}_1)^T \\ \vdots \\ \boldsymbol{k}(\boldsymbol{x}_N)^T \end{pmatrix} \frac{\partial}{\partial t}\boldsymbol{c}(t) = \begin{pmatrix} \mathcal{L}\boldsymbol{k}(\boldsymbol{x}_1)^T \\ \vdots \\ \mathcal{L}\boldsymbol{k}(\boldsymbol{x}_N)^T \end{pmatrix} \boldsymbol{c}(t). \tag{20.8}$$

The PDE has now been partially discretized to create a system of N ODEs for the expansion coefficients $\boldsymbol{c}$ which can be solved with whatever method seems appropriate.

Thus far we have discussed only the PDE, but any BVP would also have initial and boundary conditions which must be incorporated into the method of lines. Suppose a full BVP takes the form

$$\begin{aligned}
\frac{\partial}{\partial t}u(\boldsymbol{x},t) &= \mathcal{L}u(\boldsymbol{x},t) + f(\boldsymbol{x}), && \boldsymbol{x}\in\Omega,\ t>0,\\
0 &= \mathcal{B}u(\boldsymbol{x},t) + g(\boldsymbol{x}), && \boldsymbol{x}\in\partial\Omega,\ t\geq 0,\\
u(\boldsymbol{x},t) &= u_0(\boldsymbol{x}), && \boldsymbol{x}\in\Omega,\ t=0.
\end{aligned}$$

The functions u_0, f and g are provided as data to the problem. As always, we will require that the kernel centers and collocation points are equal, and choose $N_{\mathcal{L}}$ and $N_{\mathcal{B}}$ centers in the interior and boundary, respectively.

Our first step to a method of lines solution will be to compute a kernel-based interpolant of the initial condition, which will require solving

$$\begin{pmatrix} \boldsymbol{k}(\boldsymbol{x}_1)^T \\ \vdots \\ \boldsymbol{k}(\boldsymbol{x}_{N_{\mathcal{L}}})^T \\ \boldsymbol{k}(\boldsymbol{x}_{N_{\mathcal{L}}+1})^T \\ \vdots \\ \boldsymbol{k}(\boldsymbol{x}_{N_{\mathcal{L}}+N_{\mathcal{B}}})^T \end{pmatrix} \boldsymbol{c}(0) = \begin{pmatrix} u_0(\boldsymbol{x}_1) \\ \vdots \\ u_0(\boldsymbol{x}_{N_{\mathcal{L}}}) \\ u_0(\boldsymbol{x}_{N_{\mathcal{L}}+1}) \\ \vdots \\ u_0(\boldsymbol{x}_{N_{\mathcal{L}}+N_{\mathcal{B}}}) \end{pmatrix}. \tag{20.9}$$

This will provide us with the N initial conditions $\boldsymbol{c}(0)$. Now, though, we hit a snag, because our BVP does not automatically generate N coupled ODEs. Instead, the presence of the boundary conditions yields N coupled *differential algebraic equations*, or DAEs.

The discretization of the boundary conditions is relatively straightforward,

$$\begin{pmatrix} 0 \\ \vdots \\ 0 \end{pmatrix} = \begin{pmatrix} \mathcal{B}\boldsymbol{k}(\boldsymbol{x}_{N_{\mathcal{L}}+1})^T \\ \vdots \\ \mathcal{B}\boldsymbol{k}(\boldsymbol{x}_{N_{\mathcal{L}}+N_{\mathcal{B}}})^T \end{pmatrix} \boldsymbol{c}(t) + \begin{pmatrix} g(\boldsymbol{x}_{N_{\mathcal{L}}+1}) \\ \vdots \\ g(\boldsymbol{x}_{N_{\mathcal{L}}+N_{\mathcal{B}}}) \end{pmatrix}, \tag{20.10}$$

although one should note that this system is $N_{\mathcal{B}} \times N$ in size. Using (20.8) as a guide, we can write the semi-discretized PDE on the interior as

$$\begin{pmatrix} \boldsymbol{k}(\boldsymbol{x}_1)^T \\ \vdots \\ \boldsymbol{k}(\boldsymbol{x}_{N_{\mathcal{L}}})^T \end{pmatrix} \frac{\partial}{\partial t}\boldsymbol{c}(t) = \begin{pmatrix} \mathcal{L}\boldsymbol{k}(\boldsymbol{x}_1)^T \\ \vdots \\ \mathcal{L}\boldsymbol{k}(\boldsymbol{x}_{N_{\mathcal{L}}})^T \end{pmatrix} \boldsymbol{c}(t) + \begin{pmatrix} f(\boldsymbol{x}_1) \\ \vdots \\ f(\boldsymbol{x}_{N_{\mathcal{L}}}) \end{pmatrix},$$

where this system is $N_{\mathcal{L}} \times N$. Therefore, our full set of DAEs is

$$\begin{pmatrix} \boldsymbol{k}(\boldsymbol{x}_1)^T \\ \vdots \\ \boldsymbol{k}(\boldsymbol{x}_{N_{\mathcal{L}}})^T \\ 0 \\ \vdots \\ 0 \end{pmatrix} \frac{\partial}{\partial t}\boldsymbol{c}(t) = \begin{pmatrix} \mathcal{L}\boldsymbol{k}(\boldsymbol{x}_1)^T \\ \vdots \\ \mathcal{L}\boldsymbol{k}(\boldsymbol{x}_{N_{\mathcal{L}}})^T \\ \mathcal{B}\boldsymbol{k}(\boldsymbol{x}_{N_{\mathcal{L}}+1})^T \\ \vdots \\ \mathcal{B}\boldsymbol{k}(\boldsymbol{x}_{N_{\mathcal{L}}+N_{\mathcal{B}}})^T \end{pmatrix} \boldsymbol{c}(t) + \begin{pmatrix} f(\boldsymbol{x}_1) \\ \vdots \\ f(\boldsymbol{x}_{N_{\mathcal{L}}}) \\ g(\boldsymbol{x}_{N_{\mathcal{L}}+1}) \\ \vdots \\ g(\boldsymbol{x}_{N_{\mathcal{L}}+N_{\mathcal{B}}}) \end{pmatrix}, \tag{20.11}$$

and (20.11) plus the initial conditions from (20.9) fully define the kernel-based method of lines solution.

These problems are somewhat more difficult to solve than time-independent problems or standard ODEs because of the missing time derivatives from some of the equations. This can produce poor scaling between the different components of the solution. The simultaneous solution of such DAEs in MATLAB is discussed in [Shampine *et al.* (1999)]. Splitting methods also exist for solving DAEs [Constantinescu and Sandu (2010)], which allow for increased speed and accuracy.

Our discussion thus far has centered around general kernel-based method of lines solutions, but, as we demonstrated in the previous section, the use of the standard basis $\boldsymbol{k}(\cdot)$ is not acceptable in the $\varepsilon \to 0$ limit. Using the same steps from Section 20.1, we use the Hilbert–Schmidt SVD to convert (20.11) to either the stable basis

$$\begin{pmatrix} \boldsymbol{\psi}(\boldsymbol{x}_1)^T \\ \vdots \\ \boldsymbol{\psi}(\boldsymbol{x}_{N_\mathcal{L}})^T \\ 0 \\ \vdots \\ 0 \end{pmatrix} \frac{\partial}{\partial t}\boldsymbol{b}(t) = \begin{pmatrix} \mathcal{L}\boldsymbol{\psi}(\boldsymbol{x}_1)^T \\ \vdots \\ \mathcal{L}\boldsymbol{\psi}(\boldsymbol{x}_{N_\mathcal{L}})^T \\ \mathcal{B}\boldsymbol{\psi}(\boldsymbol{x}_{N_\mathcal{L}+1})^T \\ \vdots \\ \mathcal{B}\boldsymbol{\psi}(\boldsymbol{x}_{N_\mathcal{L}+N_\mathcal{B}})^T \end{pmatrix} \boldsymbol{b}(t) + \begin{pmatrix} f(\boldsymbol{x}_1) \\ \vdots \\ f(\boldsymbol{x}_{N_\mathcal{L}}) \\ g(\boldsymbol{x}_{N_\mathcal{L}+1}) \\ \vdots \\ g(\boldsymbol{x}_{N_\mathcal{L}+N_\mathcal{B}}) \end{pmatrix}, \tag{20.12}$$

or the eigenfunction basis if the low-rank approach is preferred. Now we present an example of using the method of lines.

Example 20.4. (Method of lines for the 1D heat equation)
Our first example is a simple 1D heat equation

$$\begin{aligned} \frac{\partial}{\partial t}u(x,t) &= 0.06\frac{\partial^2}{\partial x^2}u(x,t) + \mathrm{e}^{\sin(3t)}, && x \in (0,1),\ t > 0, && (20.13) \\ u(x,t) &= 1 - x, && x \in \{0,1\},\ t \geq 0, \\ u(x,0) &= (1-x)^2, && x \in (0,1),\ t = 0. \end{aligned}$$

We must decide what kernels to use for this BVP. The iterated Brownian bridge kernels are a logical choice for this problem because they automatically satisfy the homogeneous boundary values. However, from an instructional perspective, they are not ideal because no boundary term in (20.11) would be present and we would only be demonstrating a special case of a differential algebraic system with no algebraic part. To emphasize the structure of (20.11), we will use the C^4 Wendland kernels.

We can, in principle, devise our own ODE solver (something like backward Euler) to discretize and solve the system of DAEs; this was done in [McCourt (2013b)] for both the standard and HS-SVD basis. However, to emphasize the functionality in MATLAB, we choose to use the builtin function `ode15s` instead. This and other ODE solvers are well-documented and versatile tools for solving ODEs and, more importantly here, DAEs of the form $\mathsf{M}\mathbf{c}' = \boldsymbol{h}(t, \mathbf{c})$, as our (20.11) is. Program 20.4 demonstrates the functionality.

Program 20.4. Excerpts from PDEMethodofLines.m

```
% Choose the C4 Wendland kernel
rbf = @(e,r) (35*(e*r).^2+18*e*r+3).*max(1-e*r,0).^6;  ep = 3;
rbfxx = @(e,r) 56*e^2*(35*(e*r).^2-4*e*r-1).*max(1-e*r,0).^4;
% Create the initial, boundary and interior functions
fic = @(x) (1-x).^2;  fbc = @(x) 1-x;
fint = @(x,t) ones(size(x,1),1)*exp(sin(3*t));
% Choose the time range for this problem, could be anything
tspan = [0,10];
% Choose some number of discretization points in the interior
% Create the boundary points and the full collocation vector
Nx = 20;  xint = pickpoints(0,1,Nx,'halt');  xbc = [0;1];
x = [xint;xbc];
% Create the necessary distance and collocation matrices
DMint = DistanceMatrix(xint,x);  DMbc = DistanceMatrix(xbc,x);
Kint = rbf(ep,DMint);  Kxxint = rbfxx(ep,DMint);  Kbc = rbf(ep,DMbc);
K = [Kint;Kbc];
% Define the ODE system function c' = odefun(t,c)
odefun = @(t,c) [.06*Kxxint;Kbc]*c + [fint(xint,t);-fbc(xbc)];
% Define the Mass matrix, zeros for the Dirichlet boundary rows
% Create the ODE options object to pass the Mass to the solver
Mass = [Kint;zeros(size(Kbc))];
odeopts = odeset('Mass',Mass,'MassSingular','yes','MStateDependence','none');
% Create the initial conditions by interpolating the initial data
cinit = K\fic(x);
% Solve the ODE
[teval,C] = ode15s(odefun,tspan,cinit,odeopts);
% Evaluate the solution
xeval = pickpoints(0,1,30);
Keval = rbf(ep,DistanceMatrix(xeval,x));
U = Keval*C';
% Plot the solution
surf(xeval,teval',U','edgecolor','none')
xlabel('x'),ylabel('t'),zlabel('u'),zlim([0,4]),view([-.3 -1 1])
```

The Wendland kernels are not defined here using sparse matrices for simplicity, but there are many zero values in the kernel matrices. The kernel matrices with derivatives are evaluated similarly to earlier examples, but line 18 introduces a new function handle `odefun` which evaluates the right-hand side of (20.11). Note that this is where the heat constant 0.06 is incorporated, and also note the negating of the boundary condition component, to match the structure of the PDE (20.13) to the structure required by `ode15s`. Although our problem here is linear, these DAEs can be nonlinear, and a nonlinear example similar to this can be found in the GaussQR library under PDEMethodOfLinesNonlinear.m.

In line 21 we define the so-called `Mass` matrix M which plays the role of the matrix on the left hand side of (20.11); in the subsequent line we provide MATLAB some additional information about this matrix. After that, we simply call `ode15s` and evaluate and plot the solution. Caution should be observed with the transposes required to evaluate the solution.

Some, perhaps most, time-dependent problems do not fit into the PDE structure defined above. The class of PDEs we are referring to here is called *hyperbolic*, and the prototypical example would be the wave equation

$$u_{tt} = c^2 u_{xx}$$

Care must be taken when trying to solve these problems with the method of lines because of the sometimes global nature of kernel-based methods. A cone of influence around a point x [LeVeque (2002)] (sometimes called a light cone) describes the subdomain of Ω which should influence the value of x. In some sense, data should only be able to move through the domain at the propagation speed c.

Globally supported kernel-based methods likely do not respect this behavior, and unphysical results may arise. Even so, kernel methods — usually enhanced with some notion of *adaptivity* — have proved useful on hyperbolic problems (see [Iske and Sonar (1996); Hon and Mao (1998); Iske (2003); Aboiyar *et al.* (2006); Driscoll and Heryudono (2007); Kansa (2007); Sarra (2008); Aboiyar *et al.* (2010); Iske (2013)]).

20.3 Method of Fundamental Solutions

The method of fundamental solutions (MFS) is a powerful technique for solving homogeneous problems (i.e., with $f(\boldsymbol{x}) = 0$) with a linear operator $\mathcal{L}$ whose fundamental solution $G(\boldsymbol{x}, \boldsymbol{z})$ is known. Its development is detailed in [Fairweather and Karageorghis (1998); Golberg and Chen (1998)], with application to other elliptic problems in [Wei *et al.* (2007)], and fundamental solutions for some common differential operators were introduced in Section 6.5.

Essentially, MFS converts a boundary value problem to an interpolation problem. In certain ways, this idea is similar to the approach used in developing the method of lines in Section 20.2. We assume that the problem of interest fits the form

$$\mathcal{L}u(\boldsymbol{x}) = 0, \qquad \boldsymbol{x} \in \Omega, \tag{20.14a}$$

$$\mathcal{B}u(\boldsymbol{x}) = g(\boldsymbol{x}), \qquad \boldsymbol{x} \in \partial\Omega. \tag{20.14b}$$

The *fundamental solution* is a Green's kernel which satisfies

$$\mathcal{L}G(\boldsymbol{x}, \boldsymbol{z}) = \delta(\boldsymbol{x}, \boldsymbol{z}),$$

where $\delta(\boldsymbol{x}, \boldsymbol{z})$ is the Dirac delta function. We know that $\mathcal{L}G(\boldsymbol{x}, \boldsymbol{z}) = 0$ for $\boldsymbol{x} \in \Omega$ if $\boldsymbol{z} \notin \Omega$, because $\delta(\boldsymbol{x}, \boldsymbol{z}) = 0$ for $x \neq z$. The assumption is therefore made that the approximate solution $\hat{u}$ is of the form

$$\begin{aligned}\hat{u}(\boldsymbol{x}) &= \sum_{j=1}^{N} a_j G(\boldsymbol{x}, \boldsymbol{z}_j), \\ &= \boldsymbol{\gamma}(\boldsymbol{x})^T \boldsymbol{a}, \end{aligned} \tag{20.15}$$

where the N kernel centers $\boldsymbol{z}_j$, $j = 1, \dots, N$, are placed *outside* $\Omega \cup \partial\Omega$. The vector

$$\boldsymbol{\gamma}(\boldsymbol{x})^T = \big(G(\boldsymbol{x}, \boldsymbol{z}_1) \ \cdots \ G(\boldsymbol{x}, \boldsymbol{z}_N)\big) \tag{20.16}$$

is defined analogously to $\boldsymbol{k}$ as in (13.7).

Remark 20.2. The use of the letter $\boldsymbol{a}$ in defining the approximate solution $\hat{u}(\boldsymbol{x}) = \boldsymbol{\gamma}(\boldsymbol{x})^T \boldsymbol{a}$ serves only to help distinguish the solution from a positive definite interpolant of the form $s(\boldsymbol{x}) = \boldsymbol{k}(\boldsymbol{x})^T \boldsymbol{c}$ or $s(\boldsymbol{x}) = \boldsymbol{v}(\boldsymbol{x})^T \boldsymbol{b}$ depending on the choice of basis. All of these functions are constructed as a linear combination of kernels, and the different choice of $\boldsymbol{a}$ for the coefficients here is meant only to distinguish the setting during this introduction to the topic. No structural difference is intended.

Automatically, the condition (20.14a) is satisfied, meaning the coefficients $\boldsymbol{a}$ must be determined by enforcing (20.14b). This is often accomplished by choosing N collocation points $\boldsymbol{x}_i$ on the boundary, and then solving the linear system

$$\begin{pmatrix} \mathcal{B}\boldsymbol{\gamma}(\boldsymbol{x}_1)^T \\ \vdots \\ \mathcal{B}\boldsymbol{\gamma}(\boldsymbol{x}_N)^T \end{pmatrix} \begin{pmatrix} a_1 \\ \vdots \\ a_N \end{pmatrix} = \begin{pmatrix} g(\boldsymbol{x}_1) \\ \vdots \\ g(\boldsymbol{x}_N) \end{pmatrix}.$$

Remark 20.3. It should be noted that we are not required to choose exactly N *source points* (i.e., centers for the basis functions/kernels/fundamental solution). Often it is preferable to choose many fewer source terms than collocation points and solve an overdetermined system. Barnett and Betcke (2008) studied — for a Helmholtz problem on an analytic domain — the effects of the choice of locations of source points on the numerical stability of the problem. This choice depends on the shape of the domain and on the boundary data. Furthermore, the actual choice of source locations is sometimes also considered a variable in the problem (see, e.g., [Alves (2009)]). For simplicity, we will only study problems with a fixed set of N sources.

In the simplest case, when $\mathcal{B} = \mathcal{I}$, i.e., Dirichlet boundary conditions, this is a kernel-based interpolation problem, using the basis $G(\cdot, \boldsymbol{z}_j)$, $j = 1, \dots, N$. More complicated boundary conditions are handled just as easily, and greater accuracy is expected than with a collocation method because of the absence of $\mathcal{L}$. Since $\mathcal{L}$ is a differential operator of higher order than $\mathcal{B}$, more accuracy is lost when approximating it [Wendland (2005)], making any solution involving both operators lower order than a solution involving only $\mathcal{B}$. This was discussed in Chapter 19.

Example 20.5. (MFS on an L-shaped domain)
One of the great benefits of the MFS is its ability to handle more complicated boundaries and boundary conditions with ease. This example takes place on an L-shaped domain, which is the square $[0,2]^2$ with the bottom left quadrant removed. Our domain is written in three pieces

$$\begin{aligned}\Omega &= (x,y) \in (0,2)^2 \text{ such that } x > 1 \text{ or } y > 1,\\ \Gamma_N &= x \in [0,2],\ y = 2,\\ \Gamma_D &= \overline{\Omega} \setminus (\Omega \cup \Gamma_N),\end{aligned}$$

where Ω is the interior, Γ_N is the top most part of the boundary, and Γ_D is the rest of the boundary. The BVP is

$$\nabla^2 u(x,y) + \nu^2 u(x,y) = 0, \qquad (x,y) \in \Omega, \tag{20.17a}$$

$$\frac{\partial}{\partial y} u(x,y) = -\pi \sin(\pi(x+y)), \qquad (x,y) \in \Gamma_N, \tag{20.17b}$$

$$u(x,y) = \cos(\pi(x+y)), \qquad (x,y) \in \Gamma_D, \tag{20.17c}$$

where $\nu = \sqrt{2}\pi$. This *Helmholtz operator* has the fundamental solution

$$G(\boldsymbol{x},\boldsymbol{z}) = \frac{\mathrm{i}}{4} H_0^{(2)}(\nu \|\boldsymbol{x} - \boldsymbol{z}\|_2),$$

where $\mathrm{i} = \sqrt{-1}$ and $H_0^{(2)}$ is the *Hankel function* of order 2 [Abramowitz and Stegun (1965, pg. 358, (9.1.4))]. Program 20.5 provides an implementation of the MFS solution, with associated plots in Figure 20.1(f)–(g).

Program 20.5. Excerpts from `MFSDemo.m`

```
nu = sqrt(2)*pi;
% Define the kernel, here the fundamental solution of the elliptic operator
Gf = @(r) 1i/4*besselh(0,2,nu*r);
Gyf = @(r,dy) -1i*nu/4*besselh(1,2,nu*r).*dy./r;
% Set up the functions for the boundary conditions
fdir = @(x) cos(pi*(x(:,1)+x(:,2)));  fneu = @(x) -pi*sin(pi*(x(:,1)+x(:,2)));
% Choose collocation points: a uniform grid with the boundary extracted
N = 15;  xbox = pick2Dpoints(0,2,N);
xsquare = xbox(any(xbox==0 | xbox==2,2),:);
% Create L shape from square shape
xtoflip = all(xsquare<=1,2);
xL = xsquare;  xL(xtoflip,:) = bsxfun(@minus,[1 1],xsquare(xtoflip,:));
% Find the Neumann and Dirichlet boundaries
xneu = xsquare(xL(:,2)==2,:);
xdir = setdiff(xL,xneu,'rows');
% Choose Green's kernel source points (the centers)
theta = pickpoints(0,2*pi,4*N);
z = bsxfun(@plus,[1 1],2*[cos(theta),sin(theta)]);
% Define the distance matrices
DMdir = DistanceMatrix(xdir,z);  DMneu = DistanceMatrix(xneu,z);
```

```
    DiffMyneu = DifferenceMatrix(xneu(:,2),z(:,2));
22  % Define the kernel matrices
    Gdir = Gf(DMdir);  Gneu = Gyf(DMneu,DiffMyneu);
24  A = [Gdir;Gneu];
    % Define the right hand side with the boundary conditions
26  rhsdir = fdir(xdir);  rhsneu = fneu(xneu);  rhs = [rhsdir;rhsneu];
    % Solve the system
28  coef = A\rhs;
    % Evaluate the solution, and NaN points outside the domain
30  Neval = 21;  xeval = pick2Dpoints(0,2,Neval);
    Geval = Gf(DistanceMatrix(xeval,z));
32  ueval = real(Geval*coef);
    ueval(all(xeval<1,2)) = NaN;
34  % Plot the results on a surface plot
    X = reshape(xeval(:,1),Neval,Neval);Y = reshape(xeval(:,2),Neval,Neval);
36  U = reshape(ueval,Neval,Neval);surf(X,Y,U),view([-.3 -2.1 7])
```

A few key points to note about the above program:

- The function $\frac{\partial}{\partial y}G(\boldsymbol{x},\boldsymbol{z})$ appears in line 4, which is needed to compute the derivatives for the Neumann boundary condition. Computing this term requires the `DifferenceMatrix` function, which is evaluated in line 21.
- Constructing the collocation points is nontrivial and requires lines 8–15. MATLAB makes the project easier by providing tools such as `all` and `any` to identify values. Practice understanding this sequence of commands will provide insight into how matrices can be efficiently manipulated in MATLAB.
- As promised would occur way back in Remark 4.3, this example uses different collocation points than kernel centers. This is required because the Hankel function has a nonremovable singularity at $\|x - z\| = 0$.
- The different boundary conditions are handled easily by creating distance matrices of the points associated with each boundary condition. The Neumann and Dirichlet components must match on both sides of the system of equations, which is ensured by the ordering of the components in lines 24 and 26.
- In line 32, we evaluate the interpolant, but take only the real part of the MFS solution. In exact arithmetic, this would be unnecessary, because the imaginary terms would cancel out; on a computer, there will always be some residual imaginary terms because of machine precision. Executing the line `norm(imag(A_eval*coef))` produces 3×10^{-10}, meaning that the leftover imaginary terms are insignificant.
- In line 33, we introduce `NaN`s into our solution. Usually, we avoid these "Not-a-number" values during computation, but for plotting they signal MATLAB to not plot in that region. This is useful here when plotting a surface that must be evaluated on a rectangular grid but only exists on an L-shaped region.

20.4 Method of Particular Solutions

The method of fundamental solutions is more efficient than most collocation schemes because it has the advantage of considering a solution only on the boundary and thus handling a lower-order differential operator ($\mathcal{B}$ rather than $\mathcal{L}$). Unfortunately, the method of fundamental solutions is only applicable to homogeneous problems, i.e., problems of the form $\mathcal{L}u = 0$. To counteract this shortcoming, the method of particular solutions (MPS) was developed to allow for an inhomogeneous differential equation [Miele and Iyer (1970); Chen *et al.* (2012, 2014b)].

In the MPS setting, the BVP will take the same general form as in Section 20.1

$$\mathcal{L}u(\boldsymbol{x}) = f(\boldsymbol{x}), \qquad \boldsymbol{x} \in \Omega,$$

$$\mathcal{B}u(\boldsymbol{x}) = g(\boldsymbol{x}), \qquad \boldsymbol{x} \in \partial\Omega.$$

We assume, as we did in Section 20.3 that the operator $\mathcal{L}$ has the Green's kernel $G(\boldsymbol{x}, \boldsymbol{z})$. For the MFS setting, $f \equiv 0$, meaning that the solution could be built with the basis $G(\cdot, \boldsymbol{z}_j)$, $j = 1, \ldots, N_F$, but now that $f \neq 0$, we assume the solution has two parts,

$$u(\boldsymbol{x}) = u_F(\boldsymbol{x}) + u_P(\boldsymbol{x}).$$

The two components solve different problems:

- $u_P(\boldsymbol{x})$ solves the ill-posed BVP $\mathcal{L}u_P(\boldsymbol{x}) = f(\boldsymbol{x})$. If collocation with the basis $K(\cdot, \boldsymbol{\xi}_j)$, $j = 1, \ldots, N_P$, is used to solve this problem, it can be thought of as an approximation problem on the interior, using the basis $\mathcal{L}K(\cdot, \boldsymbol{\xi}_j)$, $j = 1, \ldots, N_P$.
- $u_F(\boldsymbol{x})$ requires the particular solution, and solves the BVP
$$\mathcal{L}u_F(\boldsymbol{x}) = 0, \qquad \boldsymbol{x} \in \Omega,$$
$$\mathcal{B}u_F(\boldsymbol{x}) = g(\boldsymbol{x}) - \mathcal{B}u_P(\boldsymbol{x}), \qquad \boldsymbol{x} \in \partial\Omega,$$
using MFS. This too is an approximation problem, only on the boundary, using the basis $G(\cdot, \boldsymbol{z}_j)$, $j = 1, \ldots, N_F$.

Because of the generally exceptional performance of the method of fundamental solutions, the main source of error for MPS is the approximation of the particular solution. For most f, we will not have access to the true particular solution, and when f is a black box, it would be impossible to do any better than an approximate particular solution. We will use our Hermite interpolation strategies from Section 19.2 to approximate particular solutions which can then be used to find the full solution.

Remark 20.4. The method of particular solutions has, historically, been associated with the use of an approximate particular solution constructed, as we demonstrate in this section, through a linear combination of basis functions (dating as far back as [Fox *et al.* (1967)] and more recently [Betcke and Trefethen (2005)]). Given that, nothing in the derivation above prevents u_P from being a single function, or even the exact particular solution. Such a situation is unlikely in many applications, but we leave that possibility available in this text should the opportunity ever arise.

We assume that our approximate particular solution is of the form $\hat{u}_P(\boldsymbol{\xi}) = \boldsymbol{k}(\boldsymbol{\xi})^T\boldsymbol{c}$, using N_P kernels centered at $\Xi = \{\boldsymbol{\xi}_1, \dots, \boldsymbol{\xi}_{N_P}\}$. This approximate particular solution is defined by the linear system

$$\begin{pmatrix} \mathcal{L}\boldsymbol{k}(\boldsymbol{\xi}_1)^T \\ \vdots \\ \mathcal{L}\boldsymbol{k}(\boldsymbol{\xi}_{N_P})^T \end{pmatrix} \boldsymbol{c} = \begin{pmatrix} f(\boldsymbol{\xi}_1) \\ \vdots \\ f(\boldsymbol{\xi}_{N_P}) \end{pmatrix}. \tag{20.18}$$

Although we could conceivably collocate (or perform nonsymmetric Hermite interpolation as discussed in Section 19.2.1, depending on how it is viewed) at any locations in Ω, we choose to collocate at the kernel centers, which produces a square matrix. If this matrix can be inverted, (20.18) uniquely defines the vector $\boldsymbol{c}$, with which we can solve for the fundamental solution.

We maintain our definition of the approximate homogeneous solution as $\hat{u}_F(\boldsymbol{x}) = \boldsymbol{\gamma}(\boldsymbol{x})^T\boldsymbol{a}$, a linear combination of N_F Green's kernels centered at locations $\mathcal{Z} = \{\boldsymbol{z}_1, \dots, \boldsymbol{z}_{N_F}\}$ which are likely outside of the domain Ω to account for a potentially singular G. Collocation points $\mathcal{X} = \{\boldsymbol{x}_1, \dots, \boldsymbol{x}_{N_F}\}$ at which the homogeneous solution will be fit are chosen on the boundary. Collectively, these choices discretize the boundary conditions of the now homogeneous BVP

$$\mathcal{B}\hat{u}_F(\boldsymbol{x}) = g(\boldsymbol{x}) - \mathcal{B}\hat{u}_P(\boldsymbol{x}),$$

and produce the linear system

$$\begin{pmatrix} \mathcal{B}\boldsymbol{\gamma}(\boldsymbol{x}_1)^T \\ \vdots \\ \mathcal{B}\boldsymbol{\gamma}(\boldsymbol{x}_{N_F})^T \end{pmatrix} \boldsymbol{a} = \begin{pmatrix} g(\boldsymbol{x}_1) \\ \vdots \\ g(\boldsymbol{x}_{N_F}) \end{pmatrix} - \begin{pmatrix} \mathcal{B}\boldsymbol{\phi}(\boldsymbol{x}_1)^T \\ \vdots \\ \mathcal{B}\boldsymbol{\phi}(\boldsymbol{x}_{N_F})^T \end{pmatrix} \boldsymbol{b}, \tag{20.19}$$

Solving for $\boldsymbol{a}$ will allow us to compute $\hat{u}_F$, and in turn the full approximate solution

$$\hat{u}(\boldsymbol{x}) = \hat{u}_F(\boldsymbol{x}) + \hat{u}_P(\boldsymbol{x}) = \boldsymbol{k}(\boldsymbol{x})^T\boldsymbol{c} + \boldsymbol{\gamma}(\boldsymbol{x})^T\boldsymbol{a}. \tag{20.20}$$

Again, following the discussion in Remark 20.3, we choose to solve square systems in (20.18) and (20.19) but this choice is not required in general. Furthermore, although the particular solution is defined using a linear system of kernels in the standard basis, any basis transformation or low-rank approximate kernel method introduced in Chapter 12 is equally applicable, and perhaps preferable in the presence of an ill-conditioned standard basis [McCourt (2013b)].

Example 20.6. (Method of particular solutions for a modified Helmholtz problem) To demonstrate the viability of this method, we will apply it to the *modified Helmholtz problem* on the domain Ω bounded by the parametric curve

$$r(\theta) = 0.8 + 0.1(\sin(6\theta) + \sin(3\theta)). \tag{20.21}$$

This domain was used in [Larsson *et al.* (2013b)] for solving Poisson equations with kernel-based finite differences. Our boundary value problem is

$$\nabla^2 u(\boldsymbol{x}) - \nu^2 u(\boldsymbol{x}) = f(\boldsymbol{x}), \qquad \boldsymbol{x} \in \Omega, \tag{20.22a}$$

$$u(\boldsymbol{x}) = g(\boldsymbol{x}), \qquad \boldsymbol{x} \in \partial\Omega, \tag{20.22b}$$

where we will use $\nu = 3$. The functions f and g are manufactured to recover the true solution $u(x,y) = e^{2x+2y}$. The fundamental solution for the operator $\mathcal{L} = \nabla^2 - \nu^2\mathcal{I}$ in $\mathbb{R}^2$ is (see Example 6.4)

$$G(\boldsymbol{x}, \boldsymbol{z}) = \frac{1}{2\pi} K_0(\nu\|\boldsymbol{x} - \boldsymbol{z}\|),$$

where K_0 is the modified Bessel function of the second kind of order zero [Abramowitz and Stegun (1965)]. Program 20.6 provides a MATLAB implementation of an MPS solution using C^4 Wendland kernels to form an approximate particular solution; associated plots are available in Figure 20.2(a)–(b).

Program 20.6. Excerpts from `PDEMPSDemo.m`

```
nu = 3;
% Define the MFS kernel, here the modified Bessel function
Gf = @(r) 1/(2*pi)*besselk(0,nu*r);
% We choose the C4 Wendland kernel to form the MPS solution
rbf = @(e,r) (35*(e*r).^2+18*e*r+3).*max(1-e*r,0).^6;  ep = .1;
rbfL = @(e,r) 56*e^2*(40*(e*r).^2-8*e*r-2).*max(1-e*r,0).^4;
% Set up the functions for the interior and boundary condition
fbc = @(x) exp(2*x(:,1)+2*x(:,2));    fint = @(x) -exp(2*x(:,1)+2*x(:,2));
% Define the number of points to use in each region
Nbc = 50;  Nint = 200;
% Choose collocation points on the boundary and centers out of the domain
theta = pickpoints(0,2*pi,Nbc);
bcdef = @(t) .8 + .1*(sin(6*t)+sin(3*t));
xbc = [bcdef(theta).*cos(theta),bcdef(theta).*sin(theta)];
zbc = 2*[cos(theta),sin(theta)];
% Choose collocation points/centers in interior
% Create a lot of Halton points and keep the first Nint
xtest = pick2Dpoints(-1,1,sqrt(3*Nint),'halt');
xallin = xtest(inpolygon(xtest(:,1),xtest(:,2),xbc(:,1),xbc(:,2)),:);
xint = xallin(1:Nint,:);
% Define the distance matrices
DMint = DistanceMatrix(xint,xint);  DMintformfs = DistanceMatrix(xbc,xint);
DMbc = DistanceMatrix(xbc,zbc);
% Define the MPS matrix and solve the MPS problem
K = rbf(ep,DMint);   KL = rbfL(ep,DMint);
A = KL - nu^2*K;   mpscoef = A\fint(xint);
% Define the MFS system and solve the MFS problem
G = Gf(DMbc);   Kformfs = rbf(ep,DMintformfs);
mfsrhs = fbc(xbc) - Kformfs*mpscoef;   mfscoef = G\mfsrhs;
% Define a function to evaluate the combined solution
usol = @(x)      Gf(DistanceMatrix(x,zbc))*mfscoef + ...
            rbf(ep,DistanceMatrix(x,xint))*mpscoef;
```

This program looks like a nearly direct combination of the nonsymmetric Hermite interpolation discussed in Section 19.2.1 and computed in line 26 and the MFS discussed in Section 20.3 and computed in line 29. Indeed the same worries accompany the computations, primarily fears about the invertibility of the matrix $\mathsf{K}_{\mathcal{D}^2} - \nu^2\mathsf{K}$ and choice of the location of the MFS centers. Even so, as seen in Figure 20.2(b), the solution is of good quality despite the peculiar shape of the boundary.

The boundary actually presents the newest opportunity to advertise one of MATLAB's builtin functions. Line 19 uses the function `inpolygon` to determine which of the possible Halton points, stored in `xtest` are actually within the curve (20.21). We create many more points than are needed, reject those that are unacceptable, and choose from among those remaining to form our MPS collocation (nonsymmetric Hermite interpolation) points.

20.5 Kernel-based Finite Differences

Standard (polynomial-based) finite difference (FD) methods originated in the first part of the 20th century, and the seminal work of Richardson (1911) and Courant, Friedrichs and Lewy (1928) are the standard references. These methods were designed to reproduce low-degree polynomials. Just as it is possible to devise kernel-based collocation methods which generalize polynomial-based pseudospectral methods (see Section 20.1), it is also possible to generalize traditional finite difference methods to the kernel setting, where one instead goes for reproduction of kernel functions. This idea was first proposed by Andrei Tolstykh at the beginning of the 21st century (see, e.g., Tolstykh and Shirobokov (2003)), and other researchers (see [Shu *et al.* (2003); Wright and Fornberg (2006)]) proposed similar, independently developed, approaches around the same time.

As in the global — pseudospectral — setting, the advantage that comes with using kernel methods[1] is the ability to work with unstructured grids of points. Therefore, one gains much geometric and topological flexibility, as evidenced by some of the recent papers such as [Flyer and Wright (2007, 2009); Shankar *et al.* (2015)].

An important advantage of kernel-based finite differences over the global collocation method, as well as MFS and MPS, is the fact that we now will be able to work with much larger systems. The system matrices will be *sparse*, and therefore we will gain in computational efficiency. On the other hand, the local nature of the approximation scheme will affect overall convergence rates and accuracy.

Example 20.7. (1D transport equation via kernel-based FD)
This example mimics the `p6.m` example from Trefethen (2000, pg. 26). A first-order wave equation

$$u_t = -c(x)u_x, \qquad u(0,t) = 0, \qquad u(x,0) = \mathrm{e}^{-100(x-1)^2} \tag{20.23}$$

[1]In fact, the entire literature up to now seems to have been using radial kernels.

is considered with wave coefficient $c(x) = 0.2 + \sin(x-1)^2$. $N = 200$ evenly spaced points between $[0, 2\pi]$ are used to discretize the domain, and at each point a finite difference stencil with the nearest $N_x = 13$ points is created based on the use of the C^4 Matérn kernel. The MATLAB ODE solver `ode15s` is used to conduct the time-stepping with a maximum time step of $\Delta t_{\max} = 0.0125$ until the final time $t = 8$ is reached. Note that $\Delta t_{\max}$, along with potentially other ODE solver relevant options, are set with the MATLAB function `odeset`. The solution is plotted in Figure 20.1(c) as viewed from above the x-t plane.

Program 20.7. Excerpts from `PDEFiniteDiff1D.m`

```
% Discretize the domain and choose a stencil size
N = 200;   x = pickpoints(0,2*pi,N);   Nx = 13;
% Choose our time span and max allowed time step
tspan = [0 8];  odeopts = odeset('MaxStep',.0125);
% Define the wave speed coefficients and initial condition
c = .2 + sin(x-1).^2;   u0 = exp(-100*(x-1).^2);
% Choose the C4 Matern kernel for derivative approximation
rbf = @(e,r) (1+(e*r)+(e*r).^2/3).*exp(-e*r);    ep = 1;
rbfx = @(e,r,dx) -e^2/3*dx.*(1+e*r).*exp(-e*r);
%%%%%%
% Create the finite difference operator
nearest = num2cell(knnsearch(x,x,'K',Nx),2);
FDcell = cellfun(@(xe,xi) ...
    rbfx(ep,DistanceMatrix(xe,x(xi,:)),DifferenceMatrix(xe,x(xi,:)))/...
    rbf(ep,DistanceMatrix(x(xi,:),x(xi,:))),...
    num2cell(x,2),nearest,'UniformOutput',0);
FDvecs = cell2mat(cellfun(@(row,cols,vals)[row*ones(1,Nx);cols;vals],...
    num2cell((1:N)',2),nearest,FDcell,'UniformOutput',0)');
LFD = sparse(FDvecs(1,:),FDvecs(2,:),FDvecs(3,:),N,N);
%%%%%%
% Change the first row to a boundary condition: u_t(0,t) = 0
LFD(1,:) = zeros(1,N);
% Define a function that acts like u_t = f(t,u) = -c(x)*u_x
odefun = @(t,u) -bsxfun(@times,c,LFD*u);
% Call the ODE solver
[t,U] = ode15s(odefun,tspan,u0,odeopts);
% Plot the results in a pleasant fashion
surf(x,t,U,'edgecolor','none'),  view([0 0 1]),  grid off
xlim([0,2*pi]),  xlabel('x'),  ylabel('t')
colormap gray,  C = colormap;  colormap(flipud(C))
h_color = colorbar('limits',[-5.5e-04,1],'ticks',[0 .5 1]);
```

Line 4 sets the maximum time step, although `ode15s` has the option to use a smaller time step as needed. This particular problem can result in unstable oscillations if a larger time step is used with a stencil size of $N_x = 13$ which produces a rather high-order finite difference approximation. The quality of the solution is quite good, with a wave height of 0.9974 (it should be 1) at time $t = 8$ and negative values no worse than -5.4×10^{-4}. Comments have been removed from the finite difference operator construction since it so closely mirrors that of Program 19.3.

20.6 Space-Time Collocation

In Section 3.7 we introduced the concept of space-time kernels, where the time dimension is included as an additional spatial dimension. Such kernels can be used to solve time-dependent PDEs through collocation in *both* the space and time dimensions (see, e.g., [Li and Mao (2011)]), as opposed to the much more common approach of just discretizing the space dimensions with kernels (see Section 20.1). This allows for more flexibility in the development of the approximate solution by not requiring a uniform march forward in time as a standard time-stepping approach would.

From a basic implementation standpoint, the space-time approach described in this section is not much different from the collocation approach for elliptic (time-independent) problems. Instead of solving a problem in d space dimensions we now solve a problem in $d+1$ space-time dimensions. Of course, one cannot in general expect that the spatial features of the solution are on the same order as the time scale, and therefore one should almost certainly be working with *anisotropic* kernels, which can be viewed either as anisotropic radial kernels (cf. Section 3.1.2 and Example 20.8) or anisotropic tensor product kernels (cf. Section 3.4.3 and Example 20.9). From a theoretical point of view, we are not aware of any publications that address this interesting approach to solving time-dependent PDEs.

Example 20.8. (Solving a heat equation with space-time kernels)
Our first space-time kernel example will compute the solution to the linear heat equation:

$$u_t - \frac{1}{4}u_{xx} = 0, \qquad u(0,t) = u(1,t) = 0, \qquad u(x,0) = x(1-x). \tag{20.24}$$

We use an anisotropic C^2 Matérn kernel which has exactly enough smoothness from which to construct a solution to this problem. Differentiation of this kernel is a bit tricky due to the use of a shape parameter *vector*. The basic isotropic version is discussed in Fasshauer (2007, Appendix D), but it needs to be appropriately adjusted for the anisotropic setting used here (see lines 12–14 of Program 20.8). Our implementation is presented in Program 20.8 and the associated plots are provided in Figure 20.2(d)–(e).

Program 20.8. Excerpts from `PDESpaceTime.m`

```
  % Define the initial and boundary condition
2 fic  = @(x) x(:,1).*(1-x(:,1));
  fint = @(x) zeros(size(x,1),1);  fbc  = @(x) zeros(size(x,1),1);
4 % Collocation points: scattered on interior, uniform on boundary
```

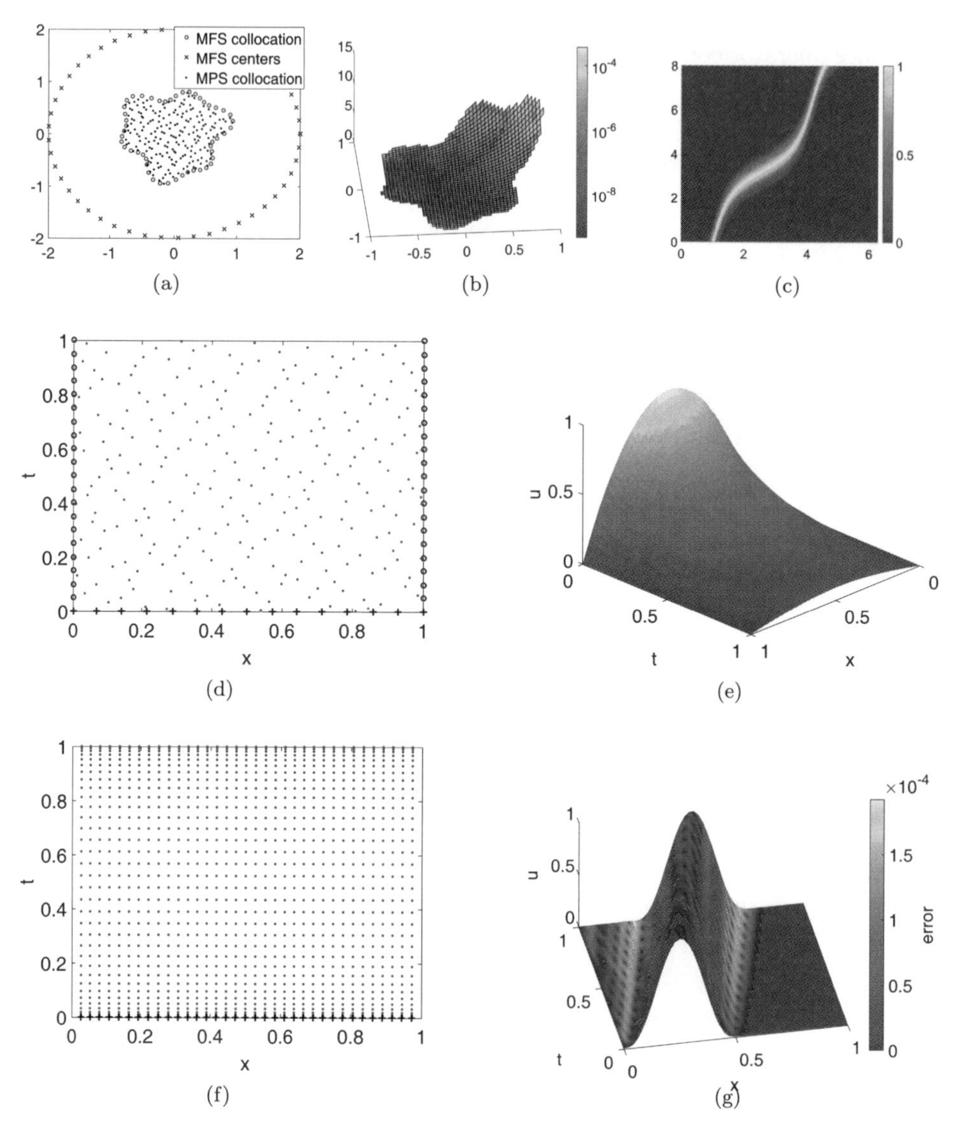

Fig. 20.2 (a) From Example 20.6. The distribution of 50 boundary points and 200 interior points. (b) From Example 20.6. The error in color as projected onto the solution surface. (c) From Example 20.7. The wave propagates at variable speeds depending on x: the yellow region represents the location of the wave as it moves in time from the bottom to the top of the plot. (d) From Example 20.8. The space-time collocation point distribution with a total of 302 points — red dot on the interior, black + for the initial condition, blue circle for the boundary condition. (e) From Example 20.8. The solution behaves as expected, with deviation from the finite difference solution of only 10^{-3}. (f) From Example 20.9. The tensor product kernel (20.26) prefers a Chebyshev/uniform grid — red dot on the interior, black + for the initial condition. Note the absence of the boundary points because the IBB kernel has builtin boundary conditions. (g) From Example 20.9. The space-time kernel collocation strategy produces a good quality solution, though error does accumulate toward the end of the simulation.

```
Nx = 15;  Nt = 20;  tmax = 1;
6   xint = pick2Dpoints([0 0],[1 tmax],[Nx-2,Nt-1],'halt');
    xbc = [kron([0;1],ones(Nt,1)),repmat(pickpoints(tmax/Nt,tmax,Nt),2,1)];
8   xic = [pickpoints(0,1,Nx),zeros(Nx,1)];
    x = [xbc;xic;xint];
10  % Choose a kernel for our problem
    % epvec(1) is the x shape parameter, epvec(2) is the t shape parameter
12  rbfM2 = @(r) (1+r).*exp(-r);    epvec = [.5,1];
    rbfM2dt = @(r,dt,ep_t) -ep_t*exp(-r).*(ep_t*dt);
14  rbfM2dxx = @(r,dx,ep_x) ep_x^2*exp(-r).*((ep_x*dx).^2./(r+eps)-1);
    % Evaluate the collocation matrix
16  Abc  = rbfM2(real(DistanceMatrix(xbc,x,epvec)));
    Aic  = rbfM2(DistanceMatrix(xic,x,epvec));
18  Aint = rbfM2dt(DistanceMatrix(xint,x,epvec), ...
                DifferenceMatrix(xint(:,2),x(:,2)),epvec(2)) - ...
20        .25*rbfM2dxx(DistanceMatrix(xint,x,epvec),...
                    DifferenceMatrix(xint(:,1),x(:,1)),epvec(1));
22  A = [Abc;Aic;Aint];
    % Evaluate the RHS with initial/boundary condition and interior components
24  rhsbc  = fbc(xbc);  rhsic  = fic(xic);  rhsint = fint(xint);
    rhs = [rhsbc;rhsic;rhsint];
26  % Solve the system for the solution coefficients
    coef = A\rhs;
28  % Choose some evaluation points and evaluate the solution at them
    xeval = pick2Dpoints([0 0],[1 tmax],50);
30  ueval = rbfM2(DistanceMatrix(xeval,x,epvec))*coef;
```

Our points are chosen scattered throughout the domain to emphasize the flexibility of space-time kernel collocation, although points are still structured on $x = 0$, $x = 1$ and $t = 0$ to ensure that the boundary and initial conditions are enforced. Figure 20.2(d) shows the structure of the points to emphasize the meshfree nature of space-time kernel collocation. We note that the collocation matrix and right-hand sides must be organized so that their rows correspond; this is why lines 22 and 25 look structurally very similar.

The solution in Figure 20.2(e) looks as expected, with decay over time and minimal incursions below zero (on the order of 10^{-3}). For comparison's sake, the `GaussQR` library version of Program 20.8 also solves the same PDE with a second-order finite difference scheme on 2001 points and a second-order BDF method with time step 6×10^{-4}. At time $t = 1$ that solution and the space-time kernel have a pointwise difference of less than 10^{-3} despite the fact that the kernel solution required inverting only a 302×302 sized, albeit dense, matrix.

Remark 20.5. Back in Remark 4.1 we mentioned that our fastest computation of distance matrices can produce rounding errors on the order of $\sqrt{\epsilon_{\text{mach}}}$. Example 20.8 is an example of one such situation where the numerical cancelation actually causes the computed distance matrix to be *complex*, which is a much more significant issue than simply being incorrect by roughly 10^{-8}. This spurious venture into complex arithmetic is quickly terminated with the `real` command in line 16, but it serves as an all too common reminder that fundamental numerical analysis is a vital area of study before any applications can be considered.

Remark 20.6. When solving a PDE with space-time collocation, properties such as positivity or conserved quantities are generally neglected. The development of such methods is still an open problem, with some progress involving other kernel-based PDE techniques (see, e.g., [Cecil *et al.* (2004); Barba *et al.* (2005); Wendland (2006); Huang *et al.* (2006)]).

Example 20.9. (Solving a transport equation with space-time kernels)
We consider a simple transport equation

$$u_t + \frac{1}{4}u_x = 0, \qquad u(x,0) = \begin{cases} 512(1-2x)^3x^3, & 0 \le x < \frac{1}{2}, \\ 0, & x \ge \frac{1}{2}, \end{cases} \tag{20.25}$$

and attempt to simulate this up to time $t = 1$ with collocation using a space-time kernel of the form

$$K((x,s),(z,t)) = K_1(x,z)K_2(s,t). \tag{20.26}$$

Because this particular problem happens to satisfy $u(0,t) = u(1,t) = 0$ for $0 < t < 1$ we choose the C^2 iterated Brownian bridge kernel (7.12) (using $\beta = 2$) as K_1. We use the C^∞ Chebyshev kernel (3.13) with $a = 1$ as K_2. Program 20.9 presents a MATLAB implementation of space-time collocation using this tensor product kernel, with an associated error plot in Figure 20.2(g).

Program 20.9. Excerpts from `PDESpaceTimeTransport.m`

```
   % Define the initial condition
2  fic  = @(x) (512*(1-2*x(:,1)).^3.*(x(:,1)).^3).*(x(:,1)<=0.5);
   % Choose kernel centers: uniform in space and Chebyshev in time
4  Nx = 35;  Nt = 36;  tmax = 1;
   x1Du = pickpoints(0,1,Nx+2); x1Du = x1Du(2:end-1);
6  x1Dc = pickpoints(0,tmax,Nt,'cheb');
   [X,T] = meshgrid(x1Du,x1Dc);  xall = [X(:),T(:)];
8  % Choose a kernel for our problem
   % KI2 - C2 IBB kernel, KCAs - scaled analytic Chebyshev
10 epvec = [10,.5];
   Kcell = {KI2,KCAs};  Kdxcell = {KI2dx,KCAs};  Kdtcell = {KI2,KCAsdt};
12 Kf = @(Kc,e,x,z) prod(cell2mat(reshape( ...
            cellfun(@(K,e,x,z) K(e,x,z), ...
14                  Kc,num2cell(e),num2cell(x,1),num2cell(z,1), ...
            'UniformOutput',0),   [1,1,length(e)])),3);
16 % Break the domain into initial condition and PDE sections
   xic = xall(xall(:,2)==0,:);  xint = xall(xall(:,2)~=0,:);  x = [xic;xint];
18 % Form the collocation matrix
   Aic = Kf(Kcell,epvec,xic,x);
20 Aint = Kf(Kdtcell,epvec,xint,x) + 0.25*Kf(Kdxcell,epvec,xint,x);
```

```
A = [Aic;Aint];
% Evaluate the RHS with initial condition and interior components
rhsic = fic(xic);  rhsint = zeros(size(xint,1),1);  rhs = [rhsic;rhsint];
% Solve the system for the solution coefficients
coef = A\rhs;
% Choose some evaluation points and evaluate the solution at them
xeval = pick2Dpoints([0 0],[1 tmax],50);
ueval = Kf(Kcell,epvec,xeval,x)*coef;
```

Because of the peculiarly structured kernel locations for this problem (Chebyshev points in time, uniform in space, no boundaries for the IBB kernel), lines 4–7 are more complicated than most point selections in this book. Lines 10–15 associate 1D kernels with the correct dimensions — `Kcell = {KI2,KCAs}` applies the C^2 IBB kernel in space and the analytic Chebyshev in time — and define the evaluation mechanism for the tensor kernel as we learned in Section 4.4. See Figure 20.2(f) for a plot of the points in use. The definitions of these 1D kernels and their derivatives are conspicuously absent but are presented in Remark 20.7 to help focus on the collocation issues here.

In a similar strategy to Example 20.8, the domain is divided into sections (`xic` for initial conditions and `xint` for the interior) and the collocation matrix and right-hand side are built in matching order. Note that the mechanism for computing derivatives with these tensor kernels is to call the same evaluation function `Kf` but pass it the tensor kernel cell array with the correct derivative, e.g., `Kdxcell` for a derivative in the x dimension. This is slightly different than the original implementation for tensor kernels from Program 4.13 because we now must consider not just a tensor kernel but also its derivatives.

For many tensor space-time kernels, different shape parameters should be used to collocate. In this particular problem, it is necessary because the IBB kernel becomes ill-conditioned for $\varepsilon < 1$ but the Chebyshev kernel requires $\varepsilon < 1$. The choices made here are arbitrary, and as was the case for interpolation and kriging problems from Chapter 14, an effective parametrization strategy would be useful for space-time kernel collocation.

Remark 20.7. Line 11 from Program 20.9 defines the kernel as `Kcell = {KI2,KCAs}`, i.e., the tensor product of the C^2 iterated Brownian bridge kernel and a scaled form of the C^∞ Chebyshev kernels. The IBB kernels are available in the `GaussQR` repository under the function `ibb`:

```
KI2 = @(e,x,z) ibb(x,z,e,2);
KI2dx = @(e,x,z) ibb(x,z,e,2,1);
```

The analytic Chebyshev kernel is also messy but is reasonable enough to be written inline as is done in the library version of Program 20.9:

```
KCA = @(b,x,z) .5 + (1-b)* ...
       (b*(1-b^2) - 2*b*bsxfun(@plus,x.^2,z.^2') + (1+3*b^2)*x*z')./ ...
```

```
            ((1-b^2)^2 + 4*b*(b*bsxfun(@plus,x.^2,z.^2')-(1+b^2)*x*z'));
KCAdx = @(b,x,z) (1-b)* ...
           (((1-b^2)^2 + 4*b*(b*bsxfun(@plus,x.^2,z.^2')-(1+b^2)*x*z')).* ...
              bsxfun(@plus,          -4*b*x,              (1+3*b^2)*z') - ...
            (b*(1-b^2) - 2*b*bsxfun(@plus,x.^2,z.^2') + (1+3*b^2)*x*z').* ...
              (4*b).*bsxfun(@minus,  2*b*x,               (1+b^2)*z'))./ ...
            ((1-b^2)^2 + 4*b*(b*bsxfun(@plus,x.^2,z.^2')-(1+b^2)*x*z')).^2;
```

Note, however, that the definition of the Chebyshev kernels from Section 3.9.2 defines them only on the domain $x, z \in [-1, 1]$. This is at odds with the domain we are using for this problem, which requires the following rescaling from [0,`tmax`] to $[-1, 1]$:

```
KCAs = @(e,s,t) KCA(e,s*2/tmax-1,t*2/tmax-1);
KCAsdt = @(e,s,t) 2/tmax*KCAdx(e,s*2/tmax-1,t*2/tmax-1);
```

The change from `@(e,s,t)` to `@(b,x,z)` is cosmetic but helps to differentiate between the time domain and the $[-1, 1]$ domain on which the Chebyshev kernels live. To account for the chain rule, `2/tmax` is applied to the derivative.

Remark 20.8. We recall from Section 7.2 that the $\beta = 2$ iterated Brownian bridge kernel requires not just zero values at $x = 0$ and $x = 1$, but zero-valued second derivatives as well. This is not specified in the PDE (20.25), leading us to worry about the effect on the solution. If the kernel boundary conditions were unacceptable in some setting, we could consider a generalized Sobolev space style modification to the kernel as discussed in Section 8.3. It turns out that, for time $0 < t < 1$ the necessary boundary conditions are satisfied so the C^2 iterated Brownian bridge kernel actually performs better than a different C^2 kernel because it automatically acts appropriately around the boundaries.

Chapter 21

Finance

This chapter introduces topics from finance, where kernel methods are either fundamental to the structure of the solution scheme or a useful tool to approximate quantities of interest. Because of the complex stochastic nature of financial problems, and the numerous strategies available for gaining insights into them, we restrict ourselves primarily to the application of previously discussed ideas in the financial setting.

Our first analysis involves the use of the Brownian motion kernel (see Example 2.4) to facilitate random walks which can be used in the computation of the high-dimensional integrals which may arise in finance. We then transition to Black–Scholes type PDEs and explain how the kernel methods described in Chapter 20 serve as useful tools for computing solutions that measure certain important financial quantities.

This chapter exists mostly as a recognition of one of the interesting applications in which kernels have found a role, and not as an introduction to finance. Most of the aspects of stochastic calculus have been removed so as to focus on the impact of positive definite kernels in this application. For those unfamiliar with stochastic calculus, [Mikosch (1998); Protter (2004)] provide excellent presentations; Shreve (2004) discusses stochastic calculus specifically for finance and we generally refer to it throughout this chapter to avoid the more complicated points.

21.1 Brownian motion

Although numerous different financial problems can be phrased and considered, we restrict ourselves to the simplest because, frankly, this is not a text on finance. Our desire here is to consider the price of an asset, which we will assume to be modeled by

$$\mathrm{d}X_t = \mu(t)X_t\,\mathrm{d}t + \sigma(t)X_t\,\mathrm{d}W_t,$$

where W_t is a Brownian motion process (or Wiener process) and X_0 is known, that is, we begin at time $t = 0$. In the rest of this chapter we discuss the Brownian motion process, its connection to kernels and its role in predicting asset prices.

To some degree, we already find ourselves somewhat out of sorts because this setting requires the use of random processes (in time) rather than random fields (in space) as we have considered throughout the rest of this book. The distinction is primarily one of verbiage and not function and we apply the same logic and theory to random processes as we have to random fields elsewhere.

21.1.1 *Brownian motion and the Brownian motion kernel*

The first appearance of Brownian motion in this text was in Example 2.4, where we simply defined $K(x, z) = \min(x, z)$ to be the Brownian motion kernel. We reconsidered it in Example 2.6 in the context of the reproducing property and in Example 5.1 we stated that the Brownian motion process had a Markov property whereby its distribution should be defined by the knowledge at the current time. To understand the connection between the Brownian motion *kernel* and the Brownian motion *process*, we must first define the Brownian motion process.

Following the framework of Shreve (2004, Chapter 3.3), we define a *Brownian motion process* W_t to be a random process with $W_0 = 0$ and increments

$$W_{t_1} - W_{t_0},\ W_{t_2} - W_{t_1}, \ldots, W_{t_N} - W_{t_{N-1}}$$

which are independent and normally distributed such that[1]

$$\begin{aligned} \mathbb{E}[W_{t_i} - W_{t_{i-1}}] &= 0, & i &= 1, \ldots, N, \\ \mathrm{Var}(W_{t_i} - W_{t_{i-1}}) &= t_i - t_{i-1}, & i &= 1, \ldots, N. \end{aligned}$$

The role of Brownian motion is to simulate a *random walk* with a path which is everywhere continuous but nowhere differentiable. Although the increments of this random walk $W_{t_i} - W_{t_{i-1}}$ are independent, the locations at each time $W_1, \ldots, W_{t_N}$ are *not* independent — they are in fact jointly normally distributed, which can be seen because each is a sum of the normally distributed increments.

If we define the length N vector

$$\boldsymbol{W} = \begin{pmatrix} W_{t_1} \\ \vdots \\ W_{t_N} \end{pmatrix},$$

we know that it has mean zero because

$$\begin{aligned} \mathbb{E}[W_{t_i}] &= \mathbb{E}[W_{t_i} - W_{t_{i-1}} + W_{t_{i-1}}] \\ &= \mathbb{E}[W_{t_i} - W_{t_{i-1}}] + \mathbb{E}[W_{t_{i-1}} - W_{t_{i-2}} + W_{t_{i-2}}] \\ &= 0 + \mathbb{E}[W_{t_{i-1}} - W_{t_{i-2}}] + \mathbb{E}[W_{t_{i-2}} - W_{t_{i-3}} + W_{t_{i-3}}] = \ldots = \mathbb{E}[W_0] = 0. \end{aligned}$$

The covariance of $\boldsymbol{W}$ can be determined using the independent increments property of Brownian motion and (5.2), which said that $\mathrm{Cov}(W_s, W_t) = \mathbb{E}[W_s W_t]$ provided

[1]This expected value, and all the rest in this chapter, are conditioned on the appropriate filtration; this is a subtlety that we ignore. See [Shreve (2004)] for more details.

$\mathbb{E}[W_s] = \mathbb{E}[W_t] = 0$ for any s and t. Begin by assuming that $s < t$, which implies that W_s and $W_t - W_s$ are independent random variables, and study

$$\begin{aligned}\mathbb{E}[W_s W_t] &= \mathbb{E}[W_s W_t - W_s^2 + W_s^2] \\ &= \mathbb{E}[W_s(W_t - W_s)] + \mathbb{E}[W_s^2] \\ &= \mathbb{E}[W_s]\mathbb{E}[W_t - W_s] + \mathrm{Var}(W_s) \\ &= (0)\mathbb{E}[W_t - W_s] + s = s, \qquad s < t,\end{aligned}$$

where we exploited $W_0 = 0$ to get $\mathrm{Var}(W_s) = s$. If, instead, we had assumed that $t < s$, we would have independence between W_t and $W_s - W_t$ to arrive at

$$\mathbb{E}[W_s W_t] = \mathbb{E}[W_t(W_s - W_t)] + \mathbb{E}[W_t^2] = t, \qquad t < s.$$

Thus $\mathrm{Cov}(W_s, W_t) = \min(s, t)$, which is why this kernel is also called the *Brownian motion kernel*: it is the covariance of a Brownian motion process.

21.1.2 *Geometric Brownian motion*

We started this section by stating that we wish to model the price of a financial asset with the stochastic differential equation

$$\mathrm{d}X_t = \mu(t) X_t \,\mathrm{d}t + \sigma(t) X_t \,\mathrm{d}W_t \tag{21.1}$$

with X_0 known. Although we now have some better understanding of the Brownian motion process W_t, we still do not know how to solve (21.1). The details are provided in [Shreve (2004, Example 4.4.8)], but it will suffice to write that the solution is

$$X_t = X_0 \exp\left(\int_0^t \sigma(s)\,\mathrm{d}W_s + \int_0^t (\mu(s) - \sigma(s)^2/2)\,\mathrm{d}s\right), \tag{21.2}$$

where $\int_0^t \sigma(s)\,\mathrm{d}W_s$ is an *Itô integral*.

Remark 21.1. The concept of Itô integrals is an interesting departure from standard Lebesgue integrals because of the so-called quadratic variation present in W_s. We can gain some intuition as to the unique aspects of this integral by approximating it with the summation[2]

$$\int_0^t \sigma(s)\,\mathrm{d}W_s = \sum_{i=1}^{N} \sigma(s_{i-1})(W_{s_i} - W_{s_{i-1}}), \quad 0 = s_0 < s_1 < \ldots < s_N = t.$$

Recall that $W_{s_i} - W_{s_{i-1}}$ is normally distributed with variance $s_i - s_{i-1}$ meaning that no matter how many increments are taken between 0 and t the magnitude of each contribution in the sum cannot be bounded. We restrict ourselves to pricing assets in a market with constant σ in this chapter, but this situation demonstrates that working in this stochastic framework is more complicated than working in a deterministic framework.

[2]Evaluating $\sigma(s_{i-1})$ and not $\sigma(s_i)$ is required because $\sigma(s_i)$ is not measurable with respect to the filtration at time s_{i-1}. This would be the equivalent of knowing the temperature tomorrow, which is impractical (or unrealistic) in most settings.

Some choices of $\mu(t)$ and $\sigma(t)$ allow for a simple form of X_t. For constant $\mu(t) = \mu_0$ and $\sigma(t) = \sigma_0$, the value of the asset is a *geometric Brownian motion*

$$X_t = X_0 \exp\left(\sigma_0 W_t + (\mu_0 - \sigma_0^2/2)t\right), \tag{21.3}$$

the expected value of which is

$$\mathbb{E}[X_t] = X_0 \mathrm{e}^{\mu_0 t} \mathbb{E}\left[\exp(\sigma_0 W_t - t\sigma_0^2/2)\right] = X_0 \mathrm{e}^{\mu_0 t}, \tag{21.4}$$

as explained in [Shreve (2004, Example 4.4.8)]. The μ_0 term here suggests that X_t has a drift in addition to any volatility imposed by σ_0.

21.1.3 *Pricing options and high-dimensional integration*

In Section 21.1.2 we introduced the use of geometric Brownian motion as a model for an asset's price X_t. While pricing an asset is an important tool, more complicated financial derivatives can be created based on the underlying value of an asset. Here we discuss the formation of options on assets. We also introduce the existence of a risk-free interest rate $r > 0$ which we assume to be constant; this allows for the opportunity to balance, or hedge, a portfolio against the risky behavior of an asset modeled by geometric Brownian motion.

A long/short option on an asset (or collection of assets) is a deal made with a fellow trader where you sign a contract at time $t = 0$ to receive/deliver at some future time. Often a time of expiry $t = T$ is set to prevent an option from being exercised years after its inception, and if an option is unexercised at expiry then it loses all value. The fair price to be paid/received for such a contract is related to its expected worth during the length of the contract $t \in [0, T]$. This would mean that the expected value of the contract would be zero which preserves the *no arbitrage* concept often underlying rational markets [Shreve (2012, Chapter 2)].

The payoff structure based on the asset X_t is encapsulated in a function g such that the fair price for the option is $\mathbb{E}[g(X_t)]$. We restrict our discussion in this section to *call options*, which are options that gain value as the value of the underlying asset increases. Each of these options relies on a fixed *exercise price*[3] (or *strike price*) E which is the price set by the contract, usually at inception, for which the asset should be exchanged at time of exercise t_E. For many options, $t_E = T$, but not for all. If the asset has value $X_{t_E} > E$, the call option can be exercised for a profit of $X_{t_E} - E$; otherwise, the call option is worthless because exercising it would cost more than the current price of the asset.

Various call options yield different forms of g, for example:

European call: $g(X_T) = \mathrm{e}^{-rT}(X_T - E)_+$,

Asian call: $g(X_{t_1}, \ldots, X_{t_p}) = \mathrm{e}^{-rT}\left(\dfrac{1}{p}\sum_{i=1}^{p} X_{t_i} - E\right)_+, \ 0 \le t_1 \le \ldots \le t_p \le T,$

[3] In most of the finance literature, the strike price is denoted with the letter K, but we use the letter E here because K already plays such a prominent role in this text.

American call: $g(X_{t_E}) = \mathrm{e}^{-rt_E}(X_{t_E} - E)_+,\ 0 < t_E \le T$,

basket European call: $g(X_T^{(1)}, \dots, X_T^{(d)}) = \mathrm{e}^{-rT}\left(\frac{1}{d}\sum_{\ell=1}^{d} X_T^{(\ell)} - E\right)_+$.

The American call option can be exercised at any time before expiry, whereas the other options fix their exercise time as $t_E = T$. In the basket European call there are d assets, the prices of which are averaged, and in the Asian option there are p times at which a single asset's price is sampled and those values are averaged in the payout[4]. Recall from our notation involving compactly supported kernels, $(x)_+ = \max(x, 0)$. Also, the e^{-rt} term is present in all of these payouts to account for the inflation caused by the risk-free interest rate $r > 0$.

To fairly price these European or Asian options, we must compute, or approximate, $\mathbb{E}[g(X_T)]$[5]; for the American option we must do even more and evaluate $g(X_{t_E})$ for the exercise time t_E chosen optimally. The European call with only one asset, which we label now as $g = C_e(X_t)$, is simple enough to consider directly. We have

$$\mathbb{E}[C_e] = n(d_1(X_0, t))X_0 - En(d_2(X_0, t))\mathrm{e}^{-rt}, \tag{21.5}$$

where $n(z)$ is the cumulative distribution function of the standard normal distribution evaluated at $z \in \mathbb{R}$ and

$$d_1(x, t) = \frac{1}{\sigma_0\sqrt{t}}\left(\log\left(\frac{x}{E}\right) + \left(r + \frac{\sigma_0^2}{2}\right)t\right),$$
$$d_2(x, t) = d_1(x, t) - \sigma_0\sqrt{t}.$$

This is determined in, e.g., [Shreve (2004, Chapter 4.5.4)]. Because this solution is determined through hedging a portfolio, it is only appropriate when $\mu_0 = r$ in the geometric Brownian motion.

For the other options above, more complicated situations arise. In the Asian call option, which we denote $C_a(X_{t_1}, \dots, X_{t_p})$, we must compute the p-dimensional expectation

$$\mathbb{E}[C_a] = \int_{[0,\infty)^p} \mathrm{e}^{-rT}\left(\frac{1}{p}\sum_{i=1}^{p} x_i - E\right)_+ f_{X_{t_1}}(x_1)\dots f_{X_{t_p}}(x_p)\,\mathrm{d}x_1\dots\mathrm{d}x_p,$$

where $f_{X_{t_i}}$ is the density of X_{t_i}. The fair price of the basket European call, denoted with $C_b(X_T^{(1)}, \dots, X_T^{(d)})$, is computed with the d-dimensional expectation

$$\mathbb{E}[C_b] = \int_{[0,\infty)^d} \mathrm{e}^{-rT}\left(\frac{1}{d}\sum_{\ell=1}^{d} x_\ell - E\right)_+ f_{X_T^{(1)}}(x_1)\dots f_{X_T^{(d)}}(x_d)\,\mathrm{d}x_1\dots\mathrm{d}x_d.$$

[4]The Asian call option is often defined with an integral, not a summation, but we restrict ourselves to the summation here.

[5]This expectation should be phrased with the risk-neutral probability measure, but we ignore this important topic here. See, e.g., [Shreve (2012, Chapter 2)].

Approximating the fair price for the American call, denoted by $C_u(X_{t_E})$, is even more complicated because it requires computing

$$\sup_{0<t_E\le T} \mathbb{E}[C_u X_{t_E}] = \sup_{0<t_E\le T} \int_0^\infty \mathrm{e}^{-rt_E}(x-E)_+ f_{X_{t_E}}(x)\,\mathrm{d}x.$$

These integrals should be phrased over $\mathbb{R}$, not $[0,\infty)$, but the asset price X_t is always positive so there is nothing lost here.

The common theme here is that pricing options accurately requires the use of numerical integration, often called quadrature or cubature, and these integrals may be in multiple dimensions. Approximating these integrals can be done using strategies as simple as the "left-hand rule" or as intricate as Gaussian quadrature, Newton–Cotes formulas, or Clenshaw–Curtis. In higher dimensions, the cost of these methods becomes unacceptable and statistical methods such as Monte Carlo and quasi-Monte Carlo become the preferred tool. The theory of quasi-Monte Carlo is another topic in which positive definite kernels play a useful role.

21.1.4 *A generic error formula for quasi-Monte Carlo integration via reproducing kernels*

To the eye of the practitioner, reproducing kernels play a mostly hidden role in high-dimensional integration. We provide a glimpse of this role by looking at an error estimate for *quasi-Monte Carlo integration*. Our discussion loosely follows the exposition from [Dick *et al.* (2013a, Section 3)].

Using a quasi-Monte Carlo rule we have

$$\int_\Omega f(\boldsymbol{x})\,\mathrm{d}\boldsymbol{x} \approx \frac{1}{N}\sum_{i=1}^N f(\boldsymbol{x}_i), \tag{21.6}$$

where the points $\boldsymbol{x}_1,\ldots,\boldsymbol{x}_N$ are for now arbitrarily chosen in the domain Ω; usually one takes $\Omega=[0,1]^d$, and all of the integrals in Section 21.1.3 can be transformed into that domain. After we have derived our error bound we will explain how it leads to criteria for choosing so-called *low-discrepancy designs* (see also Appendix B.1).

The key to obtaining error bounds for quasi-Monte Carlo rules is to assume that the integrand f lies in some reproducing kernel Hilbert space $\mathcal{H}_K(\Omega)$ with reproducing kernel K such that $\int_\Omega\int_\Omega K(\boldsymbol{x},\boldsymbol{z})\,\mathrm{d}\boldsymbol{x}\,\mathrm{d}\boldsymbol{z}<\infty$. This assumption goes back to [Hickernell (1998)] and is used only to obtain error bounds. The fact that $f\in\mathcal{H}_K(\Omega)$ does not play any role in any of the algorithms for evaluation of the integral.

Using the reproducing property, i.e., $f(\boldsymbol{x}) = \langle f, K(\boldsymbol{x},\cdot)\rangle_{\mathcal{H}_K(\Omega)}$, we can rewrite both sides of (21.6), respectively, as

$$\int_\Omega f(\boldsymbol{x})\,\mathrm{d}\boldsymbol{x} = \int_\Omega \langle f, K(\boldsymbol{x},\cdot)\rangle_{\mathcal{H}_K(\Omega)}\,\mathrm{d}\boldsymbol{x} = \left\langle f, \int_\Omega K(\boldsymbol{x},\cdot)\,\mathrm{d}\boldsymbol{x}\right\rangle_{\mathcal{H}_K(\Omega)},$$

$$\frac{1}{N}\sum_{i=1}^N f(\boldsymbol{x}_i) = \frac{1}{N}\sum_{i=1}^N \langle f, K(\boldsymbol{x}_i,\cdot)\rangle_{\mathcal{H}_K(\Omega)} = \left\langle f, \frac{1}{N}\sum_{i=1}^N K(\boldsymbol{x}_i,\cdot)\right\rangle_{\mathcal{H}_K(\Omega)}.$$

Therefore, the quadrature error is given by

$$\int_\Omega f(\boldsymbol{x})\,\mathrm{d}\boldsymbol{x} - \frac{1}{N}\sum_{i=1}^N f(\boldsymbol{x}_i) = \left\langle f, \int_\Omega K(\boldsymbol{x},\cdot)\,\mathrm{d}\boldsymbol{x} - \frac{1}{N}\sum_{i=1}^N K(\boldsymbol{x}_i,\cdot)\right\rangle_{\mathcal{H}_K(\Omega)}$$
$$= \langle f, \xi\rangle_{\mathcal{H}_K(\Omega)},$$

where

$$\xi(\boldsymbol{z}) = \int_\Omega K(\boldsymbol{x},\boldsymbol{z})\,\mathrm{d}\boldsymbol{x} - \frac{1}{N}\sum_{i=1}^N K(\boldsymbol{x}_i,\boldsymbol{z}), \qquad \boldsymbol{z}\in\Omega,$$

is referred to as the *Riesz representer of the quadrature error*.

By taking advantage of the symmetry of the kernel and its reproducing property we can obtain the *squared worst case error* as

$$\|\xi\|^2_{\mathcal{H}_K(\Omega)} = \left\langle \int_\Omega K(\boldsymbol{x},\cdot)\,\mathrm{d}\boldsymbol{x} - \frac{1}{N}\sum_{i=1}^N K(\boldsymbol{x}_i,\cdot), \int_\Omega K(\cdot,\boldsymbol{z})\,\mathrm{d}\boldsymbol{z} - \frac{1}{N}\sum_{j=1}^N K(\cdot,\boldsymbol{x}_j)\right\rangle_{\mathcal{H}_K(\Omega)}$$
$$= \int_\Omega\int_\Omega K(\boldsymbol{x},\boldsymbol{z})\,\mathrm{d}\boldsymbol{x}\,\mathrm{d}\boldsymbol{z} - \frac{2}{N}\sum_{i=1}^N\int_\Omega K(\boldsymbol{x}_i,\boldsymbol{z})\,\mathrm{d}\boldsymbol{z} + \frac{1}{N^2}\sum_{i=1}^N\sum_{j=1}^N K(\boldsymbol{x}_i,\boldsymbol{x}_j).$$

As explained in [Dick *et al.* (2013a)], the mixed first partial derivatives of the error representer ξ give rise to the *local discrepancy function*, and low-discrepancy points are obtained as (approximate) minimizers of the supremum of the local discrepancy function in Ω (which is also known as the *star discrepancy*, see Appendix B.1).

High-dimensional integration rules are usually discussed on $\Omega = [0,1]^d$, and that is why *anisotropic tensor product kernels* play a prominent role in the theory of quasi-Monte Carlo methods.

By choosing a specific kernel — and thereby a class of (smooth) integrands — one can obtain specific rates of convergence for the integration error. For example, one gets convergence of order $\mathcal{O}(N^{-1+\delta})$ with arbitrarily small $\delta > 0$ for integrands $f \in H_1([0,1])$, an anchored weighted Sobolev space with kernel $K(x,z) = 2\max(x,z)$, which is similar to the anchored weighted Sobolev space $H_{1,\varepsilon}([0,1])$ discussed in Example 8.2. The integrands appearing in Section 21.1.3 all fall in the Hilbert space of the Brownian motion kernel because they all take the value zero at zero.

Many more details are provided in the excellent survey article [Dick *et al.* (2013a)].

21.1.5 *Example of asset pricing through quasi-Monte Carlo*

To demonstrate the use of quasi-Monte Carlo in this setting we compute the value of a European call option because the true solution (21.5) is available to us. See [Liu (2008)] as a reference for standard Monte Carlo methods. For us to use (21.5) we must compute the integral

$$\mathbb{E}[C_e] = \int_0^\infty \mathrm{e}^{-rT}(x-E)_+ f_{X_T}(x)\,\mathrm{d}x$$

which we do by converting it into an integral over $[0,1]$. To do this, we need $F_{X_T}^{-1}$ which is the inverse cumulative distribution function for X_T. Fortunately, we know from (21.3) that

$$X_t = X_0 \exp\left(\sigma_0 W_t + (\mu_0 - \sigma_0^2/2)t\right),$$

and, because we assumed in Section 21.1.1 that $W_0 = 0$, we know that W_t is normally distributed with mean 0 and variance t. This means that X_t has a lognormal distribution with mean $X_0 e^{\mu_0 t}$ and variance $X_0^2 e^{2\mu_0 t}\left(e^{\sigma_0^2 t} - 1\right)$ [Øksendal (2003)]. This is sufficient to evaluate the inverse CDF $F_{X_T}^{-1}(x)$ using the `icdf` function in MATLAB's Statistics and Machine Learning Toolbox.

Using this strategy, the integral can be rephrased with essentially a u-substitution from calculus. If we state that $x = F_{X_T}^{-1}(u)$ then this integral becomes

$$\begin{aligned}\mathbb{E}[C_e] = \int_0^\infty C_e(x) f_{X_T}(x)\,\mathrm{d}x &= \int_0^\infty C_e(x)\mathrm{d}F_{X_T}(x) \\ &= \int_0^1 C_e(F_{X_T}^{-1}(u))\mathrm{d}u.\end{aligned}$$

This is also called *inverse transform sampling* in the Monte Carlo community. A sample MATLAB code which executes this integral is presented in Program 21.1 with numerical results provided in Figure 21.1.

Program 21.1. Excerpts from `FinanceEurCallQMC.m`

```
% Choose the parameters for the option
E = 1;  T = 1;  r = .05;  X_0 = 2;
% Choose the parameters for the asset
s = .3;  mu = r;
% Define the payout for the European call option at expiry
payout = @(x) exp(-r*T)*max(x-E,0);
% Define the true solution
d1 = @(x,t) 1./(E*sqrt(t)).*(log(x/E)+(r+E^2/2)*t);
d2 = @(x,t) d1(x,t) - E*sqrt(t);
ptrue = normcdf(d1(X_0,T)).*X_0 - E*normcdf(d2(X_0,T)).*exp(-r*T);
% Test the quality for different random walk lengths
Nvec = round(logspace(2,5,15));
k = 1;  MCvec = zeros(size(Nvec));  QMCvec = zeros(size(Nvec));
for N=Nvec
    % Perform the quasi-Monte Carlo method
    u = pickpoints(0,1,N,'halt');
    Finvu = icdf('Lognormal',u,log(X_0)+(mu-s^2/2)*T,s*sqrt(T));
    QMCvec(k) = abs(ptrue - mean(payout(Finvu)));
    % Perform several standard Monte Carlo integrals for comparison
    W_T = sqrt(T)*randn(N,20);
    X_T = X_0*exp((r-s^2/2)*T + s*W_T);
    MCvec(k) = mean(abs(ptrue - mean(payout(X_T))));
    k = k + 1;
end
```

Note that we have chosen $\mu_0 = r$ in line 4 which is required to use the analytic solution (21.5). That solution is defined in lines 8–10. Within the loop we begin by computing our quasi-Monte Carlo integral which requires a low-discrepancy sequence that we form in line 16. Here we use the Halton points but for this one-dimensional problem we could have just as easily used uniformly spaced points.

The computation of $F_{X_T}^{-1}$ takes place in line 17. Note that the values passed as parameters to `icdf` are *not* the mean and standard deviation of X_t; they are instead related to the normal distribution underlying the lognormal distribution and are explained in MATLAB's documentation. In line 18 we compute $\mathbb{E}[C_e]$ with `mean(payout(Finvu))` and then determine the error in that computation.

An analogous computation takes place in lines 20–22 for standard Monte Carlo methods using pseudorandom draws from the normal distribution of W_T. The results of this computation are compared to the quasi-Monte Carlo computation in Figure 21.3 where there is clearly a faster convergence of the quasi-Monte Carlo scheme, at least for this one-dimensional example. We can check this empirically by having MATLAB compute a line of best fit

```
polyfit(log(Nvec),log(QMCvec),1)
ans =
   -0.8174    0.3425
polyfit(log(Nvec),log(MCvec),1)
ans =
   -0.4954   -0.7715
```

which shows the expected $N^{-1/2}$ behavior often associated with Monte Carlo methods and a steeper rate of convergence for the quasi-Monte Carlo method.

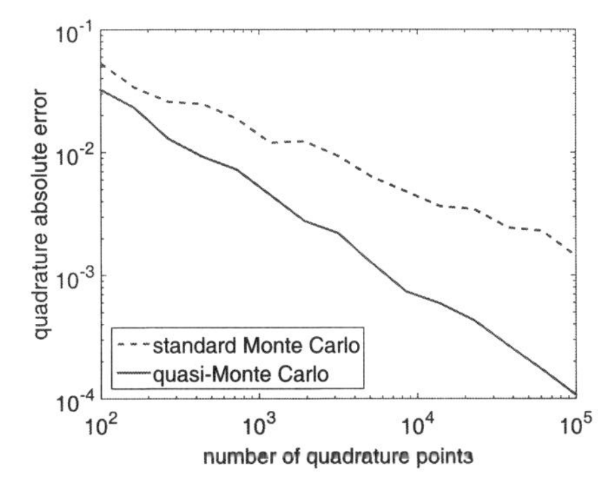

Fig. 21.1 For this simple example, the quasi-Monte Carlo method converges faster than the standard Monte Carlo method.

21.2 Black–Scholes PDEs

Another approach to option pricing is via the *Black–Scholes equation*. This linear parabolic partial differential equation can be obtained from the stochastic differential equation (21.1) via the Feynman–Kac formula (see, e.g., [Shreve (2004, Chapter 6)]). Namely, the solution of the Black–Scholes PDE represents the expected payoff modeled by (21.1) together with an appropriate payoff function. In the Black–Scholes framework this payoff function takes on the role of a terminal condition. The boundary condition for European options is straightforward to implement in the PDE setting, and therefore we will again consider this example.

The d-dimensional Black–Scholes equation that models the fair price $P(\boldsymbol{x},t)$ of an option based on a *basket* of d underlying assets $\boldsymbol{x} = \begin{pmatrix} x_1 & \cdots & x_d \end{pmatrix}^T$ at time t is of the form

$$\begin{aligned}
\frac{\partial P}{\partial t}(\boldsymbol{x},t) &= -\frac{1}{2}\sum_{i=1}^{d}\sum_{j=1}^{d}\left(\mathsf{S}\mathsf{S}^T\right)_{i,j} x_i x_j \frac{\partial^2 P}{\partial x_i x_j}(\boldsymbol{x},t) - r\sum_{i=1}^{d} x_i \frac{\partial P}{\partial x_i}(\boldsymbol{x},t) + rP(\boldsymbol{x},t), \\
&\qquad \boldsymbol{x}\in\Omega,\ 0<t\le T, \qquad\qquad (21.7)\\
P(\boldsymbol{x},0) &= F(\boldsymbol{x}), \qquad \boldsymbol{x}\in\Omega,\\
P(\boldsymbol{x},t) &= G(\boldsymbol{x},t), \qquad \boldsymbol{x}\in\partial\Omega,\ 0<t\le T,
\end{aligned}$$

where T is the time of expiry, S is the *volatility matrix*, r is the risk-free interest rate. F is the *payoff function* which lets one distinguish between call and put options, and between European, American, Asian, and other more exotic options. This model can be made significantly more realistic by including, e.g., variable r or S values, but we consider only the simplest case to focus on the role kernels play in this setting.

Following [Pettersson *et al.* (2008); Larsson *et al.* (2008, 2013a)], we use a d-dimensional simplex as our computational domain given by

$$\Omega = \left\{ \boldsymbol{x} \mid \frac{1}{d}\sum_{i=1}^{d} x_i \le 4E,\ x_i \ge 0 \right\}.$$

The boundary condition is specified via the function G and discussed in more detail for the specific examples that follow.

Remark 21.2. The formulation of (21.7) may at first seem a bit confusing, since it presupposes initial condition F when in reality we only know the *terminal condition* of how much the option would be worth at time of expiry. To reach the PDE above, one must "reverse time," i.e., $t \to T - t$, and as a result the terminal condition turns into an initial condition. We use the terms *time of inception* to be $t = T$ in this PDE setting and *time of expiry* to be $t = 0$.

21.2.1 *Single-asset European option through Black–Scholes PDEs*

We revisit the European call option, introduced in Section 21.1.5, for a single asset, i.e., $d = 1$. The associated payoff function is

$$F(x) = \max(x - E, 0), \tag{21.8}$$

where, again, E is the exercise price at which the asset may be exchanged at expiry. Unlike in Section 21.1, where we studied the use of random walks and high-dimensional integration to compute the value of the option, now our goal is to compute a numerical solution to (21.7) using the strategies developed in Chapter 20.

For our following numerical experiments, the parameters in the Black–Scholes equation are chosen as $r = 0.05$, $\mathsf{S} = 0.3$, $T = 1$ and $E = 1$. For this particular payoff function F, a closed form solution is still given in (21.5), albeit with the necessary time reversal and using $\sigma_0 = \mathsf{S}$.

To implement a numerical solution, we must impose boundary and initial conditions for the differential equation. Note that the function F, which we have defined as the payout (theoretically coming at the end of the contract), was in (21.7) phrased as the initial condition to the problem through the reversal of time. Boundary conditions in space must be chosen, and here some flexibility is available.

The boundary condition at $x = 0$ must obviously be $G(0, t) = 0$ because, no matter the exercise price, there is no value in paying E for an asset worth 0. Our other boundary was, somewhat arbitrarily, chosen to be at $x = 4$. This decision was made to place the boundary far enough from the exercise price that the asymptotic boundary condition

$$G(x, t) = x - E\mathrm{e}^{-2rt}, \qquad x = 4,$$

is acceptable [Pettersson *et al.* (2008)]. Taking all this into account, we can write a more digestible form of the BVP as

$$\begin{aligned}
\frac{\partial P}{\partial t}(x,t) &= -\frac{1}{2}(0.3)^2 x^2 \frac{\partial^2 P}{\partial x^2}(x,t) - 0.05x\frac{\partial P}{\partial x}(x,t) + 0.05P(x,t), \\
&\qquad\qquad 0 \le x \le 4,\ 0 < t \le 1, \\
P(x,0) &= (x-1)_+, \qquad 0 \le x \le 4, \\
P(0,t) &= 0,\ P(4,t) = 4 - \mathrm{e}^{-0.05t}, \qquad 0 < t \le T.
\end{aligned} \tag{21.9}$$

Example 21.1. (European call option)
In this example we solve (21.7) using kernel-based collocation with various Matérn kernels and study the convergence toward the true solution (21.5) as a function of the number of collocation points N. Program 21.2 demonstrates this in MATLAB and associated plots are provided in Figure 21.2.

Program 21.2. Excerpts from FinanceEurCallPDE.m

```
% Define parameters of the Black-Scholes PDE
E = 1;  T = 1;  S = .3;  r = .05;
% Define the payout, used as the initial condition at time of expiry
payout = @(x) max(x-E,0);
% Define the boundary conditions (for both x=0 and x=4)
bc = @(x,t) E*(4-exp(-r*t))*(x==4*E);
% Define the true solution
d1 = @(x,t) 1./(E*sqrt(t)).*(log(x/E)+(r+E^2/2)*t);
d2 = @(x,t) d1(x,t) - E*sqrt(t);
Ptrue = @(x,t) normcdf(d1(x,t)).*x - E*normcdf(d2(x,t)).*exp(-r*t);
% Define the C2 Matern kernel for this problem
rbf = @(e,r) (1+e*r).*exp(-e*r);  rbfx = @(e,r,dx) -e^2*exp(-e*r).*dx;
rbfxx = @(e,r) e^2*exp(-e*r).*(e*r-1);     ep = 2;
% Loop through a range of N values
Nvec = ceil(logspace(1,2,12));  errvec = zeros(size(Nvec));  k = 1;
for N=Nvec
    % Choose the collocation points
    xall = pickpoints(0,4*E,N);
    xbc = xall(xall==0 | xall==4*E);     xint = xall(xall~=0 & xall~=4*E);
    Nbc = length(xbc);                   Nint = length(xint);
    ibc = Nint+1:Nint+Nbc;               iint = 1:Nint;
    x = [xint;xbc];
    % Create the necessary differentiation matrices
    DM = DistanceMatrix(x,x);              V = rbf(ep,DM);
    DMint = DistanceMatrix(xint,x);        Vxxint = rbfxx(ep,DMint);
    DiffMint = DifferenceMatrix(xint,x);  Vxint = rbfx(ep,DMint,DiffMint);
    VxintVinv = Vxint/V;                   VxxintVinv = Vxxint/V;
    % Form the functions for the ODE solver
    odeint = @(u)   .5*S^2*xint.^2.*(VxxintVinv*u) + ...
                              r*xint.*(VxintVinv*u) - r*u(iint);
    odebc  = @(t,u) u(ibc)-bc(xbc,t);
    odefun = @(t,u) [odeint(u);odebc(t,u)];
    % Prepare the ODE solver
    Mass = sparse([eye(Nint),zeros(Nint,Nbc);zeros(Nbc,N)]);
    Jac = [.5*S^2*bsxfun(@times,xint.^2,VxxintVinv) + ...
                r*bsxfun(@times,xint,VxintVinv) - ...
                   [r*eye(Nint),zeros(Nint,Nbc)]; ...
           [zeros(Nbc,Nint),eye(Nbc)]                      ];
    odeopt = odeset('Jacobian',Jac,'Mass',Mass,...
        'MStateDependence','none','MassSingular','yes');
    [tsol,Psol] = ode15s(odefun,[0,T],payout(x),odeopt);
    errvec(k) = errcompute(Psol(end,:)',Ptrue(x,T));
end
```

The structure of this program draws elements from the PDE solvers explained in Chapter 20. One notable difference is the use of global differentiation matrices to propagate the solution. We previously used finite difference differentiation matrices in a method of lines framework in Example 20.7. Another difference is the addition of the Jacobian (defined on lines 35–38) matrix to the ODE solver which, although not required as we saw in other time-stepping examples, can help the ODE solver perform better. We use the lengthy notation in line 27 to describe the action of the differentiation matrices (recall Remark 19.3).

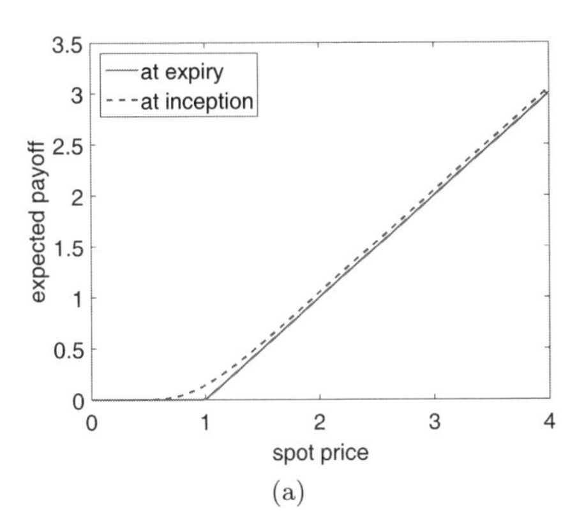

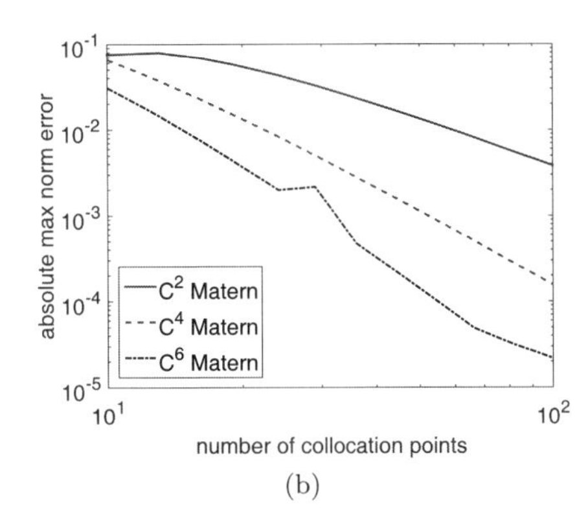

Fig. 21.2 (a) The initial condition (21.8) at time of expiry has a discontinuous first derivative; although, for $t > 0$ the solution is analytic. The expected payoff at time of inception can be used to fairly price the option. (b) The accuracy in the numerical solution at time of inception decreases as more collocation points are used. As predicted, smoother basis functions see faster convergence. All experiments used $\varepsilon = 2$.

Figure 21.2(a) displays the initial condition, as well as the expected payoff at inception. The solution begins with a discontinuity in the first derivative at spot price $x = 1$, but that discontinuity is immediately smoothed over by the diffusion part of the PDE. The solution "at inception" denotes the expected payoff of purchasing a call option given that spot price of the asset when the contract is signed. Because the value of the computed solution at time $t = 1$ with spot price $x = 2$ is 1.0497, the fair price for the call option with exercise price $E = 1$ is 0.0497. The quality of the computation seems to improve as a smoother kernel is used which matches our expectation.

Example 21.2. (Dealing with the derivative discontinuity in the initial condition) In Figure 21.2(a) we recognize the lack of smoothness in the initial condition of the PDE. In a sense, this should pose some problems when using kernels that expect more smoothness than actually exists in the solution. If we reconsider those results but monitor the accuracy of the solution at all times, rather than just the final time, we arrive at Figure 21.3; we also present the results for the Chebyshev nodes rather than uniformly spaced points for comparison.

In these figures, we see a rather frightening source of error accumulating rather far from the point $x = 1$ where the derivative is discontinuous. While this accumulation did not prevent the convergence of even the C^6 Matérn, as we showed in Figure 21.2(b), it may be possible to phrase this BVP in such a way that we actively fight this behavior. This example will demonstrate the use of coupling, as

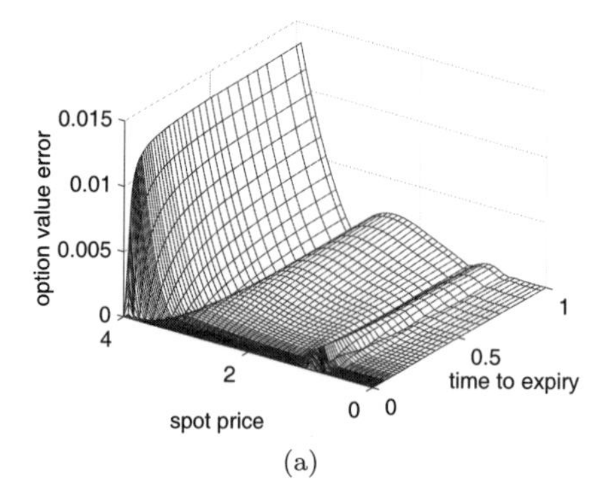

(a)

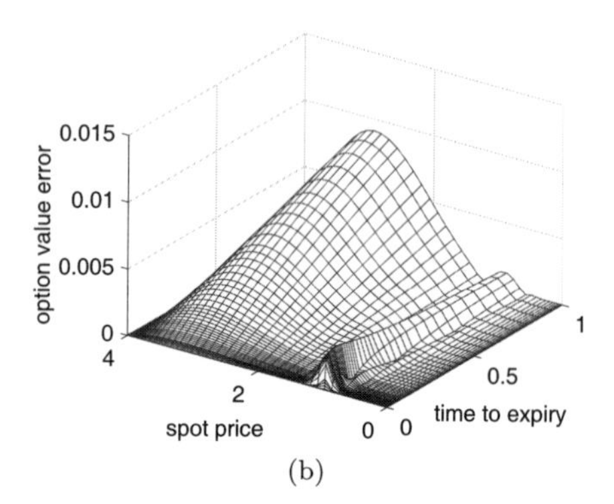

(b)

Fig. 21.3 While the error around the initial discontinuity at $x = 1$ is not terrible, errors grow in the rest of the domain. These experiments used C^2 Matérn kernels centered at $N = 50$ locations. (a) Uniformly spaced collocation points. (b) Chebyshev spaced collocation points.

introduced in Section 19.4, to describe the BVP as two coupled BVPs.

Our goal is to decompose the domain into two pieces on which the initial condition is analytic and impose the jump discontinuity in the coupling between these two domains. The two regions of interest are

$$\Omega_1 = [0, 1], \qquad \Omega_2 = [1, 4],$$

and their solutions are labeled P_1 and P_2, respectively. The same PDE and initial conditions from (21.9) are used, and boundary conditions at $x = 0$ and $x = 4$ will still be imposed.

At the interface $x = E$ we impose equality of the values and first derivatives but require that, at time $t = 0$, there is a discontinuous derivative. We do this in a rather ad hoc way, by adding a very quickly decaying term to the derivative coupling, leaving us with coupling conditions

$$\lim_{x\to 1^-} P_1(x,t) = \lim_{x\to 1^+} P_2(x,t),$$

$$\lim_{x\to 1^-} \frac{\partial}{\partial x} P_1(x,t) = \lim_{x\to 1^+} \frac{\partial}{\partial x} P_2(x,t) + \mathrm{e}^{-100000t}.$$

To better associate collocation points with this coupling strategy, we choose to cluster points near the interface $x = E$; the new point selection is demonstrated in Figure 21.4(a). This idea of clustering kernel collocation points near the exercise price was discussed in [Larsson *et al.* (2013a)] along with other, more interesting, point selection strategies. As was the case in Section 19.4, the coupling code here is omitted to spare the reader the tedious details of implementation; it is available in the `GaussQR` library under `FinanceEurCallPDECouple.m`.

By both coupling the solution across the exercise price and clustering collocation points around the exercise price we have created a more accurate solution. What's more, the Jacobian of the associated system has large off-diagonal zero blocks because the two domains Ω_1 and Ω_2 only interact through the interface. In doing this, the solution can be computed more efficiently without adversely affecting the accuracy.

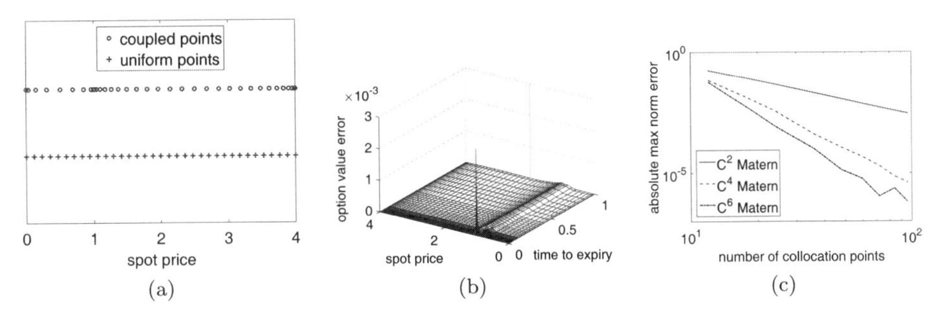

Fig. 21.4 A combination of repositioning the collocation points and coupling across the initial discontinuity have yielded better results for the smooth kernels. These experiments all used Matérn kernels with $\varepsilon = 2$. (a) An $N = 36$ coupled point distribution using Chebyshev nodes on each component domain. (b) Minor errors initially in the $N = 48$ C^6 Matérn solution fade quickly as time progresses. (c) The smoother solvers maintain their convergence rates better with the coupling and with points clustered around the exercise price. These errors were measured against the true solution at time of inception.

Remark 21.3. In Figure 21.3 and Figure 21.4(b), the lines perpendicular to the "time to expiry" axis are clustered in an irregular manner. This is the result of `ode15s` taking adaptively spaced steps to help maximize the quality of the time stepping scheme without losing accuracy. MATLAB will allow the output to be recorded at specifically chosen times by changing the `[0,T]` argument in Program 21.2 at line 41 to be a vector of times at which the solution should be evaluated during the time stepping process.

21.2.2 *Pricing American options*

As described in Section 21.1.3, an American option allows for exercise at any $t = t_E$ prior to expiry at $t = T$; the option value is found by determining the optimal exercise strategy, i.e., for a put option[6] we have

$$P_u(X_{t_E}) = \sup_{0<t_E\leq T} \mathbb{E}\left[\mathrm{e}^{-rt_E}(E - X_{t_E})_+\right],$$

where we have discounted the exercise price with the risk-free interest rate $r > 0$, as usual.

The optimal exercise strategy corresponds to finding an optimal stopping time t^* of the form

$$t^* = \inf\{t \geq 0 : x(t) \leq x^*(t)\},$$

where $x^*(t)$ denotes the optimal exercise boundary (which moves with t), i.e., if $x(t) < x^*(t)$ then the option is "in the money" and should be exercised. Thus American option pricing represents an optimal stopping problem (see, e.g., [Glasserman (2004); Kwok (2008)]), and is therefore sometimes formulated via a stochastic

[6]We discuss a put option here since it can be shown that the value of an American call option is the same as that of a European call (see, e.g., [Brandimarte (2003)]).

dynamic programming or linear complementarity approach (see, e.g., [Glasserman (2004); Tavella (2003)]). In many practical algorithms the value of an American option is computed on a discrete time grid. This simplification actually changes the nature of the option from an American option (with continuous exercise time) to a so-called *Bermudan* option (with discrete exercise times, i.e., allowing early exercise at a finite set of discrete times specified by the contract, see, e.g., [Shreve (2004)]). The binomial lattice (see, e.g., [Brandimarte (2003)]), or least squares Monte–Carlo simulation of [Longstaff and Schwartz (2001)] are two examples where American option prices are determined via the Bermudan mode.

One strategy for pricing these options with the Black–Scholes PDE approach involves the use of a *moving boundary*, which is a natural tool for allowing exercise at any time before expiry[7]. This moving boundary condition is of the form

$$P(x,t) = \max\left(E - x(T), P(x,t)\right),$$

which uniquely determines the exercise boundary together with the so-called *smooth pasting condition*

$$\frac{\partial P}{\partial x}(x^*(t), t) = -1.$$

This latter condition ensures that we have a smooth transition across the exercise boundary (see, e.g., [Tavella (2003); Kwok (2008)]).

In order to deal with that challenge, a number of alternate — somewhat heuristic — approaches have been suggested in the literature. The simplest approach — taken by [Hon and Mao (1999); Brandimarte (2003); Wu and Hon (2003); Hon and Wu (2007); Chan and Hubbert (2011)] — is to use a reasonably fine time-price grid and then to directly implement the payoff condition at each time step, i.e.,

$$P(x_i,t_j) = \max\left(E - x(T), P(x_i,t_j)\right), \qquad i = 1,\dots,N$$

where the points x_i, $i = 1,\dots,N$, are our collocation points and t_j denotes an arbitrary time step within the time discretization of $[0,T]$. The formula above indicates that the option value is updated at each time step by comparing the computed values $P(x_i,t_j)$ on the right-hand side (which are nothing but the value of a European put) against the value of the option at the time of expiry. Chan and Hubbert (2011) and the notes by Richardson (2009) provide a good motivation/argumentation for why this strategy is appropriate when computing the value of the American put.

Similarly, Ballestra and Pacelli (2013) suggest viewing an American option as the result of applying Richardson extrapolation to a Bermudan option. Alternatively, based on the work in [Nielsen *et al.* (2002, 2008)], Fasshauer *et al.* (2004); Belova *et al.* (2012); Safdari-Vaighani *et al.* (2014) use a penalty approach which converts the moving boundary problem to a standard fixed boundary problem. We present an example of pricing American options in the `GaussQR` library under `FinanceAmerPut.m`.

[7]Of course, any numerical PDE solver also discretizes the PDE so only finitely many possible exercise times can be considered.

Appendix A

Collection of Positive Definite Kernels and Their Known Mercer Series

Here we collect several positive definite kernels $K : \Omega \times \Omega \to \mathbb{R}$, some of which appeared in the text and some which did not, together with their eigenvalues λ_n and eigenfunctions φ_n so that the kernel can be represented using Mercer's Theorem 2.2 as

$$K(\boldsymbol{x}, \boldsymbol{z}) = \sum_{n=1}^{\infty} \lambda_n \varphi_n(\boldsymbol{x}) \varphi_n(\boldsymbol{z})$$

and therefore can be implemented using the Hilbert–Schmidt SVD of Chapter 13 — even if the closed form of the kernel is unknown (as is indeed the case in some of the examples listed below), or if the kernel matrix is not ill-conditioned. In this latter case, however, the series will usually converge very slowly and therefore a closed form expression for the kernel and direct implementation are preferable.

A.1 Piecewise Linear Kernels

A.1.1 *Brownian bridge kernel*

The Brownian bridge kernel is discussed, e.g., by Ritter (2000). The closed form and Hilbert–Schmidt eigenvalues and eigenfunctions are given by

$$\begin{aligned} K(x, z) &= \min(x, z) - xz, \\ \lambda_n &= \frac{1}{(n\pi)^2}, \qquad n = 1, 2, \ldots, \\ \varphi_n(x) &= \sqrt{2} \sin(n\pi x) = \sqrt{2} \sin\left(\frac{x}{\sqrt{\lambda_n}}\right). \end{aligned}$$

This kernel can also be obtained as the Green's function for the differential operator $\mathcal{L} = -\frac{\mathrm{d}^2}{\mathrm{d}x^2}$ with boundary conditions $u(0) = u(1) = 0$ (see also Example 6.6).

A.1.2 *Brownian motion kernel*

The Hilbert–Schmidt eigenvalue problem (2.5) for the Brownian motion kernel on the domain $\Omega = [0,1]$ was solved by Kailath (1966). This yields

$$\begin{aligned} K(x,z) &= \min(x,z), \\ \lambda_n &= \frac{4}{(2n-1)^2\pi^2}, \qquad n = 1,2,\ldots, \\ \varphi_n(x) &= \sqrt{2}\sin\left((2n-1)\frac{\pi x}{2}\right) = \sqrt{2}\sin\left(\frac{x}{\sqrt{\lambda_n}}\right). \end{aligned}$$

The corresponding differential operator and boundary conditions for this kernel are $\mathcal{L} = -\frac{\mathrm{d}^2}{\mathrm{d}x^2}$ with $u(0) = u'(1) = 0$ (see also Example 2.6).

A.1.3 *Another piecewise linear kernel*

If one again uses the differential operator $\mathcal{L} = -\frac{\mathrm{d}^2}{\mathrm{d}x^2}$, but changes the boundary conditions to $u'(0) = u(1) = 0$, then one obtains another piecewise linear kernel. This was done in [Stakgold (1979, Section 6.2, Ex. 2)] resulting in

$$\begin{aligned} K(x,z) &= 1 - \max(x,z), \\ \lambda_n &= \frac{4}{(2n-1)^2\pi^2}, \qquad n = 1,2,\ldots, \\ \varphi_n(x) &= \sqrt{2}\cos\left((2n-1)\frac{\pi x}{2}\right) = \sqrt{2}\cos\left(\frac{x}{\sqrt{\lambda_n}}\right). \end{aligned}$$

A.2 Exponential Kernel

The exponential kernel (see Section 3.1) is also known as the *Ornstein–Uhlenbeck kernel* or C^0 *Matérn kernel.* We give several different Mercer series representations for this kernel, depending on different choices of the (univariate) domain and boundary conditions. Like the piecewise linear kernels of Appendix A.1, all versions of this kernel are continuous but not differentiable at $x = z$. Eigenvalues and eigenfunctions in the multivariate setting are not known to us. Of course, one can create tensor products of the one-dimensional kernels. However, this will not generate kernels that are equivalent to the radial exponential kernel.

A.2.1 *Domain:* $[0, 1]$

This version is due to Hawkins (1989) and can be found in [Su and Cambanis (1993)]. It results in

$$
\begin{aligned}
K(x, z) &= \mathrm{e}^{-|x-z|}, \\
\lambda_n &= \frac{2}{1 + v_n^2}, \qquad n = 0, 1, 2, \ldots, \\
\varphi_n(x) &= \sqrt{\frac{2}{3 + v_n^2}} \left(\sin(v_n x) + v_n \cos(v_n x) \right),
\end{aligned}
$$

where the v_n are solutions of

$$
\tan v_n = \frac{2 v_n}{v_n^2 - 1}, \qquad v_n \in \left((2n-1)\frac{\pi}{2}, (2n+1)\frac{\pi}{2} \right), \qquad n = 0, 1, 2, \ldots,
$$

which need to be found numerically.

A.2.2 *Domain:* $[-L, L]$

This version was derived by Van Trees (2001, Chapter 3) and can also be found in [Xiu (2010)]. We have

$$
\begin{aligned}
K(x, z) &= \mathrm{e}^{-\varepsilon |x-z|}, \\
\lambda_n &= \begin{cases} \dfrac{2\varepsilon}{\varepsilon^2 + v_n^2}, & \text{if } n \text{ even}, \\ \dfrac{2\varepsilon}{\varepsilon^2 + w_n^2}, & \text{if } n \text{ odd}, \end{cases} \\
\varphi_n(x) &= \begin{cases} \sqrt{\dfrac{2 v_n}{2 L v_n - \sin(2 L v_n)}} \sin(v_n x), & \text{if } n \text{ even}, \\ \sqrt{\dfrac{2 w_n}{2 L w_n + \sin(2 L w_n)}} \cos(w_n x), & \text{if } n \text{ odd}, \end{cases}
\end{aligned}
$$

where v_n and w_n are solutions of

$$
\begin{aligned}
v_n + \varepsilon \tan(L v_n) &= 0, \qquad v_n \in \left((2n-1)\frac{\pi}{2L}, (2n+1)\frac{\pi}{2L} \right), \qquad n \text{ even}, \\
\varepsilon - w_n \tan(L w_n) &= 0, \qquad w_n \in \left((2n-1)\frac{\pi}{2L}, (2n+1)\frac{\pi}{2L} \right), \qquad n \text{ odd}.
\end{aligned}
$$

A.2.3 *Domain:* $[0, \infty)$

On this semi-infinite domain we require a weight function, namely $\rho(x) = 2\mathrm{e}^{-2x}$ in the Hilbert–Schmidt integral operator. The solution of an integral equation that corresponds to the associated eigenvalue problem can be found in [Juncosa (1945)].

It results in

$$K(x,z) = \mathrm{e}^{-\varepsilon|x-z|},$$
$$\lambda_n = \frac{4\varepsilon}{v_n^2}, \qquad n = 1,2,\ldots,$$
$$\varphi_n(x) = J_\varepsilon\left(v_n \mathrm{e}^{-x}\right),$$

where v_n are the positive zeros of the Bessel functions of the first kind of order $\varepsilon - 1$, i.e., $J_{\varepsilon-1}(v_n) = 0$, $n = 1,2,\ldots$.

In Example 2.5 we used the weight function $\rho(x) = \alpha\mathrm{e}^{-\alpha x}$, and then specialized to $\alpha = 4\varepsilon$. This led to

$$K(x,z) = \mathrm{e}^{-\varepsilon|x-z|},$$
$$\lambda_n = \frac{8}{(2n-1)^2\pi^2}, \qquad n = 1,2,\ldots,$$
$$\varphi_n(x) = \mathrm{e}^{\varepsilon x}\sin\left(\frac{(2n-1)\pi}{2}\mathrm{e}^{-2\varepsilon x}\right).$$

A.3 Other Continuous Kernels

A.3.1 *Tension spline kernel*

This is a special member of the family of iterated Brownian bridge kernels listed in Appendix A.5.1. The eigenfunctions satisfy the Sturm–Liouville differential eigenvalue problem

$$-\varphi''(x) + \varepsilon^2\varphi(x) = \frac{1}{\lambda}\varphi, \qquad \varphi(0) = \varphi(1) = 0, \tag{A.1}$$

where ε is a shape parameter or *tension parameter*. We get

$$K(x,z) = \begin{cases} \dfrac{\sinh(\varepsilon x)\sinh\left(\varepsilon(1-z)\right)}{\varepsilon\sinh(\varepsilon)}, & x < z, \\ \dfrac{\sinh(\varepsilon z)\sinh\left(\varepsilon(1-x)\right)}{\varepsilon\sinh(\varepsilon)}, & x > z, \end{cases}$$
$$\lambda_n = \frac{1}{n^2\pi^2 + \varepsilon^2}, \qquad n = 1,2,\ldots,$$
$$\varphi_n(x) = \sin(n\pi x).$$

These kernels minimize the norm associated with the inner product

$$\langle f,g\rangle_{\mathcal{H}_K([0,1])} = \int_0^1 f'(x)g'(x)\,\mathrm{d}x + \varepsilon^2\int_0^1 f(x)g(x)\,\mathrm{d}x.$$

The original "spline in tension" of Schweikert (1966) minimizes the norm associated with the inner product

$$\langle f,g\rangle_{\mathcal{H}_K([0,1])} = \int_0^1 f''(x)g''(x)\,\mathrm{d}x + \varepsilon^2\int_0^1 f'(x)g'(x)\,\mathrm{d}x$$

and is therefore analogous to cubic splines and the first iterate of the Brownian bridge kernel whose closed form is given in (7.12). Use of a similar inner product was proposed by Bouhamidi (2001), leading to a family of "splines in tension" minimizing the norm associated with the inner product

$$\langle f, g\rangle_{\mathcal{H}_K(\mathbb{R})} = \int_{\mathbb{R}} f^{(\beta)}(x)g^{(\beta)}(x)\,\mathrm{d}x + \varepsilon^2 \int_{\mathbb{R}} f^{(\beta-1)}(x)g^{(\beta-1)}(x)\,\mathrm{d}x.$$

Note how this kernel has a certain tension specified by the choice of the shape parameter ε.

A.3.2 *Relaxation spline kernel*

By *subtracting* the shift ε^2 in (A.1) instead of adding it we obtain a different set of eigenvalues, and therefore also a different kernel, namely

$$K(x,z) = \begin{cases} \dfrac{\sin(\varepsilon x)\sin(\varepsilon(1-z))}{\varepsilon\sin(\varepsilon)}, & x < z, \\ \dfrac{\sin(\varepsilon z)\sin(\varepsilon(1-x))}{\varepsilon\sin(\varepsilon)}, & x > z, \end{cases}$$

$$\lambda_n = \frac{1}{n^2\pi^2 - \varepsilon^2}, \qquad n = 1, 2, \ldots,$$

$$\varphi_n(x) = \sin(n\pi x).$$

Since the effects of the shape parameter here amount to a relaxation instead of a tension we choose to call this kernel a *relaxation spline* kernel.

A.3.3 *Legendre kernel*

Another kernel that is only continuous on its domain, $\Omega = [-1, 1]$, is obtained by considering the differential operator defined by $\mathcal{L}u(x) = -\frac{\mathrm{d}}{\mathrm{d}x}\left[(1-x^2)\frac{\mathrm{d}}{\mathrm{d}x}u(x)\right]$ on $(-1, 1)$ with (regular singular) boundary condition $\lim_{|x|\to 1} u(x) < \infty$. As shown in [Courant and Hilbert (1953, Chapter V, Section 15)] (see also Example 3.2) one gets

$$K(x,z) = -\frac{1}{2}\log(1-\min(x,z))(1+\max(x,z)) + \log 2 - \frac{1}{2},$$

$$\lambda_n = \frac{1}{n(n+1)}, \qquad n = 1, 2, \ldots,$$

$$\varphi_n(x) = \sqrt{\frac{2n+1}{2}}P_n(x),$$

where the P_n are Legendre polynomials of degree n.

A.4 Modified Exponential Kernel

In [Spanos *et al.* (2007)] the Hilbert–Schmidt integral eigenvalue problem for the C^2 Matérn kernel $K(x,z) = (1+\varepsilon|x-z|)\mathrm{e}^{-\varepsilon|x-z|}$ on $[-L, L]$ is solved by differentiating the integral equation four times and introducing appropriate boundary conditions. The structure of the eigenvalues and eigenfunctions is similar to those in Appendix A.2.2, but significantly more technical. We therefore refer the interested reader to the original paper for more details.

A.5 Families of Iterated Kernels

Here we list several families of kernels obtained either as iterated kernels of a Hilbert–Schmidt integral operator as as solution of a differential eigenvalue problem with iterated differential operator as discussed in detail in Chapter 7 for the iterated Brownian bridge kernels listed in Appendix A.5.1.

A.5.1 *Iterated Brownian bridge kernels*

A detailed derivation of this family of kernels was given in Chapter 7. The eigenvalues and eigenfunctions on the domain $[0,1]$ are given by

$$\lambda_n = \frac{1}{(n^2\pi^2 + \varepsilon^2)^\beta},$$
$$\varphi_n(x) = \sqrt{2}\sin(n\pi x).$$

A closed form for these kernels is, however, known only in special cases. For $\varepsilon = 0$ we get piecewise polynomial kernels whose closed form is given in terms of Bernoulli polynomials by (cf. (7.7))

$$K_{\beta,0}(x,z) = (-1)^{\beta-1}\frac{2^{2\beta-1}}{(2\beta)!}\left(B_{2\beta}\left(\frac{|x-z|}{2}\right) - B_{2\beta}\left(\frac{x+z}{2}\right)\right),$$

and when $\varepsilon > 0$ we have the tension spline kernel of Appendix A.3.1 for $\beta = 1$, and the kernel (7.12) for $\beta = 2$.

A.5.2 *Periodic spline kernels*

These trigonometric splines were discussed in [Wahba (1990, Chapter 2)]. The kernels, given in closed form in terms of Bernoulli polynomials and as a Fourier cosine series, along with its eigenvalues and eigenfunctions are (cf. (3.2))

$$K_\beta(x,z) = \frac{(-1)^{\beta-1}}{(2\beta)!} B_{2\beta}(|x-z|)$$
$$= \sum_{n=1}^{\infty} \frac{2}{(2n\pi)^{2\beta}} \cos(2n\pi(x-z)),$$
$$\lambda_n = \begin{cases} (2j\pi)^{-2\beta}, & n = 2j-1, \\ (2j\pi)^{-2\beta}, & n = 2j, \end{cases} \qquad j = 1,2,\ldots,$$
$$\varphi_n(x) = \begin{cases} \sqrt{2}\sin(2j\pi x), & n = 2j-1, \\ \sqrt{2}\cos(2j\pi x), & n = 2j, \end{cases} \qquad j = 1,2,\ldots.$$

We can see that here each eigenvalue has two eigenfunctions associated with it. Such an occurrence of multiple eigenvalues is typical in the multivariate setting, and we can interpret the periodic spline kernels as restrictions of bivariate Euclidean kernels to the circle (or "sphere") $\mathbb{S}^1$ (see also the discussion of zonal kernels in Section 3.4.2).

A.5.3 *Periodic kernels*

We can generalize the periodic spline kernels from Appendix A.5.2 by adding a shift in the eigenvalues (analogous to the shift that generated the tension spline kernel from the Brownian bridge kernel in Appendix A.3.1). This results in (see also (3.9))

$$K_{\beta,\varepsilon}(x,z) = \sum_{n=1}^{\infty} \frac{2}{(4n^2\pi^2+\varepsilon^2)^\beta} 2\cos\left(2n\pi(x-z)\right), \quad x,z \in [0,1],\ \varepsilon > 0, \beta \in \mathbb{N},$$
$$= \sum_{n=1}^{\infty} \frac{2}{(4n^2\pi^2+\varepsilon^2)^\beta} \left(\cos(2n\pi x)\cos(2n\pi z) + \sin(2n\pi x)\sin(2n\pi z)\right),$$
$$\lambda_n = \begin{cases} (4j^2\pi^2+\varepsilon^2)^{-\beta}, & n = 2j-1, \\ (4j^2\pi^2+\varepsilon^2)^{-\beta}, & n = 2j, \end{cases} \qquad j = 1,2,\ldots,$$
$$\varphi_n(x) = \begin{cases} \sqrt{2}\sin(2j\pi x), & n = 2j-1, \\ \sqrt{2}\cos(2j\pi x), & n = 2j, \end{cases} \qquad j = 1,2,\ldots.$$

Due to the shift in the eigenvalues these kernels are no longer piecewise polynomial splines, and we do not readily have a closed form expression available for these kernels. This situation is similar to what happens for the iterated Brownian bridge kernels in Chapter 7.

A.5.4 *Chebyshev kernels*

In Section 3.9.2 we introduced two families of designer kernels on $[-1,1]$ based on eigenfunctions specified in terms of appropriately normalized Chebyshev polynomials and eigenvalues with either geometric or algebraic decay. Note that for these two examples the Mercer series is indexed beginning with $n = 0$.

For the geometrically decaying eigenvalues we had for $a \in (0,1]$ and $b \in (0,1)$ the family

$$\begin{aligned}
K_{a,b}(x,z) &= 1 - a + 2a(1-b)\frac{b(1-b^2) - 2b(x^2+z^2) + (1+3b^2)xz}{(1-b^2)^2 + 4b\left(b(x^2+z^2) - (1+b^2)xz\right)}, \\
\lambda_0 &= 1 - a, \\
\lambda_n &= \frac{a(1-b)b^n}{b}, \qquad n = 1, 2, \ldots, \\
\varphi_n(x) &= \sqrt{2 - \delta_{n0}}T_n(x),
\end{aligned}$$

where the T_n are Chebyshev polynomials of degree n (so that a weight $\rho(x) = \frac{1}{\pi\sqrt{1-x^2}}$ is employed in the underlying Hilbert–Schmidt integral operator).

For algebraically decaying eigenvalues, on the other hand, we got with $a \in (0,1]$ and $\beta \in \mathbb{N}$ that

$$\begin{aligned}
K_{a,\beta}(x,z) &= 1 - a + \frac{a(-1)^{\beta+1}(2\pi)^{2\beta}}{2(2\beta)!\zeta(2\beta)}\left(B_{2\beta}\left(\frac{|\cos^{-1}(x) + \cos^{-1}(z)|}{2\pi}\right)\right. \\
&\qquad\qquad \left. + B_{2\beta}\left(\frac{|\cos^{-1}(x) - \cos^{-1}(z)|}{2\pi}\right)\right), \\
\lambda_0 &= 1 - a, \\
\lambda_n &= \frac{a}{\zeta(2\beta)n^{2\beta}}, \qquad n = 1, 2, \ldots, \\
\varphi_n(x) &= \sqrt{2 - \delta_{n0}}T_n(x),
\end{aligned}$$

where the T_n are again Chebyshev polynomials of degree n, ζ is the Riemann zeta function, and B_n are Bernoulli polynomials of degree n.

A.6 Kernel for the First Weighted Sobolev Space

Three weighted Sobolev spaces were discussed in [Novak and Woźniakowski (2008, Appendix A.2.1)] (see also Example 8.2). These kernels were introduced by Thomas-Agnan (1996) (as well as [Duc-Jacquet (1973)] for the $\varepsilon = 1$ case) and satisfy

$$\begin{aligned}
K(x,z) &= \frac{\cosh(\varepsilon \min(x,z))\cosh(\varepsilon(1 - \max(x,z)))}{\varepsilon \sinh(\varepsilon)}, \\
\lambda_1 &= \frac{1}{\varepsilon^2}, \\
\lambda_n &= \frac{1}{\varepsilon^2 + ((n-1)\pi)^2}, \qquad n > 1, \\
\varphi_1(x) &= 1, \\
\varphi_n(x) &= \sqrt{2}\cos\left((n-1)\pi x\right),
\end{aligned}$$

where ε is a shape parameter.

Similar kernels for other weighted Sobolev spaces can be found, e.g., in [Wasilkowski and Woźniakowski (1999); Werschulz and Woźniakowski (2009); Dick *et al.* (2013b)].

A.7 Gaussian Kernel

In Section 12.2.1 we discussed the eigenvalues and eigenfunctions of the Gaussian kernel. They are at the heart of our implementation of the GaussQR algorithm and on $(-\infty, \infty)$ are given by

$$K(x,z) = \mathrm{e}^{-\varepsilon^2|x-z|^2},$$
$$\lambda_n = \sqrt{\frac{\alpha^2}{\alpha^2+\delta^2+\varepsilon^2}}\left(\frac{\varepsilon^2}{\alpha^2+\delta^2+\varepsilon^2}\right)^{n-1}, \qquad n = 1,2,\ldots,$$
$$\varphi_n(x) = \gamma_n \mathrm{e}^{-\delta^2 x^2} H_{n-1}(\alpha\beta x),$$

where the H_n are Hermite polynomials of degree n, and

$$\beta = \left(1 + \left(\frac{2\varepsilon}{\alpha}\right)^2\right)^{\frac{1}{4}}, \quad \gamma_n = \sqrt{\frac{\beta}{2^{n-1}\Gamma(n)}}, \quad \delta^2 = \frac{\alpha^2}{2}\left(\beta^2 - 1\right),$$

are constants defined in terms of the shape parameter ε and the parameter α in the weight function $\rho(x) = \frac{\alpha}{\sqrt{\pi}}\mathrm{e}^{-\alpha^2 x^2}$ of the Hilbert–Schmidt integral operator.

On $\mathbb{R}^d$ one uses the product form as discussed in Section 12.2.1.

A.8 Sinc Kernel

In [Slepian and Pollak (1961)] the Hilbert–Schmidt eigenfunctions of the sinc kernel are given in terms of scaled *prolate angular spheroidal wave functions* Ps_n^0 (see also [DLMF, Eq. (30.15.1)], i.e.,

$$\int_{-1}^{1} \frac{\varepsilon}{\pi}\frac{\sin\varepsilon(x-z)}{\varepsilon(x-z)}\varphi_n(z)\,\mathrm{d}z = \lambda_n\varphi_n(x),$$

so that

$$\varphi_n(x) = \sqrt{\frac{2n+1}{2}}\sqrt{\lambda_n}Ps_n^0(x,\varepsilon^2), \qquad n = 0,1,2,\ldots,$$
$$\lambda_n = \frac{2\varepsilon}{\pi}\left(K_n^0(\varepsilon)A_n^0(\varepsilon^2)\right)^2,$$

where the coefficients $K_n^0(\varepsilon)$ and $A_n^0(\varepsilon^2)$ are rather complicated and defined in [DLMF, Section 30.11(v)]. It may be interesting to note that for $\varepsilon \to 0$ the angular spheroidal wave functions Ps_n^0 converge to the Legendre polynomials P_n of degree n.

A.9 Zonal Kernels

A.9.1 *Spherical inverse multiquadric*

The spherical inverse multiquadric kernel on the sphere $\mathbb{S}^2$ was discussed in Section 3.4.2. It is given by

$$\begin{aligned} K(\boldsymbol{x},\boldsymbol{z}) &= \frac{1}{\sqrt{1+\gamma^2-2\gamma\boldsymbol{x}^T\boldsymbol{z}}}, \qquad \boldsymbol{x},\boldsymbol{z}\in\mathbb{S}^2,\ \gamma\in(0,1), \\ &= \sum_{n=0}^{\infty}\frac{4\pi\gamma^n}{2n+1}\sum_{\ell=1}^{2n+1} Y_{n,\ell}(\boldsymbol{x})Y_{n,\ell}(\boldsymbol{z}), \end{aligned}$$

where the $Y_{n,\ell}$, $\ell=1,\ldots,2n+1$, are spherical harmonics of degree n and order ℓ.

A.9.2 *Abel–Poisson kernel*

In addition to the spherical inverse multiquadric kernel of Appendix A.9.1, another popular kernel on $\Omega=\mathbb{S}^2$ is given by the *Abel–Poisson kernel* [Michel (2013, Eq. (6.88))]

$$\begin{aligned} K(\boldsymbol{x},\boldsymbol{z}) &= \frac{1-\gamma^2}{(1+\gamma^2-2\gamma\boldsymbol{x}^T\boldsymbol{z})^{3/2}}, \qquad \boldsymbol{x},\boldsymbol{z}\in\mathbb{S}^2,\ \gamma\in(-1,1), \\ &= \sum_{n=0}^{\infty} 4\pi\gamma^n\sum_{\ell=1}^{2n+1} Y_{n,\ell}(\boldsymbol{x})Y_{n,\ell}(\boldsymbol{z}), \end{aligned}$$

where $Y_{n,\ell}$ is again a spherical harmonic of degree n and order ℓ.

Appendix B

How To Choose the Data Sites

The distribution of data points throughout the domain plays an important role in determining the accuracy of kernel-based approximation methods. Some applications — such as surrogate modeling, computer experiments or the numerical solution of PDEs — allow the user to choose the location of the data sites, while in other settings — such as the fitting of given data measurements or in machine learning applications — the data locations are usually fixed. We mentioned this issue in a few places in the book, such as in Chapters 7, 12 and 16.

Generally speaking, one may distinguish between data-independent and data-dependent strategies, and between strategies for regression vs. those for interpolation. Data-independent strategies are more universal and also more generic, while data-dependent strategies often involve iterative and adaptive algorithms. Since we did not focus on the latter in the main part of this book we similarly avoid it here, and instead summarize some of the ideas for data-independent point selection strategies.

The main difference between optimal point distributions for regression and those for interpolation — especially on bounded domains — is that the former tend to favor a uniform distribution, while the latter usually push points toward the boundary of the domain. This is well known in the numerical analysis community where points for interpolation should be clustered near the boundary of the domain if one works with smooth methods such as polynomials or many types of kernels. The oscillations that otherwise occur near the boundary are known as the *Runge phenomenon* (see, e.g., [Fornberg and Zuev (2007)]) and tend to be eliminated with Chebyshev-like point distributions. In higher dimensions, however, such point distributions are impractical due to the increasing proportion of space taken up by the boundary as the dimension increases. Therefore — especially when working in higher dimensions — most people seem to agree that generic, near-uniform, distribution of the data sites is a good strategy since this provides a way to deal with the *curse of dimensionality*. In the literature on experimental design and on quasi-Monte Carlo integration a specific set of data sites is usually referred to as a *design*, and we now also follow this convention.

B.1 Low Discrepancy Designs

Error bounds in quasi-Monte Carlo integration are often expressed in terms of the so-called *star discrepancy* of the design $\mathcal{X}$ (see our discussion is Section 21.1.4 and also [Dick *et al.* (2013a)]). Since one wants to choose a design which optimizes the accuracy of an integration method one will then look for so-called *low discrepancy designs.* Such designs will yield a low star discrepancy $D_N(\mathcal{X}) = \sup_{\boldsymbol{x}\in[0,1]^d} |\Delta_{\mathcal{X}}(\boldsymbol{x})|$, where $\Delta_{\mathcal{X}}(\boldsymbol{x})$ denotes the *local discrepancy* of a design $\mathcal{X}$ at $\boldsymbol{x}$ and measures the difference between the proportion of design points in the open box $[\mathbf{0}, \boldsymbol{x}) = [0, x_1) \times \cdots \times [0, x_d) \subset [0,1]^d$ and the "ideal" proportion[8], i.e., the volume of the box $[\mathbf{0}, \boldsymbol{x})$ (which can be computed as an integral).

Typical low discrepancy designs include *Halton sequences* and *Sobol′ sequences.* We focus on these two types of designs since they are available in MATLAB's Statistics and Machine Learning Toolbox as `haltonset` and `sobolset`, respectively. The former were introduced by Halton (1960) and the latter by Sobol′ (1967). The construction of both of these sequences is of a number-theoretical nature and described in detail in [Dick *et al.* (2013a)]. One difference between Halton sequences and Sobol′ sequences is that the former generate nested designs, while the latter do not. This means that a set of $M < N$ Halton points is a subset of the set of N points, while for Sobol′ points this is generally not the case. Figure B.1 shows 100 Halton points and 100 Sobol′ points in the unit square $[0, 1]$. The points are generated with the MATLAB code `points = net(haltonset(2),100)` and `points = 100*net(sobolset(2),100)/99`, respectively and plotted with `plot([points(:,1),points(:,2)],'ko')`.

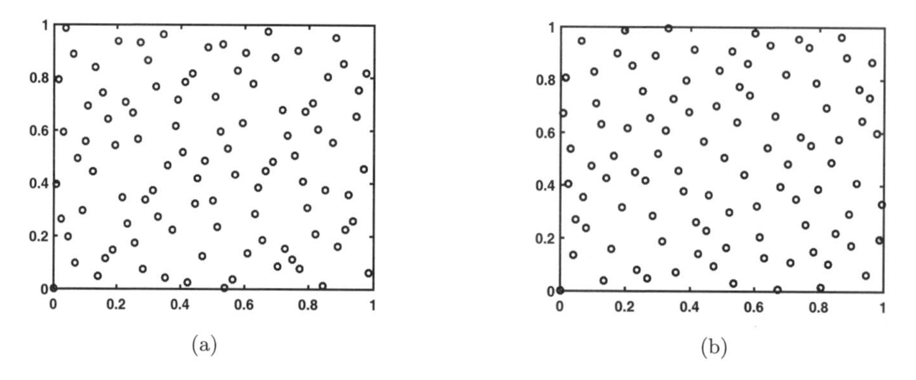

Fig. B.1 (a) 100 Halton points; (b) 100 Sobol′ points in $[0, 1]^2$.

Other low discrepancy designs found in the literature include Hammersley points, Faure and Niederreiter sequences, and more generally lattice rules and digital nets (see, e.g., [Dick *et al.* (2013a)] for more details). Another type of design

[8] An algebraic interpretation of local discrepancy as the derivative of the Riesz representer of the integration error was provided in Section 21.1.4.

which sometimes is used for high-dimensional integration and also numerical solution of PDEs are the so-called *sparse grids* of Bungartz and Griebel (2004). We refer the reader to that survey paper or the book [Garcke and Griebel (2013)] for more details.

Despite their asymptotic properties, the simplest implementations of low-discrepancy sequences suffer for small numbers of points during the trek toward high dimensions. Standard Halton points generated in 22 dimensions see a very nice distribution between the first and second dimensions but a much less useful distribution between the twenty-first and twenty-second dimensions, as seen in Figure B.2. Strategies for fighting this includes scrambled sequences [Braaten and Weller (1979)], randomized sequences [Wang and Hickernell (2000); Hickernell and Hong (2002)] and shuffled sequences [Hess and Polak (2003)]. This does not play a role in the experiments in this texts, but readers should be aware when conducting their own experiments.

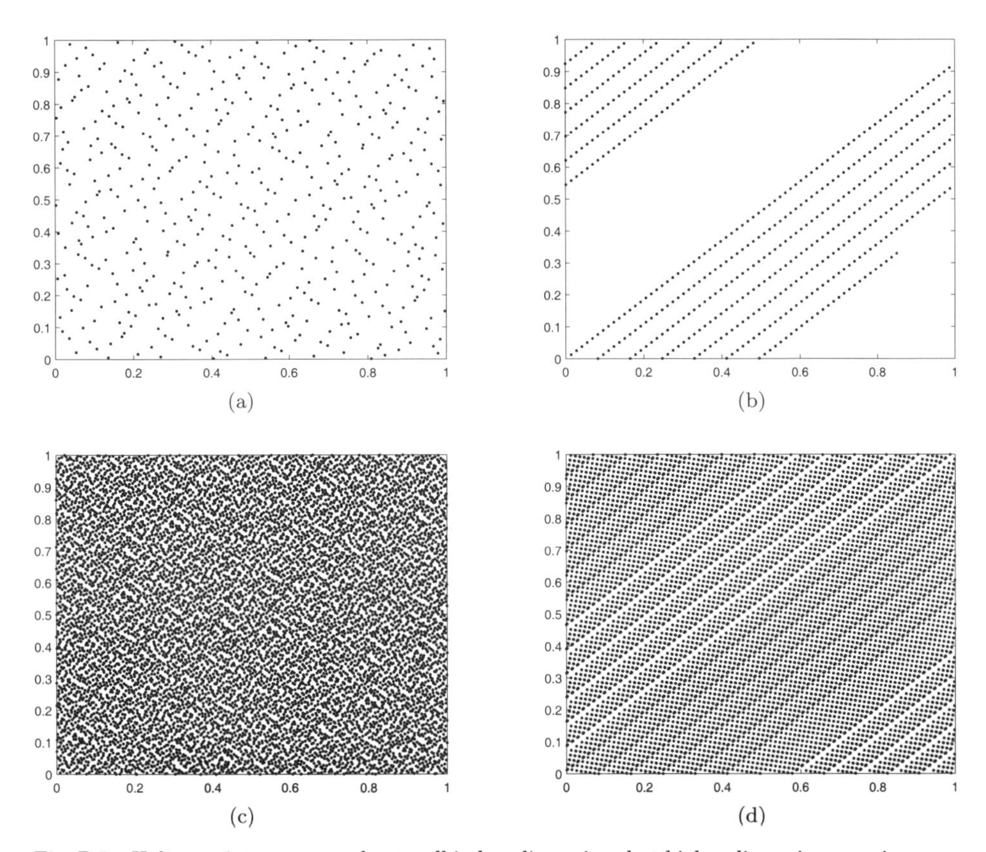

Fig. B.2 Halton points can spread out well in low dimensions but higher dimensions require more points to cover the entire domain. This example was provided by Jessica Garrett at the University of Colorado Denver. (a) 500 points comparing 1st and 2nd dimensions, (b) 500 points comparing 21st and 22nd dimensions, (c) 10000 points comparing 1st and 2nd dimensions, (d) 10000 points comparing 21st and 22nd dimensions.

B.2 Optimal Designs in Statistics

In the literature on surrogate modeling, kriging and experimental design one can find other types of "optimal" designs. Some good references are, e.g., [Sacks *et al.* (1989); Morris *et al.* (1993); Fang *et al.* (2006)]. If we again focus on those designs available via MATLAB's Statistics and Machine Learning Toolbox, then we are led to the *Latin hypercube designs* of McKay *et al.* (1979), which generalize Latin squares and simply ensure that each (multidimensional) row and column, i.e., axis-parallel hypercube, of the unit cube is occupied by a design point, and to *D-optimal* designs, which maximize the determinant of the Gram matrix (or information matrix) $\mathsf{K}^T\mathsf{K}$. However, the routines provided in MATLAB do this for regression with *polynomial*-based K-matrices (such as Vandermonde matrices), not kernel-based ones and therefore lead to rather regular designs. Figure B.3 shows 100 Latin hypercube points and 9 D-optimal points for a quadratic model in the unit square $[0, 1]$. The points are generated with the MATLAB code `points = lhsdesign(100,2)` and `points = cordexch(2,9,'quadratic','bounds',[0 0;1 1])`, respectively and plotted with `plot([points(:,1),points(:,2)],'ko')`. As Figure B.3 shows, these designs are well distributed and therefore might still be useful for kernel-based methods. While uniform distribution alone is not optimal, D-optimality and similar criteria aim at minimizing statistical error bounds based on, e.g., the variance of the estimator. Other popular designs in the statistics literature are so-called *A-optimal* or *E-optimal* designs. They are constructed as to minimize the trace of $\mathsf{K}^T\mathsf{K}$ or the smallest eigenvalue of $(\mathsf{K}^T\mathsf{K})^{-1}$, respectively (see, e.g., [Morris *et al.* (1993); Fang *et al.* (2006)]).

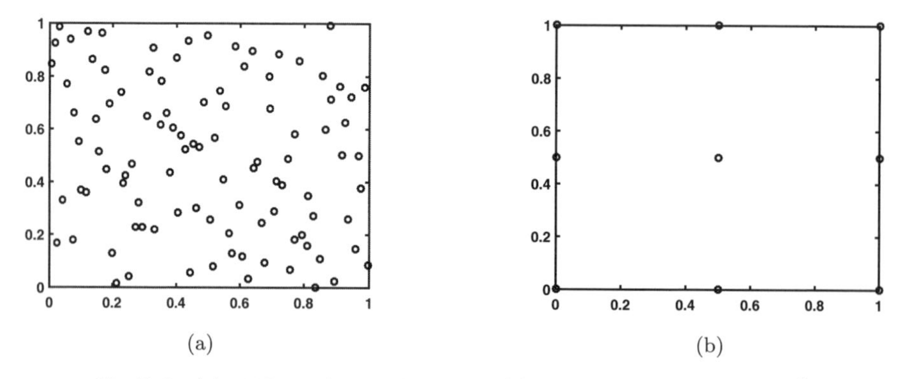

Fig. B.3 (a) 100 Latin hypercube points; (b) 9 D-optimal points in $[0, 1]^2$.

B.3 Optimal Points in Approximation Theory

The study of optimal designs is much less prominent in the numerical analysis and approximation theory literature. One optimal design in 1D, selected by minimizing the Lebesgue constant, are the Chebyshev-Lobatto points. For integer $N > 1$, the N Chebyshev–Lobatto points [Trefethen (2013)] (also called the Chebyshev nodes, Chebyshev points or Chebyshev–Gauss–Lobatto points) are defined as the roots of the degree N Chebyshev polynomial, as discussed in Section 3.9.2. These points are most naturally defined on $[-1, 1]$, though they can be linearly scaled as needed. They are accessed by passing the `'cheb'` option in the `pickpoints` function in the `gaussqr` library.

The concept of (approximate) *Fekete points* is analogous to D-optimal designs in statistics, i.e., they maximize the determinant of the interpolation matrix K (see, e.g., [De Marchi (2003); Briani *et al.* (2012)]). This idea is natural in approximation theory since Fekete points also minimize the Lebesgue constant, i.e., the norm of the interpolation operator, and therefore improve the accuracy of the interpolation method. Other optimal point sets — mostly aimed at keeping the Lebesgue constant under control — are, e.g, (approximate) *Leja points* (see, e.g., [De Marchi (2003)]), *Padua points* [Caliari *et al.* (2008)] and *weakly admissible meshes* [Bos *et al.* (2011b)], which are often just Chebyshev–Lobatto points (as in Figure B.4). As already mentioned for the D-optimal points in the previous section, these strategies which were designed for use with polynomial approximation methods might also be good choices for kernel methods — especially since they focus on interpolation and quadrature rather than regression.

Other approaches in the literature on kernels include the greedy points of De Marchi *et al.* (2005), which are obtained by iteratively placing points at the maximum (monitored on a very fine grid) of the power function. This results in data-independent near-uniform point distributions. Iske (2000) uses a similar strategy. And, finally, CVT points, a design based on centroidal Voronoi tesselations, was proposed in [Du *et al.* (1999, 2002, 2010)].

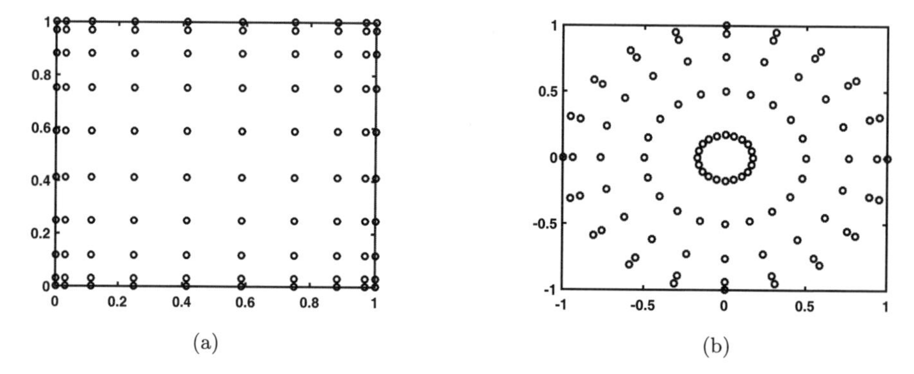

Fig. B.4 Weakly admissible mesh with (a) 100 points in $[0, 1]^2$ and (b) 100 points in $\mathbb{S}^1$.

Appendix C

A Few Facts from Analysis and Probability

Some good references for the following are, e.g., [Cheney (2001); Hunter and Nachtergaele (2001); Atkinson and Han (2009)] or [Steinwart and Christmann (2008, Appendix)].

Definition C.1 (Bounded linear operator). *Let X, Y be normed linear spaces. A linear operator (or linear transformation) $\mathcal{T}: X \to Y$ is* bounded *if there exists a constant $M \geq 0$ such that $\|\mathcal{T}f\|_Y = M\|f\|_X$ for all $f \in X$.*

A bounded linear operator is also referred to as a continuous linear operator. In particular, on a finite-dimensional linear space, every linear operator is bounded. Integral and differential operators are special kinds of linear operators. The former are often bounded, the latter usually not. For example, the integral operator

$$\mathcal{K}f(\boldsymbol{x}) = \int_\Omega K(\boldsymbol{x}, \boldsymbol{z}) f(\boldsymbol{z})\,\mathrm{d}\boldsymbol{z}, \qquad \boldsymbol{x} \in \Omega,$$

is bounded provided the kernel K is continuous on $\Omega \times \Omega$. The differential operator $\mathcal{D}f = f'$, on the other hand, is an unbounded linear operator on $C^1([0,1])$ since $f(x) = \mathrm{e}^{\lambda x}$ is an eigenfunction of $\mathcal{D}$ for any $\lambda \in \mathbb{R}$, i.e., $\mathcal{D}f = \lambda f$, which shows that $\frac{\|\mathcal{D}f\|}{\|f\|} = |\lambda|$, which is unbounded.

The classical definition of the adjoint of an operator is the following one which applies in the case of *bounded* operators.

Definition C.2 (Adjoint of a bounded operator). *Let $\mathcal{T}$ be a bounded operator on a Hilbert space $\mathcal{H}$ with inner product $\langle \cdot, \cdot \rangle$. The* adjoint *of $\mathcal{T}$, $\mathcal{T}^*$, which is also a bounded operator on $\mathcal{H}$, is defined via*

$$\langle f, \mathcal{T}g \rangle = \langle \mathcal{T}^* f, g \rangle \qquad \text{for all } f, g \in \mathcal{H}.$$

A typical example is given by a Hilbert–Schmidt integral operator whose adjoint is an operator of the same type. If $\mathcal{T}^* = \mathcal{T}$ then we call $\mathcal{T}$ a *self-adjoint* operator. Note that in the finite-dimensional matrix setting the adjoint corresponds to the transpose, and self-adjointness to symmetry.

Differential operators, however, are usually not bounded. Therefore we need

Definition C.3 (Adjoint of an unbounded operator). *Suppose that* $\mathcal{L} : D(\mathcal{L}) \subset \mathcal{H} \to \mathcal{H}$ *is a densely defined unbounded linear operator on a Hilbert space* $\mathcal{H}$. *The* adjoint $\mathcal{L}^* : D(\mathcal{L}^*) \subset \mathcal{H} \to \mathcal{H}$ *is the operator with domain*

$$D(\mathcal{L}^*) = \{g \in \mathcal{H} \mid \text{there is an } h \in \mathcal{H} \text{ with } \langle \mathcal{L}f, g\rangle = \langle f, h\rangle \text{ for all } f \in D(\mathcal{L})\}.$$

If $g \in D(\mathcal{L}^*)$, *then we define* $\mathcal{L}^* g = h$, *where* h *is the unique element such that* $\langle \mathcal{L}f, g\rangle = \langle f, h\rangle$ *for all* $f \in D(\mathcal{L})$.

In particular, the adjoint of a differential operator is another differential operator which can be obtained via integration by parts. The domain $D(\mathcal{L})$ determines the boundary conditions for $\mathcal{L}$, and the domain $D(\mathcal{L}^*)$ determines the boundary conditions for $\mathcal{L}^*$. These boundary conditions are chosen in such a way that the boundary terms for the integration by parts procedure vanish.

Definition C.4 (Precompact set). *A subset* A *of a metric space* X *is* precompact *if its closure in* X *is compact.*

Definition C.5 (Compact operator). *Let* X *and* Y *be normed linear spaces. A linear operator* $\mathcal{T} : X \to Y$ *is* compact *if* $\mathcal{T}(B)$ *is a precompact subset of* Y *for every bounded subset* B *of* X.

Equivalently, $\mathcal{T}$ is compact if and only if every bounded sequence $\{x_n\} \subset X$ has a subsequence $\{x_{n_k}\}$ such that $\{\mathcal{T}x_{n_k}\}$ converges in Y.

Definition C.6 (Haussdorff space). *A topological space* X *is called a* Hausdorff space *if every pair of distinct points* $x, y \in X$ *has a pair of nonintersecting neighborhoods. Moreover,* X *is* locally compact *if every point* $x \in X$ *has a compact neighborhood.*

Any metric space is a Hausdorff space. In particular, the Euclidean space $\mathbb{R}^d$ is a locally compact (but not compact) Hausdorff space.

Theorem C.1 (Spectral theorem). *If* $\mathcal{T}$ *is a compact self-adjoint operator defined on a Hilbert space* $\mathcal{H}$, *then* $\mathcal{T}$ *is of the form* $\mathcal{T}f = \sum \lambda_n \langle f, e_n\rangle_{\mathcal{H}} e_n$ *for an appropriate (possibly finite) orthonormal sequence* $\{e_n\}$ *and appropriate real numbers* λ_n *satisfying* $\lim \lambda_n = 0$. *Furthermore, we have* $\mathcal{T}e_n = \lambda_n e_n$.

Definition C.7 (σ-algebra and measurable space). *A* σ-algebra $\mathcal{A}$ *on a set* X *is a collection of subsets of* X *such that*

(1) the empty set $\emptyset$ *is in* $\mathcal{A}$,
(2) if $A \in \mathcal{A}$ *then the complement* A^c *of* A *is in* $\mathcal{A}$, *i.e.,* $A^c = X \setminus A \in \mathcal{A}$,
(3) if $\{A_1, A_2, \ldots\}$ *is a countable family of sets in* $\mathcal{A}$, *then* $\bigcup_{i=1}^{\infty} A_i \in \mathcal{A}$.

Moreover, a measurable space $(X, \mathcal{A})$ *is a set* X *and a* σ*-algebra* $\mathcal{A}$ *on* X. *The elements of* $\mathcal{A}$ *are called* measurable sets.

Definition C.8 (Measurable function). *Let* $(X, \mathcal{A})$ *and* $(Y, \mathcal{B})$ *be measurable spaces. A* measurable function *is a mapping* $f : X \to Y$ *such that* $f^{-1}(B) \in \mathcal{A}$ *for every* $B \in \mathcal{B}$.

Definition C.9 (Measure). *A* measure μ *on a set* X *is a mapping* $\mu : \mathcal{A} \to [0, \infty)$ *on a* σ*-algebra* $\mathcal{A}$ *of* X *such that*

(1) $\mu(\emptyset) = 0$,
(2) if $\{A_1, A_2, \ldots\}$ *is a countable family of mutually disjoint sets in* $\mathcal{A}$, *i.e.,* $A_i \cap A_j = \emptyset$ *for* $i \neq j$, *then* $\mu\left(\bigcup_{i=1}^{\infty} A_i\right) = \sum_{i=1}^{\infty} \mu(A_i)$.

Definition C.10 (Borel measure). *Let* X *be a locally compact Hausdorff space and let* $\mathcal{B}(X)$ *be the* σ*-algebra of Borel sets on* X, *i.e., the smallest* σ*-algebra that contains all the open sets of* X. *Any measure* μ *on* $\mathcal{B}(X)$ *is called a* Borel measure.

Definition C.11 (Mean-square continuous random field). *A random field* Y *is called* mean-square continuous *(or second-order) at* $\boldsymbol{x}$ *if* $\mathbb{E}\left[|Y_{\boldsymbol{x}}|^2\right]$ *is finite and* $\lim_{\boldsymbol{z} \to \boldsymbol{x}} \mathbb{E}\left[|Y_{\boldsymbol{x}} - Y_{\boldsymbol{z}}|^2\right] = 0$.

Appendix D

The GaussQR Repository in MATLAB

One of the goals of this book is to provide demonstrations of how kernels can be used in computation. To that end, we have created a library called GaussQR where relevant kernel computations are performed in MATLAB. In this appendix, we introduce this library and show how it facilitates computation with kernels.

Because both MATLAB and the GaussQR library are constantly evolving to provide users with the best software possible, the discussion here serves primarily to facilitate an understanding of the programs provided in the text. MATLAB's and GaussQR's documentation are more up-to-date and explain topics more thoroughly, so the content in this appendix should only be used as a reference if a MATLAB installation is inaccessible.

D.1 Accessing GaussQR

The GaussQR library was developed at the Illinois Institute of Technology and is still housed at

http://math.iit.edu/~mccomic/gaussqr

From this location, the most recent stable release can be downloaded. Releases are made roughly once every 12–18 months, although there is no set deadline. For those users who want to use the most up-to-date version of the library, there is a Mercurial repository which is accessible to everyone available at the same site.

After acquiring the library, the directions for installation and usage should be provided in the home directory (or else we have not been doing our job). Basically though, all that is needed is to point MATLAB in the right direction by adding GaussQR to its path and start using the repository. The script rbfsetup will automatically make the necessary additions to the path and create the GAUSSQR_PARAMETERS global object that stores certain commonly used quantities such as the coefficients of some Hermite polynomials.

D.2 Common functions in `GaussQR`

Within this text we have, at times, used certain `GaussQR` functions without going into great detail about them. This serves two purposes: first, it allows us to focus on the mathematics of the situation without getting distracted with tedium, and second, it allows the text to remain as accurate as possible even as the `GaussQR` library continues to evolve. In this section we list some functions that appear in the book as a reference when MATLAB is not available; when using MATLAB to execute code or run experiments presented in the book, it is still recommended to use the documentation available in the `GaussQR` library as this section could quickly become out-of-date.

Our first discussion involves common functions:

- `rbfsetup` — The script that allows for access to the rest of the repository. Also used to define user preferences and tolerances.
- `DistanceMatrix` — Useful for computing matrices of $\|\boldsymbol{x} - \boldsymbol{z}\|$ values which appear in radial kernels. Can be used for evaluating sparse kernel matrices if a compactly supported kernel is passed.
- `DifferenceMatrix` — Useful for computing matrices of $x_i - z_i$ values for differentiating radial kernels.
- `pickpoints/pick2Dpoints` — Used to create points distributed in one or two dimensions according to various different distribution strategies.
- `pickfunc` — Access to a selection of test functions that are too complicated for an inline definition.
- `errcompute` — Used for computing the error between two vectors as judged in various different ways. The choice of error can be set in `GAUSSQR_PARAMETERS`.
- `gqr_downloaddata` — Downloads content from the internet, most likely from the `GaussQR` data directory. Allows us to avoid binary files in the `GaussQR` repository.
- `rescale_data` — Converts data from its input domain to $[-1, 1]^d$.

These are functions specific to the Gaussian kernel:

- `gqr_phi` — Evaluates the Gaussian eigenfunctions.
- `gqr_solveprep` — Takes the steps to form the HS-SVD basis: computing β and δ^2, choosing the necessary eigenfunctions, and forming $\bar{\mathsf{C}}$.
- `gqr_solve` — Solves for the HS-SVD basis interpolation coefficients of a Gaussian interpolant.
- `gqr_rsolve` — Creates a low-rank eigenfunction approximate Gaussian interpolant.
- `gqr_eval` — Evaluates a Gaussian interpolant (either low-rank or HS-SVD basis) at selected locations.

Other functions that appear in this text include:

- `gssfunc` — An overloaded function. Can be called to access optimization criteria for the generalized Sobolev space example in Program 8.2 or to evaluate the residual left during the search for the desired GSS kernel.
- `sphHarm` — Evaluates the spherical harmonics. Adapted from [Wright (2014)].
- `haltonseq` — A poor implementation of a Halton point generator for those users without MATLAB's Statistics and Machine Learning Toolbox.
- `gqr_formMarr` — Creates an array of multi-indices in graded lexicographic order (discussed in Section 12.2).
- `ibb` — Evaluates the iterated Brownian bridge kernel and derivatives as requested.
- `HSSVD_IBBSolve` — See Appendix D.3.

D.3 Full Hilbert–Schmidt SVD sample solver

This section displays a single function for implementing the Hilbert–Schmidt SVD of an iterated Brownian bridge kernel in MATLAB. We do this to help present some of the topics in this text in one single place and also to present some MATLAB programming strategies. Relevant points to note:

- This function allows for overloading, which is to say that some of the inputs are optional and the function will act differently depending on how it is called.
- To ensure that the user has passed viable arguments, the inputs are checked and errors are thrown as needed.
- We use the `persistent` variable `IBB` to store computations that have taken place and would need to be repeated unnecessarily during subsequent calls. In this example, the persistent variable stores the interpolation coefficients after the solve for future evaluation. This is the same structure as `polyfit` in MATLAB and `gqr_solve` in the `GaussQR` library, though both of those functions return the object that is computed to allow it to be used elsewhere.
- Within an `M`-file, other functions can be defined that are only accessible within that file. To access a function from outside an `M`-file, that file must bear the same name as the function.
- This function can accept multiple $\boldsymbol{y}$ vectors as `y`= $\begin{pmatrix}\boldsymbol{y}_1 & \cdots & \boldsymbol{y}_M\end{pmatrix}$. In doing so, the various coefficients will be stored in the same order, i.e., `c`= $\begin{pmatrix}\boldsymbol{c}_1 & \cdots & \boldsymbol{c}_M\end{pmatrix}$. This allows for the interpolation of different functions on the same points simultaneously, as was required in Program 7.2.

Program D.3. `HSSVD_IBBSolve_Full.m`

```
% HSSVD_IBBSolve_Full
% This function allows the user to solve an interpolation problem with the
% Iterated Brownian Bridge kernel using the HSSVD basis
```

```
4   function yeval = HSSVD_IBBSolve_Full(ep,beta,x,y,xeval)
    % Evaluate the HS-SVD basis IBB interpolant
6   %
    % Requires a positive ep shape parameter and positive integer beta
8   % smoothness parameter
    % Data locations x and data values y should be equal length column vectors
10  %
    % function yeval = HSSVD_IBBSolve_Full(ep,beta,x,y,xeval)
12  %   Inputs:  ep    - shape parameter
    %            beta  - smoothness parameter
14  %            x     - data locations
    %            y     - data values
16  %            xeval - locations at which to evaluate the interpolant
    %   Outputs: yeval - interpolant values at xeval
18  %
    % Note that multiple sets of data values can be passed for simultaneous
20  % computation, e.g., y = [y1 y2 y3], seval = [seval1 seval2 seval3]
    %
22  %
    %
24  % This function can be called a second way, depending on if the user has
    % already solved the interpolation problem once and just wants to evaluate
26  % at a different set of points
    %
28  % function yeval = HSSVD_IBBSolve_Full(xeval)
    %   Inputs:  xeval - locations at which to evaluate the interpolant
30  %   Outputs: yeval - interpolant values at xeval
    %
32  % You can only call the function in this form if you have already performed
    % a solve.  To solve the linear system but not perform evaluations:
34  %
    % function HSSVD_IBBSolve_Full(ep,beta,x,y)
36
    % Define an IBB data structure which will store the important solve content
38  % in it for use at a later time
    % This is overwritten if the user passes new data to the problem
40  persistent IBB

42  % Perform some basic error testing to confirm things are acceptable
    % Unpack the arguments passed to the function
44  % Solve for the IBB object if the appropriate data is passed
    % Can also use varargin to manage input arguments if preferred
46  switch nargin
        case 1
48          if isempty(IBB)
                error('Cannot evaluate because no coefficients have been stored')
50          else
                % Rename the first info passed to the function
52              xeval = ep;
            end
```

```
    case 4,5
        if nargin==4
            if nargout>0
                error('No evaluation points requested ... nothing can be returned')
            end
        end
        if numel(ep)>1 || numel(beta)>1
            error('Must have scalar ep and beta; check calling sequence')
        end
        if ep~=abs(ep)
            error('Shape parameter ep=%g unacceptable',ep)
        end
        if beta~=abs(floor(beta))
            error('Smoothness parameter beta=%g unacceptable',beta)
        end
        if size(x,1)~=size(y,1)
            error('Input data must be the same size')
        end
        if size(x,2)~=1
            error('The IBB kernel can only handle 1D problems')
        end

        % If we have reached this point, we can form the HSSVD basis
        % The try block and temporary storage prevents the IBB object from
        % being overwritten in the event of a faulty call
        try
            IBB_new = HSSVD_IBBForm(ep,beta,x,y);
        catch err
            rethrow(err);
        end
        IBB = IBB_new;
    otherwise
        error('Unacceptable inputs, nargin=%d',nargin)
end

% If there are points to be evaluated, do so
if exist('xeval','var')
    yeval = HSSVD_IBBEval(IBB,xeval);
end

end

%%%%%%%%%%%%%%%%%%%%%%%%%%%
% These functions can only be called within this Matlab file because they
% have been declared within a file of a different name

function IBB = HSSVD_IBBForm(ep,beta,x,y)
N = size(x,1);

% Define the eigenfunction and eigenvalue handles
```

```
phifunc = @(n,x) sqrt(2)*sin(pi*x*n);
lamfunc = @(b,e,n) ((pi*n).^2+e^2).^(-b);

% Determine how many eigenfunctions will be needed for accuracy
M = ceil(1/pi*sqrt(eps^(-1/beta)*(N^2*pi^2+ep^2)-ep^2));
narr = 1:M;

% Evaluate the Phi and lamvec
Phi1 = phifunc(narr(1:N),x);
Phi2 = phifunc(narr(N+1:end),x);
lamvec1 = lamfunc(beta,ep,narr(1:N));
lamvec2 = lamfunc(beta,ep,narr(N+1:end));

% Create the CbarT object
CbarT = bsxfun(@rdivide,lamvec2',lamvec1).*(Phi2'/Phi1');

% Form the Psi matrix and solve for the interpolation coefficients
Psi = Phi1 + Phi2*CbarT;
c = Psi\y;

% Pack the IBB object
IBB.CbarT = CbarT;
IBB.c = c;
IBB.narr = narr;
IBB.phifunc = phifunc;
end

function yeval = HSSVD_IBBEval(IBB,xeval)
% Unpack the IBB object
CbarT = IBB.CbarT;
c = IBB.c;
narr = IBB.narr;
phifunc = IBB.phifunc;
N = size(CbarT,2);

% Evaluate the Phieval components
Phieval1 = phifunc(narr(1:N),xeval);
Phieval2 = phifunc(narr(N+1:end),xeval);
yeval = Phieval1*c + Phieval2*(CbarT*c);
end
```

Bibliography

Aboiyar, T., Georgoulis, E., and Iske, A. (2010). Adaptive ADER methods using kernel-based polyharmonic spline WENO reconstruction, *SIAM J. Sci. Comput.* **32**, 6, pp. 3251–3277.

Aboiyar, T., Georgoulis, E. H., and Iske, A. (2006). High order WENO finite volume schemes using polyharmonic spline reconstruction, in *Proceedings of the International Conference on Numerical Analysis and Approximation Theory NAAT 2006* (Cluj-Napoca (Romania)), pp. 113–126.

Abrahamsen, P. (1997). A review of Gaussian random fields and correlation functions, Technical Report 917, Norwegian Computing Center, Oslo.

Abramowitz, M. and Stegun, I. A. (1965). *Handbook of Mathematical Functions: with Formulas, Graphs, and Mathematical Tables* (Dover Publications, New York).

Adler, R. J. (2009). *The Geometry of Random Fields (reprint of 1981 edition)* (Society for Industrial and Applied Mathematics, Philadelphia, PA).

Agler, J. and McCarthy, J. E. (2002). *Pick Interpolation and Hilbert Function Spaces* (American Mathematical Society, Providence, RI).

Ahlberg, J. H., Nilson, E. N., and Walsh, J. L. (1967). *The Theory of Splines and Their Applications* (Academic Press, New York).

Ahrem, R., Beckert, A., and Wendland, H. (2005). A new multivariate interpolation method for large-scale spatial coupling problems in aeroelasticity, in *The Proceedings to Int. Forum on Aeroelasticity and Structural Dynamics*, DGLR-Bericht 2005-04.

Akhiezer, N. I. and Glazman, I. M. (1993). *Theory of Linear Operators in Hilbert Space* (Dover Publications, New York).

Alfeld, P., Neamtu, M., and Schumaker, L. L. (1996). Fitting scattered data on sphere-like surfaces using spherical splines, *J. Comput. Appl. Math.* **73**, 1–2, pp. 5–43.

Ali, S. T., Antoine, J.-P., and Gazeau, J.-P. (2000). *Coherent States, Wavelets and Their Generalizations*, Graduate Texts in Contemporary Physics (Springer, Berlin; New York).

Allasia, G., Cavoretto, R., and De Rossi, A. (2013a). Lobachevsky spline functions and interpolation to scattered data, *Comput. Appl. Math.* **32**, 1, pp. 71–87.

Allasia, G., Cavoretto, R., and De Rossi, A. (2013b). Numerical integration on multivariate scattered data by Lobachevsky splines, *Int. J. Comput. Math.* **90**, 9, pp. 2003–2018.

Allen, D. M. (1974). The relationship between variable selection and data agumentation and a method for prediction, *Technometrics* **16**, 1, pp. 125–127.

Alpaydin, E. (2009). *Introduction to Machine Learning*, 2nd edn. (MIT Press, Cambridge, MA).

Alves, C. J. (2009). On the choice of source points in the method of fundamental solutions,

Eng. Anal. Bound. Elem. **33**, 12, pp. 1348–1361.

An, C., Chen, X., Sloan, I., and Womersley, R. (2010). Well conditioned spherical designs for integration and interpolation on the two-sphere, *SIAM J. Numer. Anal.* **48**, 6, pp. 2135–2157.

An, J. and Owen, A. (2001). Quasi-regression, *J. Complexity* **17**, 4, pp. 588–607.

Anitescu, M., Chen, J., and Wang, L. (2012). A matrix-free approach for solving the parametric Gaussian process maximum likelihood problem, *SIAM J. Sci. Comput.* **34**, 1, pp. A240–A262.

Arbogast, T., Estep, D., Sheehan, B., and Tavener, S. (2014). A posteriori error estimates for mixed finite element and finite volume methods for problems coupled through a boundary with nonmatching grids, *IMA J. Numer. Anal.* **34**, 4, pp. 1625–1653.

Arcangéli, R. (1974). *Études de problèmes de type elliptique ou parabolique avec conditions ponctuelles*, Ph.D. thesis, University of Toulouse.

Arcangéli, R., López de Silanes, M. C., and Torrens, J. J. (2007). An extension of a bound for functions in Sobolev spaces, with applications to (m, s)-spline interpolation and smoothing, *Numer. Math.* **107**, 2, pp. 181–211.

Arcangéli, R., López de Silanes, M. C., and Torrens, J. J. (2009). Estimates for functions in Sobolev spaces defined on unbounded domains, *J. Approx. Theory* **161**, 1, pp. 198–212.

Arcangéli, R., López de Silanes, M. C., and Torrens, J. J. (2012). Extension of sampling inequalities to Sobolev semi-norms of fractional order and derivative data, *Numer. Math.* **121**, 3, pp. 587–608.

Arlot, S. and Celisse, A. (2010). A survey of cross-validation procedures for model selection, *Statistics Surveys* **4**, pp. 40–79.

Aronszajn, N. (1943). La théorie des noyaux reproduisants et ses applications, Prèmiere Partie, *Math. Proc. Cambridge* **39**, 3, pp. 133–153.

Aronszajn, N. (1950). Theory of reproducing kernels, *T. Am. Math. Soc.* **68**, 3, pp. 337–404.

Aronszajn, N., Creese, T. M., and Lipkin, L. J. (1983). *Polyharmonic Functions*, Oxford Mathematical Monographs (Clarendon Press, Oxford).

Aronszajn, N. and Smith, K. T. (1957). Characterization of positive reproducing kernels. Applications to Green's functions, *Amer. J. Math.* **79**, 3, pp. 611–622.

Atkinson, K. and Han, W. (2009). *Theoretical Numerical Analysis: A Functional Analysis Framework*, 3rd edn. (Springer, Berlin; New York).

Atkinson, K. E. (1997). *The Numerical Solution of Integral Equations of the Second Kind*, Cambridge Monographs on Applied and Computational Mathematics (Cambridge University Press, Cambridge).

Atluri, S. N. (2004). *The Meshless Method (MLPG) for Domain & BIE Discretizations* (Tech Science Press, Encino, CA).

Atluri, S. N. and Shen, S. (2002). *The Meshless Local Petrov-Galerkin (MLPG) Method* (Tech Science Press, Encino, CA).

Atluri, S. N. and Zhu, T. (1998). A new meshless local Petrov–Galerkin (MLPG) approach in computational mechanics, *Comput. Mech.* **22**, pp. 117–127.

Atteia, M. (1966). Existence et détermination des fonctions "spline" à plusieurs variables, *C. R. Acad. Sci. Paris* **262**, pp. 575—578.

Aubin, J.-P. (2000). *Applied Functional Analysis*, 2nd edn. (Wiley–Interscience, New York).

Auffray, Y., Barbillon, P., and Marin, J.-M. (2012). Maximin design on non hypercube domains and kernel interpolation, *Stat. Comput.* **22**, 3, pp. 703–712.

Bach, F. R. and Jordan, M. I. (2003). Kernel independent component analysis, *J. Mach.*

Learn. Res. **3**, pp. 1–48.

Bai, Z., Fahey, G., and Golub, G. (1996). Some large-scale matrix computation problems, *J. Comput. Appl. Math.* **74**, 1, pp. 71–89.

Baldi Antognini, A. and Zagoraiou, M. (2010). Exact optimal designs for computer experiments via Kriging metamodelling, *J. Stat. Plan. Infer.* **140**, 9, pp. 2607–2617.

Ballestra, L. V. and Pacelli, G. (2013). Pricing European and American options with two stochastic factors: A highly efficient radial basis function approach, *J. Econ. Dyn. Control* **37**, 6, pp. 1142–1167.

Barba, L. A., Leonard, A., and Allen, C. B. (2005). Advances in viscous vortex methods — meshless spatial adaption based on radial basis function interpolation, *Int. J. Numer. Meth. Fluids* **47**, 5, pp. 387–421.

Barnett, A. H. and Betcke, T. (2008). Stability and convergence of the method of fundamental solutions for Helmholtz problems on analytic domains, *J. Comput. Phys.* **227**, 14, pp. 7003–7026.

Barry, R. P. and Pace, R. K. (1999). Monte Carlo estimates of the log determinant of large sparse matrices, *Linear Algebra Appl.* **289**, 1, pp. 41–54.

Baxter, B. J. C. (2006). Scaling radial basis functions via Euclidean distance matrices, *Comput. Math. Appl.* **51**, 8, pp. 1163–1170.

Baxter, B. J. C. and Sivakumar, N. (1996). On shifted cardinal interpolation by Gaussians and multiquadrics, *J. Approx. Theory* **87**, 1, pp. 36–59.

Bayona, V., Moscoso, M., Carretero, M., and Kindelan, M. (2010). RBF-FD formulas and convergence properties, *J. Comput. Phys.* **229**, 22, pp. 8281–8295.

Bayona, V., Moscoso, M., and Kindelan, M. (2011). Optimal constant shape parameter for multiquadric based RBF-FD method, *J. Comput. Phys.* **230**, 19, pp. 7384–7399.

Bayona, V., Moscoso, M., and Kindelan, M. (2012a). Gaussian RBF-FD weights and its corresponding local truncation errors, *Eng. Anal. Bound. Elem.* **36**, 9, pp. 1361–1369.

Bayona, V., Moscoso, M., and Kindelan, M. (2012b). Optimal variable shape parameter for multiquadric based RBF-FD method, *J. Comput. Phys.* **231**, 6, pp. 2466–2481.

Beatson, R. K., Cherrie, J. B., and Mouat, C. T. (1999). Fast fitting of radial basis functions: methods based on preconditioned GMRES iteration, *Adv. Comput. Math.* **11**, pp. 253–270.

Beatson, R. K., Davydov, O., and Levesley, J. (2010). Error bounds for anisotropic RBF interpolation. *J. Approx. Theory* **162**, pp. 512–527.

Beatson, R. K., Levesley, J., and Mouat, C. T. (2011). Better bases for radial basis function interpolation problems, *J. Comput. Appl. Math.* **236**, 4, pp. 434–446.

Beatson, R. K., Light, W. A., and Billings, S. (2001). Fast solution of the radial basis function interpolation equations: domain decomposition methods, *SIAM J. Sci. Comput.* **22**, 5, pp. 1717–1740.

Beatson, R. K. and Newsam, G. N. (1992). Fast evaluation of radial basis functions: I, *Comput. Math. Appl.* **2**, pp. 7–19.

Beatson, R. K. and Newsam, G. N. (1998). Fast evaluation of radial basis functions: moment-based methods, *SIAM J. Sci. Comput.* **19**, pp. 1428–1449.

Beckert, A. and Wendland, H. (2001). Multivariate interpolation for fluid-structure-interaction problems using radial basis functions, *Aerospace Science and Technology* **5**, 2, pp. 125–134.

Bejancu, A. and Hubbert, S. (2012). A study of the uniform accuracy of univariate thin plate spline interpolation, *Appl. Numer. Math.* **62**, 12, pp. 1781–1789.

Belova, A., Ehrhardt, M., and Shmidt, T. (2012). Meshfree methods in option pricing, in *Advances and Challenges in Embedded Computing*, Vol. 1, pp. 242–245.

Belytschko, T. and Chen, J. S. (2007). *Meshfree and Particle Methods* (Wiley, New York).

Belytschko, T., Krongauz, Y., Organ, D., Fleming, M., and Krysl, P. (1996). Meshless methods: An overview and recent developments, *Comput. Methods Appl. Mech. Eng.* **139**, 1–4, pp. 3–47.

Belytschko, T., Lu, Y. Y., and Gu, L. (1994). Element-free Galerkin methods, *Int. J. Numer. Meth. Eng.* **37**, 2, pp. 229–256.

Ben-Ari, E. N. and Steinberg, D. M. (2007). Modeling data from computer experiments: An empirical comparison of kriging with MARS and projection pursuit regression, *Quality Engineering* **19**, 4, pp. 327–338.

Berg, C., Christensen, J. P. R., and Ressel, P. (1984). *Harmonic Analysis on Semigroups: Theory of Positive Definite and Related Functions*, Graduate Texts in Mathematics (Springer, Berlin; New York).

Berger, J. O., De Oliveira, V., and Sansó, B. (2001). Objective Bayesian analysis of spatially correlated data, *J. Am. Stat. Assoc.* **96**, 456, pp. 1361–1374.

Bergman, S. (1950). *The Kernel Function and Conformal Mapping*, 2nd edn. (American Mathematical Society, Providence, RI).

Bergman, S. and Schiffer, M. (1947). On Green's and Neumann's functions in the theory of partial differential equations, *B. Am. Math. Soc.* **53**, 12, pp. 1141–1151.

Bergman, S. and Schiffer, M. (1953). *Kernel Functions and Elliptic Differential Equations in Mathematical Physics* (Academic Press, New York).

Bergmann, S. (1921). *Über die Entwicklung der harmonischen Funktionen der Ebene und des Raumes nach Orthogonalfunktionen*, Ph.D. thesis, University of Berlin.

Bergmann, S. (1922). Über die Entwicklung der harmonischen Funktionen der Ebene und des Raumes nach Orthogonalfunktionen, *Math. Ann.* **86**, 3–4, pp. 238–271.

Berlinet, A. and Thomas-Agnan, C. (2004). *Reproducing Kernel Hilbert Spaces in Probability and Statistics* (Kluwer, Dordrecht).

Bernardi, C., Maday, Y., and Patera, A. T. (1994). A new nonconforming approach to domain decomposition: the mortar element method, in H. Brezis and J. Lions (eds.), *Nonlinear PDEs and Their Applications* (Longman, London).

Bernstein, D. S. (2009). *Matrix Mathematics: Theory, Facts, and Formulas*, 2nd edn. (Princeton University Press, Princeton, N.J.).

Berrut, J.-P. and Trefethen, L. N. (2004). Barycentric Lagrange interpolation, *SIAM Rev.* **46**, 3, pp. 501–517.

Betcke, T. and Trefethen, L. N. (2005). Reviving the method of particular solutions, *SIAM Rev.* **47**, 3, pp. 469–491.

Birkhoff, G. and de Boor, C. (1965). Piecewise polynomial interpolation and approximation, in H. L. Garabedian (ed.), *Approximation of Functions* (Elsevier, Amsterdam), pp. 164–190.

Bochev, P. and Shashkov, M. (2005). Constrained interpolation (remap) of divergence-free fields, *Comput. Methods Appl. Mech. Eng.* **194**, 2–5, pp. 511–530.

Bochner, S. (1932). *Vorlesungen über Fouriersche Integrale, Mathematik und ihre Anwendungen*, Vol. 12 (Akademische Verlagsgesellschaft, Leipzig).

Bochner, S. (1933). Monotone Funktionen, Stieltjes Integrale und harmonische Analyse, *Math. Ann.* **108**, pp. 378–410.

Bollig, E. F., Flyer, N., and Erlebacher, G. (2012). Solution to PDEs using radial basis function finite-differences (RBF-FD) on multiple GPUs, *J. Comput. Phys.* **231**, 21, pp. 7133–7151.

de Boor, C. (2001). *A Practical Guide to Splines*, revised edn. (Springer, Berlin; New York).

de Boor, C. (2006). On interpolation by radial polynomials, *Adv. Comput. Math.* **24**, 1–4, pp. 143–153.

de Boor, C. and Ron, A. (1990). On multivariate polynomial interpolation, *Constr. Approx.* **6**, pp. 287–302.
de Boor, C. and Ron, A. (1992a). Computational aspects of polynomial interpolation in several variables, *Math. Comp.* **58**, 198, pp. 705–727.
de Boor, C. and Ron, A. (1992b). The least solution for the polynomial interpolation problem, *Math. Z.* **210**, pp. 347–378.
Bos, L., Calvi, J.-P., Levenberg, N., Sommariva, A., and Vianello, M. (2011a). Geometric weakly admissible meshes, discrete least squares approximations and approximate Fekete points, *Math. Comp.* **80**, pp. 1623–1638.
Bos, L., De Marchi, S., Sommariva, A., and Vianello, M. (2011b). Weakly admissible meshes and discrete extremal sets, *Numer. Math. Theor. Meth. Appl.* **4**, pp. 1–12.
Bos, L. P., De Marchi, S., Sommariva, A., and Vianello, M. (2011c). On multivariate Newton interpolation at discrete Leja points, *Dolomites Res. Notes Approx.* **4**, Special Issue on Kernel Functions and Meshless Methods, pp. 15–20.
Boser, B. E., Guyon, I. M., and Vapnik, V. N. (1992). A training algorithm for optimal margin classifiers, in *Proceedings of the Fifth Annual Workshop on Computational Learning Theory*, COLT '92 (ACM, New York), pp. 144–152.
Bouhamidi, A. (2001). Hilbertian approach for univariate spline with tension, *Approx. Theory Applic.* **17**, 4, pp. 36–57.
Boutry, G., Elad, M., Golub, G. H., and Milanfar, P. (2005). The generalized eigenvalue problem for nonsquare pencils using a minimal perturbation approach, *SIAM J. Matrix Anal. Appl.* **27**, 2, pp. 582–601.
Box, G. E. P. and Draper, N. R. (2007). *Response Surfaces, Mixtures, and Ridge Analyses*, 2nd edn. (Wiley–Interscience, Hoboken, NJ).
Box, G. E. P. and Wilson, K. B. (1951). On the experimental attainment of optimum conditions, *J. Roy. Stat. Soc. B Met.* **13**, 1, pp. 1–45.
Boyd, J. P. (2001). *Chebyshev and Fourier Spectral Methods*, 2nd edn. (Dover Publications, New York).
Boyd, J. P. (2010). The uselessness of the Fast Gauss Transform for summing Gaussian radial basis function series, *J. Comput. Phys.* **229**, 4, pp. 1311–1326.
Boyd, J. P. and Wang, L. (2009). An analytic approximation to the cardinal functions of Gaussian radial basis functions on an infinite lattice, *Appl. Math. Comput.* **215**, 6, pp. 2215–2223.
Bozzini, M., Lenarduzzi, L., Rossini, M., and Schaback, R. (2015). Interpolation with variably scaled kernels, *IMA J. Numer. Anal.* **35**, 1, pp. 199–219.
Bozzini, M., Lenarduzzi, L., and Schaback, R. (2002). Adaptive interpolation by scaled multiquadrics, *Adv. Comput. Math.* **16**, pp. 375–387.
Bozzini, M., Rossini, M., and Schaback, R. (2013). Generalized Whittle–Matérn and polyharmonic kernels, *Adv. Comput. Math.* **39**, 1, pp. 129–141.
Braaten, E. and Weller, G. (1979). An improved low-discrepancy sequence for multidimensional quasi-Monte Carlo integration, *J. Comput. Phys.* **33**, 2, pp. 249–258.
Brandimarte, P. (2003). *Numerical Methods in Finance: A MATLAB-Based Introduction* (John Wiley & Sons, New York).
Brenner, S. C. and Scott, L. R. (1994). *The Mathematical Theory of Finite Element Methods* (Springer, Berlin; New York).
Brezis, H. (2010). *Functional Analysis, Sobolev Spaces and Partial Differential Equations* (Springer, Berlin; New York).
Briani, M., Sommariva, A., and Vianello, M. (2012). Computing Fekete and Lebesgue points: Simplex, square, disk, *J. Comput. Appl. Math.* **236**, 9, pp. 2477–2486.
Buhmann, M. D. (1990). Multivariate cardinal interpolation with radial-basis functions,

Constr. Approx. **6**, 3, pp. 225–255.

Buhmann, M. D. (2000). Radial basis functions, *Acta Numerica* **9**, pp. 1–38.

Buhmann, M. D. (2003). *Radial Basis Functions: Theory and Implementations*, Cambridge Monographs on Applied and Computational Mathematics (Cambridge University Press, Cambridge).

Bump, D. (1998). *Automorphic Forms and Representations* (Cambridge University Press, Cambridge).

Bungartz, H.-J. and Griebel, M. (2004). Sparse grids, *Acta Numerica* **13**, pp. 1–123.

Bylund, C. and Mayner, W. (2012). A two-parameter family of compact Matérn kernels for data interpolation on the unit interval, Draft manuscript.

Caflisch, R. E., Morokoff, W., and Owen, A. (1997). Valuation of mortgage backed securities using Brownian bridges to reduce effective dimension, *J. Comput. Finance* **1**, pp. 27–46.

Cakmakci, O., Fasshauer, G. E., Foroosh, H., Thompson, K. P., and Rolland, J. P. (2008). Meshfree approximation methods for free-form surface representation in optical design with applications to head-worn displays, *Proc. SPIE* **7061**, 1, pp. 70610D–70610D–15.

Cakmakci, O., Kaya, I., Fasshauer, G. E., Thompson, K. P., and Rolland, J. P. (2010). Application of radial basis functions to represent optical freeform surfaces, *Proc. SPIE* **7652**, 1, pp. 76520A–76520A–8.

Caliari, M., De Marchi, S., and Vianello, M. (2008). Algorithm 886: Padua2D: Lagrange interpolation at Padua points on bivariate domains, *ACM T. Math. Software* **35**, 3, pp. 21:1–21:11.

Campbell, S. L. and Meyer, C. D. (2009). *Generalized Inverses of Linear Transformations* (Society for Industrial and Applied Mathematics, Philadelphia, PA).

Candès, E. J., Romberg, J. K., and Tao, T. (2006). Stable signal recovery from incomplete and inaccurate measurements, *Commun. Pure Appl. Math.* **59**, 8, pp. 1207–1223.

Carey, V., Estep, D., and Tavener, S. (2009). A posteriori analysis and adaptive error control for multiscale operator decomposition solution of elliptic systems I: triangular systems, *SIAM J. Numer. Anal.* **47**, 1, pp. 740–761.

Carlson, R. E. and Foley, T. A. (1992). Interpolation of track data with radial basis methods, *Comput. Math. Appl.* **24**, pp. 27–34.

Carlson, R. E. and Natarajan, B. K. (1994). Sparse approximate multiquadric interpolation, *Comput. Math. Appl.* **27**, pp. 99–108.

Carr, J. C., Beatson, R. K., McCallum, B. C., Fright, W. R., McLennan, T. J., and Mitchell, T. J. (2003). Smooth surface reconstruction from noisy range data, in *Proceedings of the 1st International Conference on Computer Graphics and Interactive Techniques in Australasia and South East Asia*, GRAPHITE '03 (ACM, New York), pp. 119–126.

Casciola, G., Lazzaro, D., Montefusco, L. B., and Morigi, S. (2006). Shape preserving surface reconstruction using locally anisotropic radial basis function interpolants, *Comput. Math. Appl.* **51**, 8, pp. 1185–1198.

Casciola, G., Montefusco, L. B., and Morigi, S. (2007). The regularizing properties of anisotropic radial basis functions, *Appl. Math. Comput.* **190**, 2, pp. 1050–1062.

Casciola, G., Montefusco, L. B., and Morigi, S. (2010). Edge-driven image interpolation using adaptive anisotropic radial basis functions, *J. Math. Imaging Vis.* **36**, 2, pp. 125–139.

zu Castell, W. and Filbir, F. (2005). Radial basis functions and corresponding zonal series expansions on the sphere, *J. Approx. Theory* **134**, 1, pp. 65–79.

Catoni, O. (2004). *Statistical Learning Theory and Stochastic Optimization* (Springer,

Berlin; New York).

Cavoretto, R. (2010). *Meshfree Approximation Methods, Algorithms and Applications*, Ph.D. thesis, University of Turin.

Cavoretto, R., Fasshauer, G. E., and McCourt, M. J. (2015). An introduction to the Hilbert–Schmidt SVD using iterated Brownian bridge kernels, *Numer. Algorithms* **68**, pp. 393–422.

Cecil, T., Qian, J., and Osher, S. (2004). Numerical methods for high dimensional Hamilton–Jacobi equations using radial basis functions, *J. Comput. Phys.* **196**, 1, pp. 327–347.

Chan, R. T. L. and Hubbert, S. (2011). A numerical study of radial basis function based methods for options pricing under the one dimension jump-diffusion model, `http://arxiv.org/abs/1011.5650`.

Chang, M.-S. and Wu, X. (2015). Transformation-based nonparametric estimation of multivariate densities, *J. Multivariate Anal.* **135**, 0, pp. 71–88.

Chen, C. S., Hon, Y. C., and Schaback, R. A. (2012). *Scientific Computing with Radial Basis Functions* (in preparation).

Chen, D., Menegatto, V., and Sun, X. (2003). A necessary and sufficient condition for strictly positive definite functions on spheres, *P. Am. Math. Soc.* **131**, 9, pp. 2733–2740.

Chen, J. (2014). Data structure and algorithms for recursively low-rank compressed matrices, Argonne National Laboratory, Preprint ANL/MCS-P5112-0314.

Chen, J., Wang, L., and Anitescu, M. (2014a). A fast summation tree code for Matérn kernel, *SIAM J. Sci. Comput.* **36**, 1, pp. A289–A309.

Chen, S., Cowan, C. F. N., and Grant, P. M. (1991). Orthogonal least squares learning algorithm for radial basis function networks, *IEEE T. Neural Networ.* **2**, 2, pp. 302–309.

Chen, W., Fu, Z.-J., and Chen, C. S. (2014b). *Recent Advances on Radial Basis Function Collocation Methods*, Springer Briefs in Applied Sciences and Technology (Springer, Berlin; New York).

Chen, Y., Lee, J., and Eskandarian, A. (2006). *Meshless Methods in Solid Mechanics* (Springer, Berlin; New York).

Cheney, E. W. and Light, W. A. (1999). *A Course in Approximation Theory* (Brooks/Cole, Pacific Grove, CA).

Cheney, W. (2001). *Analysis for Applied Mathematics* (Springer, Berlin; New York).

Cheng, A.-D. (2012). Multiquadric and its shape parameter — A numerical investigation of error estimate, condition number, and round-off error by arbitrary precision computation, *Eng. Anal. Bound. Elem.* **36**, 2, pp. 220–239.

Chernih, A. and Hubbert, S. (2014). Closed form representations and properties of the generalised Wendland functions, *J. Approx. Theory* **177**, pp. 17–33.

Chernih, A., Sloan, I. H., and Womersley, R. S. (2014). Wendland functions with increasing smoothness converge to a Gaussian, *Adv. Comput. Math.* **40**, 1, pp. 185–200.

Cherrie, J. B., Beatson, R. K., and Newsam, G. N. (2002). Fast evaluation of radial basis functions: methods for generalized multiquadrics in $\mathbb{R}^n$, *SIAM J. Sci. Comput.* **23**, pp. 1549–1571.

Chiles, J.-P. and Delfiner, P. (2012). *Geostatistics Modeling Spatial Uncertainty*, 2nd edn. (John Wiley & Sons, Hoboken, N.J.).

Chu, D. and Golub, G. H. (2006). On a generalized eigenvalue problem for nonsquare pencils, *SIAM J. Matrix Anal. Appl.* **28**, 3, pp. 770–787.

Cialenco, I., Fasshauer, G. E., and Ye, Q. (2012). Approximation of stochastic partial differential equations by a kernel-based collocation method, *Int. J. Comput. Math.*

89, 18, pp. 2543–2561.

Cline, D. B. H. and Hart, J. D. (1991). Kernel estimation of densities with discontinuities or discontinuous derivatives, *Statistics* **22**, 1, pp. 69–84.

Cochran, J. A. (1972). *The Analysis of Linear Integral Equations* (McGraw–Hill, New York).

Coles, S. (2001). *An Introduction to Statistical Modeling of Extreme Values*, Lecture Notes in Control and Information Sciences (Springer).

Constantine, P. and Gleich, D. (2014). Computing active subspaces, `http://arxiv.org/abs/1408.0545`.

Constantine, P. G., Dow, E., and Wang, Q. (2014a). Active subspace methods in theory and practice: Applications to Kriging surfaces, *SIAM J. Sci. Comput.* **36**, 4, pp. A1500–A1524.

Constantine, P. G., Dow, E., and Wang, Q. (2014b). Erratum: Active subspace methods in theory and practice: Applications to Kriging surfaces, *SIAM J. Sci. Comput.* **36**, 6, pp. A3030–A3031.

Constantinescu, E. M. and Sandu, A. (2010). Extrapolated implicit-explicit time stepping, *SIAM J. Sci. Comput.* **31**, 6, pp. 4452–4477.

Courant, R., Friedrichs, K., and Lewy, H. (1928). Über die partiellen Differenzengleichungen der mathematischen Physik, *Math. Ann.* **100**, 1, pp. 32–74.

Courant, R. and Hilbert, D. (1953). *Methods of Mathematical Physics, Vol. 1* (Wiley, New York).

Cover, T. M. (1965). Geometrical and statistical properties of systems of linear inequalities with applications in pattern recognition, *IEEE Trans. Electron.* **14**, pp. 326–334.

Craven, P. and Wahba, G. (1979). Smoothing noisy data with spline functions, *Numer. Math.* **31**, 4, pp. 377–403.

Cressie, N. (1993). *Statistics for Spatial Data*, revised edn. (Wiley–Interscience, New York).

Cressie, N. and Huang, H.-C. (1999). Classes of nonseparable, spatio-temporal stationary covariance functions, *J. Am. Stat. Assoc.* **94**, pp. 1330–1340.

Cressie, N. and Johannesson, G. (2008). Fixed rank kriging for very large spatial data sets, *J. Roy. Stat. Soc. B Met.* **70**, 1, pp. 209–226.

Cristianini, N. and Shawe-Taylor, J. (2000). *An Introduction to Support Vector Machines and Other Kernel-based Learning Methods* (Cambridge University Press, Cambridge).

Cucker, F. and Smale, S. (2002). On the mathematical foundations of learning, *B. Am. Math. Soc.* **39**, 1, pp. 1–49.

Cucker, F. and Zhou, D. X. (2007). *Learning Theory: An Approximation Theory Viewpoint*, Cambridge Monographs on Applied and Computational Mathematics (Cambridge University Press, Cambridge).

Curtis, P. C., Jr (1959). n-parameter families and best approximation, *Pacific J. Math.* **9**, 4, pp. 1013–1027.

Dai, F. and Xu, Y. (2013). *Approximation Theory and Harmonic Analysis on Spheres and Balls*, Springer Monographs in Mathematics (Springer, Berlin; New York).

Dalzell, C. and Ramsay, J. (1993). Computing reproducing kernels with arbitrary boundary constraints, *SIAM J. Sci. Comput.* **14**, 3, pp. 511–518.

Das, S. and Neumaier, A. (2013). Solving overdetermined eigenvalue problems, *SIAM J. Sci. Comput.* **35**, 2, pp. A541–A560.

Davies, B. and Martin, B. (1979). Numerical inversion of the Laplace transform: a survey and comparison of methods, *J. Comput. Phys.* **33**, 1, pp. 1–32.

Davis, P. J. (1963). *Interpolation and Approximation.* (Blaisdell Pub. Co., New York).

Davydov, O. (2014). *TSFIT: A Software Package for Two-Stage Scattered Data Fitting,* https://www.staff.uni-giessen.de/odavydov/tsfit/.

Davydov, O. and Oanh, D. T. (2011). On the optimal shape parameter for Gaussian radial basis function finite difference approximation of the Poisson equation, *Comput. Math. Appl.* **62**, 5, pp. 2143–2161.

Davydov, O. and Schaback, R. (2015). Error bounds for kernel-based numerical differentiation, http://www.staff.uni-giessen.de/odavydov/rbf_consistency.html.

Deng, J., Li, K., and Irwin, G. W. (2012). Locally regularised two-stage learning algorithm for RBF network centre selection, *Int. J. Syst. Sci.* **43**, 6, pp. 1157–1170.

Deng, Q. and Driscoll, T. (2012). A fast treecode for multiquadric interpolation with varying shape parameters, *SIAM J. Sci. Comput.* **34**, 2, pp. A1126–A1140.

De Marchi, S. (2003). On optimal center locations for radial basis function interpolation: computational aspects, *Rend. Sem. Mat. Univ. Pol. Torino* **61**, 3, pp. 343–358.

De Marchi, S. and Santin, G. (2013). A new stable basis for radial basis function interpolation, *J. Comput. Appl. Math.* **253**, pp. 1–13.

De Marchi, S. and Schaback, R. (2010). Stability of kernel-based interpolation, *Adv. Comput. Math.* **32**, 2, pp. 155–161.

De Marchi, S., Schaback, R., and Wendland, H. (2005). Near-optimal data-independent point locations for radial basis function interpolation, *Adv. Comput. Math.* **23**, 3, pp. 317–330.

DeVore, R. A. and Ron, A. (2010). Approximation using scattered shifts of a multivariate function, *T. Am. Math. Soc.* **362**, 12, pp. 6205–6229.

Diaconis, P. (1988). Bayesian numerical analysis, in S. S. Gupta and J. O. Berger (eds.), *Statistical Decision Theory and Related Topics IV, Papers from the 4th Purdue Symp., West Lafayette/Indiana 1986*, Vol. 1 (Springer, Berlin; New York), pp. 163–175.

Dick, J., Kuo, F. Y., and Sloan, I. H. (2013a). High-dimensional integration: the quasi-Monte Carlo way, *Acta Numerica* **22**, pp. 133–288.

Dick, J., Nuyens, D., and Pillichshammer, F. (2013b). Lattice rules for nonperiodic smooth integrands, *Numer. Math.* **126**, 2, pp. 259–291.

Dick, J. and Pillichshammer, F. (2010). *Digital Nets and Sequences: Discrepancy Theory and Quasi-Monte Carlo Integration* (Cambridge University Press, Cambridge).

DLMF (2012). NIST Digital Library of Mathematical Functions, http://dlmf.nist.gov/, Release 1.0.5 of 2012-10-01.

Doane, D. P. and Seward, L. E. (2011). Measuring skewness: a forgotten statistic, *J. Stat. Ed.* **19**, 2, pp. 1–18.

Driscoll, T. A. and Fornberg, B. (2002). Interpolation in the limit of increasingly flat radial basis functions, *Comput. Math. Appl.* **43**, 3–5, pp. 413–422.

Driscoll, T. A. and Heryudono, A. (2007). Adaptive residual subsampling methods for radial basis function interpolation and collocation problems, *Comput. Math. Appl.* **53**, 6, pp. 927–939.

Du, D., Li, X., Fei, M., and Irwin, G. W. (2012). A novel locally regularized automatic construction method for RBF neural models, *Neurocomputing* **98**, pp. 4–11.

Du, Q., Faber, V., and Gunzburger, M. (1999). Centroidal Voronoi tessellations: applications and algorithms, *SIAM Rev.* **41**, 4, pp. 637–676.

Du, Q., Gunzburger, M., and Ju, L. (2002). Meshfree, probabilistic determination of point sets and support regions for meshless computing, *Comput. Methods Appl. Mech. Eng.* **191**, 13–14, pp. 1349–1366.

Du, Q., Gunzburger, M., and Ju, L. (2010). Advances in studies and applications of centroidal Voronoi tessellations, *Numer. Math. Theor. Meth. Appl.* **3**, pp. 119–142.

Duarte, C. A. and Oden, J. T. (1996a). An *h-p* adaptive method using clouds, *Comput.*

Methods Appl. Mech. Eng. **139**, 1, pp. 237–262.

Duarte, C. A. and Oden, J. T. (1996b). *H-p* clouds — an *h-p* meshless method, *Numer. Meth. Part. D. E.* **12**, pp. 673–705.

Duc-Jacquet, M. (1973). *Approximation des fonctionnelles lineaires sur les espaces hilbertiens autoreproduisants*, Ph.D. thesis, University of Grenoble.

Duchon, J. (1976). Interpolation des fonctions de deux variables suivant le principe de la flexion des plaques minces, *RAIRO Analyse Numerique* **10**, pp. 5–12.

Duchon, J. (1977). Splines minimizing rotation-invariant semi-norms in Sobolev spaces, in W. Schempp and K. Zeller (eds.), *Constructive Theory of Functions of Several Variables, Oberwolfach 1976* (Springer, Berlin; New York), pp. 85–100.

Duchon, J. (1978). Sur l'erreur d'interpolation des fonctions de plusieurs variables par les d^m-splines, *RAIRO Analyse Numerique* **12**, 4, pp. 325–334.

Duffy, D. G. (2001). *Green's Functions with Applications* (Chapman and Hall/CRC, Boca Raton, FL).

Dyn, N. (1987). Interpolation of scattered data by radial functions, in C. K. Chui, L. L. Schumaker, and F. Utreras (eds.), *Topics in Multivariate Approximation* (Academic Press, New York), pp. 47–61.

Dyn, N. (1989). Interpolation and approximation by radial and related functions, in C. Chui, L. Schumaker, and J. Ward (eds.), *Approximation Theory VI* (Academic Press, New York), pp. 211–234.

Dyn, N. and Levin, D. (1981). Bell shaped basis functions for surface fitting, in Z. Ziegler (ed.), *Approximation Theory and Applications* (Academic Press, New York), pp. 113–129.

Dyn, N. and Levin, D. (1983). Iterative solution of systems originating from integral equations and surface interpolation, *SIAM J. Numer. Anal.* **20**, 2, pp. 377–390.

Dyn, N., Levin, D., and Rippa, S. (1986). Numerical procedures for surface fitting of scattered data by radial functions, *SIAM J. Sci. Stat. Comp.* **7**, pp. 639–659.

Dyn, N., Narcowich, F. J., and Ward, J. D. (1997). A framework for interpolation and approximation on Riemannian manifolds, in *Approximation Theory and Optimization (Cambridge, 1996)* (Cambridge University Press, Cambridge), pp. 133–144.

Dyn, N., Narcowich, F. J., and Ward, J. D. (1999). Variational principles and Sobolev-type estimates for generalized interpolation on a Riemannian manifold, *Constr. Approx.* **15**, 2, pp. 175–208.

Estep, D., Ginting, V., Ropp, D., Shadid, J. N., and Tavener, S. (2008). An a posteriori-a priori analysis of multiscale operator splitting, *SIAM J. Numer. Anal.* **46**, pp. 1116–1146.

Euler, L. (1755). *Institutiones calculi differentialis*, Vol. II (Petersburg).

Everitt, B. and Skrondal, A. (2010). *The Cambridge Dictionary of Statistics* (Cambridge University Press).

Evgeniou, T., Pontil, M., and Poggio, T. (2000). Regularization networks and support vector machines, *Adv. Comput. Math.* **13**, 1, pp. 1–50.

Fairweather, G. and Karageorghis, A. (1998). The method of fundamental solutions for elliptic boundary value problems, *Adv. Comput. Math.* **9**, pp. 69–95.

Fang, K. T., Li, R., and Sudjianto, A. (2006). *Design and Modeling for Computer Experiments*, Computer Science and Data Analysis (Chapman & Hall, New York).

Fasshauer, G. E. (1997). Solving partial differential equations by collocation with radial basis functions, in A. Le Méhauté, C. Rabut, and L. L. Schumaker (eds.), *Surface Fitting and Multiresolution Methods* (Vanderbilt University Press, Nashville, TN), pp. 131–138.

Fasshauer, G. E. (1999). Hermite interpolation with radial basis functions on spheres, *Adv.*

Comput. Math. **10**, 1, pp. 81–96.

Fasshauer, G. E. (2007). *Meshfree Approximation Methods with* MATLAB, *Interdisciplinary Mathematical Sciences*, Vol. 6 (World Scientific Publishing Co., Singapore).

Fasshauer, G. E. (2011a). Green's functions: taking another look at kernel approximation, radial basis functions and splines, in M. Neamtu and L. L. Schumaker (eds.), *Approximation Theory XIII: San Antonio 2010*, *Springer Proceedings in Mathematics*, Vol. 13 (Springer, Berlin; New York), pp. 37–63.

Fasshauer, G. E. (2011b). Positive definite kernels: past, present and future, *Dolomites Res. Notes Approx.* **4**, Special Issue on Kernel Functions and Meshless Methods, pp. 21–63.

Fasshauer, G. E., Hickernell, F. J., and Woźniakowski, H. (2012a). Average case approximation: convergence and tractability of Gaussian kernels, in H. Woźniakowski and L. Plaskota (eds.), *Monte Carlo and Quasi-Monte Carlo Methods 2010*, no. 23 in Springer Proceedings in Mathematics and Statistics (Springer, Berlin; New York), pp. 309–324.

Fasshauer, G. E., Hickernell, F. J., and Woźniakowski, H. (2012b). On dimension-independent rates of convergence for function approximation with Gaussian kernels, *SIAM J. Numer. Anal.* **50**, 1, pp. 247–271.

Fasshauer, G. E., Khaliq, A. Q. M., and Voss, D. A. (2004). Using meshfree approximation for multi-asset American option problems, *J. Chinese Institute Engineers* **27**, pp. 563–571.

Fasshauer, G. E. and McCourt, M. J. (2012). Stable evaluation of Gaussian radial basis function interpolants, *SIAM J. Sci. Comput.* **34**, 2, pp. A737–A762.

Fasshauer, G. E. and Schumaker, L. L. (1998). Scattered data fitting on the sphere, in M. Dæhlen, T. Lyche, and L. L. Schumaker (eds.), *Mathematical Methods for Curves and Surfaces II* (Vanderbilt University Press, Nashville, TN), pp. 117–166.

Fasshauer, G. E. and Ye, Q. (2011). Reproducing kernels of generalized Sobolev spaces via a Green function approach with distributional operators, *Numer. Math.* **119**, pp. 585–611.

Fasshauer, G. E. and Ye, Q. (2013a). Kernel-based collocation methods versus Galerkin finite element methods for approximating elliptic stochastic partial differential equations, in M. A. Schweitzer (ed.), *Meshfree Methods for Partial Differential Equations VI* (Springer, Berlin), pp. 155–170.

Fasshauer, G. E. and Ye, Q. (2013b). Reproducing kernels of Sobolev spaces via a Green kernel approach with differential operators and boundary operators, *Adv. Comput. Math.* **38**, 4, pp. 891–921.

Fasshauer, G. E. and Ye, Q. (2014). A kernel-based collocation method for elliptic partial differential equations with random coefficients, in *MCQMC 2012*, *Springer Proceedings in Mathematics and Statistics*, Vol. 65 (Springer, Berlin; New York), pp. 331–347.

Fasshauer, G. E. and Zhang, J. G. (2007a). Iterated approximate moving least squares approximation, in V. M. A. Leitão, C. Alves, and C. A. Duarte (eds.), *Advances in Meshfree Techniques* (Springer, Berlin; New York), pp. 221–240.

Fasshauer, G. E. and Zhang, J. G. (2007b). On choosing "optimal" shape parameters for RBF approximation, *Numer. Algorithms* **45**, 1–4, pp. 345–368.

Fasshauer, G. E. and Zhang, J. G. (2009). Preconditioning of radial basis function interpolation systems via accelerated iterated approximate moving least squares approximation, in A. J. M. Ferreira, E. J. Kansa, G. E. Fasshauer, and V. M. A. Leitão (eds.), *Progress on Meshless Methods* (Springer, Berlin; New York), pp. 57–75.

Faul, A. C., Goodsell, G., and Powell, M. J. D. (2005). A Krylov subspace algorithm for multiquadric interpolation in many dimensions, *IMA J. Numer. Anal.* **25**, 1, pp. 1–24.

Faul, A. C. and Powell, M. J. D. (1999). Proof of convergence of an iterative technique for thin plate spline interpolation in two dimensions, *Adv. Comput. Math.* **11**, pp. 183–192.

Faul, A. C. and Powell, M. J. D. (2000). Krylov subspace methods for radial basis function interpolation, in *Numerical Analysis 1999 (Dundee)* (Chapman & Hall/CRC, Boca Raton, FL), pp. 115–141.

Fedoseyev, A. I., Friedman, M. J., and Kansa, E. J. (2002). Improved multiquadric method for elliptic partial differential equations via PDE collocation on the boundary, *Comput. Math. Appl.* **43**, pp. 439–455.

Fernández-Navarro, F., Hervás-Martínez, C., Gutiérrez, P. A., and Carbonero-Ruz, M. (2011). Evolutionary q-Gaussian radial basis function neural networks for multiclassification, *Neural Networks* **24**, 7, pp. 779–784.

Ferreira, J. C. and Menegatto, V. A. (2009). Eigenvalues of integral operators defined by smooth positive definite kernels, *Integr. Equ. Oper. Theory* **64**, 1, pp. 61–81.

Ferreira, J. C. and Menegatto, V. A. (2013). Positive definiteness, reproducing kernel Hilbert spaces and beyond, *Ann. Funct. Anal.* **4**, 1, pp. 64–88.

Fine, S. and Scheinberg, K. (2002). Efficient SVM training using low-rank kernel representations, *J. Mach. Learn. Res.* **2**, pp. 243–264.

Flake, G. W. and Lawrence, S. (2002). Efficient SVM regression training with SMO, *Mach. Learn.* **46**, 1-3, pp. 271–290.

Floater, M. S. (2003). Mean value coordinates, *Comput. Aided Geom. D.* **20**, 1, pp. 19–27.

Flyer, N. and Lehto, E. (2010). Rotational transport on a sphere: Local node refinement with radial basis functions, *J. Comput. Phys.* **229**, 6, pp. 1954–1969.

Flyer, N., Lehto, E., Blaise, S., Wright, G. B., and St-Cyr, A. (2012). A guide to RBF-generated finite differences for nonlinear transport: Shallow water simulations on a sphere, *J. Comput. Phys.* **231**, 11, pp. 4078–4095.

Flyer, N. and Wright, G. B. (2007). Transport schemes on a sphere using radial basis functions, *J. Comput. Phys.* **226**, 1, pp. 1059–1084.

Flyer, N. and Wright, G. B. (2009). A radial basis function method for the shallow water equations on a sphere, *P. Roy. Soc. A – Math. Phy.* **465**, 2106, pp. 1949 –1976.

Foley, T. A. (1994). Near optimal parameter selection for multiquadric interpolation, *J. Appl. Sc. Comp.* **1**, pp. 54–69.

Folland, G. B. (1992). *Fourier Analysis and Its Applications* (American Mathematical Society, Providence, RI).

Forbes, G. W. (2010). Robust and fast computation for the polynomials of optics, *Optics Express* **18**, 13, pp. 13851–13862.

Forbes, G. W. (2012). Characterizing the shape of freeform optics, *Optics Express* **20**, 3, pp. 2483–2499.

Fornberg, B. (1998). *A Practical Guide to Pseudospectral Methods*, Cambridge Monographs on Applied and Computational Mathematics (Cambridge University Press).

Fornberg, B. and Flyer, N. (2005). Accuracy of radial basis function interpolation and derivative approximations on 1-D infinite grids, *Adv. Comput. Math.* **23**, 1–2, pp. 5–20.

Fornberg, B. and Flyer, N. (2015). Solving PDEs with radial basis functions, *Acta Numerica* **24**, pp. 215–258.

Fornberg, B., Flyer, N., and Russell, J. (2010). Comparisons between pseudospectral and radial basis function derivative approximations, *IMA J. Numer. Anal.* **30**, 1, pp. 149–172.

Fornberg, B., Larsson, E., and Flyer, N. (2011). Stable computations with Gaussian radial basis functions, *SIAM J. Sci. Comput.* **33**, 2, pp. 869–892.

Fornberg, B., Larsson, E., and Wright, G. (2006). A new class of oscillatory radial basis functions, *Comput. Math. Appl.* **51**, 8, pp. 1209–1222.

Fornberg, B., Lehto, E., and Powell, C. (2013). Stable calculation of Gaussian-based RBF-FD stencils, *Comput. Math. Appl.* **65**, 4, pp. 627–637.

Fornberg, B. and Piret, C. (2008a). On choosing a radial basis function and a shape parameter when solving a convective PDE on a sphere, *J. Comput. Phys.* **227**, 5, pp. 2758–2780.

Fornberg, B. and Piret, C. (2008b). A stable algorithm for flat radial basis functions on a sphere, *SIAM J. Sci. Comput.* **30**, 1, pp. 60–80.

Fornberg, B. and Wright, G. (2004). Stable computation of multiquadric interpolants for all values of the shape parameter, *Comput. Math. Appl.* **48**, 5–6, pp. 853–867.

Fornberg, B., Wright, G., and Larsson, E. (2004). Some observations regarding interpolants in the limit of flat radial basis functions, *Comput. Math. Appl.* **47**, pp. 37–55.

Fornberg, B. and Zuev, J. (2007). The Runge phenomenon and spatially variable shape parameters in RBF interpolation, *Comput. Math. Appl.* **54**, 3, pp. 379–398.

Forrester, A. I. J., Sobester, A., and Keane, A. J. (2008). *Engineering Design via Surrogate Modelling* (Wiley, Chichester).

Fox, L., Henrici, P., and Moler, C. (1967). Approximations and bounds for eigenvalues of elliptic operators, *SIAM J. Numer. Anal.* **4**, 1, pp. 89–102.

Franke, C. and Schaback, R. (1998a). Convergence order estimates of meshless collocation methods using radial basis functions, *Adv. Comput. Math.* **8**, pp. 381–399.

Franke, C. and Schaback, R. (1998b). Solving partial differential equations by collocation using radial basis functions, *Appl. Math. Comput.* **93**, 1, pp. 73–82.

Franke, R. (1979). A critical comparison of some methods for interpolation of scattered data, NPS-53-79-003, Naval Postgraduate School, Monterey, CA.

Franke, R. (1982). Scattered data interpolation: Tests of some method, *Math. Comp.* **38**, 157, pp. 181–200.

Fredholm, I. (1903). Sur une classe d'équations fonctionnelles, *Acta Mathematica* **27**, 1, pp. 365–390.

Freeden, W., Gervens, T., and Schreiner, M. (1998). *Constructive Approximation on the Sphere: With Applications to Geomathematics* (Oxford University Press, Oxford).

Frieze, A., Kannan, R., and Vempala, S. (2004). Fast Monte-Carlo algorithms for finding low-rank approximations, *J. ACM* **51**, 6, pp. 1025–1041.

Fuselier, E. (2008a). Improved stability estimates and a characterization of the native space for matrix-valued RBFs, *Adv. Comput. Math.* **29**, 3, pp. 311–313.

Fuselier, E. (2008b). Sobolev-type approximation rates for divergence-free and curl-free RBF interpolants, *Math. Comp.* **77**, 263, pp. 1407–1423.

Fuselier, E. (2015). RBF Vector Decomposition, Retrieved February 25, 2015, from `http://math.highpoint.edu/~efuselier/RBFVectorDecomposition/`.

Fuselier, E., Hangelbroek, T., Narcowich, F. J., Ward, J. D., and Wright, G. B. (2013). Localized bases for kernel spaces on the unit sphere, *SIAM J. Numer. Anal.* **51**, 5, pp. 2538–2562.

Fuselier, E., Narcowich, F., Ward, J., and Wright, G. (2008). Error and stability estimates for surface-divergence free RBF interpolants on the sphere, *Math. Comp.* **78**, 268, pp. 2157–2186.

Fuselier, E. and Wright, G. B. (2012). Scattered data interpolation on embedded submanifolds with restricted positive definite kernels: Sobolev error estimates, *SIAM J. Numer. Anal.* **50**, 3, pp. 1753–1776.

Fuselier, E. and Wright, G. B. (2015). Order-preserving derivative approximation with periodic radial basis functions, *Adv. Comput. Math.* **41**, 1, pp. 23–53.

Fuselier, E. J. and Wright, G. B. (2009). Stability and error estimates for vector field interpolation and decomposition on the sphere with RBFs, *SIAM J. Numer. Anal.* **47**, 5, pp. 3213–3239.

Fuselier, E. J. and Wright, G. B. (2013). A high-order kernel method for diffusion and reaction-diffusion equations on surfaces, *J. Sci. Comput.* **56**, 3, pp. 535–565.

Gandin, L. S. (1963). *Objective Analysis of Meteorological Fields* (Gidrometeorologicheskoe Izdatel'stvo (GIMIZ), Leningrad).

Garcke, J. and Griebel, M. (2013). *Sparse Grids and Applications, Lecture Notes in Computational Science and Engineering*, Vol. 88 (Springer, Berlin; New York).

Gasca, M. and Sauer, T. (2000). Polynomial interpolation in several variables, *Adv. Comput. Math.* **12**, 4, pp. 377–410.

Gauß, C. F. (1809). *Theoria motus corporum coelestium in sectionibus conicis solem ambientium* (Friedrich Perthes and I.H. Besser, Hamburg).

Gautschi, W. (2001). Barycentric formulae for cardinal (SINC-) interpolants by Jean-Paul Berrut, *Numer. Math.* **87**, 4, pp. 791–792.

Genton, M. G. (2002). Classes of kernels for machine learning: a statistics perspective, *J. Mach. Learn. Res.* **2**, pp. 299–312.

Ghanem, R. G. and Spanos, P. D. (2003). *Stochastic Finite Elements: A Spectral Approach* (Courier Dover Publications, New York).

Gingold, R. A. and Monaghan, J. J. (1977). Smoothed particle hydrodynamics — Theory and application to non-spherical stars, *Mon. Not. R. Astron. Soc.* **181**, pp. 375–389.

Glasserman, P. (2004). *Monte Carlo Methods in Financial Engineering, Applications of Mathematics*, Vol. 53 (Springer, Berlin; New York).

Gneiting, T., Genton, M. G., and Guttorp, P. (2007). Geostatistical space-time models, stationarity, separability and full symmetry, in B. Finkenstaedt, L. Held, and V. Isham (eds.), *Statistics of Spatio-Temporal Systems* (Chapman & Hall/CRC, Boca Raton, FL), pp. 151–175.

Golberg, M. A. and Chen, C. S. (1998). *The Method of Fundamental Solutions for Potential, Helmholtz and Diffusion Problems*, Computational Engineering (Computational Mechanics Publications, WIT Press).

Golberg, M. A., Chen, C. S., and Karur, S. R. (1996). Improved multiquadric approximation for partial differential equations, *Eng. Anal. Bound. Elem.* **18**, pp. 9–17.

Golomb, M. and Weinberger, H. F. (1959). Optimal approximation and error bounds, in R. E. Langer (ed.), *On Numerical Approximation* (University of Wisconsin Press), pp. 117–190.

Golub, G. H., Heath, M., and Wahba, G. (1979). Generalized cross-validation as a method for choosing a good ridge parameter, *Technometrics* **21**, 2, pp. 215–223.

Golub, G. H. and Van Loan, C. F. (2012). *Matrix Computations*, 4th edn. (Johns Hopkins University Press, Baltimore, MD).

Golub, G. H. and Von Matt, U. (1997). Generalized cross-validation for large-scale problems, *J. Comput. Graph. Stat.* **6**, 1, pp. 1–34.

Gönen, M. and Alpaydın, E. (2011). Multiple kernel learning algorithms, *J. Mach. Learn. Res.* **12**, pp. 2211–2268.

Good, I. J. and Gaskins, R. A. (1980). Density estimation and bump-hunting by the penalized likelihood method exemplified by scattering and meteorite data, *J. Am. Stat. Assoc.* **75**, 369, pp. 42–56.

Green, G. (1828). An essay on the application of mathematical analysis to the theories of electricity and magnetism, Nottingham.

Green, P. J. and Silverman, B. W. (1993). *Nonparametric Regression and Generalized Linear Models: A roughness penalty approach* (CRC Press, Boca Raton, FL).

Griebel, M., Rieger, C., and Zwicknagl, B. (2015). Multiscale approximation and reproducing kernel Hilbert space methods, *SIAM J. Numer. Anal.* **53**, 2, pp. 852–873.

Griewank, A. and Walther, A. (2008). *Evaluating Derivatives*, Other Titles in Applied Mathematics (Society for Industrial and Applied Mathematics, Philadelphia, PA).

Gu, C. (2013). *Smoothing Spline ANOVA Models*, 2nd edn. (Springer, Berlin; New York).

Gu, C., Jeon, Y., and Lin, Y. (2013). Nonparametric density estimation in high-dimensions, *Statistica Sinica* **23**, pp. 1131–1153.

Gumerov, N. A. and Duraiswami, R. (2007). Fast radial basis function interpolation via preconditioned Krylov iteration, *SIAM J. Sci. Comput.* **29**, 5, pp. 1876–1899.

Gutmann, H.-M. (2001). A radial basis function method for global optimization, *J. Global Optim.* **19**, pp. 201–227.

Guttorp, P. and Gneiting, T. (2005). On the Whittle–Matérn correlation family, Tech. Rep. NRCSE-TRS No. 080, University of Washington.

Gutzmer, T. (1996). Interpolation by positive definite functions on locally compact groups with application to SO(3), *Results Math.* **29**, pp. 69–77.

Gutzmer, T. and Melenk, J. (2001). Approximation orders for natural splines in arbitrary dimensions, *Math. Comp.* **70**, 234, pp. 699–703.

Haaland, B. and Qian, P. Z. G. (2011). Accurate emulators for large-scale computer experiments, *Ann. Stat.* **39**, 6, pp. 2974–3002.

Haar, A. (1918). Die Minkowskische Geometrie und die Annäherung an stetige Funktionen, *Math. Ann.* **18**, pp. 294–311.

Haberman, R. (2013). *Applied Partial Differential Equations*, 5th edn. (Pearson Prentice Hall, Upper Saddle River, NJ).

Hackbusch, W. (1989). *Integralgleichungen: Theorie und Numerik* (B.G. Teubner, Leipzig).

Hackbusch, W. (1999). A sparse matrix arithmetic based on $\mathcal{H}$-matrices. part I: Introduction to $\mathcal{H}$-matrices, *Computing* **62**, 2, pp. 89–108.

Halton, J. H. (1960). On the efficiency of certain quasi-random sequences of points in evaluating multi-dimensional integrals, *Numer. Math.* **2**, pp. 84–90.

Handcock, M. S. and Stein, M. L. (1993). A Bayesian analysis of kriging, *Technometrics* **35**, 4, pp. 403–410.

Hangelbroek, T. (2008). Error estimates for thin plate spline approximation in the disk, *Constr. Approx.* **28**, pp. 27–59.

Hangelbroek, T., Madych, W., Narcowich, F., and Ward, J. D. (2012). Cardinal interpolation with Gaussian kernels, *J. Fourier Anal. Appl.* **18**, 1, pp. 67–86.

Hangelbroek, T., Narcowich, F. J., Rieger, C., and Ward, J. D. (2014). An inverse theorem on bounded domains for meshless methods using localized bases, `http://arxiv.org/abs/1406.1435`.

Hangelbroek, T., Narcowich, F. J., Sun, X., and Ward, J. D. (2011). Kernel approximation on manifolds II: the L_∞ norm of the L_2 projector, *SIAM J. Math. Anal.* **43**, 2, pp. 662–684.

Hangelbroek, T., Narcowich, F. J., and Ward, J. D. (2010). Kernel approximation on manifolds I: Bounding the Lebesgue constant, *SIAM J. Math. Anal.* **42**, 4, pp. 1732–1760.

Hangelbroek, T. and Ron, A. (2010). Nonlinear approximation using Gaussian kernels, *J. Funct. Anal.* **259**, 1, pp. 203–219.

Hangelbroek, T. and Schmid, D. (2011). Surface spline approximation on SO(3), *Appl. Comput. Harmon. A.* **31**, 2, pp. 169–184.

Hansen, P. C. (1988). Computation of the singular value expansion, *Computing* **40**, 3, pp. 185–199.

Harder, R. L. and Desmarais, R. N. (1972). Interpolation using surface splines, *J. Aircraft* **9**, 2, pp. 189–191.

Hardy, R. L. (1971). Multiquadric equations of topography and other irregular surfaces, *J. Geophys. Res.* **76**, 8, pp. 1905–1915.

Hardy, R. L. and Göpfert, W. M. (1975). Least squares prediction of gravity anomalies, geodial undulations, and deflections of the vertical with multiquadric harmonic functions, *Geophys. Res. Letters* **2**, pp. 423–426.

Hastie, T., Tibshirani, R., and Friedman, J. (2009). *Elements of Statistical Learning: Data Mining, Inference, and Prediction*, 2nd edn., Springer Series in Statistics (Springer, Berlin; New York).

Hawkins, D. L. (1989). Some practical problems in implementing a certain sieve estimator of the Gaussian mean function, *Commun. Stat. Simulat.* **18**, 2, pp. 481–500.

Herbrich, R. (2002). *Learning Kernel Classifiers: Theory and Algorithms* (MIT Press, Cambridge, MA).

Hess, S. and Polak, J. (2003). An alternative method to the scrambled Halton sequence for removing correlation between standard Halton sequences in high dimensions, in *ERSA conference papers*, ersa03p406 (European Regional Science Association).

Hickernell, F. (1998). A generalized discrepancy and quadrature error bound, *Math. Comp.* **67**, 221, pp. 299–322.

Hickernell, F. (2009). Shape parameter problem, Private communication.

Hickernell, F. (2014). Positive definite kernels based on Chebychev functions, Private communication.

Hickernell, F. and Hong, R. H. S. (2002). Quasi-Monte Carlo methods and their randomizations, in R. Chan, Y.-K. Kwok, D. Yao, and Q. Zhang (eds.), *Applied Probability*, Vol. 26 (American Mathematical Society, Providence, RI), pp. 59–78.

Hickernell, F. J. and Hon, Y. C. (1999). Radial basis function approximations as smoothing splines, *Appl. Math. Comput.* **102**, 1, pp. 1–24.

Hilbert, D. (1904). Grundzüge einer allgemeinen Theorie der linearen Integralgleichungen I, *Göttinger Nachrichten, Math.-Phys. Kl.* , pp. 49–91.

Hille, E. (1972). Introduction to general theory of reproducing kernels, *Rocky Mt. J. Math.* **2**, 3, pp. 321–368.

Hochstadt, H. (1973). *Integral Equations* (Wiley, New York).

Holladay, J. C. (1957). A smoothest curve approximation, *Mathematical Tables and Other Aids to Computation* **11**, 60, pp. 233–243.

Holmes, M., Gray, A., and Isbell, C. (2007). Fast nonparametric conditional density estimation, in *Proceedings of the Twenty-Third Conference Annual Conference on Uncertainty in Artificial Intelligence (UAI-07)* (AUAI Press, Corvallis, Oregon), pp. 175–182.

Hon, Y. C. and Mao, X. Z. (1998). An efficient numerical scheme for Burgers' equation, *Appl. Math. Comput.* **95**, 1, pp. 37–50.

Hon, Y. C. and Mao, X. Z. (1999). A radial basis function method for solving options pricing model, *Financial Engineering* **8**, pp. 31–49.

Hon, Y. C. and Schaback, R. (2001). On unsymmetric collocation by radial basis functions, *Appl. Math. Comput.* **119**, 2–3, pp. 177–186.

Hon, Y. C. and Wei, T. (2004). A fundamental solution method for inverse heat conduction problem, *Eng. Anal. Bound. Elem.* **28**, 5, pp. 489–495.

Hon, Y. C. and Wu, Z. (2007). Error estimation on using radial basis functions for solving option pricing models, *Wilmott* **9040286**, pp. 1–21.

Horn, R. A. and Johnson, C. R. (2013). *Matrix Analysis*, 2nd edn. (Cambridge University Press, Cambridge).

Huang, C.-S., Lee, C.-F., and Cheng, A. (2007). Error estimate, optimal shape factor, and high precision computation of multiquadric collocation method, *Eng. Anal. Bound. Elem.* **31**, 7, pp. 614–623.

Huang, C.-S., Wang, S., Chen, C., and Li, Z.-C. (2006). A radial basis collocation method for Hamilton–Jacobi–Bellman equations, *Automatica* **42**, 12, pp. 2201–2207.

Hubbert, S. (2012). Closed form representations for a class of compactly supported radial basis functions, *Adv. Comput. Math.* **36**, 1, pp. 115–136.

Hubbert, S. and Baxter, B. J. C. (2001). Radial basis functions for the sphere, in W. Haussman, K. Jetter, and M. Reimer (eds.), *Recent Progress in Multivariate Approximation, International Series of Numerical Mathematics*, Vol. 137 (Birkhäuser, Basel), pp. 33–47.

Hubbert, S. and Müller, S. (2007). Thin plate spline interpolation on the unit interval, *Numer. Algorithms* **45**, pp. 167–177.

Hunter, J. K. and Nachtergaele, B. (2001). *Applied Analysis* (World Scientific Publishing Company, Singapore).

Hur, J. and Baldick, R. (2012). Spatial prediction of wind farm outputs using the augmented kriging-based model, in *IEEE Power and Energy Society General Meeting*, pp. 1–7.

Hutchinson, M. (1990). A stochastic estimator of the trace of the influence matrix for laplacian smoothing splines, *Comm. Stat. Simul. Comput.* **19**, 2, pp. 433–450.

Iske, A. (1999). Perfect centre placement for radial basis function methods, Tech. rep., Technische Universität München.

Iske, A. (2000). Optimal distribution of centers for radial basis function methods, Technical report m0004, Technische Universität München.

Iske, A. (2003). Radial basis functions: basics, advanced topics and meshfree methods for transport problems, *Rend. Sem. Mat. Univ. Pol. Torino* **61**, pp. 247–285.

Iske, A. (2004). *Multiresolution Methods in Scattered Data Modelling, Lecture Notes in Computational Science and Engineering*, Vol. 37 (Springer, Berlin; New York).

Iske, A. (2013). On the construction of kernel-based adaptive particle methods in numerical flow simulation, in R. Ansorge, B. Hester, A. Meister, and T. Sonar (eds.), *Recent Developments in the Numerics of Nonlinear Hyperbolic Conservation*, Notes on Numerical Fluid Mechanics and Multidisciplinary Design (NNFM) (Springer, Berlin), pp. 197–221.

Iske, A. and Sonar, T. (1996). On the structure of function spaces in optimal recovery of point functionals for ENO-schemes by radial basis functions, *Numer. Math.* **74**, 2, pp. 177–201.

Jester, P., Menke, C., and Urban, K. (2011). B-spline representation of optical surfaces and its accuracy in a ray trace algorithm, *Applied Optics* **50**, 6, pp. 822–828.

Jetter, K., Stöckler, J., and Ward, J. D. (1999). Error estimates for scattered data interpolation on spheres, *Math. Comp.* **68**, 226, pp. 733–747.

Jiao, X. and Heath, M. T. (2004). Common-refinement-based data transfer between non-matching meshes in multiphysics simulations, *Int. J. Numer. Methods Eng.* **61**, 14, pp. 2402–2427.

Joachims, T. (2002). *Learning to Classify Text Using Support Vector Machines: Methods, Theory and Algorithms* (Springer, Berlin; New York).

Johnson, M. J. (2012). Compactly supported, piecewise polyharmonic radial functions with prescribed regularity, *Constr. Approx.* **35**, 2, pp. 201–223.

Jones, D. R. (2001). A taxonomy of global optimization methods based on response surfaces, *J. Global Optim.* **21**, 4, pp. 345–383.

Jordão, T. and Menegatto, V. A. (2010). Integral operators generated by multi-scale kernels, *J. Complexity* **26**, 2, pp. 187–199.

Juncosa, M. L. (1945). An integral equation related to Bessel functions, *Duke Math. J.* **12**, 3, pp. 465–471.

Kac, M. and Siegert, A. J. F. (1947). On the theory of noise in radio receivers with square law detectors, *J. Appl. Phys.* **18**, 4, pp. 383–397.

Kailath, T. (1966). Some integral equations with 'nonrational' kernels, *IEEE T. Inform. Theory* **12**, 4, pp. 442–447.

Kang, L. and Joseph, V. R. (2014). Kernel approximation: from regression to interpolation, Submitted.

Kansa, E. J. (1986). Application of Hardy's multiquadric interpolation to hydrodynamics, in *Proc. 1986 Simul. Conf.*, pp. 111–117.

Kansa, E. J. (1990a). Multiquadrics — A scattered data approximation scheme with applications to computational fluid-dynamics — I: Surface approximations and partial derivative estimates, *Comput. Math. Appl.* **19**, 8-9, pp. 127–145.

Kansa, E. J. (1990b). Multiquadrics — A scattered data approximation scheme with applications to computational fluid-dynamics — II: Solutions to parabolic, hyperbolic and elliptic partial differential equations, *Comput. Math. Appl.* **19**, 8–9, pp. 147–161.

Kansa, E. J. (2007). Exact explicit time integration of hyperbolic partial differential equations with mesh free radial basis functions, *Eng. Anal. Bound. Elem.* **31**, 7, pp. 577–585.

Kansa, E. J. and Carlson, R. E. (1992). Improved accuracy of multiquadric interpolation using variable shape parameters, *Comput. Math. Appl.* **24**, pp. 99–120.

Kansa, E. J. and Hon, Y. C. (2000). Circumventing the ill-conditioning problem with multiquadric radial basis functions: Applications to elliptic partial differential equations, *Comput. Math. Applic.* **39**, pp. 123–137.

Karhunen, K. (1947). Über lineare Methoden in der Wahrscheinlichkeitsrechnung, *Ann. Acad. Sci. Fennicae. Ser. A. I. Math.-Phys.* **37**, pp. 1–79.

Karlin, S. (1968). *Total Positivity* (Stanford University Press, Stanford).

Kaya, I. and Rolland, J. P. (2010). A radial basis function method for freeform optics surfaces, in *Frontiers in Optics 2010/Laser Science XXVI*, OSA Technical Digest (CD) (Optical Society of America), p. FThX1.

Kaya, I. and Rolland, J. P. (2013). Hybrid RBF and local ϕ-polynomial freeform surfaces, *Advanced Optical Technologies* **2**, 1, pp. 81–88.

Kenett, R., Zacks, S., and Amberti, D. (2014). *Modern Industrial Statistics: With Applications in R, MINITAB and JMP*, 2nd edn. (Wiley, Hoboken, NJ).

Keyes, D. E., McInnes, L. C., Woodward, C., Gropp, W. D., Myra, E., Pernice, M., Bell, J., Brown, J., Clo, A., Connors, J., Constantinescu, E., Estep, D., Evans, K., Farhat, C., Hakim, A., Hammond, G., Hansen, G., Hill, J., Isaac, T., Jiao, X., Jordan, K., Kaushik, D., Kaxiras, E., Koniges, A., Lee, K., Lott, A., Lu, Q., Magerlein, J., Maxwell, R., McCourt, M., Mehl, M., Pawlowski, R., Peters, A., Reynolds, D., Riviere, B., Rüde, U., Scheibe, T., Shadid, J., Sheehan, B., Shephard, M., Siegel, A., Smith, B., Tang, X., Wilson, C., and Wohlmuth, B. (2013). Multiphysics simulations: Challenges and opportunities, *Int. J. High Perform. Comput. Applic.* **27**, 1, pp. 4–83.

Khaliq, A. Q. M., Voss, D. A., and Fasshauer, G. E. (2008). A parallel time stepping approach using mesh-free approximations for pricing options with non-smooth payoffs, *J. Risk* **10**, 4, pp. 135–142.

Khintchine, A. (1934). Korrelationstheorie der stationären stochastischen Prozesse, *Math. Ann.* **109**, 1, pp. 604–615.

Kimeldorf, G. and Wahba, G. (1971). Some results on Tchebycheffian spline functions, *J. Math. Anal. Appl.* **33**, pp. 82–95.

Kincaid, D. and Cheney, W. (2002). *Numerical Analysis: Mathematics of Scientific Computing*, 3rd edn. (Brooks/Cole, Pacific Grove, CA).

Kitanidis, P. K. (1997). *Introduction to Geostatistics: Applications in Hydrogeology* (Cambridge University Press, Cambridge).

Knoll, D. A. and Keyes, D. E. (2004). Jacobian-free Newton–Krylov methods: a survey of approaches and applications, *J. Comput. Phys.* **193**, pp. 357–397.

Kolmogorov, A. N. (1962). *Interpolation and Extrapolation of Stationary Random Sequences (Translated from the original 1941 Russian edition)* (Rand Corp., Santa Monica, CA).

König, H. (1986). *Eigenvalue Distribution of Compact Operators* (Birkhäuser, Basel).

Kounchev, O. (2001). *Multivariate Polysplines: Applications to Numerical and Wavelet Analysis* (Academic Press, San Diego, CA).

Krasny, R. and Wang, L. (2011). Fast evaluation of multiquadric RBF sums by a Cartesian treecode, *SIAM J. Sci. Comput.* **33**, 5, pp. 2341–2355.

Kress, R. (1999). *Linear Integral Equations* (Springer, Berlin; New York).

Krige, D. G. (1951). A statistical approach to some basic mine valuation problems on the Witwatersrand, *J. Chem. Met. & Mining Soc., S. Africa* **52**, 6, pp. 119–139.

Kryżyak, A. (2011). Radial basis function networks with optimal kernels, in L. Devroye (ed.), *2011 IEEE International Symposium on Information Theory Proceedings (ISIT)*, pp. 860–863.

Kupradze, V. D. and Aleksidze, M. A. (1964). The method of functional equations for the approximate solution of certain boundary value problems, *USSR Comp. Math. Math.* **4**, 4, pp. 82–126.

Kwok, Y.-K. (2008). *Mathematical Models of Financial Derivatives*, 2nd edn. (Springer, Berlin; New York).

Kybic, J., Blu, T., and Unser, M. (2002). Generalized sampling: A variational approach — part I: Theory, *IEEE T. Signal Proces.* **50**, pp. 1965–1976.

Lancaster, P. and Šalkauskas, K. (1981). Surfaces generated by moving least squares methods, *Math. Comp.* **37**, pp. 141–158.

Lanckriet, G. R. G., Cristianini, N., Bartlett, P., El Ghaoui, L., and Jordan, M. I. (2004). Learning the kernel matrix with semidefinite programming, *J. Mach. Learn. Res.* **5**, pp. 27–72.

de Laplace, P.-S. (1778). *Oeuvres complètes de Laplace. Tome 10 / publiées sous les auspices de l'Académie des sciences, par MM. les secrétaires perpétuels* (Gauthier–Villars, Paris).

Larkin, F. M. (1970). Optimal approximation in Hilbert spaces with reproducing kernel functions, *Math. Comp.* **24**, 112, pp. 911–921.

Larsson, E., Åhlander, K., and Hall, A. (2008). Multi-dimensional option pricing using radial basis functions and the generalized Fourier transform, *J. Comput. Appl. Math.* **222**, 1, pp. 175–192.

Larsson, E. and Fornberg, B. (2003). A numerical study of some radial basis function based solution methods for elliptic PDEs, *Comput. Math. Appl.* **46**, 5–6, pp. 891–902.

Larsson, E. and Fornberg, B. (2005). Theoretical and computational aspects of multivariate interpolation with increasingly flat radial basis functions, *Comput. Math. Appl.* **49**, 1, pp. 103–130.

Larsson, E., Gomes, S. M., Heryudono, A., and Safdari-Vaighani, A. (2013a). Radial basis function methods in computational finance, in *Proceedings of the 13th International Conference on Computational and Mathematical Methods in Science and Engineer-*

ing, CMMSE 2013.

Larsson, E., Lehto, E., Heryudono, A., and Fornberg, B. (2013b). Stable computation of differentiation matrices and scattered node stencils based on Gaussian radial basis functions, *SIAM J. Sci. Comput.* **35**, 4, pp. A2096–A2119.

LeCun, Y. A., Bottou, L., Orr, G. B., and Müller, K.-R. (2012). Efficient backprop, in *Neural networks: Tricks of the trade* (Springer, Berlin; New York), pp. 9–48.

Lee, C.-F., Ling, L., and Schaback, R. (2009). On convergent numerical algorithms for unsymmetric collocation, *Adv. Comput. Math.* **30**, 4, pp. 339–354.

Lee, Y. J. and Micchelli, C. A. (2013). On collocation matrices for interpolation and approximation, *J. Approx. Theory* **174**, pp. 148–181.

Lee, Y. J., Micchelli, C. A., and Yoon, J. (2014). On convergence of flat multivariate interpolation by translation kernels with finite smoothness, *Constr. Approx.* **40**, 1, pp. 37–60.

Lee, Y. J., Yoon, G. J., and Yoon, J. (2007). Convergence of increasingly flat radial basis interpolants to polynomial interpolants, *SIAM J. Math. Anal.* **39**, 2, pp. 537–553.

LeVeque, R. J. (2002). *Finite Volume Methods for Hyperbolic Problems*, Cambridge Texts in Applied Mathematics (Cambridge University Press, Cambridge).

Levesley, J. and Ragozin, D. (2002). Positive definite kernel interpolation on manifolds: convergence rates, in C. K. Chui, L. L. Schumaker, and J. Stöckler (eds.), *Approximation Theory X: Abstract and Classical Analysis* (Vanderbilt University Press, Nashville, TN), pp. 277–285.

Levesley, J. and Ragozin, D. L. (2007). Radial basis interpolation on homogeneous manifolds: convergence rates, *Adv. Comput. Math.* **27**, 2, pp. 237–246.

Li, S., Gu, M., Wu, C. J., and Xia, J. (2012). New efficient and robust HSS Cholesky factorization of SPD matrices, *SIAM J. Matrix Anal. Appl.* **33**, 3, pp. 886–904.

Li, S. and Liu, W. K. (2007). *Meshfree Particle Methods* (Springer, Berlin; New York).

Li, Z. and Mao, X.-Z. (2011). Global space-time multiquadric method for inverse heat conduction problem, *Int. J. Numer. Meth. Eng.* **85**, 3, pp. 355–379.

Lin, Y., Lv, F., Zhu, S., Yang, M., Cour, T., Yu, K., Cao, L., and Huang, T. (2011). Large-scale image classification: fast feature extraction and SVM training, in *Computer Vision and Pattern Recognition (CVPR), 2011 IEEE Conference on* (IEEE), pp. 1689–1696.

Ling, L. (2006). Finding numerical derivatives for unstructured and noisy data by multiscale kernels, *SIAM J. Numer. Anal.* **44**, 4, pp. 1780–1800.

Ling, L., Opfer, R., and Schaback, R. (2006). Results on meshless collocation techniques, *Eng. Anal. Bound. Elem.* **30**, 4, pp. 247–253.

Little, G. and Reade, J. B. (1984). Eigenvalues of analytic kernels, *SIAM J. Math. Anal.* **15**, 1, pp. 133–136.

Liu, G. R. (2002). *Mesh Free Methods: Moving Beyond the Finite Element Method* (CRC Press, Boca Raton, FL).

Liu, G. R. and Liu, M. B. (2003). *Smoothed Particle Hydrodynamics: A Meshfree Particle Method* (World Scientific Publishing Company, Singapore).

Liu, G. R. and Trung, N. T. (2010). *Smoothed Finite Element Methods* (CRC Press, Boca Raton, FL).

Liu, G. R. and Zhang, G. Y. (2013). *Smoothed Point Interpolation Methods: G Space Theory and Weakened Weakforms* (World Scientific Publishing Company, Singapore).

Liu, J. S. (2008). *Monte Carlo Strategies in Scientific Computing* (Springer, Berlin; New York).

Liu, W. K., Jun, S., and Zhang, Y. F. (1995). Reproducing kernel particle methods, *Int. J. Numer. Meth. Fluids* **20**, 8–9, pp. 1081–1106.

Lockwood, B. A. and Anitescu, M. (2011). Gradient-enhanced universal kriging for uncertainty propagation in nuclear engineering, *Trans. Am. Nucl. Soc.* **104**, pp. 168–195.

Lockwood, B. A. and Anitescu, M. (2012). Gradient-enhanced universal kriging for uncertainty propagation, *Nucl. Sci. Eng.* **170**, 2, pp. 168–195.

Loève, M. (1977). *Probability Theory I* (Springer, Berlin; New York).

Logan, B. F. and Shepp, L. A. (1975). Optimal reconstruction of a function from its projections, *Duke Math. J.* **42**, 4, pp. 645–659.

Longstaff, F. A. and Schwartz, E. S. (2001). Valuing American options by simulation: a simple least-squares approach, *Rev. Financ. Stud.* **14**, pp. 113–147.

Lowitzsch, S. (2005). Matrix-valued radial basis functions: stability estimates and applications, *Adv. Comput. Math.* **23**, 3, pp. 299–315.

Lucy, L. B. (1977). A numerical approach to the testing of the fission hypothesis, *Astron. J.* **82**, pp. 1013–1024.

Luh, L.-T. (2012). The shape parameter in the shifted surface spline III, *Eng. Anal. Bound. Elem.* **36**, 11, pp. 1604–1617.

Lukić, M. N. and Beder, J. H. (2001). Stochastic processes with sample paths in reproducing kernel Hilbert spaces, *T. Am. Math. Soc.* **353**, 10, pp. 3945–3969.

Lyche, T. and Schumaker, L. L. (1973). On the convergence of cubic interpolating splines, in *Spline functions and approximation theory (Proc. Sympos., Univ. Alberta, Edmonton, Alta., 1972), Internat. Ser. Numer. Math.*, Vol. 21 (Birkhäuser, Basel), pp. 169–189.

Macedo, I., Gois, J., and Velho, L. (2009). Hermite interpolation of implicit surfaces with radial basis functions, in *2009 XXII Brazilian Symposium on Computer Graphics and Image Processing (SIBGRAPI)*, pp. 1–8.

Madych, W. R. (2006). An estimate for multivariate interpolation II, *J. Approx. Theory* **142**, 2, pp. 116–128.

Madych, W. R. and Nelson, S. A. (1983). *Multivariate interpolation: a variational theory* (Unpublished manuscript).

Madych, W. R. and Nelson, S. A. (1990). Polyharmonic cardinal splines, *J. Approx. Theory* **60**, 2, pp. 141–156.

Madych, W. R. and Potter, E. H. (1985). An estimate for multivariate interpolation, *J. Approx. Theory* **43**, 2, pp. 132–139.

Mairhuber, J. C. (1956). On Haar's theorem concerning Chebychev approximation problems having unique solutions, *P. Am. Math. Soc.* **7**, 4, pp. 609–615.

Mao, K. Z. (2002). RBF neural network center selection based on Fisher ratio class separability measure, *IEEE T. Neural Networ.* **13**, 5, pp. 1211–1217.

Mardia, K. V., Kent, J. T., and Bibby, J. M. (1979). *Multivariate Analysis* (Academic Press, London; New York).

Mardia, K. V., Kent, J. T., Goodall, C. R., and Little, J. A. (1996). Kriging and splines with derivative information, *Biometrika* **83**, 1, pp. 207–221.

Marsden, M. (1974). Cubic spline interpolation of continuous functions, *J. Approx. Theory* **10**, 2, pp. 103–111.

Matérn, B. (1986). *Spatial Variation, Lecture Notes in Statistics*, Vol. 36, 2nd edn. (Springer, Berlin; New York).

Matheron, G. (1962). *Traité de géostatistique appliquée. Tome I* (Technip, Paris).

Matheron, G. (1965). *Les Variables Régionalisées et Leur Estimation* (Masson, Paris).

Matheron, G. (1973). The intrinsic random functions and their applications, *Adv. Appl. Probab.* **5**, 3, pp. 439–468.

Mathias, M. (1923). Über positive Fourier-Integrale, *Math. Zeit.* **16**, pp. 103–125.

Matías, J. M., Vaamonde, A., Taboada, J., and González-Manteiga, W. (2004). Comparison of Kriging and neural networks with application to the exploitation of a slate mine, *Math. Geol.* **36**, 4, pp. 463–486.

Maz'ya, V. and Schmidt, G. (2007). *Approximate Approximations, Mathematical Surveys and Monographs*, Vol. 141 (American Mathematical Society, Providence, RI).

McCafferty, A. E., Horton, R. J., Stanton, M. R., McDougal, R. R., and Fey, D. L. (2011). Geophysical, geochemical, mineralogical, and environmental data for rock samples collected in a mineralized volcanic environment, upper Animas River watershed, Colorado, Tech. Rep. 595, U.S. Geological Survey Data Series.

McCourt, M. (2013a). *Building Infrastructure for Multiphysics Simulations*, Ph.D. thesis, Cornell University.

McCourt, M. (2013b). Using Gaussian eigenfunctions to solve boundary value problems, *Adv. Appl. Math. Mech.* **5**, pp. 569–594.

McCourt, M. and Fasshauer, G. E. (2014). Stable likelihood computation for Gaussian random fields, Submitted.

McCourt, M., Smith, B., and Zhang, H. (2015). Sparse matrix-matrix products executed through coloring, *SIAM J. Matrix Anal. Appl.* **36**, 1, pp. 90–109.

McCourt, M. J. (2013c). A fast QR method for Gaussian eigenfunction approximation, Submitted.

McDonald, D. B., Grantham, W. J., Tabor, W. L., and Murphy, M. J. (2007). Global and local optimization using radial basis function response surface models, *Appl. Math. Model.* **31**, 10, pp. 2095–2110.

McKay, M. D., Conover, W. J., and Beckman, R. J. (1979). A comparison of three methods for selecting values of input variables in the analysis of output from computer code, *Technometrics* **21**, pp. 239–245.

McNamee, J., Stenger, F., and Whitney, E. L. (1971). Whittaker's cardinal function in retrospect, *Math. Comp.* **25**, 113, pp. 141–154.

Meinguet, J. (1979). Multivariate interpolation at arbitrary points made simple, *Z. Angew. Math. Phys.* **30**, 2, pp. 292–304.

Melenk, J. M. and Babuška, I. (1996). The partition of unity finite element method: basic theory and applications, *Comput. Methods Appl. Mech. Eng.* **139**, pp. 289–314.

Menegatto, V. A. (1999). Strict positive definiteness on spheres, *Analysis* **19**, 3, pp. 217–233.

Mercer, J. (1909). Functions of positive and negative type, and their connection with the theory of integral equations, *Philos. T. R. Soc. Lond.* **209**, pp. 415–446.

Meschkowski, H. (1962). *Hilbertsche Räume mit Kernfunktion* (Springer, Berlin; New York).

Mhaskar, H. N., Narcowich, F. J., Prestin, J., and Ward, J. D. (2010). l^p Bernstein estimates and approximation by spherical basis functions, *Math. Comp.* **79**, pp. 1647–1679.

Micchelli, C. A. (1986). Interpolation of scattered data: distance matrices and conditionally positive definite functions, *Constr. Approx.* **2**, 1, pp. 11–22.

Micchelli, C. A. and Pontil, M. (2005). Learning the kernel function via regularization, *J. Mach. Learn. Res.* **6**, pp. 1099–1125.

Michel, V. (2013). *Lectures on Constructive Approximation: Fourier, Spline, and Wavelet Methods on the Real Line, the Sphere, and the Ball*, Applied and Numerical Harmonic Analysis (Birkhäuser, Basel).

Miele, A. and Iyer, R. R. (1970). General technique for solving nonlinear, two-point boundary-value problems via the method of particular solutions, *J. Optimiz. Theory App.* **5**, pp. 382–399.

Mikosch, T. (1998). *Elementary Stochastic Calculus with Finance in View*, Advanced Series on Statistical Science & Applied Probability (World Scientific Publishing, Singapore).

Moës, N., Dolbow, J., and Belytschko, T. (1999). A finite element method for crack growth without remeshing, *Int. J. Numer. Meth. Eng.* **46**, 1, pp. 131–150.

Moguerza, J. M. and Muñoz, A. (2006). Support vector machines with applications, *Stat. Sci.* **21**, 3, pp. 322–336.

Moler, C. B. (2008). *Numerical Computing with MATLAB* (Society for Industrial and Applied Mathematics, Philadelphia, PA).

Mongillo, M. (2011). Choosing basis functions and shape parameters for radial basis function methods, *SIAM Undergraduate Research Online* **4**, pp. 190–209.

Moore, E. H. (1916). On properly positive Hermitian matrices, *B. Am. Math. Soc.* **23**, 2, p. 59.

Moore, E. H. (1935). *General Analysis, Part I, Memoirs*, Vol. 1 (Amer. Philos. Soc., Philadelphia).

Moridis, G. J. and Kansa, E. J. (1994). The Laplace transform multiquadric method: A highly accurate scheme for the numerical solution of linear partial differential equations, *J. Appl. Sc. Comp.* **1**, pp. 375–407.

Morris, M. D., Mitchell, T. J., and Ylvisaker, D. (1993). Bayesian design and analysis of computer experiments: use of derivatives in surface prediction, *Technometrics* **35**, 3, pp. 243–255.

Morse, P. M. and Feshbach, H. (1953). *Methods of Theoretical Physics* (McGraw–Hill, New York).

Mouat, C. T. (2001). *Fast algorithms and preconditioning techniques for fitting radial basis functions*, Ph.D. thesis, University of Canterbury, NZ.

Müller, C. (1966). *Spherical Harmonics, Lecture Notes in Mathematics*, Vol. 17 (Springer, Berlin; New York).

Müller, S. (2009). *Komplexität und Stabilität von kernbasierten Rekonstruktionsmethoden*, Ph.D. thesis, Universität Göttingen.

Müller, S. and Schaback, R. (2009). A Newton basis for kernel spaces, *J. Approx. Theory* **161**, 2, pp. 645–655.

Nabney, I. T. (2004). Efficient training of RBF networks for classification, *Int. J. Neural Syst.* **14**, 03, pp. 201–208.

Nagy, B., Loeppky, J. L., and Welch, W. J. (2007). Fast Bayesian inference for Gaussian process models, Tech. rep., The University of British Columbia, Department of Statistics.

Narayan, A. and Xiu, D. (2012). Stochastic collocation methods on unstructured grids in high dimensions via interpolation, *SIAM J. Sci. Comput.* **34**, 3, pp. A1729–A1752.

Narcowich, F. J. (1995). Generalized Hermite interpolation and positive definite kernels on a Riemannian manifold, *J. Math. Anal. Appl.,* **190**, 1, pp. 165–193.

Narcowich, F. J., Sivakumar, N., and Ward, J. D. (1994). On condition numbers associated with radial-function interpolation, *J. Math. Anal. Appl.,* **186**, pp. 457–485.

Narcowich, F. J. and Ward, J. D. (1991). Norms of inverses and condition numbers for matrices associated with scattered data, *J. Approx. Theory* **64**, pp. 69–94.

Narcowich, F. J. and Ward, J. D. (1992). Norm estimates for the inverses of a general class of scattered-data radial-function interpolation matrices, *J. Approx. Theory* **69**, pp. 84–109.

Narcowich, F. J. and Ward, J. D. (1994). Generalized Hermite interpolation via matrix-valued conditionally positive definite functions, *Math. Comp.* **63**, 208, pp. 661–687.

Narcowich, F. J., Ward, J. D., and Wendland, H. (2005). Sobolev bounds on functions with scattered zeros, with applications to radial basis function surface fitting, *Math.*

Comp. **74**, pp. 743–763.

Narcowich, F. J., Ward, J. D., and Wendland, H. (2006). Sobolev error estimates and a Bernstein inequality for scattered data interpolation via radial basis functions, *Constr. Approx.* **24**, 2, pp. 175–186.

Nayroles, B., Touzot, G., and Villon, P. (1992). Generalizing the finite element method: diffuse approximation and diffuse elements, *Comput. Mech.* **10**, 5, pp. 307–318.

Neumaier, A. (1998). Solving ill-conditioned and singular linear systems: A tutorial on regularization, *SIAM Rev.* **40**, 3, pp. 636–666.

Nguyen, V. P., Rabczuk, T., Bordas, S., and Duflot, M. (2008). Meshless methods: a review and computer implementation aspects, *Math. Comput. Simulat.* **79**, 3, pp. 763–813.

Nielsen, B. F., Skavhaug, O., and Tveito, A. (2002). Penalty and front-fixing methods for the numerical solution of American option problems, *J. Comput. Finance* **5**, 4, pp. 69–97.

Nielsen, B. F., Skavhaug, O., and Tveito, A. (2008). Penalty methods for the numerical solution of American multi-asset option problems, *J. Comput. Appl. Math.* **222**, 1, pp. 3–16.

Novak, E. and Woźniakowski, H. (2001). Intractability results for integration and discrepancy, *J. Complexity* **17**, 2, pp. 388–441.

Novak, E. and Woźniakowski, H. (2008). *Tractability of Multivariate Problems Volume 1: Linear Information*, no. 6 in EMS Tracts in Mathematics (European Mathematical Society).

Oeuvray, R. and Bierlaire, M. (2009). Boosters: a derivative-free algorithm based on radial basis functions, *Int. J. Model. Simul.* **29**, 1, pp. 26–36.

Øksendal, B. K. (2003). *Stochastic Differential Equations: An Introduction with Applications*, 6th edn. (Springer, Berlin; New York).

Olver, F. W. J., Lozier, D. W., Boisvert, R. F., and Clark, C. W. (eds.) (2010). *NIST Handbook of Mathematical Functions* (Cambridge University Press, New York).

Omre, H. (1987). Bayesian kriging — merging observations and qualified guesses in kriging, *Math. Geol.* **19**, 1, pp. 25–39.

Opfer, R. (2006). Multiscale kernels, *Adv. Comput. Math.* **25**, 4, pp. 357–380.

Orr, M. J. L. (1995). Regularization in the selection of radial basis function centers, *Neural Comput.* **7**, 3, pp. 606–623.

Orr, M. J. L. (1996). Introduction to radial basis function networks, Tech. rep., University of Edinburgh, Centre for Cognitive Sciences.

Pace, R. K. and LeSage, J. P. (2004). Chebyshev approximation of log-determinants of spatial weight matrices, *Computational Statistics & Data Analysis* **45**, 2, pp. 179–196.

Parzen, E. (1961). An approach to time series analysis, *Ann. Math. Stat.* **32**, 4, pp. 951–989.

Parzen, E. (1962). *Stochastic Processes* (Holden Day, San Francisco, CA).

Parzen, E. (1970). Statistical inference on time series by RKHS methods, in *Proceedings 12th Biennial Seminar* (Canadian Mathematical Congress), pp. 1–37.

Pazouki, M. and Schaback, R. (2011). Bases for kernel-based spaces, *J. Comput. Appl. Math.* **236**, 4, pp. 575–588.

Pazouki, M. and Schaback, R. (2013). Bases for conditionally positive definite kernels, *J. Comput. Appl. Math.* **243**, pp. 152–163.

Pearson, J. W. (2013). A radial basis function method for solving PDE-constrained optimization problems, *Numer. Algorithms* **64**, pp. 481–506.

Pesenson, I. (2004). Variational splines on Riemannian manifolds with applications to

integral geometry, *Adv. Appl. Math.* **33**, 3, pp. 548–572.
Pettersson, U., Larsson, E., Marcusson, G., and Persson, J. (2008). Improved radial basis function methods for multi-dimensional option pricing, *J. Comput. Appl. Math.* **222**, 1, pp. 82–93.
Pinkus, A. (2013). On ridge functions, *Azerbaijan J. Math.* **3**, 2, pp. 122–130.
Platt, J. C. (1999). Fast training of support vector machines using sequential minimal optimization, in B. Schölkopf, C. J. C. Burges, and A. J. Smola (eds.), *Advances in Kernel Methods* (MIT Press, Cambridge, MA), pp. 185–208.
Platte, R. B. and Driscoll, T. A. (2005). Polynomials and potential theory for Gaussian radial basis function interpolation, *SIAM J. Numer. Anal.* **43**, 2, pp. 750–766.
Poggio, T., Mukherjee, S., Rifkin, R., Rakhlin, A., and Verri, A. (2001). b, Tech. rep., MIT AI Memo 2001-011.
Pogorzelski, W. (1966). *Integral Equations and their Applications* (Pergamon Press, Tarrytown, NY, USA).
Poincaré, H. (1896). *Calcul des Probabilités* (George Carré, Paris).
Porcu, E., Mateu, J., and Bevilacqua, M. (2007). Covariance functions that are stationary or nonstationary in space and stationary in time, *Stat. Neerl.* **61**, 3, pp. 358–382.
Powell, M. J. D. (1987). Radial basis functions for multivariable interpolation: a review, in J. C. Mason and M. G. Cox (eds.), *Algorithms for the Approximation of Functions and Data* (Oxford University Press, Oxford), pp. 143–167.
Powell, M. J. D. (1992). The theory of radial basis functions in 1990, in W. Light (ed.), *Advances in Numerical Analysis II: Wavelets, Subdivision, and Radial Basis Functions* (Oxford University Press, Oxford), pp. 105–210.
Powell, M. J. D. (1999). Recent research at Cambridge on radial basis functions, in M. W. Müller, M. D. Buhmann, D. H. Mache, and M. Felten (eds.), *New Developments in Approximation Theory* (Birkhäuser, Basel), pp. 215–232.
Protter, P. E. (2004). *Stochastic Integration and Differential Equations: Version 2.1*, Vol. 21 (Springer Science & Business Media).
R Core Team (2014). *R: A Language and Environment for Statistical Computing*, R Foundation for Statistical Computing, Vienna, Austria.
Ramsay, J. and Silverman, B. W. (2005). *Functional Data Analysis*, 2nd edn. (Springer, Berlin; New York).
Rasmussen, C. E. and Williams, C. (2006). *Gaussian Processes for Machine Learning* (MIT Press, Cambridge, MA).
Reade, J. B. (1983). Eigenvalues of positive definite kernels, *SIAM J. Math. Anal.* **14**, 1, pp. 152–157.
Reade, J. B. (1984). Eigenvalues of positive definite kernels II, *SIAM J. Math. Anal.* **15**, 1, pp. 137–142.
Renka, R. J. (1987). Interpolatory tension splines with automatic selection of tension factors, *SIAM J. Sci. Stat. Comp.* **8**, 3, pp. 393–415.
Richardson, L. F. (1911). The approximate arithmetical solution by finite differences of physical problems involving differential equations, with an application to the stresses in a masonry dam, *Philos. T. Roy. Soc. A* **210**, 459–470, pp. 307–357.
Richardson, M. (2009). Numerical methods for option pricing, M.Sc. Mathematical Modelling and Scientific Computing "Special Topic", Oxford University.
Rieger, C. (2008). *Sampling Inequalities and Applications*, Ph.D. thesis, Universität Göttingen.
Rieger, C., Schaback, R., and Zwicknagl, B. (2010). Sampling and stability, in M. Dæhlen, M. Floater, T. Lyche, J.-L. Merrien, K. Mørken, and L. L. Schumaker (eds.), *Mathematical Methods for Curves and Surfaces, Lecture Notes in Computer Science*, Vol.

5862 (Springer, Berlin; New York), pp. 347–369.

Rieger, C. and Zwicknagl, B. (2008). Sampling inequalities for infinitely smooth functions, with applications to interpolation and machine learning, *Adv. Comput. Math.* **32**, 1, pp. 103–129.

Rieger, C. and Zwicknagl, B. (2014). Improved exponential convergence rates by oversampling near the boundary, *Constr. Approx.* **39**, 2, pp. 323–341.

Riesz, F. and Sz.-Nagy, B. (1955). *Functional Analysis* (Dover Publications, New York).

Rippa, S. (1999). An algorithm for selecting a good value for the parameter c in radial basis function interpolation, *Adv. Comput. Math.* **11**, 2–3, pp. 193–210.

Ritter, K. (2000). *Average-Case Analysis of Numerical Problems*, *Lecture Notes in Mathematics*, Vol. 1733 (Springer, Berlin; New York).

Ron, A. and Sun, X. (1996). Strictly positive definite functions on spheres, *Math. Comp.* **65**, 216, pp. 1513–1530.

Runge, C. (1901). Über empirische Funktionen und die Interpolation zwischen äquidistanten Ordinaten, *Z. Math. Phys.* **46**, pp. 224–243.

Rynne, B. P. and Youngson, M. A. (2008). *Linear Functional Analysis*, Springer Undergraduate Mathematics Series (Springer, Berlin; New York).

Sacks, J., Welch, W. J., Mitchell, T. J., and Wynn, H. P. (1989). Design and analysis of computer experiments, *Stat. Sci.* **4**, 4, pp. 409–423.

Safdari-Vaighani, A., Heryudono, A., and Larsson, E. (2014). A radial basis function partition of unity collocation method for convection–diffusion equations arising in financial applications, `http://link.springer.com/article/10.1007/s10915-014-9935-9`.

Saitoh, S. (1988). *Theory of Reproducing Kernels and Its Applications* (Longman Higher Education, London).

Saitoh, S. (1997). *Integral Transforms, Reproducing Kernels and Their Applications* (Chapman and Hall/CRC, Boca Raton, FL).

Salemi, P. L. (2014). *Gaussian Markov Random Fields and Moving Least Squares for Metamodeling and Optimization in Stochastic Simulation*, Ph.d. thesis, Northwestern University, Evanston, IL.

Santner, T. J., Williams, B. J., and Notz, W. I. (2003). *The Design and Analysis of Computer Experiments* (Springer, Berlin; New York).

Sapidis, N. S. (ed.) (1987). *Designing Fair Curves and Surfaces: Shape Quality in Geometric Modeling and Computer-Aided Design* (Society for Industrial and Applied Mathematics, Philadelphia, PA).

Sarra, S. A. (2008). A numerical study of the accuracy and stability of symmetric and asymmetric RBF collocation methods for hyperbolic PDEs, *Numer. Meth. Part. D. E.* **24**, 2, pp. 670–686.

Sarra, S. A. and Kansa, E. J. (2009). *Multiquadric Radial Basis Function Approximation Methods for the Numerical Solution of Partial Differential Equations*, *Advances in Computational Mechanics*, Vol. 2 (Tech Science Press, Encino, CA).

Sarra, S. A. and Sturgill, D. (2009). A random variable shape parameter strategy for radial basis function approximation methods, *Eng. Anal. Bound. Elem.* **33**, 11, pp. 1239–1245.

Schaback, R. (1993). Comparison of radial basis function interpolants, in K. Jetter and F. Utreras (eds.), *Multivariate Approximation: From CAGD to Wavelets* (World Scientific Publishing, Singapore), pp. 293–305.

Schaback, R. (1995a). Error estimates and condition numbers for radial basis function interpolation, *Adv. Comput. Math.* **3**, 3, pp. 251–264.

Schaback, R. (1995b). Multivariate interpolation and approximation by translates of a basis function, in C. K. Chui and L. L. Schumaker (eds.), *Approximation Theory VIII, Vol. 1: Approximation and Interpolation* (World Scientific Publishing, Singapore), pp. 491–514.

Schaback, R. (1999). Native Hilbert spaces for radial basis functions I, in M. W. Müller, M. D. Buhmann, D. H. Mache, and M. Felten (eds.), *New Developments in Approximation Theory* (Birkhäuser, Basel), pp. 255–282.

Schaback, R. (2000). A unified theory of radial basis functions: native Hilbert spaces for radial basis functions II, *J. Comput. Appl. Math.* **121**, 1–2, pp. 165–177.

Schaback, R. (2005). Multivariate interpolation by polynomials and radial basis functions, *Constr. Approx.* **21**, pp. 293–317.

Schaback, R. (2007). Convergence of unsymmetric kernel-based meshless collocation methods, *SIAM J. Numer. Anal.* **45**, pp. 333–351.

Schaback, R. (2008). Limit problems for interpolation by analytic radial basis functions, *J. Comput. Appl. Math.* **212**, pp. 127–149.

Schaback, R. (2010). Unsymmetric meshless methods for operator equations, *Numer. Math.* **114**, 4, pp. 629–651.

Schaback, R. (2011a). *Kernel-Based Meshless Methods* (Manuscript).

Schaback, R. (2011b). The missing Wendland functions, *Adv. Comput. Math.* **34**, 1, pp. 67–81.

Schaback, R. (2013). Greedy sparse linear approximations of functionals from nodal data, *Numer. Algorithms* **63**, 3, pp. 531–547.

Schaback, R. (2014). All well-posed problems have uniformly stable and convergent discretizations, Preprint, Universität Göttingen.

Schaback, R. and Wendland, H. (2000). Adaptive greedy techniques for approximate solution of large RBF systems, *Numer. Algorithms* **24**, pp. 239–254.

Schaback, R. and Wendland, H. (2006). Kernel techniques: From machine learning to meshless methods, *Acta Numerica* **15**, pp. 543–639.

Scheuerer, M. (2011). An alternative procedure for selecting a good value for the parameter c in RBF-interpolation, *Adv. Comput. Math.* **34**, 1, pp. 105–126.

Scheuerer, M., Schaback, R., and Schlather, M. (2013). Interpolation of spatial data — a stochastic or a deterministic problem? *Eur. J. Appl. Math.* **24**, 4, pp. 601–629.

Schmidt, E. (1907). Zur Theorie der linearen und nichtlinearen Integralgleichungen. I. Teil: Entwicklung willkürlicher Funktionen nach Systemen vorgeschriebener, *Math. Ann.* **63**, pp. 433–476.

Schmidt, E. (1908). Über die Auflösung linearer Gleichungen mit unendlich vielen Unbekannten, *Rend. Circ. Mat. Palermo* **25**, 1, pp. 53–77.

Schoenberg, I. J. (1938). Metric spaces and completely monotone functions, *Ann. Math.* **39**, 4, pp. 811–841.

Schoenberg, I. J. (1942). Positive definite functions on spheres, *Duke Math. J.* **9**, pp. 96–108.

Schoenberg, I. J. (1946a). Contributions to the problem of approximation of equidistant data by analytic functions, Part A: On the problem of smoothing or graduation, a first class of analytic approximation formulas, *Quart. Appl. Math.* **4**, pp. 45–99.

Schoenberg, I. J. (1946b). Contributions to the problem of approximation of equidistant data by analytic functions, Part B: On the problem of osculatory interpolation, a second class of analytic approximation formulae, *Quart. Appl. Math.* **4**, pp. 112–141.

Schoenberg, I. J. (1969). Cardinal interpolation and spline functions, *J. Approx. Theory* **2**, 2, pp. 167–206.

Schoenberg, I. J. (1973). *Cardinal Spline Interpolation* (Society of Industrial and Applied Mathematics, Philadelphia, PA).

Schölkopf, B. and Smola, A. J. (2002). *Learning with Kernels: Support Vector Machines, Regularization, Optimization, and Beyond* (MIT Press, Cambridge, MA).

Schräder, D. and Wendland, H. (2011). A high-order, analytically divergence-free discretization method for Darcy's problem, *Math. Comp.* **80**, pp. 263–277.

Schultz, M. and Varga, R. (1967). *L*-splines, *Numer. Math.* **10**, pp. 319–345.

Schumaker, L. L. (1981). *Spline Functions: Basic Theory* (John Wiley & Sons, New York).

Schweikert, D. G. (1966). An interpolation curve using a spline in tension, *J. Math. Phys.* **45**, pp. 312–317.

Scott, D. W. (2009). *Multivariate Density Estimation: Theory, Practice, and Visualization*, Vol. 383 (John Wiley & Sons, Chichester).

Scott, D. W. (2012). Multivariate density estimation and visualization, in J. E. Gentle, W. K. H ardle, and Y. Mori (eds.), *Handbook of Computational Statistics*, Springer Handbooks of Computational Statistics (Springer, Berlin; New York), pp. 549–569.

Seeger, M. (2004). Gaussian processes for machine learning, *Int. J. Neural Syst.* **14**, 2, pp. 69–106.

Seguret, S. and Huchon, P. (1990). Trigonometric kriging: A new method for removing the diurnal variation from geomagnetic data, *J. Geophys. Res.–Sol. Ea.* **95**, B13, pp. 21383–21397.

Shampine, L., Reichelt, M., and Kierzenka, J. (1999). Solving index-1 DAEs in MATLAB and Simulink, *SIAM Rev.* **41**, 3, pp. 538–552.

Shankar, V., Wright, G. B., Kirby, R. M., and Fogelson, A. L. (2015). A radial basis function (RBF)-finite difference (FD) method for diffusion and reaction-diffusion equations on surfaces, *J. Sci. Comput.* **63**, 3, pp. 745–768.

Shapiro, H. S. (1971). *Topics in Approximation Theory* (Springer, Berlin; New York).

Shawe-Taylor, J. and Cristianini, N. (2004). *Kernel Methods for Pattern Analysis* (Cambridge University Press, Cambridge).

Shawe-Taylor, J., Williams, C., Cristianini, N., and Kandola, J. (2002). On the eigenspectrum of the Gram matrix and its relationship to the operator eigenspectrum, in N. Cesa-Bianchi, M. Numao, and R. Reischuk (eds.), *Algorithmic Learning Theory*, no. 2533 in Lecture Notes in Computer Science (Springer, Berlin; New York), pp. 23–40.

Shreve, S. (2012). *Stochastic Calculus for Finance I: The Binomial Asset Pricing Model* (Springer, Berlin; New York).

Shreve, S. E. (2004). *Stochastic Calculus for Finance II: Continuous-time Models*, Vol. 11 (Springer, Berlin; New York).

Shu, C., Ding, H., and Yeo, K. S. (2003). Local radial basis function-based differential quadrature method and its application to solve two-dimensional incompressible Navier-Stokes equations, *Comput. Methods Appl. Mech. Eng.* **192**, pp. 941–954.

Sibson, R. (1981). A brief description of natural neighbor interpolation, in V. Barnett (ed.), *Interpreting Multivariate Data* (Wiley, Chichester), pp. 21–36.

Silverman, B. W. (1982). On the estimation of a probability density function by the maximum penalized likelihood method, *Ann. Stat.* **10**, 3, pp. 795–810.

Silverman, B. W. (1986). *Density estimation for statistics and data analysis*, Vol. 26 (CRC press).

Singham, D. I., Royset, J. O., and Wets, R. J.-B. (2013). Density estimation of simulation output using exponential EPI-splines, in *Proceedings of the 2013 Winter Simulation Conference: Simulation: Making Decisions in a Complex World*, WSC '13 (IEEE Press), pp. 755–765.

Slepian, D. and Pollak, H. O. (1961). Prolate spheroidal wave functions, Fourier analysis and uncertainty. I, *Bell Syst. Tech. J.* **40**, pp. 43–63.

Smirnov, O. and Anselin, L. (2001). Fast maximum likelihood estimation of very large spatial autoregressive models: a characteristic polynomial approach, *Computational Statistics & Data Analysis* **35**, 3, pp. 301–319.

Smithies, F. (1958). *Integral Equations* (Cambridge University Press, Cambridge).

Sobol′, I. M. (1967). The distribution of points in a cube and the approximate evaluation of integrals, *U.S.S.R. Comput. Math. and Math. Phys.* **7**, pp. 86–112.

Sobolev, S. L. (1963). *Applications of Functional Analysis in Mathematical Physics* (American Mathematical Society, Rhode Island).

Solak, E., Murray-Smith, R., Leithead, W. E., Leith, D. J., and Rasmussen, C. E. (2003). Derivative observations in Gaussian process models of dynamic systems, in S. Becker, S. Thrun, and K. Obermayer (eds.), *Advances in Neural Information Processing Systems 15* (MIT Press, Cambridge, MA), pp. 1057–1064.

Song, G., Riddle, J., Fasshauer, G. E., and Hickernell, F. J. (2012). Multivariate interpolation with increasingly flat radial basis functions of finite smoothness, *Adv. Comput. Math.* **36**, 3, pp. 485–501.

Sonnenburg, S., Rätsch, G., Schäfer, C., and Schölkopf, B. (2006). Large scale multiple kernel learning, *J. Mach. Learn. Res.* **7**, pp. 1531–1565.

Spanos, P. D., Beer, M., and Red-Horse, J. (2007). Karhunen–Loéve expansion of stochastic processes with a modified exponential covariance kernel, *J. Eng. Mech.* **133**, 7, pp. 773–779.

Stakgold, I. (1979). *Green's Functions and Boundary Value Problems* (Wiley, New York).

Stehfest, H. (1970a). Algorithm 368: Numerical inversion of Laplace transforms, *Commun. ACM* **13**, 1, pp. 47–49.

Stehfest, H. (1970b). Remark on algorithm 368: Numerical inversion of Laplace transforms, *Commun. ACM* **13**, 10, p. 624.

Stein, M. L. (1990). A comparison of generalized cross validation and modified maximum likelihood for estimating the parameters of a stochastic process, *Ann. Stat.* **18**, 3, pp. 1139–1157.

Stein, M. L. (1999). *Interpolation of Spatial Data: Some Theory for Kriging* (Springer, Berlin; New York).

Stein, M. L. (2005a). Nonstationary spatial covariance functions, Technical report, University of Chicago.

Stein, M. L. (2005b). Space-time covariance functions, *J. Am. Stat. Assoc.* **100**, 469, pp. 310–321.

Stein, M. L. (2013). On a class of space-time intrinsic random functions, *Bernoulli* **19**, 2, pp. 387–408.

Steinwart, I. and Christmann, A. (2008). *Support Vector Machines*, Information Science and Statistics (Springer, Berlin; New York).

Steinwart, I. and Scovel, C. (2012). Mercer's theorem on general domains: On the interaction between measures, kernels, and RKHSs, *Constr. Approx.* **35**, pp. 363–417.

Stewart, G. (1994). Perturbation theory for rectangular matrix pencils, *Linear Algebra Appl.* **208–209**, pp. 297–301.

Stewart, G. W. (2011). Fredholm, Hilbert, Schmidt: Three fundamental papers on integral equations, Translated with commentary by G. W. Stewart, University of Maryland.

Stewart, J. (1976). Positive definite functions and generalizations, an historical survey, *Rocky Mt. J. Math.* **6**, 3, pp. 409–434.

Stone, M. (1977). Asymptotics for and against cross-validation, *Biometrika* **64**, 1, pp. 29–35.

Su, Y. and Cambanis, S. (1993). Sampling designs for estimation of a random process, *Stoch. Proc. Appl.* **46**, 1, pp. 47–89.

Sukumar, N., Moran, B., and Belytschko, T. (1998). The natural element method in solid mechanics, *Int. J. Numer. Meth. Eng.* **43**, 5, pp. 839–887.

Sun, H. (2005). Mercer theorem for RKHS on noncompact sets, *J. Complexity* **21**, 3, pp. 337–349.

Sun, X. (1994). Scattered Hermite interpolation using radial basis functions, *Linear Algebra Appl.* **207**, pp. 135–146.

Surjanovic, S. and Bingham, D. (2014). Virtual library of simulation experiments: Test functions and datasets, Retrieved November 22, 2014, from `http://www.sfu.ca/~ssurjano`.

Suykens, J. A. K., Van Gestel, T., De Brabanter, J., De Moor, B., and Vandewalle, J. (2002). *Least Squares Support Vector Machines* (World Scientific Publishing Company, Singapore).

Swartz, B. K. and Varga, R. S. (1972). Error bounds for spline and L-spline interpolation, *J. Approx. Theory* **6**, 1, pp. 6–49.

Szegő, G. (1939). *Orthogonal Polynomials (Colloquium Publications)*, 4th edn. (American Mathematical Society, Providence, RI).

Tavella, D. (2003). *Quantitative Methods in Derivatives Pricing: An Introduction to Computational Finance* (Wiley, New York).

Thomas-Agnan, C. (1996). Computing a family of reproducing kernels for statistical applications, *Numer. Algorithms* **13**, 1, pp. 21–32.

Tolstykh, A. I. and Shirobokov, D. A. (2003). On using radial basis functions in a finite difference mode with applications to elasticity problems, *Comput. Mech.* **33**, 1, pp. 68–79.

Tongarlak, M., Ankenman, B., Nelson, B., Borne, L., and Wolfe, K. (2008). Using simulation early in the design of a fuel injector production line, in *Simulation Conference, 2008. WSC 2008. Winter*, pp. 471–478.

Trahan, C. J. and Wyatt, R. E. (2003). Radial basis function interpolation in the quantum trajectory method: optimization of the multi-quadric shape parameter, *J. Comput. Phys.* **185**, 1, pp. 27–49.

Van Trees, H. L. (2001). *Detection, Estimation, and Modulation Theory, Part I* (Wiley–Interscience, New York).

Trefethen, L. N. (2000). *Spectral Methods in MATLAB*, Software, Environments, Tools (Society for Industrial and Applied Mathematics, Philadelphia, PA).

Trefethen, L. N. (2013). *Approximation Theory and Approximation Practice* (Society for Industrial and Applied Mathematics, Philadelphia, PA).

Trefethen, L. N. and Weideman, J. A. C. (1991). Two results on polynomial interpolation in equally spaced points, *J. Approx. Theory* **65**, 3, pp. 247–260.

Unser, M., Aldroubi, A., and Eden, M. (1992). On the asymptotic convergence of B-spline wavelets to Gabor functions, *IEEE T. Inform. Theory* **38**, 2, pp. 864–872.

Van Loan, C. F. and Fan, K.-Y. D. (2010). *Insight Through Computing: A MATLAB Introduction to Computational Science and Engineering* (Society for Industrial and Applied Mathematics, Philadelphia, PA).

Vandebril, R., Van Barel, M., and Mastronardi, N. (2008). *Matrix Computations and Semiseparable Matrices: Linear Systems*, Matrix Computations and Semiseparable Matrices (Johns Hopkins University Press).

Vapnik, V. and Lerner, A. (1963). Pattern recognition using generalized portrait method, *Automat. Rem. Contr.* **24**, 6, pp. 774–780.

Vapnik, V. N. (1998). *Statistical Learning Theory* (Wiley–Interscience, New York).

Varga, R. S. (1987). *Functional Analysis and Approximation Theory in Numerical Analysis* (Society for Industrial and Applied Mathematics, Philadelphia, PA).

Wackernagel, H. (2003). *Multivariate Geostatistics: An Introduction with Applications*, 3rd edn. (Springer, Berlin; New York).

Wahba, G. (1975). Smoothing noisy data with spline functions, *Numer. Math.* **24**, pp. 383–393.

Wahba, G. (1990). *Spline Models for Observational Data* (Society for Industrial and Applied Mathematics, Philadelphia, PA).

Wang, X. and Hickernell, F. J. (2000). Randomized Halton sequences, *Math. Comput. Model.* **32**, 7–8, pp. 887–899.

Wang, Y. H. and Zhang, K. (2013). Adaptive genetic algorithm for multilayer RBF network and its application on real function approximation, *Appl. Mech. Mater.* **380**, pp. 1166–1169.

Wasilkowski, G. W. and Woźniakowski, H. (1999). Weighted tensor product algorithms for linear multivariate problems, *J. Complexity* **15**, 3, pp. 402–447.

Wei, T., Hon, Y. C., and Ling, L. (2007). Method of fundamental solutions with regularization techniques for Cauchy problems of elliptic operators, *Eng. Anal. Bound. Elem.* **31**, 4, pp. 373–385.

Wells, J. H. and Williams, L. R. (1976). *Embeddings and Extensions in Analysis*, Ergebnisse der Mathematik und ihrer Grenzgebiete (Springer, Berlin; New York).

Wendland, H. (1995). Piecewise polynomial, positive definite and compactly supported radial functions of minimal degree, *Adv. Comput. Math.* **4**, 1, pp. 389–396.

Wendland, H. (2005). *Scattered Data Approximation, Cambridge Monographs on Applied and Computational Mathematics*, Vol. 17 (Cambridge University Press, Cambridge).

Wendland, H. (2006). Spatial coupling in aeroelasticity by meshless kernel-based methods, in P. Wesseling, E. Oñate, and J. Périaux (eds.), *European Conference on Computational Fluid Dynamics — ECCOMAS CFD 2006* (TU Delft).

Wendland, H. (2008). Hybrid methods for fluid-structure-interaction problems in aeroelasticity, in M. Griebel and M. A. Schweitzer (eds.), *Meshfree Methods for Partial Differential Equations IV, Lecture Notes in Computational Science and Engineering*, Vol. 65 (Springer, Berlin), pp. 335–358.

Wendland, H. and Rieger, C. (2005). Approximate interpolation with applications to selecting smoothing parameters, *Numer. Math.* **101**, 4, pp. 729–748.

Werschulz, A. G. and Woźniakowski, H. (2009). Tractability of multivariate approximation over a weighted unanchored Sobolev space, *Constr Approx* **30**, 3, pp. 395–421.

Whittaker, J. M. (1915). On the functions which are represented by expansions of the interpolation theory, *Proc. Roy. Soc. Edinburgh* **35**, pp. 181–194.

Wiener, N. (1938). The homogeneous chaos, *Amer. J. Math.* **60**, pp. 897–936.

Wiener, N. (1949). *Extrapolation, Interpolation, and Smoothing of Stationary Time Series* (MIT Press, Cambridge, MA).

Williams, C. and Seeger, M. (2000). The effect of the input density distribution on kernel-based classifiers, in *Proceedings of the 17th International Conference on Machine Learning* (Morgan Kaufmann), pp. 1159–1166.

Williams, C. and Seeger, M. (2001). Using the Nyström method to speed up kernel machines, in *Advances in Neural Information Processing Systems 13* (MIT Press, Cambridge, MA), pp. 682–688.

Wold, H. (1938). *A Study in the Analysis of Stationary Time Series* (Almqvist & Wiksells, Uppsala).

Womersley, R. (2007). Interpolation and cubature on the sphere, Retrieved December 5, 2014, from `http://web.maths.unsw.edu.au/~rsw/Sphere/`.

Womersley, R. S. and Sloan, I. H. (2001). How good can polynomial interpolation on the sphere be? *Adv. Comput. Math.* **14**, 3, pp. 195–226.

Wright, G. (2014). Radial basis functions for scientific computing, Retrieved December 5, 2014, from `http://math.boisestate.edu/~wright/montestigliano/`.

Wright, G. B. and Fornberg, B. (2006). Scattered node compact finite difference-type formulas generated from radial basis functions, *J. Comput. Phys.* **212**, 1, pp. 99–123.

Wu, Z. (1992). Hermite–Birkhoff interpolation of scattered data by radial basis functions, *Approx. Theory Applic.* **8**, 2, pp. 1–10.

Wu, Z. and Hon, Y. C. (2003). Convergence error estimate in solving free boundary diffusion problem by radial basis functions method, *Eng. Anal. Bound. Elem.* **27**, 1, pp. 73–79.

Wu, Z.-M. and Schaback, R. (1993). Local error estimates for radial basis function interpolation of scattered data, *IMA J. Numer. Anal.* **13**, 1, pp. 13–27.

Xia, J., Chandrasekaran, S., Gu, M., and Li, X. S. (2010). Fast algorithms for hierarchically semiseparable matrices, *Numer. Linear Algebr.* **17**, 6, pp. 953–976.

Xia, J., Xi, Y., and Gu, M. (2012). A superfast structured solver for Toeplitz linear systems via randomized sampling, *SIAM J. Matrix Anal. Appl.* **33**, 3, pp. 837–858.

Xiu, D. (2010). *Numerical Methods for Stochastic Computations: A Spectral Method Approach* (Princeton University Press, Princeton, NJ).

Xu, Y. and Cheney, E. W. (1992). Strictly positive definite functions on spheres, *P. Am. Math. Soc.* **116**, pp. 977–981.

Yang, C., Duraiswami, R., and Davis, L. S. (2004). Efficient kernel machines using the improved fast Gauss transform, in *Adv. Neur. In.*, pp. 1561–1568.

Yokota, R., Barba, L. A., and Knepley, M. G. (2010). PetRBF — A parallel O(N) algorithm for radial basis function interpolation with Gaussians, *Comput. Methods Appl. Mech. Eng.* **199**, 25–28, pp. 1793–1804.

Yuille, A. L. and Grzywacz, N. M. (1989). A mathematical analysis of the motion coherence theory, *Int. J. Comput. Vision* **3**, 2, pp. 155–175.

Zaremba, S. (1907). L'équation biharmonique et une class remarquable de functions fondamentales harmoniques, *Bull. Int. Acad. Sci. Cracovie* **3**, pp. 147–196.

Zaremba, S. (1909). Sur le calcul numérique des fonctions demandées dans le probléme de Dirichlet et le probléme hydrodynamique, *Bull. Int. Acad. Sci. Cracovie* **2**, pp. 125–195.

Zhang, K., Tsang, I. W., and Kwok, J. T. (2008). Improved Nyström low-rank approximation and error analysis, in *Proceedings of the 25th International Conference on Machine Learning* (ACM), pp. 1232–1239.

Zhou, P. and Yang, Z. (2013). A novel backward elimination algorithm for construction of RBF neural networks, *Proc. SPIE* **8921**, pp. 892115–892115–7.

Zhu, H., Williams, C. K., Rohwer, R. J., and Morciniec, M. (1998). Gaussian regression and optimal finite dimensional linear models, in C. M. Bishop (ed.), *Neural Networks and Machine Learning* (Springer, Berlin; New York).

Zwicknagl, B. (2008). Power series kernels, *Constr. Approx.* **29**, 1, pp. 61–84.

Zwicknagl, B. and Schaback, R. (2013). Interpolation and approximation in Taylor spaces, *J. Approx. Theory* **171**, pp. 65–83.

Index